Recent Trends on Renewable Energy, Smart Grid, and Electric Vehicle Technologies (RESGEVT 2020)

IOP Conference Series: Materials Science and Engineering
Volume 937

Vellore, India
9 July 2020

Editors:

K. Palanisamy
R. Sitharthan
Umashankar Subrmaniam
Jens Bo Holm-Nielsen
Sanjeevikumar Padmanaban
Akhtar Kalam
Konstantin Gomonov

ISBN: 978-1-7138-4773-1
ISSN: 1757-8981

Printed from e-media with permission by:

Curran Associates, Inc.
57 Morehouse Lane
Red Hook, NY 12571

Some format issues inherent in the e-media version may also appear in this print version.

Institute of Physics
Dirac House, Temple Back
Bristol BS1 6BE UK

Phone: 44 1 17 929 7481
Fax: 44 1 17 920 0979

techtracking@iop.org

Additional copies of this publication are available from:

Curran Associates, Inc.
57 Morehouse Lane
Red Hook, NY 12571 USA
Phone: 845-758-0400
Fax: 845-758-2633
Email: curran@proceedings.com
Web: www.proceedings.com

TABLE OF CONTENTS

IOP Conf. Series: Materials Science and Engineering **937** (2020) 011001 doi:10.1088/1757-899X/937/1/011001

PREFACE

The Virtual Conference on Recent trends on Renewable Energy, Smart Grid, and Electric Vehicle Charging is a platform for researchers, academics as well as professionals from all over the world to present, discuss and promote advances in knowledge, research, and practice in the field of Smartgrid, Renewable Energy Integration, Energy Storage System, Power Quality in Smart Grid, Transactive Energy Systems. RESGEVT-20 offered fantastic opportunity to all the global researchers to share their view through digital platform without any restrictions due to COVID-19.

RESGEVT-20 focuses on renewable energy, smart grid, and electric vehicle charging technology. Renewable energy, and smart grid technology has a way for a solution for better generation of electric power and an efficient way for transmission and distribution of this power. Due to its versatility it can be more easily installed and required less space as compared to traditional grids. Concept of Smart Grid design is aimed for grid observability, create controllability of assets, enhance performance and security of power system and specially the economic aspects of operations, maintenance and planning. That's why it is also consider that smart grid technology can be used to micro-grid level which eventually connect to all other micro-grids to form a large network of Smart Grid. These smart grids have a huge potential and could be a solution of reliability of power transmission and distribution in developing countries which lack infrastructure. In US only 20% of the all carbon dioxide is been emitted by transportation while generation of electricity has 40% of the carbon dioxide emitting share in it. This is due to the high demand rise of electricity. Smart Grids are been considering as a key role to address this problem by distributing electric power in an efficient way and ultimately reduce greenhouse gasses and pollutants like NOx and Sox.

The virtual conference RESGEVT 20 was organized by Centre for Smart grid Technologies, School of Electrical Engineering, Vellore Institute of Technology, Vellore, India. In Association with CEMR-RUDN University, Russia; Victoria University, Australia; Aalborg University, Denmark; and Prince Sultan University, Riyadh.

Editorial Board

IOP Conf. Series: Materials Science and Engineering **937** (2020) 011002 doi:10.1088/1757-899X/937/1/011002

Peer review declaration

All papers published in this volume of *IOP Conference Series: Materials Science and Engineering* have been peer reviewed through processes administered by the Editors. Reviews were conducted by expert referees to the professional and scientific standards expected of a proceedings journal published by IOP Publishing.

- **Type of peer review:** Single-blind
- **Conference submission management system:** University server
- **Number of submissions received:** 121
- **Number of submissions sent for review:** 121
- **Number of submissions accepted:** 58
- **Acceptance Rate (Number of Submissions Accepted / Number of Submissions Received X 100):** 31/60= 47%
- **Average number of reviews per paper:** 2 times (Plagiarism check, language check, Major Technical review, minor review and formatting)
- **Total number of reviewers involved:** 32 reviewers (21 National and 11 International)
- **Any additional info on review process:** Turnitin and Grammarly software's has been used
- **Contact person for queries:** Dr. K. Palanisamy, School of Electrical Engineering, Vellore Institute of Technology, Vellore, Tamil Nadu, India. (kpalanisamy@vit.ac.in)

IOP Conf. Series: Materials Science and Engineering **937** (2020) 012001 doi:10.1088/1757-899X/937/1/012001

Cyber Security Issues for IoT based Smart Grid Infrastructure

Mariya Ouaissa[1], Mariyam Ouaissa[1]

[1]ENSAM, Moulay Ismail University, Meknes, Morocco

E-mail: mariya.ouaissa@edu.umi.ac.ma

Abstract. We have been witnessing for several years the increasing networking of objects of all kinds that were not previously used to communicate. These newly connected objects together form the concept of Internet of Things (IoT). The energy sector is currently exploring the possibilities offered by the Internet of Things. However, for this sector, network intelligence is not a completely new phenomenon. The electrical networks already have information retrieval capacities for their monitoring and control, but these are still limited. The meeting of IoT and Smart Grids opens the way to data collection from all points of the network and in real time. Certain data previously collected in an aggregated manner over large geographic areas could now be collected on a local scale, near production or transmission points, and at a reasonable cost. It is also the massive accumulation of this data and its processing that is possible today. The smart grid is a network for the transport and distribution of electricity that is enhanced by capacities in terms of digital control, surveillance and telecommunications. It provides a two-way, real-time exchange of energy and information between the various players in the electricity chain, from the production site to commercial, industrial and residential users. The energy transition and the new smart grids raise many challenges in terms of industrial cybersecurity. Digital exchanges are intensifying (production, distribution and consumption are interconnected) and the use of connected objects (facilitating remote monitoring, observation and control of the entire value chain) increases the degree of vulnerability of information systems industrial. The cybersecurity of smart grids is today a decisive element in ensuring global security of our global energy systems. The purpose of this work is to present a comprehensive overview of challenges, threats and some solutions of security in IoT based smart grid. We focus in our study in network vulnerabilities, security requirements and cyber-attacks in smart grid in order to determine their impact in the network and give a guide on future research directions for cyber-security in smart grid applications.

1. Introduction

Currently the Internet knows a remarkable growth due to the development of connected objects. Internet is defined as a concept of communication of people anytime and anywhere, with the appearance of connected objects, the physical world can now communicate, whether for person-to-person, person-to-object or object-to-object relationships [1]. The Internet of Things (IoT) is defined as a global network of interconnected services and intelligent objects of all kinds designed to support humans in activities of daily living thanks to their detection, calculation and communication capabilities. Their ability to observe the physical world and provide information for decision making will be an integral part of the architecture of the Internet of the future. These objects must integrate into a more global system than the digital world and adapt to it [2].

The IoT relies on the implementation of a multitude of new sensors, actuators and intelligent systems, which communicate with each other. The energy sector is currently exploring the possibilities offered by the internet of things [3]. Network intelligence is not a completely new phenomenon for this sector. The Smart Grid already have information retrieval capacities for their monitoring and control, but these are still limited.

Smart grids are electrical grids that adjust the flow of electricity between providers and consumers using information technology. These networks provide a balance of production, distribution and consumption by gathering information on the state of the network. Smart grids improve the balance between production and consumption, the efficiency and security of grids, promote the integration of renewable energy sources throughout the network and, thanks to better system management, generate energy savings and lower costs (for both production and consumption) [4].

The advent of the internet of things requires an increased need for cybersecurity. The increase in the number of points of interaction with the network potentially exposes smart grids to modern threats targeting information systems [5]. Various factors are directly implicated; the lack of consideration of safety during the design phases, the risk of obsolescence in the face of rapid technological developments, the absence of applicable and followed standards in terms of security and confidentiality, or the routing, storage and management of an exponential amount of data [6] .

In this paper, we present the cyber security issues and challenges of the IoT based smart grid infrastructure. The next section describes the system network, features and standard of smart grid. In section 3 we discuss briefly the IoT concept and the link between IoT and smart grid. We present the security requirements of smart grid in section 4. We investigate security issues and challenges in the IoT based smart grid in section 5.

2. The Smart Grid Network

National and international energy saving needs make the modernization of electricity grids an absolute necessity. The smart grid constitutes a solution in term of management, metering, integration of renewable energies in the network. In this section, we start by briefly presenting the smart grid network and citing its advantages. Next, we will describe the most important standards in this field.

2.1. Smart Grid

For a long time, electricity was produced mainly from nonrenewable energy resources (nuclear, coal, natural gas, oil) centrally. These resources were becoming more and more expensive and scarce. Problems with the traditional power grid have accelerated the creation of a new concept called the smart grid, which is an electricity distribution network that uses information technologies to optimize the production, consumption and distribution of electricity [7]. It seeks to balance supply and demand for electrical energy by smoothing consumption peaks, which are costly and polluting. In practice, smart grids are presented as electrical networks to which a digital system of bidirectional communication between the supplier and the consumer has been added, an intelligent measurement system and a control system, the intelligent measurement system generally being an integral part (Figure1).

2.2. The Features of the Smart Grid Network

The features can be classified into four classes: operational efficiency, energy efficiency, customer satisfaction and environmental benefits [8].

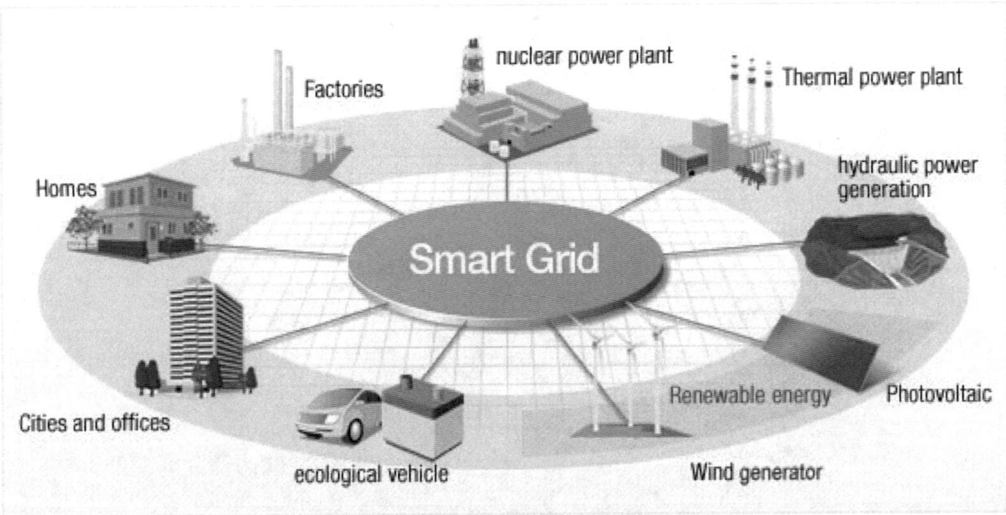

Figure 1. Smart Grid

2.2.1. Operational Efficiency. The smart grid increases the operational efficiency, first, it integrates the decentralized production of Distributed Energy Resources (DER). Each user can produce energy using energy resources renewable. There is an evolution from the model of one producer and several consumers to a model of several producers and several consumers. Then, it enables remote monitoring by using the Supervisory Control and Data Acquisition (SCADA) system and diagnostics by integrating sensors that are installed across the entire network. These sensors instantly indicate electrical flows and consumption levels. Still, it improves the reliability of public service power, operational performance and overall productivity.

2.2.2. Energy Efficiency. The smart grid helps reduce losses on the energy transmission and distribution network by optimizing electricity flows. It allows you to know the network status in real time, to anticipate incidents and to facilitate decision-making.

2.2.3. Customer Satisfaction. The smart grid offers several advantages to consumers. First, it reduces the duration and frequency of outages through remote maintenance with the Outage Management System (OMS). Then it improves the quality of the power supply with the voltage setting. Second, it allows consumers to reduce energy costs since they can decide when and how they want to use electricity. On the other hand, it helps empower the consumer to manage their energy consumption, to save money and overcome certain constraints in the power system, without compromising their lifestyle.

2.2.4. Environmental Benefits. The smart grid has several environmental benefits. First, it allows the integration of renewable energy resources. Then, it activates the wide adoption of electric vehicles. These vehicles reduce carbon emissions. Still, it ensures and enhances reliability and security of environment in order to resist to disruptions, attacks and natural disasters, anticipating and responding to system disruptions and strengthening security supply through capabilities of improved transfers.

2.3. Security Standards for the Smart Grid

Several standardization organizations have defined standards for the deployment of the smart grid network. The American organization National International Standard Technology (NIST) has defined an architecture for the smart grid. The European organization Institute of Electrical and Electronics Engineers (IEEE) defined the three layers of the network. He added the concept of centralized management of renewable energy [9].

IOP Conf. Series: Materials Science and Engineering **937** (2020) 012001 doi:10.1088/1757-899X/937/1/012001

2.3.1. The Architecture Proposed by NIST. NIST proposed the basic architecture for smart grids. This architecture has seven domains namely: distribution, transmission, customer, markets, operations, bulk Generation, service provider as shown in Figure 2.

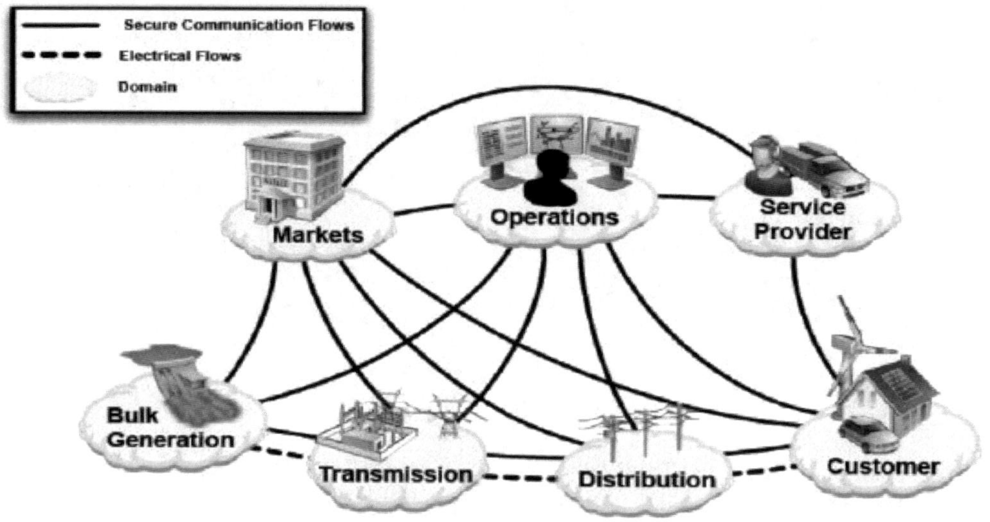

Figure 2. NIST Smart Grid Architecture

2.3.2. The Architecture Proposed by IEEE. In the architecture of NIST, each consumer can generate electricity using renewable resources. The surplus of produced energy is managed by the consumer himself, the management of the surplus of energy is decentralized. Given the complexity of the management task, the IEEE proposed an architecture based on that of NIST, but defined a new field called DER, which allows to manage the surplus of energy produced, the management surplus energy is centralized. In addition, IEEE has provided guidelines for understanding and defining the interoperability of smart grids. It defines three integrated architectural perspectives: the energy layer, the communication layer and the information layer (Figure 3).

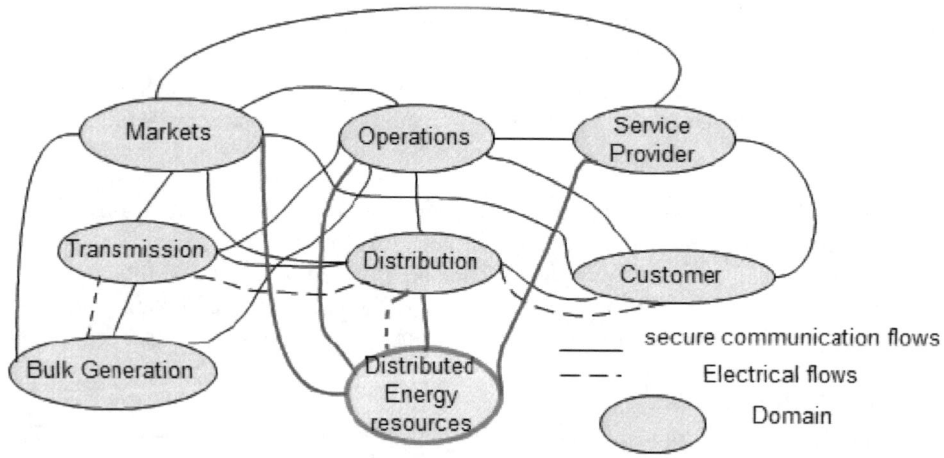

Figure 3. European Extension of the NIST Model

3. IoT based Smart Grids

The internet of things is a major development in the field of Information Technology (IT) which increasingly dominates and reigns in the market for computer systems. IoT is the result of the development and combination of different technologies. It encompasses almost all areas of information technology today such as smart cities, Machine to Machine systems, connected vehicles, sensor networks, etc. IoT, also known as Web 3.0, represents an extension of the Internet to things and places in the physical world. Each physical object that can be a person, a computer, a smartphone, a car, a sensor, a smart house, is associated with a virtual entity, which behaves like an active entity in the system. The integration of IoT with a smart environment extends the capabilities of smart objects by enabling the user to monitor the environment from remote sites. IoT can be integrated with different smart environments and can generally be classified into the following areas: smart cities, smart homes, smart grid, smart buildings, smart transportation, smart health, and smart industry. Figure 4 illustrates the IoT-based smart environments [10].

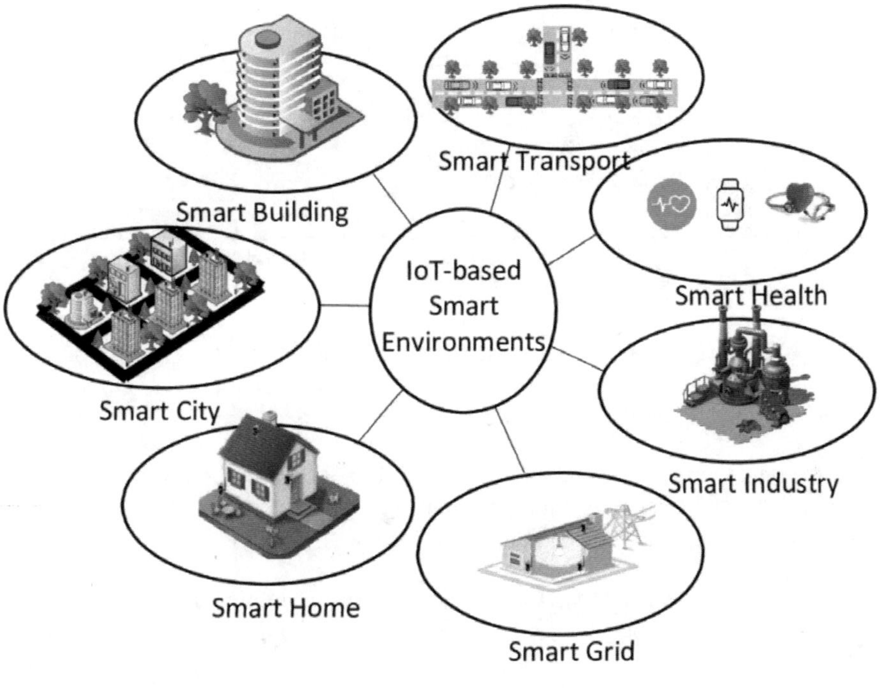

Figure 4. IoT based Smart Environment

The Internet of Things (IoT) has recently emerged as enabling technology for the smart gird. Each device in the grid can be considered as an object. Utilizing the concept of IoT, each device can have a unique IP address that can upload its status and download control commands via the Internet. IoT allows devices on the smart grid infrastructure to be sensed and controlled remotely via a scalable communication network, which allows for easier integration between the physical world grid devices and computer based control systems thereby leading to improved efficiency and accuracy and enabling the grid to meet the energy needs of current and future generations (Figure 5) [11].

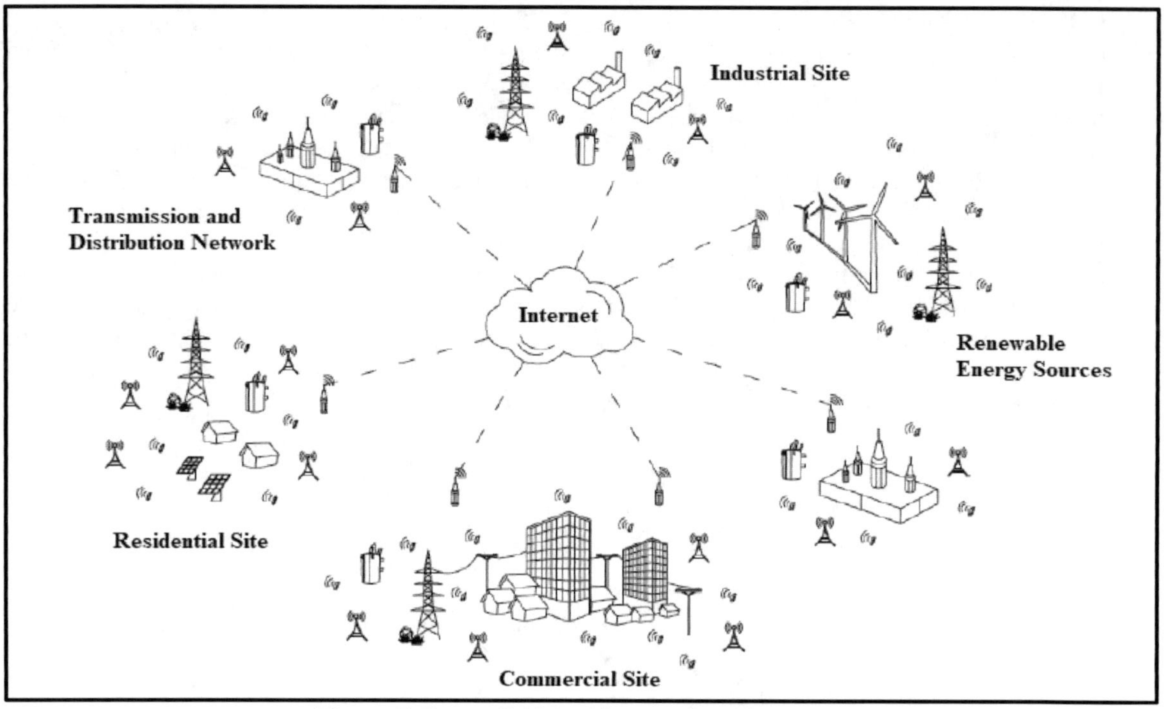

Figure 5. Smart Grid Architecture based IoT

4. Security Requirements of IoT based Smart Grid

A certain number of security services are necessary to secure the exchanges of the objects based smart grid with the control center and to avoid attacks. The main challenges [12] of securing these exchanges can be summed up in the implementation of:

- An authentication service to prevent spoofing and fake message injection attacks. The injection of false disconnect commands, for example, can deprive electricity of an entire neighborhood or even sensitive public buildings such as hospitals, police stations, etc.

- An integrity service to avoid modification attacks. For example, the messages containing the measurements of energy consumed at the Home Area Network (HAN) level are of great importance in the smart grid network and their modification can have serious consequences on consumption costs.

- A privacy service to counter eavesdropping and invasion of privacy attacks. This service is of primary importance, since it will make the messages exchanged between the objects and the control center incomprehensible to any attacker. The use of the privacy service should meet the needs of applications in the smart grid network.

- Communications between the smart grid and the control center require the implementation of anti-replay mechanisms to avoid the replay of certain messages or orders such as; fault messages, consumption rate messages, invoice, and cut-off commands.

- Regarding the availability service, a malfunction of objects can have serious consequences, hence the need for the implementation of an availability service to avoid denial of service attacks. The same is true for the control center and the collector.

5. Cyber Security Challenges and Issues

Several equipment's of smart grid network exchange a huge amount of data o daily basis to optimize the production, consumption and distribution of energy [13]. Securing these data flows is essential. A single failure or attack could compromise the security of the entire power grid. In this section, we will describe in detail the attacks that can affect the intelligent network by classifying them [14].

5.1. Cyber Attacks on the Smart Grid Architecture

The deployment of information and communication technologies on power grids raises more concerns about the security of the power system and the protection of consumer data than with traditional power grids. We review some attacks that can be carried out on smart grids. We have chosen the classification of which makes it easier for us to identify smart grid attacks, since it is concerned with the various components of the smart grid network, namely:

- Devices
- Systems
- Networks

5.1.1. Attacks on Smart Grid Network Devices. The different devices of a smart grid network can be affected by several types of attack. In this part, we will present these attacks.

- Smart Meter

A malicious node can disrupt the functioning of smart meters by making several types of attacks:

- ✓ The jamming attack can be launched to prevent the Smart Meter from communicating with other nodes in the smart grid network.
- ✓ The wiretapping attack can be performed to detect sensitive information about the customer's energy consumption. Similarly, this attack can lead to an attack on the privacy of consumers. In effect, the attacker can recover several private information. This information is critical since it can be exploited by thieves.
- ✓ Stealthy attacks: the attacker can inject false information about electricity prices, which can significantly increase consumers' bills.
- ✓ The application called Remote Connect Disconnect (RCD) can be used by attackers to perform a remote disconnection attack on the smart meter, thus depriving the customer of electricity. Still, attackers can use this application to connect a smart meter and take advantage of illegal power.
- ✓ Replay attack: the attacker can use the smart meters out of use by injecting incorrect data into the system, which can lead to incorrect energy prices or inaccurate predictions
- ✓ Smart Meters are placed outside houses. An attacker can have physical access to at least one smart counter. It can launch a compromising attack via the Test Action Group (JTAG) interface to steal keys and other information.
- ✓ The energy requirements of certain devices are based on the energy consumption data supplied by the smart meter to plan their load according to demand and prices. An attacker can spoof the identity of a smart meter and send a false response to these devices in order to cause a great demand for energy, to increase the electricity bill or to cause the equipment to shut down.

- Phasor Measurements Units (PMU)

PMUs are placed on the transmission lines. They help create a snapshot of the state of the power grid (voltage measurement). They are able to collect measurements on electrical voltages and quantities and send them to the Phasor Data Concentrator (PDC). The PDC reads data from multiple PMUs and then merges it into a single message. This message will be communicated to the area of operations. PMUs transmit their measurements to the PDC hierarchically. Hierarchical architectures suffer from drawbacks such as delay in messages. This drawback can be exploited by attackers. Package destinations of data to be sent is preconfigured. Manipulation of the destination lists of a (damaged) PMU by an attacker can cause attacks to spread to damage other PMUs. A malicious node can perform a spoofing attack on a PMU, modify PMU messages that contain energy measurement data, and can also read messages in transit between the PMU and the PDC. These attacks affect critical decision operations such as detection and the location of the event. For example, when an attacker re-plays an old PMU message that contains line failure information or energy loss measures, Grid systems can make a decision to cut power to an area. The availability of real-time PMU data is a critical issue; any inaccessibility of real-time data can affect the functionality of Wide Area Monitoring, Protection and Control (WAMPAC).

- Plug-in Hybrid Electric Vehicle (PHEV)

Although a two-way communications infrastructure can bring many advantages for the smart grid, it can introduce new vulnerabilities. An attacker can manipulate the real-time pricing information that is communicated by the utility company for vehicles. The attacker can disrupt the transmission of electricity price information to the PHEV owner, resulting in the loss of the pricing information, which is, in fact, one of the possible attacks Denial of Service (DoS) on the smart grid. Likewise, it is possible for the attacker to manipulate pricing information by injecting incorrect price values in order to mislead owners of plug-in hybrid vehicles.

- Remote Terminal Unit (RTU)

Remote Terminal Units are traditionally used to configure and troubleshoot devices in the remote intelligent network. This remote access functionality can give rise to attacks that allow malicious nodes to take control of devices. A denial of service attack on a smart grid device can saturate the computing power of the CPU, memory or bandwidth and will result in delay in real-time data exchange. As a result, control center operators do not have a complete picture of the state of the power grid, leading to incorrect decisions.

- Sensors or Intelligent Electronic Devices (IEDs)

Switches are used to protect the power infrastructure in substations, when an IED detects an anomaly, it sends a message to open/close to the switches to balance the power load. If an attacker impersonates a surveillance IED, he could send false close/open messages to the switches and damage the protection system, resulting in potential loss of power to clients.

5.1.2. Attacks on Smart Grid Systems. In this part, we will study the security of some smart grid systems.

- Control and Management Systems

Control and management systems (For example: Distribution Management System (DMS), Energy Management System (EMS), Supervisory Control and Data Acquisition (SCADA)) play a key role in the smart grid network. It is necessary to protect them from attack. A malicious node can carry out DoS attacks on control systems and make them unavailable.

- Wide Area Monitoring, Protection and Control (WAMPAC)

The Wide Area Monitoring, Protection and Control system exchanges transmission data with other control systems to provide real-time monitoring and alarm functions and ensure efficient power transmission, production and aggregation in the electric grid. The WAMPAC system is also vulnerable to DoS attacks while these services are provided in real time. Denial of service can affect the different layers of communication.

- Advanced Metering Infrastructure (AMI)

AMI is the communication infrastructure for smart meters. It is used to provide two-way communication between customers and suppliers. The communication channels used by AMI to communicate data between smart meters and utilities are also vulnerable to cyber attacks. Data transit through these channels can be intercepted. The Man In The Middle (MITM) attack can be launched. Energy consumption data can be changed before sending messages. Also, by listening to the wireless communication channel, an attacker could obtain the information exchanged between smart meters and the control center. Demand Response (DR) is an essential part of automatic load management and builds on the ability of the AMI communications infrastructure to send load reduction requests for smart meters and other devices, for dynamically manage the overall system load.

- Outage Management System (OMS)

Automated outage management requires smart meters to send out fault information. The utility uses the information to restore power in reduced time. A disruption of this application directly affects by delaying detection and the correction of faults. Crash management is resilient if the utility can still identify and recover from crashes within a limited time, where time is dependent on the specific requirements of the utilities. An attacker can spoof the identity of a smart meter and send a crash message, however he can modify the message sent to influence the resilience of the grid. On a larger scale, several attackers can impersonate several smart meters in the same geographic area and send almost identical messages to learn about a disaster. The control center can make the decision to cut the power in this geographic area.

5.1.3. Attacks on Smart Grid Networks. Smart grids are connected and controlled by communication networks. We have chosen to classify the attacks that can be carried out on networks, in two categories.

- Attacks used on the routing protocols
- Attacks used on the communication protocols

- Attacks on Routing Protocols

Among the routing protocols that can be used in NAN networks, we cite:

✓ Routing Protocol for Low-Power and Lossy Networks (RPL) defined in RFC6553 by the Internet Engineering Task Force (IETF). It has been designed to support the specific requirements of these networks. RPL is a proactive distance vector routing protocol that builds a Destination Oriented Directed Acyclic Graph (DODAG). The DODAG built allows each node to transmit the data it has collected to the DODAG root. Each node in the DODAG selects a parent according to a given routing metric and an objective function. The collected data is forwarded from child to parent to the root. RPL routing protocol for IoT can be affected by Selective Forwarding Attacks: malicious nodes try to stop packets in the network by refusing to forward or delete the messages that send them. crosses. With these attacks, it is possible to launch DoS attacks where malicious nodes selectively transmit some packages. This attack is mainly intended to disrupt the routing paths.

✓ The Energy Minimum Transmission Protocol (MTE) takes the basic Dynamic Source Routing (DSR) protocol and assigns a weight to each of the links which is a function of the energy required to transmit a packet on this route. Routing takes place along lighter routes, aggregating all the links making up a path. The attacks of the wireless sensor network are applicable to the MTE routing protocol (DoS, ...).

- Attacks on Communication Protocols

The communication protocols used within a smart grid are also a major source of vulnerabilities. Certain protocols can be affected by attacks: DoS, listening, modifying.

- ✓ ZigBee: ZigBee has a major drawback, all passwords are stored in clear text in the storage space. If the attacker obtains physical access to the device, he can copy the memory of the device to the computer, and then he can find the key. ZigBee can be affected by DoS attacks.
- ✓ Protocol C37.118: is an IEEE standard for the use of synchrophasors in power systems. The C37.118 protocol does not encrypt the messages exchanged between the PDC and PMU. An attacker can perform a listening attack. Still, it is vulnerable to MITM attacks because it does not verify the source of the messages it receives from the PMU.
- ✓ DNP3: Distributed Network Protocol (DNP3) is a communication protocol widely used by public services of electricity. They used iperf (a network traffic generator) to occupy the communication channel, which reduces network availability. They showed that the packages Long DNP3 are more vulnerable to DoS attacks than short DNP3 packets. The DNP3 protocol is a documented protocol, it may be subject to reverse engineering.

5.2. Countermeasures against Cyber Attacks

To protect system from multiple attack vectors, we find that the countermeasures, which until now were almost exclusively software, are no longer sufficient and that a good knowledge of the hardware platform becomes more and more important for a better mastery of the security of a computer system. Thus, it is essential today to know not only the functioning of the programs, but also to ensure the good behavior of the entire processing chain, including lesser known hardware components, such as the chipset, inputs-outputs or peripherals. Many challenges are encountered to secure intelligent grid systems from many attacks and to achieve the security demands and objectives of the intelligent grid, since intelligent grid assets are distributed across large areas. The cyber defense solutions should protect all aspects of smart grid systems. The defense solution integrate multiple defense techniques including proactive real-time intrusion prevention/detection systems (IPS/IDS) using machine learning and artificial intelligence, network segmentation, controlled wireless propagation, authentication, authorization, certification [15]. The proposed solutions should comprise of scalable, resilient, and adaptive cyber security/defense techniques for smart grid operation without affecting any legitimate smart grid operations.

6. Conclusion

The smart grid is an energy distribution network that uses communication technologies to optimize the production, consumption and distribution of electricity. IoT is the next step towards a globally and pervasive connection to any communication and computation enabled objects. The smart grid is the biggest deployment of IoT technology, where smart devices are deployed along the energy path, all the way from the generation plant to the end-customer. In the literature, several researchers have looked into the security of the smart grid network and have identified a number of attacks that can be carried out, namely identity theft attacks by a smart meter, one of the essential components of the smart grid network. Similarly, the attacker can replay energy consumption messages that aim to disrupt the billing system of users. Some attacks can even deprive the customer of electricity. The integration of communications technologies into the electrical grid presents a source of vulnerability. In this paper, we presented a comprehensive survey of the main security issues and challenges for the IoT based smart grid infrastructure. In this context the use of the IoT provide a better control and monitoring of the power grid in smart grid network, so it is necessary to concentrate on the security challenges in order to develop to implementation and integration of the IoT devices in the smart grid.

References

[1] Atzori L, Iera A, and Morabito G, The Internet of Things: A survey, *Computer Networks* **54(15)**, 2787–2805, 2010.
[2] Yacchirema DC and Palau C, Smart IoT Gateway For Heterogeneous Devices Interoperability, *IEEE Latin America Transactions* **14(8),** 3900-3906, 2016.
[3] Ouaissa M, Ouaissa M, and Rhattoy A, An efficient and secure authentication and key agreement protocol of LTE mobile network for an IoT system, *International Journal of Intelligent Engineering and Systems*, **12(4)**, 212-222, 2019.
[4] Fang X, Misra S, Xue G, ans Yang D, Smart grid—The new and improved power grid: A survey, *IEEE communications surveys & tutorials*, **14(4)**, 944-980, 2011.
[5] El Mrabet Z, Kaabouch N, El Ghazi H, and El Ghazi H, Cyber-security in smart grid: Survey and challenges, *Computers & Electrical Engineering*, **67**, 469-482, 2018.
[6] Kimani K, Oduol V, ans Langat K, Cyber security challenges for IoT-based smart grid networks, *International Journal of Critical Infrastructure Protection*, **25**, 36-49, 2019.
[7] Chen SY, Song SF, Li LX, ans Shen J, Survey on smart grid technology, *Power system technology*, **33(8)**, 1-7, 2009.
[8] Yan Y, Qian Y, Sharif H, ans Tipper D, A survey on smart grid communication infrastructures: Motivations, requirements and challenges, *IEEE communications surveys & tutorials*, **15(1)**, 5-20, 2012.
[9] Gungor VC, Sahin D, Kocak T, Ergut S, Buccella C, Cecati C, and Hancke GP, Smart grid technologies: Communication technologies and standards, *IEEE transactions on Industrial informatics*, **7(4)**, 529-539, 2011.
[10] Ahmed E, Yaqoob I, Gani A, Imran M, ans Guizani M, Internet-of-things-based smart environments: state of the art, taxonomy, and open research challenges, *IEEE Wireless Communications*, **23(5)**, 10-16, 2016.
[11] Mugunthan SR and Vijayakumar T, Review on IoT based Smart Grid Architecture Implementations, *Journal of Electrical Engineering and Automation (EEA)*, **1(1)**, 12-20, 2019.
[12] Rawat DB ans Bajracharya C, Cyber security for smart grid systems: Status, challenges and perspectives. In SoutheastCon, IEEE, 1-6, 2015.
[13] Bekara C, Security issues and challenges for the IoT-based smart grid. In FNC/MobiSPC, 532-537, 2014
[14] Gunduz MZ and Das R, Cyber-security on smart grid: Threats and potential solutions, *Computer Networks*, **169**, 107094, 2020.
[15] Wang W and Lu Z, Cyber security in the smart grid: Survey and challenges, *Computer networks*, **57(5)**, 1344-1371, 2013.

IOP Conf. Series: Materials Science and Engineering **937** (2020) 012002 doi:10.1088/1757-899X/937/1/012002

Three-Level Topology for Shunt Active Power Filters

S.P.Richard [1], R.Manivasagam [2] , P.Vigneshwaran [3]

[1,2,3] K. Ramakrishnan College of Engineering, Trichy – India.

sprichard23@gmail.com, manivasagammn3@gmail.com, vigneshgemini@gmail.com

Abstract. The development in the field of power electronics has led to its optimum and efficient use in domestic and industrial arena especially in control and automation processes. In turn this has created a threat in terms of power quality issues. To some extent solution to power quality issues is obtained again using power electronic devices in the form of active filtering. A shunt active power filter (SAPF) is used for harmonic filtering. The performance of such system is enhanced by using multilevel inverters. A novel three level inverter topology is proposed here. A three phase transformer fed from both ends at its primary terminals with two independent voltage source inverters is used to replace a filter inductance of a conventional system. The benefits of multilevel inverter are utilized in SAPF and also system power factor is kept near unity in all conditions.

Keywords: Active powers filter (APF), current controlled voltage source inverter, dual inverter.

1. Introduction

The rising uncertainties in AC system and increasing dynamics and non-linearity of loads have attracted more attention of engineers/researchers to evolve improved methods. This leads to evolution of universal control technique of compensation having substantially improved transient performance under unbalanced load conditions (Ali, 2011). Here we have used an indirect method of reference current generation as given in (Patidar and Singh, 2010). Section 2 describes the system used in the paper. The indirect control method is explained in section 3. System modeling is covered in section 4 while section 5 covers the results and discussions. A brief conclusion on the work is covered in section 6. (Baiju, M.R., Shivakumar et al., 2002; Somasekhar et al., 2004).

2. System Descrıptıon

Two DC capacitors are connected at the two inputs of the VSI's. An elevated stage of manufacturing proficiency must necessities the appropriate analysis of PQ troubles. Moreover, the complexity of PQ analysis is necessities that the professional awareness in several region of electric power, e.g., electric drives, sensors, revolving machines, transformers, and common power systems function etc,. (Manivasagam.R et al. 2020). The observation of utilities, utensils producer, and clients might be entirely dissimilar in endeavor to describe PQ. Utilities observe PQ from the system consistency point of observation. Furthermore, utensils producer deem PQ as being that stage allocate for suitable function of their utensils, but clients deem superior PQ that make sure the constant operation of progression, function, and trade. A PQ trouble could be distinct as some power difficulty evident in voltage, current, or frequency variation that consequence in breakdown or malfunction of client utensils.

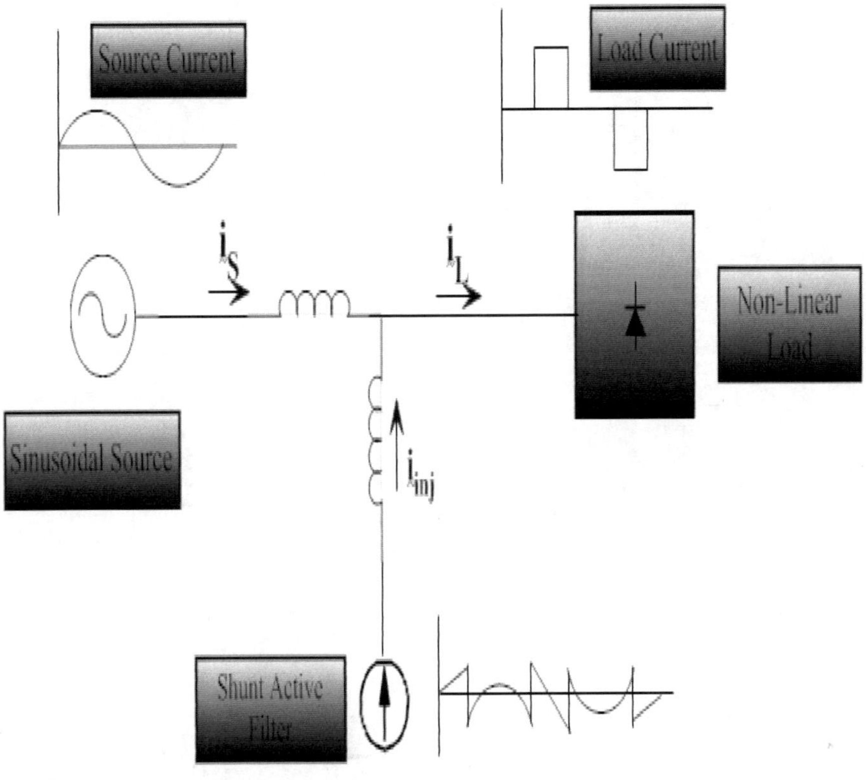

Figure.1. Basic shunt compensation principle

On the other hand, the requirement of electric power contains persistent to produce quickly. Immediate power recompense procedures are employed to maintain the grid presentation at a superior eminence. Distribution system undergoes current with voltage associated PQ troubles, which comprise deprived power feature, imprecise basis current, voltage conflict. The significant PQ troubles are voltage dissimilarity, voltage sag, load derange, harmonic deformation, disruption, voltage swells, and voltage flicker, that can origin disruption in the dispensation plants and economic fatalities. These conflicts can origin troubles, such as overheating, motor breakdown, imprecise metering, and disoperation of defensive utensils. Voltage swell and sag can take place because of lightning, capacitor controlling, motor opening, close to circuit error, or misfortune, and can also direct to power disruption. There are frequent processes to moderate harmonics, swells, voltage sags, and disturb in the power allocation format. Nonetheless, there are abundant methods to diminish the consequence of harmonics unaccompanied in the format.

3. Control Method

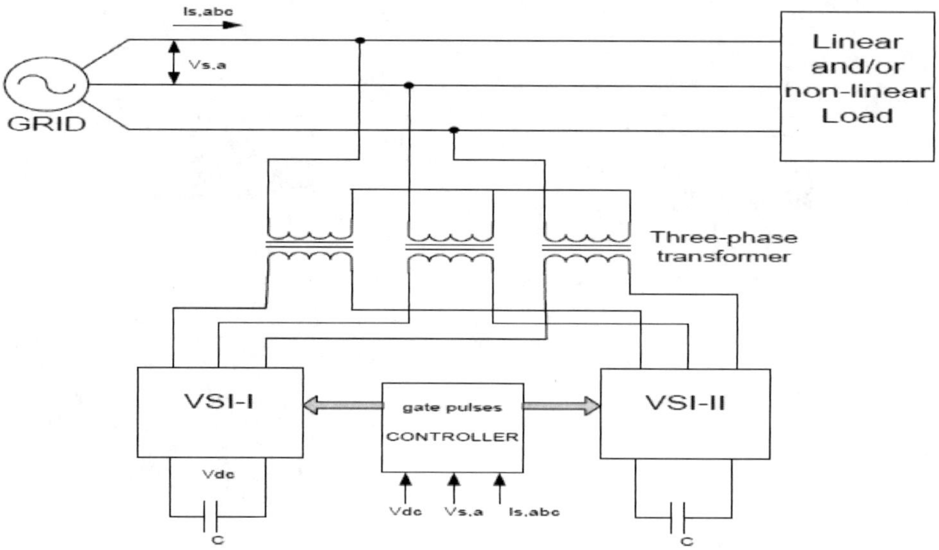

Figure.2 Three–phase system with dual inverter fed SAPF

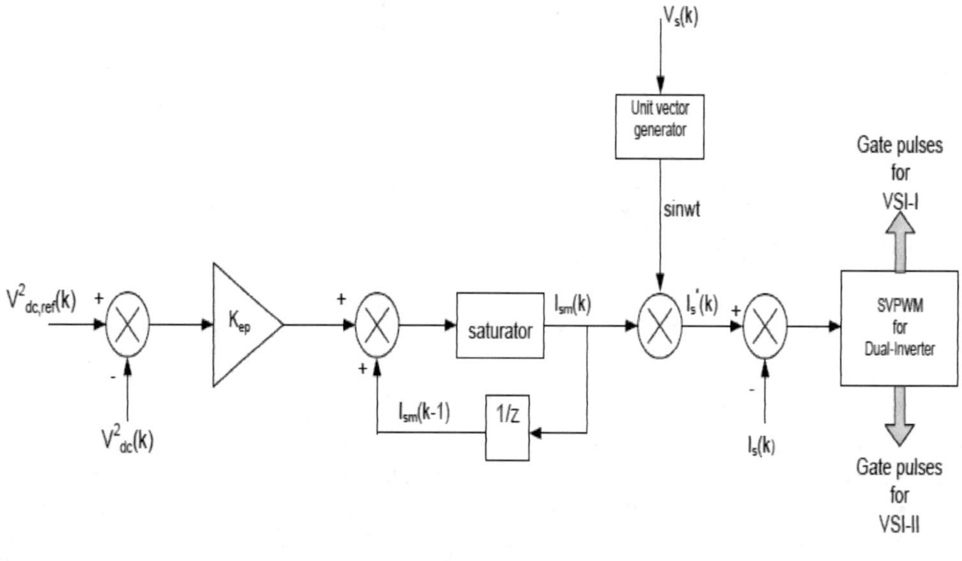

Figure.3.Control Circuit

This method is to employ active power filter that capitulate harmonic current of identical magnitude and contradictory division to that of the harmonic current perverted in the format such that it extract the harmonic current in the format. APF is exploited its remuneration of elevated speed reaction and suppleness in function as it encompass the power electronic devices. Furthermore the identical device is exploited to incorporate the power from the RES to the allocation format. Subsequently the necessity of supplementary utensils is escaped. Thus, the APF substance is the grid interfacing inverter for interfacing RES to the electrical grid devoid of further power stipulation utensils. These active power

IOP Conf. Series: Materials Science and Engineering **937** (2020) 012002 doi:10.1088/1757-899X/937/1/012002

filters are habitually exploited to diminish only the THD in load region and source region of some format but not accomplished of explanatory foremost PQ troubles like voltage sag, disturb and voltage swell etc. The convention of a practice power device is considered to be the significant powerful procedure.

4. SV-PWM For Dual Inverter

A new multilevel topology is being used to drive induction motor (Baiju et al., 2004). Based on similar lines a dual fed three phase transformer is used here in place of an induction motor (Manivasagam et al 2014,2019) . Out of six free terminals of primary winding of this transformer, three ends are fed from VSI-I and other three free ends are fed from VSI-II as shown in figure 4. The state phasor locations from individual inverters are shown in figure 5 (Ragavi et al 2015) .

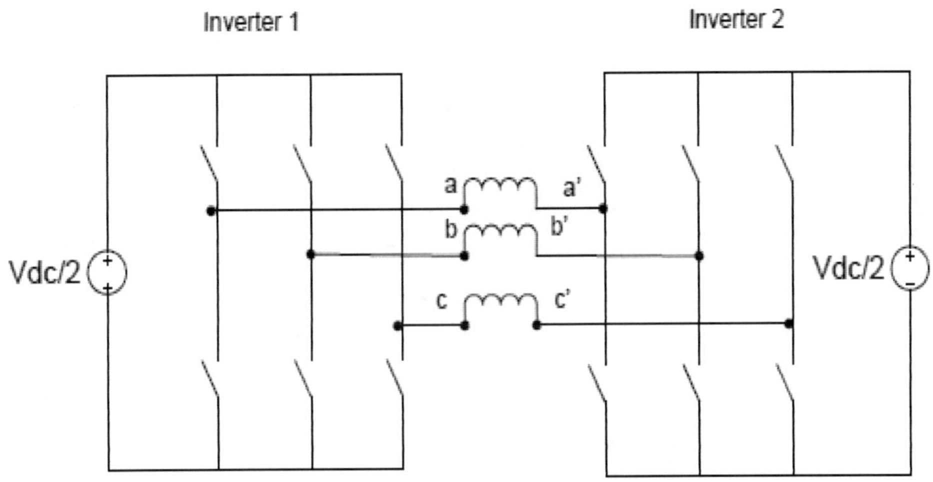

Figure.4. Open end fed primary winding of transformer

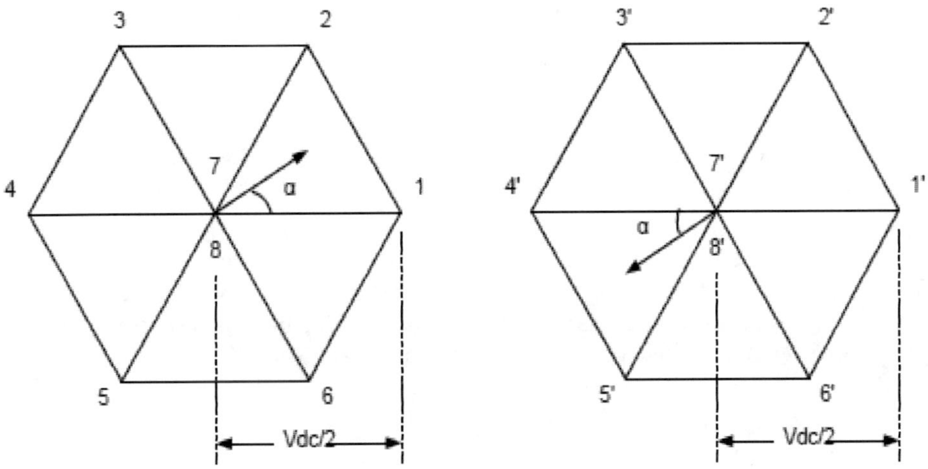

IOP Conf. Series: Materials Science and Engineering **937** (2020) 012002 doi:10.1088/1757-899X/937/1/012002

Figure. 5. Voltage vector locations for inverter –I and II

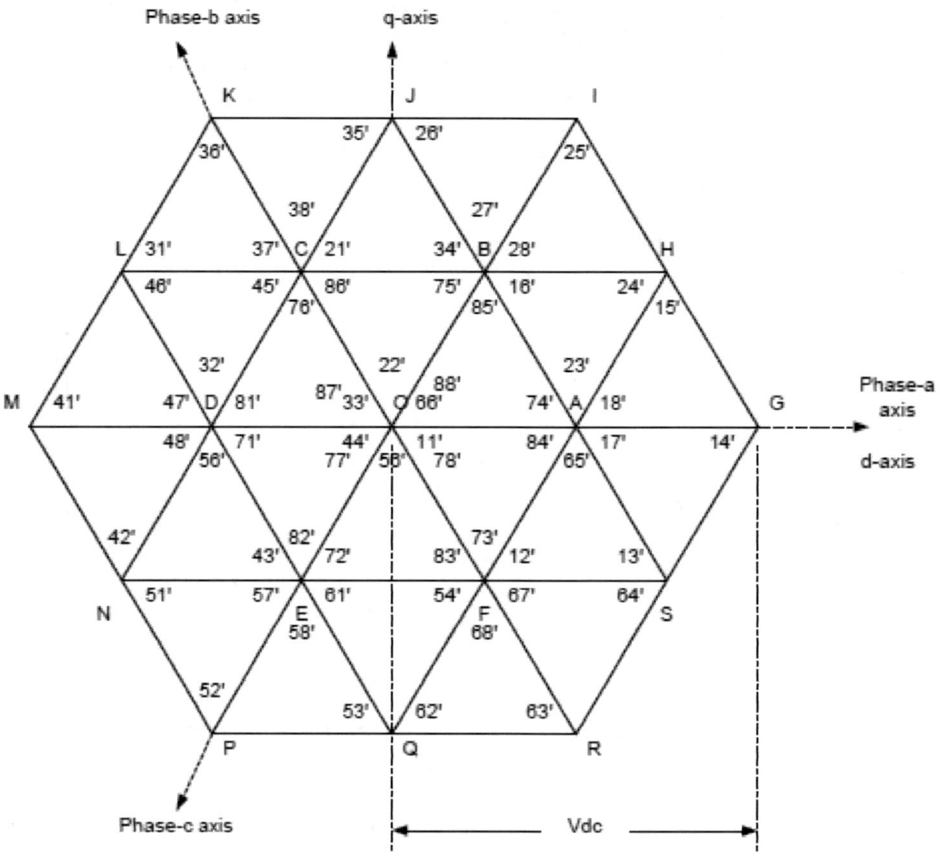

Figure.6. Voltage vector positions in three-level inverter.

The control strategy of these two inverters is selected such that three level output voltage is generated across each phase of the transformer primary. The space phasor combinations (23X23 = 26 = 64) as each inverter is assumed with 8 states independently of each other) from the 2 inverters is shown in figure 6. The resultant space vector combination locations shown in figure 6 are obtained.

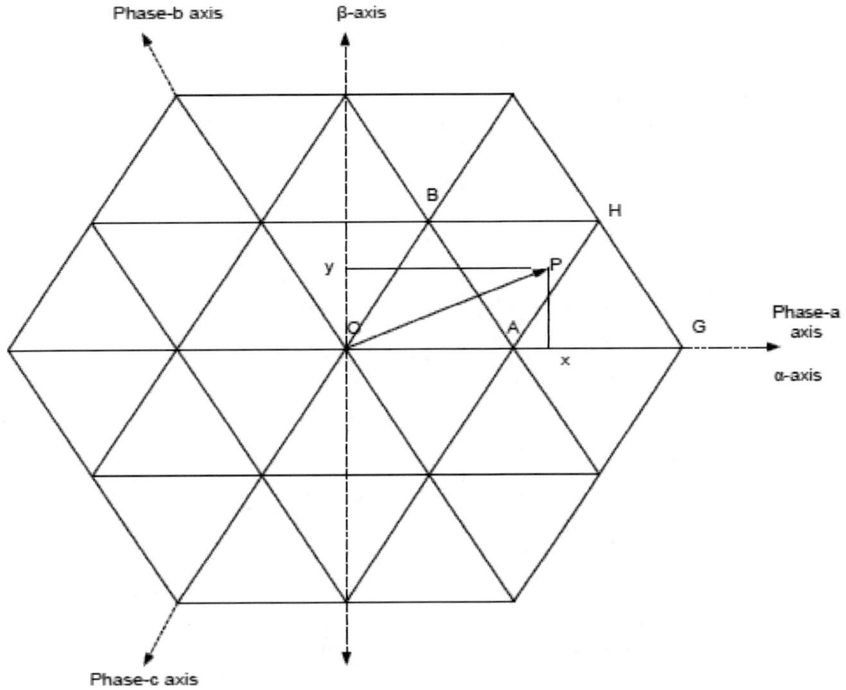

Figure. 7. Switching vector components along d and q axis

In order to have equal switching losses in the two inverters the role of each inverter (clamping or switching) is altered every 600 (D,Aarthi et al 2017).

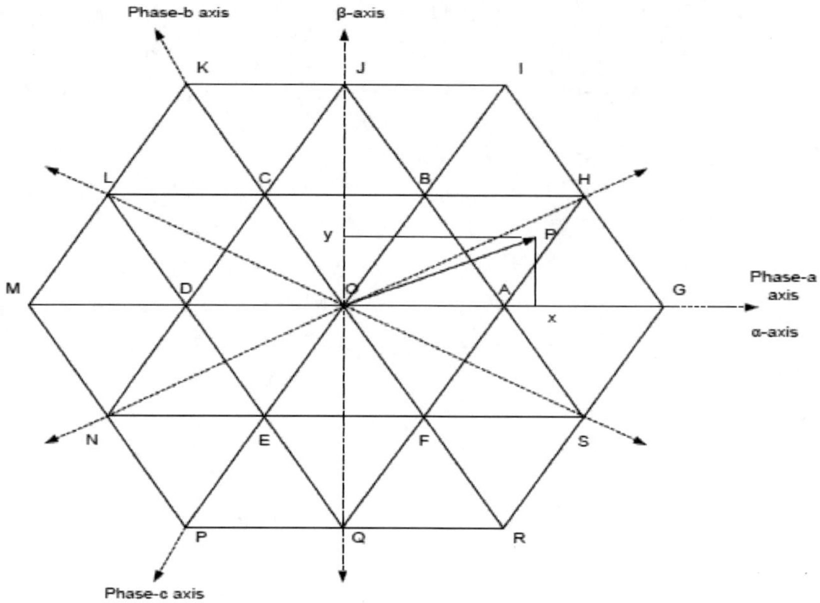

Figure.8. Principle of NSH switching scheme

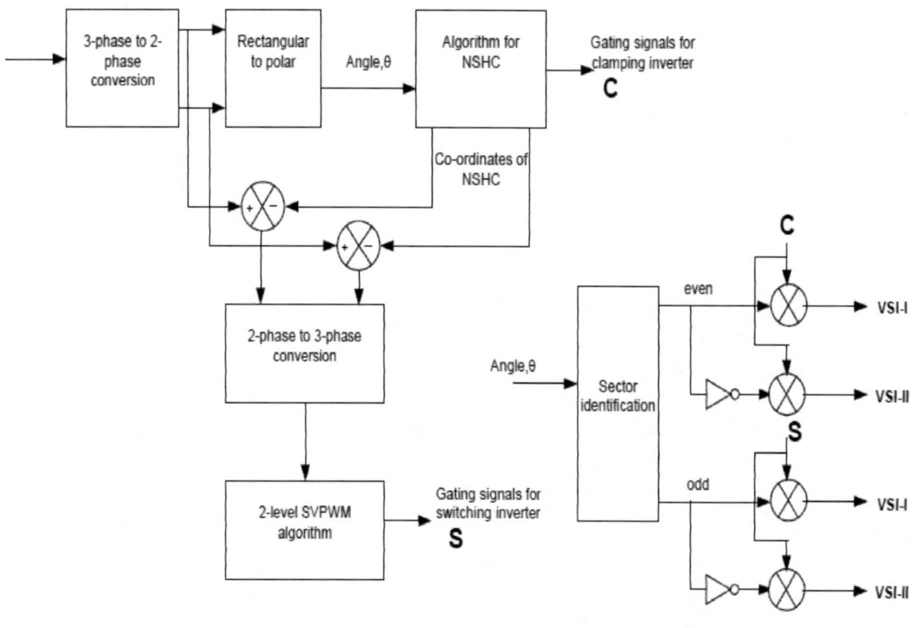

Figure.9. Dual inverter SVPWM functional diagram

5. Results And Discussions

Table 1. Sımulatıon Parameters

Location	Parameter	Value
Supply	Supply voltage	(i) 400 V, 3 phase, 50 Hz (ii)400V+5thand7thharmonic components
Active Power Filter	1. open end 3-phase transformer 2.DC-bus capacitance	10kVA, 800V/800V, 0.002 pu resistance and 0.045 pu reactance1000 μF

Case (a): sinusoidal supply and non-linear load. (Figure 10 (a) to (d))Here supply voltage is assumed to be balanced sinusoidal voltage. Initially the system runs without SAPF. The source current is equal to the load current. At t=0.1 sec, SAPF is added to the circuit. Now the source current becomes sinusoidal as shown in fig 10(b) and the source current lies in phase with the source voltage as shown in fig 10(c). Fig 10(d) shows the dual inverter phase voltage. Case (b): sinusoidal voltage and change in load. (Figure 11(a) to (d))Here too supply voltage is assumed to be balanced sinusoidal voltage. Initially the system runs without SAPF. The source current is equal to the load current. At t=0.1 sec, SAPF is added to the circuit. Now the source current becomes sinusoidal as shown in fig 11(b) and the source current lies in phase with the source voltage as shown in fig 11(c). At t=0.4 sec additional load is connected and the effect is seen in fig 11(b) and (c). Still the system maintains unity pf. Fig 11(d) shows the dual inverter phase voltage.

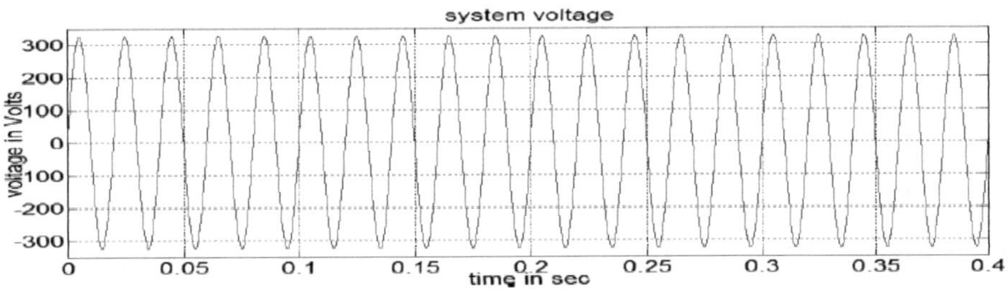

Figure. 10(a). Source phase voltage

IOP Conf. Series: Materials Science and Engineering **937** (2020) 012002 doi:10.1088/1757-899X/937/1/012002

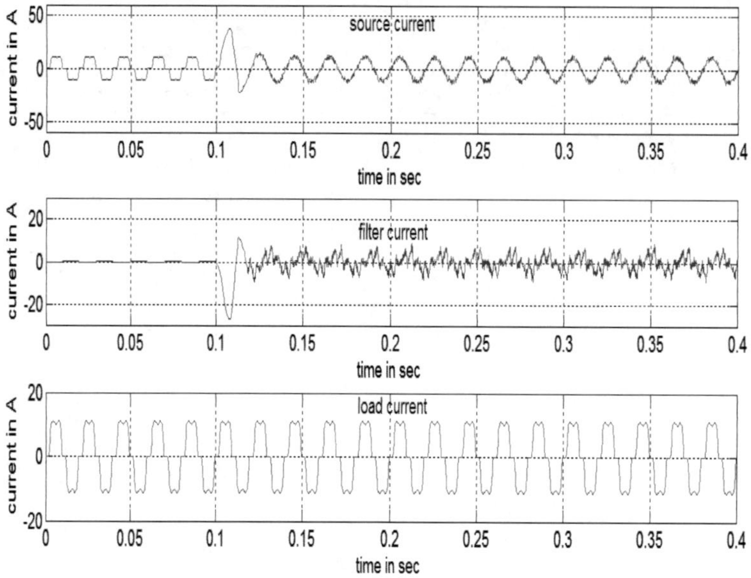

Figure. 10(b). Source, filter and load current

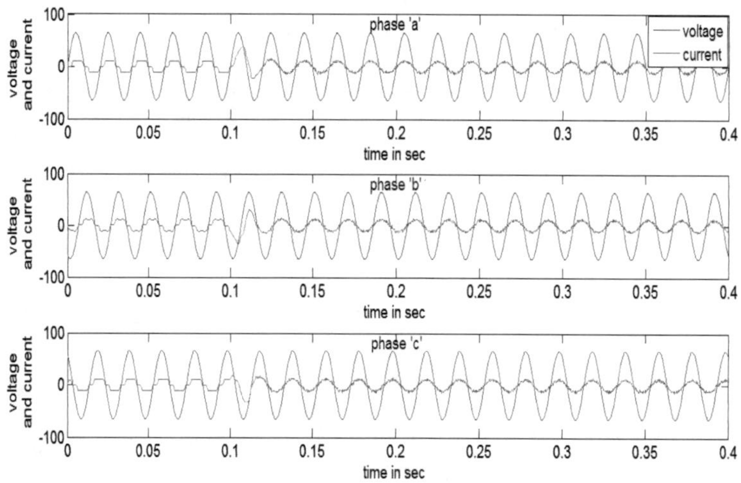

Figure. 10(c). Power factor

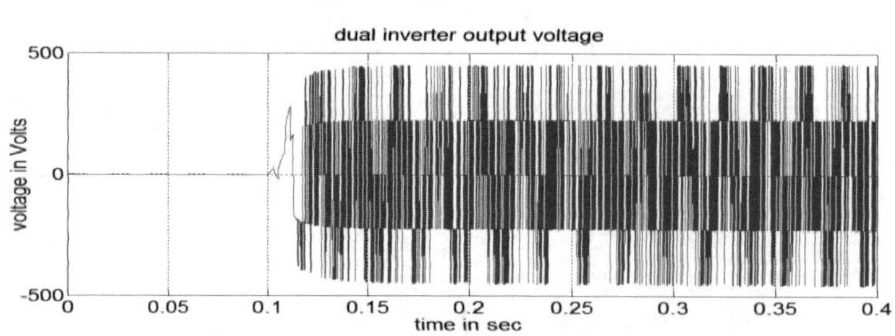

Figure. 10(d). Dual inverter phase voltage

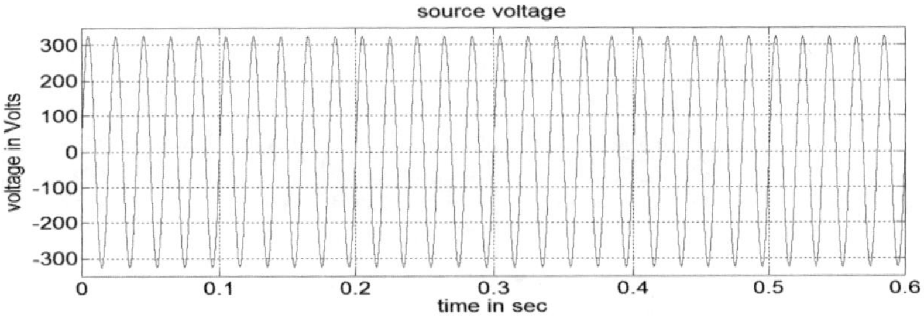

Figure. 11(a). Source phase voltage

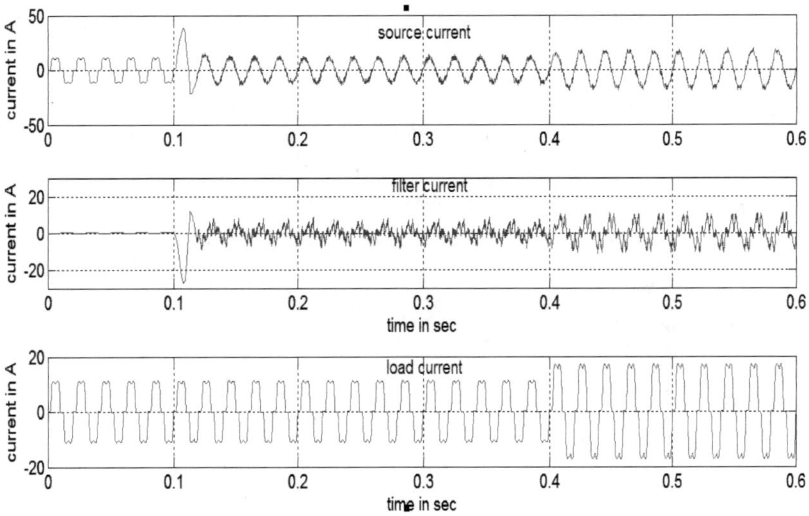

Figure. 11(b). Source, filter and load current

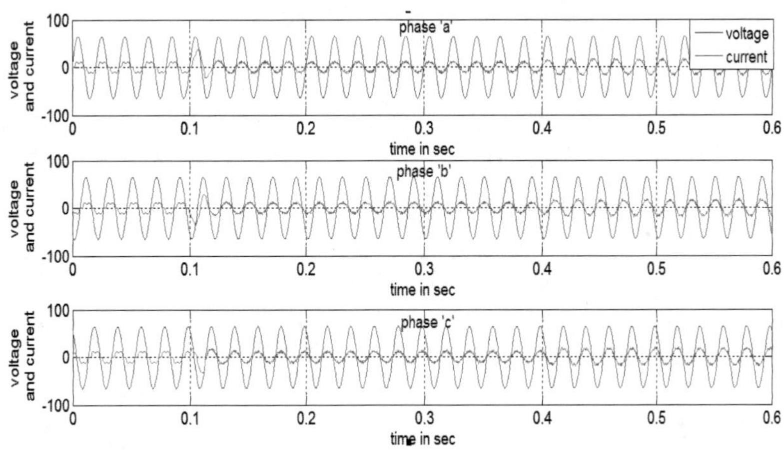

Figure. 11(c). Power factor

IOP Conf. Series: Materials Science and Engineering **937** (2020) 012002 doi:10.1088/1757-899X/937/1/012002

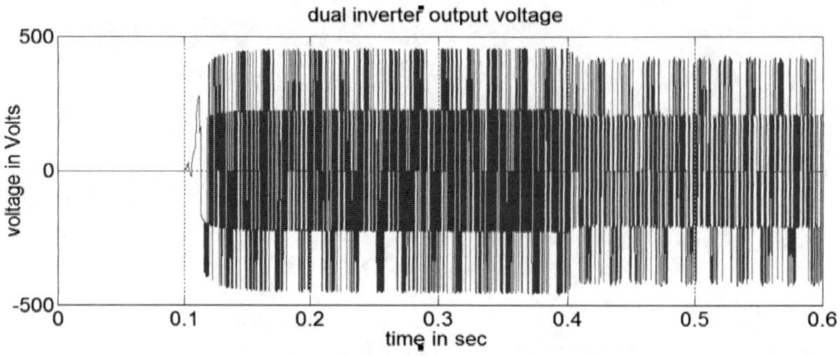

Figure. 11.(d). Dual inverter phase voltage

Case (c): non-sinusoidal supply and non-linear load. (Figure 12(a) to (d)).Here supply voltage is distorted with 5thand 7thharmonic components of voltages added to the fundamental. Initially the system runs without SAPF. The source current is equal to the load current. At t=0.1 sec, SAPF is added to the circuit. Now the source current becomes sinusoidal as shown in fig 12(b) and the source current lies in phase with the source voltage as shown in fig 12(c). Fig 12(d) shows the dual inverter phase voltage.

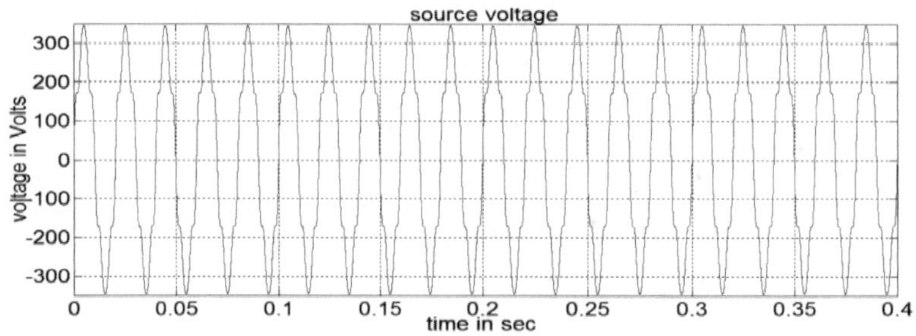

Figure. 12.(a). Source phase voltage

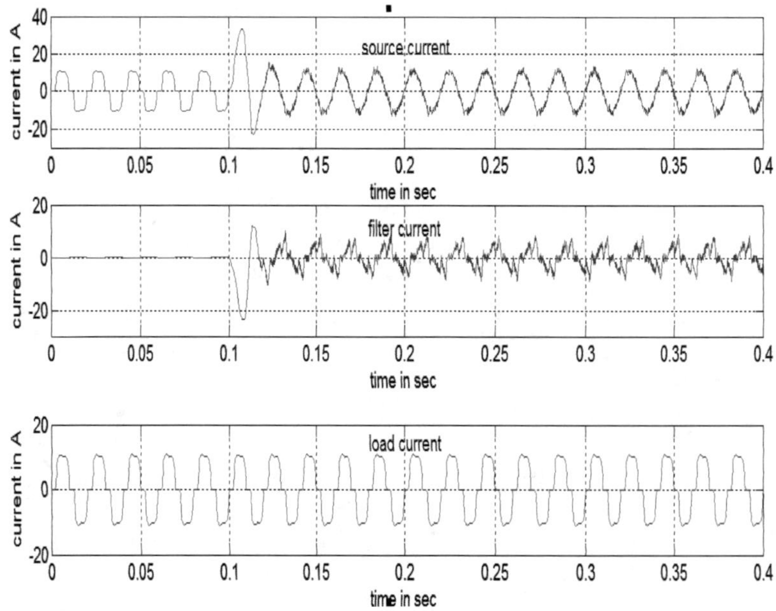

IOP Conf. Series: Materials Science and Engineering **937** (2020) 012002 doi:10.1088/1757-899X/937/1/012002

Figure. 12.(b). Source, filter and load current

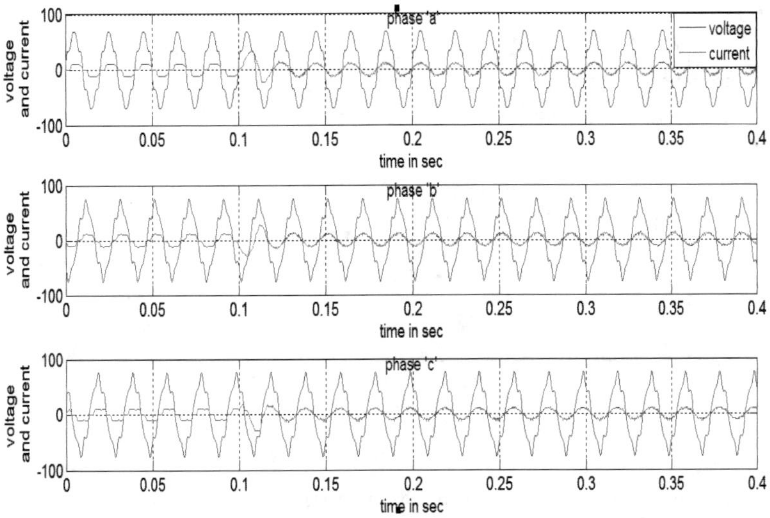

Figure. 12.(c). Power factor

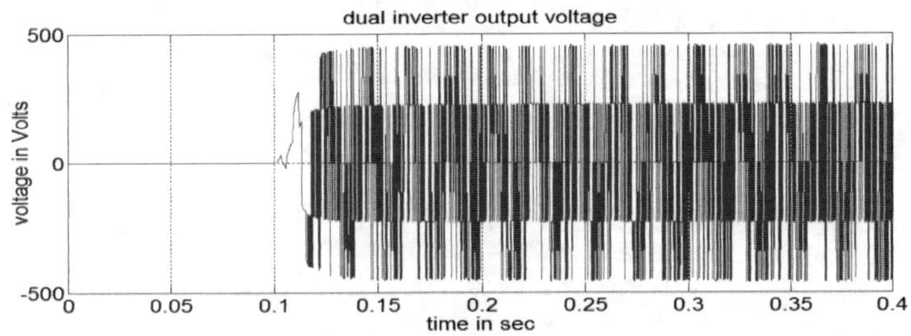

Figure. 12.(d). dual inverter phase voltage

Figures 13(a) and (b) shows the load current and source current THD respectively under balanced supply conditions.

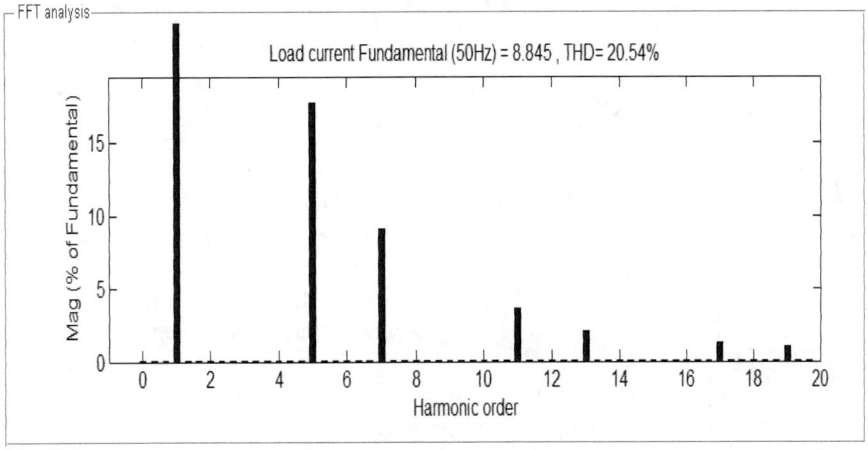

Figure. 13.(a). Load current THD under balanced supply voltage

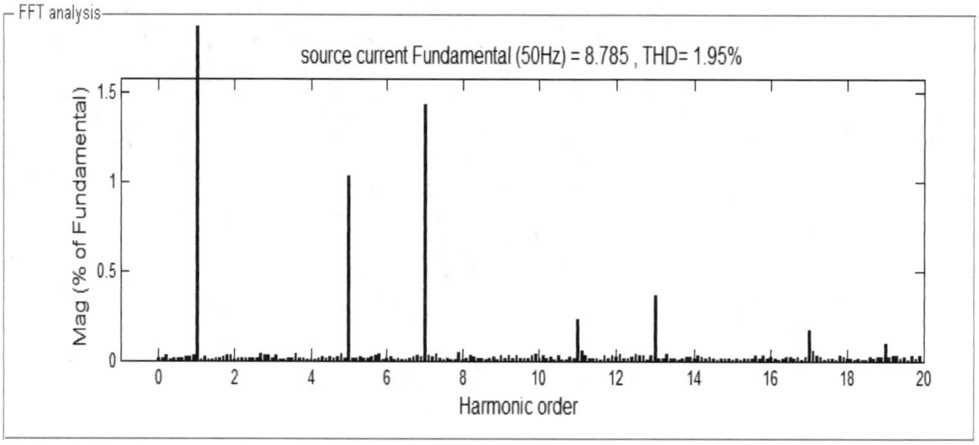

Figure. 13.(b). Source current THD under balanced supply voltage

6. Conclusion

The system power factor is kept near unity in all conditions. The system performance is also verified for different load conditions to check the dynamic behavior of the system. The implementation through digital controller is simple as SVPWM technique is used to drive the dual inverter.

References

[1] Ali, A., 2011, Active power filter based on cascaded transformer multilevel inverter. (IJAEST) international journal of advanced engineering sciences and technologies Vol. No. 7, Issue No. 2, 313 –318.

[2] R. Manivasagam and V. Dharmalingam, 2014, Power quality problem mitigation by unified power quality conditioner: An adaptive hysteresis control technique. International Journal of Power Electronics, vol. 6, pp. 403-425, Available at: https://doi.org/10.1504/ijpelec.2014.067442.

[3] Baiju, M.R., Mohapatra, K.K., Kanchan, R.S., and Gopakumar, K., 2004, A dual two-level inverter scheme with common mode voltage elimination for an induction motor drive. IEEE Trans. Power Electronics, 19, (3), pp. 794–805.

[4] Raghavi, R ,Manivasagam.R,2015, Modeling of a Grid Connected New Energy Vehicle Charging Station. International Journal of Applied Engineering Research (IJAER), Volume 10, Number 20 , Special Issues, pp. 15870- 15875.

[5] Jalal, A., Nasser, T. and Ali A., 2011, Damping of low frequency oscillations in power system with Neuro-Fuzzy UPFC controller. 10th International Conference on, Environment and Electrical Engineering (EEEIC), pp. 1-5.

[6] R. Manivasagam, P. Parthasarathy, and R. Anbumozhi, 2019, Robust analysis of T-S fuzzy controller for nonlinear system using H-infinity. Advances in Intelligent Systems and Computing, vol. 949, pp. 643-651, Available at: https://doi.org/10.1007/978-981-13-8196-6_56.

[7] Patidar R. D. and Singh, S. P., 2010, Digital signal processor based shunt active filter controller for customer-generated harmonics and reactive power compensation. Electric Power Components

and Systems, Vol. 38, pp. 937-959.
[8] D. Aarthi and R. Manivasagam , 2017, Design of UPFC using ten switch converter with switch reduction.International Journal for Scientific Research & Evelopment, vol. 5, pp. 490-493.
[9] Santiago, A. and Verneet, 2009, 35th Annual Conference of IEEE Industrial Electronics, IECON '09, pp. 3211 – 3216.
[10] D Aarthi and R.Manivasagam,2017, "Design and Implementation of UPFC Using Ten Switch Converter With Switch Reduction", International Journal Of Engineering And Computer Science, vol.6,Issue.6,pp. 21861- 21869.
[11] Baiju, M.R., Somasekhar, V.T and Gopakumar, K., 2004, Dual two level inverter scheme for an open end winding induction motor drive with a single DC power supply and improved DC bus utilization. IEE Proc., Electonics. Power Appl., 151, (2), pp. 230–238.
[12] Reddy, Satheesh, G., B. T. and sai babu, C.H., 2011, Three-level voltage generation for Dual Inverter fed open end winding Induction motor drive. International Journal of Engineering Science and Technology (IJEST), Vol. 3 No. 5, pp. 3892-3991.
[13] Manivasagam, R. and Prabakaran, R. 2020 , Power quality improvement by UPQC using ANFIS-based hysteresis controller, Int. J. Operational Research, Vol. 37,No. 2, pp.174–197.

IOP Conf. Series: Materials Science and Engineering **937** (2020) 012003 doi:10.1088/1757-899X/937/1/012003

The influence of energy production from renewable energy sources on energy efficiency and total factor productivity

S. A. Balashova, K. G. Gomonov, V. M. Matyushok

Department of economic and mathematical modelling, International Center for Emerging Markets Research, Peoples' Friendship University of Russia (RUDN University). 6 Miklukho-Maklaya Street, Moscow, 117198, Russian Federation

Abstract. The article analyzes the international experience of improving energy efficiency due to increasing the level of energy production from renewable sources and technological changes. Empirical verification of the related hypothesis has been carried out for the USA, China, India, Germany and Russia using econometric tools. The necessity of the transition to a new energy structure based on the active introduction of energy-efficient technologies, renewable energy sources, integrated systems and intellectual energy services based on an open grid architecture is substantiated. The posed hypotheses confirm the following: the production of energy from renewable sources is increasing in all countries, not only in the countries with the lack of fossil energy sources (Germany, India, China), but also in the USA and Russia; the growth of energy consumption caused by the growth of the economy and the population is humped due to the introduction of energy-efficient technologies and due to an increase in the share of renewable energy; growth in the share of renewable sources in energy consumption and total factor productivity increases energy efficiency in terms of GDP per unit of energy used in Germany, India and Russia.

1. Introduction

With an increase in the population of the Earth to 10 billion by 2050 and further urbanization, the problems of providing the population not only with food and water but also with electricity become more acute. Currently, cities consume about 75% of all energy resources with a specific gravity of the urban population of about 53%. According to Schneider Electric forecasts, the share of the urban population will increase to 86% by 2050, which will lead to an increase in energy consumption by 56-58%. Besides, almost 1.3 billion people who did not have access to electricity before will show an additional increase in demand for electricity. At the same time, modern economies and traditional energy are faced with the problem of low energy efficiency, which is basically of a different nature: depreciation of equipment and infrastructure, high losses in grids and environmental pollution.

A key role in improving energy efficiency is played by modern concepts of the development of the electric power industries, which involve greater automation, the development of smart grids and the introduction of renewable energy sources. The emergence of new energy development vectors is dictated by the inextricable link of the industry with economic growth opportunities. Today, the traditional way of developing the energy industry by increasing capacities, expanding the quantitative composition of energy equipment and increasing resource extraction is inferior to an intensive, innovative and breakthrough development path through the active introduction of energy-efficient equipment, advanced energy and information and communication technologies, renewable energy sources, integrated systems and smart energy services built on open network architecture and marks the beginning of the transition to the next energy structure.

IOP Conf. Series: Materials Science and Engineering **937** (2020) 012003 doi:10.1088/1757-899X/937/1/012003

New technologies are already yielding tangible results. For example, Germany's GDP for 1990-2016 increased in constant prices by 48.4%, and annual electricity consumption - only by 10.3%. The share of electricity production from primary energy resources (hydrocarbons) decreased from 68.6% to 53.9%, while the energy intensity of GDP decreased by 1.6 times (World Bank, 2020). The indicators for other countries are not so visible, but the trend towards increasing energy efficiency through the introduction of new technologies and renewable energy has been emerging in the last decade. In Russia, the Presidential Council for Modernization and Innovative Development of the Economy in early 2017 approved the EnergyNet roadmap, which provides for specific measures to develop the smart energy market until 2035.

Productivity is a measure of output divided by a measure of input. Following the neoclassical approach, aggregate output (measured as GDP at country-level) growth can be decomposed into contributions from aggregate capital input (K), aggregate labour input (L) and aggregate total factor productivity (TFP) growth as:

$$\Delta\, Log\, GDP = \nu_K \Delta\, Log\, K + \nu_L \Delta\, Log\, L + \Delta\, Log\, TFP \qquad (1)$$

where ν_K is the output elasticity of capital and ν_L is the output elasticity of labour. Under the assumption of constant returns to scale $\nu_K + \nu_L = 1$. TFP is derived as a residual after the contributions of *K* and *L* are deducted from GDP growth.

TFP can be regarded as the measure of productivity, which considers not only labour as an input but also physical, human, and other intangible capital. TFP is driven by non-input factors such as technological progress that are not tied to explicit input usage (Solow, 1957). In theory, TFP growth captures technical change and overall efficiency. A review of the literature and TFP growth data analysis suggest there are different key drivers of TFP growth in different groups of countries. Public policy toward R&D, to boost innovation in the private sector, is found to be among these drivers in OECD countries (Duverger & van Pottelsberghe de la Potterie, 2011; Guellec & Van Pottelsberghe De La Potterie, 2003). The effect of technological changes on TFP growth is examined in (Heshmati & Kumbhakar, 2014). The impact of ICT on TFP through an integrative framework of ICT-induced externalities and ICT-leveraged innovations is assessed in (Chou, Hao-Chun Chuang, & Shao, 2014). For emerging economies and oil-exporting countries, such factors as FDI also affect economic growth and productivity (Balashova, Revinova, & Lazanyuk, 2016; Bozkurt, Erdem, & Eroğlu, 2015).

Another measure of productivity is GDP per unit energy use. This indicator is reported by the World Bank and used as a measure of energy efficiency of the economy. The inverse ratio (energy consumption per unit of GDP) is called energy intensity. It tends to decline in nearly all regions of the world (*International Energy Outlook 2016*, 2016)

We assume that TFP growth can affect the energy efficiency of the economy. A possible channel is the use of more energy efficiency technologies. On the other hand, for the fast-growing economies such as India and China, the growing market boost TFP growth and technological changes, which are not necessarily connected with energy-saving technologies. Manufacturing-focused economies tend to use more energy per dollar of GDP than service-focused economies. Climate and distance between urban areas also influence the energy efficiency of the economy.

The purpose of the analysis is to search for empirical evidence of the contribution of energy from renewable resources and productivity growth to improving energy efficiency in several countries using econometric technique. The answer to this question is not apparent, since energy efficiency is growing unevenly and, in addition to technical innovations and investments, an active program at the state level is required (Allcott & Greenstone, 2013).

The following countries were selected for analysis: Germany as the economic leader of the EU and one of the leaders in developing renewable energy; the USA and China being the locomotives of the global economy and significant investors in Smart Grid technologies; India as one of the fastest-growing developing countries, one of the world leaders in wind energy (Dent, 2013), actively increasing investments in wind and solar generation. Assessments were carried out for Russia as one of the global leaders among oil and gas producers.

2. Methodology and data

IOP Conf. Series: Materials Science and Engineering **937** (2020) 012003 doi:10.1088/1757-899X/937/1/012003

To assess the impact of renewable energy on energy efficiency in different countries, we put forward the following hypotheses:

2.1. Hypothesis 1.
The production of energy from renewable sources not taking into account hydropower has been multiplying over the past 25 years. However, the growth rate varies in different countries depending on whether the country is an importer or exporter of energy resources and the level of support for "green energy" at the government level.
A trend model is evaluated to obtain empirical confirmation of hypothesis 1.

$$RNWX\_PROD_{it} = A_i e^{\gamma_i t} v_{it} \tag{2}$$

Here $RNWX\_PROD_{it}$ is the production of electricity from renewable sources (except for hydropower) in the country i for a period of time t. A_i and γ_i - model parameters specific to each of the countries in question. v_{it} - is a random component of the model, meeting the OLS regression assumptions. The parameter γ_i (for small values) is interpreted as the growth rate of renewable energy. To obtain estimates of the parameters, equation (2) is transformed to a linear form. The assessment is made over a pool of countries for the whole period 1990-2017, assuming the parameters of equation (2) to be specific for each country. We assume that the growth rate can be expressed as a linear functions of dummy variables Dexp and Dsup

$$y_{it} = y_0 + \delta_1 Dexp_{it} + \delta_2 Dsup_{it} \tag{3`}$$

Where $\mathrm{Dexp_{it}} = 1$ to energy-rich countries (such as Russia and USA) and 0 otherwise. $\mathrm{Dsup_{it}} = 1$ for countries with intensive government support for low-carbon energy supply (such as Germany and India).

2.2. Hypothesis 2.
Energy use per capita tends to decrease with the increase of renewable energy production.
A model is estimated of the dependence of the amount of energy consumed on the population, GDP and the share of electricity received from renewable energy sources to test this hypothesis

$$ENG\_USE\_CAP_{it} = \alpha_{1i} RNEWX_{it} + \alpha_{2i} GDP\_CAP_{it} + \alpha_{3i} + \varepsilon_{it} \tag{4}$$

Here, ENG_USE_CAP is the natural logarithm of energy used from all sources (kg of oil equivalent) per capita, GDP_CAP is the natural logarithm of GDP per capita (in constant US dollars 2010). $RNEWX$ is the share of energy from renewable sources in the total energy consumption from all sources. ε - random component. The parameters α_{2i} is the elasticities of energy consumption by GDP per capita in country i, α_{1i} shows the growth rate of energy consumption with an increase of 1 pp in $RNEWX$ while controlling other variables (Ayvazyan, 2014).

2.3. Hypothesis 3.
An increase in total factor productivity impacts on energy efficiency, due to the implementation of modern energy-saving technologies, complementing the impact of renewable energy consumption (% of total final energy consumption).
The depended variable is GDP per unit of energy use (constant 2011 PPP $ per kg of oil equivalent). The explanatory variables are the share of renewable energy in total energy consumption $RNWZ_{it}$ and the total factor productivity TFP.
Thus, to test hypothesis 3, we evaluate the model

$$GDP\_ENG_{it} = \beta_{i1} + \beta_{i2} RNWZ_{it} + \beta_{i3} TFP + \varepsilon_{it} \tag{5}$$

The estimation is carried out for each country individually by the least-squares method with the correction of standard errors according to the Newey-West method due to autocorrelation in the residuals.
Here GDP_ENG_{it} is the natural logarithm of GDP per unit of energy spent in the current period t for the country i, ε - random component, subject to autocorrelation.

We retrieve data for TFP growth from the Total Economy Database (TED). TED is a comprehensive database with annual data covering growth accounting variables for 123 countries in the world ("Total Economy Database," 2019). Data for the other time series are retrieved from the World Bank database and OECD.

3. Empirical Analysis Results

3.1. Renewable energy growth

Renewable energy sources include the following types: solar energy, geothermal energy, wind energy, hydropower, ocean energy (thermal gradient, wave power and tidal power), biomass, draft animal power, firewood, peat, shale and tar sands. The most widely used for technological purposes is hydropower. According to Renewables 2017 Global Status Report for 2016, the share of renewable energy sources in the total final consumption in the world is more than 19.3%, and according to the most optimistic forecasts, by 2050 it can reach 35%. Renewable energy plays a vital role in the concept of the 17 Sustainable Development Goals for the next 15 years. One of these goals is "Sustainable Energy for All", which implies increasing efficiency and expanding the use of renewable energy sources.

Among the countries, the United States accounts for the enormous amount of electricity from renewable sources (excluding hydropower). Germany leads the EU in the share of electricity generated from renewable energy sources. Since 2000, China and India have been increasing their volumes and increasing the share of renewable energy sources in electricity production. In Russia, the volume and share of renewable energy sources are tiny, although, start growing since 2000s (Figure 1).

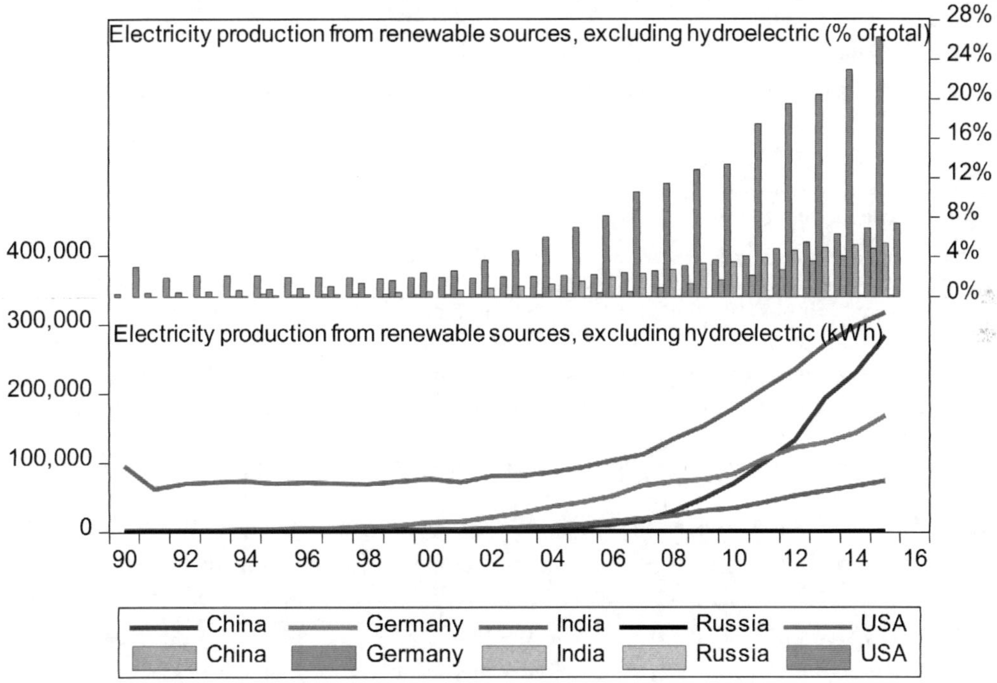

Figure 1. Production of electricity from renewable sources (excluding hydropower) (left axis, billion kW), the share of RES from the generation of total electricity (right axis, %). Source: Compiled by the authors based on the World Bank statistics.

Table 1. The estimated growth rate of electricity production from renewable sources

	Germany	USA	China	India	Russia
			1990-2015		
γ	0.20***	0.06***	0.31***	0.32***	0.13***
R^2			0.977		
			2000-2015		
γ	0.17***	0.10***	0.35***	0.22***	0.12***
R^2			0.99		

We conclude from table 1 a rather steady growth rate of electricity production from renewable sources, excluding hydroelectric, for all considered countries. The low growth rate for the USA is related to a relatively high production already in early 90-es and the effect of shale revolution (Huntington et al., 2020).
The estimation of the equation (2')

$$y_i = 0.30 - 0.2 * Dexp_{it}$$

supports the hypothesis that the growth rate of electricity production from renewable sources is very high for countries with insufficient fossil resources and significantly low for energy-rich countries. However, for the considered group of countries we cannot verify the assumption that government support has a particular impact on the growth rate of the production, because according to the International Energy Agency, all the countries considered have government programs to support green energy (International Energy Agency, 2017).
It should be noted that according to New Energy Outlook 2017 forecasts, more than 70% of investments in the energy sector will go to renewable energy production, and the amount will increase to $ 400 billion annually by2040. Among the leaders in this segment of the industry will be Germany, the USA, China and India.
Russia has also joined the race for the leadership in renewable energy issues. Investments allocated to R&D in the energy sector are yielding results. In addition to equation (1), we estimated the ADL (1,1) model (5) for Russia

$$Log(RNWX\_PROD)_t = \alpha_1 + \alpha_2 Log(RNWX\_PROD)_{t-1} + \alpha_3 \text{Log}(RD\_EG)_{t-1} + \varepsilon_t \quad (6)$$

Here RD_EG are the R&D expenditures in the energy sector (source - OECD database[1]). The estimation was carried out using the least-squares method with correction of autocorrelation in the residues according to the Newey-West method (Greene, 2008) for the period from 1995 to 2014.

Estimation of the parameter $\alpha_3 = 0.7$ (significant at all levels) suggests that in the short term, an increase of 1% in R&D expenditures in the energy sector corresponds to a 0.7% increase in electricity production from renewable sources in the next period. In the long run, the return on growth in R&D expenditures by 1% increases to 2.5% (parameter α_2 can be interpreted as 1 minus the speed of adjustment, the estimation $\alpha_2 = 0.715$ is significant at all levels). It should be recognized that in Russia, R&D expenditures in the energy sector are small compared to the other countries under consideration. For example, in 2012, according to OECD estimates, R&D expenditures in the energy sector from all sources of financing were only $1,551 million (constant 2010 international $) in Russia (of which $360 million were the state-funded), and in Germany, only from state sources the amount of 1,184 million dollars (international $²) was spent on R&D in the energy sector. Nevertheless, R&D expenditures in this area indicate attention to the problems of energy, both from the side of the business and from the government.

3.2. Energy consumption and renewable energy

In the world, energy consumption is growing with an increase in world GDP and a growing world population. According to the World Energy Agency, the demand for electricity will increase by 58%, or the average growth will be 2% per year by 2040.

The dynamics of energy consumption are multidirectional in the observed countries (Fig. 2). Consumption growth, both gross and per capita, is observed in China, although the growth rate of per capita consumption is declining. A similar situation is for India, though the growth rate is lower than in China. The amount of energy consumed per capita in India is four times less than in China and ten times less than in the United States. In Germany and the United States, there has been a trend towards lower energy consumption (Figure 2).

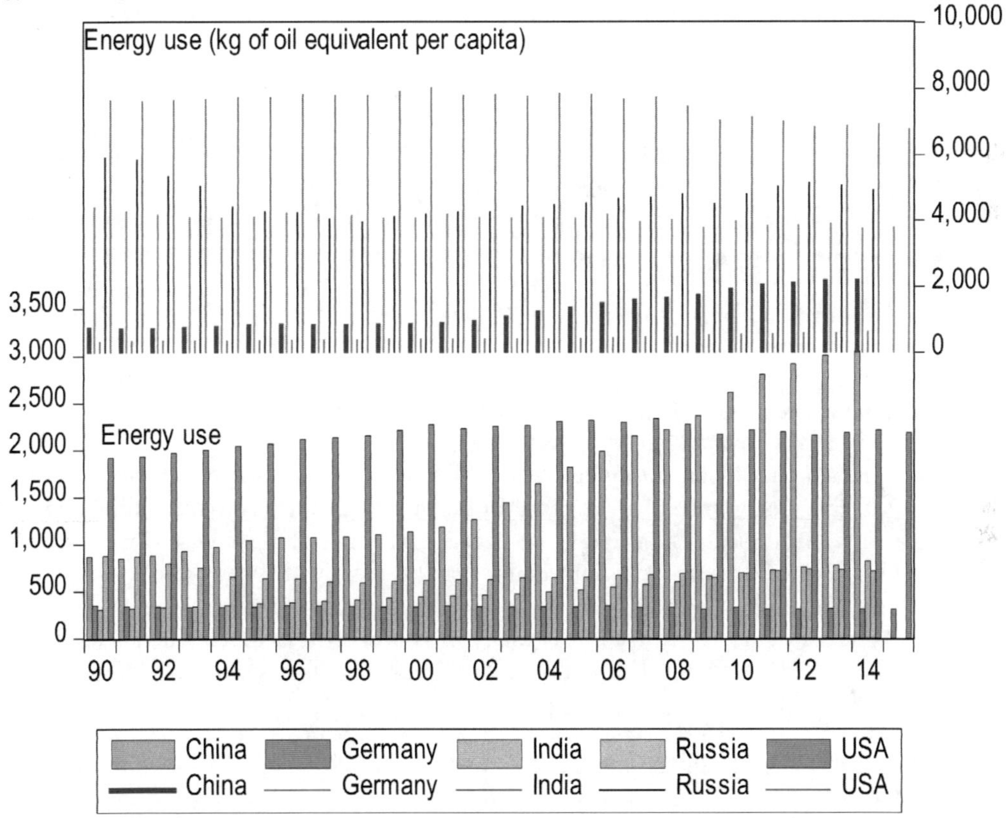

Figure 2. Energy use (kg of oil equivalent), left axis, and energy use (kg of oil equivalent per capita), right axis. Source: Compiled by the authors based on World Bank statistics.

The time series in equation (3) is proved to be I(1)), based on the ADF test (Dickey & Fuller, 1981). Thus, we assess parameters of equation (3), using first differences of the logs to eliminate linear trends (Fig.2). Consequently, we estimate the impact of changes in the share of renewable sources in electricity production on the growth rate of energy use per capita, controlling for the growth rate of GDP per capita (Table 2).

IOP Conf. Series: Materials Science and Engineering **937** (2020) 012003 doi:10.1088/1757-899X/937/1/012003

Table 2. The estimated impact of changes in the share of renewable sources in electricity production on the growth rate of energy use per capita based on equation (3)

	Germany	USA	China	India	Russia
Dependent variable $\Delta Log(ENG\_USE\_CAP)$					
$\Delta(RNEWX)$	-0.01**	-0.0005	0.0004	-0.005*	-0.01**
$\Delta Log(GDP\_CAP)$	0.42*	0.82***	0.99**	0.5***	0.45***
R^2	0.22	0.57	0.20	0.35	0.80

Variables are the log differences of the initial time series except for RNEWX, which is the difference of the share of renewable sources in electricity production.

From the estimation results, we can conclude that an increase in the renewable electricity output (% of total electricity output) decreases energy use per capita, controlling for GDP per capita in Germany, India and Russia. In the case of the USA, the sign of the coefficient is what is expected; however, the estimation is insignificant. As can be seen from Fig.2, the energy consumption is very high in the USA, and tend to decrease only since late 2000-s.

In China, energy consumption per capita increases during the whole considered period despite the growing rate of renewable energy production. The share of renewable sources in China is rather small (Fig.1) and do not impact on overall energy savings.

In the ranking of the American Council for an energy-efficient economy in terms of energy efficiency, the United States is given the thirteenth place in the world[3]. In terms of energy consumption per capita, the United States (like Australia and Canada) is one of the most wasteful in the world - 6.8 TOE against 3.3 TOE for the EU. The US federal government allocates less money in comparable terms for energy-saving events than the EU[4]. However, the US has great potential for energy efficiency growth. According to a McKinsey study, the installed capacity of energy-efficient technologies in the United States will be able to reduce energy consumption by 23% by 2020, according to Rhodium Group estimates, the widespread introduction of energy-saving technologies and equipment can reduce energy consumption in the USA by 42-59%. Note that since 2009, a large-scale program for the creation of "smart energy grids of the 21st century" has been carried out, which provides for a series of measures to modernize trunk transmission lines and distribution networks using advanced electronic control and measuring equipment, communications and computer technologies. According to the calculations of the National Laboratory of Energy Technologies, the implementation of the program involves attracting investments of $ 17-24 billion annually for 20 years (Dmitriev, 2014).

3.3. Energy efficiency and total factor productivity

Various transmission channels exist through which total factor productivity may have an impact on energy efficiency. TFP reflects the spillover of technological changes and the use of modern technologies and efficient business processes in the production of goods and services. The link between renewable energy and energy efficiency is not so strong. However, the use of solar panels by

IOP Conf. Series: Materials Science and Engineering **937** (2020) 012003 doi:10.1088/1757-899X/937/1/012003

householders proved to decrease energy consumption in USA (Balashova et al. 2020, Ratner, S., & Ratner, P. 2017). The results, presented in Table 2, also speaks in favour of the link between the share of renewable energy and energy use, controlling for GDP. So, we can expect, that controlling for technological changes an increase in the share of renewable sources in energy consumption helps to increase the energy efficiency (in terms of GDP per unit of energy use).

To estimate the equation (4) we first check the time series for weak stationarity using the ADF test. To avoid the spurious regression, we use the first differences of the natural logarithm of the series GDP_ENG and the first difference of the series RNWZ. Note, that data on TFP is retrieved from TED as a difference between GDP growth and capital and labour growth and thus measure the TFP growth rate. Results of estimation of modified equation (4) are reported in Table 3.

As seen from Table 3, hypothesis 3 is supported for Germany and India. India is fighting to decrease its colossal energy losses, which drops from 28% in 2000 to 19% in 2014, but still very large. TFP growth is associated with the energy efficiency increase in this country. In Germany, we also see the upward trend in energy efficiency, and the impact of TFP growth and renewable energy consumption (% of total final energy consumption) on energy efficiency is positive and significant.

Table 3. The estimated impact of changes in the share of renewable sources in energy consumption and total factor productivity on the growth rate on GDP per unit of energy use based on equation (4)

	Germany	**USA**	**China**	**India**	**Russia**
Dependent variable $\Delta Log(GDP\_ENG)$					
$\Delta(RNWZ)$	0.023**	0.002	0.03***	0.02***	0.035
$\Delta Log(TFP)$	0.68**	0.02	0.32	0.87***	0.63***
R^2	0.33	0.005	0.43	0.70	0.68

GDP per unit of energy use is increasing in China, and it is in line with the increase in electricity production from renewable sources, excluding hydroelectric (% of total). However, the TFP growth is negative in China after the global financial crisis, and so its impact on energy efficiency for the period 1990-2017 is not significant (Shen, Y., 2020).

The growth rate of energy efficiency in the USA is very volatile and cannot be associated with renewable energy consumption (as well as production) or TFP growth rate.

The estimates obtained accurately capture the trend in energy efficiency over the past 25 years. In Fig. 3 shows the actual and estimated values of this indicator.

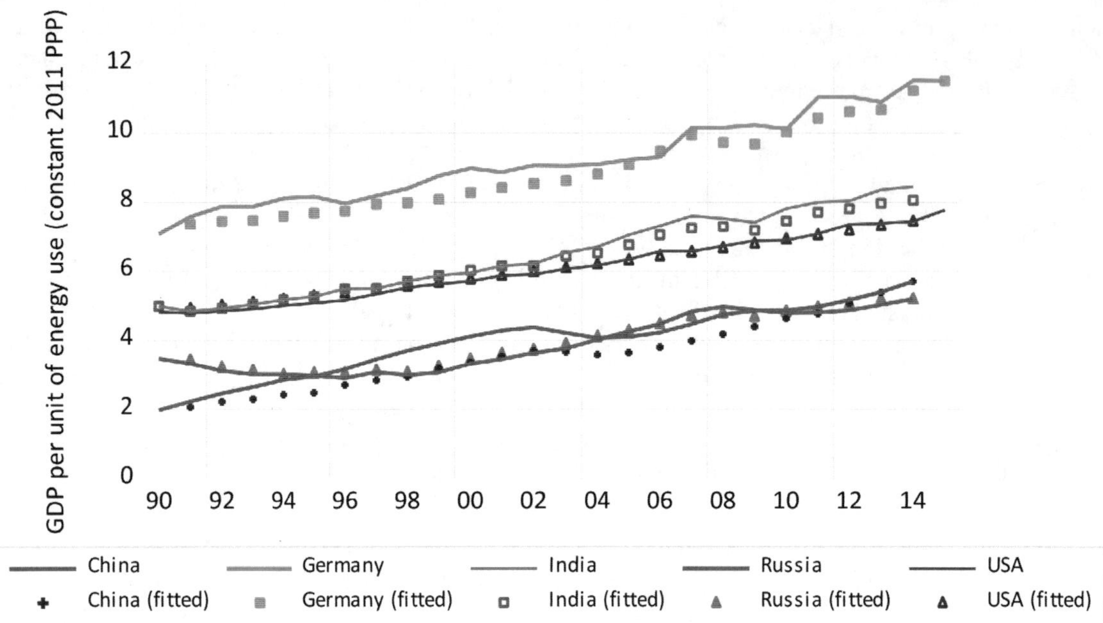

Figure 3. GDP per unit of energy use (in 2011 prices for PPP): observed and fitted values.

Source: Compiled by the authors based on World Bank statistics and estimates.

According to Fig. 3, Germany has the highest energy efficiency rate among the countries in question. Germany is actively increasing the share of energy from renewable energy sources in its final consumption. In 1990, this value was only 2% and increased by more than seven times to 14.6% in 2015 (according to Eurostat). Note that according to Strategy 2020, the average target for EU countries is 20%. In the framework of this strategy, each country has its target indicator depending on the initial values, the potential for using renewable energy sources and the level of economic development; for Germany, the target indicator of the share of energy received from renewable energy sources in the final consumption is 18% by 2020.

In the USA, energy efficiency is not as high as in Germany, and at comparable prices, it is slightly lower than in India. The increase in energy efficiency in the United States only in the last decade can be attributed to an increase in the share of renewable energy sources, which reached almost 8% in production (excluding hydropower, Fig. 1) and almost 9% in total energy consumption. In India and China, energy efficiency tends to increase in the period under review, which can be primarily associated with the high growth rates of the economies of these countries. However, the contribution of RES to energy efficiency growth has not been recorded, despite the increased share of these sources in electricity production (Fig. 1). In the final energy consumption in China and India, the share of fossil fuel energy consumption increased significantly (88% in 2014 versus 76% in 1990 in China, in India growth to 73% in 2014 from 54% in 1990), respectively, the share of renewable energy decreased.

For Russia, the energy efficiency indicator has a dynamics similar to the dynamics of GDP (energy consumption varied to a much lesser extent than GDP, and its ratio changed mainly due to changes in the numerator {Formatting Citation}. The share of energy from renewable energy sources in the total consumption is tiny (about 4%), practically does not vary and does not affect the dynamics of the studied indicator.

4. Conclusion

The production of energy from renewable sources is increasing in all countries, not only in the countries with the lack of fossil energy sources (Germany, India, China), but also in the USA and Russia. The performed analysis based on the available statistical data allows us to conclude that energy production from renewable sources in the countries is growing at different rates due to the level of state support.

The growth of energy consumption caused by the growth of the economy and the population is humped due to the introduction of energy-efficient technologies and due to an increase in the share of renewable energy. An increase in the renewable electricity output (% of total electricity output) decreases energy use per capita, controlling for GDP per capita in Germany, India and Russia. However, it is not the case for the USA and China for different reasons.
Growth in the share of renewable sources in energy consumption and total factor productivity increases energy efficiency in terms of GDP per unit of energy used in Germany, India and Russia. This hypothesis is not supported for China and the USA.

Funding
This article has been prepared with the support of the Economic Faculty of RUDN University.

References

[1] Aivazyan S.A. Metody ekonometriki [Methods of Econometrics]. - Infra-M, 2014. ISBN 978-5-9776-0153-5
[2] Allcott, H., & Greenstone, M. (2013). Is There an Energy Efficiency Gap?. Energy Efficiency, 133-161. doi: 10.1016/b978-0-12-397879-0.00005-0
[3] Balashova, S., Revinova, S. and Lazanyuk, I. (2016) 'Total Factor Productivity Growth and Innovation Policy Tools: Empirical Study', in SGEM International Multidisciplinary Scientific Conference on Social Sciences and Arts, pp. 917–922. doi: 10.5593/sgemsocial2016B25
[4] Bozkurt, Ö. G., Erdem, C., & Eroğlu, L. (2015). Identifying the factors affecting the economic growth of oil-producing countries. International Journal of Trade and Global Markets, 8(2), 97–111. https://doi.org/10.1504/IJTGM.2015.069421
[5] Bozkurt, Ö., Erdem, C., & Eroğlu, l. (2015). Identifying the factors affecting the economic growth of oil-producing countries. International Journal Of Trade And Global Markets, 8(2), 97. doi: 10.1504/ijtgm.2015.069421
[6] Chou, Y., Hao-Chun Chuang, H., & Shao, B. (2014). The impacts of information technology on total factor productivity: A look at externalities and innovations. International Journal Of Production Economics, 158, 290-299. doi: 10.1016/j.ijpe.2014.08.003
[7] Chou, Y.-C., Hao-Chun Chuang, H., & Shao, B. B. M. (2014). The impacts of information technology on total factor productivity: A look at externalities and innovations. International Journal of Production Economics, 158, 290–299. https://doi.org/10.1016/J.IJPE.2014.08.003
[8] Dent, C. (2013). Wind energy development in East Asia and Europe. Asia Europe Journal, 11(3), 211-230. doi: 10.1007/s10308-013-0360-8
[9] Dickey, D. A., & Fuller, W. A. (1981). Likelihood Ratio Statistics for Autoregressive Time Series with a Unit Root. Econometrica, 49(4), 1057. https://doi.org/10.2307/1912517
[10] Dmitriev S.S. Energeticheskaya strategiya B. Obamy: opora na innovatsii i tekhnologicheskoe liderstvo [B. Obama's energy strategy: reliance on innovation and technological leadership]. – M.: IMEMO RAN, 2014. 162
[11] Duverger, C., & van Pottelsberghe de la Potterie, B. (2011). Determinants of productivity growth: Science and technology policies and the contribution of R&D. EIB Papers, 16(1), 52–60. Retrieved from http://www.eib.org/attachments/efs/eibpapers/eibpapers_2011_v16_n01_en.pdf#page=54
[12] Gomonov, K., Balashova, S., Matyushok, V., Balashova, S., & Gomonov, K. (2019). ELECTRICITY CONSUMPTION FOR THE RUSSIAN ECONOMY: DOES SMART GRID MATTER? International Journal of Economic Policy in Emerging Economies, 12(4), 407–430. https://doi.org/10.1504/IJEPEE.2019.10024443
[13] Greene, W. H. (2008). Econometric Analysis. (P. Education, Ed.), Journal of the American Statistical Association (6th ed., Vol. 97). Pearson Prentice Hall. https://doi.org/10.1198/jasa.2002.s458
[14] Guellec, D., & Van Pottelsberghe De La Potterie, B. (2003). The impact of public R&D expenditure on business R&D*. Economics Of Innovation And New Technology, 12(3), 225-243. doi: 10.1080/10438590290004555

[15] Guellec, D., & Van Pottelsberghe De La Potterie, B. (2003). The impact of public R&D expenditure on business R&D. Economics of Innovation and New Technology.
[16] Heshmati, A., & Kumbhakar, S. (2013). A general model of technical change with an application to the OECD countries. Economics Of Innovation And New Technology, 23(1), 25-48. doi: 10.1080/10438599.2013.805918
[17] Heshmati, A., & Kumbhakar, S. C. (2014). A general model of technical change with an application to the OECD countries. Economics of Innovation and New Technology, 23(1), 25–48. https://doi.org/10.1080/10438599.2013.805918
[18] Huntington, H. G., Bhargava, A., Daniels, D., Weyant, J. P., Avraam, C., Bistline, J., … Victor, N. (2020). Key findings from the core North American scenarios in the EMF34 intermodel comparison. Energy Policy, 144, 111599. https://doi.org/10.1016/j.enpol.2020.111599
[19] International Energy Outlook 2016. (2016). Retrieved from www.eia.gov/outlooks/ieo/pdf/industrial.pdf
[20] Loschi, H., Leon, J., Iano, Y., Filho, E., Conte, F., Lustosa, T., & Freitas, P. (2015). Energy Efficiency in Smart Grid: A Prospective Study on Energy Management Systems. Smart Grid And Renewable Energy, 06(08), 250-259. doi: 10.4236/sgre.2015.68021
[21] Nosko V.P. Vvedenie v analiz vremennykh ryadov[Introduction to the analysis of time series]. - Moskva, 2002.
[22] Ratner, S., & Ratner, P. (2017). Developing a Strategy of Environmental Management for Electric Generating Companies Using DEA-Methodology. Advances in Systems Science and Applications, 17(4), 78-92. https://doi.org/10.25728/assa.2017.17.4.521
[23] Shen, Y., Yue, S., Sun, S., & Guo, M. (2020). Sustainable total factor productivity growth: The case of China. Journal of Cleaner Production, 256, 120727. https://doi.org/10.1016/j.jclepro.2020.120727
[24] Solow, R. M. (1957) 'Technical Change and the Aggregate Production Function', The Review of Economics and Statistics, 39(3), pp. 312–320.
[25] Solow, R. M. (1957). Technical Change and the Aggregate Production Function. The Review of Economics and Statistics, 39(3), 312–320.
[26] Svetlana Balashova, Svetlana Ratner, Konstantin Gomonov, Andrey Berezin. Modeling Consumer and Industry Reaction to Renewable Support Schemes: Empirical Evidence from the USA and Applications for Russia International Journal of Energy Economics and Policy, 2020, 10(3), 158-167.
[27] Balashova, S., Revinova, S., & Lazanyuk, I. (2016). Total Factor Productivity Growth and Innovation Policy Tools: Empirical Study. In SGEM International Multidisciplinary Scientific Conference on Social Sciences and Arts (pp. 917–922). https://doi.org/10.5593/sgemsocial2016B25
[28] Total Economy Database. (2019). Retrieved March 1, 2020, from https://www.conference-board.org/data/economydatabase/
[29] World Bank Open Data | Data. (2020). Retrieved 26 June 2020, from https://data.worldbank.org
[30] World Energy Outlook 2017 – Analysis - IEA. (2020). Retrieved 26 June 2020, from https://www.iea.org/reports/world-energy-outlook-2017

Fast and Integrated Chargers for Lithium Ion Batteries

M Sabarimuthu[1], N Senthilnathan[2], M Arun Kumar[3], M S Kamalesh[4], V Jayaprakash[5], R Srinivasan[6]

[1,4]AP, Department of Electrical and Electronics Engineering, Kongu Engineering College Perundurai, India
[2] Professor, Department of Electrical and Electronics Engineering, Kongu Engineering College Perundurai, India
[3] Assistant Professor, Department f EECE, GITAM University, Bengaluru campus, Bengaluru.
[5,6]UG Scholar, Department of Electrical and Electronics Engineering, Kongu Engineering College, Perundurai, India

eeesabari@gmail.com

Abstract. Electric vehicles were introduced in the year of 1908, where they have larger battery for power supply. Now- a- days the technique for charging e-bikes is done either by using grid supply or solar power. This paper says about charging circuit for batteries which uses both solar power and power from the grid to charge the electric vehicles. The charging with solar helps to reduce the emission from power grid. Ever-increasing demand for fuel supply, rising fuel prices, and increased environmental awareness among masses are paving the way for electric vehicles (EVs). Although, in recent years EV market has seen an exponential growth, one of the major challenges faced by this automotive/vehicular market. Now the Indian government had ordered to Indian Vehicle Association to launch the electric vehicles with in 2023. But the charging of electric Batteries has taken more time, which compare to fossil fuel and gas. To overcome this issue use of turbo charger, to reduce the charging time as well as self charging by solar during running time is proposed. The charger uses the source from EB grid and solar as well as at parking time. The results proves that proposed work have reduced the charging time of the lithium ion battery compared to the conventional charging technique and all under controlled temperature raise.

Key words: Electric Vehicles, Turbo charger, Lithium ion battery charger Start your abstract here…

1. Introduction

In the present scenario the usage of non-conventional fuels for vehicles are in great demand and increasing in price. To reduce the non-conventional fuel usage, electric vehicles are the trending technology for the conventional vehicle replacement, by using e-bikes emission problems will be reduced. Photovoltaic systems may be installed on rooftops to build charging stations. Solar canopies installed on parking lots make an excellent choice or solar-powered EV charging stations. Photovoltaic cells that convert are solar energy into required voltage to charge the battery. As far as electric vehicles are concerned lead acid and lithium ion batteries were most commonly used, Lead acid batteries are lower cost and have higher current carrying capacity [1]. They are heavier in weight and have only smaller life time while lithium ion batteries have lower weight, whereas they are higher cost. In this proposed method uses lithium ion battery for electric vehicles. Generally, grid supply is used to charge the batteries of the electric vehicle. The charging time is longer. This method using both grid supply and solar. Where supply for charging the battery is transmitted using both dc and ac converters. Where battery is charged either by grid or solar power [2-4]. The availability of both sources for a long duration is insufficient. So the modification technique is done by comparing both the grid supply and solar supply, the efficient supply is given to the battery. In this process only DC supply given to the components. The 230V AC supply from the power grid is rectified and given to the step down transformer which is converted into 12V. The proposed system is designed in such a way that it is easy to charge the battery. The rectifier is used to convert the AC into DC which is fed to the controller. Now calculating state of charge, depth of discharge and input temperature or measure. By using Arduino uno the outputs are displayed in voltage level indicator. Normally charging time for battery is above 7hours [5-7], where in this technique charging time is less than 1hours 30 minutes. Below charging circuit is designed, then the smart way methodology is proposed and the analysis of battery charging stated is done. In previous paper A three phase, 6 pulse SCR based rectifier topology is used to charge the battery [8] and constant current, It charges the battery upto 5.7% higher soc in short time compared to CC charging [9-20]. Constant voltage, constant current method is used to charge the battery [21]. But In this method constant voltage, different current is used to charge the battery and also determining soc and temperature of the battery [21-30]. It explore the optimal charging frequency in AC impedance analysis. The SRC proposed the optimal charging performance indicates the minimum-ac-impedance frequency fZ min [31-33]. Realistic battery module is used to charging the battery for minimize the charging time. Two types of system is used to charge the battery, one is on-board and other is off board [33].

2. Existing methods

In existing system, the battery has charging time with minimum of 6 to 7hrs. It causes major drawback for electric vehicle during emergency. And also in IC engine the diesel or petrol can fill with in 2 or 3 minutes during maximum of period [18-20].

Their major drawback in Electric vehicle has slow charging. Some methods are proposed as constant current, different voltage method used to charge the battery but life of the battery is reduced. Constant voltage, constant current method is used to charge the battery but charging time is high.

3. Problems identified

- The charging of battery is very slower
- The petrol engine causes air pollution

3.1. Objective

To reduce the charging time of batteries where used in electric vehicle by using fast charger to charge the battery and detecting the percentage and temperature of the battery and then input power will be change by using controller.

4. Block diagram

Figure 1 explains the block diagram of proposed work and it consists of EB source, Renewable energy, Battery, Battery level detector, Arduino, relay, Current controller. The controller detects the SOC and Temperature for the battery to charge. The SOC of the battery is detected using battery level detector and temperature of the battery to be measured using temperature sensor. Arduino plays a major role of the proposed work. It is used as controller of the work. In this paper lithium ion battery is used because lithium ion battery is familiar in EVs now days. During night hours renewable energy will not available, hence controller does the change of input power to EB source.

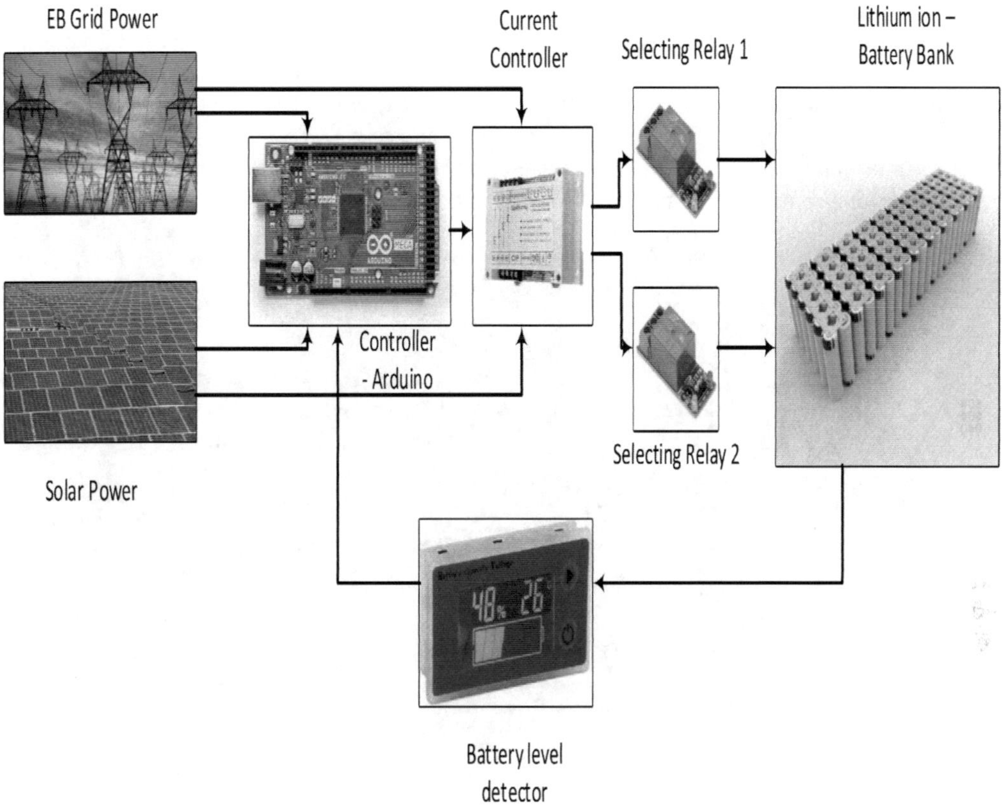

Figure 1. Block Diagram.

5. Components used

Table 1: components used in proposed work.

APPARATUS	RATING	QUANTITY
Solar panel, EB grid Source	(36V,8AH),(230v)	1,1
relay, battery level detector	(5v dc input),(12v-70v)	3,1
Step down Transformer	230v/12v-5A	5
Dc converter (or) Rectifier	AC-DC(output 13.6v,5A)	1
Lithium ion battery	36v-6A	1
Micro controller (ARDUINO)	Atmega	1
Current controller	5A, 3A	Each 1

The above table 1 gives the details of the components used in proposed work which includes source side and battery side specifications.

6. Battery specifications

Figure 2 shows three 36-volt lithium ion battery. Specification of the battery used as shown in table 2. The important things as, cells are connected in series connection to get the battery pack voltage as to meet the EV rated voltage. In this example the resulting pack voltage is 36 volts. In this paper 36V, 6Ah lithium ion battery is used. This assumes that individual batteries are the same capacity of battery pack. Do not mix and match different size of battery pack in the same batteries. Portable equipment needing higher voltages use battery pack with two or more cells connected in series.

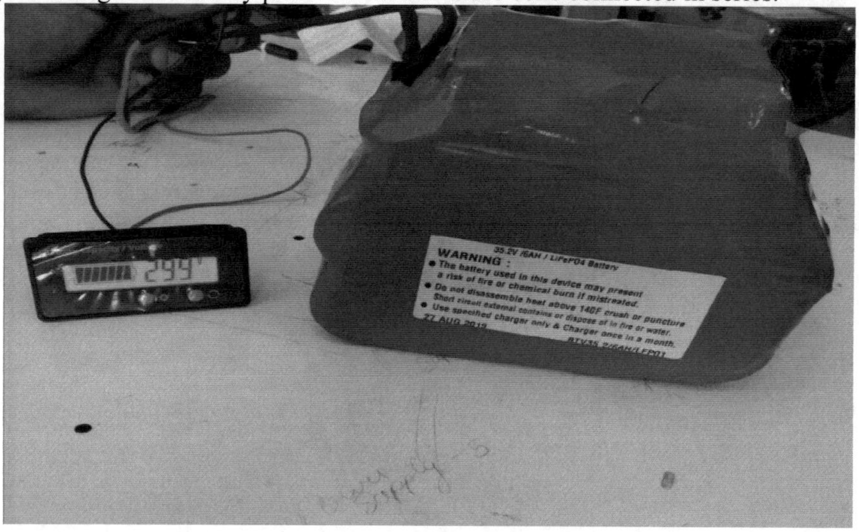

Figure 2. 36V Lithium Ion Battery.

6.1. Specification of battery

Table 2: Specification of battery.

Terminology	Specification
Nominal Voltage	36v
Type	Lithium ion battery
Initial Current	3c
Capacity	6Ah
Charging Hour	1hour 30mins
Cycle use	36v,6A
Charging Instruction	45 °C
Battery	36v
Initial current	1A

7. Hardware Setup

The 230V AC supply from the grid is given to the step down transformers which is converted in series for 12v-36v and also parallel connection for 6A. The rectifier circuit is used to convert the AC supply into DC supply which is fed to the bridge rectifier. In fast charger, the Arduino plays a major role of control the battery charging process.

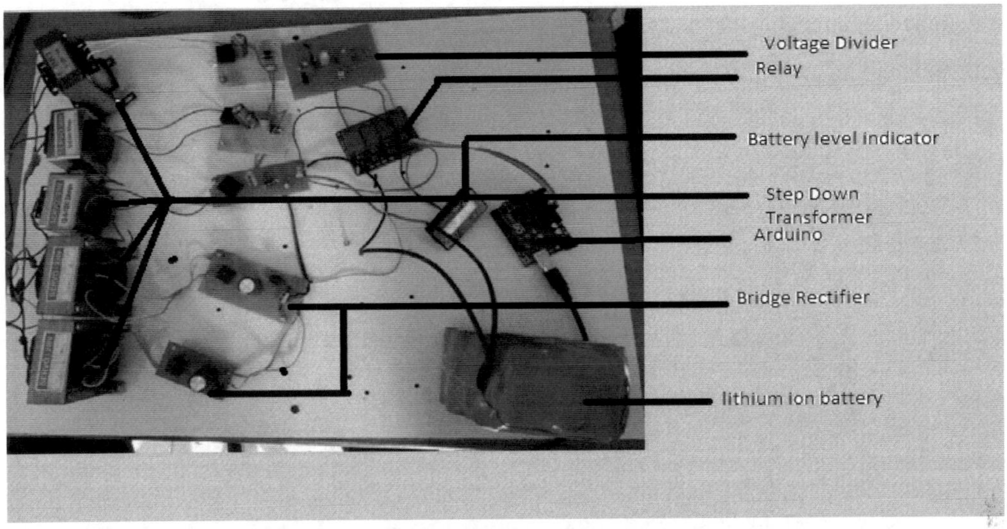

Figure 3. Hardware Setup of Proposed System.

For charging, the power supply is taken from Renewable energy and electricity board is fed to the controller. The input power supply of Renewable energy is compared with EB grid and close the circuit to the battery. The state of charge is measured by battery level detector which gives signal to the controller. If the state of charge is less than 30% the input power of battery is 37.6V,5A, If the state of charge is greater than 30% the input power is 37.6V,3A which is control by current controller. During the high power input, the battery temperature is detected by temperature sensor. If this exceeds the set level, controller stops charging. The temperature range of lithium ion battery is 45'c. If the battery temperature is greater than 37°c The input power is reduce to 37.6V,1.5A by controller. Finally the charger is designed as based on temperature and state of charge of battery

8. Results and Discussion

Table 3: Battery voltage and its corresponding state of charge percentage.

State of Charge in percentage	Lithium ion battery voltage (36v Battery)	Normal / Conventional charger (time taken)	Proposed Fast charger (time taken)
100%	36+	5h	1h 35m
75%	34.5+	3h 30m	55m
50%	33+	2h 45m	45m
25%	31.5+	1h 45m	30m
0%	29+	Nil	Nil

In existing method, the normal mode charging has take more than 5 hours to charge the battery up-to 100%. It's the major drawback of electric vehicle. To overcome this drawback, fast charger has introduced. It can charge the battery with 1hour 30 minutes, but life cycle of battery has reduced. And also temperature also increase, due to high temperature battery can result in explosion. And also internal resistance of battery has been increase. Thus proposed fast charger is used to overcome this above drawback and hence it charge the battery based on temperature and percentage of battery as shown in table 3 and its graphical representation as in figure 4.

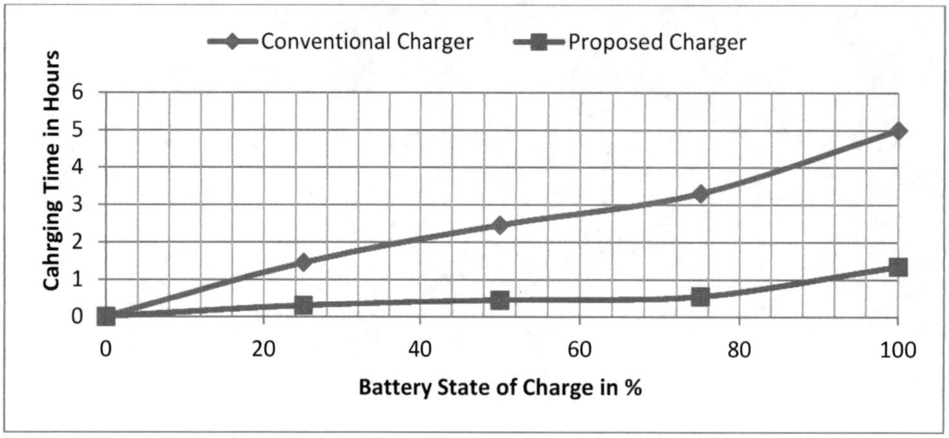

Figure 4. Result comparison of conventional charger with proposed charger.

9. Conclusion

In this paper, the development of charging circuit for a battery with performance characteristics analysis and also taking into consideration the supply of both the electricity and renewable energy (solar power) was done. The proposed work, utilizes both solar energy and grid electricity for charging lithium ion battery. The voltage from these two sources is compared and which has the high value is used to charge the battery. Electricity is preferred during night hours and also during insufficient solar energy. Arduino is used as controller of the battery charging in different ways. The result table proves that the proposed charger charges the battery with 1h 35 minutes., where the conventional charger takes more than 5 hours to reach the same level. At last, this method concluding that this approach reduces the pollution and increase the usage of EVs as a result creating pollution free environment and reduces the charging time of the lithium ion battery.

References

[1]. Sitharthan R, Geethanjali M and Pandy TKS 2016 Adaptive protection scheme for smart microgrid with electronically coupled distributed generations *Alexandria Engineering Journal* **55(3)** 2539-2550

[2]. Fathima AH, and Palanisamy K 2014 Battery energy storage applications in wind integrated systems—a review *IEEE International Conference on Smart Electric Grid* 1-8

[3]. Prabaharan N and Palanisamy K 2015 Investigation of single-phase reduced switch count asymmetric multilevel inverter using advanced pulse width modulation technique *International Journal of Renewable Energy Research* **5(3)** 879-890.

[4]. Jerin ARA, Kaliannan P and Subramaniam U 2017 Improved fault ride through capability of DFIG based wind turbines using synchronous reference frame control based dynamic voltage restorer. *ISA transactions* **70** 465-474

[5]. Sitharthan, R, Sundarabalan CK, Devabalaji KR, Nataraj SK and Karthikeyan M 2018 Improved fault ride through capability of DFIG-wind turbines using customized dynamic voltage restorer *Sustainable cities and society* **39** 114-125

[6]. Prabaharan N and Palanisamy K 2016 A single-phase grid connected hybrid multilevel inverter for interfacing photo-voltaic system *Energy Procedia* **103** 250-255

[7]. Palanisamy K, Mishra JS, Raglend IJ and Kothari DP 2010 Instantaneous power theory based unified power quality conditioner (UPQC) *IEEE Joint International Conference on Power Electronics, Drives and Energy Systems* 1-5

[8]. Sitharthan R and Geethanjali M 2017 An adaptive Elman neural network with C-PSO learning algorithm-based pitch angle controller for DFIG based WECS Journal of Vibration and Control **23(5)** 716-730

[9]. Sitharthan R and Geethanjali M 2015 Application of the superconducting fault current limiter strategy to improve the fault ride-through capability of a doubly-fed induction generator–based wind energy conversion system *Simulation* **91(12)** 1081-1087

[10]. Sitharthan R, Karthikeyan M, Sundar DS and Rajasekaran S 2020 Adaptive hybrid intelligent MPPT controller to approximate effectual wind speed and optimal rotor speed of variable speed wind turbine *ISA transactions* **96** 479-489

[11]. Sitharthan R, Devabalaji KR and Jees A 2017 An Levenberg–Marquardt trained feed-forward back-propagation based intelligent pitch angle controller for wind generation system *Renewable Energy Focus* **22** 24-32

[12]. Sitharthan R, Sundarabalan CK, Devabalaji KR, Yuvaraj T and Mohamed Imran A 2019 Automated power management strategy for wind power generation system using pitch angle controller *Measurement and Control* **52(3-4)** 169-182

[13]. Sundar DS, Umamaheswari C, Sridarshini T, Karthikeyan M, Sitharthan R, Raja AS and Carrasco MF 2019 Compact four-port circulator based on 2D photonic crystals with a 90° rotation of the light wave for photonic integrated circuits applications *Laser Physics* **29(6)** 066201

[14]. Sitharthan R, Parthasarathy T, Sheeba Rani S and Ramya KC 2019. An improved radial basis function neural network control strategy-based maximum power point tracking controller for wind power generation system *Transactions of the Institute of Measurement and Control* **41(11)** 3158-3170

[15]. Rajesh M and Gnanasekar JM 2017 Path observation based physical routing protocol for wireless ad hoc networks *Wireless Personal Communications* **97(1)** 1267-1289

[16]. Palanisamy K, Varghese LJ, Raglend IJ and Kothari DP 2009. Comparison of intelligent techniques to solve economic load dispatch problem with line flow constraints *IEEE International Advance Computing Conference* 446-452

[17]. Sitharthan R, Ponnusamy M, Karthikeyan M and Sundar DS 2019 Analysis on smart material suitable for autogenous microelectronic application *Materials Research Express* **6(10)** 105709

[18]. Rajaram R, Palanisamy K, Ramasamy S and Ramanathan P 2014 Selective harmonic elimination in PWM inverter using fire fly and fireworks algorithm *International Journal of Innovative Research in Advanced Engineering* **1(8)** 55-62

[19]. Sitharthan R, Swaminathan JN and Parthasarathy T 2018 March. Exploration of wind energy in India: A short review *IEEE National Power Engineering Conference* 1-5

[20]. Karthikeyan M, Sitharthan R, Ali T and Roy B 2020 Compact multiband CPW fed monopole antenna with square ring and T-shaped strips *Microwave and Optical Technology Letters* **62(2)** 926-932

[21]. Sundar D Sridarshini T, Sitharthan R, Madurakavi Karthikeyan, Sivanantha Raja A, and Marcos Flores Carrasco 2019 Performance investigation of 16/32-channel DWDM PON and long-reach PON systems using an ASE noise source *In Advances in Optoelectronic Technology and Industry Development: Proceedings of the 12th International Symposium on Photonics and Optoelectronics* 93

[22]. Sitharthan R and Geethanjali M 2014 Wind Energy Utilization in India: A Review *Middle-East J. Sci. Res.* **22** 796–801 doi:10.5829/idosi.mejsr.2014.22.06.21944

[23]. Sitharthan R and Geethanjali M 2014 ANFIS based wind speed sensor-less MPPT controller for variable speed wind energy conversion systems *Australian Journal of Basic and Applied Sciences* **8**14-23

[24]. Jerin ARA, Kaliannan P, Subramaniam U and El Moursi MS 2018 Review on FRT solutions for improving transient stability in DFIG-WTs *IET Renewable Power Generation* **12(15)** 1786-1799

[25]. Prabaharan N, Jerin ARA, Palanisamy K and Umashankar S 2017 Integration of single-phase reduced switch multilevel inverter topology for grid connected photovoltaic system *Energy Procedia* **138** 1177-1183

[26]. Rameshkumar K, Indragandhi V, Palanisamy K and Arunkumari T 2017 Model predictive current control of single phase shunt active power filter *Energy Procedia* **117** 658-665

[27]. Fathima AH and Palanisamy K 2016 Energy storage systems for energy management of renewables in distributed generation systems *Energy Management of Distributed Generation Systems* 157

[28]. Rajesh M 2020 Streamlining Radio Network Organizing Enlargement Towards Microcellular Frameworks *Wireless Personal Communications* 1-13

[29]. Subbiah B, Obaidat MS, Sriram S, Manoharn R and Chandrasekaran SK 2020 Selection of intermediate routes for secure data communication systems using graph theory application and grey wolf optimisation algorithm in MANETs IET *Networks* doi:10.1049/iet-net.2020.0051

[30]. Singh RR and Chelliah TR 2017 Enforcement of cost-effective energy conservation on single-fed asynchronous machine using a novel switching strategy *Energy* **126** 179-191

[31]. Amalorpavaraj RAJ, Palanisamy K, Umashankar S and Thirumoorthy AD 2016 Power quality improvement of grid connected wind farms through voltage restoration using dynamic voltage restorer *International Journal of Renewable Energy Research* **6(1)** 53-60

[32]. Singh RR, Chelliah TR, Khare D and Ramesh US 2016 November. Energy saving strategy on electric propulsion system integrated with doubly fed asynchronous motors *IEEE Power India International Conference* 1-6

[33]. Singh RR, Mohan H and Chelliah TR 2016 November. Performance of doubly fed machines influenced to electrical perturbation in pumped storage plant-a comparative electromechanical analysis *IEEE 7th India International Conference on Power Electronics* 1-6

IOP Conf. Series: Materials Science and Engineering **937** (2020) 012005 doi:10.1088/1757-899X/937/1/012005

Simultaneous and Incremental Approach for Optimal Placement and Sizing of Multiple Distributed Generations in Distribution System for Minimization of Power Losses

Nisha R Godha (Dagade) [1,2], Vishram N Bapat[3], Iranna Korachagaon[4]

[1]Research Scholar, VTU, Belgavi, Karnataka, India
[2]Asst. Prof., NBN Sinhgad School of Engineering, Pune, Maharashtra, India
[3]Corporate member, Institution of Engineers, Kolkata, West Bengal, India
[4]Professor, Tontadarya College of Engineering, Gadag, Karnataka, India

Email: dagade.nisha@gmail.com

Abstract. Several benefits have been obtained through integration of distributed generation in distribution systems. These advantages include reducing power losses, improving voltage profiles, and many more. These advantages can be accomplished and improved by optimizing the size and location of dispersed generations in the system. This paper introduces simultaneous and incremental approach using genetic algorithm for placement of multiple distributed generations to reduce power losses while keeping the network voltage profiles within specified limits. The algorithm is tested on IEEE 33-bus system and an Indian 85-bus system. It is clearly observed that, the proposed approaches provide optimal deployment of multiple distributed generations which reduces power losses and improves voltage profile.

1. Introduction

The definition of distributed generation (DG) is changing with time. In the literature, there is no unique definition of DG and it has been defined in many ways. DGs are the small power generators that can be connected to grid at any point close to the load. DGs which mainly include wind power, solar power, small hydropower and miniature gas turbine etc. is a supplement to centralized power generation. The combination of DGs and centralized generation can increase the advantages. Appropriate placement of DG in a distribution network can reduce power losses, distributed power investments, improve voltage profile and power quality, reduce interruption cost for network enforcing, total harmonic distortion etc. [1].

The extent to which DG reduces power losses and improves voltage profile depends on mode of operation, size and location of DGs [2]. Hence to maximize the benefits, it is very important to find the appropriate size of DG and the location in the network, as improper size and location can worsen the system operation and may lead to poor power quality and reliability, increase in system losses and costs. Many techniques have been proposed in the literature to solve the problem of placement of distributed generation. These methods can be classified into different groups: Analytical, Numerical, Heuristic, Meta-heuristic and some hybrid methods. [3] provides a survey of the state of the art of heuristic and metaheuristic techniques for optimal placement of DG.

Simultaneous placement of multiple DGs is proposed in [4], [5], [6], [7], [8] and incremental approach is proposed in [9], [10]. GA based approach is applied to obtain optimal location and size of

IOP Conf. Series: Materials Science and Engineering **937** (2020) 012005 doi:10.1088/1757-899X/937/1/012005

DGs in radial and networked system considering single and multiple DG placement with different load patterns [4]. Multi-location DG placement problem is solved by loss sensitivity factor and GA for minimizing loss and increasing the overall economic benefits [5]. A method based on genetic algorithm is used for optimally allocating distributed generation for profit, loss reduction and voltage profile improvement [6]. GA based approach has been used for allocation of DG in the power system with the mixed load models. The objective function is designed to minimize active and reactive power losses and to improve voltage profile [7]. Multi objective function is developed [8] considering power losses, voltage and reliability index and shift factor indices. It is solved by genetic algorithm and particle swarm optimization for various systems taking different type of loads. GA is utilized to select appropriate DG technology with optimal location and size for optimal operation of power system to minimize power loss in the network [9]. Nonlinear-constrained with single objective GA technique is employed for planning of renewable energy-based DGs [10]. GA is applied for reconfiguration and placement of DG [11]. Effect of distributed generation installation on power loss is presented in [12]- [13] using genetic algorithm method. A GA based on multi objective performance index is proposed [14] for determination of size and location of DG in distribution systems with different load models. Many parameters like loss reduction, voltage profile improvement, environmental effects, exploitation and maintenance expenses and cost of load prediction are considered in objective function and genetic algorithm based method is used to solve the problem of placement of DGs [15]. In [16] authors presented a structure to locate and size various types of DGs to improve the voltage profile, boost the voltage stability index, and reduce the energy losses. Efforts made to solve DG placement problem using genetic algorithm on various theoretical and practical system are summarized in table 1.

In this paper, simultaneous and incremental approach using Genetic algorithm (GA) is proposed to minimize the power losses by selecting the proper location and size of multiple distributed generations. The problem solved by genetic algorithm gives better results. The characteristics of genetic algorithm make it suitable to ill-structured optimization problems [17]. There can be various objective functions for enhancing the system such as minimization of operation cost, maximization of benefit/cost ratio, minimization of network loss, voltage profile improvement etc..

Table 1. Summary of literature on genetic algorithm used for placement of DG

Ref.	Objectives function	Number of DGs	Design Variable	Load model	System used
[4]	Minimize the power loss	Multiple	Location, size	Constant power	11 bus radial feeder, 6 bus system and 30 bus system
[5]	Maximization of benefits	Multiple	Location, size	Variable load model	33 and 69 bus system
[6]	Multi objective	Multiple	Location, size	Constant power	12-Bus radial rural distribution network
[7]	Minimize the power loss and improve voltage profile	Multiple	Location, size	Constant power	38- bus system
[8]	Multi objective	Multiple	Location, size	Different load model-constant power	33, 69 and 54 bus radial system
[9]	Minimize the power loss	Multiple	Type, Location, size	Constant power	14 bus system

[10]	Minimize the power loss	Multiple	Location, size	Constant power	33, 69 and 119 bus system
[11]	Minimize the power loss	Multiple	Reconfiguration Location, size	Constant power	33 bus system
[12]	Minimize the power loss	Single	Location, size	Constant power	30- bus system
[13]	Minimize the power loss	Single	Location, size	Constant power	33- bus system
[14]	Multi objective	Single	Location, size	Different load model-variable power	16 and 37- bus system
[15]	Multi objective	Single	Location, size	Constant power	13- bus network
[16]	Multi objective	Single	Location, size	Variable load model	33 bus system

IEEE bus systems: 13, 30, 33, 69, 119

Most of the work in the literature presented simultaneous or incremental approach alone for placement of multiple DGs. In this paper the proposed method considers both the approaches for placement of multiple DGs. The proposed algorithm is analyzed on IEEE 33-bus system and an Indian 85-bus system. This paper is arranged as follows. Section II includes the problem definition and objective function considered in the work. Section III introduces the simultaneous and incremental approach and presents the methodology for solving the problem of optimal placement of multiple DGs. In Sections IV and V, the simulation results are discussed and conclusions are drawn respectively.

2. Problem Formulation

As the installation of distributed generation increases in the system, it is very important to consider the size and location of DGs. Inadequate position and size of DGs will increase power loss and degrade reliability. High-capacity DGs are not suggested to install in the network. Moreover, the complete capacity of DGs should be consumed within the limit of the network. Any effort to install DG with high capacity to export energy beyond the substation leads to very high loss due to inverse power flow through substation. Hence, the network configuration and the load on the distribution system play an important role while selecting the optimal size of DG for minimum system loss.

The main purpose of the paper is to employ the proposed approach to minimize power losses without violating the voltage limits and other constraints in the power system. The proposed simultaneous and incremental approach using Genetic algorithm is utilized to determine the optimal sizes and locations of the placement of multiple DGs at which the system loss becomes minimum. Figure 1 shows the optimization model used. In this paper, DG injecting real power with a fixed generator output is considered.

2.1 Objective Function

Consider the power flow from the bus i to bus j is S_{ij} and from bus j to bus i is S_{ji}. The power loss in the line can be given by,

$$S_{Lij} = S_{ij} + S_{ji} \quad (1)$$

$$S_{ij} = V_i I_{ij}^{*} \quad (2)$$

$$S_{ji} = V_j I_{ji}^{*} \quad (3)$$

Where, the current I and the voltage V at the corresponding bus are found from the power flow. S is complex power flow with its real and imaginary part corresponding to the real and imaginary power loss on the line respectively. The amount of energy injections at all the nodes in the network represents losses. The total loss in the network can be calculated as:

$$P_{loss} = \sum_{k=1}^{nbr} real(S_{Lk}) \quad (4)$$

Where, 'nbr' is number of transmission lines in the network. The objective function is to reduce P_{loss} according to the constraints of power flow. DG is regarded as negative loads on buses; therefore, the equation of the power balance has been modified as $P_{di}=P_{di}-P_{dgi}$, where P_{di} - Power demand at bus '*i*' and P_{dgi} – Power from DG connected at bus '*i*'. There is no change in reactive power balance equations as DG is producing only real power. In this paper, loss minimization has been used as a fitness function, and the GA has been used to optimize the fitness function.

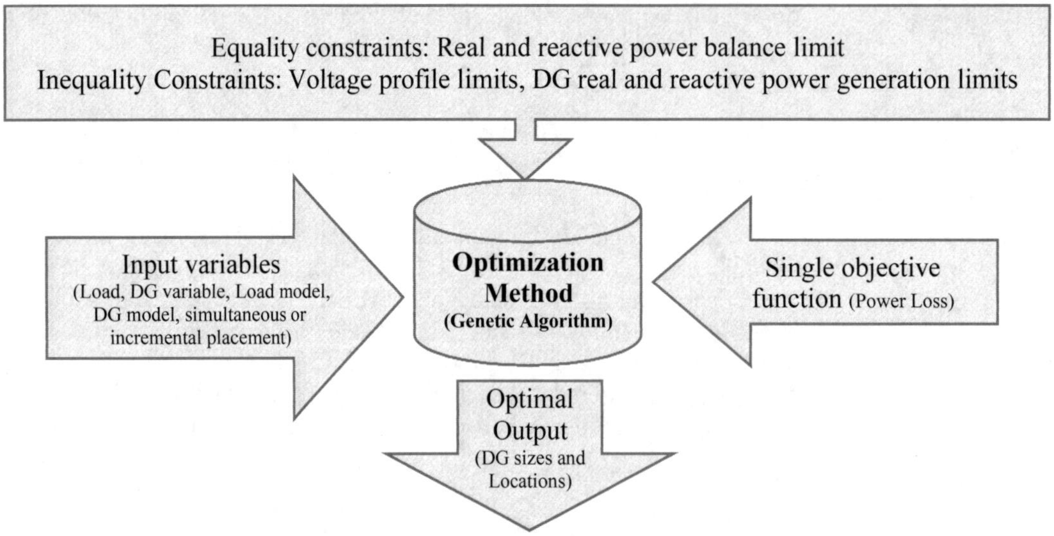

Figure 1. Optimization Model

2.2 Operating Constraints
The operating constraints of the problem are:

2.2.1 Equality constraints: Summation of all incoming powers including power generated by 'm' DG units should be equal to outgoing powers and power losses for distribution system.

$$P_{gs} + \sum_{i=1}^{m} P_{DGi} = P_{load} + P_{loss} \quad (5)$$

Where, P_{gs} – Power available at substation from central generation, P_{load} – Total load of on the substation.

2.2.2 Inequality constraints:
The injected power by increasing DG unit is limited by their maximum and minimum limits as:

$$P_{minDGi} \leq P_{DGi} \leq P_{maxDGi} \quad (6)$$

Bus voltage limits (as per Indian standard ±5% margin)

$$0.95 \text{ pu} \leq V_i \leq 1.05 \text{ pu} \quad (7)$$

3. Simultaneous and incremental approach using Genetic Algorithm

To reduce the losses with high value and to compensate the demand in a power system having no DG, many DGs can be placed simultaneously as shown in figure 2(a). It is also possible that the system may have some penetration of DG, and to improve the system, it is needed to deploy more DGs. In that case incremental approach is suitable to solve the problem as shown in figure 2(b).

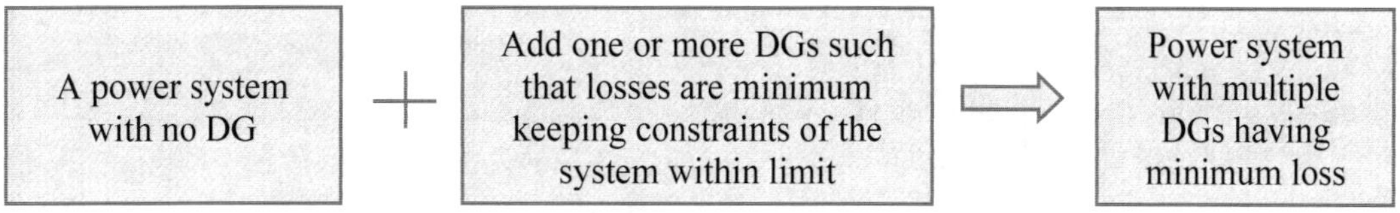

Figure 2(a). Simultaneous approach of placement of DGs

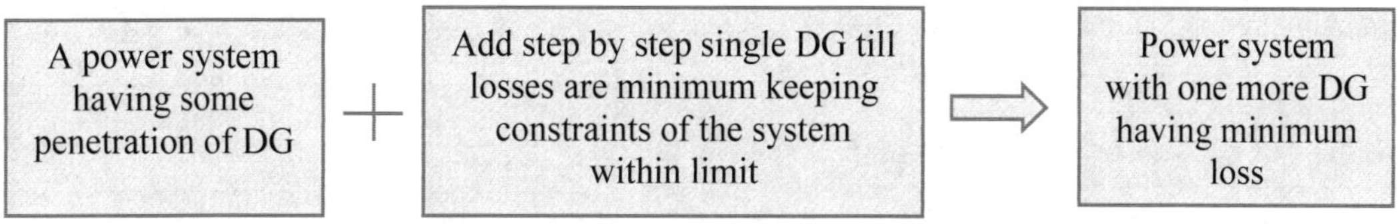

Figure 2(b). Incremental approach of placement of DGs

If integration of nDG in the distribution network doesn't violate the network regulations and sufficient funding is available then system planner can add appropriate number of DGs all at once i.e. simultaneously. But, sometime it is not possible to invest large amount at the same time and adding more DG in the system may have regulation problem then incremental approach is suitable. In incremental approach it is important to consider system data with DG to place next DG.

Genetic algorithm is a metaheuristic approach belonging to a larger class of evolutionary algorithms. It is used to solve constrained and unconstrained search and optimization problems. This uses natural selection controlled by reproductive biological process. GA requires: genetic representation of solution domain and fitness function. The most important components in a GA consist of initialization, population generation, evaluation, selection, crossover and mutation.

As initial approach, a group of chromosomes called population are used. The quality of initial solutions is evaluated through the fitness function extracted from the objective function. Three GA operators generate a new array of candidate solutions with an improved fitness value over each generation (i.e. reproduction, crossover and mutation). The GA framework is explained below:

Step 1: Initialization- Randomly generates an initial population. The chromosome consists of DG location ('loc') and DG size ('P_{DG}'). The proposed chromosome will be:

$$X_i = [loc_1, loc_2, \dots\dots, loc_n, P_{DG1}, P_{DG2}, \dots\dots\dots, P_{DGn}]$$

Where i =1,2,3,.....m. The variable m indicates the number of locations and DGs. 'n' is number of buses in the radial or networked system.

Step 2: Evaluation- evaluate fitness function after calculating the power loss

Step 3: Selection- select the best population from the current population.

Step 4: Crossover- generate offspring using crossover function with crossover probability Pc = 0.8

Step 5: Mutation- Apply the mutation function for each population generated in step 3 with the mutation probability Pm = 0.01.

Step 6: Termination- if the termination criterion (Maximum iterations and constraints) is fulfilled then halt. Otherwise update iterations and return to step 2.

Algorithmic flowcharts for the proposed simultaneous and incremental approach are shown in figure 3 and figure 4.

IOP Conf. Series: Materials Science and Engineering **937** (2020) 012005 doi:10.1088/1757-899X/937/1/012005

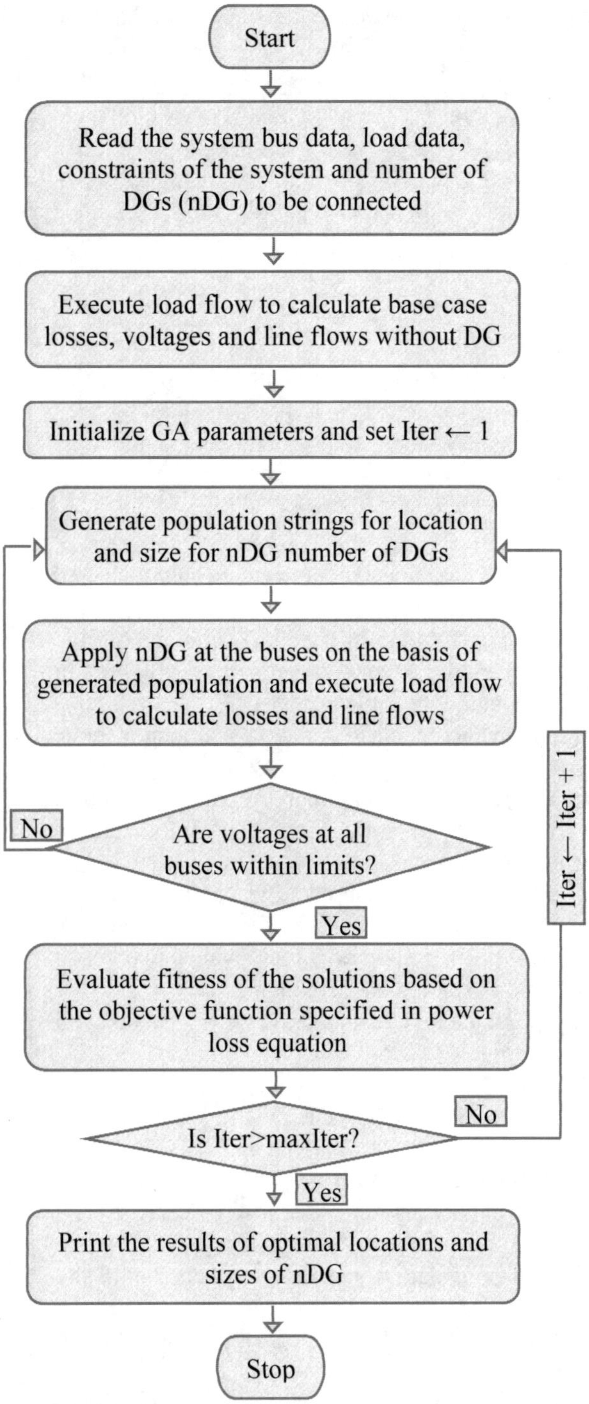

Figure 3. Flow chart for simultaneous approach

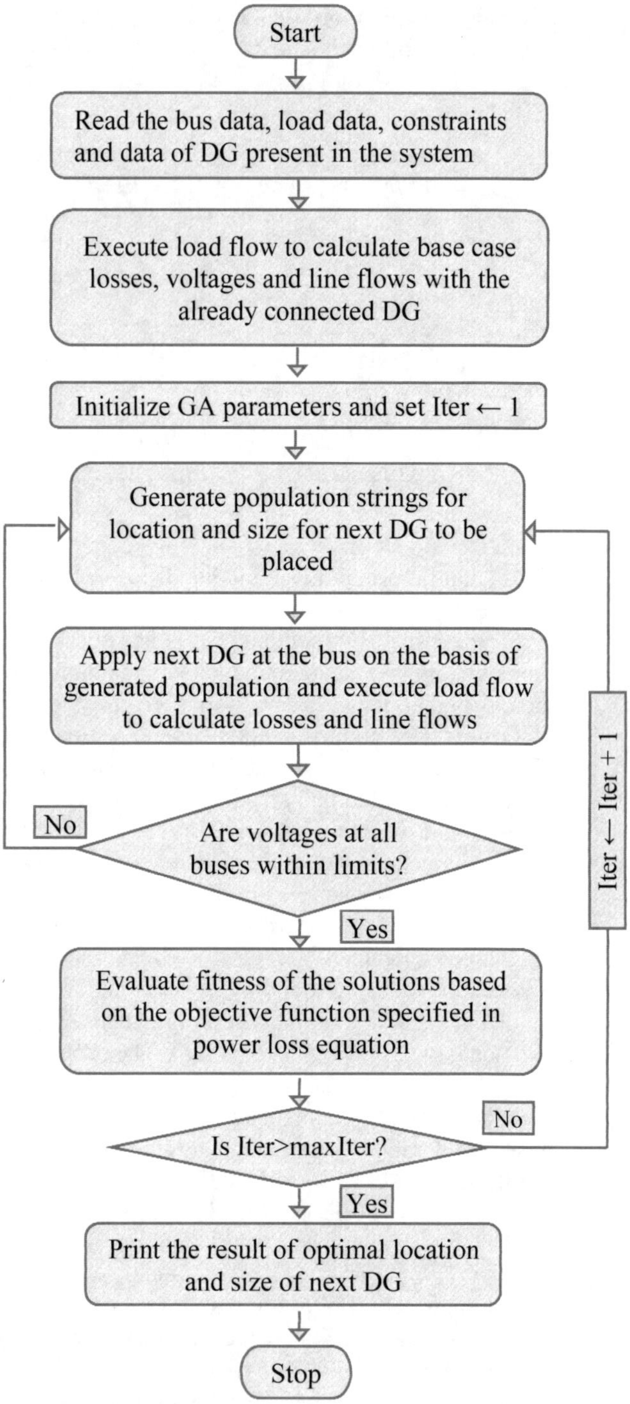

Figure 4. Flow chart for incremental approach

4. Results and Discussion

The optimal locations and sizes for the placement of multiple DGs are obtained by simultaneous and incremental approach. The proposed approach is implemented in MATLAB. The line flows are computed using Newton-Raphson method. The success of the method is tested on IEEE 33-bus [18] and Indian 85-bus test system [19].

4.1 IEEE 33 - bus system

Figure 5 represents single line diagram of IEEE 33-bus test system and its line and bus data are taken from [18]. The system consists of 32 sectionalizing branches at nominal voltage of 12.66 kV and a total system load of 3.72 MW and 2.3 MVAR

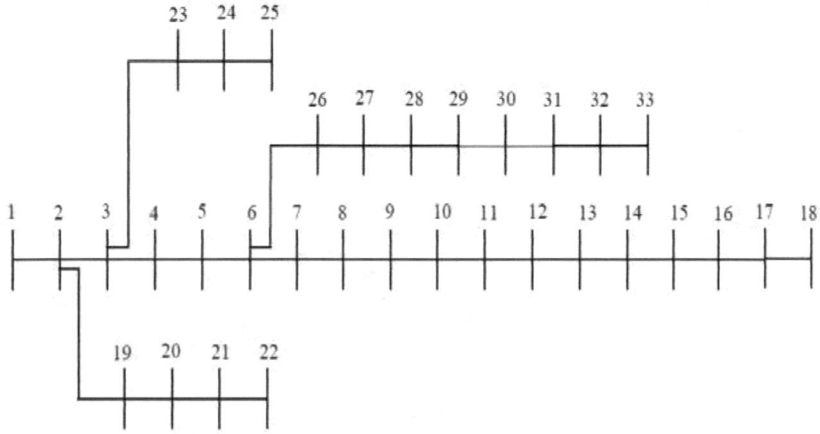

Figure 5. Single line diagram of IEEE 33-bus system

4.1.1 Case I: Simultaneous approach for placement of multiple DGs

It is considered that initially there is no DG in the system. Appropriate location and size of single DG, two DG and three DG is obtained using the algorithm as shown in algorithmic figure 3 and figure 4.

Table 2 shows the power losses, % reduction in losses and minimum and maximum voltage values before and after installation of single DG, two DGs and three DGs simultaneously. Before DG installation, real losses of the system are 212.2 kW.

After installation of single DG injecting real power only, there is reduction in losses by 47.42% with the value of loss as 111.57 kW and the optimal location is 6 with the size 2641 kW. In two DG case the optimal locations are bus 13 and bus 30 with the size 895.8 kW and 1277.6 kW respectively. The optimal locations for three DGs are obtained as 25, 16 and 31 giving minimum losses. Losses are reduced by 62.24%. Further increase in number of DGs does not reduce the losses by large amount. Figure 6 gives the voltage profile at different buses when the single and multiple DGs are placed at optimal location with appropriate size for case I. Due to placement of DG the voltage variations are reduced. With addition of 4 and more number of DGs, the losses get reduced, but the difference in loss reduction is very small.

Table 2. Simulation results of case I for IEEE 33-bus system

Options	Without DG	Single DG	Two DG	Three DG
Optimum bus locations for placement of DG	-	6	13 30	25 16 31
Optimal size of DG (kW)	-	2641.0	895.8 1277.6	533.3 685.3 999.2
Total active power loss (kW)	212.2	111.57	88.364	80.112
Reduction in power loss (%)	-	47.42	58.36	62.24
Minimum voltage in p.u	0.903	0.943	0.972	0.966
Maximum voltage in p.u	1.00	1.00	1.00	1.00

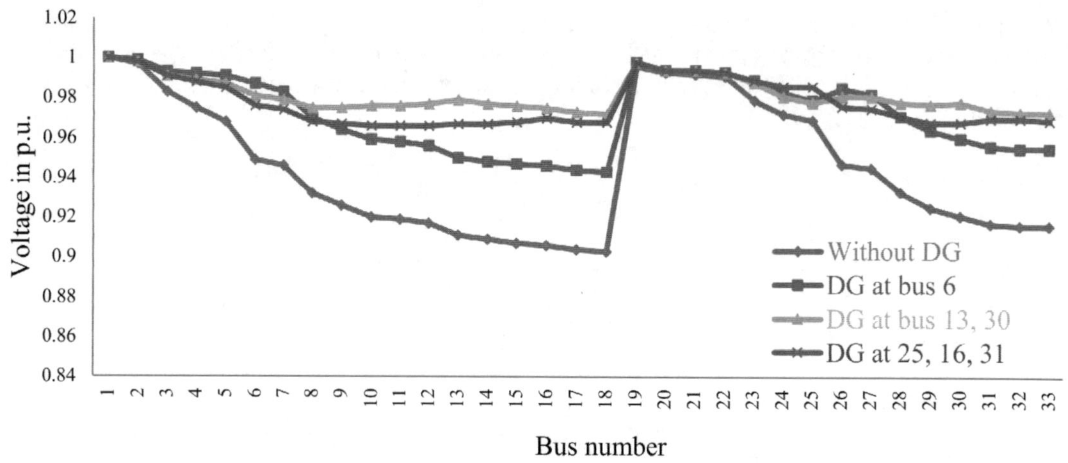

Figure 6. Voltage magnitudes at each bus for Case I for IEEE 33-bus system after placement of DG

4.1.2 Case II: Incremental approach for placement of multiple DGs

In this case it is assumed that already there is some penetration of DG in the system. Optimal location and size for placement of next DG is decided on the basis of previously connected DGs.

Table 3 shows the power losses, % reduction in losses and minimum and maximum voltage values before and after installation of single DG, two DGs and three DGs for case II. The system has DG of size 2641 kW already connected at bus 6. Now to add new DG in the system, genetic algorithm is applied by generating the population for only two variables – one for next DG location and DG size.

The next optimal DG size is 472.9 kW at bus 15 which will give minimum losses. Further, for placement of next DG, it is assumed that two DGs are connected at bus 6 and 15 of size 2641 kW and 472.9 kW respectively. Only two variables are generated and GA is executed and the optimal location for third DG is at bus 25 with its optimal size 623.2 kW. By adding one more DG there is small reduction in losses but the cost associated with DG increases with large amount. Figure 7 shows the voltage profile of each bus after placement of DGs at proper location.

The placements of one, two and three DG have improved the voltage profile of the distribution system. From figure 6 and 7, it is observed that the voltages are below the standard values at buses 6-18 and 26-33 when DG is not placed in the network. With the DG, voltage values are within acceptable range and it is improved at all buses. The highest percentage improvement of voltage is at buses 13-18 while buses 2-6 and 19-22 has lowest voltage improvement. This voltage profile improvement depends on the location of DG whether it is further connected to branch or not.

Table 3. Simulation results of case II for IEEE 33-bus system

Options	Without DG	Single DG	Two DG	Three DG
Bus number at which DG already present in the System	-	-	6	6 & 15
Next optimum location for placement of DG	-	6	15	25
Optimal size of DG (kW)	-	2641.0	2641.0 472.9	2641.0 472.9 623.2
Total active power loss (kW)	212.2	111.57	97.301	90.02
Reduction in power loss (%)	-	47.42	54.15	57.49
Minimum voltage in p.u	0.903	0.943	0.962	0.964
Maximum voltage in p.u	1.00	1.00	1.00	1.00

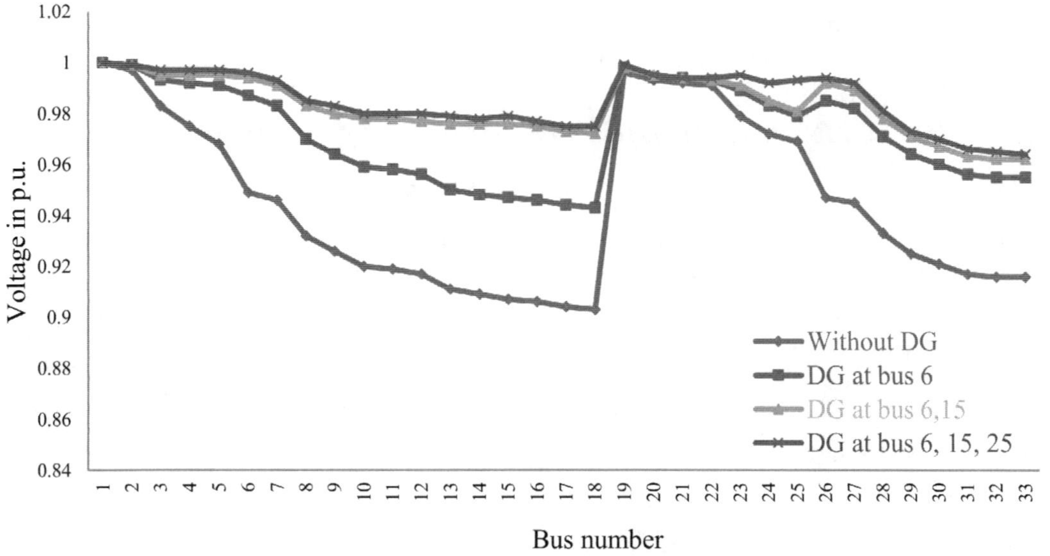

Figure 7. Voltage magnitudes at each bus for Case II for IEEE 33-bus system after placement of DG

4.2 Indian-85 bus system

This system consists of 85 buses and 84 sectionalising branches shown in figure 8. The bus, line and generator data have been taken from [19]. The total real power demand of the system is 2.549 MW and the reactive power demand is 2.601 MVAR. The base values are 100MVA, 11 kV. The base case power loss without DG is 349.1 kW. This RDS is solved for P-type DG units i.e injecting only real power, using the proposed GA algorithm. The minimum voltage is at the 54th bus.

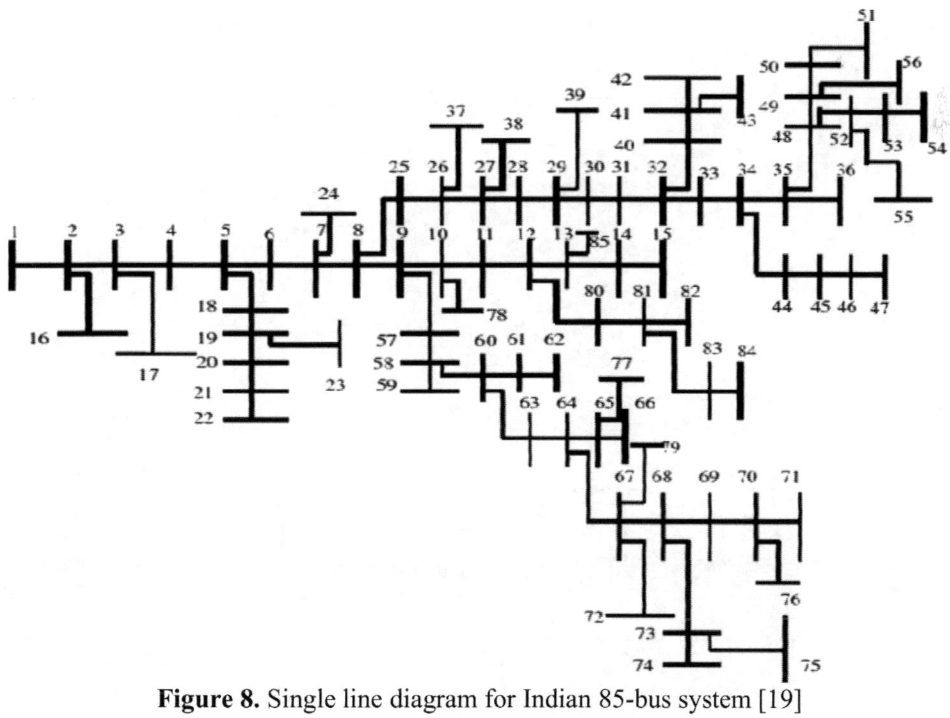

Figure 8. Single line diagram for Indian 85-bus system [19]

4.2.1 Case I: Simultaneous approach for placement of multiple DGs

Table 4 shows the power losses, percent reduction in losses and minimum and maximum voltage values before and after installation of single DG, two DGs and three DGs. Before DG installation real losses of the system were 349.1kW.

Table 4. Simulation results of case I for Indian 85-bus system

Options	Without DG	Single DG	Two DG	Three DG
Optimum bus locations for placement of DG	-	8	36 9	69 33 9
Optimal size of DG (kW)	-	2474.0	591.2 1597.5	321.1 851.2 744.3
Total active power loss (kW)	349.1	192.2	170.0	166.44
Reduction in power loss (%)	-	44.93	51.30	52.32
Minimum voltage in p.u	0.899	0.969	0.985	0.985
Maximum voltage in p.u	1.05	1.05	1.05	1.05

After installation of single DG injecting real power only, there is reduction in losses by 44.93 percent with the value of loss as 192.2 kW and the optimal location 8 with the size 2474.0kW. For multiple DG installation, it is considered that there is no DG in the system and two or three DGs are required to be connected. For two DGs, population for four variables are generated- two for location and two for size. In two DG case the optimal locations are bus 36 and bus 9 with the size 591.2 kW and 1597.5 kW respectively.

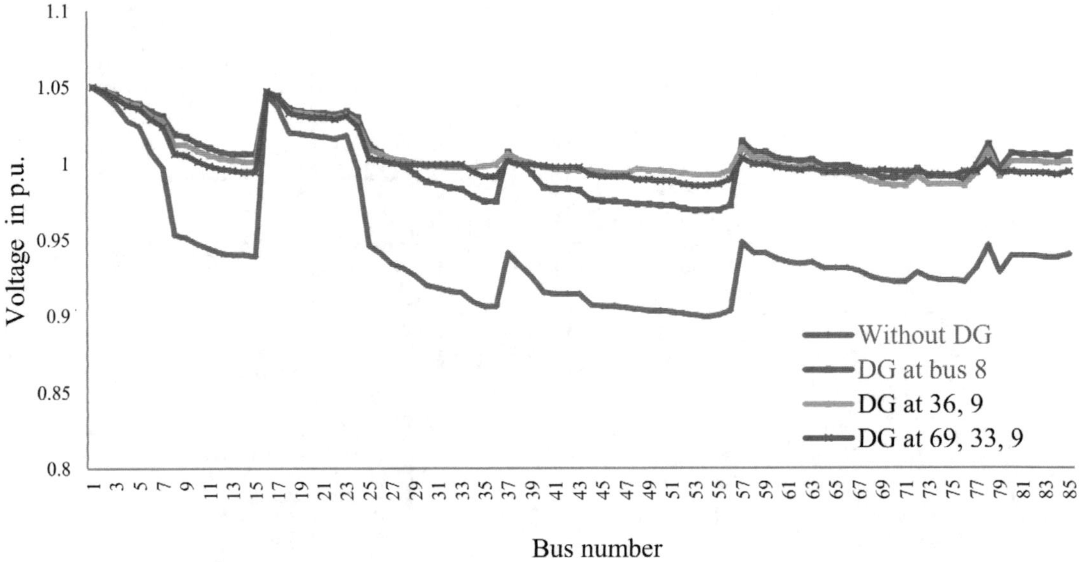

Figure 9. Voltage magnitudes at each bus for Case I for Indian 85-bus system after placement of DG

IOP Conf. Series: Materials Science and Engineering **937** (2020) 012005 doi:10.1088/1757-899X/937/1/012005

In three DG case, the population for 6 variables- three for each location and size is generated. The optimal locations for three DGs are obtained as 69, 33 and 9 giving minimum losses. Losses are reduced by 52.32%. Further increase in number of DGs does not reduce the losses significantly.

Figure 9 illustrates the voltage magnitudes for Indian 85-bus system when single DG, two DG and three DG are connected with their appropriate size at suitable location.

4.2.2 Case II: Incremental approach for placement of multiple DGs

Optimal location and size for placement of next DG is decided on the basis of previously connected DGs. The value of power losses, % reduction in losses and minimum and maximum voltage values before and after installation of single, two and three DGs are included in table 5.

Table 5. Simulation results of case II for Indian 85-bus system

Options	Without DG	Single DG	Two DG
Bus number at which DG already present in the System	-	-	8
Next optimum location for placement of DG	-	8	48
DG optimal size (kW)	-	2474.0	2474.0 371.4
Total active power loss (kW)	349.1	192.2	179.8
Reduction in losses (%)	-	44.93	48.49
Minimum voltage in p.u	0.899	0.969	0.999
Maximum voltage in p.u	1.05	1.05	1.05

The system has DG of size 2474.0 kW is already connected at bus 8. The next optimal DG size is 371.4 kW at bus 48 which will give minimum losses. Figure 10 illustrates the voltage magnitudes for Indian 85-bus system when one or multiple DGs are connected with their appropriate size at suitable location.

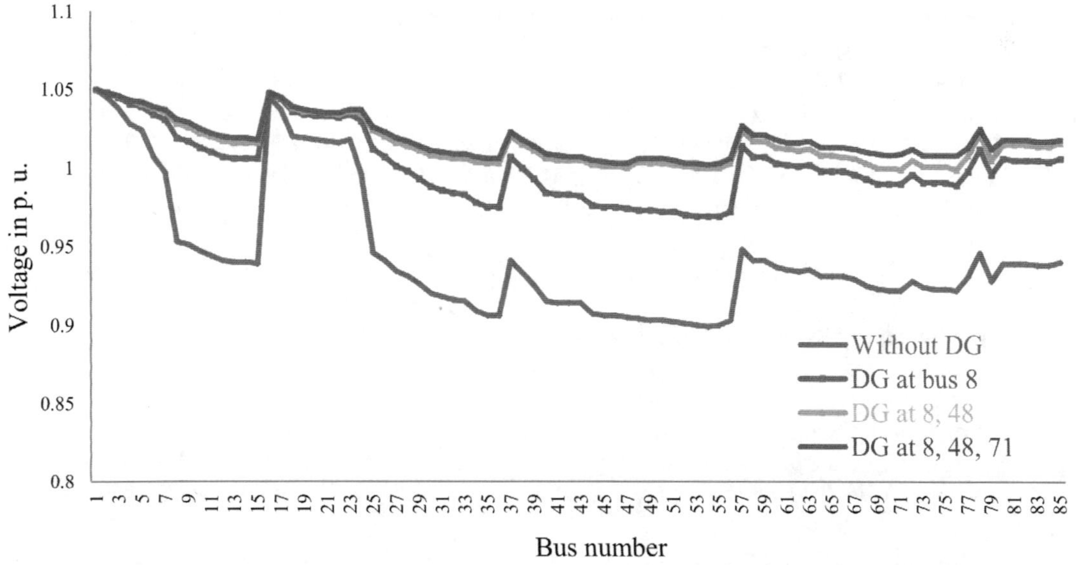

Figure 10. Voltage magnitudes at each bus for Case II for Indian 85-bus system after placement of DG

With single DG placed at bus 8 the loss is 192.2 kW with reduction of 42.93 % from its base loss. After addition of two DGs the loss reduction is 48.49 %.

From the case II for 85-bus system it is observed that even though DG is previously connected in the system, the losses reduced by 12.40 kW, but it violates the power balance constraints of the system and it is not the feasible solution as it does not satisfy the standards of the grid. From figure 9 and 10, it is observed that the voltages are below the standard values at buses 10-14 and 25-85 when DG is not placed in the network. With the DG voltage values are within acceptable range and it is improved at all buses. The highest percentage improvement of voltage is at buses 30-36 and 40-56 which is around 15 to 21 percent while bus 22 and 20 has lowest voltage improvement for one or more DG placement. The voltage profile improvement is dependent on the location of DG whether it is further connected to branch or not.

5. Conclusion

Inappropriate allocation of DG sources in the power system can result in increased power losses, as well as inappropriate operation of the system. To achieve the maximum environmental, economic and technical benefits, DG should be installed in the optimal position. This paper presented simultaneous and incremental approaches using GA algorithm to obtain optimum locations and sizes of multiple DGs in radial distribution system to minimize total power loss. The use of simultaneous or incremental approach depends on the availability of sufficient amount of funding and constraints in technical and environmental regulation. Loss reduction and voltage profile improvement is observed with the increase in penetration of DG. However, this trend reverses after a certain number of locations and size of DG.

References

[1] Ali Y, Sedighizadeh M, Sadighi M. A *particle swarm optimization for sitting and sizing of Distributed Generation in distribution network to improve voltage profile and reduce THD and losses*. 43rd International Universities Power Engineering Conference, IEEE; 2008. p. 1–5.

[2] Ochoa LF, Padilha-Feltrin A, Harrison GP. *Evaluating Distributed Generation Impacts With a Multi-objective Index*. IEEE Trans Power Delivery. 2006 Jul; **21**(3):1452–8.

[3] Dagade NR, Bapat VN, Korachagaon I. *Placement of Distributed Generation in Distribution Networks: A Survey on Different Heuristic Methods.* 2nd International Conference on Advanced Technologies for Societal Applications - vol. **1**. Springer International Publishing, 2019. p.693–707.

[4] Singh RK, Goswami SK. *Optimum Siting and Sizing of Distributed Generations in Radial and Networked Systems.* Electr Power Components System. 2009 Jan 28; **37**(2): 127–45.

[5] Shukla TN, Singh SP, Srinivasarao V, Naik KB. *Optimal Sizing of Distributed Generation Placed on Radial Distribution Systems*. Electr Power Components System. 2010 Jan 29; **38**(3): 260–74.

[6] Singh RK, Goswami SK. *Optimum allocation of distributed generations based on nodal pricing for profit, loss reduction, and voltage improvement including voltage rise issue*. Int J. Electr Power Energy System. 2010 Jul; **32**(6):637–44.

[7] Kalantari M, Kazemi A, Zakerinia MS. *Optimization of Distributed Generation Placement for Minimizing Power Losses and Voltage Profile Improvement Using Genetic Algorithm*. Int J Adv Comput Sci. 2012; **2**(10): 376–81.

[8] Bohre AK, Agnihotri G, Dubey M. *Optimal sizing and sitting of DG with load models using soft computing techniques in practical distribution system.* IET Gener Transm Distrib. 2016; 1–16.

[9] Ayodele TR, Ogunjuyigbe ASO, Akinola OO. *Optimal Location, Sizing and Appropriate Technology Selection of Distributed Generators for Minimizing Power Loss Using Genetic Algorithm.* J Renew Energy. 2015; 1–9.

[10] Hassan AA, Fahmy FH, Nafeh AESA, Abu-elmagd MA. *Genetic single objective optimisation*

for sizing and allocation of renewable DG systems. Int J Sustain Energy. 2015;

[11] Dahalan WM, Mokhlis H. *Simultaneous network reconfiguration and sizing of distributed generation*. Electric Distribution Network Planning, Power Systems,. Springer Verlag; 2018. p. 279–98.

[12] Hasibuan A, Masri S, Othman WAFWB. *Effect of distributed generation installation on power loss using genetic algorithm method.* IOP Conf Ser Mater Sci Eng. 2018 Feb; **308**(1): 012034.

[13] Mohan Kashyap AM and SK . *Optimal Placement of Distributed Generation Using Genetic Algorithm Approach,*. Proceeding of the Second International Conference on Microelectronics, Computing & Communication Systems (MCCS 2017). 2019. vol. **476**.

[14] Singh D, Singh D, Verma KS. *Multi-objective Optimization for DG Planning with Load Models.* IEEE Trans Power Syst. 2009 Feb; **24**(1): 427–36.

[15] Aref A. *Optimal placement and estimation of DG capacity in distribution network's using genetic algorithm-based method.* 2012; **5**(3): 2281–6.

[16] Abd HA, Azmy AM, Saad MA. *Optimal DG Deployment Based on Technical and Economic Considerations with Daily Load Variation*. Int J Eng Res Africa. 2019;45:115–31.

[17] Srinivasan D, Fushuan Wen, Chang CS, Liew AC. *A survey of applications of evolutionary computing to power systems.* In: Proceedings of International Conference on Intelligent System Application to Power Systems. IEEE; 1996. p. 35–41.

[18] Sahoo NC, Prasad K. *A fuzzy genetic approach for network reconfiguration to enhance voltage stability in radial distribution systems.* Energy Convers. Manag. 2006 Nov; **47**(18–19): 3288–306.

[19] Shuaib YM, Kalavathi MS, Asir CC. *Optimal capacitor placement in radial distribution system using Gravitational Search Algorithm.* Int J Electr Power Energy Syst. 2015; **64**:384–97.

Implementation of market methods of price forming in the field of power industry with the application of the consolidated balance-market model of economic systems

I D Grachev[1], S A Nekrasov[2], S N Larin[3], N V Noack[4] and N M Baranova[5]

[1]Dr.sc.oec., Chief Researcher, e-mail: idg19@mail.ru, ORCID: 0000-0003-1815-5898
[2]PhD in Technical Sciences, Senior Researcher, e-mail: san693@mail.ru, ORCID: 0000-0002-7649-0515
[3]PhD in Technical Sciences, Senior Researcher, e-mail: sergey77707@rambler.ru; ORCID: 0000-0001-5296-5865
[4]PhD in Psychological Sciences, Leading Researcher, e-mail: n.noack@mail.ru; ORCID: 0000-0001-8696-5767

Central Economics and Mathematics Institute RAS, 47, Nakhimovsky prospect, 117418 Moscow, Russia

[5]PhD in Pedagogic Sciences, Associate professor, Department of Economic and Mathematical modeling, Peoples' Friendship University of Russia (RUDN University), 6, Miklukho-Maklaya Str., 117198 Moscow, Russia, e-mail: baranova_nm@pfur.ru; ORCID: 0000-0002-7201-9435

Abstract.
Object. The power industry of Russia is being developing with the use of its natural resources, advanced ideas, implementation of existing opportunities. So, the key task of the energy strategy of Russia till 2035 is the transfer from the commodity-heavy to the resource-innovative development of both the power industry and the economics of the country. All conditions have been created for its solution as of now – the power industry is expected to increase by 20-25% if demand for the hydrocarbon raw materials falls on the immediate horizon.
In this juncture such branch as the power industry seems to be a crutial and sophisticated object of the research on the dynamics of economic systems infrastructural transformations. Anyway, its implementation in the field of the power industry involved quasi-market mechanisms of the preservation of existing ones and the building of new non-competitive powers with the use of such mechanisms as "adhesion contract" and "marginal price forming".
The work shows the inhibiting impact of these mechanisms on the economic system. The offer for its replacement to increase the pace of the power industry development and of the Russian economics substantiates.
The research purpose is to substantiate the application of the consolidated balance-market model of economic systems for the analysis of the dynamics of the development of the technologically heterogenic branch, interacting with the other economic system, provided Leontief correlations. The power industry was selected as the studied one. Peculiarities of the reforming of power price rates, combining both market and regulating (quasi-market) mechanisms, were considered.
Methods. The random balance-market model became the methodologic basis for mixed economic systems in the option of the self-consistent field of Leontief matrices. It make it possible to solve practical issues of the development of most important multi-agent branches in its interaction with the other economic system. In the work this model is applied in the quasitensor form for the evaluation of mixed mechanisms of the regulation of price rates in the power industry.
Results. It was revealed that the common use of such mechanisms as "marginal price forming" and "adhesion contracts" fully suppresses the economic growth in all branches of economics. In order to find the way out was offered the consolidated random model of mixed economic

systems in the quasitensor form. Its application will make it possible to obtain the new distribution of the capital by goods and agents with the better use of its technical advantages with such level of power rates.

Conclusions. The possibility of the subtle analysis of the dynamics of the specialization and localization of agents was shown for the following development of distributed centralized energoinformational systems in the Russian Federation. It is expedient to use offered approach and developed algorithms of modeling for the evaluation of the accelerated economic growth of other key branches of the Russian economics.

Key words: mixed economic systems, power branch, market and quasi-market mechanisms, regulation of rates, random balance-market model, Leontief matrices, self-consistent field

1. Introduction

The authors developed a probabilistic model of mixed economic systems. In it, the market is modeled by a self-consistent field of market values or other quasi-tensor parameters characterizing the economic system in a multidimensional economic space [1, 2].

It allowed us to effectively solve a wide range of problems: fundamental issues of modeling the innovative generation of Kondratieff cycles, the bitcoin course and the prospects of cryptocurrencies.

In this juncture the special case of the model with the self-consistent field of Leontief matrices, allowing to analyze the development of branches with the mixed competitively regulated management in its interaction with the other part of the economic system is of considerable interest [3].

The electricity field appears an important and comfortable object for studies of economic systems [4-5], where within frames of "market" reforms are implemented quasi-market mechanisms for the preservation of existing and building of new non-competitive capacities with use of mechanisms of so called "marginal price forming" and "adhesion contracts" [6].

In the practical aspect the timeliness of terms of reference and of the solving of this task is referred to the digitalization of Russian economics, which principally allow to form any types of Leontief matrices, namely, specific two good matrices of "power energy-everything else" type. Anyway, the expedience and efficiency of such use of the digital information requires both the theoretic and model processing of the problem for the branch, which is of interest to us. Certain approaches to the solution of mentioned problems will be disclosed in this article.

2. Research purpose

The purpose of this work is to substantiate the possibility of the application of the consolidated balance-market model of economic systems for the analysis of the dynamics of the development of the technologically heterogenic branch, interacting with the remaining economic system, provided Leontief correlations at the example of the reformation of the electric power industry, combining both market and regulating mechanisms of such type as marginal price forming and adhesion contracts.

3. Materials and methods

Nowadays, reforms of the electric power industry enter its final lap. So, we can be confident enough that along with positive results of its implementation there are also obvious unsatisfactory results [7]. To most significant of latter ones following ones can be referred.

1. Number of employees has been doubled at the expense of the number of different kinds of intermediates with annual generation volumes, equivalent to the former RSFSR epoch.

2. The number of hours of use of capacities has been considerably reduced.

In this context the modern Russia has lost its leading positions. Before the dissolution of the Soviet Union, for example, stations of its energetic system have been working better than any ones in the world, regularly showing the use of capacities up to 6 thousand and more hours. Anyway, for the electric power reform Russia has followed the example of tough US states, at which power stations the use of capacities was not exceeding 4.5 thousand hours. As a result today we have got the opposite pattern: best countries in the world near by the number of hours of capacities to Soviet results, and the Russia's index is about 4.5 thousand hours, as of underdeveloped countries [8].

IOP Conf. Series: Materials Science and Engineering **937** (2020) 012006 doi:10.1088/1757-899X/937/1/012006

3. With different kinds of calculation by kWh, the accidence in the Russian electricity industry has grown in 1.5-2 times.

All these facts are well known and obvious, which should be purposively worked out in order to obtain its improvement.

As to use of mechanisms of the so called "marginal pricing" (MP) and «capacity connection agreement» (CCA) everything is much more difficult. These mechanisms are oriented to the support of the functioning of existing and the building of new non-competitive stations for safety reasons and the whole range of other factors due to the shift of the electric power price above its equilibrium value [9].

For clarity let's show the simple example, when within CCA frames 1.7 GW of the regeneration for the amount of $857 mn. In comparison with the reserve foundation of the country and of amounts, annually going to the budget proficit, this figure is not comparable marginally. Anyway, from the point of view of the theory and practice even this minor amount exactly shows the amount of the mechanism itself. The explication of all provisions of the theory and of its practical interpretation will take a lot of place, but the essence of the mechanism will not change in this juncture and will be very simple – the minor amount will be equivalent to the price shift, which will be significant enough.

So what is the essence of the CCA mechanism? To be simple, its use will cause the increase of the electricity price in comparison with its average value. For example, let's take the building of about 4 GW, which have been distributed in conformity with CCA mechanism within a year. Let's multiply this figure on the average cost of the building of one GW and let's compare it with total volumes of power sales at the national wholesale market. As a result of these simple calculations it will be found out, that the power price shift will be about 10% upwards from its average value. At the first sight we can see the increase of the visible power efficiency.

But the main problem is that the power is used almost in all branches. Of course, there are export hydrocarbon branches, which do not strongly depend from power prices. But we hardly can highlight many such branches, as most of it strongly depend from power prices. So it happens that the power price increase stipulates the growth of prices for the production, manufactured in other branches, where it is being used. So, possibilities of the growth of these areas are being drastically downshifting.

We can analyze this situation in more details, while using the concept of self-consistent Leontief matrices. Anyway, as it will require new mathematic formulations, first of all we will build evaluations, reflecting the essence of the studied problem with lowest terms of the option, provided endless resources and identic agents, which we have already successfully applied in digital economic-epidemic models [11-15]. That's how it looks at the graph below (figure 1).

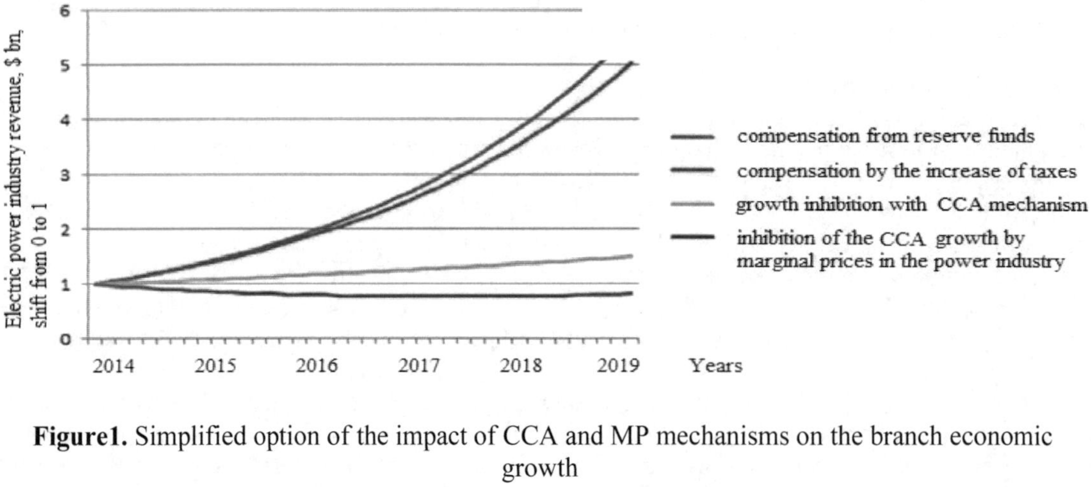

Figure1. Simplified option of the impact of CCA and MP mechanisms on the branch economic growth

Source: estimated by the authors

The first upper curve is typical for objects of Zainskaya State District Power Station (SDPS) like objects, which really needs modernization with use of reserve funds. The modernization of such kind infrastructure objects of the power system of the country requires about 4 bn USD. Provided that annually about 100 bn USD are forwarded to the sovereign welfare fund of Russia, which is the successor of the former stabilization fund, as well as the partially unused budget proficit, the mentioned figure seems rather minor. But with that it can provide for the economics growth of about 4-5% [15-19].

The second curve shows the case, when the modernization of sites like Zainskaya SDPS will not require the sovereign wealth fund, which is placed in other countries and creates certain losses for the country, as well as will be used funds, received at the expense of the increase of taxes in all areas, including exporters of hydrocarbon raw materials. Let's increase by above mentioned 4 bn USD the amount, required for the modernization of the branch infrastructure and minor for the whole branch due to the tax return. The result is that this approach reduces the economic growth by a negligible margin [20-35].

The third curve shows how the CCA mechanism goes, when the power price grows a little, for about 10%. Anyway, the result is the cardinal suppression of the economic growth.

The lower curve shows what is going on with the simultaneous use of the whole scope of both CCA and marginal price forming mechanisms. The main difference here is that the use of the CCA mechanism assumes the building of new stations and the use of the marginal price forming mechanism assumes the increase of the price in order that some unprofitable, but already existing stations continue working. The joint use of these mechanisms fully suppresses the initial five per cent economic growth. So, is evident the impact of one of most powerful mechanisms of the suppression of the economic growth in the country. Accordingly, the combination of such mechanisms (CCA and MP) together with the regulation of the key rate, overestimated taxes and the indefinitely reinforced ruble, that's what the Central Bank of Russia is being doing, what is rather sufficient in order to fully suppress not only the economic growth of 3-5%, but also 8-10% growth, which has been recently demonstrated by China [12].

Intermediate result: we need to correct the current situation with the electric industry on a timely basis, as CCA and MP mechanisms considerably hinder the activity of all consumers of the electric power [13]. In fact, in order to principally improve the position of all consumers of the electric power it is enough to use the direct budget support or to directly involve national welfare funds.

4. Results and discussion

For the more full analysis, which will have practical prospects, as the digitalization provides the necessary experimental information, we will use the consolidated balance-market model of mixed economic systems, which has been developed by us [1-2] and which contains both fundamental, derived from physical, chemical and other correlation laws (ore-coal-iron), as well as situational technologic correlations between branches, which looks like this in the simplest price record:

$$\bar{P} = L \times \bar{P} \qquad (1)$$

$\bar{P}$ - is the vector of prices for m-goods

L - Leontief matrix in the price record

The work [2] offers the idea of averaged Leontief balances, allowing to in-build the market chaos into Leontief balances by choosing a certain averaging mechanism, corresponding to the two hundred year practice of evaluation activities.

In fact, all real economics are mixed [14, 15], it really simultaneously contains a variety of cardinally different Leontief balances. In its logical end this uncontested fact makes it possible for each agent *(j)* to assign the individual Leontief matrix *L(j)*, corresponding to the idealized situation of the replacement of all agents, acting with equivalent goods, by *j*-type agents with its peculiar individual technologies.

The work [2] shows, that if the consolidate Leontief matrix shows up as "capitalist" averaging out of individual *L(j)* – Leontief matrices of agents, then it corresponds to the standard determinist one. It also shows that goods' prices, equal for it averaged Leontief matrix, coincide with market costs in its classic determination by practitioners and theoreticians of evaluating activities [16, 17, 18].

So, we have schematically in-built the statistic ensemble of agents with considerably different technologies of activities in scheduled Leontief balances within a fixed time period *(i)*. In order to obtain a working dynamic model we need to describe, how the interaction of agents *(j)* inside the Leontief cycle $i \rightarrow i{+}1$ causes the redistribution of the capital between them, consequently, to the change of coefficients in the capitalist averaging, what is equivalent to the endogenous change of the most averaged Leontief matrix etc.

The complexity of the front solution of this task is obvious for everyone, who is familiar with the problem of many bodies in physics and mathematics, which, starting from already three bodies, first of all, has not got a common analytic solution, and, secondly, for multitudinal systems it causes such problems as instability of computer computations [19]. If we take into consideration, that the interaction of agents in economics is much more sophisticated than in physics, there are evident complications, arising with the choice of such an approach to its solution.

Anyway, physics itself suggests the possible outcome from this complication in the form of the concentration of the self-consistent field [20]. Without giving specifics, let's highlight, that the concept of the self-consistent field determines the dynamics in the *i*-moment for the *j*-particles by its interaction with the summary (self-consistent) field, created by all other particles in the point of the *j*-particle, getting rid of the reliably inaccurate count of numerous pairwise interactions [21].

This concept is quite natural for real economics. In fact, when you comes to the market to order to purchase apples, you really do interact at a certain counter not with a certain vendor, who, when coming to the market, becomes unfree in its transactions, is included in his "cooperative effects", but with the market itself and with its averaged evaluations of the demand and offer.

The key dimension of the market, around which can be built the initial concept of the self-consistent field, is the market cost, which in two hundred years theory and practice of the evaluation activities [16, 18] is determined as the most probable price of the object and is determined by the practically weighed, by involved capitals, averaging by transactions with equivalent sites.

It should also be pointed out, that independent professional evaluators, while carrying out the total evaluation of the real estate in cities, while carrying out equivalent transactions, are, basically, using the field methodology, while "weighing" objects, geographically remote from the center of the city and of it other "poles", while introducing the dependence of the market price from the set of distances to "poles".

According to Leontief, at the first step of the model building let's limit oneself by market prices, homogenous in the geographic and other kinds of market prices. While synchronizing following actions with the object (goods) by the special case of the "release of goods", we will obtain the result of the interaction of *j*-agents at the *i*-cycle in the one goods approach as [22, 23]:

$$\bar{A}_{i+1} = \bar{A}_i - diag\left(\bar{\xi}_i\right) \times \bar{A}_i + \frac{(\bar{A}_i^T \cdot \bar{\xi}_i)}{(\bar{A}_i^T \cdot \bar{I})} \times \bar{A}_i \qquad (2)$$

$\bar{A}$ – vector of the capital distribution by *j*-agents at *i*-cycle
$\bar{I}$ – single vector
$\bar{\xi}_i$ – normalized vector of the self-cost of manufacturing of goods by *j*-agents
$\frac{(\bar{A}_i^T \cdot \bar{\xi}_i)}{(\bar{A}_i^T \cdot \bar{I})} = < \bar{\xi} >_A$ – «capitalist» averaging, i.e. the weighing of capital transactions

The mathematic kind of the formula (2) is the simplest rule of any market – «if you work worse than the average market – you lose the capital, if you work better – you purchase the capital» at the each $i \rightarrow i{+}1$ cycle.

The scheme of the combination of Leontief balances (1) and of the market redistribution of capitals by (2) in the self-consistent field of market costs is offered in [2]. Below is given the scheme of the quasitensor consolidate model, consolidated for the case "every agent manufactures any goods", which form is convenient for the following computer modeling (3).

$$A(i{+}1,j,k){=}A(i,j,k)-\xi(i,j,k)\times A(i,j,k)+\frac{\sum_{j}^{m} A\,(i,j,k)\times\xi(i,j,k)}{\sum_{j=1}^{m} A\,(i,j,k)}\times A\,(i,j,k) \quad (3)$$

Current individual Leontief matrices (technologic receipts) give individual Leontief prices (self-cost) of the *j*-agent in conformity with the common Leontief scheme (4).

$$P(i,j,k) = \sum_{l=1}^{m} L(i,j,k,l)\times P_0(i,l) \quad (4)$$

For formulas (3), (4):
i – time (cycles) index, *j* – index of agents, $\{k,l\}$ – branch indices of Leontief matrices *L(i, j)*;
L(i, j, k, l) – individual of the Leontief matrix of the *j*-agent at the *i*-cycle, i.e. this is the number of units of goods (e), necessary for the production of goods to (*k*) (*j*)-agent at the (*i*)-cycle;
$P_0(i,k)$ – starting prices for the *i*-cycle of all *k*-goods (possible and non-equilibrium);
A(i, j, k) – starting for the *i* distribution of the capital by the *j*-agent for the release of *k*-goods;
$\xi(i,j,k)$ – relative deviation of the "self cost" from the capitalist averaging, which detailed option will be determined below;
$P_0(i,l)$ – starting market prices for goods at the entry to the *i*-cycle.

The formula (3) is immediately limited to the standard Leontief one (1) by the removal of the dependence from *i* and *j*, i.e. the transfer to technologically identical agents. The transfer to the each certain case (each *j*-agent is dealing with the release of one good) is carried out by the resetting of all other positions in the quasitensor *A(i,j,e)*.

The vector of the self-consistent field of market costs $< P(i,k) >_A$ (5) is, as earlier, the "capitalist" averaging, i.e. the convolution by *j*-index with *A(i,j,k)*:

$$< P(i,k) >_A = \sum_{j=1}^{m} A(i,j,k)\times P(i,j,k)\times\left(\sum_{j=1}^{m} A(i,j,k)\right)^{-1} \quad (5)$$

And after each technologically individual agent interacts with the vector self-consistent field of market costs in conformity with the same rule «worse than the market» – you lose the capital, «better than the market» – you acquire the capital.

Technically, for the use of worked out algorithms let's use the three index matrix (6) of relative deviations of individual prices from self-consistent ones.

$$\xi(i,j,k) = \frac{P(i,j,k)-P_s(i,k)}{P_s(i,k)};\quad P_s(i,k) \cong < P(i,k) > \quad (6)$$

And when the Leontief autoprogress cycle is reduced to the redistribution of the capital by all agents *j* and goods *k* provided its individual technologic Leontief matrices.

We got new version of capital allocation by goods and agents that increases its technological benefit under existing price level.

We introduce external data for nonzero growth of the sum $\sum_{j,k}^{n,m} A(j,i,k)$) instead of *P(0,l)* [16]. Then we enter update on current market prices $P_s(i,l)$ into equation (1) instead of *P(0,l)* and repeat the equation (3) for *A(i +1,j,k,e)*.

A two-product model of "power - everything else" is presented to illustrate the algorithm [21]. The actual spread in own needs of power and remaining items producers is used as basis when constructing the model of two-industry Leontief matrices. In its simplest version it leads to the model matrix (7)-(9):

$$L(j) = LE\times j + LV(1\text{-}j\,/\,N) \quad (7)$$

$$LE = \begin{bmatrix} 0{,}1 & 0{,}03 \\ 0{,}9 & 0{,}97 \end{bmatrix} \quad (8)$$

$$LV = \begin{bmatrix} 0{,}2 & 0{,}1 \\ 0{,}8 & 0{,}9 \end{bmatrix} \quad (9)$$

Approximated matrices *LE, LV* are built on the basis of evaluations of the dispersion of power costs for own needs of Russia generation, as well as shares of the electrical energy industry in the GDP of Russia [24]. Its detailing requires the collection of the digital information from the whole generation of Russia.

IOP Conf. Series: Materials Science and Engineering **937** (2020) 012006 doi:10.1088/1757-899X/937/1/012006

The absence of training of agents in the course of Leontief cycles as well as the uniform start distribution of capitals by agents and goods.

The figure 2 shows the logarithmical scale of the growth dynamics, as well as "specializations" of electric power manufacturers (left wing) and of other goods (right wing) with its link only through Leontief correlations.

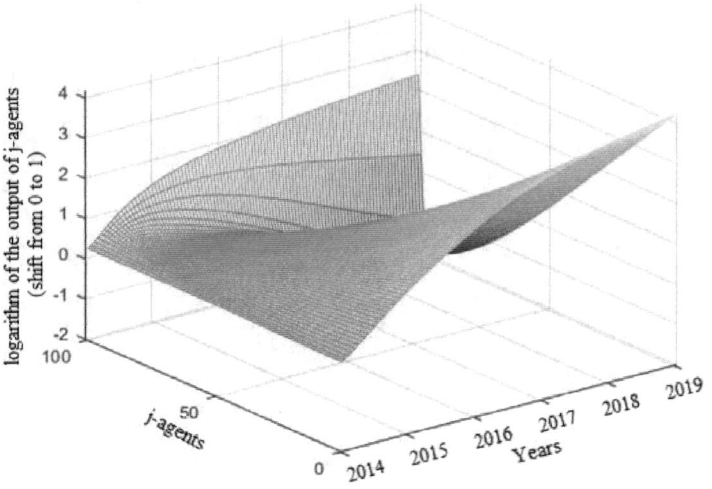

Figure 2. Growth dynamics and "specializations" of manufacturers of the power energy (left wing) and of other goods (right wing) with its connection exclusively through Leontief correlations.
Source: estimated by the authors [25]

The figure 3 shows the logarithmic scale of the integral dynamics of the economic growth of such a system. The same model with the included marginal price forming is shown at figures 4 and 5. Figures 2-5 show the obvious and unjustified redistribution of the capital in favor of agents – manufacturers of the electric power with the suppression of the economic growth of all other branches and of the whole economic system of the country.

5. Conclusions

The digitalization of the Russian economics provides the opportunity for the application of the consolidate balance-market theory of economic systems, developed by authors, provided its systemic and random internal and external interaction. In the work such an opportunity is practically implemented for the "two goods" system – "electric power – everything else".

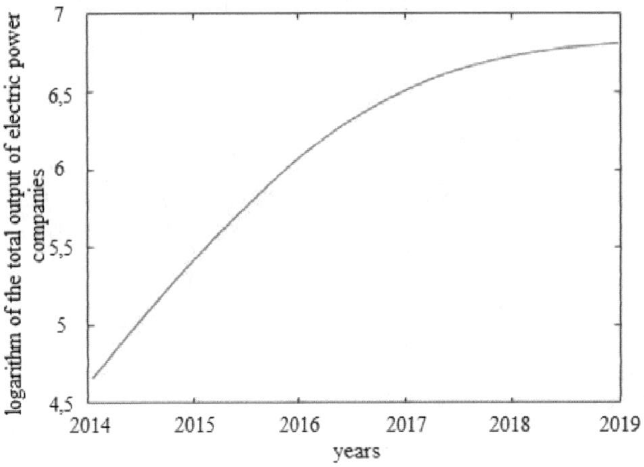

Figure 3. Integral dynamics of the economic growth of the system, represented at figure 2.
Source: estimated by the authors [25]

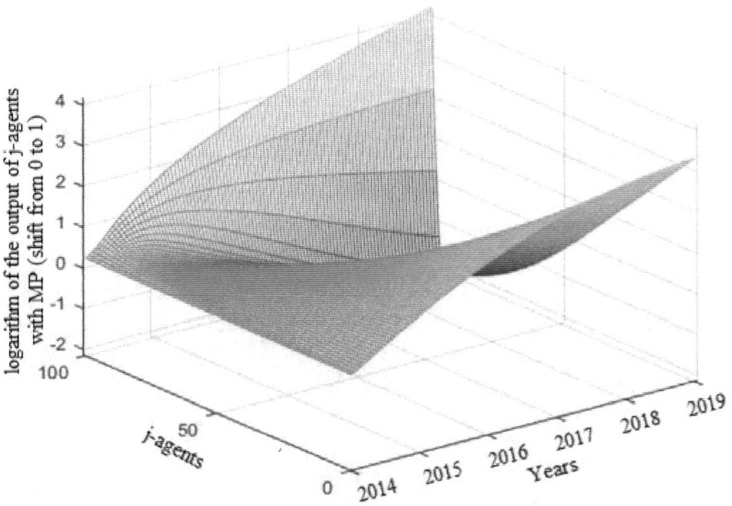

Figure 4. Growth dynamics and "specializations" of manufacturers of the power energy (left wing) and of other goods (right wing) with its connection exclusively through Leontief correlations with the inclusion of the marginal price forming. **Source**: estimated by the authors [25]

Was confirmed the evaluation of the suppression influence of mechanisms of the marginal price forming and of adhesion contracts for the whole economic system.

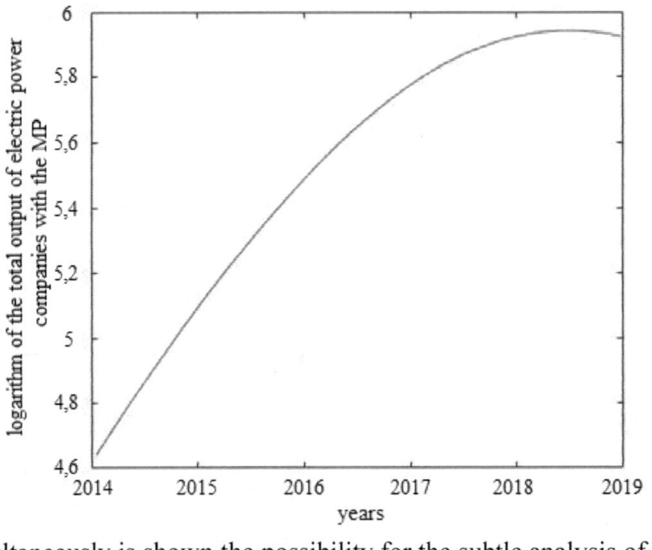

Figure 5. Integral dynamics of the economic growth of the system, represented at the figure 4 with the inclusion of the marginal price forming. **Source:** estimated by the authors [25]

Simultaneously is shown the possibility for the subtle analysis of the dynamics of the specialization and localization of agents, which will be rather important, provided the following development of centralized and distributed power information systems in Russia.

It is expedient to use developed and tested algorithms for the evaluation of the accelerated development of following significant branches, namely, economics of the knowledge.

6. References

[1]. Grachev I D 2010 *Methodology and Econophysical Tools for Economic Progress Modeling* (Moscow: CEMI) p 409

[2]. Grachev I D 2017 *Innovation economy models and techniques* vol 11, ed E Y Khrustalyov and K A Bagrinovsky (Moscow: CEMI) pp 42–47

[3]. Anikeenok O A and Eremin M V 2019 *Guidance of Self-consistent Field Method with Hartree-Fock Approximation* (Kazan: Kazan State University) p 30

[4]. Sabir A and Ibrir S 2020 Robust disturbance rejection control of grid-connected fuel cell converters with grid support ability under unbalanced faults *Energy Syst.* **11** 673–698 https://doi.org/10.1007/s12667-019-00332-4
[5]. Choudhury T R, Nayak B, De A et al 2020 A comprehensive review and feasibility study of DC–DC converters for different PV applications: ESS, future residential purpose, EV charging *Energy Syst.* **11** 641–671 https://doi.org/10.1007/s12667-019-00331-5
[6]. Voropay N I 2017 *Tenth Int. Conf. Management of Large-Scale System Development (MLSD)* (Moscow: National Research University – Higher School of Economics)
[7]. Koutovoy G P 2019 *Regional energy and energy conservation* vol 3, ed E Zholudeva et al (Moscow: System Consulting) pp 80–84
[8]. Fedorov Ya P 2019 Financial and Economic Aspects of Power: 10 Years After Liquidations of RAO UES *Russian Journal of Entrepreneurship* vol 20(1), ed O D Protsenko et al (Moscow: Creative Economy) pp 173–184 DOI: 10.18334/rp.20.1.39753
[9]. Gushchin N and Gubaidullina A 2019 *Realnoe Vremya* **21**
[10]. Edelstein-Keshet L 2005 *Mathematical Models in Biology* (Vancouver: University of British Columbia) p 586
[11]. Brondbo H K, Storebo A, Boomsma T K et al 2020 A real options approach to generation capacity expansion in imperfectly competitive power markets *Energy Syst.* **11** 515–550 https://doi.org/10.1007/s12667-019-00325-3
[12]. Makarov A A and Veselov V F 2018 *Eleventh Int. Conf. Management of Large-Scale System Development (MLSD)* (Moscow: National Research University – Higher School of Economics)
[13]. Glaziev S Y 2016 National economy structures in the global economic development *Economics and Mathematic Methods* vol 52(2), ed V L Makarov et al (Moscow: Science) 3–29
[14]. Glaziev S Y 2016 Applied results in the theory of world economic structures *Economics and Mathematic Methods* vol 52(3), ed V L Makarov et al (Moscow: Science) 3–21
[15]. Belyaeva A V and Grebenyuk E A 2014 Building of models of the mass evaluation of real estate objects provided the spatial dependence *Control Sciences* vol 1, ed F T Aleskerov et al (Moscow: Sensidat-Plus) 45–52
[16]. Federal Act No 135 of 29 July 1998 On Concerning Valuation Activities in the Russian Federation *Russian Federation Code*
[17]. Krzysztof G et al 2000 *The European Group of Valuers' Associations* (Belgium – Gillis nv/sa) p 378
[18]. Weiss Pierre 1907 *J. Phys. Theor. Appl.* **6** 661-690 https://doi.org/10.1051/jphystap:019070060066100
[19]. Boudec J Y L et al 2007 *Fourth Int. Conf. on the Quantitative Evaluation of Systems (QEST)* (UK – Edinburgh) pp 3–18 doi: 10.1109/QEST.2007.8 https://dx.doi.org/10.1007/s11537-007-0657-8
[20]. Grachev I D and Nekrasov S A 2013 *National Interests Priorities and Security* vol 4(193), ed S F Vikulov et al (Moscow: Finance and Credit) 2–6
[21]. Stridbeck U et al 2005 Electricity Market Experience: Lessons from liberalised electricity markets *International Energy Agency* (France: Cedex) 274
[22]. Sitharthan R, Geethanjali M and Pandy TKS 2016 Adaptive protection scheme for smart microgrid with electronically coupled distributed generations *Alexandria Engineering Journal* **55(3)** 2539-2550
[23]. Fathima AH, and Palanisamy K 2014 Battery energy storage applications in wind integrated systems—a review *IEEE International Conference on Smart Electric Grid* 1-8

[24]. Prabaharan N and Palanisamy K 2015 Investigation of single-phase reduced switch count asymmetric multilevel inverter using advanced pulse width modulation technique *International Journal of Renewable Energy Research* **5(3)** 879-890.

[25]. Jerin ARA, Kaliannan P and Subramaniam U 2017 Improved fault ride through capability of DFIG based wind turbines using synchronous reference frame control based dynamic voltage restorer. *ISA transactions* **70** 465-474

[26]. Sitharthan, R, Sundarabalan CK, Devabalaji KR, Nataraj SK and Karthikeyan M 2018 Improved fault ride through capability of DFIG-wind turbines using customized dynamic voltage restorer *Sustainable cities and society* **39** 114-125

[27]. Prabaharan N and Palanisamy K 2016 A single-phase grid connected hybrid multilevel inverter for interfacing photo-voltaic system *Energy Procedia* **103** 250-255

[28]. Palanisamy K, Mishra JS, Raglend IJ and Kothari DP 2010 Instantaneous power theory based unified power quality conditioner (UPQC) *IEEE Joint International Conference on Power Electronics, Drives and Energy Systems* 1-5

[29]. Sitharthan R and Geethanjali M 2017 An adaptive Elman neural network with C-PSO learning algorithm-based pitch angle controller for DFIG based WECS Journal of Vibration and Control **23(5)** 716-730

[30]. Sitharthan R and Geethanjali M 2015 Application of the superconducting fault current limiter strategy to improve the fault ride-through capability of a doubly-fed induction generator–based wind energy conversion system *Simulation* **91(12)** 1081-1087

[31]. Sitharthan R, Karthikeyan M, Sundar DS and Rajasekaran S 2020 Adaptive hybrid intelligent MPPT controller to approximate effectual wind speed and optimal rotor speed of variable speed wind turbine *ISA transactions* **96** 479-489

[32]. Sitharthan R, Devabalaji KR and Jees A 2017 An Levenberg–Marquardt trained feed-forward back-propagation based intelligent pitch angle controller for wind generation system *Renewable Energy Focus* **22** 24-32

[33]. Sitharthan R, Sundarabalan CK, Devabalaji KR, Yuvaraj T and Mohamed Imran A 2019 Automated power management strategy for wind power generation system using pitch angle controller *Measurement and Control* **52(3-4)** 169-182

[34]. Sundar DS, Umamaheswari C, Sridarshini T, Karthikeyan M, Sitharthan R, Raja AS and Carrasco MF 2019 Compact four-port circulator based on 2D photonic crystals with a 90° rotation of the light wave for photonic integrated circuits applications *Laser Physics* **29(6)** 066201

[35]. Sitharthan R, Parthasarathy T, Sheeba Rani S and Ramya KC 2019. An improved radial basis function neural network control strategy-based maximum power point tracking controller for wind power generation system *Transactions of the Institute of Measurement and Control* **41(11)** 3158-3170

Design and testing of proton exchange membrane fuel cell (PEMFC) power pack for platform vehicle

G Anilkumar[1], N Puneetha[2], G V Brahmendra Kumar[3], K Palanisamy[4*] and Amrish Gupta[5]

[1,2] School of Renewable Energy, JNT University, Kakinada, 533003, India.
[3,4]Centre for Smart Grid Technologies, Vellore Institute of Technology, Vellore, 632014, India.
[5]BHEL R&D Division, Hyderabad, 500093, India
Email: gorantlaanilkumar@gmail.com, brahmendrakumar.g@gmail.com
*Corresponding author: kpalanisamy@vit.ac.in

Abstract. During the last few years and in this context of constant innovation, several alternatives for internal combustion engines arose with the aim to find feasible options to fossil fuel consuming devices. As one may recall, the main alternative for internal combustion cars is without a doubt BEVs (battery electric vehicle). Those vehicles rely on electricity in order to function, storing the electricity in a battery pack and using it to feed a motor in order to give motion. However, some issues are still to solve in order to market them as appealing as internal combustion cars. Two of the most important issues to tackle are mileage range and charging time. On the other hand, fuel cell technology which began in the late sixties is beginning to take part in the automotive sector. This paper presents design evaluation and testing of 100 kW PEM fuel cell power pack for battery operated platform vehicle application developed to drive PMSM motor with requisite energy range that can be adapted by the power conditioning elements of the system under implementation to attain desired speed under Matlab/Simulink environment and the overall system dynamic performance was found to be coherent by analyzing basic operating parameters.

1. Introduction

Nowadays there are many technologies that could actually partially or even totally substitute fossil fuels in energy matters, all related in some way to electricity production such as hydro power generation, wind power, photovoltaic, solar thermal energy, etc. [1]. Just as energy industry, automotive industry which is always in a context of constant innovation has been inflicting change trying to find feasible alternatives to internal combustion engines which are the main cause of fossil fuel consumption and pollution in mobility currently [2]. Recently companies like Toyota, Hyundai, Honda or Nikola Motors have been making progress in this technology, and even launching commercial models such as the Toyota Mirai. This type of technology takes advantage of hydrogen being the most abundant element in nature and tackles the problem of autonomy range and recharge time. [3] discussed about comparative analysis of the plugin fuel cell vehicles (PFCV) regarding different topologies, drive cycles and control strategies for PFCV systems. Then the proposed strategy is able to reduce hydrogen consumption for the high-speed drive cycles with the less braking operation. As well as the on-off switching frequency of the PEMFC system to design and development in this presentation presented. An optimization strategy is proposed at first by regulating the power distribution between the battery and the PEMFC system. Then a direct multiple shooting (DMS) algorithms are used to solve this nonlinear programming problem (NLP) over the spatial domain. The PEMFC in portable applications and simulation results was presented in [4]. A detailed analysis of the model-independent tracking techniques to extract maximum power (MPP) and maximum efficiency point (MEP). PEMFC exhibits nonlinear output characteristics which may result in low efficiency and low power operation of the PEMFC. However, the output and efficiency of the PEMFC should be high for a portable application but they are low at MEP and MPP respectively in the paper. The main advantage is a portable application mobile in nature like a small vehicle truck utilizing fuel cell electric power. In [5], Stated that for vehicle applications to be qualified as high efficiency electrical system, it should consist of a fuel cell, a battery and superconductor. A precise power management unit (PMU) was discussed to propose a hybrid energy storage (HES) to improve its performance by combining various promising devices in [6-10]. These devices are used for storage purpose because of the sudden vehicle hydrogen gas will empty the power supply from the battery to run the vehicle. Here, the hybrid vehicle design consideration is predominantly energy storage. In [11-16] carried a study with an emphasis on reducing fossil fuel consumption and environmental protection. A delivery van design was proposed based on the fuel cell and battery size, to reduce emissions in vehicular trucks using world operational data and duty cycles from GPS technique. However, cost is critical to the success and adoption of these zero emissions trucks. In [17-28] studied the influence of the physical and geometrical gas flow parameters on the consumption of reagents in PEMFC and identified the performance of the fuel cell, a wider channel improve, chemical reaction in the catalyst layer, voltage drop, performance increase in saturation temperature, performance of the PEMFC rates, effect of rib/channels width which showed greatest sensitivity, activation and diffusion, highest power densities are made with speeds of flow low due to the smaller activation [29-36].

2. Motivation for using PEMFC

A fuel cell of polymer electrolyte membrane also called proton exchange membrane the fuel cells applications, including transportation and automotive industry applications, are currently the most used type of fuel cell for most of PEM as well as offering low weight and volume when put aside other types of fuel cells [7]. This type of cell is capable to deliver high power density (nowadays 0.7 Wcm^{-2}). In PEMFCs use a polymer as an electrolyte, carbon electrodes and gas diffusion layers and platinum-based catalysts. The main advantage of this type of fuel cell is that they feed from hydrogen and oxygen from the air. The typical PEMFC is presented in figure 1.

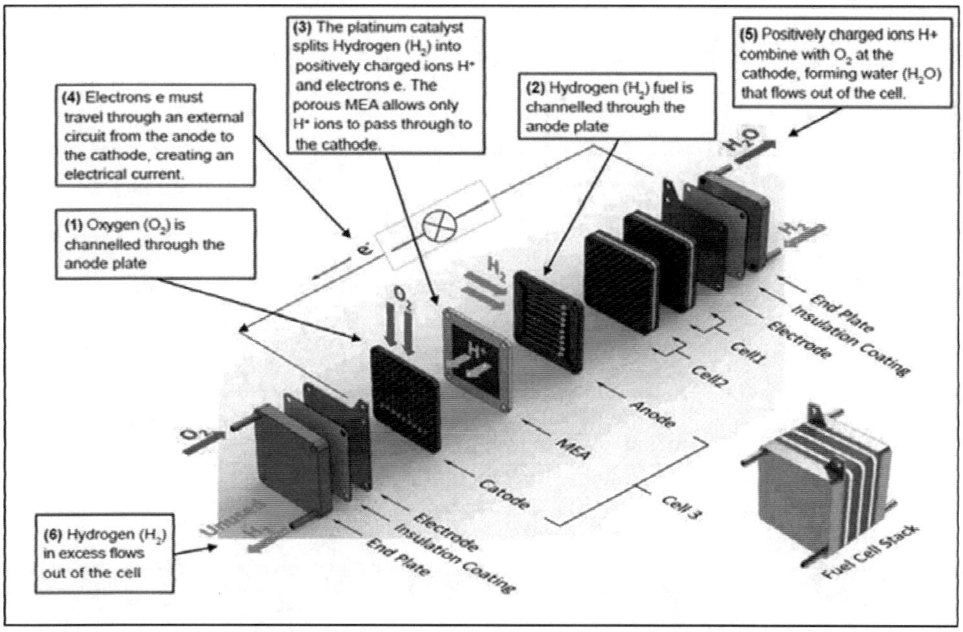

Figure 1. Layout and operation of PEMFC.

Normal operating temperature for this type of fuel cell are around 80°C (considered relatively low temperatures). However, it has been proved that cold starts for fuel cell vehicles are possible even in below 0°C situations [37]. As they are able to start in low temperatures, they have a quick start and that entail less wear of the components of the system and therefore, gives more durability to the whole set. Nevertheless, the use of platinum as catalyst increases the cost of the fuel cell stack drastically [38]. In all, PEMFCs are a very suitable and feasible alternative for mass producing hydrogen powered vehicles due to how technologically advanced and reliable they are, as well as their good power to weight ratio [39-46].

3. Modeling and testing of PEMFC power pack for platform truck

To achieve high current density optimal operating conditions, need to be identified for a fuel cell by system thickness, catalysts, and alloys particle size, catalyst quantity, catalyst layer thickness to analyze in this research. The most promising system among several different kinds of fuel cells due to their various advantages such an easy startup and operating temperature range from 30^{0}C to 80^{0}C, no liquid electrolyte, the optimal thickness of the membrane. Membranes are usually lightweight and their thickness is in microns (look like a very thin sheet around 20 microns which enables necessary ions only to pass between anode and cathode). In short it consists of a polymer electrolyte membrane along with catalyst and diffusion layers to achieve the desired purpose.

3.1 Balance of Parts in PEM Fuel Cell for Platform Truck

The following figure 2 displays fuel cell power conversion system components that are utilized in the development of PEMFC based platform truck and the following sections brief description about them. The following are the subcomponents of the fuel cell system that are used in the development of PEMFC based platform truck which are discussed as follows.

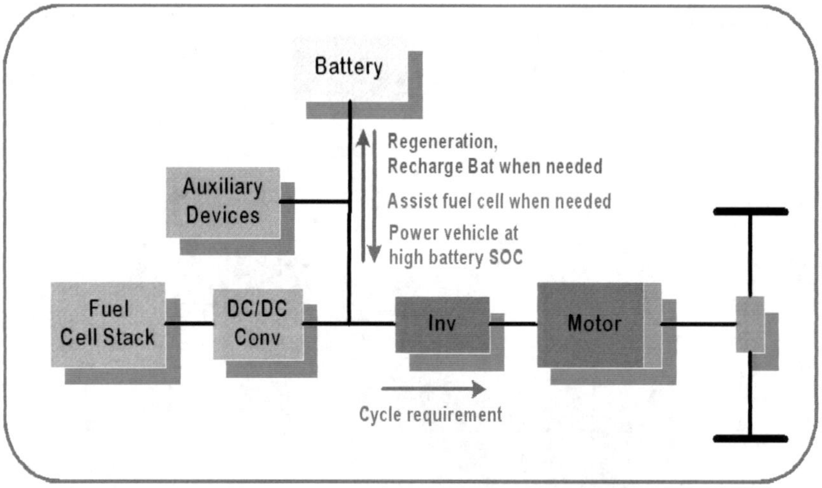

Figure 2. Fuel cell power conversion system components.

3.1.1 Fuel cell stack

A single fuel cell consists of membrane electrode assembly (MEA), flow fields for reactant gas flow delivering the cell voltage about 0.7-0.9 V. The output of the fuel cell depends upon the size of the stack. Increasing the number of cells increases the voltage. To obtain the maximum power of 3 kW (2.5 kW net power), 75 cells are used and assembled, called a stack. Considering stack reference parameters, it is intended to develop a new stack in Matlab/Simulink environment, to achieve 100 kW by increasing area as well as number of cells along with their dimensions.

3.1.2 Power Conditioners

According to the polarization curve [7], unregulated DC voltage generated in fuel cell drops down when the current increases. As a tradeoff, power is usually conditioned before supplying to load with the use of DC/DC regulators and/or inverter regulators to raise the voltage between battery and inverter.

3.1.3 DC/DC converter

It should be ensured that overall fuel cell stack output voltage must be equal to the grid voltage while designing the number of cells and their output interconnections either in series or parallel. If for some reason the number of cells is chosen differently, a DC/DC converter can be used to achieve the right output voltage level. The voltage over the stack is normally not constant and when the current increases there will be a voltage drop. In fuel cells this voltage drop is greater than in normal electrical power generators. In this case there is also need for a DC/DC converter to solve this problem [9]. This problem can be rectified by developing the new fuel cell stack in Matlab environment by modifying (either by increasing or decreasing the number of cells) to achieve the desired voltage (ex: 400 cells for 400 V assuming 1 V output from each fuel cell).

3.1.4 DC/AC inverter

The inverter or DC/AC converter is the device responsible for transforming the DC current obtained in the boost converter after the fuel cell stack to AC inverter current in order to feed the electric motor in the drive train system. In order to do those parameters such as output voltage, output frequency, and total delivered power have to be specified, and the final design of the converter will vary if those parameters vary. When decelerating, the inverter has to be capable to obtain the AC current generated in the motor and transform it in DC current in order to feed the battery. For that reason, it is needed a three-phase inverter, scheme of which is shown below.

3.1.5 Battery

Demand for extra power during driving tasks for longer periods of time should be available at high speeds such as towing a load up an incline. This extra energy required for a hard acceleration can generally be provided by batteries for less cost and mass. Further, it is also useful when the vehicle is operating in a very low-efficiency region. During low-power demands, the control method adopted by battery management system can be used to improve the overall efficiency of the vehicle [10][11].

3.1.6 Motor

As seen before, fuel cell vehicles obtain electricity to feed an electric motor. Electric motors are divided in two types according to whether they work with direct (PMDC) or alternating current (PMAC). For PMAC motors we can find two different types; Brushless DC Motors (BLDCM) and Permanent Magnet Synchronous Motor (PMSM). But the most used type of motor in automotive solutions due to power/weight ratio, durability and technology knowledge is the Permanent Magnets Synchronous Motor (working with alternate current). While brushless DC motors generate a trapezoidal counter electromotive force when the rotor spins, PMSM generate a sinusoidal force.

4. Configuration of the proposed PEMFC vehicle system

The configuration of an entire driving system formed by the fuel cell stack, the DC/DC boost converter, DC/AC converter and PMSM motor has been chosen. The Matlab/Simulink tool is used to implement an integrated design level sub system structures such as energy management, electrical and fuel cell vehicle (FCV) dynamics to model complex systems to analyze the desired scenario as shown in figure 3.

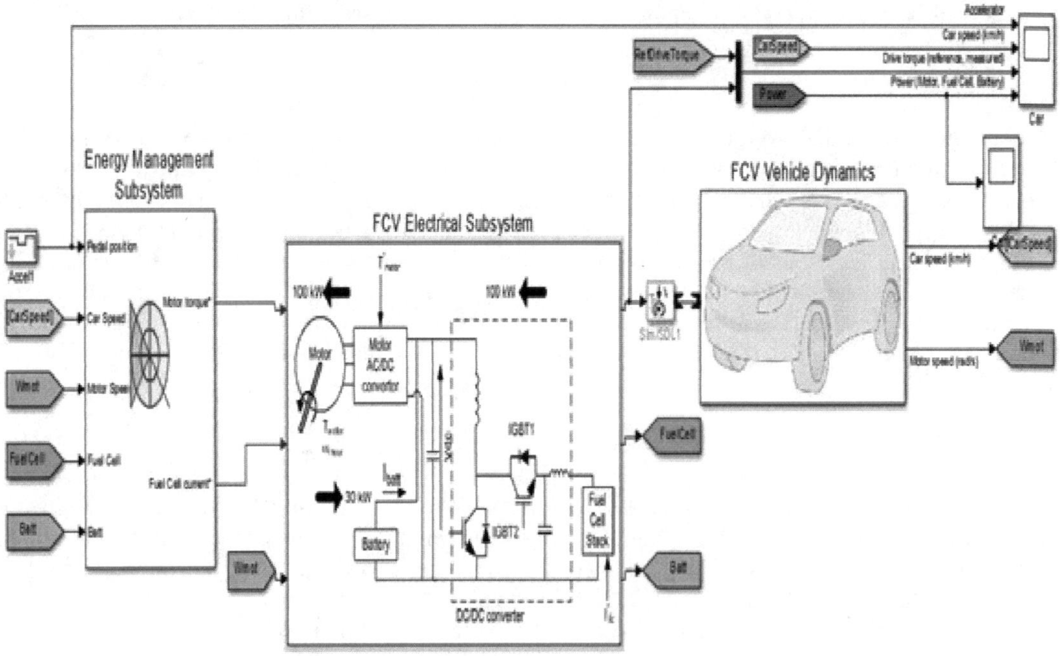

Figure 3. Proposed system of the 100-kW fuel cell vehicle.

4.1 Energy management subsystem

It is essential to implement energy management strategy to ensure dynamic response as well as reliable operation of a PEM fuel cell generator to enhance its lifetime. The following figure 3 shows the integration of energy management, electrical and fuel cell dynamics simulation subsystems with its ancillaries connected via DC electrical network.

For a conventional PEMFC fuel cell, the output voltage can obtain through the Nernst equation in order to attain reversible potential of the fuel cell [12]. This equation only corresponds to the behavior for one single fuel cell. Every fuel cell within the fuel cell stack is driven by this same equation. The corresponding equation in a proper electrical model for a fuel cell can be designed. However, this model will depend on the inputs of both temperature and current to work properly.

$$\Delta G = nFE \tag{1}$$

Where ΔG = Gibbs free energy, n is the number of electrons transferred in the reaction (from balanced reaction), F is the Faraday constant (96,500 C/mol), and E is potential difference.

Figure 3 exhibits the simplified electrical modelling diagram of fuel cell system and its energy management strategy using Simulink platform. The proposed energy management strategy must be able to respond to dynamics requested by fuel cell electrical quantities. The cooling circuit must maintain fuel cell temperature in its operating range and the fuel cell voltage must be monitored to avoid gas diffusion limitation problems. The fuel cell Stack is supplied with hydrogen by high pressure cylinders (From 300 to 700 bars) aboard vehicle and its flow is regulated by pressure reducing valve. The determination of air flow reference depends on fuel cell voltage. Indeed, by increasing air flow over its real consumption in the stack, partial pressure of oxygen will increase in cathode compartment. This overpressure will then increase requisite fuel cell voltage. The following table 1 shows the list of simulation parameters along with their specification that are considered for project execution.

Table 1. System parameters and specifications.

Parameter	Specifications
Power of Stack	
Nominal of stack power	85500 W
Maximal of stack power	100022.4 W
Resistance of FC	1.1 Ω
Voltage (En)	1.1 V
Utilization of nominal	
Nominal of Hydrogen (H_2)	95.24%
Nominal of Oxidant (O_2)	50.03%
Consumption of nominal	
Nominal of fuel flow	794 LPM
Nominal of air flow	1891 LPM
Current (i_o)of exchange	0.024152 A
Coefficient of exchange	1.1912
FC signal variation parameter	
Composition Fuel (x_H_2)	99.95%
Composition Oxidant (y_O_2)	21%
At nominal hydrogen fuel flow rate utilization	
Nominal of litre per minute (LPM)	374.8 LPM
Maximum of litre per minute	456.77 LPM
At nominal oxidant utilization: Air flow rate	
Nominal	1698 LPM
Maximum	2069 LPM
Temperature (T)	368 K
Pressure of fuel supply	3 bars
Pressure of air supply	3 bars

4.2 FCV electrical system

The power systems of a fuel cell vehicle are the battery (Li-ion), the inverter and the motor. The battery is set after the boost converter in order to take advantage of the risen voltage, and manage energy charge and discharge of the system. Therefore, the battery is set in parallel between the boost converter and the inverter. Nominal voltage of the battery is set as the value obtained in the converter of output boost. Battery capacity set this nominal voltage and the power that is to be needed. When implementing the motor-inverter structure, a motor model is selected.

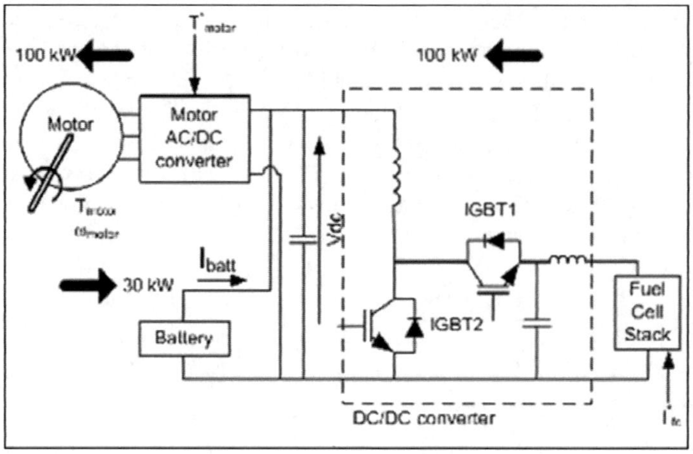

Figure 4. FCV electrical subsystem.

In this case just a 100-kW peak power motor is required and motor parameters can be subtracted. These parameters such as stator resistance, motor inductances and flux linkage can be set into the inverter model in order to link those two elements. However, these elements need to be controlled and for that, a speed and current control has to be implemented. As a speed and vector controller is implemented, taking the motor speed, obtaining the required torque and then obtaining the current control in order to drive the inverter.

4.3 FCV vehicle dynamics

In order to test the system in real conditions once it has all been modelled, it has been decided to put this system through a driving cycle consisting in an acceleration ramp, a velocity maintaining phase and a deceleration ramp between 0 and 25 km/h, as the object of study is a vehicle with small power output, such a scooter or a truck. In order to do that, several equations have to be implemented in what is called the dynamic model of the vehicle. That means that the force balance is implemented in order to get the force required by the wheel and eventually, both torque and angular speed in order to be able to control the electrical system of the vehicle.

When implementing the dynamic model of the vehicle, several assumptions have been made, such as considering the vehicle a rigid body with only one axis affected by the forces that intervene in the system. Hence, the which the model will be based is law of motion which says that the acceleration of an object as produced by a net force is mass times proportional in the same direction to this same force. In the case of the system in study, the resulting force can be written as the forces that help the vehicle accelerate (tractive forces) and the forces that make the vehicle decelerate (resistive forces) [13]. Several forces can be described as resistive, in which the aerodynamic drag, the rolling force and the grading force can be included. The following table 2 list out the dynamic model parameters that are considered for the execution of this project.

Table 2. dynamic model parameters.

Parameter	**Specifications**
Mass (kg)	1625 kg
Ground from the system height	0.5 m
Front axle from horizontal distance	1.4 m
Area of frontal	2.711 m^2
Atmospheric density	1.2 kgm^{-3}

IOP Conf. Series: Materials Science and Engineering **937** (2020) 012007 doi:10.1088/1757-899X/937/1/012007

Drag coefficient (C_d)	0.26
Effective rolling radius	0.25 m
Load vertical rated	3000 N
Rated load longitudinal peak force rated	3500 N
Rated load (%) slip at peak force	10%

5. Results and discussion

To ensure proper functioning of every part of the proposed PEM fuel cell system, the simulation results are carried out on completion of the desired system modelling. Simulation of each component and their relevant functional testing results are outlined in the figure 4.

Figure 4 shows the simulation results obtained by integrating the above discussed subsystems. From figure 4 one can clearly examine time in seconds on x-axis and parameters of concern such as acceleration, vehicle speed, driving torque and power delivered on y-axis. A discussion related to the analysis of these four important parameters with respect to time is presented below. Figure 4 shows the simulation results of dynamically modelled fuel cell vehicle exhibiting system characteristics for a time duration of 16s under four different operation modes i.e. acceleration, cruising, fast acceleration and deceleration.

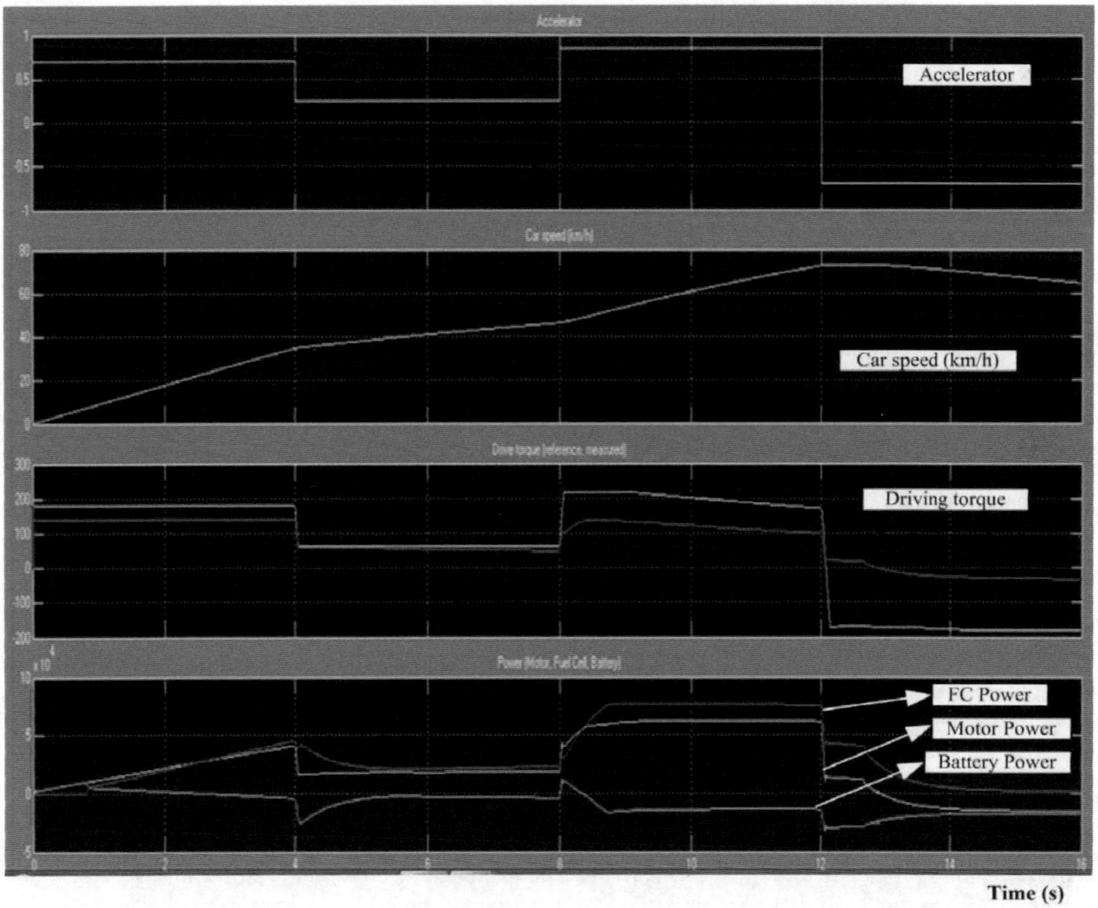

Figure 5. Simulation results and power outputs throughout the full driving cycle.

At t = 0s, the FCV is stopped and pushes the accelerator pedal to 60%. Since the fuel cell cannot start as quickly as required, the battery provides the motor power until the fuel cell starts. In fact, in real operating conditions, this is mainly due to the inertia exhibited by fuel cell stack to attain required chemical reaction rate between hydrogen and oxygen.

At t = 0.7s, the fuel cell begins to provide power but is not able to reach the reference power due to its large time constant. Therefore, the battery continues to provide electrical power to the motor.

In real operation condition, the fuel cell stack initially starts up small power will be required because of the auxiliary power systems operates fuel cell staring initial condition. Like pressure gauge and safety valves temperature controller operating at the initial stage. So that power is taken from the battery. Therefore, the battery continues to provide electrical power to the motor. Real operation of the fuel cell stack therefore the battery continues to provide electrical power to the motor. because of continuous hydrogen react with a membrane so unwanted particle enters into the pipeline reaction rate will be suddenly reduced at the stack power produces also decrease so suddenly the speed of car will slow that problem will be rectified. Therefore, the battery continues to provide electrical power to the motor.

At t = 4s, the accelerator pedal is released to 25%. The fuel cell cannot decrease its power instantaneously. Therefore, the battery absorbs excess of the fuel cell power in order to maintain the required power.

At t = 6s, the fuel cell power is equal to the reference power. The battery is no more needed. Once the fuel cell generates the electrical power, the battery is no more needed.

At t = 8s, the accelerator pedal is pushed to 85%. The battery helps the fuel cell by providing an extra power of 50 kW.

At t = 12s, the accelerator pedal is set to -25%, corresponding to a regenerative power of 50 kW. The motor acts as a generator driven by the vehicle's wheels. The kinetic energy of the FCV is transformed into electrical energy which is stored in the battery. For this pedal position, the battery absorbs the regenerative power and the residual fuel cell power.

The improvement of the EMS as well as the dimensioning of the components can be obtained when the simulation is done with great exactitude. For example, so as not to exceed the maximum power of the battery, it is possible to directly control the electric motor in order to limit its current. These current peaks can be determined with precision through dynamic simulation. This model is thus the central point of the electric vehicular components since all other systems depend on the battery behavior. A systematic approach has to be followed while switching ON and OFF of the system to ensure reliable operation of the above-mentioned fuel cell system. During ON condition, the battery should be turned ON at first followed by fuel cell, DC/DC converter and inverter in a sequential manner otherwise due to current and voltage fluctuations, inverter leads to failure operation. During OFF state, initially the inverter has to be switched OFF followed by DC/DC converter, fuel cell and then the battery to curtail current and voltage fluctuations.

6. Conclusion

Regarding fuel cell technologies, it is clear from that PEMFC is the technology offering better performance when it comes to applying those to the automotive sector. Other technologies might be better when it comes to stationary energy procurement purposes. Nevertheless, PEM fuel cells still have a long way in terms of material improvement and cost. The major breakthrough in this type of technology will come when manufacturing is cheap enough to compete with conventional modes of transportation, as other characteristics have been already matched, such as autonomy range and recharge time. In the proposed system, it can be seen that the results of the overall system are coherent and that the system responds as it should in the proposed situation. The stack of cells is able to deliver the range of energy for which it was conceived and the power conditioning elements are able to adapt and control this power in order to give the electric motor the power to move the vehicle to the desired speed and much improvement can be done in order to perfect this model.

***Acknowledgments**:* The authors express their gratitude to the BHEL R&D Division, Hyderabad; JNT University, Kakinada, and Centre for Smart Grid Technologies, Vellore Institute of Technology, Vellore, India for their entire support towards this research work.

References

[1] Brahmendra Kumar G. V. and Palanisamy K. (2019) "A Review on Microgrids with Distributed Energy Resources," *2019 Innovations in Power and Advanced Computing Technologies (i-PACT)*, Vellore, India, 1-6, 2019, doi: 10.1109/i-PACT44901.2019.8960189.

[2] Kumar, G.V.B.; Sarojini, R.K.; Palanisamy, K.; Padmanaban, S.; Holm-Nielsen, J.B. (2019) "Large Scale Renewable Energy Integration: Issues and Solutions," *Energies*, *12*, 1996, 2019.

[3] Yang B., Rajashekara K., Hongfei X., and Yaoyu L. (2017) "A comparative analysis of fuel cell vehicle architectures," *2017 36th Chinese Control Conference (CCC)*, Dalian, 2017, 9226-9231, doi: 10.23919/ChiCC.2017.8028826.

[4] Kumaraswamy V. K. and Quaicoe J. E. (2016) "Tracking techniques for the PEMFC in portable applications," *2016 IEEE Electrical Power and Energy Conference (EPEC)*, Ottawa, 1-6, doi: 10.1109/EPEC.2016.7771720.

[5] Sami, B.S., Imene, B.Z., Yahyaoui, I., Nasri, S., Adnane, C., & Moussa, M.F. (2017) "Study of dynamic performance of a Hybrid Electric System combining FC/Battery/SC," *2017 8th International Renewable Energy Congress (IREC)*, 1-6, 2017.

[6] Kumar, G.V.B.; Kaliannan, P.; Padmanaban, S.; Holm-Nielsen, J.B.; Blaabjerg, F.; (2020) "Effective Management System for Solar PV Using Real-Time Data with Hybrid Energy Storage System. *Appl. Sci.* 2020, *10*, 1108, 2020.

[7] Hoogers, G. (Ed.). (2003) "Fuel Cell Technology Handbook" Boca Raton: CRC Press, 2003. https://doi.org/10.1201/9781420041552

[8] Boucetta, A., Ghodbane, H., Ayad, M.Y., Bahri, M. (2016) "A review on the performance and modelling of proton exchange membrane fuel cells," AIP Conf. Proc. 1758, 030019, 2016.

[9] Kumar G. V. B. and Palanisamy K., (2019) "Interleaved Boost Converter for Renewable Energy Application with Energy Storage System," *2019 IEEE 1st International Conference on Energy, Systems and Information Processing (ICESIP)*, Chennai, India, 2019, pp. 1-5, doi: 10.1109/ICESIP46348.2019.8938306.

[10] Brahmendra Kumar G. V., Kumar G. A., Eswararao S. and Gehlot D. (2018) "Modelling and Control of BESS for Solar Integration for PV Ramp Rate Control," *2018 International Conference on Computation of Power, Energy, Information and Communication (ICCPEIC)*, 368-374, 2018. doi: 10.1109/ICCPEIC.2018.8525173.

[11] Gundumalla V. B. K. and Eswararao S., (2018) "Ramp Rate Control Strategy for an Islanded DC Microgrid with Hybrid Energy Storage System," *2018 4th International Conference on Electrical Energy Systems (ICEES)*, Chennai, 2018, pp. 82-87, doi: 10.1109/ICEES.2018.8442363.

[12] Frederick J. Barclay., (2006) "Fuel Cells, Engines and Hydrogen: An Exergy Approach," 2006. doi:10.1002/9780470030264.

[13] Petrinić M. and Jakopović Ž. (2018) "Modeling and simulation of PEM fuel cell – power converter system," 2018**.** https://bib.irb.hr/datoteka/308658.MIPRO2007_Petrinic.pdf.

[14] Sitharthan R, Geethanjali M and Pandy TKS 2016 Adaptive protection scheme for smart microgrid with electronically coupled distributed generations *Alexandria Engineering Journal* **55(3)** 2539-2550

[15] Fathima AH, and Palanisamy K 2014 Battery energy storage applications in wind integrated systems—a review *IEEE International Conference on Smart Electric Grid* 1-8

[16] Prabaharan N and Palanisamy K 2015 Investigation of single-phase reduced switch count asymmetric multilevel inverter using advanced pulse width modulation technique *International Journal of Renewable Energy Research* **5(3)** 879-890.

[17] Jerin ARA, Kaliannan P and Subramaniam U 2017 Improved fault ride through capability of DFIG based wind turbines using synchronous reference frame control based dynamic voltage restorer. *ISA transactions* **70** 465-474
[18] Sitharthan, R, Sundarabalan CK, Devabalaji KR, Nataraj SK and Karthikeyan M 2018 Improved fault ride through capability of DFIG-wind turbines using customized dynamic voltage restorer *Sustainable cities and society* **39** 114-125
[19] Prabaharan N and Palanisamy K 2016 A single-phase grid connected hybrid multilevel inverter for interfacing photo-voltaic system *Energy Procedia* **103** 250-255
[20] Palanisamy K, Mishra JS, Raglend IJ and Kothari DP 2010 Instantaneous power theory based unified power quality conditioner (UPQC) *IEEE Joint International Conference on Power Electronics, Drives and Energy Systems* 1-5
[21] Sitharthan R and Geethanjali M 2017 An adaptive Elman neural network with C-PSO learning algorithm-based pitch angle controller for DFIG based WECS Journal of Vibration and Control **23(5)** 716-730
[22] Sitharthan R and Geethanjali M 2015 Application of the superconducting fault current limiter strategy to improve the fault ride-through capability of a doubly-fed induction generator–based wind energy conversion system *Simulation* **91(12)** 1081-1087
[23] Sitharthan R, Karthikeyan M, Sundar DS and Rajasekaran S 2020 Adaptive hybrid intelligent MPPT controller to approximate effectual wind speed and optimal rotor speed of variable speed wind turbine *ISA transactions* **96** 479-489
[24] Sitharthan R, Devabalaji KR and Jees A 2017 An Levenberg–Marquardt trained feed-forward back-propagation based intelligent pitch angle controller for wind generation system *Renewable Energy Focus* **22** 24-32
[25] Sitharthan R, Sundarabalan CK, Devabalaji KR, Yuvaraj T and Mohamed Imran A 2019 Automated power management strategy for wind power generation system using pitch angle controller *Measurement and Control* **52(3-4)** 169-182
[26] Sundar DS, Umamaheswari C, Sridarshini T, Karthikeyan M, Sitharthan R, Raja AS and Carrasco MF 2019 Compact four-port circulator based on 2D photonic crystals with a 90° rotation of the light wave for photonic integrated circuits applications *Laser Physics* **29(6)** 066201
[27] Sitharthan R, Parthasarathy T, Sheeba Rani S and Ramya KC 2019. An improved radial basis function neural network control strategy-based maximum power point tracking controller for wind power generation system *Transactions of the Institute of Measurement and Control* **41(11)** 3158-3170
[28] Rajesh M and Gnanasekar JM 2017 Path observation based physical routing protocol for wireless ad hoc networks *Wireless Personal Communications* **97(1)** 1267-1289
[29] Palanisamy K, Varghese LJ, Raglend IJ and Kothari DP 2009. Comparison of intelligent techniques to solve economic load dispatch problem with line flow constraints *IEEE International Advance Computing Conference* 446-452
[30] Sitharthan R, Ponnusamy M, Karthikeyan M and Sundar DS 2019 Analysis on smart material suitable for autogenous microelectronic application *Materials Research Express* **6(10)** 105709
[31] Rajaram R, Palanisamy K, Ramasamy S and Ramanathan P 2014 Selective harmonic elimination in PWM inverter using fire fly and fireworks algorithm *International Journal of Innovative Research in Advanced Engineering* **1(8)** 55-62
[32] Sitharthan R, Swaminathan JN and Parthasarathy T 2018 March. Exploration of wind energy in India: A short review *IEEE National Power Engineering Conference* 1-5
[33] Karthikeyan M, Sitharthan R, Ali T and Roy B 2020 Compact multiband CPW fed monopole antenna with square ring and T-shaped strips *Microwave and Optical Technology Letters* **62(2)** 926-932

[34] Sundar D Sridarshini T, Sitharthan R, Madurakavi Karthikeyan, Sivanantha Raja A, and Marcos Flores Carrasco 2019 Performance investigation of 16/32-channel DWDM PON and long-reach PON systems using an ASE noise source *In Advances in Optoelectronic Technology and Industry Development: Proceedings of the 12th International Symposium on Photonics and Optoelectronics* 93

[35] Sitharthan R and Geethanjali M 2014 Wind Energy Utilization in India: A Review *Middle-East J. Sci. Res.* **22** 796–801 doi:10.5829/idosi.mejsr.2014.22.06.21944

[36] Sitharthan R and Geethanjali M 2014 ANFIS based wind speed sensor-less MPPT controller for variable speed wind energy conversion systems *Australian Journal of Basic and Applied Sciences* **8**14-23

[37] Jerin ARA, Kaliannan P, Subramaniam U and El Moursi MS 2018 Review on FRT solutions for improving transient stability in DFIG-WTs *IET Renewable Power Generation* **12(15)** 1786-1799

[38] Prabaharan N, Jerin ARA, Palanisamy K and Umashankar S 2017 Integration of single-phase reduced switch multilevel inverter topology for grid connected photovoltaic system *Energy Procedia* **138** 1177-1183

[39] Rameshkumar K, Indragandhi V, Palanisamy K and Arunkumari T 2017 Model predictive current control of single phase shunt active power filter *Energy Procedia* **117** 658-665

[40] Fathima AH and Palanisamy K 2016 Energy storage systems for energy management of renewables in distributed generation systems *Energy Management of Distributed Generation Systems* 157

[41] Rajesh M 2020 Streamlining Radio Network Organizing Enlargement Towards Microcellular Frameworks *Wireless Personal Communications* 1-13

[42] Subbiah B, Obaidat MS, Sriram S, Manoharn R and Chandrasekaran SK 2020 Selection of intermediate routes for secure data communication systems using graph theory application and grey wolf optimisation algorithm in MANETs IET *Networks* doi:10.1049/iet-net.2020.0051

[43] Singh RR and Chelliah TR 2017 Enforcement of cost-effective energy conservation on single-fed asynchronous machine using a novel switching strategy *Energy* **126** 179-191

[44] Amalorpavaraj RAJ, Palanisamy K, Umashankar S and Thirumoorthy AD 2016 Power quality improvement of grid connected wind farms through voltage restoration using dynamic voltage restorer *International Journal of Renewable Energy Research* **6(1)** 53-60

[45] Singh RR, Chelliah TR, Khare D and Ramesh US 2016 November. Energy saving strategy on electric propulsion system integrated with doubly fed asynchronous motors *IEEE Power India International Conference* 1-6

[46] Singh RR, Mohan H and Chelliah TR 2016 November. Performance of doubly fed machines influenced to electrical perturbation in pumped storage plant-a comparative electromechanical analysis *IEEE 7th India International Conference on Power Electronics* 1-6

Real time Implementation of Home appliance control using ALEXA

S Somesh[1], N Senthilnathan[2], M Sabarimuthu[3], A Santhosh Kumar[4], R Rishikeshanan[5], V Bala[6]

1- UG Scholar, Department of Electrical and Electronics Engineering, Kongu Engineering College, Perundurai, India

2- Professor, Department of Electrical and Electronics Engineering, Kongu Engineering College, Perundurai, India

3- Assistant Professor, Department of Electrical and Electronics Engineering, Kongu Engineering College, Perundurai, India

4-6- UG Scholar, Department of Electrical and Electronics Engineering, Kongu Engineering College, Perundurai, India

someshsaravanankeceee@gmail.com

Abstract. In recent years, Internet of Things (IOT) is being playing a major role in automation of Industries. Smart home systems are being employed in buildings to improve energy efficiency and reduce power wastage. Voice recognition is the current trend in automation. Regular methods of switching systems are difficult for paraplegic persons and blind persons since blind people can't use the SMS method of automation. So voice recognition methods are preferred. Voice recognition systems involve calibration and programming to accept the user commands and operate the appliance. Amazon Echo dot, a smart speaker has an internally programmed Artificial Intelligence known as Alexa. It offers dual interaction and can be used to connect to Smart devices. To remedy the high cost of smart devices, we register a name in Alexa server as a smart device which comprises a Wi-Fi module and a relay. The Smart speaker and the Wi-Fi module form a single way communication which helps in controlling of appliance.

1. Introduction

The Machine Interaction has led to voice recognition. Speech recognition makes use of inbuilt algorithms and programs to interface with the user command to perform the specific. Home automation makes of lighting and other appliance control and it can be even extended to industries if calibrated properly. Automation calibration and setting need not be same for all kinds of applications. Engineers use complex algorithms and calibrations to meet up with the specific demands. Automation is essential for smart city construction. It involves in many kinds of process such as Traffic monitoring, water monitoring, sewage control etc... Smart home system involves appliance control, security control which also includes metering of light, moisture and other parameters of the home. The household appliance must be interconnected through IOT any other means to act as an intelligent smart device. Amazon developed a smart speaker known as Amazon Echo dot. It has an interactive artificial Intelligence programmed inside it which is known as "Alexa". The device is able to play music, podcasts, weather forecasting, interaction etc… It can be used in Smart home systems to be used for not only appliance control but also to provide home monitoring. In India, automation systems for home automation systems are very costly so low cost systems. Alexa integrated systems offer low cost automation to homes and buildings. Through Alexa app and Server, we can register the load device names as smart devices and enable as home automation.

2. Literature

The There are several studies which deals with the pros and cons of various automation methods. Several methods of home automation systems are available such as Blue tooth control, Wi-Fi control, IOT systems, SMS and through Remote control. Automation greatly helps in reduction of labour, improves energy efficiency, faster communication etc… One of the earliest interpersonal intelligent Artificial Intelligence developed was "Siri". It was developed by Siri Inc. Later when apple acquired Siri, it was introduced on iPhone 4S. It was also intended to be released for Blackberry and Android devices. Ana et al. [1] proposed the use of Siri, Artificial Intelligence to be used for Home automation. It was designed to recognize the voice recognition commands through interconnected Zigbee networks. But the problem was that it was uneconomical and it can't be used to indicate the status of appliance owing to the absence of sensors. Also, it was available only for IOS users. Piyare&Tazil [2] et al., Mr.Vaibhav Malav[3] proposed the use of Bluetooth for short range control of appliances. It makes use of Cell phone to transfer the commands to the Arduino controller wirelessly. But the disadvantage is its short range of operation and its low data rate. S. Hidayat et al. [4] proposed the control of Home automation system using Bluetooth technology. Raspberry is attached to a relay system and by using the Voice command, the device is controlled. But the device proves to be costly and the wired medium is needed to be used. Also, there was another problem which greatly reduced its performance. Its operating time was more than a minute which was highly undesirable since the appliance require faster turn on and off process. P.S.N.Priyanka et.al [5], Shreya Ghosh et.al [6] proposed Home automation control system employing DTMF. Even though DTMF systems are simple, they produce call tariff problem and a manual is needed for the key identification. Mobile communication based control of appliance employing GSM technology is discussed by Rozita Teymourzadeh et.al [7] in their paper. The sharing of same bandwidth by multiple users can create transmission interference. Ma Naing et.al [8] proposed automation employing Arduino. Even though programming of Arduino is simple, it's structure and building takes a huge cost when developed on a large scale. Also the systems mentioned above are not suitable for blind persons. So voice recognition is a good way to address these problems. But the voice recognitions need to have the ability to understand the commands correctly and to give the proper responses. They should even be able to interpret our commands in the noisy environments. One such system is the AI "Alexa" which is incorporated in a device known as "Echo dot" [9-20]. This device built on Particle photon connects to the voice controlled intelligent personal assistant service Alexa, which responds to the word "Alexa". Alexa Voice Service (AVS) is Amazon's intelligence voice recognition and natural language understanding service. Similar to Alexa, Google Assistant is an artificial intelligence-powered virtual assistant developed by Google that is primarily available on mobile and smart home devices. It can engage in two-way conversations. The full-bodied sound of Amazon makes it better suited to your audio tastes [21-30].

3. Problems found

Paraplegic persons and blind persons at home find difficult to operate appliance manually. Even though remote control of appliance is available, blind persons find it difficult to operate it. So voice recognition systems greatly help for such people. Voice recognition systems on advanced level are

IOP Conf. Series: Materials Science and Engineering **937** (2020) 012008 doi:10.1088/1757-899X/937/1/012008

costly to afford which is unnecessary for simple control of home appliance. So we make use of commercially available AI such as Google Assistant, Alexa etc... Even though Alexa integrated systems are available, the commercial systems need to reconfigure the Wi-Fi every time when the load gets shut down [31-36]. This leads to unnecessary time consumption. Also the commercially available smart devices compatible to the Alexa systems are costly which makes it difficult for large scale installation.

4. Proposed System

The proposed system makes use of Alexa integrated with a Wi-Fi network which is configured in an ESP8266 Wi-Fi module. The load (device) name which is to be addressed in the voice command is registered in the ALEXA Server. This eliminates the need to frequent reconfiguring of Wi-Fi using the ALEXA app. This is done by registering a Wi-Fi name in the ESP node and the Alexa cloud server so that it automatically connects itself to the Wi-Fi.

5. Block Diagram:

The working of the block diagram is shown in figure 1. The 230 V supply is rectified, filtered and made it to operate the relay. The Echo dot is always ON waiting for the wake word. The voice command gets processed through Alexa cloud server. Alexa cloud server has an inbuilt Voice recognition system known as Alexa Voice Service. The commands from the cloud are received by the ESP8266 which processes the command and activates the relay. Every device is given a name inside the Node MCU through programming. Every time the command is passed, the data is recognised from the registered server through the internet and the proper relay is activated.

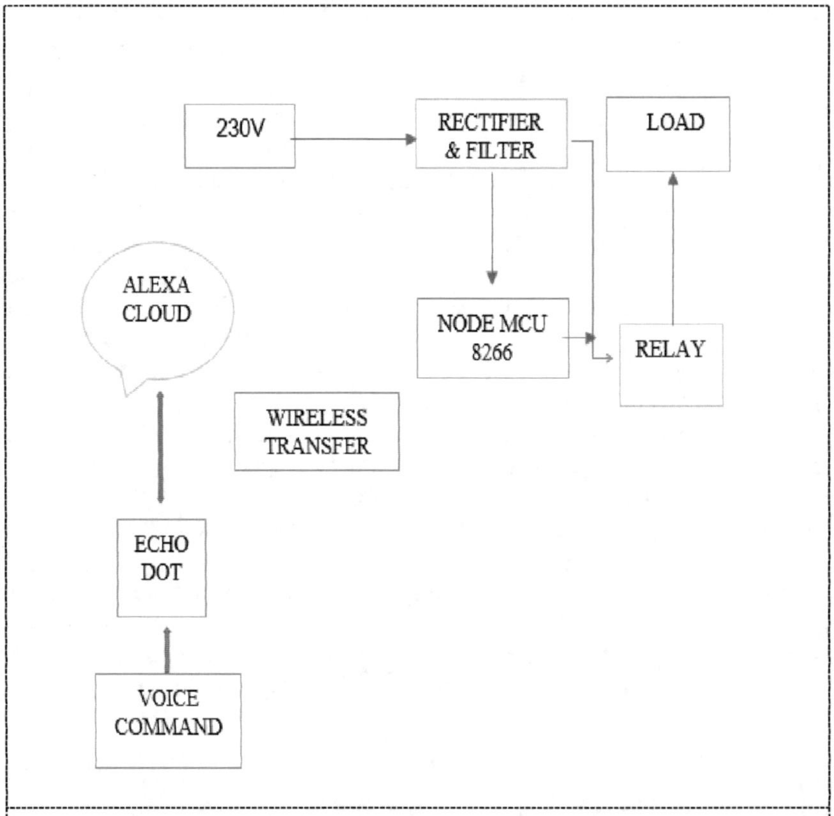

Figure 1. Block Diagram

IOP Conf. Series: Materials Science and Engineering **937** (2020) 012008 doi:10.1088/1757-899X/937/1/012008

6. Components required

6.1. Amazon Echo dot

Figure 2. Echo dot

Alexa is an interactive artificial system developed by Amazon. This core is incorporated in a device known as Amazon echo dot. It can be used to listen to music, control home appliance etc… Different types of Echo dots are available which we can use based upon our needs. The figure 2 is the second generation model of Echo dot. Its added advantage is its application in smart home monitoring systems such as temperature control etc… They are suitable for many devices such as Philips Hue etc… It can be called with a wake word "Alexa"... The internet is connected from the Echo dot through Wi-Fi. Echo dot has an added advantage of recognising voice commands even among the noisy environments and can be operated even from a certain distance. The commands need not be perfectly grammatical. The voice recognition system in the cloud is very good at understanding even the simplistic commands.

6.2. ESP8266 Node MCU:

Figure 3. ESP8266 Node MCU

IOP Conf. Series: Materials Science and Engineering **937** (2020) 012008 doi:10.1088/1757-899X/937/1/012008

It is a low cost device which enables to apply many IOT applications. It can act as both the access point (create hotspot) and can act as Wi-Fi connectivity. It is easy to program with Arduino Uno. Its features include I2S, I2C interfaces, GPIO pins, 10 bit ADC etc... It works on IEEE 801.11 Wi-Fi configuration. The Processor is L106 32-bit RISC microprocessor core based on the Tensilica Xtensa Diamond Standard 106Micro running at 80 MHz and 32 Kb instruction RAM (period) at the end.

6.3. Load Relays:

Figure 4. Relay

The load is connected by means of a 5V Relay module. The figure 4 is an electromagnetic relay consisting of three pins namely VCC, INI, and GND. VCC and GND acts as polarity whereas the IN acts as a pulse for the operation of relay. Based on the configuration of NC or NO, the relay can be made to either ON or OFF when the pulse is given. The electromagnetic relays are very faster in action thus avoiding unnecessary time delay. Relays are faster in approach and based on the number of devices used to operate, we can use the appropriate relays for example such as two channel, three channel and multi-channel relays. Different appliances have different ratings. So before using a relay, the current and voltage rating of the appliance should be considered otherwise the relay might not withstand the excess voltage and current parameters. The ESP8266 operates with the DC 5V of Input and output and so it becomes much easier to integrate with the Relay.

6.4. Rectifier:

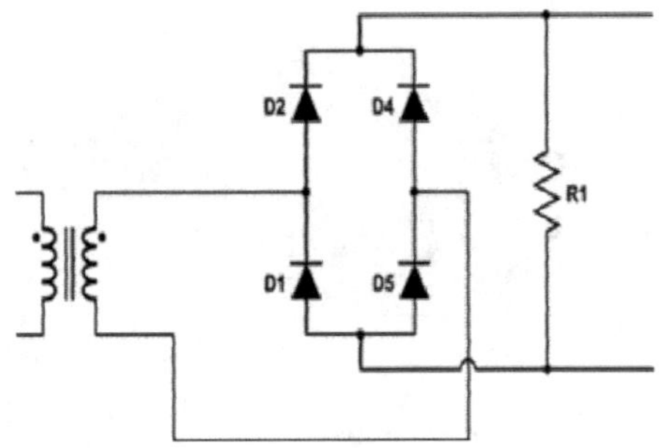

Figure 5. Bridge Rectifier

IOP Conf. Series: Materials Science and Engineering **937** (2020) 012008 doi:10.1088/1757-899X/937/1/012008

The load is connected by means of a 5V Relay module. But the supply is AC. So it is necessary to convert to AC to DC.Figure.5 shows the bridge rectifier circuit. The two diodes D2 and D5 conducts during one half cycle and the another sets of diodes D1 and D4 conducts during another half cycle. This together forms an output of pulsating DC. But the output is not pure DC. It is pulsating because of the unidirectional ripples. This can be eliminated by employing capacitive filter circuits at the end of Rectifier output which eliminates the ripples.

7. Circuit Diagram

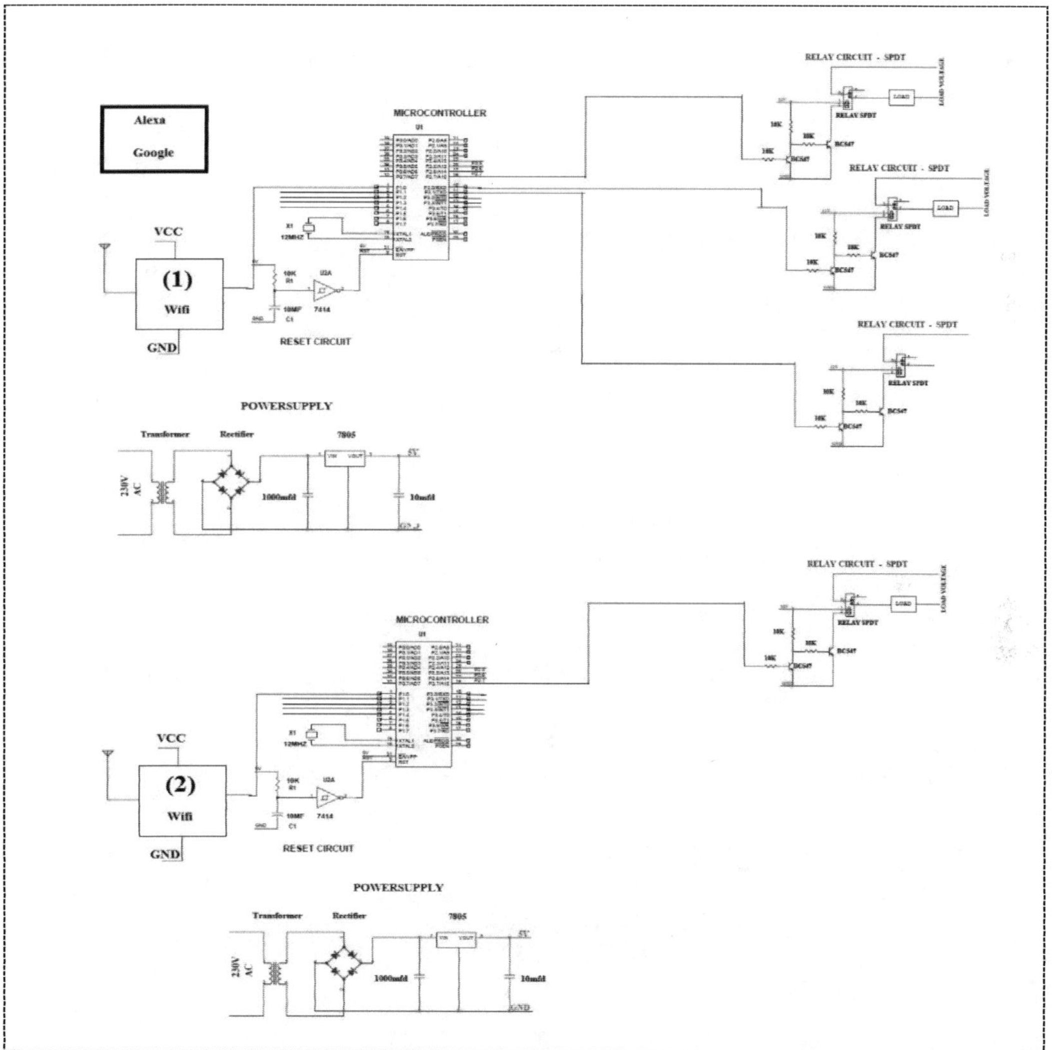

Figure 6. Circuit Diagram

8. Flowchart:

IOP Conf. Series: Materials Science and Engineering **937** (2020) 012008 doi:10.1088/1757-899X/937/1/012008

The below Figure.6 explains the step by step process of working of Alexa integrated automated system as a flowchart.

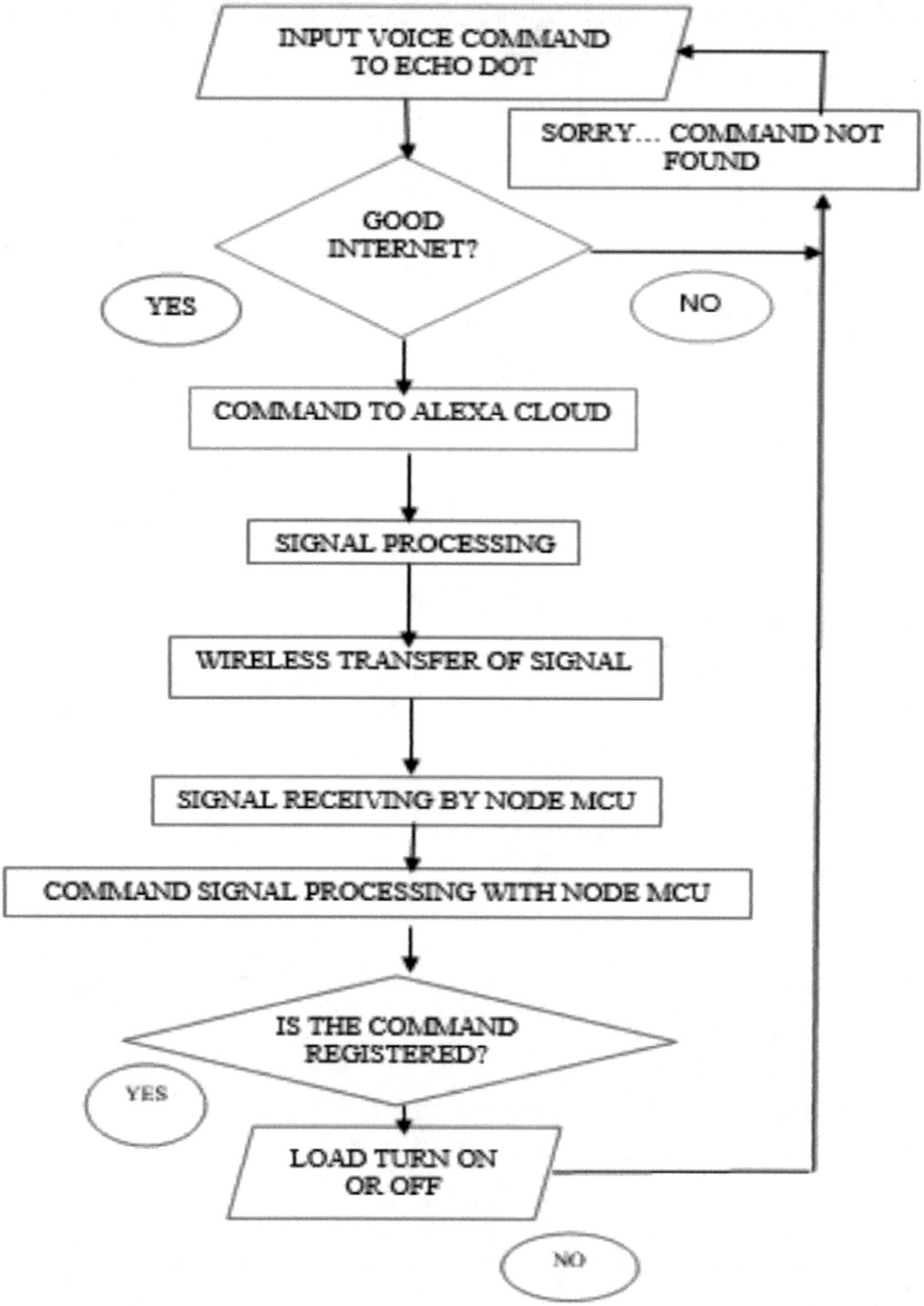

Figure 7. Flow Chart

9. Working:

The device name which has to be controlled is pre-programmed and registered in the ALEXA server. So every time, we call upon the device name to be turned on and off, ALEXA gets its information from its Server through the internet connection. When we give the Wake word "ALEXA", the device (Echo dot) gets ready to accept the command which is indicated by the blue circular light. When we give the command such as "Alexa, turn on the light" or in some similar manner which is recognised by its advanced AI system, the command gets processed through the Server and the processed signal is transferred to the ESP8266.The ESP8266 is registered with the APP KEY to enable connecting to the phone and the Wi-Fi and to help in auto connection to the Wi-Fi, a SSID name and its password is registered in the ESP module. Each ESP is distinguished from one another by its APPKEY. The command gets received by the ESP8266 and the command gets interfaced with the programming of the ESP module. Based on this, the relay gets turned ON. The Command signal irrespective of the load device activates the load. They operate irrespective of the voice of the user and it is greatly useful in Automation system in offices. By simply connecting to a Wi-Fi network whose name can be configured in the ALEXA cloud server, automation of appliance becomes easier.

10. Conclusion

Voice Automated system greatly helps in automation and for helping paraplegic persons since they can't use remote or any other means. Voice automation is extremely helpful for blind people since normal remote controlled home appliance systems make use of IR remotes which has to be showed towards the appliances for control. Voice command systems can be done even by illiterate person by simply mentioning the device name and telling "ON" or "OFF". These systems are highly reliable and it can be operated even by integrating with Mobile phones since they operate with Wi-Fi. Hence they can be operated from anywhere. Systems such as Blue tooth or Wi-Fi employed remote control systems even though they are efficient, only the registered smart phones can be used to operate the appliance. The voice automated system eliminates the disadvantage and makes the system accessible to everyone.

References

[1] Marie, A., Benedict, I., Zandrae, A., Neil, A., Gustilo, R. 2015 *Home Automation Using Raspberry Pi through Siri Enabled Mobile Devices.*

[2] Piyare, R., and Tazil, M. 2011 *Bluetooth based Home Automation System using Cell Phone*

[3] Mr. Vaibhav Malav, Mr. Raushan Kumar Bhagat, Mr. Rahul Saini , Mr. Udit Mamodiya Conference 2019, *Research paper on bluetooth based home automation using Arduino*

[4] Sitharthan R, Geethanjali M and Pandy TKS 2016 Adaptive protection scheme for smart microgrid with electronically coupled distributed generations *Alexandria Engineering Journal* **55(3)** 2539-2550

[5] Fathima AH, and Palanisamy K 2014 Battery energy storage applications in wind integrated systems—a review *IEEE International Conference on Smart Electric Grid* 1-8

[6] Prabaharan N and Palanisamy K 2015 Investigation of single-phase reduced switch count asymmetric multilevel inverter using advanced pulse width modulation technique *International Journal of Renewable Energy Research* **5(3)** 879-890.

[7] Jerin ARA, Kaliannan P and Subramaniam U 2017 Improved fault ride through capability of DFIG based wind turbines using synchronous reference frame control based dynamic voltage restorer. *ISA transactions* **70** 465-474

[8] Sitharthan, R, Sundarabalan CK, Devabalaji KR, Nataraj SK and Karthikeyan M 2018 Improved fault ride through capability of DFIG-wind turbines using customized dynamic voltage restorer *Sustainable cities and society* **39** 114-125

[9] Prabaharan N and Palanisamy K 2016 A single-phase grid connected hybrid multilevel inverter for interfacing photo-voltaic system *Energy Procedia* **103** 250-255

[10] Palanisamy K, Mishra JS, Raglend IJ and Kothari DP 2010 Instantaneous power theory based unified power quality conditioner (UPQC) *IEEE Joint International Conference on Power Electronics, Drives and Energy Systems* 1-5

[11] Sitharthan R and Geethanjali M 2017 An adaptive Elman neural network with C-PSO learning algorithm-based pitch angle controller for DFIG based WECS Journal of Vibration and Control **23(5)** 716-730

[12] Sitharthan R and Geethanjali M 2015 Application of the superconducting fault current limiter strategy to improve the fault ride-through capability of a doubly-fed induction generator–based wind energy conversion system *Simulation* **91(12)** 1081-1087

[13] Sitharthan R, Karthikeyan M, Sundar DS and Rajasekaran S 2020 Adaptive hybrid intelligent MPPT controller to approximate effectual wind speed and optimal rotor speed of variable speed wind turbine *ISA transactions* **96** 479-489

[14] Sitharthan R, Devabalaji KR and Jees A 2017 An Levenberg–Marquardt trained feed-forward back-propagation based intelligent pitch angle controller for wind generation system *Renewable Energy Focus* **22** 24-32

[15] Sitharthan R, Sundarabalan CK, Devabalaji KR, Yuvaraj T and Mohamed Imran A 2019 Automated power management strategy for wind power generation system using pitch angle controller *Measurement and Control* **52(3-4)** 169-182

[16] Sundar DS, Umamaheswari C, Sridarshini T, Karthikeyan M, Sitharthan R, Raja AS and Carrasco MF 2019 Compact four-port circulator based on 2D photonic crystals with a 90° rotation of the light wave for photonic integrated circuits applications *Laser Physics* **29(6)** 066201

[17] Sitharthan R, Parthasarathy T, Sheeba Rani S and Ramya KC 2019. An improved radial basis function neural network control strategy-based maximum power point tracking controller for wind power generation system *Transactions of the Institute of Measurement and Control* **41(11)** 3158-3170

[18] Rajesh M and Gnanasekar JM 2017 Path observation based physical routing protocol for wireless ad hoc networks *Wireless Personal Communications* **97(1)** 1267-1289

[19] Palanisamy K, Varghese LJ, Raglend IJ and Kothari DP 2009. Comparison of intelligent techniques to solve economic load dispatch problem with line flow constraints *IEEE International Advance Computing Conference* 446-452

[20] Sitharthan R, Ponnusamy M, Karthikeyan M and Sundar DS 2019 Analysis on smart material suitable for autogenous microelectronic application *Materials Research Express* **6(10)** 105709

[21] Rajaram R, Palanisamy K, Ramasamy S and Ramanathan P 2014 Selective harmonic elimination in PWM inverter using fire fly and fireworks algorithm *International Journal of Innovative Research in Advanced Engineering* **1(8)** 55-62

[22] Sitharthan R, Swaminathan JN and Parthasarathy T 2018 March. Exploration of wind energy in India: A short review *IEEE National Power Engineering Conference* 1-5

[23] Karthikeyan M, Sitharthan R, Ali T and Roy B 2020 Compact multiband CPW fed monopole antenna with square ring and T-shaped strips *Microwave and Optical Technology Letters* **62(2)** 926-932

[24] Sundar D Sridarshini T, Sitharthan R, Madurakavi Karthikeyan, Sivanantha Raja A, and Marcos Flores Carrasco 2019 Performance investigation of 16/32-channel DWDM PON and long-reach PON systems using an ASE noise source *In Advances in Optoelectronic Technology and Industry Development: Proceedings of the 12th International Symposium on Photonics and Optoelectronics* 93

[25] Sitharthan R and Geethanjali M 2014 Wind Energy Utilization in India: A Review *Middle-East J. Sci. Res.* **22** 796–801 doi:10.5829/idosi.mejsr.2014.22.06.21944

[26] Sitharthan R and Geethanjali M 2014 ANFIS based wind speed sensor-less MPPT controller for variable speed wind energy conversion systems *Australian Journal of Basic and Applied Sciences* **8**14-23

[27] Jerin ARA, Kaliannan P, Subramaniam U and El Moursi MS 2018 Review on FRT solutions for improving transient stability in DFIG-WTs *IET Renewable Power Generation* **12(15)** 1786-1799

[28] Prabaharan N, Jerin ARA, Palanisamy K and Umashankar S 2017 Integration of single-phase reduced switch multilevel inverter topology for grid connected photovoltaic system *Energy Procedia* **138** 1177-1183

[29] Rameshkumar K, Indragandhi V, Palanisamy K and Arunkumari T 2017 Model predictive current control of single phase shunt active power filter *Energy Procedia* **117** 658-665

[30] Fathima AH and Palanisamy K 2016 Energy storage systems for energy management of renewables in distributed generation systems *Energy Management of Distributed Generation Systems* 157

[31] Rajesh M 2020 Streamlining Radio Network Organizing Enlargement Towards Microcellular Frameworks *Wireless Personal Communications* 1-13

[32] Subbiah B, Obaidat MS, Sriram S, Manoharn R and Chandrasekaran SK 2020 Selection of intermediate routes for secure data communication systems using graph theory application and grey wolf optimisation algorithm in MANETs IET *Networks* doi:10.1049/iet-net.2020.0051

[33] Singh RR and Chelliah TR 2017 Enforcement of cost-effective energy conservation on single-fed asynchronous machine using a novel switching strategy *Energy* **126** 179-191

[34] Amalorpavaraj RAJ, Palanisamy K, Umashankar S and Thirumoorthy AD 2016 Power quality improvement of grid connected wind farms through voltage restoration using dynamic voltage restorer *International Journal of Renewable Energy Research* **6(1)** 53-60

[35] Singh RR, Chelliah TR, Khare D and Ramesh US 2016 November. Energy saving strategy on electric propulsion system integrated with doubly fed asynchronous motors *IEEE Power India International Conference* 1-6

[36] Singh RR, Mohan H and Chelliah TR 2016 November. Performance of doubly fed machines influenced to electrical perturbation in pumped storage plant-a comparative electromechanical analysis *IEEE 7th India International Conference on Power Electronics* 1-6

IOP Conf. Series: Materials Science and Engineering **937** (2020) 012009 doi:10.1088/1757-899X/937/1/012009

Global MPP Tracking for Partial shaded PV System using Fractional Order Extreme Seeking controller

R. Ramaprabha[1]and G. Ramya[2]

[1] Associate Professor, Department of Electrical and Electronics Engineering, SSN College of Engineering, Rajiv Gandhi Salai, Kalavakkam-603110, Chennai, Tamil Nadu, India1,3

[2] Assistant Professor, Department of Electrical and Electronics Engineering, Saveetha School of Engineering, Thandalam-602105, Chennai, Tamil Nadu, India2

ramaprabhar@ssn.edu.in (corresponding author), ramyamrose@gmail.com

Abstract. Renewable energy tends to replace natural resources like coal and fuel, nowadays research work is in progress on solar power utilization. To overcome the photovoltaic (PV) installation issue, building-integrated PV (BiPV) system has been introduced. Partial shading is major issue in BiPV system. In order to prevent this issue, an anti-parallel bypass diode is added across each panel/cell. But it results in multiple peaks p-v characteristics with respect to varying irradiations. To track global peak power among all other local peaks, a special control algorithm based on extreme seeking (ES) strategy with fractional order controller (FOC) is developed. The proposed control method increases the power generation and efficient tracking of maximum power at MPP.

Keywords –Maximum Power Point Tracking (MPPT); Integer Order Extreme Seeking (IOES); Fractional Order Extreme Seeking (FOES); DC-DC Converter; Photovoltaic Module.

1. Introduction

Due to swelling inhabitants and industries, electricity demand is increasing day by day. Hence, renewable energy plays a major role to satisfy this energy demand and also it tends to replace natural resources like coal and fuel, strong research work is being concentrated on power generation by solar because it is abundant in nature and inexhaustible. So, PV system with maximum energy conversion efficiency is gaining more responsibility [1] A major issue in PV system is a large space requirement for the installation of photovoltaic (PV) system. Building-integrated PV (BiPV) system is used to overcome this issue, which becomes popular nowadays. However, BiPV model experience partial shading of PV array due to irradiation change [2]. The damage of PV cells/panels due to partial shading leads to power loss, which can be prevented by anti-parallel diode connected on sequence PV cells/panel [3]. This will prevent the panel damage, but causes multiple power curves in panel characteristics of proposed PV system, which makes critical problem to track maximum power point [4]. There are some special algorithms present in the literature to track global peak among all other local peaks [5-15]. Different techniques were implemented to improve the power generation efficiency.

In this work, to track extreme power of PV, FOES scheme is introduced. The mainstream developments in linear, nonlinear, and adaptive control seek to achieve the objectives of control issues to regulate the output PV system. However, in many control applications, the desired set point is unknown, it must be determined so that the output of the dynamic system reaches its maximum or minimum extreme point [16-20]. This type of control scheme is known as extreme seeking control (ESC).

Fractional order controller (FOC) is introduced in extreme seeking control to increase the conversion efficiency and stability of the system under variation in external conditions [21-30]. It is a non-model based gradient search scheme that dynamically seeks the optimal point of the output mapping of a system, which makes these methods robust to model [31-33]. By using maximum power point tracking algorithms (MPPT), the converter can adaptively control the voltage or current and consequently power, based on their control logic conditions. Figure 1 presents the blocks of proposed system with source and controller with load.

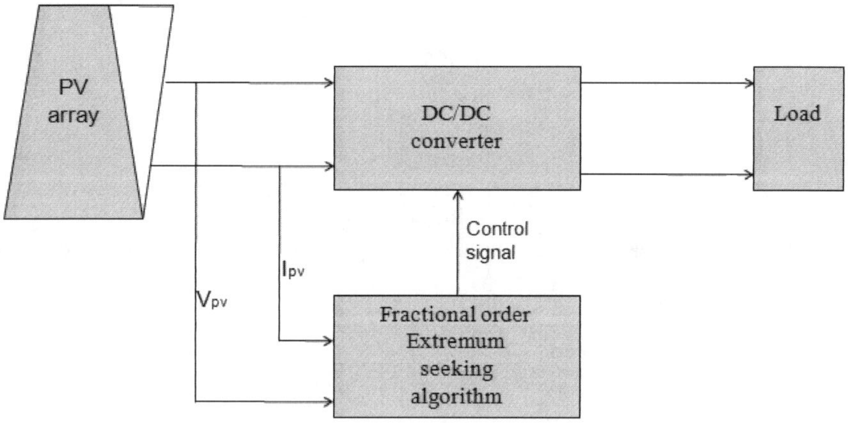

Figure 1. Block diagram of partial shaded PV with FOES controller

2. Fractional order controller

Nowadays in many applications fractional control technique is used rapidly. The main motive to use this controller is due to its increased performance. In this work, extreme seeking strategy based on a novel proposed control is established to enhance the output to improve the system performance.

2.1. Fractional Calculus

Integral and differential to non -integer order is represented as

$$ {}_aD_t^m = \begin{cases} \dfrac{d^m}{dt^m} & Re(m)>0 \\ 1 & Re(m)=0 \\ \int_a^t d\tau^{-m} & Re(m)<0 \end{cases} \quad (1)$$

Where m is the derivative or integral order and Re (m) represents the real part. Many schemes are introduced with fractional order controller. In this paper, singularity function method proposed which is used as a fraction order controller.

2.2. Charef method

Various ideas can be carried out to implement fractional order operators in both time or frequency domain. Charef singularity function [8] is similar to Oustaloup's method which allows the fractional order transfer function with the balanced one as given in Eq. (2)

$$H(s) = s^m, m \in R^+ \quad (2)$$

where m is the fractional number. Fractal dimension is represented in s function as in Eq. (3)

IOP Conf. Series: Materials Science and Engineering **937** (2020) 012009 doi:10.1088/1757-899X/937/1/012009

$$H(s)=\frac{\prod_{i=0}^{N-1}\left(1+\frac{s}{z_i}\right)}{\prod_{i=0}^{N}\left(1+\frac{s}{p_i}\right)} \tag{3}$$

The coefficient of zeros and poles are calculated in time domain of y in terms if dB as in Eqs. (4) – (6).

$$a = 10^{\left[\frac{y}{10(1-m)}\right]} \tag{4}$$

$$b = 10^{\left[\frac{y}{10(m)}\right]} \tag{5}$$

$$ab = 10^{\left[\frac{y}{10m(1-m)}\right]} \tag{6}$$

The poles and zeros are derived by,

$$p_0 = p_t 10^{\left[\frac{y}{20m}\right]} \tag{7}$$

$$p_i = (ab)^i p_0 \tag{8}$$

$$z_i = (ab)^i a p_0 \tag{9}$$

The number of poles and zeros can also be represented by band width ω_{max} as shown in Eq. (10)

$$N = \left(\frac{ln\left(\frac{\omega_{max}}{p_0}\right)}{ln(ab)}\right) + 1 \tag{10}$$

3. Fractional Order Extremum Seeking Control

An MPP seeker is also required to supply the controller an appropriate reference input. Fields in which the fractional order extreme seeking technique have been proved to be extremely powerful in the optimization process. The main issue on seeking the global optimum of P (•): R → R , where the input is varied by disturbance d(t) is shown in Figure 2.

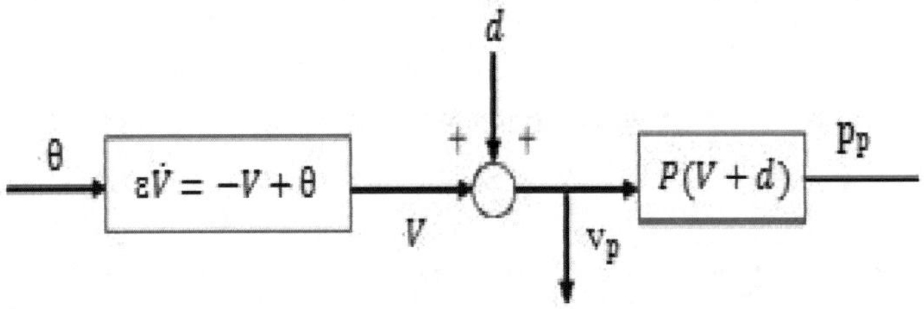

Figure 2. Schematic of the controller circuit

As in [7], we consider the following assumptions.

Assumption 1: P (•): R → R is assumed to be Lipchitz and bounded. In extreme seeking problem it is assumed to achieve the probing signal θ (•) which is controlled as the function of P (•) through a dynamic system by Eq. (11).

$$\varepsilon \dot{V} = -V + \theta \tag{11}$$

Where ε is a scalar. The unknown map signal is P (•) and shown in Fig. 2.

$$\left.\begin{aligned} v_p &= V(t) + d(t) \\ p_p &= P(V(t) + d(t)) \end{aligned}\right\} \tag{12}$$

Assumption 2: The disturbance d (•) is implicit to be bounded and contains first and second order time derivatives, $\bar{d}$ and $\bar{d}_d$ as $|d(t)| \leq \bar{d}$, $|d_d| \leq \bar{d}_d$ for all values of $t \geq 0$. Let Eq. (11) is assumed to be available. The following Eq. (13) is known.

$$\left.\begin{aligned} z_1(t) &= \dot{v}_p(t) = \dot{V}(t) + \dot{d}(t) \\ z_2(t) &= \dot{p}_p(t) = \frac{\partial P(V(t)+d(t))}{\partial P}\left(\dot{V}(t)+\dot{d}(t)\right) \end{aligned}\right\} \quad (13)$$

The signals in Eq. (13) are approximated to use in actual implementation. By introducing an integer order integral 1/s we obtain the controller strategy, which is explained in [9]. Block diagram of extremum seeking controller is represented in Fig. 3. P (•) is an input in which the signals are V and d. Voltage and power measured from source is given as input to FOES algorithm. The schematic of closed-loop dynamic system is seen in Figure 3.

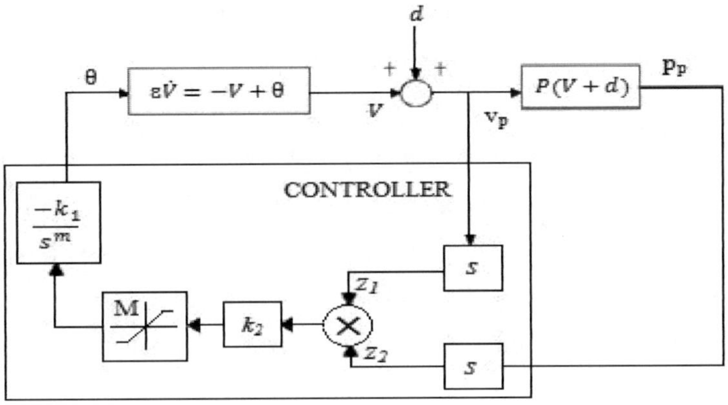

Figure 3. Dynamic extremum seeking control blocks

Convergence speed of V is set to θ by a parameter ε>0, where ε static gain. The signal $k_2 * z_1(t) * z_2(t)$ is fed with the saturation and is integrated by $1/s^m$ and then product to k_1, which results in plant reference θ along k_1 and k_2.

FO integration function Eq. (14) is obtained.

$$\frac{d^m\theta}{dt^m} = -k_1 sat\left(k_2 z_1(t) z_2(t)\right) \quad (14)$$

Where sat(_) is the saturation function, the above function can be written as

$$s^m\theta = -k_1 sat\left(k_2 z_1(t) z_2(t)\right) \quad (15)$$

the θ becomes with the control law as in Eq. (16)

$$\theta = \frac{-k_1 sat\left(k_2 z_1(t) z_2(t)\right)}{s^m} \quad (16)$$

with gain $k_1 \,\&\, k_2 > 0$. Eq. (16) involves iteration to solve.

The main objective to get the better stability control is done by regulating parameter k_2 sufficiently. The expression of θ in Eq. (16) assures that $\theta \leq k_1/s^m$, as it avoids high frequency dynamics. It is helpful in controlling the saturation rate constraints of the system.

The flow diagram shown in Figure 4 explains the operation of FOES algorithm to seek the global peak of photovoltaic array.

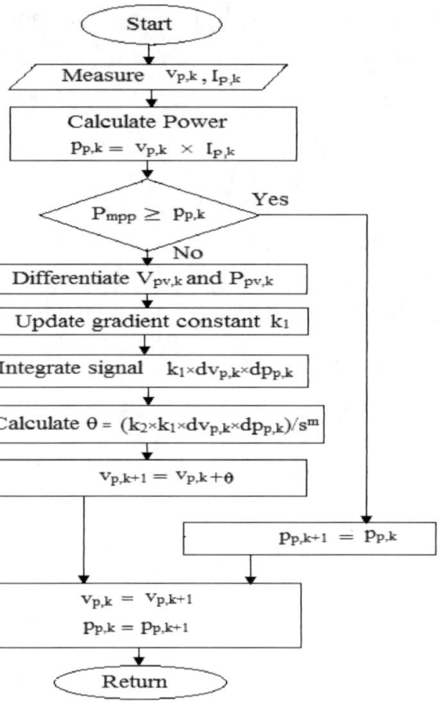

Figure 4. Fractional order extremum seeking operation flow

4. Photovoltaic System

The photovoltaic system is designed as in [6], [11]. Reference module for simulation is taken by SOLKAR 36 W peak power PV rated at 37.08 W_P; 16.56 V and 2.25 A at peak power. Three series connected PV panels are used in this work which results in necessary output current rating and voltage rating under ordinary conditions. The PV cell structure is represented in Figure 5.

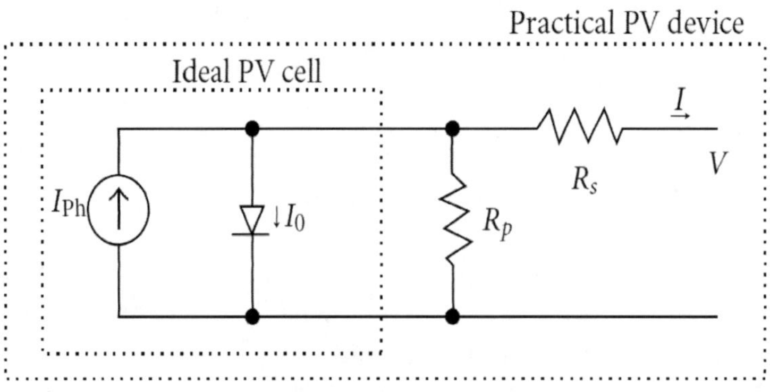

Figure 5. PV cell block

The photo generated current I_p of the photovoltaic module depends on the solar irradiation linearly and is also affected by the temperature as shown in the Eq. (17).

$$I_p = \left[I_{SC} + K_i\left(T_k - T_{ref}\right)\right] \times \frac{\lambda}{1000} \tag{17}$$

Where I_{SC} is the short circuit current in A taken from the datasheet, K_i is the temperature coefficient (0.0017A/K), T_k and T_{ref} are the actual and reference temperatures in K and λ is the insolation in W/m^2. Module reverse saturation current is expressed by Eq. (18)

$$I_{rs} = \frac{I_{SC}}{exp\left(\frac{qV_{oc}}{N_s AKT}\right) - 1} \tag{18}$$

Where V_{oc} is the open circuit voltage, q (=1.6 × 10^{-19}C) is the charge of electron, K (=1.3805 × 10^{-23} J/K) is the Boltzmann constant and A (=1.6) is the ideality factor. Module saturation current varies with temperature of cell shown by Eq. (19)

$$I_o = I_{rs}\left[\frac{T}{T_r}\right]^3 exp\left[\frac{qE_{go}}{AK}\left\{\frac{1}{T_r} - \frac{1}{T}\right\}\right] \tag{19}$$

Where E_{g0} (= 1.1eV) is the semiconductor band-gap energy. The module current (I_{pv}) is given in Eq. (20).

$$I_{pv} = N_p I_p - N_p I_o\left[exp\left(\frac{q(V_{pv} + I_{vv} R_s)}{N_s AKT}\right) - 1\right] - V_{pv} + \left(\frac{I_{pv} R_s}{R_{sh}}\right) \tag{20}$$

Where N_p and N_s are the parallel and series connected cells numbers, R_s and R_{sh} is the equivalent series and shunt connected resistance. Generally, the value of R_{sh} is high and hence neglected for model simplicity as in Eq. (21).

$$I_{pv} = N_p I_p - N_p I_o\left[exp\left(\frac{q(V_{pv} + I_{pv} R_s)}{N_s AKT}\right) - 1\right] \tag{21}$$

5. Partial Shaded Model

Because of neighbour building, dirt, aging effect of PV module partial shading condition is taking place. PV module output power is affected by shaded cells and it is difficult to consider the shading of each and every cell, so the effect of shading is also considered in the module.

We have considered three modules with different illumination. When one cell is shaded, it becomes reverse bias, breakdown voltage will occur in these situations, which cause damage in the cell. So anti parallel diode connected in series to bypass the current.

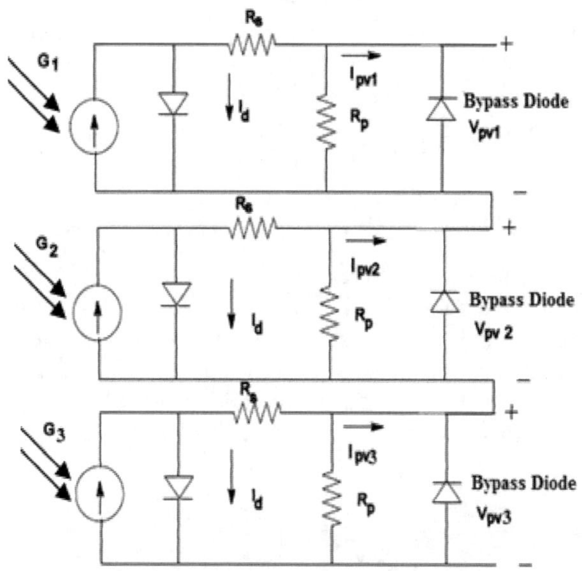

Figure 6. Partial shaded PV module

PV array in partial shading condition is in need to simulate with MPPT controller because P-V curve had multiple peaks, so MPPT should be good enough to track the global peak among all other local peaks. The circuit diagram of partial shaded PV module is shown in Figure 6.

IOP Conf. Series: Materials Science and Engineering **937** (2020) 012009 doi:10.1088/1757-899X/937/1/012009

6. Converter Modelling

The MPPT is applied to regulate the voltage and the duty cycle of converter. The PV module is designed to supply desired maximum power. The basic blocks of the dc-dc converter are illustrated in Figure 7.

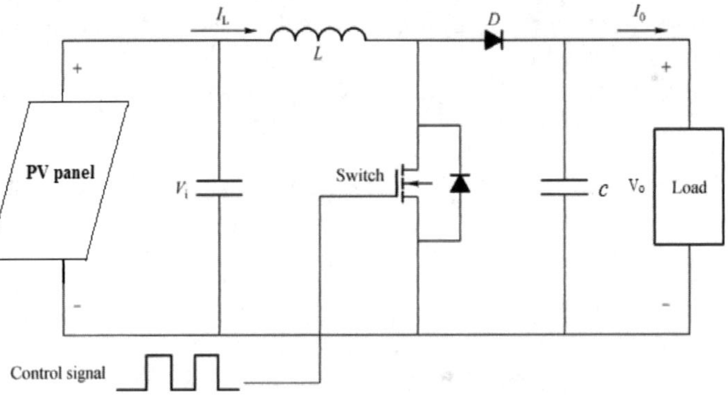

Figure 7. Boost converter structure

The output of designed converter is controlled by switching ON/OFF the controllable switch with constant frequency. Signal produced by the MPPT is compared with constant switching frequency triangular signal which will produce a square wave and is fed to DC-DC converter. The current flow in the inductor L is given in Eq. (22), as in [12].

$$\begin{cases} \text{Turn on time : L} = \left(\dfrac{V_i}{I_{\max} - I_{\min}} \right) t_{on} \\ \text{Turn off time : L} = \left(\dfrac{V_o - V_i}{I_{\max} - I_{\min}} \right) t_{off} \end{cases} \quad (22)$$

The control signal duty ratio is expressed as in Eq. (23).

$$D = \frac{T_s}{t_{on} - t_{off}} \quad (23)$$

By neglecting the difference of I_{max} - I_{min}, the output voltage of the converter can be obtained as

$$V_o = V_i \left(\frac{t_{on} + t_{off}}{t_{off}} \right) = \frac{V_i}{1 - D} \quad (24)$$

7. Simulation Results

The specifications of the various parameters of the partial shaded connected PV model are as follows. Irradiation levels at three panels are considered as, G_1 = 1000 W/m^2, G_2 = 200 W/m^2, G_3 = 600 W/m^2 and temperature is kept at T = 25°C. The MatLab schematics for varying irradiation condition are shown in Fig. 8. The power loss due to partial shaded PV is more. To overcome from this issue, bypass diodes are added across each module in series connected system as shown in Figure 8 and the output characteristics of the partial shaded PV system is shown in the below Figure 9.

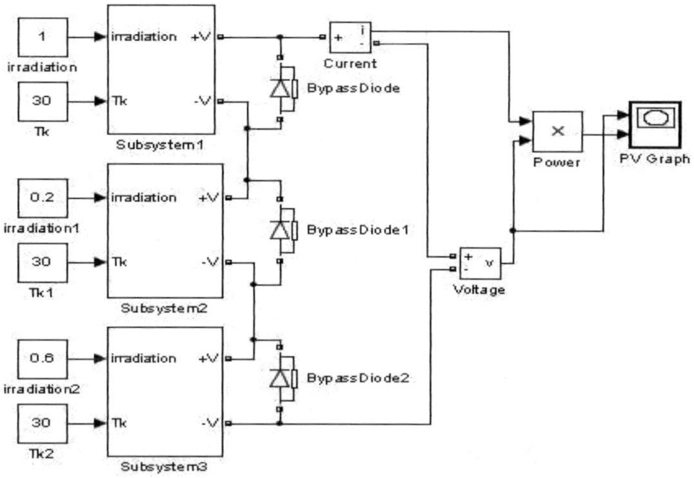

Figure 8. Partial shaded PV array with bypass diodes

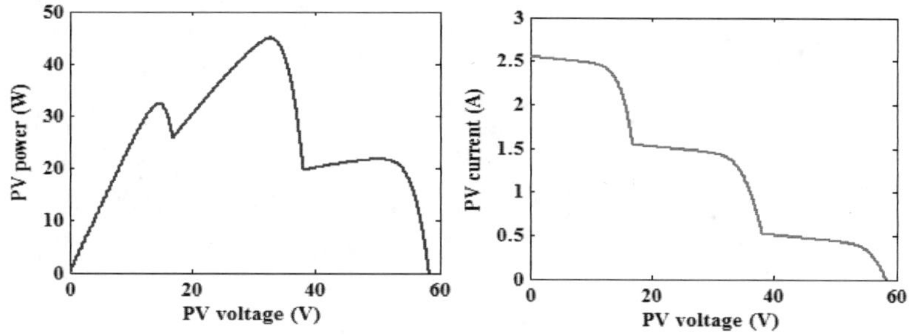

Figure 9. V-I and P-V characteristics of partial shaded PV

From the Figure 9, it is observed that the bypass diode introduced each series connected panel causes multiple peaks in characteristics, however it improves the power. To track the global peak among all other peaks, the developed FOES algorithm is used and also the convergence time is increased.

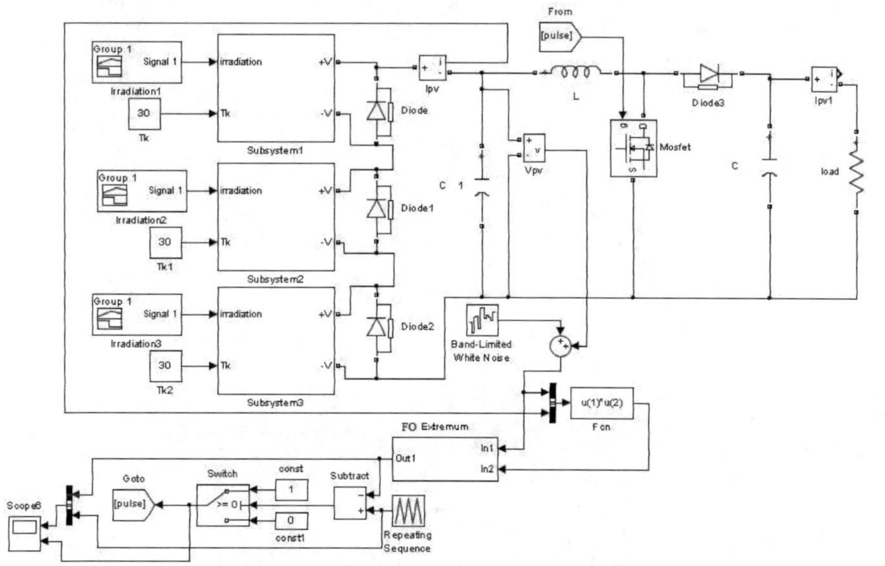

Figure 10. Developed system in MatLab/Simulink

The developed FOES algorithm tracking global peak among all other local peaks can be seen from the characteristic Figure 11.

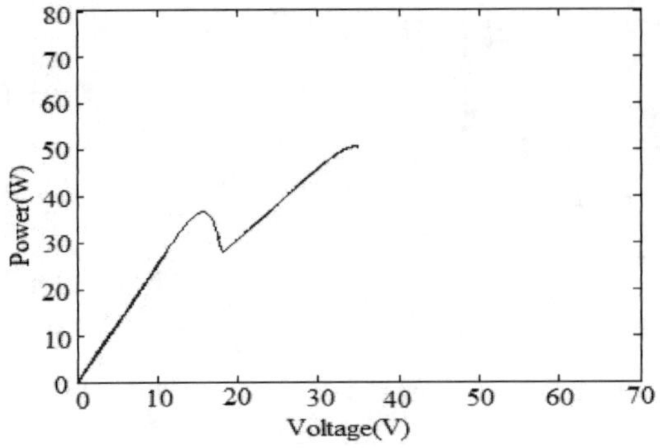

Figure 11. MPPT at global maximum

Maximum power point curve for three different level of irradiation at particular instance is shown in following Figure 12.

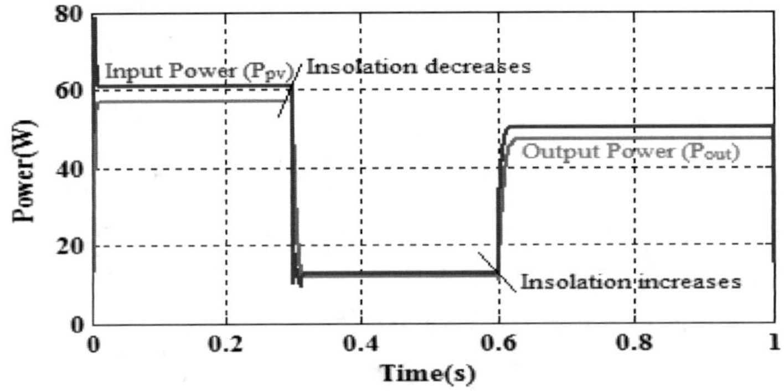

Figure 12. MPP curve for varying irradiation

Converter output power for different fractional values are inferred in Figure 13, in which the convergence period is comparatively high at fractional value m = 0.7. For fractional value m = 1 FOES controller acts similar like normal Integer Order ES controller.

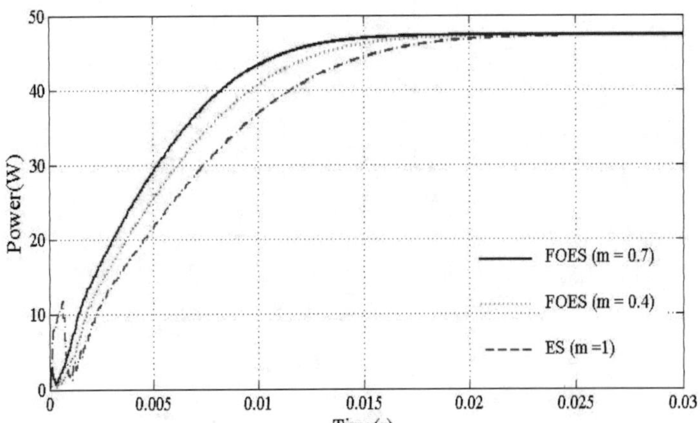

Figure13. Convergence time comparison plot

8. Hardware Results

An electronic load is modelled to study the characteristics of PV module shown in Figure 14.

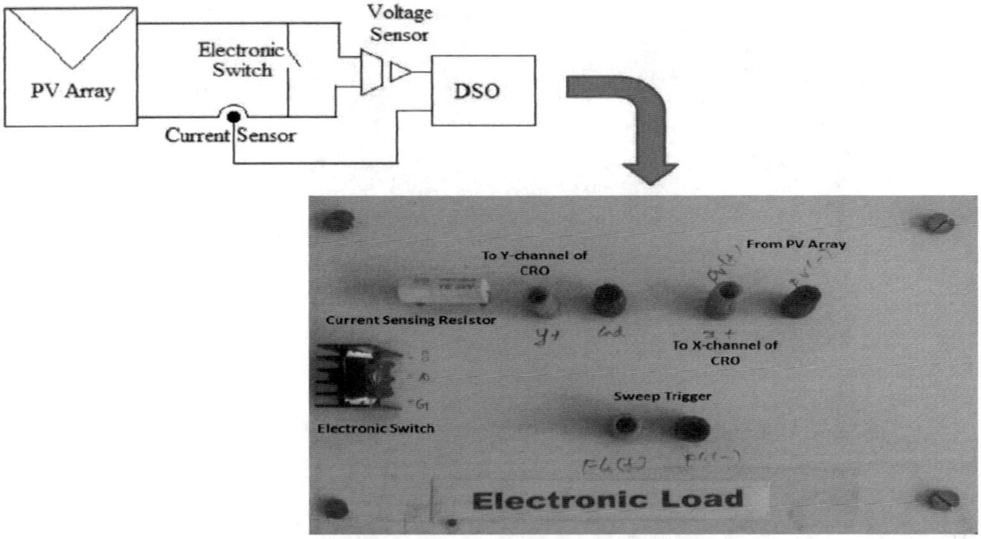

Figure 14. Electronic load schematic

The circuit is connected such that the IPV is proportional to voltage at the amplifier terminal. The MOSFET gate terminal is triggered by triangular signal such that the current swings between zero and short circuit value. When the IPV varies from zero to maximum and vice versa, the characteristic curve is drawn. DSO has been used to trace the characteristics of without shading effect and also under partial shaded conditions. The traced characteristics under shading conditions are shown in Figure 15. The hardware prototype of the partial shaded PV interfaced with conventional boost converter with the developed MPPT algorithm is shown in Figure 16. The input and output voltage for boost converter with developed algorithm, gate pulse for MOSFET switch and load current waveforms for un shaded and under shaded conditions are shown in Figure 17 and Figure 18

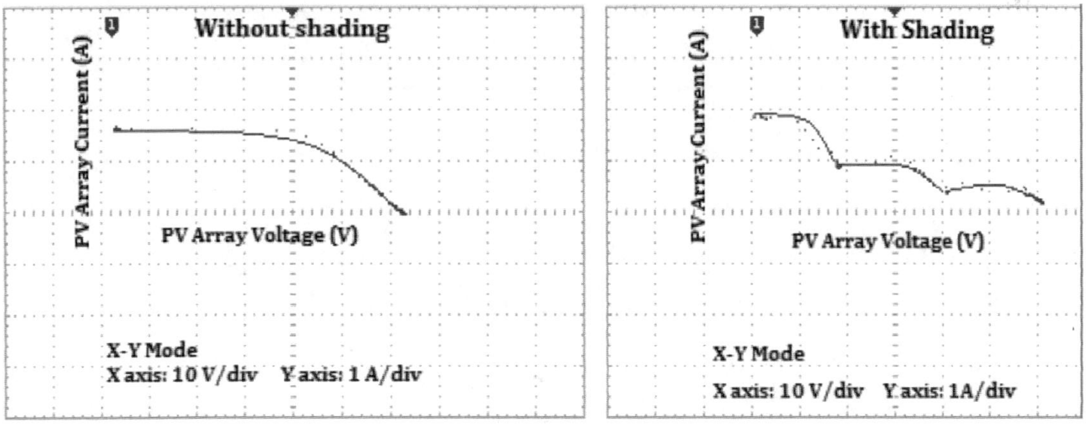

Figure 15. VI characteristics of PV array with and without shading Conditions

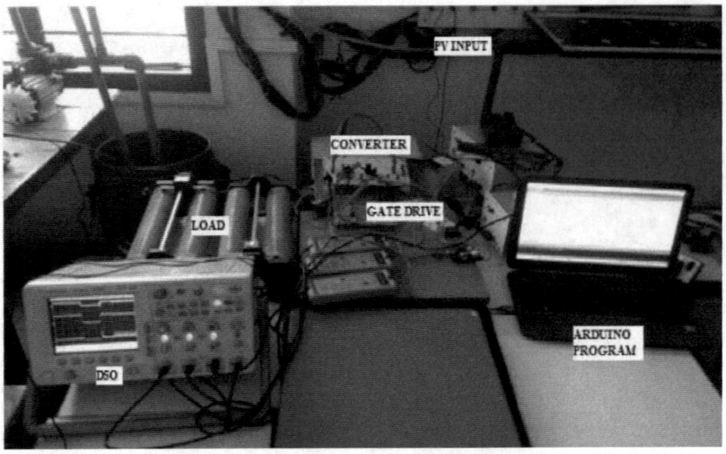

Figure 16. Overall system prototype

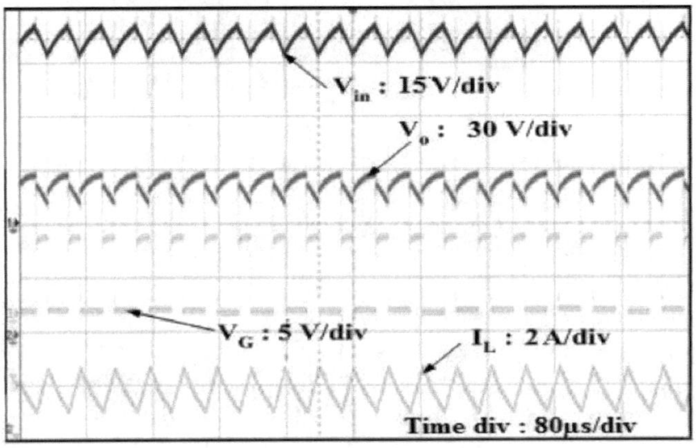

Figure 17. Input & output voltage of converter, gate pulse voltage and load current waveform under unshaded condition

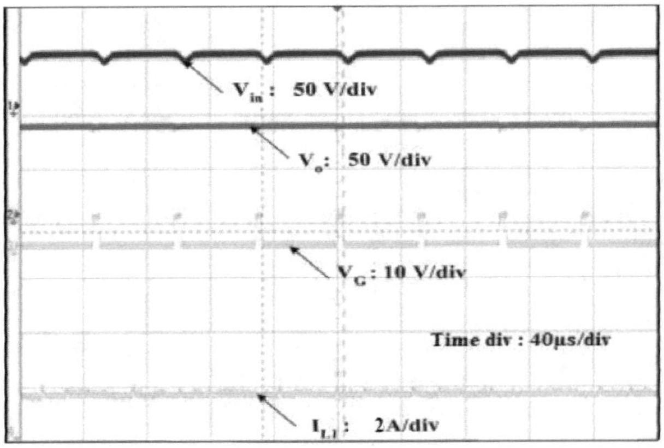

Figure 18. Input & output voltage of converter, gate pulse voltage and load current waveform under partial shaded condition

9. Conclusion

In this research, the proposed control method is developed with abusing external disturbance produced in the system. Three PV panels connected in series have been characterized under partial shaded condition with bypass diodes. The developed FOES controller results in efficient performance and robustness on tracking maximum global peak of the PV system. Performance analysis has been carried out with developed FOES algorithm for partial shaded PV system. FOES controller increases the time of convergence which is proven by comparison plot for different fractional values. The proposed system has also been validated experimentally.

References

[1] Sitharthan R, Geethanjali M and Pandy TKS 2016 Adaptive protection scheme for smart microgrid with electronically coupled distributed generations Alexandria Engineering Journal 55(3) 2539-2550

[2] Fathima AH, and Palanisamy K 2014 Battery energy storage applications in wind integrated systems—a review IEEE International Conference on Smart Electric Grid 1-8

[3] Prabaharan N and Palanisamy K 2015 Investigation of single-phase reduced switch count asymmetric multilevel inverter using advanced pulse width modulation technique International Journal of Renewable Energy Research 5(3) 879-890.

[4] Jerin ARA, Kaliannan P and Subramaniam U 2017 Improved fault ride through capability of DFIG based wind turbines using synchronous reference frame control based dynamic voltage restorer. ISA transactions 70 465-474

[5] Sitharthan, R, Sundarabalan CK, Devabalaji KR, Nataraj SK and Karthikeyan M 2018 Improved fault ride through capability of DFIG-wind turbines using customized dynamic voltage restorer Sustainable cities and society 39 114-125

[6] Prabaharan N and Palanisamy K 2016 A single-phase grid connected hybrid multilevel inverter for interfacing photo-voltaic system Energy Procedia 103 250-255

[7] Palanisamy K, Mishra JS, Raglend IJ and Kothari DP 2010 Instantaneous power theory based unified power quality conditioner (UPQC) IEEE Joint International Conference on Power Electronics, Drives and Energy Systems 1-5

[8] Sitharthan R and Geethanjali M 2017 An adaptive Elman neural network with C-PSO learning algorithm-based pitch angle controller for DFIG based WECS Journal of Vibration and Control 23(5) 716-730

[9] Sitharthan R and Geethanjali M 2015 Application of the superconducting fault current limiter strategy to improve the fault ride-through capability of a doubly-fed induction generator–based wind energy conversion system Simulation 91(12) 1081-1087

[10] Sitharthan R, Karthikeyan M, Sundar DS and Rajasekaran S 2020 Adaptive hybrid intelligent MPPT controller to approximate effectual wind speed and optimal rotor speed of variable speed wind turbine ISA transactions 96 479-489

[11] Sitharthan R, Devabalaji KR and Jees A 2017 An Levenberg–Marquardt trained feed-forward back-propagation based intelligent pitch angle controller for wind generation system Renewable Energy Focus 22 24-32

[12] Sitharthan R, Sundarabalan CK, Devabalaji KR, Yuvaraj T and Mohamed Imran A 2019 Automated power management strategy for wind power generation system using pitch angle controller Measurement and Control 52(3-4) 169-182

[13] Sundar DS, Umamaheswari C, Sridarshini T, Karthikeyan M, Sitharthan R, Raja AS and Carrasco MF 2019 Compact four-port circulator based on 2D photonic crystals with a 90° rotation of the light wave for photonic integrated circuits applications Laser Physics 29(6) 066201

[14] Sitharthan R, Parthasarathy T, Sheeba Rani S and Ramya KC 2019. An improved radial basis function neural network control strategy-based maximum power point tracking controller for wind power generation system Transactions of the Institute of Measurement and Control 41(11) 3158-3170

[15] Rajesh M and Gnanasekar JM 2017 Path observation based physical routing protocol for

wireless ad hoc networks Wireless Personal Communications 97(1) 1267-1289
[16] Palanisamy K, Varghese LJ, Raglend IJ and Kothari DP 2009. Comparison of intelligent techniques to solve economic load dispatch problem with line flow constraints IEEE International Advance Computing Conference 446-452
[17] Sitharthan R, Ponnusamy M, Karthikeyan M and Sundar DS 2019 Analysis on smart material suitable for autogenous microelectronic application Materials Research Express 6(10) 105709
[18] Rajaram R, Palanisamy K, Ramasamy S and Ramanathan P 2014 Selective harmonic elimination in PWM inverter using fire fly and fireworks algorithm International Journal of Innovative Research in Advanced Engineering 1(8) 55-62
[19] Sitharthan R, Swaminathan JN and Parthasarathy T 2018 March. Exploration of wind energy in India: A short review IEEE National Power Engineering Conference 1-5
[20] Karthikeyan M, Sitharthan R, Ali T and Roy B 2020 Compact multiband CPW fed monopole antenna with square ring and T-shaped strips Microwave and Optical Technology Letters 62(2) 926-932
[21] Sundar D Sridarshini T, Sitharthan R, Madurakavi Karthikeyan, Sivanantha Raja A, and Marcos Flores Carrasco 2019 Performance investigation of 16/32-channel DWDM PON and long-reach PON systems using an ASE noise source In Advances in Optoelectronic Technology and Industry Development: Proceedings of the 12th International Symposium on Photonics and Optoelectronics 93
[22] Sitharthan R and Geethanjali M 2014 Wind Energy Utilization in India: A Review Middle-East J. Sci. Res. 22 796–801 doi:10.5829/idosi.mejsr.2014.22.06.21944
[23] Sitharthan R and Geethanjali M 2014 ANFIS based wind speed sensor-less MPPT controller for variable speed wind energy conversion systems Australian Journal of Basic and Applied Sciences 814-23
[24] Jerin ARA, Kaliannan P, Subramaniam U and El Moursi MS 2018 Review on FRT solutions for improving transient stability in DFIG-WTs IET Renewable Power Generation 12(15) 1786-1799
[25] Prabaharan N, Jerin ARA, Palanisamy K and Umashankar S 2017 Integration of single-phase reduced switch multilevel inverter topology for grid connected photovoltaic system Energy Procedia 138 1177-1183
[26] Rameshkumar K, Indragandhi V, Palanisamy K and Arunkumari T 2017 Model predictive current control of single phase shunt active power filter Energy Procedia 117 658-665
[27] Fathima AH and Palanisamy K 2016 Energy storage systems for energy management of renewables in distributed generation systems Energy Management of Distributed Generation Systems 157
[28] Rajesh M 2020 Streamlining Radio Network Organizing Enlargement Towards Microcellular Frameworks Wireless Personal Communications 1-13
[29] Subbiah B, Obaidat MS, Sriram S, Manoharn R and Chandrasekaran SK 2020 Selection of intermediate routes for secure data communication systems using graph theory application and grey wolf optimisation algorithm in MANETs IET Networks doi:10.1049/iet-net.2020.0051
[30] Singh RR and Chelliah TR 2017 Enforcement of cost-effective energy conservation on single-fed asynchronous machine using a novel switching strategy Energy 126 179-191
[31] Amalorpavaraj RAJ, Palanisamy K, Umashankar S and Thirumoorthy AD 2016 Power quality improvement of grid connected wind farms through voltage restoration using dynamic voltage restorer International Journal of Renewable Energy Research 6(1) 53-60
[32] Singh RR, Chelliah TR, Khare D and Ramesh US 2016 November. Energy saving strategy on electric propulsion system integrated with doubly fed asynchronous motors IEEE Power India International Conference 1-6
[33] Singh RR, Mohan H and Chelliah TR 2016 November. Performance of doubly fed machines

influenced to electrical perturbation in pumped storage plant-a comparative electromechanical analysis IEEE 7th India International Conference on Power Electronics 1-6

IOP Conf. Series: Materials Science and Engineering **937** (2020) 012010 doi:10.1088/1757-899X/937/1/012010

Hybrid DPSO based MPPT control of high static gain converter in photovoltaic system for DC microgrid applications

J Sevugan Rajesh[1], R Karthikeyan[2] and R Revathi[3]

[1]Assistant Professor, EEE Department, SVS College of Engineering, Coimbatore, India.
[2]Professor, EEE Department, M.Kumarasamy College of Engineering, Karur, India.
[3]Assistant Professor, EEE Department, KPR Institute of Engineering and Technology, Coimbatore, India.

Abstract. This paper deals with efficiency enhancement of photovoltaic renewable system under Partial Shaded Conditions (PSC) using Hybrid DPSO (Deterministic Particle Swarm Optimization) algorithm based maximum power point tracker embedded in a high gain DC/DC SEPIC converter. The proposed high static gain converter with Hybrid DPSO algorithm eliminates the multiple peaks under varying insolation conditions; speedup the tracking process and also dynamic responses of the converter. The performance parameters of the proposed Hybrid DPSO method are analyzed and comparing it against the PSO and classical Perturb and Observe (P&O) techniques in MATLAB simulink. The evaluation result proves that proposed method shows better tracking performance under fast varying insolation conditions.

1. INTRODUCTION

Fast development in green energy generation investments and government subsidized policies quite significant in recent years has led to a sustained development in the power sector. Renewable power consumption grew day by day and creates great impact in the world's electricity generation. This scenario induces the researchers to attract interest towards green energy generation in the form of distributed generation (DG). Solar, wind, hydroelectricity, geothermal energy and wave power, and combustible renewable and renewable waste (solid biomass, landfill gas, liquid biofuels, and waste incineration) are the vital renewable energy among them few sources produces DC power, examples: fuel cells, PV (photovoltaic) etc. Energy storage devices such as super capacitor, lead acid, sodium ion, nickel cadmium and Li-ion secondary batteries etc, are supplying DC power. DG systems employed in DC micro-grid guarantees reliable power to consumers by optimal control of flow of power from the generation to consumers load. One more imperative feature is majority of the end user appliances are moreover DC adaptable loads (e.g. LED, BLDC fan, CCTV cameras, computers etc.). PV panel have the drawbacks of lower cell voltages and non-linear PV, IV characteristics with respect to irradiance and temperature leads to poor energy conversion efficiency. To overcome the above said limitations a power conditioning units such as Boost converter and maximum power point tracker (MPPT) are used in linking the DC bus and PV Panels. MPPT as its name signify, allows tracking of power over a period of time and detecting the peak power point (MPP) that the panel is customized to provide under varying temperature and solar insolation [4]. The process of MPP tracking is to get optimizes and superior exploit of PV systems thus maximize the PV array efficiency which depends upon the particular MPPT technique. The MPPT algorithms that are mainly categorized into two groups, namely conventional and nonconventional algorithms in several literatures. The conventional techniques includes incremental conductance method or hill climbing algorithm, P&O (perturb and observe), ripple correlation method, short circuit current, open circuit voltage etc [7-10]. The MPPT control using P&O method is most preferred method in the literatures because of its simplicity and easy implementation in real time. But under varying environmental conditions a shaded cell would act as a load and draws additional current from the main load causes tracking PV power get diminished, process of tracking get oscillates and leads to wastage of power [11]. Thus fast varying insolation conditions in P&O method alter the PV power output and distress the reliability of the system. Non-conventional techniques are biologically inspired algorithms has ability to handle multiple peaks in PV characteristics curve and faster convergence to the actual peak point [12-18]. Particle swarm optimization algorithm (PSO), ant colony systems (ACS), genetic algorithm, and flashing firefly algorithm (FA) are few accepted and recent evolutionary algorithms (EA) [19-25]. Each algorithm has its own merits and limitations, in this paper, swarm optimization algorithms are utilized because of its simple construction and easy implementation. Particle seeker optimization efficiently maximizes the efficiency of solar array by tracking optimum global peak point (GPP) [26-30]. But the random numbers makes the algorithm little slow, when tracking the peak power with complex computations at more number of iterations which causes some substantial amount of power loss. To overcome the losses occurred and speed up the tracking process, hybrid DPSO (Deterministic Particle Swarm Optimization) technique is initiated [31-33]. Hybrid DPSO algorithm tracks the global peak point in less convergence time as compared to other soft computing techniques.

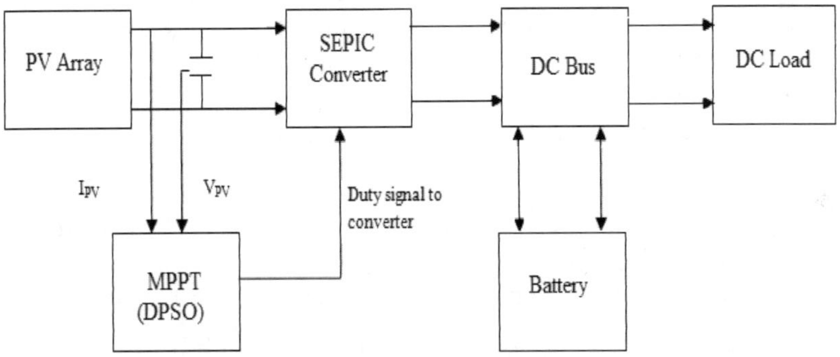

Figure 1. General Block Diagram of Photovoltaic system for DC Microgrid

The proposed controller is able to track optimum global peak operating point with an optimal speed, and it indicates a very good dynamic response under exclusive partial shadowing conditions. The figure 1 indicates the blocks of proposed PV system for DC microgrid applications. The option of an appropriate converter is another imperative subject of matter when employing a PV-MPPT system. Several literatures reveal that the custom of midway converters for power conditioning and tracking applications, which are: classical boost, buck, buck-boost, SEPIC and Cuk converter topologies [9] and [13]. The above said converter topologies suffers with the problems like electromagnetic interference and reverse recovery, when it is subjected to extreme high duty cycle value D>0.8, to get higher voltage gain. The SEPIC, conventional buck-boost and Cuk topologies were ability to either step up and down the voltage applied to it. In this work, a SEPIC converter has been preferred with an intention to use the system for high static gain and high voltage applications [2]. Rest of the work is structured as follows. The hybrid DPSO technique will discuss in section 2. Section 3 and section 4 illustrates the high static gain converter. The comparison results of P&O, PSO and the DPSO hybrid techniques are disused in section 5. At last, the paper is concluded in section 6.

IOP Conf. Series: Materials Science and Engineering **937** (2020) 012010 doi:10.1088/1757-899X/937/1/012010

2. Implementation of Hybrid DPSO Based MPPT

In this paper, hybrid DPSO-P&O MPPT technique is proposed for PV systems under PSC to get maximum efficiency, fast convergence time and low complexity. In PSO based MPPT control to the DC-DC converter be capable of react to one command at an instance, the particles are initialized and assessed in a consecutive way. The successive particles intervals should be larger than the settling time (t_s) of the overall system in order to get acceptable voltage and current samples. Each particle takes ample amounts of time to carry out entire aforesaid steps, hence the convergence time of particles converge to the Global Peak Point (GPP) is greater and it causes some considerable power loss. To reduce convergence time, loss of power and get peak power (Ppv), Hybrid - DPSO technique is initiated. The implementation of hybrid P&O-DPSO algorithm in MPPT controller reduces the search space and hence, the time of convergence gets decreased. Moreover, each particle's velocity is controlled to a meticulous value to remain the space surrounded by the peaks. Therefore the classical PSO composed into the deterministic construction. Figure 2 represents the flow of proposed P&O algorithm. In conventional PSO algorithm, the particles position is defined by PV array voltage as reference and all the particles should converge to GPP by updating the P_{best} and G_{best} using equation 1 and 2. However, this procedure involves large number of iterations and huge variation in the velocity component may cause the particle to move out as of the vicinity of global peak point [3].

$$D_i^{k+1} = D_i^k + \varphi_i^{k+1} \tag{1}$$

$$\varphi_i^{k+1} = W\varphi_i^k + C_1 r_1 \{P_{best} - x_i^k\} + C_2 r_2 \{G_{best} - x_i^k\} \tag{2}$$

Thus, by eliminating the random number r_1 and r_2 in (2) and restraining the velocity factor (V_{max}) hence the conventional PSO is changed to a new deterministic construction. The basic concept is removing the random number from PSO acceleration factor in the equation 2 and modified velocity equation of PSO can therefore be written as:

$$\varphi_i^{k+1} = W\varphi_i^k + \{P_{best} - x_i^k\} + \{G_{best} - x_i^k\} \tag{3}$$

Hybrid DPSO-P&O MPPT algorithm is proposed by modifying the velocity represent in equation 3. The velocity perturbation with respect to $V_{P\&O}$ using flowchart shown in figure 2 helps the fast tracking ability to reach the GPP. The modified velocity equation as depicted below:

$$\varphi_i^{k+1} = W\varphi_i^k + \{P_{best} - x_i^k\} + \{G_{best} - x_i^k\} + V_{P\&O} \tag{4}$$

Where $V_{P\&O}$ is the variation of voltage offer by the classical P&O algorithm illustrated on figure 2. The controlling the value of φ_i^{k+1} the values of duty cycles to SEPIC converter is decrease or increase gradually in two succeeding MPPT cycles. By realizing DPSO method, subsequent parameters are utilized: particle size is about 15. One parameter i.e. the inertia weight (W) required to tune so efforts for parameter regulation is very much reduced. Figure 3 represents the overall hybrid DPSO-P&O MPPT algorithm. Hybrid DPSO-P&O method is employed to extract maximum power from PV array with reduced number of computation and iteration. Power losses tracking speed and dynamic response are better as compared to other intelligent control techniques. Here, soft computing technique decides the duty signal of converter thus proposed circuit tracks peak power point effectively.

IOP Conf. Series: Materials Science and Engineering **937** (2020) 012010 doi:10.1088/1757-899X/937/1/012010

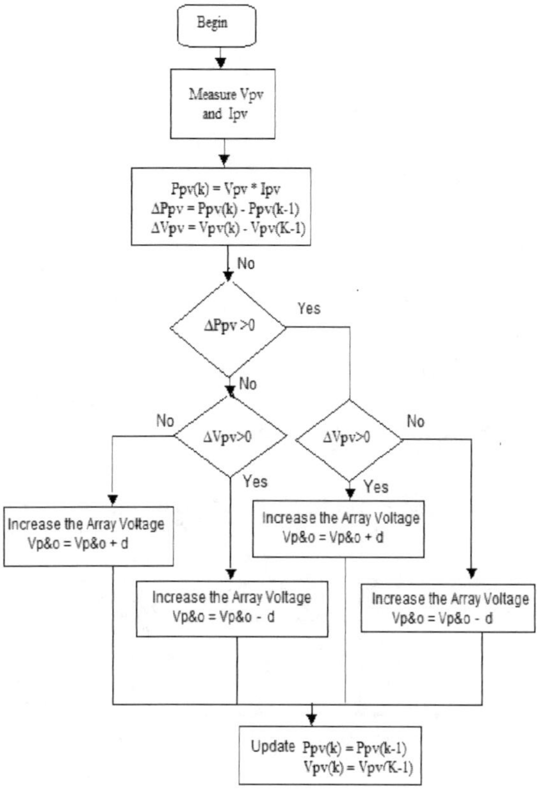

Figure 2. Flowchart of P&O algorithm method

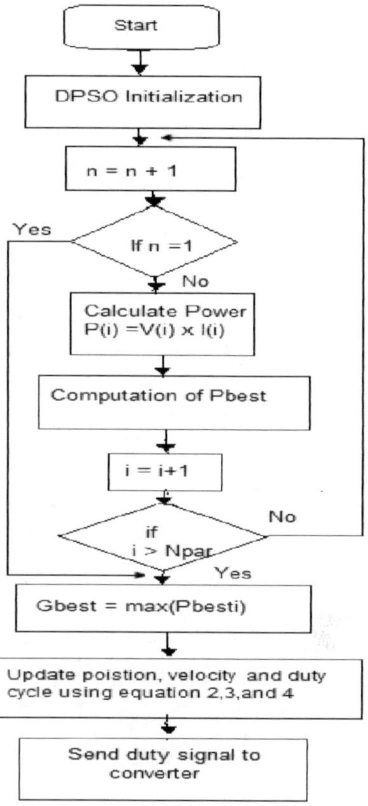

Figure 3. Flowchart of proposed hybrid DPSO-P&O MPPT algorithm

3. DESIGN OF SEPIC CONVERTER

The DC to DC conversion using DC/DC converter is heart of MPPT circuit which is generally used to regulating the voltages of PV array at the Maximum Power Point (MPP) and providing maximum power transfer by load matching. SEPIC is a DC/DC converter similar to buck-boost and Cuk converter, that has the ability to produce the output either may lesser or greater, otherwise equal to the voltage at the input terminal, but has added benefit of have non-inverted polarity output. The SEPIC converter voltages obtained at the output terminal are depending upon the switching rate of the power electronic switch. The basic structure of the classical SEPIC converter circuit is point up in figure 4. The converter construction to be modified by including two components capacitor C_M, and diode D_M in existing SEPIC converter as presented in figure 5.

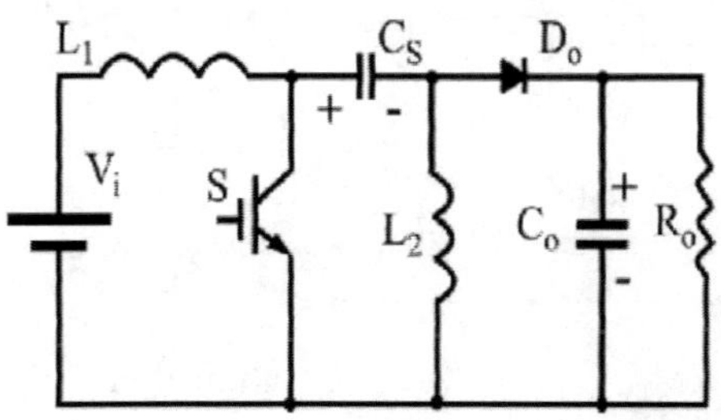

Figure 4. Classical SEPIC converter

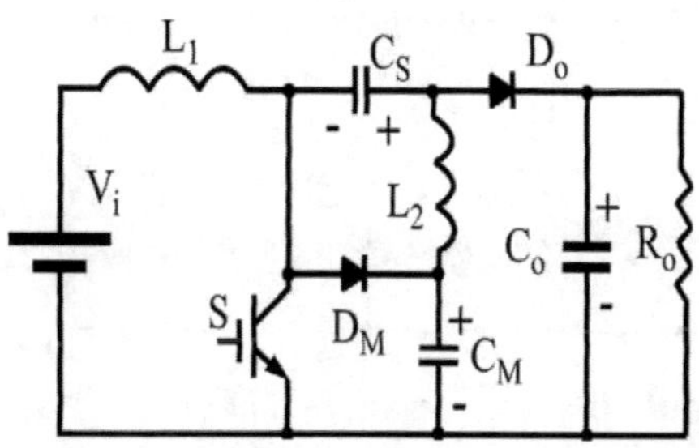

Figure 5. SEPIC converter with voltage multiplier

IOP Conf. Series: Materials Science and Engineering **937** (2020) 012010 doi:10.1088/1757-899X/937/1/012010

By the aforementioned alteration numerous functional features of the classical SEPIC converter are transformed and get twice the time of classical boost converter gain for a duty-cycle near to D = 0.85. However few applications needs larger static gain at lower duty cycle less than 0.8 (D < 0.8). So the customized converter to be tuned to get the higher static gain at duty cycle smaller than 0.8.

4. PROPOSED DC-DC CONVERTER WITH MAGNETIC COUPLING

A simple alteration made in modified SEPIC converter to build higher the static gain at the smaller duty cycle D < 0.8 is by inserting a secondary coil with the L_2 inductor. The magnetic coupling between the winding inductor L_{2p} and L_{2s} will increase the voltage at output and working like a fly-back transformer. Figure 6 demonstrates the modified arrangement of the proposed high gain circuit. The insertion of coil L_{2s} drives the over voltages across output diode D_0 terminals. This enormous voltage at diode D_0 terminals will not suppress with traditional snubber circuit. Hence a clamping circuit consist of capacitor C_{S2} and diode D_{M2} are added with the secondary coil L_{2S} as provided in figure 7. This clamping circuit also act as a voltage multiplier for the output diode D_0.

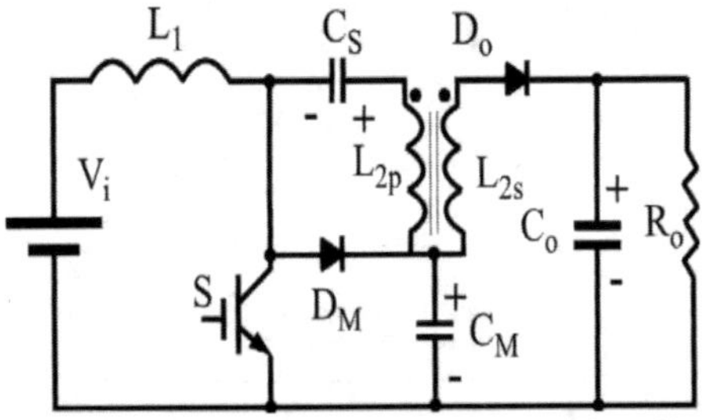

Figure 6. SEPIC converter with coil L_{2s} as secondary

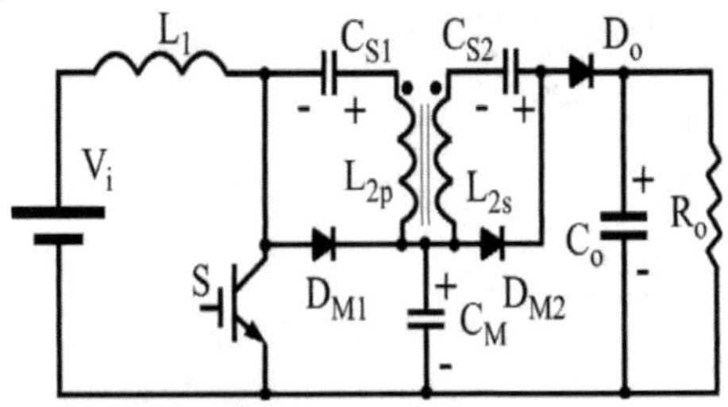

Figure 7. Proposed high static gain DC-DC converter

The voltage multiplier circuit helps transfer of energy from the leakage reactance to the output terminals and subsequently the stress on the diode D_0 gets suppressed and the voltage across the output diode is reduced than the voltage at output which guides to increase the overall converter static gain.

The high static gain SEPIC converter operates in continuous conduction mode (CCM) illustrates in two operation stages.

4.1 First Stage of operation

When the MOSFET switch S is in ON state, the input voltage charges inductor coil L_1. The inductor L_{2S} discharges their stored energy to capacitor C_{S2} through the diode D_{M2} which forms a closed path and block the output diode D_0 as shown in figure 8. Once the charging of capacitor C_{S2} is completed the diode D_{M2} becomes reverse biased which opens the path as illustrated in figure 9.

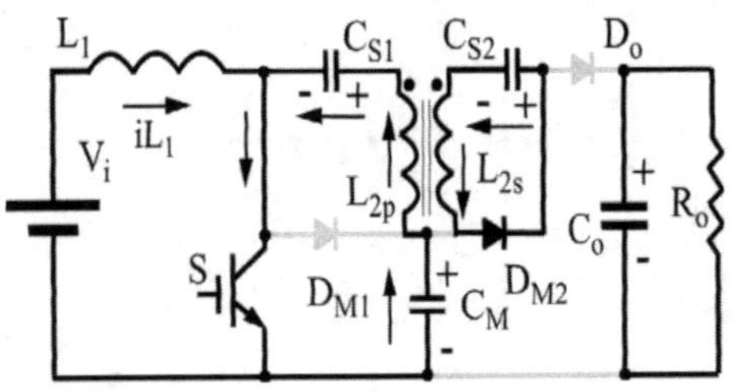

Figure 8. First stage of operation

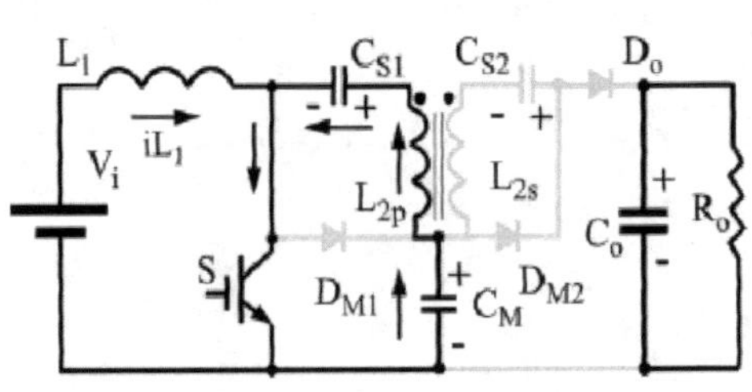

Figure 9. First stages of operation after Diode D_{M2} open

IOP Conf. Series: Materials Science and Engineering **937** (2020) 012010 doi:10.1088/1757-899X/937/1/012010

4.2 Second Stage of operation

When the control switch S is in OFF state, the stored energy in the inductor L_1 discharging through D_{M1} diode to the capacitor C_M illustrated in figure 10. Once the energy transfer from Inductor L_1 makes the capacitor C_M charged which reverse biased diode D_{M1}. The charges stored in inductor L_2, capacitors C_{S1}, and C_{S2}, charges the capacitor C_0 and energize the load R_L through output diode D_0 as shown in figure 11 until the switch S get ON.

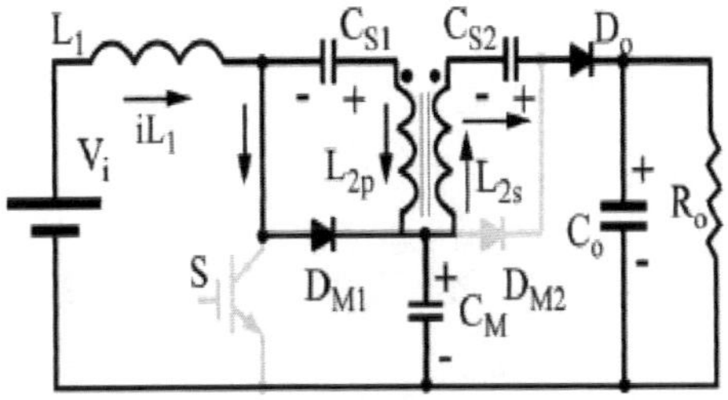

Figure 10. Operation at second stage

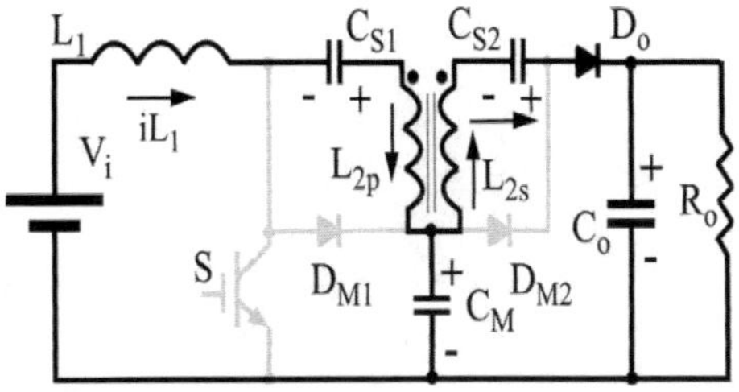

Figure 11. Diode D_{M1} blocked at end of second stage of operation

5. RESULTS AND DISCUSSION

The Matlab simulink model of overall PV system for DC microgrid environment is presented in the figure 12. The outputs of PV arrays are directly fed to the proposed high gain SEPIC converter in which the hybrid P&O-DPSO MPPT algorithm is embedded. The figure 13, 14, and 15 shows hybrid P&O-DPSO based MPPT provides better results than conventional P&O and PSO methods and tracks global maximum power point under PSC. The proposed converter can operate with high static gain ($q > 10$) and employs fewer quantity of switching devices results less switching losses. Consequently the overall PV system effectiveness and efficiency gets increased.

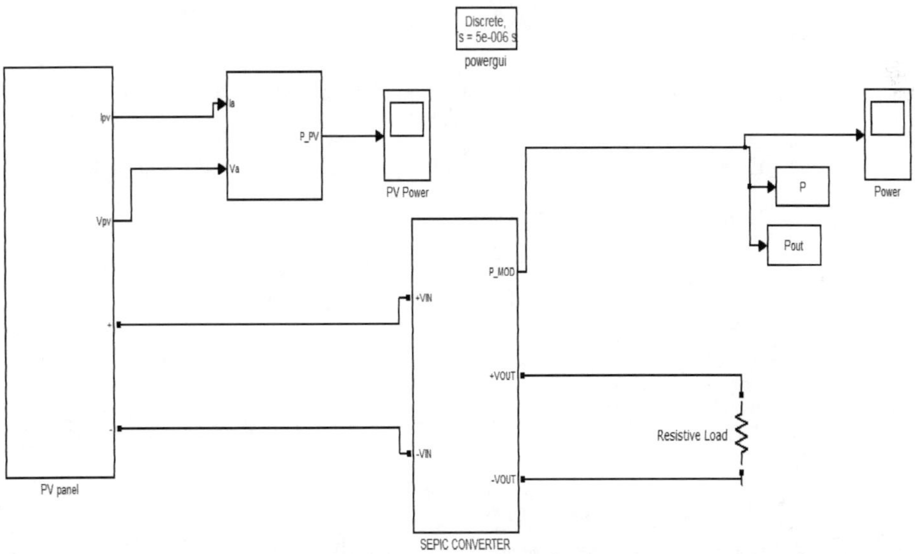

Figure 12. Simulink model of Hybrid P&O-DPSO MPPT control of proposed high static gain converter

IOP Conf. Series: Materials Science and Engineering **937** (2020) 012010 doi:10.1088/1757-899X/937/1/012010

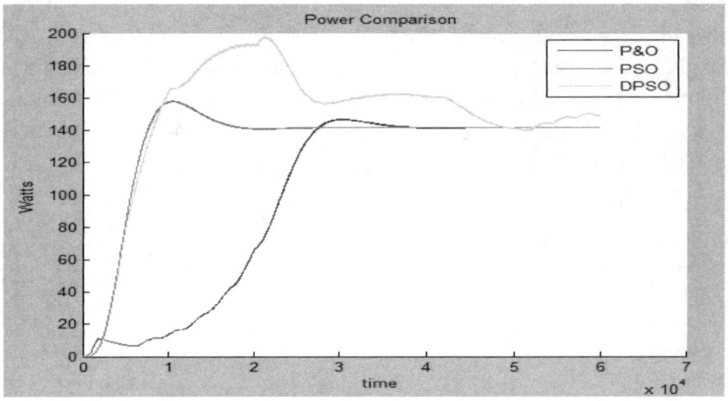

Figure 13. Power variation between P&O, PSO and hybrid DPSO techniques

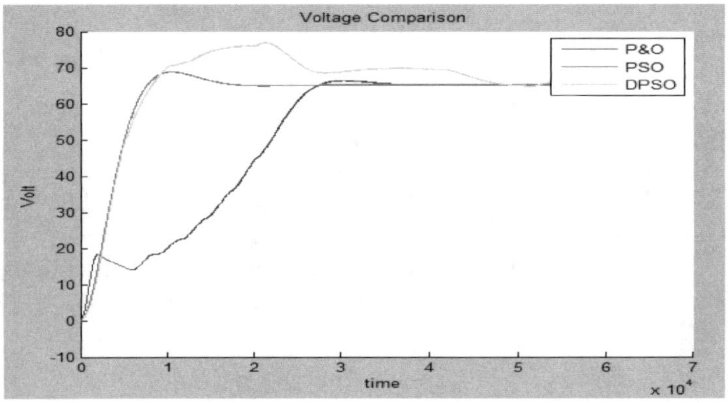

Figure 14. Voltage variation between P&O, PSO and hybrid DPSO techniques

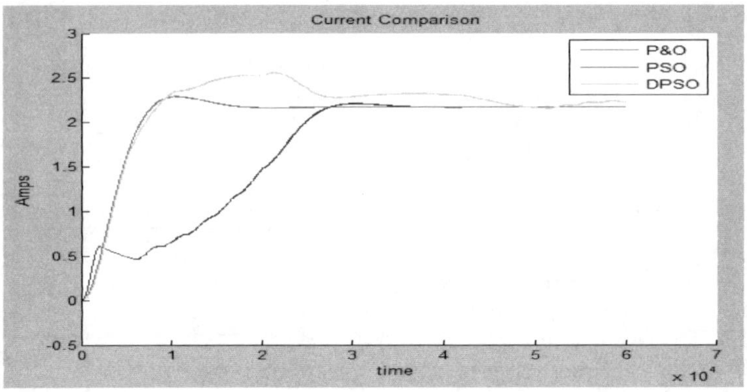

Figure 15. Current variation between P&O, PSO and hybrid DPSO techniques

6. CONCLUSION

This paper gives a hybrid P&O-DPSO based MPPT, and a high static gain SEPIC converter. The proposed hybrid control algorithms gives better results and efficiently track the optimum peak point for the PV array underneath partial shading circumstance. In this proposed technique, the output voltage has very high static gain and constant under normal or shaded conditions of solar panel using hybrid P&O-DPSO based MPPT algorithm implemented on modified SEPIC converter using MATLAB / SIMULINK. The proposed high static gain converter working with low voltage input is able to produce better output voltage and efficiency at duty cycle D < 0.8. So the control mechanism is competent to utilize in grid connected renewable system using various multilevel inverter topologies.

7. REFERENCES

[1]. Sitharthan R, Geethanjali M and Pandy TKS 2016 Adaptive protection scheme for smart microgrid with electronically coupled distributed generations *Alexandria Engineering Journal* **55(3)** 2539-2550

[2]. Fathima AH, and Palanisamy K 2014 Battery energy storage applications in wind integrated systems—a review *IEEE International Conference on Smart Electric Grid* 1-8

[3]. Prabaharan N and Palanisamy K 2015 Investigation of single-phase reduced switch count asymmetric multilevel inverter using advanced pulse width modulation technique *International Journal of Renewable Energy Research* **5(3)** 879-890.

[4]. Jerin ARA, Kaliannan P and Subramaniam U 2017 Improved fault ride through capability of DFIG based wind turbines using synchronous reference frame control based dynamic voltage restorer. *ISA transactions* **70** 465-474

[5]. Sitharthan, R, Sundarabalan CK, Devabalaji KR, Nataraj SK and Karthikeyan M 2018 Improved fault ride through capability of DFIG-wind turbines using customized dynamic voltage restorer *Sustainable cities and society* **39** 114-125

[6]. Prabaharan N and Palanisamy K 2016 A single-phase grid connected hybrid multilevel inverter for interfacing photo-voltaic system *Energy Procedia* **103** 250-255

[7]. Palanisamy K, Mishra JS, Raglend IJ and Kothari DP 2010 Instantaneous power theory based unified power quality conditioner (UPQC) *IEEE Joint International Conference on Power Electronics, Drives and Energy Systems* 1-5

[8]. Sitharthan R and Geethanjali M 2017 An adaptive Elman neural network with C-PSO learning algorithm-based pitch angle controller for DFIG based WECS Journal of Vibration and Control **23(5)** 716-730

[9]. Sitharthan R and Geethanjali M 2015 Application of the superconducting fault current limiter strategy to improve the fault ride-through capability of a doubly-fed induction generator–based wind energy conversion system *Simulation* **91(12)** 1081-1087

[10]. Sitharthan R, Karthikeyan M, Sundar DS and Rajasekaran S 2020 Adaptive hybrid intelligent MPPT controller to approximate effectual wind speed and optimal rotor speed of variable speed wind turbine *ISA transactions* **96** 479-489

[11]. Sitharthan R, Devabalaji KR and Jees A 2017 An Levenberg–Marquardt trained feed-forward back-propagation based intelligent pitch angle controller for wind generation system *Renewable Energy Focus* **22** 24-32

[12]. Sitharthan R, Sundarabalan CK, Devabalaji KR, Yuvaraj T and Mohamed Imran A 2019 Automated power management strategy for wind power generation system using pitch angle controller *Measurement and Control* **52(3-4)** 169-182

[13]. Sundar DS, Umamaheswari C, Sridarshini T, Karthikeyan M, Sitharthan R, Raja AS and Carrasco MF 2019 Compact four-port circulator based on 2D photonic crystals with a 90° rotation of the light wave for photonic integrated circuits applications *Laser Physics* **29(6)** 066201

[14]. Sitharthan R, Parthasarathy T, Sheeba Rani S and Ramya KC 2019. An improved radial basis function neural network control strategy-based maximum power point tracking controller for wind power generation system *Transactions of the Institute of Measurement and Control* **41(11)** 3158-3170

[15]. Rajesh M and Gnanasekar JM 2017 Path observation based physical routing protocol for wireless ad hoc networks *Wireless Personal Communications* **97(1)** 1267-1289

[16]. Palanisamy K, Varghese LJ, Raglend IJ and Kothari DP 2009. Comparison of intelligent techniques to solve economic load dispatch problem with line flow constraints *IEEE International Advance Computing Conference* 446-452

[17]. Sitharthan R, Ponnusamy M, Karthikeyan M and Sundar DS 2019 Analysis on smart material suitable for autogenous microelectronic application *Materials Research Express* **6(10)** 105709

[18]. Rajaram R, Palanisamy K, Ramasamy S and Ramanathan P 2014 Selective harmonic elimination in PWM inverter using fire fly and fireworks algorithm *International Journal of Innovative Research in Advanced Engineering* **1(8)** 55-62

[19]. Sitharthan R, Swaminathan JN and Parthasarathy T 2018 March. Exploration of wind energy in India: A short review *IEEE National Power Engineering Conference* 1-5

[20]. Karthikeyan M, Sitharthan R, Ali T and Roy B 2020 Compact multiband CPW fed monopole antenna with square ring and T-shaped strips *Microwave and Optical Technology Letters* **62(2)** 926-932

[21]. Sundar D Sridarshini T, Sitharthan R, Madurakavi Karthikeyan, Sivanantha Raja A, and Marcos Flores Carrasco 2019 Performance investigation of 16/32-channel DWDM PON and long-reach PON systems using an ASE noise source *In Advances in Optoelectronic Technology and Industry Development: Proceedings of the 12th International Symposium on Photonics and Optoelectronics* 93

[22]. Sitharthan R and Geethanjali M 2014 Wind Energy Utilization in India: A Review *Middle-East J. Sci. Res.* **22** 796–801 doi:10.5829/idosi.mejsr.2014.22.06.21944

[23]. Sitharthan R and Geethanjali M 2014 ANFIS based wind speed sensor-less MPPT controller for variable speed wind energy conversion systems *Australian Journal of Basic and Applied Sciences* **8**14-23

[24]. Jerin ARA, Kaliannan P, Subramaniam U and El Moursi MS 2018 Review on FRT solutions for improving transient stability in DFIG-WTs *IET Renewable Power Generation* **12(15)** 1786-1799

[25]. Prabaharan N, Jerin ARA, Palanisamy K and Umashankar S 2017 Integration of single-phase reduced switch multilevel inverter topology for grid connected photovoltaic system *Energy Procedia* **138** 1177-1183

[26]. Rameshkumar K, Indragandhi V, Palanisamy K and Arunkumari T 2017 Model predictive current control of single phase shunt active power filter *Energy Procedia* **117** 658-665

[27]. Fathima AH and Palanisamy K 2016 Energy storage systems for energy management of renewables in distributed generation systems *Energy Management of Distributed Generation Systems* 157

[28]. Rajesh M 2020 Streamlining Radio Network Organizing Enlargement Towards Microcellular Frameworks *Wireless Personal Communications* 1-13

[29]. Subbiah B, Obaidat MS, Sriram S, Manoharn R and Chandrasekaran SK 2020 Selection of intermediate routes for secure data communication systems using graph theory application and grey wolf optimisation algorithm in MANETs IET *Networks* doi:10.1049/iet-net.2020.0051

[30]. Singh RR and Chelliah TR 2017 Enforcement of cost-effective energy conservation on single-fed asynchronous machine using a novel switching strategy *Energy* **126** 179-191

[31]. Amalorpavaraj RAJ, Palanisamy K, Umashankar S and Thirumoorthy AD 2016 Power quality improvement of grid connected wind farms through voltage restoration using dynamic voltage restorer *International Journal of Renewable Energy Research* **6(1)** 53-60

[32]. Singh RR, Chelliah TR, Khare D and Ramesh US 2016 November. Energy saving strategy on electric propulsion system integrated with doubly fed asynchronous motors *IEEE Power India International Conference* 1-6

[33]. Singh RR, Mohan H and Chelliah TR 2016 November. Performance of doubly fed machines influenced to electrical perturbation in pumped storage plant-a comparative electromechanical analysis *IEEE 7th India International Conference on Power Electronics* 1-6

Minimization of Power Factor Penalty Charges for Non-Linear Domestic Loads with IOT Technology

A.Anton Amala Praveen[1],M.Muthu Kumaran ,Dr.A.Nazar ali[3],Dr.K.Premkumar[4]

[1,2,3,4] Department of Electrical and Electronics Engineering

[1]K.Ramakrishnan college of Technology,Trichy.

[2]Sree Sowdambika college of Engineering, Aruppukottai.

[3,4]Rajalakshmi Engineering College, Chennai

E.mail : praveenantoneee@gmail.com

Abstract: This Paper is dealt about the Power factor plays a major role for improving the quality of the power supply and also used to measure the system's power efficiency. Existing solution is for automatic correction of power factor and it does not provide an information about load causing poor power factor. In proposed system, the status of domestic load are monitored and it intimate the power factor range of the concern load utilization. By using Microcontroller Arduino IDE, Developing a program for calculation action for the project is to dumped in it. The simulation results were obtained using the Proteus software and the penalty charge for the different power factor values was obtained using the program dumped in the Arduino Uno controller. The power factor range of the concern loads utilization was intimated to the consumer through IoT .

1.INTRODUCTION

The Ratio of valuable true power in terms of Kilowatts to reactive power in terms of Kilo Volt-ampere obtained from the electrical supply as Hardware otherwise forming of full step up of electrical supply in it. Resultant work is to be obtained from the electrical energy in better manner. The power factor is to be mandatory for the electrical circuit which is operated in the AC Supply. Extra Power is required to finish the things in better manner.

The path which produces and concentrate about the supply and circulation in it. Power factor with the unity brings about the productive stacking in the electrical supply. Power factor must be maintained in the Indutries. If the Power factor didn't maintain in the industries. Industries must pay the penalty according to their reduction in the power factor.

At the point when the force factor is short of what one. Inductive load which affects the power factor. To improve the power factor, capacitor bank is to be added in it. It will enhance the power factor in the load side. It will avoid the penalty which is paid by the industries. Reactive power can be useless power which added with the real power and it affects the efficiency of the equipment.

It will add the extra amount to the customer bill. A poor power factor will affects not only that equipment and also affects the line which connects to that equipment. Non linear loads, Apart from the fundamental frequency, other order frequency will add in it. The waveform becomes non sinusoidal waveform. We have to maintain the unity power factor in the industries.

If the power factor is not maintained, penalty will be collected from the industries.When we use large inductive loads , it will draws more current in the non linear load. It will decrease the power factor in it. An equipment is added to the circuit which enhances the power factor in the circuit.Normally inverters are having the power factor in the range of 50% to 75%. But some type of inverters has the best power factor and the value is more than 95%.It make to act like resistive load type.

The power factor of the circuit is 0.95 .The irregular waveform is to be obtained and along these lines adds to expand misfortunes. A similarly little better performance in power factor would realize a considerable decrement in misfortunes while misfortunes is proportional to the squares of the flow of electrons. The correction in the Power factor is to be the way toward repaying a main current with the slacking current, through the inventor interfacing with the capacitance.

Power factor produced in th circuit. It obtained with the help of angle between the current and voltage. Capacitors contained in most power factor revision framework draws current that drives voltage and produces a main power factor. Hypothetically, capacitors could give 100% of required reactive power , adjusting the power factor much closer to unity may bring about consonant bending.

A circuit which consists of capacitor which reduces the inductive nature and it will boosted the power factor range in the circuit. It will improve the efficiency and decrease the angle between the voltage and current in it. The magnetizing current is to be reduce with the help of Power factor amendment and useful to kill. It is to reduce misfortunes in the dispersion structure. It offers numerous advantages to the business electrical buyer, remembering diminished service charges by disposing of charges for reactive power, decreased misfortunes making more reactive power in the circuit.

Consequently it improves the strength effectiveness. IoT is new processing and correspondence worldview in which the objects of regular day to day existence have furnished with sensor, microcontroller and handset to detect the encompassing ecological parameters. What's more, correspondence of the detected information with each other or client, turning into a basic piece of internet framework Spaces.

2.EXISTING MODEL

In the existing model, main drawbacks are to be that power factor is risen to 0.97 and it is difficult to determine the reactive power for the circuit. Another paper, the drawbacks are to be failed in triggering the switches for compensating the power factor. Normally, early days power factor will consider and also it affects the circuit.

It consists of Arduino Uno , power supply unit, voltage sensor unit, current sensor unit, current transformer, voltage transformer , LCD display, capacitor bank. It mainly used in the areas of industries to maintain the power factor with the satisfactory level.

With the help of sensing unit of current and voltage , current and voltage is to be sensed .It give to microcontroller input. The current, voltage is to be calculated with the help of Micro controller. The useful power, reactive power for the load also calculated with the help of Micro controller. By using if condition , range of the power factor are to be corrected with the help of program used in it .

.If power factor range will not be greater than 0.9 no capacitive load will be added .The system will operate with an unity power factor does not cause any problem to the electrical system . If power factor range will be less than 0.9 an capacitive load will be added to maintain the range of the power factor as a unity .The power factor value is to corrected with the help of proposed system.The drawbacks of this existing system is to be Power factor range cannot be displayed to consumer, No penalty charge reduction and Consumer cannot able to identify load causing poor power factor.

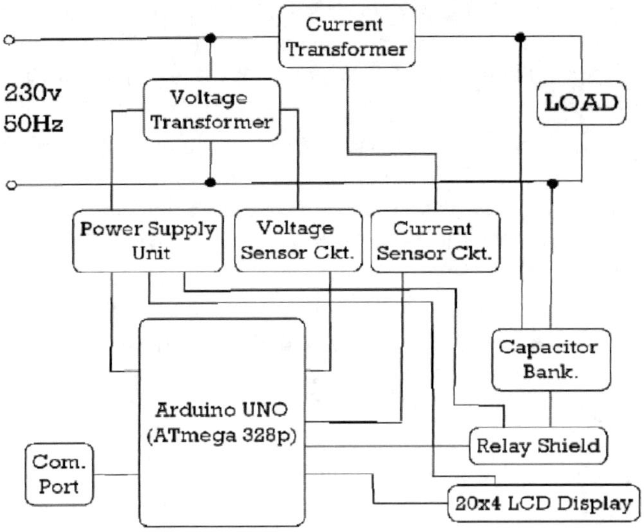

Figure1.Block Diagram of Automated Power Factor correction

3.PROPOSED SYSTEM DESIGN

Power factor correction is proposes in the proposed system with the help of circuit which shown in the figure 2 . Power factor is to be calculated and corrected .It will used to calculate the Penalty charge for the loads.

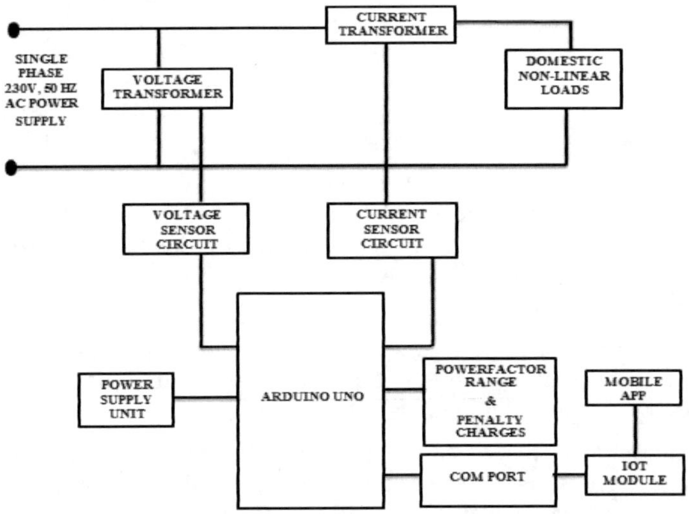

Figure 2. Proposed model

The proposed work is planned to intimate power factor range for concern utilization of the load. It builds penalty charge for the obtained power factor. To calculate power & power factor range of the individual non-linear loads, current and voltage which is flowing in the circuit measured with the help of sensing of current and voltage.. The sensors are interfaced with the ARDUINO UNO

IOP Conf. Series: Materials Science and Engineering **937** (2020) 012011 doi:10.1088/1757-899X/937/1/012011

controller thereby monitoring the power factor.

The controllers continuously update the measured power factor value in the IOT. The customer will receive information about the load causing poor power factor through IOT and they able to reduce the poor power factor causing electrical equipment. It will reduce the penalty amount for the respective poor power factor in their loads.
The Arduino UNO based on the ATmega 328 type. It is a Microcontroller. It consists of input/output pins of 14. 6 out of 14, used for output pin. Power factor is to be improved In increase the current carrying level.
By reducing the losses in the circuit, It obtain lower electric bills. The most straight forward method of improving the power factor by adding the capacitor to the electrical framework. Reactive current generators acted as a Power factor correction.
From above table 1, I mentioned the power factor range and penalty charges for not maintaining the power factor.Here, in my paper, I set a power factor range of .9 and more as a good P.F. For the power factor range goes down in the range of 0.9 to 0.85.If power factor will reduce 10% from unity power factor upto 15%, 1% charges will collected extra form normal current consumption charges.

Table1. POWER FACTOR SURCHARGE

RANGE OF POWER FACTOR	PENALTY CHARGES
< 0.75	Reduce in 0.01 power factor from 90% , 2% charges for current consumption will collected along with normal payment.
0.75 >= P.F > 0.85	Reduce in 0.01 in power factor form 90%, 1.5% charges for current consumption will collected along with normal payment.
0.85>= P.F > 0.90	Reduce in 0.01 in power factor from 90%, 1% charges for current consumption will collected along with normal payment

Power factor range goes between the range of 0.85 to 0.75, Additional 1.5% will collected extra along with normal consumption charges for reduction 1% from the power factor for this range. Power factor goes less than 0.75, Additional 2% will collected extra along with normal consumption charges for reduction of 1% from the power factor for this range. Power factor must be maintained in the Industries.
Dynamic Power factor correction, in some cases alluded to as ongoing power factor amendment, utilized for electrical adjustment in instances of quick burden changes. Dynamic power factor correction is valuable when standard power factor modification would cause over or under correction. Dynamic power factor correction utilizes semiconductor switches , regularly thyristors, to rapidly associate and separate capacitors or inductors to improve power factor.
From the figure 3, we noticed the power factor in the range of 0.9 in it. The current and voltage rating of the circuit displayed in the figure 3. The power factor will intimated to the customer. If the power factor obtained in the circuit below the nominal value, it will message to the customer

automatically. If the power factor is low, customer cut the circuit from the supply. Otherwise it will affects the circuit which connect to this equipment also.

Figure3. Hardware model for Power factor Correction

4. SIMULATION

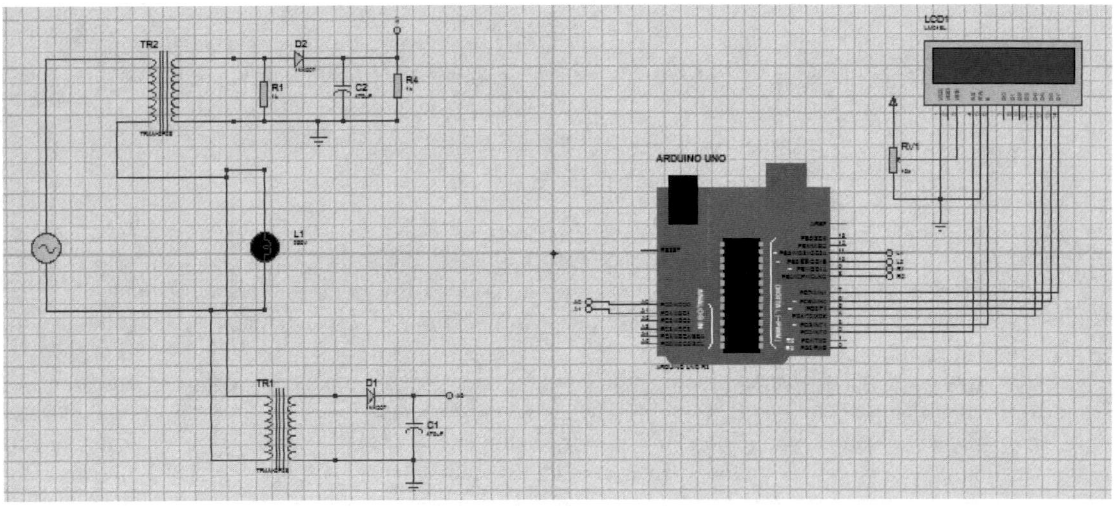

Figure 4. Proteus Model for proposed system

In this proposed paper, Model of proteus type in the proposed system are shown in the Figure 4.In this circuit, Power factor is to calculated with the help of Arduino .The customer will receive information about the load causing poor power factor through IOT and they able to reduce the poor power factor causing electrical equipment.It will reduce the penalty amount for the respective poor power factor in their loads.

By using current and voltage sensor, we can sensed the current and voltage in the circuit.Non-linear loads. Power and power factor calculated with the help of sensor mentioned above. Sensors are connected with the controller of ARDUINO UNO. It will update the power factor value through the IoT. Customer obtain the result of power factor causing electrical equipment.

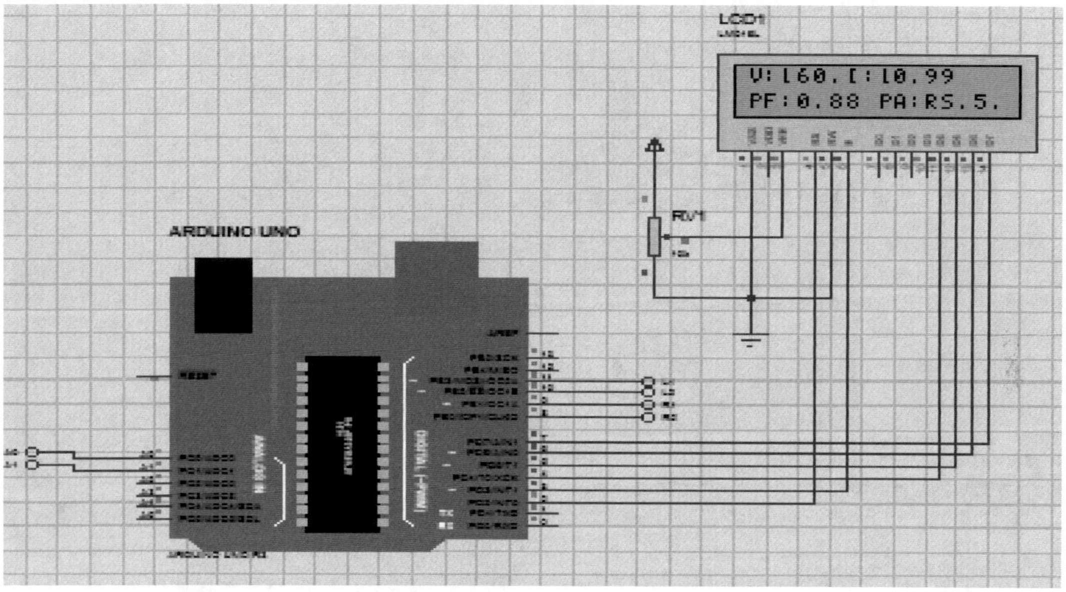

Figure 5 Range of power factor between 0.85 and 0.9

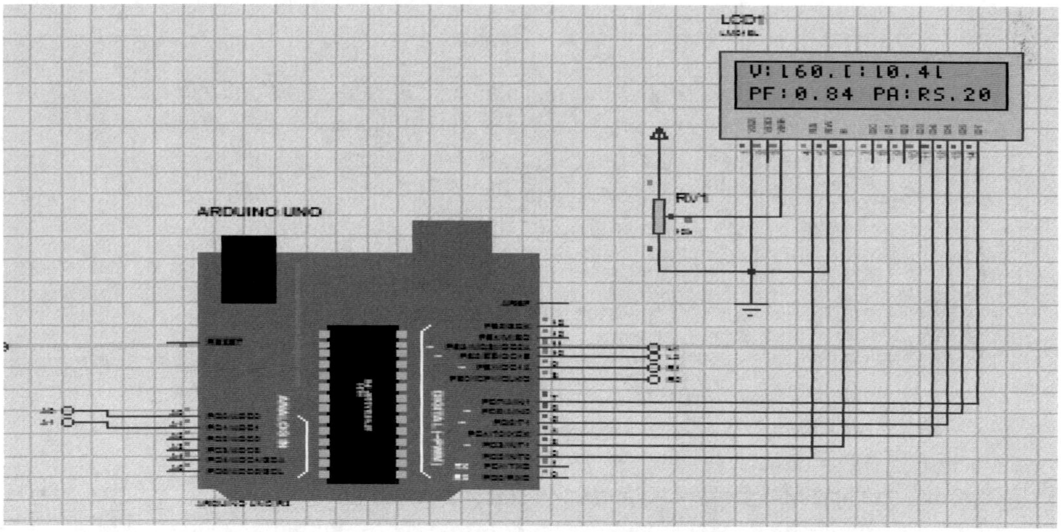

Figure 6 Range of power factor between 0.8 and 0.85

IOP Conf. Series: Materials Science and Engineering **937** (2020) 012011 doi:10.1088/1757-899X/937/1/012011

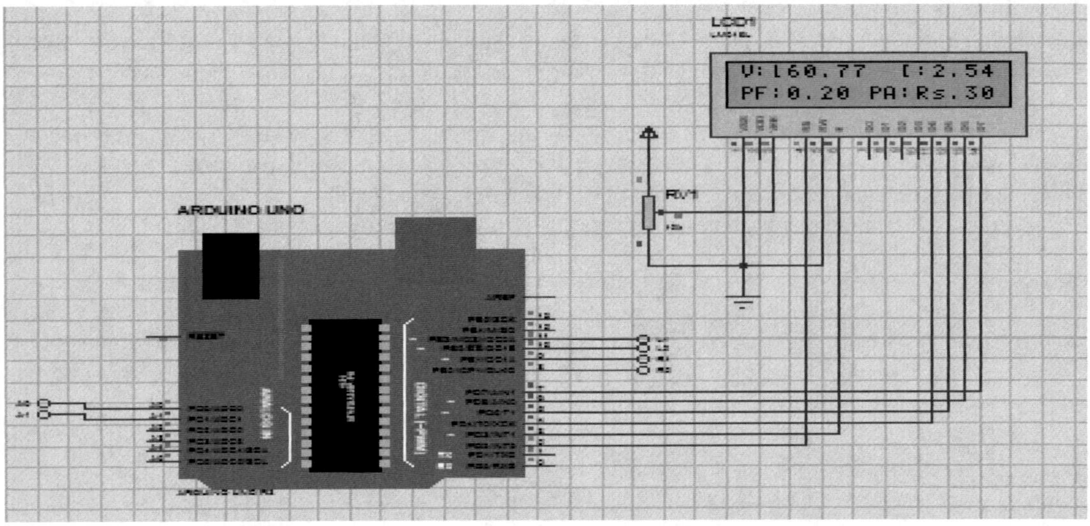

Figure 7 Range of Power factor between 0.75 and 0.85

5.CONCLUSION

Minimization of power factor penalty charges for non-linear domestic loads using IOT technology has been presented in this project. With the help of microcontroller using Arduino IDE, calculations made for the project with help of programs developed in it.. With the help of program, power factor is to be calculated here. penalty charges of different power factor values are calculated. The simulation results were obtained using the PROTEUS software .The prototypes for minimization of power factor correction are shown in it. The extracted result indicates that the poor power factor is to be identified for the load in the proposed method. Through IOT, loads utilization was intimated to the customer and also range of power factor also intimated to the customer.

6 .REFERENCES

[1] A.Zuazua-Ros, J.C.Ramos and C.Mart in- G'omez. "*System simulation of a linear concentration of energy meter Billing system*" National Science and Technology Council, June 2015, last accessed: June 18, 2017.

[2] M.Chandrasekar, S.Suresh and T.Senthilkumar, "*Energy balance model of billing system in real time*". International Journal of Research technology.

[3] R.Jai Ganesh,S.Koodeswaran,M.Kavitha,T.Ramkumar," *Performance analysis of piezoelectric harvesting system employing bridgeless power factor correction boost rectifier*"

[4] A.Anton Amala Praveen,S.Murugesan," *Integration of Hybrid Energy sources for Grid connected AC Districution System*", proceedings of the 4th International Conference on Advances in Electrical Electronics Information Communication and Bio-informatics 2018.

[5] S.Murugesan and M.V. Suganyadevi , " *Hybrid renewable Energy parameters Monitoring and control of smart street Light using IoT*", International Journal of scientific and Technology Research, ISSN:2277-8616,volume 8,issue 10,pp.645-651,October 2019.

[6] Subramanian, A.T sankar, P. sabarish and A. Nazar Ali , "*A Power factor correction based canonical switching cell converter for VSI fed BLDC motor by using voltage follower technique.*"Electrical, Instrumentation and communication Engineering (ICEICE),2017 IEEE International conference on IEEE , 2017.

[7] S.Vijayalakshmi, P. sabarish , S.R.Paveethra, Dr.P.R.Sivaraman , Dr. V.Venkatesh, " Exploration and application of Electronic Balance for High Power Discharge Lamps at High Frequency through power factor modification", International Journal of Scientific & Technology Research voluime 9, Issue 02,February 2020 ISSN 2277-8616.

[8] M.Kavitha, R.jai Ganesh & A.Rajkumar, "*Facilities Navigation and Patient Monitoring system using IBEACON Technology*", International Journal of Mechanical and production Engineering Research and development ISSN(P): 2249-6890; ISSN(E): 2249-8001 Vol. 9, special Issue 1, Jan 2019,562-570.

IOP Conf. Series: Materials Science and Engineering **937** (2020) 012012 doi:10.1088/1757-899X/937/1/012012

Torque Ripple Reduction in PMSG for Standalone Wind Energy System Using Three Level DTC with A Dither Signal

Dattatraya Kalel[1], Umashankar Subramaniam[2], Santoshkumar Hunachal[3]

1. SKN, Sinhgad College of Engineering, Korti, Pandharpur (MH)
2. Renewable Energy Lab, College of Engineering, Prince Sultan University Riyadh, Saudi Arabia
3. Research Scholar, SELECT, VIT, Vellore (TN)

Corresponding mail: ddkalel183@gmail.com

Abstract: In this paper, focused on direct torque control (DTC) for permanent magnet synchronous generator direct driven standalone variable speed wind turbine. The machine side converter (MSC) employed three-level DTC with dither signal strategy proposed for minimizes the torque & flux ripple. The triangular signal used as a dither signal & injected in a hysteresis band controller of torque & flux comparator. The optimum magnitude of the triangular wave is 3% of its rated torque & flux used in order to reduce acoustic noise. The proposed strategy minimizes torque & flux ripple as compared to conventional two-level DTC technique. Three-level SVPWM strategy employed for a load-side converter (LSC) in order to reduce harmonic distortion.

Keywords: Three level DTC; Dither Signal; Three level SVPWM; PMSG.

Introduction

The development of new technologies in the region renewable energy for generating clean and pollution free energy, special attention is given worldwide. In recent years, Renewable energy technologies such as solar, tidal, biomass, geothermal & wind energy are gaining attention in the energy market globally. The wind energy technology is one of the most emerging renewable energy technologies[1]. Nowadays, back-to-back Voltage Source converters for direct driven wind generation system with Permanent Magnet Synchronous Generator (PMSG) have been widely applied [5].

The advantages of the direct driven PMSG generator compared to a generator with a gearbox are increased reliability and the reduction of the weight of the whole system. The PMSG constructed with a large number of poles and operated at low speed. The results of directly coupling the generator with a wind turbine allow for the elimination of a gearbox installation and a reduction of maintenance costs [2], [7]. Fixed-speed wind turbine generator system WTGS and variable-speed WTGS are the type of the wind turbine generator system. In the beginning wind power development the fixed -speed WTGS has been applied. However, it has disadvantages like low power quality & less efficiency [3]. These disadvantages overcome variable speed generation systems. Variable wind speed generator have advantages such as improved wind generation energy, exact maximum power from the turbine [14].

In [3] PMSG directly coupled to wind turbine & it is controlled vector control (Field Orientation Control) using PI controller. However, in vector control d-q coordinate transformation required [11]. Without d-q coordinate transformation, direct torque control (DTC) with switching table & DTC with space vector pulse width modulation (DTC-SVPWM) has high dynamic performance. In DTC- SVPWM has expense of increased complexity is the highlights of implementing a DTC [13]. It is to control the torque and flux linkage by selecting one of the voltage vectors generated by a VSI [9], [12].

DTC worked in either torque control method or speed control method. In speed controller, difference between reference speed & generator speed which is error given to Proportional & integrative (PI) controller which generates reference torque command. This reference torque compared with actual estimated torque in hysteresis controller. On this the torque error, appropriate voltage vectors selected so that actual torque tracks the reference torque and thus generator rotor rotate its reference speed. In variable Wind speed, to extract more power, to maintain the Tip speed ratio (λi). Thus, reference speed variation accordingly wind speed, this can be conclude that speed error results in ripples in torque, thus deteriorating the steady-state performance. Torque ripples is directly proportionally to stator current, resulting in an increment in harmonic distortion of stator current [8].This drawback reduced by multilevel converter such as three level neutral clamped converter (3-L NPC). In [14] five-level inverter-fed to load employed for reduce the harmonic reduction.

In this paper focused on three level NPC based DTC with dither signal based PMSG direct drive wind turbine & three-level Space vector pulse width modulation (3-L SVPVM) has been praposed. Typical configuration of Wind Energy Conversion System illustrated in figure 1. The stator winding of PMSG connected to a full-scale, back-to-back Voltage Source Converter system, which consist of the Machine Side Converter (MSC) and the load Side Converter (GSC). This model simulated in MATLAB/ Simulink 2016a & its quick response comparatively Induction generator.

This paper is structered as follows. In the section II, presents the modeling of wind turbine, Direct drive, PMSG & power converter. The Section III, presents the proposed control methodology of PMSG i.e. DTC with dither Signal, modelling of DC bus, Pitch angle controller, & 3- Level SVPWM finally; the last section presents the simulation waveform results & conclusion.

SECTION II:

Figure 1 shows the model of WECS consists of horizontal axis wind turbine direct driven PMSG through mechanical coupling shaft, rectifier connected to capacitor, & output of capacitor to inverter fed to load.

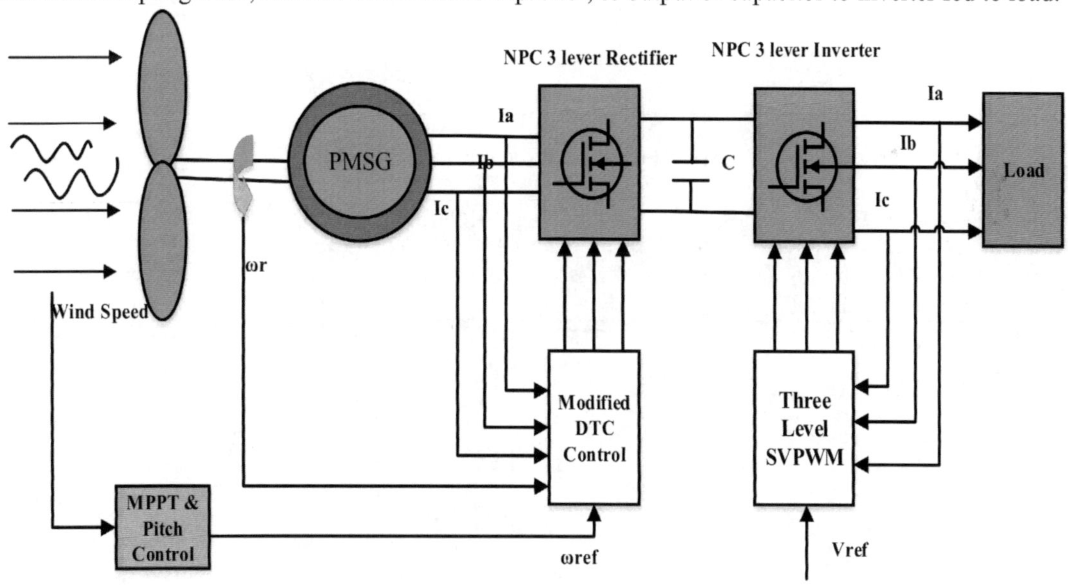

Figure 1 Block diagram of the Standalone WECS structure

2.1 Wind turbine model

The power developed by wind in the turbine blades, which written in the form according the following equation:

IOP Conf. Series: Materials Science and Engineering **937** (2020) 012012 doi:10.1088/1757-899X/937/1/012012

$$P_w = \frac{1}{2}\cdot\rho\cdot\pi\cdot r^2\cdot V_w^3 \qquad (1)$$

Where, ρ= Air density 1.225 Kg/m^3, Vw=Wind speed m/s, r= radius of blade m.
Wind turbine Power captured from the wind is depend on bentz limit Cp(λ, ß).

$$P_t = \frac{1}{2}\cdot\rho\cdot\pi . r^2\cdot V_w^3\cdot \mathrm{Cp}(\lambda, \beta) \qquad (2)$$

Where, λ= Tip Speed ratio & ß=Pitch Angle

$$\lambda = \frac{(\omega_m \cdot R)}{V_w} \qquad (3)$$

Where ωm Generator speed in rad/sec
Several models for the power coefficient has been developed such as in [3]which approximates Cp, by:

$$C_p(\lambda, \beta) = 0.517\left(\frac{116}{\lambda_i} - 0.4\beta - 5\right)\exp\left(-\frac{21}{\lambda_i}\right) + 0.006795\cdot\lambda_i \qquad (4)$$

Where, λi expressed as;

$$\lambda_i = \frac{1}{\frac{1}{\lambda + 0.008 * \beta} - \frac{0.03}{\beta^3 + 1}} \qquad (5)$$

The relationship between power coefficient Cp and output power in p.u.(in Matlab/ Simulink) for different value of wind speed (Vw) in p.u. is clearly shows in figure 2

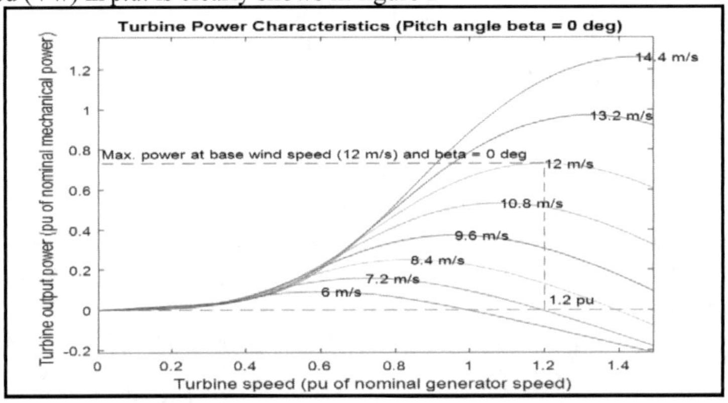

Figure 2 Power coefficient Cp(λ, ß)curves

Figure 3 shows the relationship between the gnerated output power (W) vs Wind speed Vw (m/s). WECS works in between cut in speed & Cutoff speed i.e.3 m/s to 25m/s [3], [20].
Figure 4 shows that the power coefficient Cp is most extreme at a specific estimation value of tip speed ratio, which is called as optimum tip speed ratio λopt.

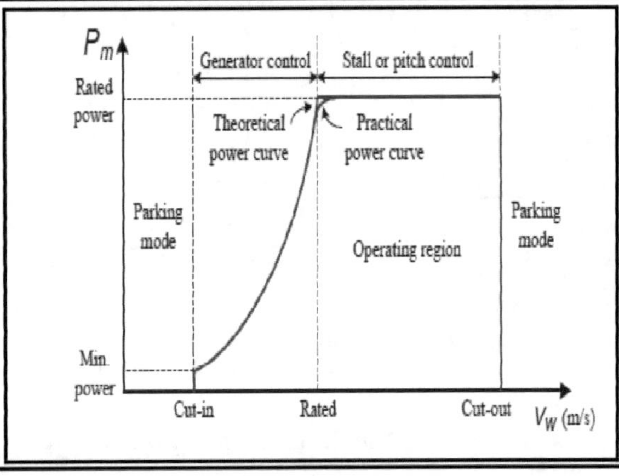

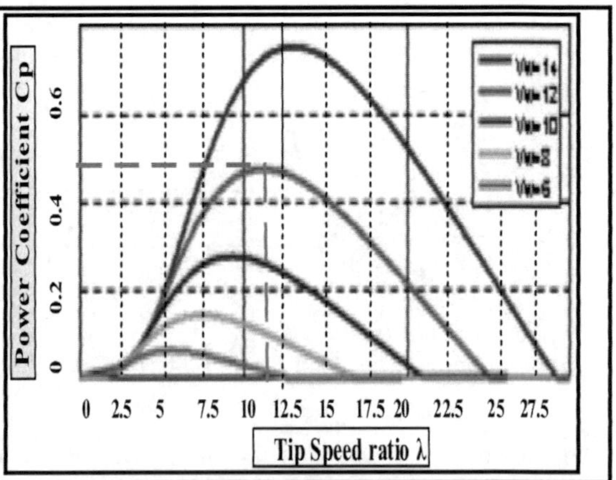

Figure 3 Relation between output power Vs wind speed

Figure 4 Power coefficient (Cp) λ curves

To extract a maximum power from wind, the Wind turbine should consistently work at λopt. By using the MPPT (Maximum Power Point Tracking) technique can be controlling the rotor speed of permanent magnet synchronous generator.

$$\omega_{opt} = \frac{\lambda_{opt} \cdot V_w}{R} \tag{6}$$

The torque equation of the turbine as;

$$T_t = \frac{P_t}{\omega_m} \tag{7}$$

Therefore, the torque equation will be,

$$T_t = \frac{0.5 \cdot \rho \cdot \pi \cdot r^2 \cdot V_w^3 \cdot C_p(\lambda, ß)}{\omega_m} \tag{8}$$

2.2 Direct Driven

Wind turbine direct driven PMSG have absent of gearbox, can avoid the problem of wear and tear of gear & less maintaince, it can help wind turbine operate more reliable. The expression is the angular speed of the PMSG rotor ωe and rotor mechanical speed ωg can written as; [2]

$$\omega_e = P_n \cdot \omega_g \tag{9}$$

Where Pn is no. of pole pair.

The dyanmic torque equation expressed as; [3]

$$T_{em} - T_t = J\frac{d\omega_m}{dt} + F\omega_m \tag{10}$$

Where, Tem is Electrimagnetic torue, J is total inertia of the PMSG shaft & F is wind turbine friction.

2.3 PMSG model

The PMSG has advantages are high reliability, good torque response & power density per volume ratio, because of DC excitation which is replaced by permanent magnet & eliminates brushes & slip ring. Thw PMSG also eliminates the rotor circuit loss due to permanent magnet [12].

The stator phase volatge equarion as;

$$V_a = R_s I_a + \frac{d\psi_a}{dt} \tag{11}$$

$$V_b = R_s I_b + \frac{d\psi_b}{dt} \tag{12}$$

$$V_c = R_s I_c + \frac{d\psi_c}{dt} \tag{13}$$

Where, Va, Vb, Vc & Ia,Ib, Ic are stator voltage & current of a,b,c respectively. R is Resistance of stator circuit, & ψa, ψb, ψc are stator flux of a, b, c axis respectively.

Figure 5 shows the park's (abc to dq) transformation. It is needed for PMSG because, this model is transform the time varient into time invarient system. With help of park transformation, the stator volages equation in the dq refernce frame as below,

The sator voltage equation as;

$$V_{ds} = R_s I_{ds} + \frac{d\psi_d}{dt} - \omega_e \psi_q \tag{14}$$

$$V_{qs} = R_s I_{qs} + \frac{d\psi_d}{dt} + \omega_e \psi_d \tag{15}$$

Where, Vds, Vqs & Ids, Ids are the volage & current of staor on d & q axixs, ωe is rotor angle with respect to a-axis.

The torque equetion of the PMSG (Tem) is,

$$T_{em} = \frac{3}{2}\frac{p_n}{2}(\psi_{ds} I_{qs} - \psi_{qs} I_{ds}) \tag{16}$$

Where, P is pair of pole.

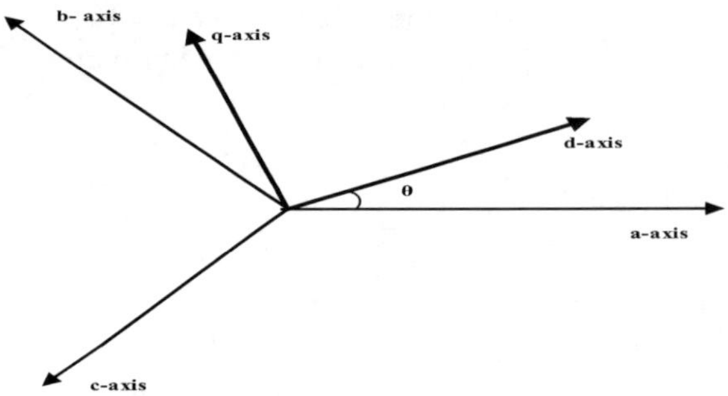

Figure 5 abc to dq refernce frame of PMSG

2.4 Power Converter

In this paper, Figure 1 shows the back-to-back converter (B2B). The B2B converter has two Voltage Source Converters (VSC) and a capacitor in between them.

These first VSC can work as a generator side converter i.e. generated ac voltage rectified to DC. The capacitor acts as a filter for the voltage ripple produced by the VSC. Next, is DC converted AC through the load side inverter. The capacitor DC-link voltage always more than the peak voltage of the load voltage [9]. The rectifier circuit of the PMSG side is controlled by hysteresis current based FOC & load side converter is controlled by three-level SVPWM technique.

SECTION III

3.1 Control of sytem

In this section, the DTC controlled strategy used for PMSG, DC bus voltage, and three level (NPC) SVPWM technique used for the load side converter.

3.1.1 Direct Torque control:

The machine side converter (MSC) controls the rotor speed of the PMSG to extract achieve max. Power from variable speed wind energy conversion system. In DTC method, Torque directly control by angle between the stator and rotor reference frame. According switching pattern the voltage vector selected for controlling the converter speed & torque. In conventional DTC, there drawback is generate torque ripple. The Eques. no.16 expressed In the stator flux reference frame, the torque of PMSG in terms of the amplitude of stator flux and torque angle.

$$T_e = \frac{3}{4}\frac{P_n|\psi_s|}{L_d L_q}[2\psi_f L_q sin(\partial) - |\psi_s|(L_q - L_q)\sin(2\partial)] \quad (17)$$

Where, ∂-torque angle (the angle between stator and rotor flux vector)

For a salient for PMSG, namely, $L_d = L_q = Ls$, the equation can be simplified as:

$$T_e = \frac{3}{4}\frac{P_n}{L_s}|\psi_s|\psi_f sin(\partial) \quad (18)$$

Eques. (18), the torque directly controlled by regulating the torque angle ∂. If the amplitude of the ψ_s is kept constant. Furthermore, the torque increases with the increase of ∂.

Control of the Amplitude of Stator Flux

The stator flux linkage in the ɤ & ß stationary reference frame expressed as;

$$\psi_s = \int(V_s - R_s I_s)dt \quad (19)$$

$$\psi_s = V_s \Delta t + \psi_{\_s|t=0} \quad (20)$$

Where, R_sI_s is fraction of voltage so it neglected. $\psi_{s|t=0}$ Is initial value of flux at t=0

3.1.2 Torque & flux estimator

The torque equetion of the PMSG (Tem) , can be obtained from Eques. No. 16

IOP Conf. Series: Materials Science and Engineering **937** (2020) 012012 doi:10.1088/1757-899X/937/1/012012

Flux equation of PMSG in dq refernce frame as;

$$\psi_{ds} = L_d I_{ds} + \psi_f \tag{21}$$

$$\psi_{qs} = L_q I_{qs} \tag{22}$$

Where, ψ_f is permanent magnet flux.
The stator flux linkage is

$$\psi_s = \sqrt{\psi_{ds}^2 + \psi_{qs}^2} \tag{23}$$

To find the sector angle θs expressed as;

$$\Theta_s = \arctan\left(\frac{\psi_{qs}}{\psi_{ds}}\right) \tag{24}$$

3.1.3 Proposed Direct Torque Control

In proposed DTC, figure 6 shows block diagram, the scheme includes three level Direct Torque Control (DTC) & two hysteresis relay controllers [4]. DTC is a stator flux oriented scheme and based on the hysteresis band control of both flux and torque.

The stator flux controller imposes the time duration of the active voltage vectors, which move the stator flux along the reference trajectory. The torque controller determines the time duration of the zero Voltage vectors, which keep the torque within hysteresis tolerance band. At every sampling time, the voltage vector selection block selects the switching state, which reduces the instantaneous flux and torque errors.

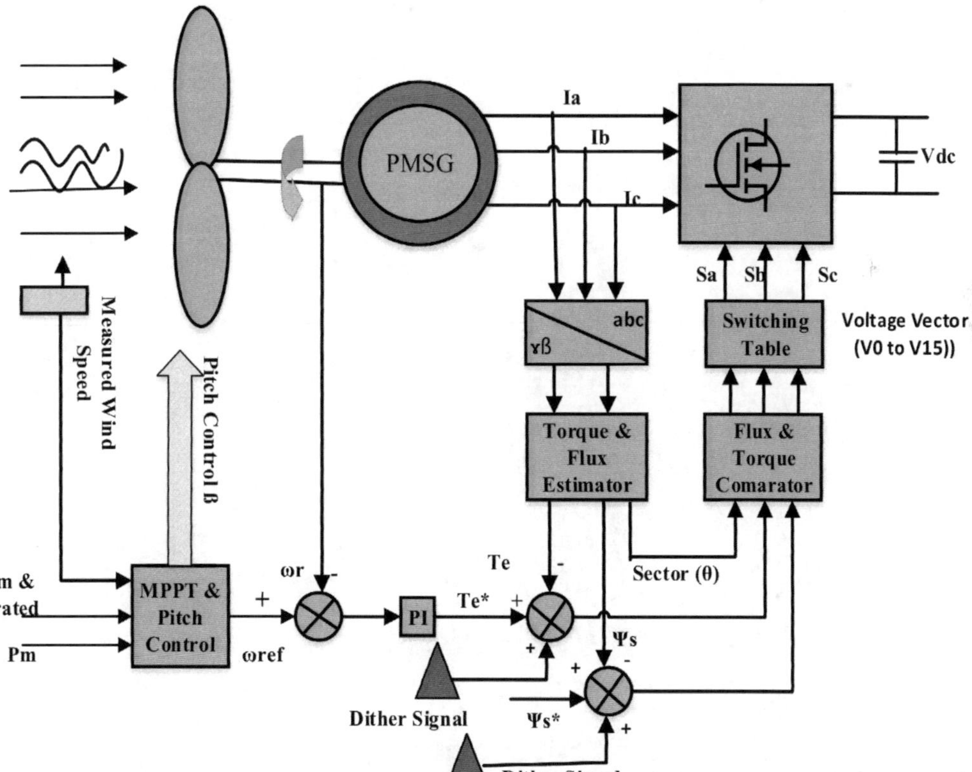

Figure 6 Detailed DTC with Dither signal for machine side converter

IOP Conf. Series: Materials Science and Engineering **937** (2020) 012012 doi:10.1088/1757-899X/937/1/012012

3.1.3 Hysteresis comparator

Five-level torque comparator as shown in figure 7a and 7b, three-level flux comparator are employed for three-level inverter based DTC.

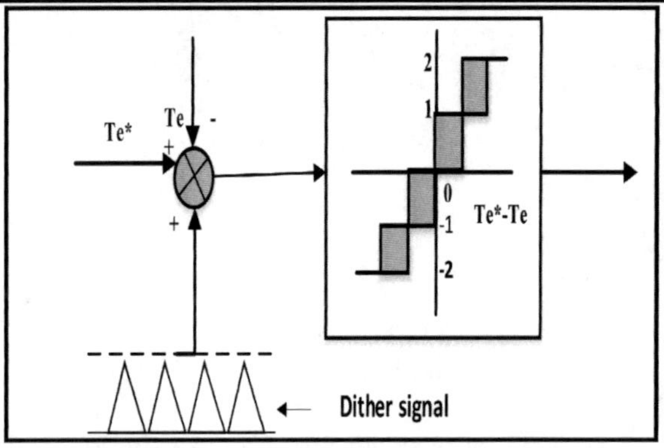

Figure 7a five level torque comparator

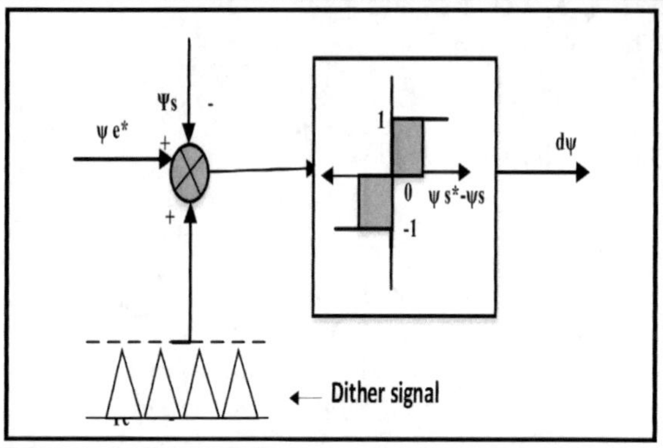

Figure 7b three level flux comparator

Figure 7a–b show the comparator of the dither signal added, where triangular waves employed as the dither signals in the error blocks. The torque error & flux error has to be reduced to adding dither signal i.e. triangular signal. In this paper, a triangular dither signal injected of minute amplitude (3% of rated torque and 3% of reference flux as hysteresis band in torque control and flux control loops respectively) in the error block.

In three level NPC converter, increases no. of sequences to employed for switching table. In the case of 3-level converter, sequence voltage 211 is similar to the sequence 100 so 221 is like 110, etc. Therefore, we have 19 combinations of vectors achievable voltages including zero sequence [6]. Figure 8 shows the six-sector division and utilization of all Voltage vectors.

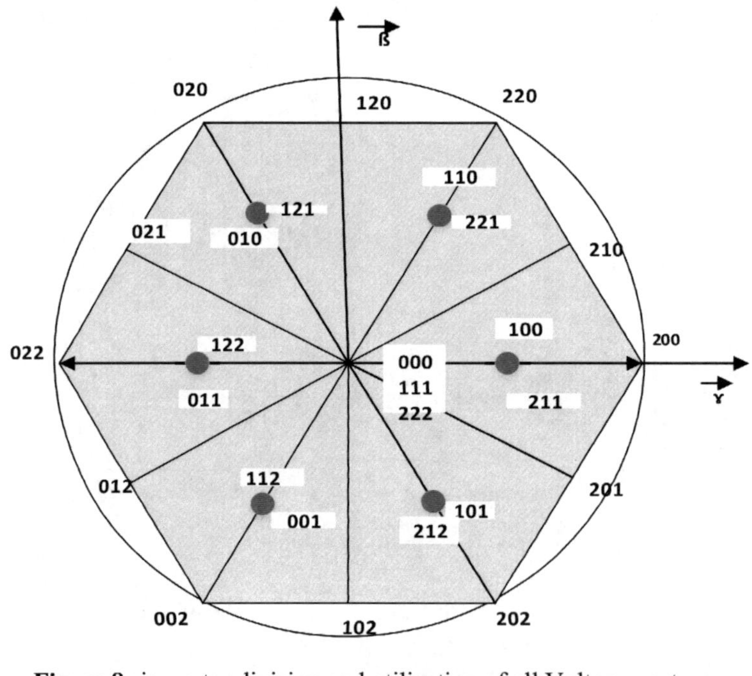

Figure 8 six-sector division and utilization of all Voltage vectors

Table 1 shows Switching table for three level DTC, it is possible to obtain six sector, twenty-four none zero vector & three zero vector.

Table 1 shows switching table for three level DTC.

Δφs	**ΔTe**	**S1**	**S2**	**S3**	**S4**	**S5**	**S6**
1	2	220	020	022	002	202	200
	1	210	120	021	012	102	201
	0	200	220	020	022	002	202
	-1	202	210	120	021	012	102
	-2	202	200	220	020	022	002
0	2	120	021	012	102	201	210
	1	120	021	012	102	201	210
	0	000	000	000	000	000	000
	-1	102	201	210	120	021	012
	-2	102	201	210	120	021	012
-1	2	020	022	002	202	200	220
	1	121	122	112	212	211	221
	0	211	221	121	122	112	212
	-1	112	212	211	221	121	122
	-2	002	202	200	220	020	022

3.1.4 Pitch angle controller

The pitch angle is kept constant at zero degree at wind speed is rated. Beyond the cut out speed, the pitch angle is proportional to the speed deviation from rated speed. The control system illustrated in the following figure [16].

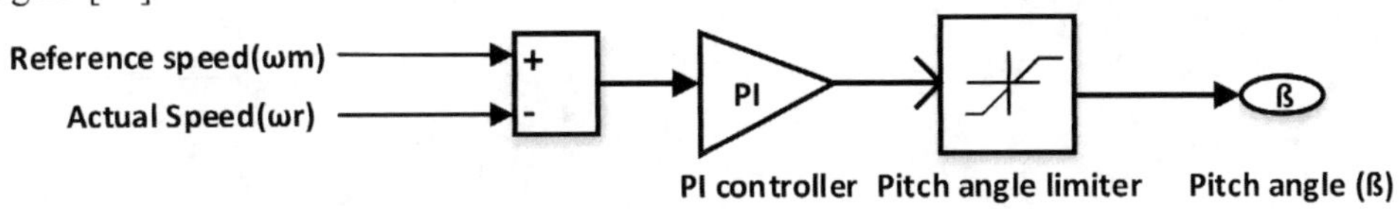

Figure 9 Pitch angle controller

3.2 SVPWM

3.2.1 3-Level Neutral Point Clamped (NPC) Inverter

Realizing SVM for more than two-level inverter (conventional six-switch inverter) called Multi level Inverter MLI [19]. Three phase three level neutral point clamped inverter employed for DTC inducton motor drives [18]. NPC multilevel inverters synthesize the stepped level of the output voltage from DC capacitor voltages, Figure 9 shows the, A 3-level NPC-MLI inverter consists of (k-1) capacitors i.e. two capacitor on the DC bus link, 2(k-1) power switching devices i.e. 4 swiches per phase and 2(k-2) clamping diodes i.e two diodes per phase [17].

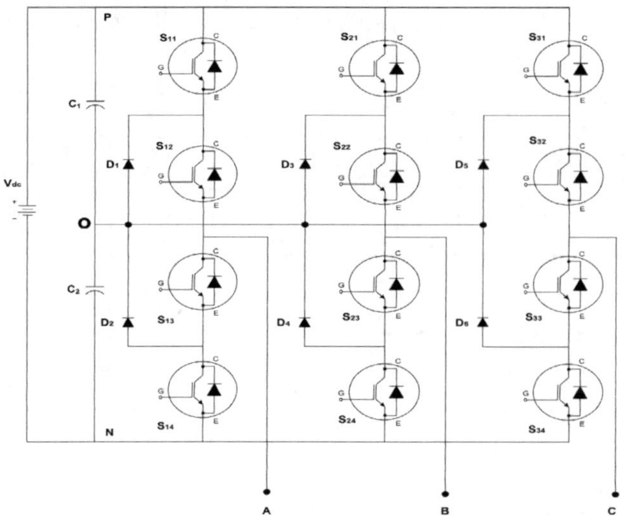

Figure 10 Three level NPC Inverter

Any of the three-phase n-level space vector diagram (SVD) can be separated into six sectors (Si), where i = 1 to 6. These sectors are further separated into $(n-1)^2$ sub triangles (△ i,j) where j = 0 to 3 and i = 1 to 6. Hence, the n-level SVPWM consists of n^3 switching states. For 3 level SVPWM has 27 switching states. shown in Figure 10 of voltage vector diagram, out of which have three null voltage vectror others are non zero vector. Each Sector have four traingles.

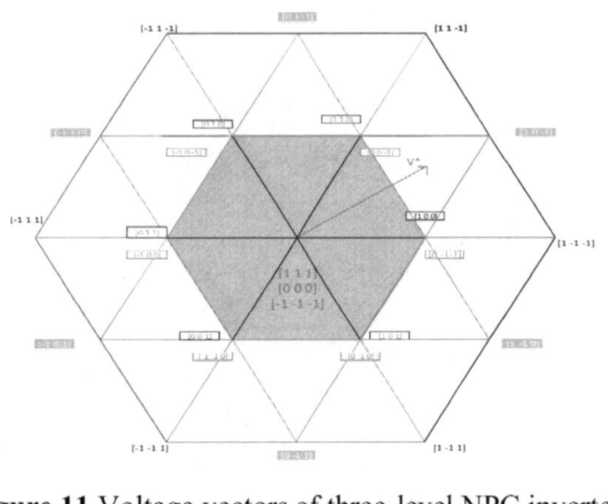

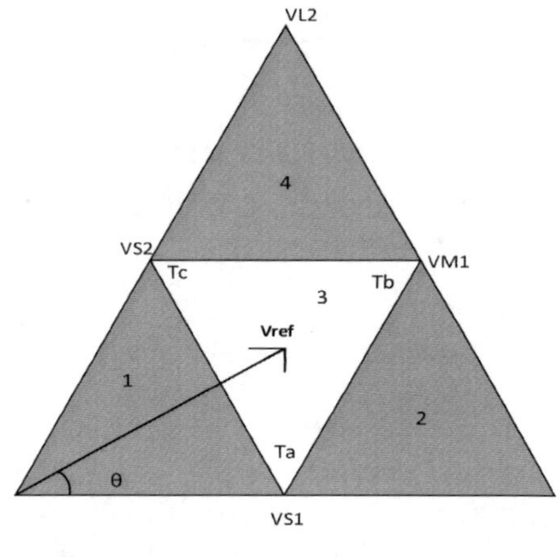

Figure 11 Voltage vectors of three-level NPC inverter **Figure 12** Voltage vectors of NPC in sector-I

According to voltage amplitude, divided them into four segment are zero voltage VZ, Medium voltage VM, Small voltage VS & Large voltage vector VL. Table 2 shows the three-level SVPWM inverter switching states.

Table 2 shows the switching states for 3 level SVPWM.

Switching Symbols	Switching Conditions Q1	Q2	Q3	Q4	Output Voltage
1	ON	ON	OFF	OFF	Vdc/2
0	OFF	ON	ON	OFF	0
-1	OFF	OFF	ON	ON	-Vdc/2

Table 3 Three-level inverter Space vector Diagram switching states.

Switching vector name	Total no. of states	Switching states	Output voltage of states
V_Z	3	[111], [000], [-1-1-1]	0
V_S	12	[100], [0-1-1], [110], [00-1], [010], [101], [011], [100], [001], [110], [101], [0-10]	$Vdc/2$
V_M	6	[10-1], [01-1], [-110], [-101], [0-11], [1-10]	$\frac{\sqrt{3}}{2}.Vdc$
V_L	6	[1-1-1], [11-1], [-11-1], [-111], [-1-11], [1-11];	Vdc

3.2.2 Voltage Vector & Time Calculation of Switching

Figure 11 displays the voltage vectors V_{S1}, V_{M1} & V_{S2} are create the triangle. This formed triangle is subdivided into four triangles; these are 1, 2, 3 and 4. In the SVPWM, assume voltage reference V_{ref} & to find its closest three vectors structure SVD to decrease the harmonic components of the output voltage and the current. The time of each vector calculated by vector calculation. Assume, reference voltage in the triangle 3 shown in Figure 12, the time of each voltage vector can be obtained by the following equations.

$$V_{S1} \cdot T_a + V_{M1} \cdot T_b + V_{S2} \cdot T_c = V_{ref} \cdot T_s \quad (25)$$

$$T_a + T_b + T_c = T_s \quad (26)$$

Where, $V_{S1} = \frac{V_{dc}}{2}, V_{M1} = \frac{\sqrt{3}}{2} V_{dc} \cdot e^{\frac{i\Pi}{6}}, V_{S2} = \frac{V_{dc}}{2} \cdot e^{\frac{i\Pi}{3}}$ (27)

Substituting Eques. (27) In Eques. (25)

$$\frac{V_{dc}}{2} \cdot T_a + \frac{\sqrt{3}}{2} V_{dc} \cdot \left(\cos\left(\frac{\Pi}{6}\right) + isin\left(\frac{\Pi}{6}\right)\right) \cdot T_b + \frac{V_{dc}}{2}\left(\cos\left(\frac{\Pi}{3}\right) + isin\left(\frac{\Pi}{3}\right)\right) \cdot T_c = V_{dc}(cos\theta + isin\theta) \cdot T_s \quad (28)$$

Separate the real and imaginary parts of the Eques. (28)

$$\frac{V_{dc}}{2} \cdot T_a + \frac{\sqrt{3}}{2} V_{dc} \cdot \left(\cos\left(\frac{\Pi}{6}\right)\right) \cdot T_b + \frac{V_{dc}}{2}\left(\cos\left(\frac{\Pi}{3}\right) +\right) \cdot T_c = V \cdot cos\theta \cdot T_s \quad (29)$$

$$\frac{\sqrt{3}}{2} V_{dc} \cdot \left(sin\left(\frac{\Pi}{6}\right)\right) \cdot T_b + \frac{V_{dc}}{2} \sin\left(\frac{\Pi}{3}\right) \cdot Tc = V_{dc}(sin\theta) \cdot \mathrm{Ts} \quad (30)$$

The values of Ta, Tb and Tc by solving Eques. (31), Eques. (32) and Eques. (33)

$$T_a = T_s[1 - 2Ksin(\theta)] \quad (31)$$

$$T_b = T_s\,[2Ksin(\theta + 60) - 1] \quad (32)$$

$$T_c = T_s\,[2Ksin(\theta - 60) + 1] \quad (33)$$

Where, $K = \frac{2}{\sqrt{3}} \cdot V_{dc}$

4. Results

This simulation results carried out in Matlab/ Simulink. In this simulation considered two cases are Variable wind speed (i.e. generator speed Around 40 RPM) & constant wind speed(generator speed Around 40 RPM) figure 13 shows the wind speed profile, Figure 14 & 15 shows the reference speed & actual speed of generator.

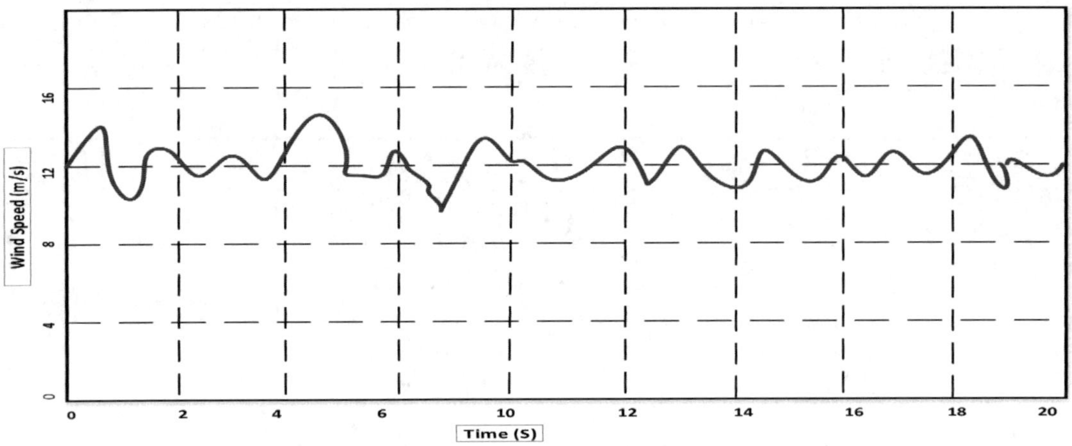

Figure 13 wind profile

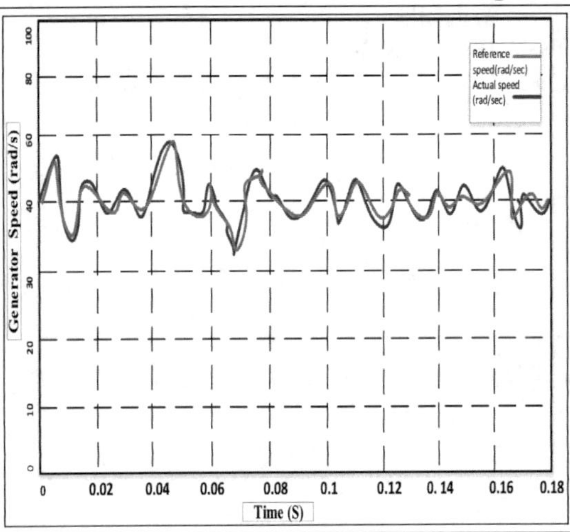

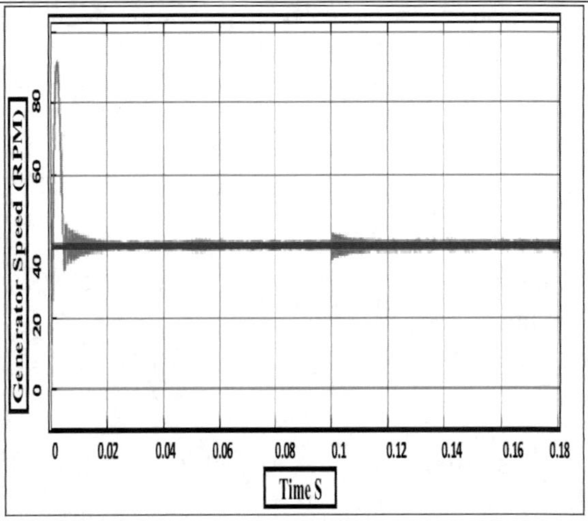

Figure 14 Generator speed RPM for variable wind speed **Figure 15** Constant generator speed at 40 RPM

The electromagnetic torque developed by PMSG for variable speed & constant wind speed. Figure 16 comparison shows that the proposed DTC controlled with dither signal having smooth torque response compared to with without dither signal.

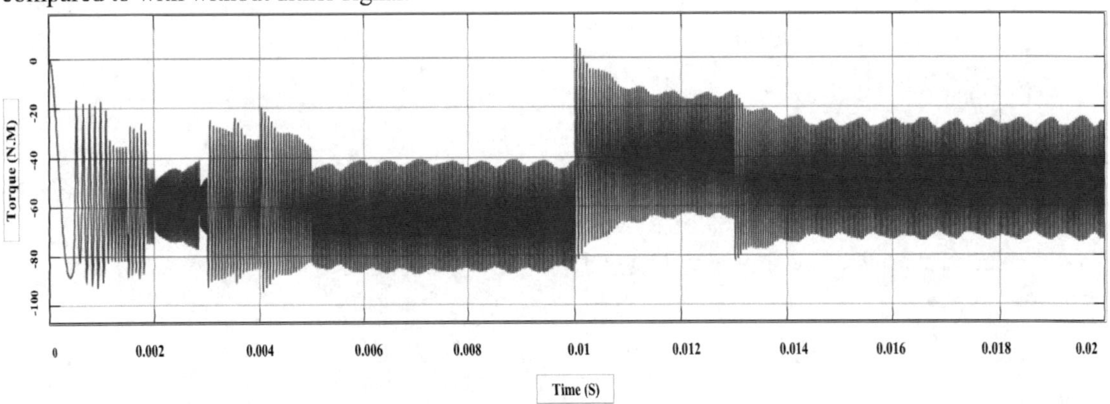

Figure 16a Torque response for variable generator speed without dither signal

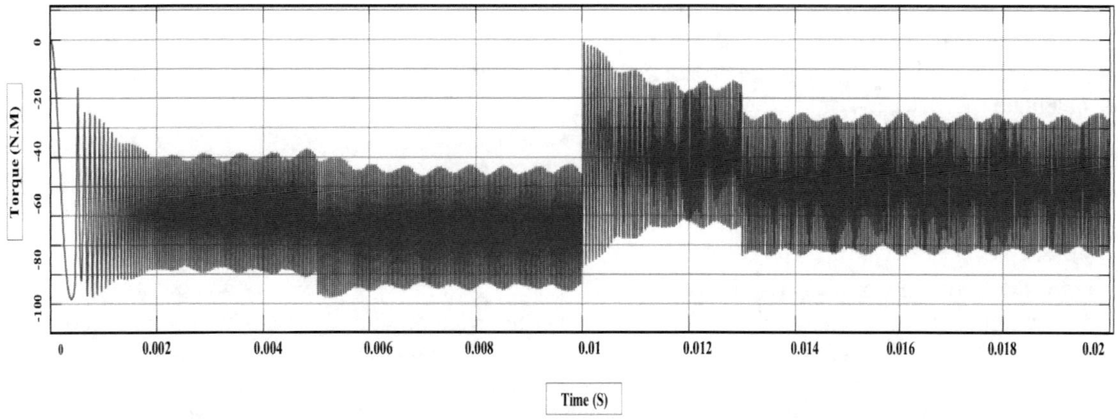

Figure 16b Torque response for constant generator speed without dither signal

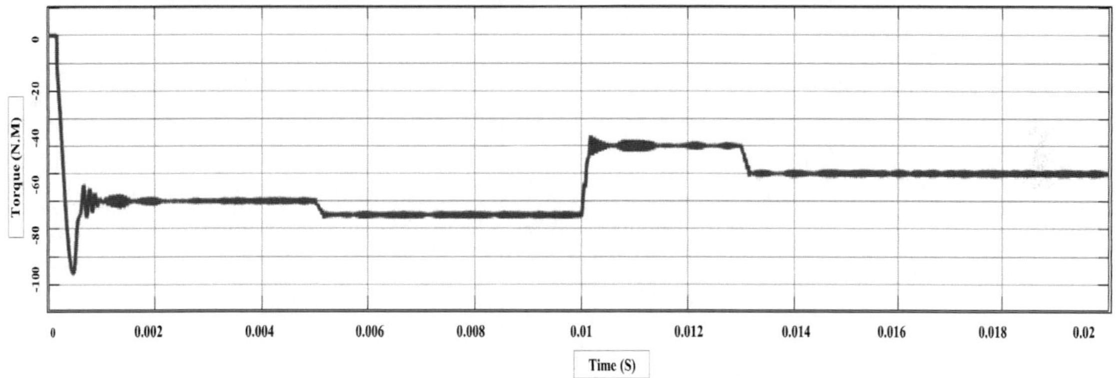

Figure 16c Torque response DTC with dither signal

Figure 17a-b shows the flux trajectory of PMSG using DTC without dither signal & with dither signal. Results compares, the DTC with dither signal has better response.

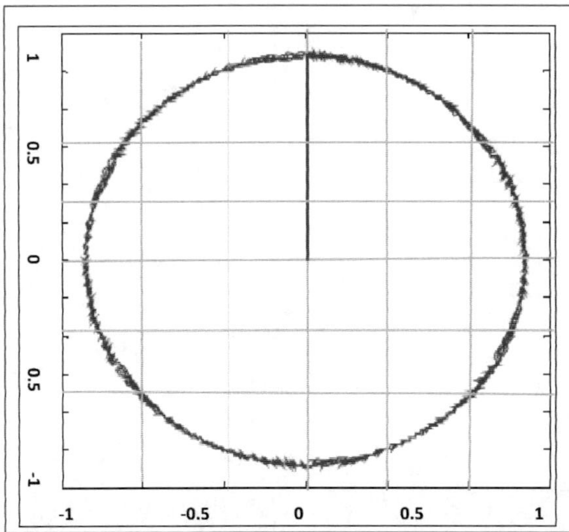

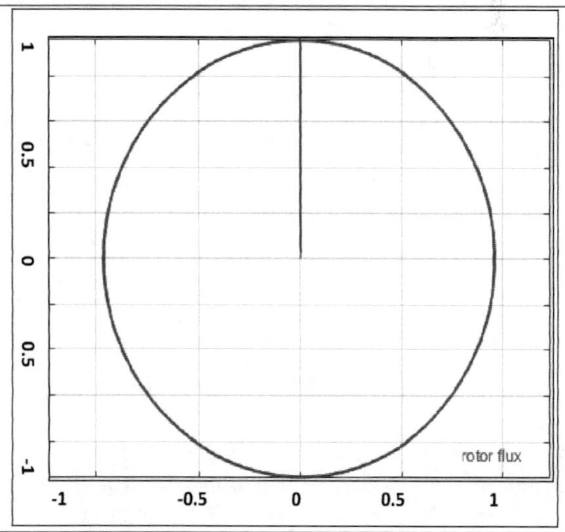

Figure 17a Flux response without dither signal **Figure 17b** Flux response with dither signal

4.1 Torque & Flux ripple analysis

The torque ripple and flux ripple are also significant performance indices when controlling the drive. The performance parameters to judge the effectiveness of the proposed method are torque and flux ripples can be calculate as Eques. (34) and Eques. (35) [10].

$$T_r = \frac{1}{N}\sum_{k=0}^{n}((T_{e^*} - Ti)^2 \quad (34)$$

$$\lambda_r = \frac{1}{N}\sum_{k=0}^{n}(\lambda_{eref} - \lambda i\,)^2 \quad (35)$$

Where, T_e* is average torque N.m., T_i is estimated torque, λ_{ref} reference flux & λ_i estimated flux.

Figure, 18 shows the torque ripple in variable speed wind turbine using DTC, constant speed wind turbine using DTC without dither signal & torques response DTC with dither signal. This comparison shows that DTC with dither signal have less ripple as compare to conventional DTC.

In conventional three level DTC, for variable & constant wind speed have peak-to-peak torque ripple is 10 N.m. & 16 N.m. respectively. In DTC with dither signal have peak-to-peak torque ripple is 2 N.m. so it can be conclude that 60% torque ripple has to be reduced in PMSG.

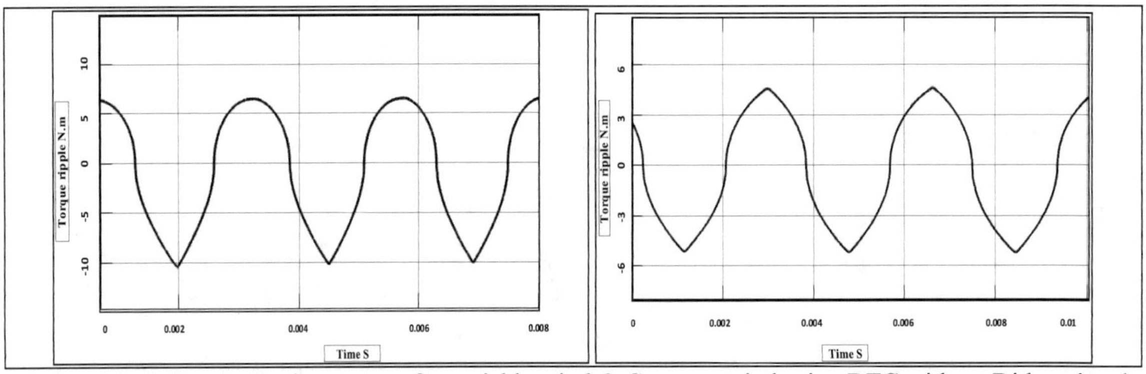

Figure 18a-b Torque ripple response for variable wind & Constant wind using DTC without Dither signal

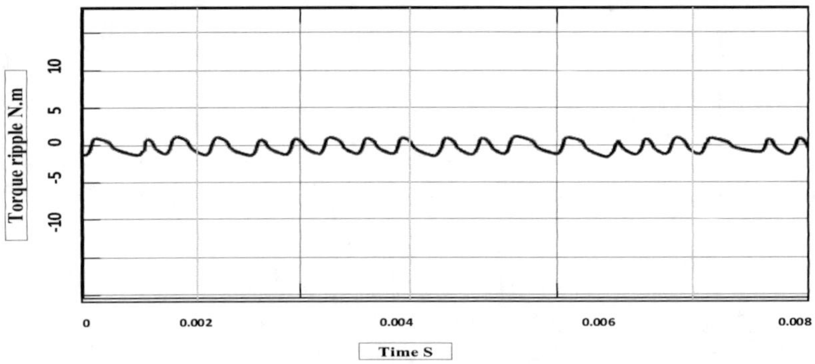

Figure 18c Torque ripple response using DTC with Dither signal.

In DTC control system, mainly focused torque control. However, in DTC due to voltage vector, flux ripple also produced. Figure 19 shows the flux ripple in PMSG using DTC without dither signal & with dither signal. In DTC, without dither signal have peak-to-peak flux ripple is 0.08 wb. & peak-to-peak flux ripple 0.01 wb in DTC with dither signal. This can conclude that 60% flux ripple has been reduces.

IOP Conf. Series: Materials Science and Engineering **937** (2020) 012012 doi:10.1088/1757-899X/937/1/012012

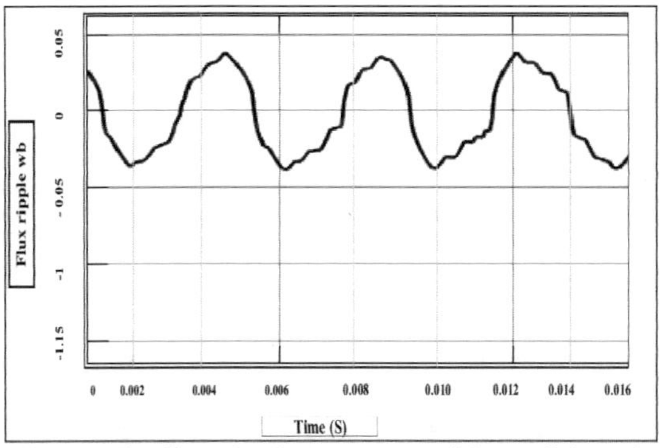

Figure 19a flux ripple response without dither signal

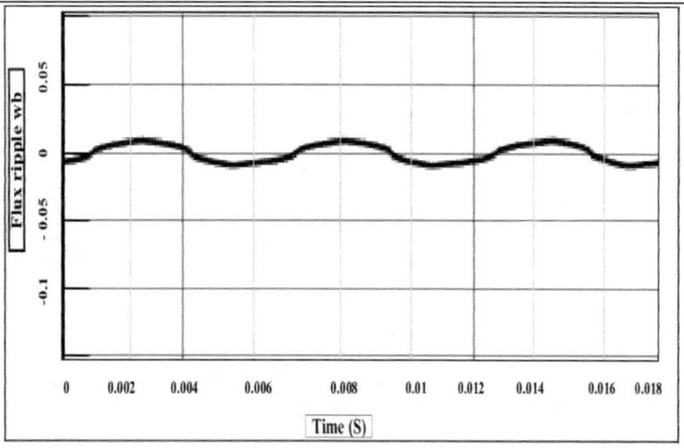

Figure 19b flux ripple response with dither signal

Figure 20 shows the Stator Current waveform. Compare the both results; it can be conclude that harmonics is reduces in current by using DTC with dither signal. Figure 21 Shows the DC bus voltage.

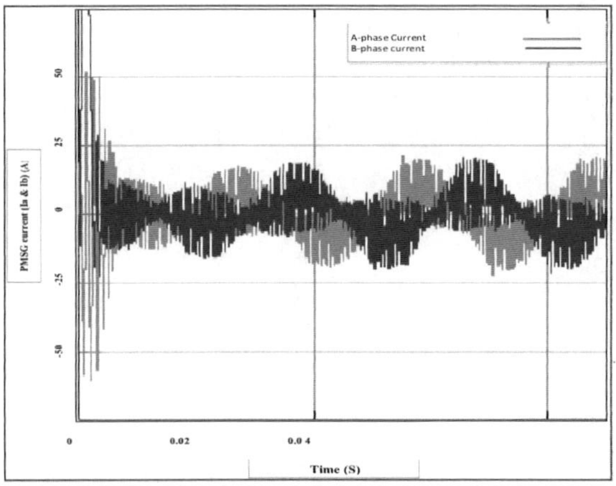

Figure 20a I_a & I_b stator current without dither signal

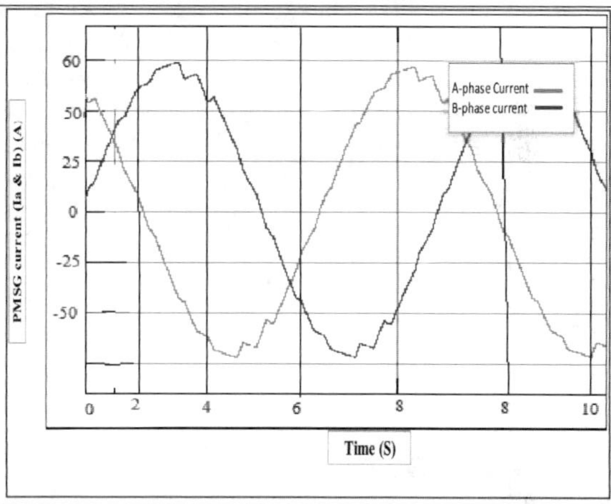

Figure 20a Ia & Ib stator current dither signal

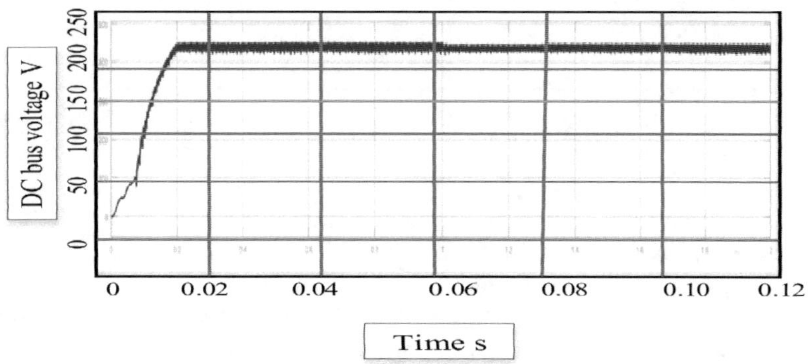

Figure 21 DC bus voltage between in back-to-back converter.

Figure 22 displays the abc voltage of NPC inverter by using SVPWM method, Figure 23 shows the load side voltage & current waveform.

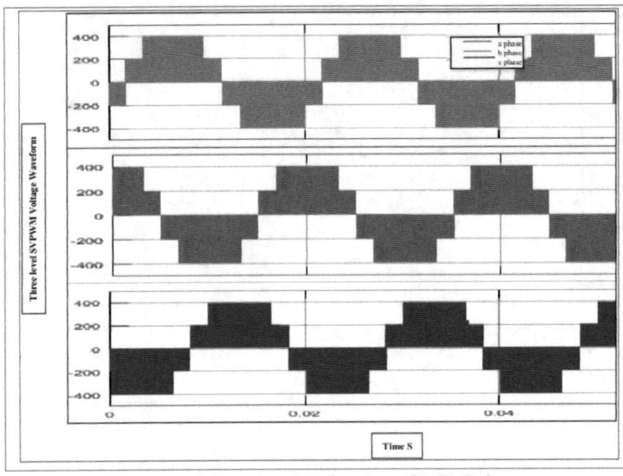

Figure 22a abc Voltage of NPC inverter

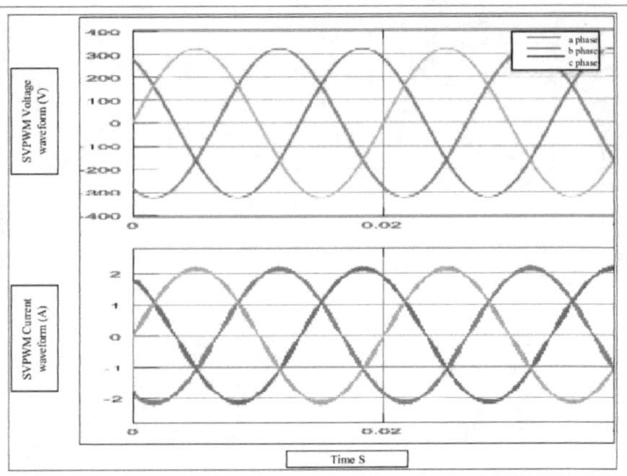

Figure 22b Load side abc voltage & current waveform

Figure 23 shows FFT analysis of PWM & three level SVPWM technique. The PWM method has Harmonic percentage is 42.63% & three level SVPWM method has 0.31%. If both the results compared, three level SVPWM technique has less harmonics compared to PWM technique.

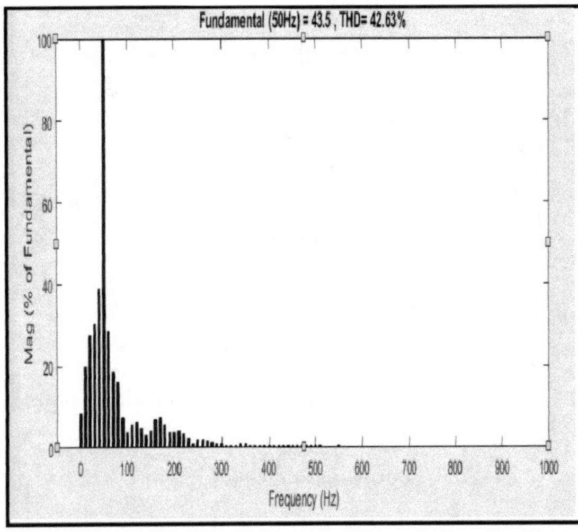

Figure 22a FFT analysis of PWM based inverter

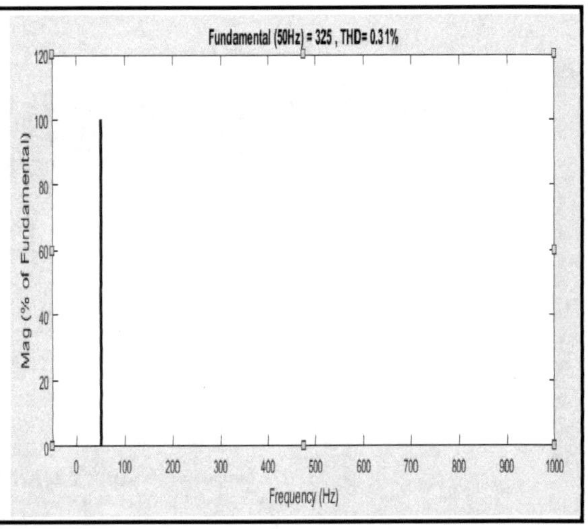

Figure 23b FFT analysis of three level SVPWM

5. Conclusion

In this paper, Dither signal injected in hysteresis band controller for three level (NPC) DTC for PMSG direct driven wind turbine. The conventional two level DTC compared on the performance basis such as torque ripple, flux ripple & stator current in Matlab Simulink software.

- The proposed technique reduces the torque, flux & stator current ripple, this controller exhibits satisfactory results in comparison with the conventional DTC.
- DTC with dither signal for PMSG is Suitable employed for wind turbine because; the application of DTC with dither signal enhances the performance of the PMSG with improved flux and torque control capability. Simulation results shows the torque & flux ripple reduced by 60% of peak-to-peak.
- In addition, load side inverter three level NPC (inverter) has less (THD 0.31%) harmonics as compared to other conventional PWM inverter method.

References

1. Yaramasu, V., Dekka, A., Durán, M. J., Kouro, S., & Wu, B. (2017). PMSG-based wind energy conversion systems: survey on power converters and controls. *IET Electric Power Applications, 11(6), 956-968*
2. Yang, S., & Zhang, L. (2013, January). Modeling and control of the PMSG wind generation system with a novel controller. *In 2013 Third International Conference on Intelligent System Design and Engineering Applications (pp. 946-949). IEEE.*
3. Torki, W., Grouz, F., & Sbita, L. (2017, March), Vector control of a PMSG direct- drive wind turbine, *In 2017 International Conference on Green Energy Conversion Systems (GECS) (pp. 1-6). IEEE.*
4. Behera, R. K., & Das, S. P. (2008). Improved direct torque control of induction motor with dither injection. *Sadhana, 33(5), 551-564.*
5. Hua, X., Yang, J., Wu, H., & Liu, J. (2014, October). Direct Torque Control of a permanent magnet synchronous generator based on space vector modulation. *In 2014 17th International Conference on Electrical Machines and Systems (ICEMS) (pp. 3520-3523). IEEE.*
6. Ameur, A., Mokhtari, B., Mokrani, L., Azoui, B., Essounbouli, N., & Hamzaoui, A. (2010). An Improved Sliding Mode Observer for Speed Sensorless Direct Torque Control of PMSM Drive with a Three-Level NPC Inverter Based Speed and Stator Resistance Estimator. *Journal of Electric Engineering, 10(4).*
7. Gajewski, P., & Pieńkowski, K. (2016). Direct Torque Control and Direct Power Control of wind turbine system with PMSG. *Przegląd Elektrotechniczny, 92(10), 249-253.*
8. Yang, S. Y., Wu, Y. K., & Lin, H. J. (2015). New application of predictive direct torque control in permanent magnet synchronous generator-based wind turbine. *Journal of Renewable and Sustainable Energy, 7(2), 023108.*
9. Romeral, L., Fabrega, A., Cusido, J., Garcia, A., & Ortega, J. A. (2008, June). Torque ripple reduction in a PMSM driven by direct torque control. *In 2008 IEEE Power Electronics Specialists Conference (pp. 4745-4751). IEEE.*
10. Shankar, V. A., Umashankar, S., Padmanaban, S., & Paramasivam, S. (2017). Adaptive neuro-fuzzy inference system (anfis) based direct torque control of pmsm driven centrifugal pump. *International Journal of Renewable Energy Research (IJRER), 7(3), 1436-1447.*
11. Madhusudan Rao. P, Rami Reddy. CH and Umashankar. S, "Real power control of wind induction generator fed by PWM converters using indirect vector control", *Second International Conference on Science, Engineering and Technology(SET), VIT University, Vellore, April 20-21, 2011.*
12. Yang, S. Y., Wu, Y. K., & Lin, H. J. (2014, November). The direct torque control of the PMSG based wind turbine with two level voltage source converter. *In 2014 CACS International Automatic Control Conference (CACS 2014) (pp. 39-44). IEEE.*
13. El Mourabit, Y., Derouich, A., El Ghzizal, A., El Ouanjli, N., & Zamzoum, O. (2017). DTC-SVM Control for permanent magnet synchronous generator based variable speed wind turbine. *International Journal of Power Electronics and Drive Systems, 8(4), 1732.*
14. Porselvi, T., Deepa, K., Subramaniam, U., Bhaskar, M. S., Sanjeevikumar, P., Leonowicz, Z., & Mitolo, M. (2020). Combined Harmonic Reduction and DC Voltage Regulation of a Single DC Source 5-Level Multilevel Inverter for Wind Electric System. *Electronics, 9(6), 979.*
15. Tiwari, R., Babu, N. R., Padmanaban, S., Martirano, L., & Siano, P. (2017, June). Coordinated DTC and VOC control for PMSG based grid connected wind energy conversion system. *In 2017 IEEE International Conference on Environment and Electrical Engineering and 2017 IEEE Industrial and Commercial Power Systems Europe (EEEIC/I&CPS Europe) (pp. 1-6). IEEE.*
16. Tiwari, R., & Babu, N. R. (2017). Comparative analysis of pitch angle controller strategies for PMSG based wind energy conversion system. *Int. J. Intell. Syst. Appl, 9(5), 62-73.*

17. Palanisamy, R., & Vijayakumar, K. (2018). Paper SVPWM for 3-phase 3-level neutral point clamped inverter fed induction motor control. *Indonesian Journal of Electrical Engineering and Computer Science (IJEECS), 9(3), 703-710.*
18. Krishna, C. H., Amarnath, J., & Kamakshaiah, S. (2012, August). Simplified SVPWM algorithm for neutral point clamped 3-level inverter fed DTC-IM drive. *In 2012 International Conference on Advances in Power Conversion and Energy Technologies (APCET) (pp. 1-6). IEEE.*
19. Madasamy P., Pongiannan, R. K., Ravichandran, S., Padmanaban, S., Chokkalingam, B., Hossain, E., & Adedayo, Y. (2019). A Simple Multilevel Space Vector Modulation Technique and MATLAB System Generator Built FPGA Implementation for Three-Level Neutral-Point Clamped Inverter. *Energies, 12(22), 4332.*
20. D. P. Kothari, S. Umashankar, "Wind Energy Systems and Applications", *Narosa Publications, Newdelhi, May 2014 (ISBN: 978-81-7319-699-7).*

An Energy Efficient Microwave Based Wireless Solar Power Transmission System

P Sabarish[1], L Hubert Tony Raj [2], G Ramprakash[3] and Dr R Karthick [4]

[1,2,3&4]Department of Electrical and Electronics Engineering,
[1]K.Ramakrishnan College of Technology, Trichy, Tamilnadu, India.
[2& 3]SRM TRP Engineering College, Trichy, Tamilnadu, India.
[4]Sethu Institute of Technology, Karriyapatti, Tamilnadu, India.

E.mail: sabarish.san@gmail.com

Abstract: Worldwide force age is estimate to increment by some 60% somewhere in the range of 2017 and 2040 to cover a fourth of essential vitality request, the IEA said in its World Energy Outlook 2018. "Power markets are experiencing a special change with more popularity brought by the advanced economy, electric vehicles" however numerous components are impacting the agreeableness of vitality innovation. The change of the vitality area is searching for clean vitality innovation that is additionally appropriate for base burden power generation. The sun based vitality is most utilized clean vitality today, but Terrestrial sun oriented vitality has several obstructions, Putting colossal sun based power satellite systems for arena addresses a lot of creative mechanical benefits than may give gigantic degree, normally clean base weight power into natural markets and it is one of the way in which satisfied essentialist demand in best in class days. Later, a symmetric concentrated with a high sun-based arrangement competence has been proposed.

1. Introduction

The present power age and transmission framework isn't extremely productive as far as vitality move. Around 25 to 30% vitality is lost during the transmission and dissemination of the power. So the force part is chipping away at the tasks to improve a definitive power supply. Researchers are searching for interchange and productive innovations to give 100% power move and age [1-2]. That way was the sunlight based remote force transmission framework .the sun based vitality is one of the most spotless and environmentally friendly power vitality; it was no expense of energy. In space there is a continuous accessibility of immense measure of sun based vitality as light andWarmth. So the usage of satellites basically made arrangements for social affair the sun based impressiveness and shaft it back to the earth is being thought of. In geo-synchronous hover, for instance of 36,000 km. In Solar Force Satellite (SFS) could have the alternative in stand up to the sun over 99% of the satellite time.

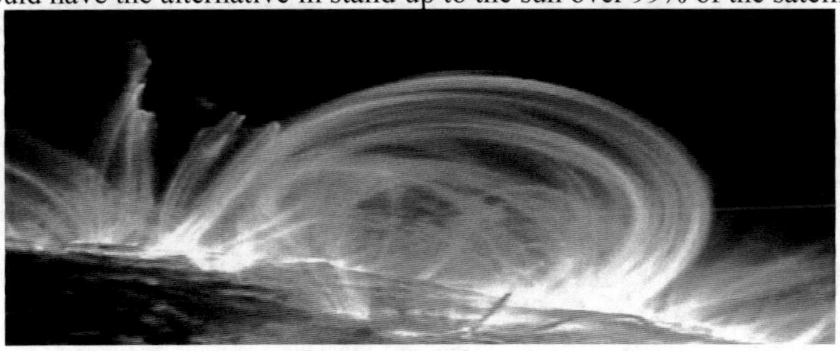

Figure 1. Solar storm on space

Force can be transmitted to the area where it is required, need not need to put resources into as extensive as a grid. In 1968, Dr. Diminish Glaser was first man the presented the idea of a large amount of sun powered force satellite arrangement of sun based authorities in high geo-synchronous circle (GEO is a circle thirty six thousands kilometer over the equator) Assortment in transforming the vitality of the sun into an electromagnetic microwave pillar to transmit some vitality to the enormous acceptance of receiving wires on earth for appropriation in the framework of the national electrical force. [3-4].

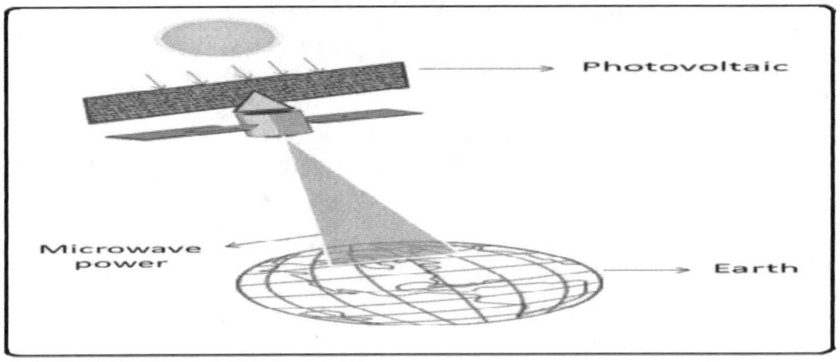

Figure 2. Solar power transfer from space

2. Solar Power Satellite System (SPS)

Sun oriented vitality was changed over into electrical vitality by utilizing the procedure of photovoltaic (PV) transformation, and sunlight based dynamic) change. Be that as it may most of the sun based force satellites have stronger than PV transformation. The Photovoltaic changing that apply of semiconductor cells in silicon or gallium arsenide to straightforwardly change over radiations into some electrical force by means of in quantum mechanism.[5]

To change over the DC power made by the sun situated satellite to microwave for the transmission the cruel radio wire towards the world's tolerant getting wire, Microwave oscillators, such as klystrones, magnetrons, can be used. The radiation glow is a high voltage structure. It uses a winding electric current, a crucial alluring field, an anode structure and a barrel of molded cathode. A daylight-based power satellite with a transmittance of 10 Wind power will transmit a flat-out force of one microwatt in a four hundred hertz station range [6-7].

Figure 3. Solar Power Satellite (SPS)

IOP Conf. Series: Materials Science and Engineering **937** (2020) 012013 doi:10.1088/1757-899X/937/1/012013

3. Transmissions of Wireless Electricity Technologies

It is a process that occurs in any kind of system in which it is transmitted. The alternating energy is passed from the power source to the electrical charge. What does this process make? Special is that no cabling of any sort used to connect the device to an energy source [8-9].

3.1 *Low-range*

It further differs from a few centimeters. Example transducer.

3.2 *Medium, range*

In this technology can be used for getting charged. The digital objects are automatic. The potential of our technology to deliver supply energy, effectively and wirelessly will enhance goods.

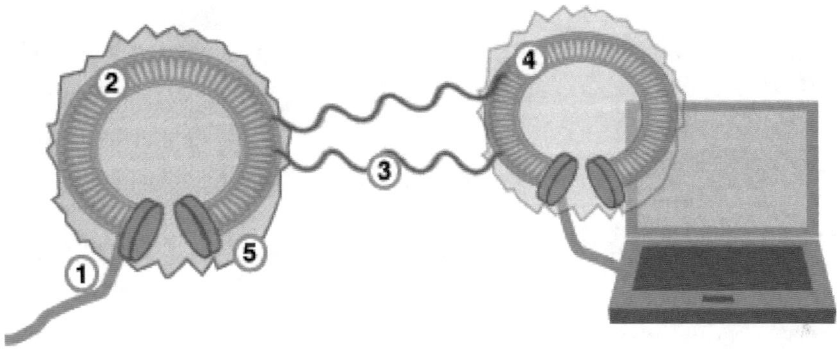

Figure 4. Working of Wireless electricity

3.3 *High Term*

Power transmission proposals involve moving electric power over miles of distance. Wide Wireless Distance energy is a technology that transfers energy to planet.

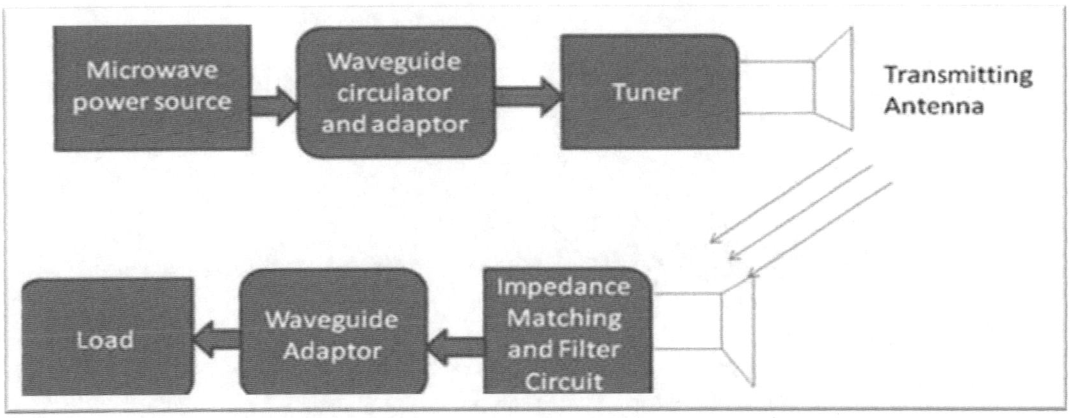

Figure 5. Wireless Microwave Power Transmission

IOP Conf. Series: Materials Science and Engineering **937** (2020) 012013 doi:10.1088/1757-899X/937/1/012013

4. Wireless transmission component details

The Microwave Generator is really the main aspect of the transmission of wireless electricity. The antenna and the reception antenna.

4.1 *Microwave Generator*

This is known as Microwave Vacuum Tubes (Magnetron, Klystron) and Microwave Power Modules (MPM)) and Microwave Transmitters and Amplifiers.

4.2 It's Rectennas

It is a rectification antenna. It is directly convert into microwave transmission to DC energy source.

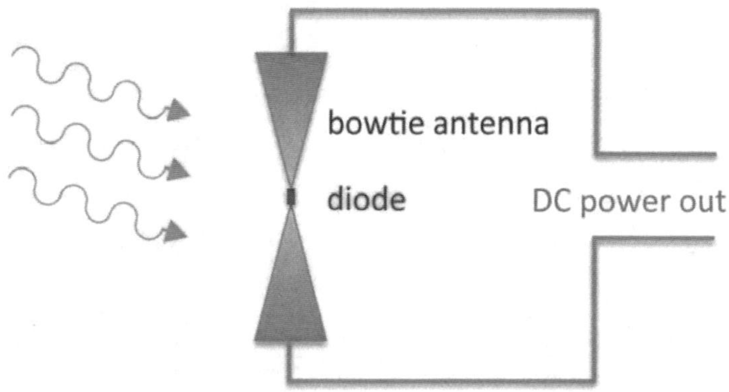

Figure 6. Rectanna

4.3 Antenna transmitting

Slotted wave-guide antenna, micro strip patch antenna and parabolic antenna. The dish antenna is the most common type of transmitting antenna. The waveguide slotted antenna is suitable for power transmission due to its high aperture efficiency (> 95%) and high power handling capability. [10-12].

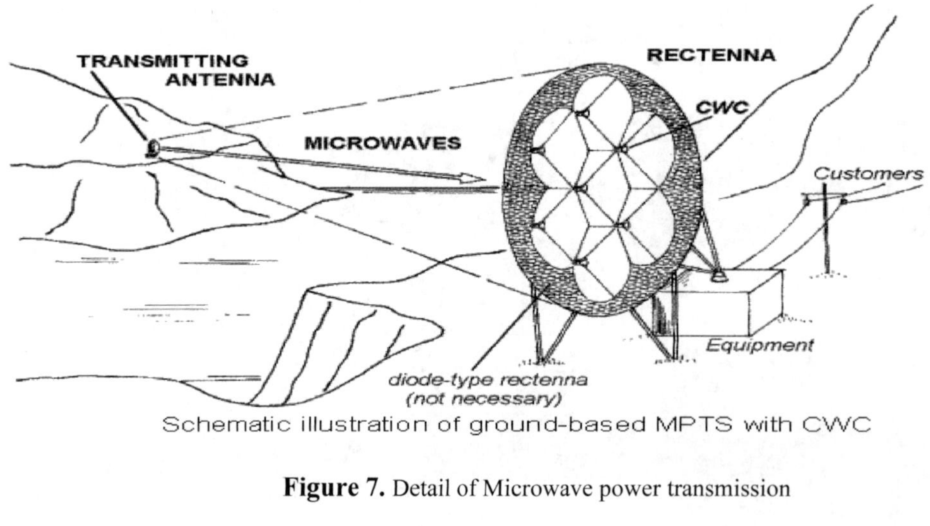

Figure 7. Detail of Microwave power transmission

5. Transmission

In this techniques for radio waves could be made constantly dimensional and it can be permitting longer segment of power transmitting with smaller frequencies of electro-magnetic radiation. Transmission using radio waves is already developed also for transmitting to essentialists through light-based force space towards Earth, and the transmission of power to transmit the departing circle has been proposed from satellite to universe. [13].

A most resonating acknowledgement methodology is the most complete fit for radiation. In this future we have to produce the existing power transmission lines and empower the spread relationship of electric age plants in a common wireless transmission from space to earth and suitable for wireless power application [14].

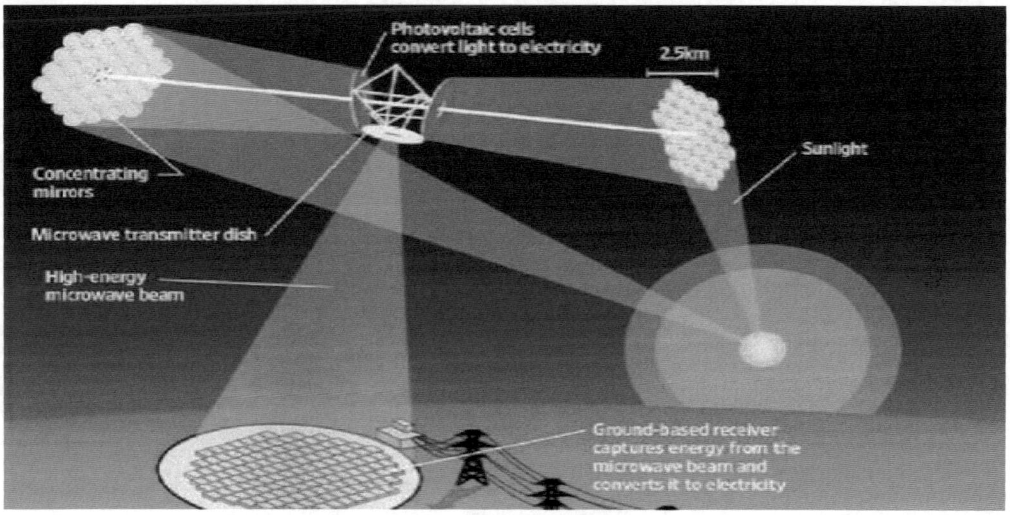

Figure 8. SPS wireless transmission

5.1 Receiving station

The framework of the satellite power system will require enormous accepting zone with the Retina cluster and the force organize associated with the current force lattices on the terrain. Since each retina component provides just a few watts, the Gig Watt (GW) has gotten the full and power energy conserved and utilized efficiently [15-16].

. A Retina could be used to transform the vitality of the microwave into power again. Retina change efficiencies in excess of 95% have been identified. "Retina" is framed as an "amending circuit."

A receptive wire called retina is modified by transmitting force and by altering the microwave ability to coordinate current (DC) source. The retina is an inactive component with a redressing diode, and is worked with no additional force origin. In between the retina low pass channel in reception apparatus and connecting diodes to smoother radiations of higher sounds in receiving side.

IOP Conf. Series: Materials Science and Engineering **937** (2020) 012013 doi:10.1088/1757-899X/937/1/012013

Figure 9. SPS Receiving Station

6. Experimental Setup

In this experimental setup of the transceiver unit has been carried out. The inductive coupling is the electromagnetic induction of the receiver. Sources of voltage for these transceivers have been supplied with solar cells [17-18]. The inductive coupling shall be used as a wireless power antenna delivered from the transmitter to the receiver input.

Transfer of wireless power is the transfer of electrical information. From the power source to the electrical charge, the driver is not made by an individual. Wireless transmission the use of wireless solar energy is challenging, not harmful, and safe.

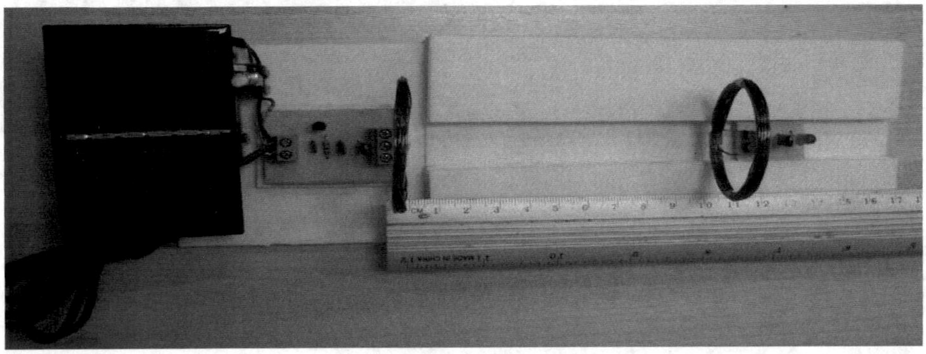

Figure 10. Experimental setup

7. Result and analysis

An experiment was performed to achieve the efficiency of the wireless transfer of power. Inductive coupling transmission was provided from the DC direct current source. The distance difference between the transmitter and the receiver is varied in order to achieve an optimal distance for wireless transmission.

Table 1. The difference information versus solar direct current output voltage and frequency Voltage reference parameters

Length (cm)	Solar source (volt)	Direct current output voltage(volt)	Frequency(MHz)	Efficiency (%)
0	10	8.69	3.33	98.99
1	10	8.46	3.13	92.05
2	10	8.02	3.05	88.00
3	10	7.86	2.82	75.00
4	10	6.93	2.61	66.02

5	10	4.88	2.51	55.33
6	10	4.16	2.30	40.02
7	10	3.06	2.22	38.89

8. Conclusion

The growing demand for global vitality is likely to continue for a long time in this paper. Be that as it may, vitality autonomy is something that the sun-powered force of Space can convey. Satellite-based Solar-power system idea is appealing in the light of the fact that it is considerably more beneficial than ground-based Sun-powered Force. It is predicted that the world will require 30TW of renewable energy by 2030, and that sun-based productivity alone will have the potential to produce about 600TW. The levels of CO2 emission may be limited and managed. In this way, the issue of a global temperature boost will be illuminated. Also, space-based force frameworks appear to have numerous critical natural points of interest when contrasted with elective ways to meet the growing earthly needs for vitality, including the need for far less land than earth-bound, sun-oriented land. In spite of the fact that the accomplishment of room sunlight based force relies upon effective advancement of key innovation.

9. References

[1] M D Udayakumar et al 2019 IOP Conf. Ser.: Mater. Sci. Eng. 623 012018.

[2] A. T. S. Subramanian, P. Sabarish and A. N. Ali, "A power factor correction based canonical switching cell converter for VSI fed BLDC motor by using voltage follower technique," 2017 IEEE International Conference on Electrical, Instrumentation and Communication Engineering (ICEICE), Karur, 2017, pp. 1-8, doi: 10.1109/ICEICE.2017.8191932.

[3] A. T. Sankara Subramanian, P. Sabarish, M. D.Udayakumar and T. Vishnu kumar," Performance Analysis of Various Photovoltaic Configurations Under Uniform Shading and Rapid Partial Shading Formations", Biosc.Biotech.Res. Comm. Special Issue Vol 13 No (3) 2020 Pp-185-192.

[4] B.Karthikeyan, R. Ramkumar, L. Nagarajan, A. Anton Amala Praveen and S. Vijayalakshmi," ANN Based MPPT Controller for PEMFC System with Interleaved Resonant PWM High Step up Converter" Biosc.Biotech.Res.Comm. Special Issue Vol 13 No (3) 2020 Pp-44-47.

[5] S.R.Paveethra, S.Vijayalakshmi, C. Kalavalli, S.Murugesan and Dr.S.Jeyasudha," MPPT based Harmonic Reduction in PV Module for Grid Connected System," Biosc.Biotech.Res.Comm. Special Issue Vol 13 No (4) 2020 Pp-43-48.

[6] P. Sabarish et al 2019 IOP Conf. Ser.: Mater. Sci. Eng. 623 012011

[7] A T Sankara Subramanian et al 2019 IOP Conf. Ser.: Mater. Sci. Eng. 623 012012

[8] Dr.A. Rajkumar, R. Jai Ganesh, V. Suresh Kumar, T. Vishnu Kumar and T. Ram Kumar," Implementation of High Efficient Single Input Triple output DC-DC Converter," Biosc.Biotech.Res.Comm. Special Issue Vol 13 No (3) 2020 Pp-48-55.

[9] R. Ramkumar, B. Karthikeyan, Dr. A. Rajkumar, V. Venkatesh and A. Anton Amala Praveen," Design and Implementation of IOT Based Smart Library Using Android Application," Biosc.Biotech.Res.Comm. Special Issue Vol 13 No (3) 2020 Pp-56-62.

[10] S.Murugesan, C.Kalavalli, S.Vijayalakshmi, S.R.Paveethra and M.V.Suganyadevi," Development of IoT Architecture for Physically Challenged People Using BLYNK Server," Biosc.Biotech.Res.Comm. Special Issue Vol 13 No (4) 2020 Pp-49-53.

[11] S.Jeyasudha, A. Anton Amala Praveen and B. Geethalakshmi," Fractional Order Controlled Multilevel Boost Hybrid Converter," Biosc.Biotech.Res.Comm. Special Issue Vol 13 No (3) 2020 Pp-202-210.

[12] R. Jaiganesh, L. Nagarajan, A. Arthi and V. Venkatesh," Anti Theft Control of Automatic Teller Machine Using Wireless Sensors," Biosc.Biotech.Res.Comm. Special Issue Vol 13 No (3) 2020 Pp-18-22.

[13] P. Sabarish, A. T. Sankara Subramanian, S. Murugesan and V. Sureshkumar,"A New Methodology of Arterial Blood Clot Removal Using Bio Molecular Devices for ATPase Nuclear Motors." Biosc.Biotech.Res.Comm. Special Issue Vol 13 No (3) 2020 Pp-197-201.

[14] A Kavitha, S Prasad Jones Christydass, J Silamboli, K Premkumar, A Nazar Ali "Metamaterial Inspired Triple Band Antenna For Wireless Communication", International Journal of Scientific & Technology Research, Vol 9, issue 04, pp. 6483-90.

[15] A Nazar Ali, D Sivamani, R Jaiganesh M Pradeep (2019), Solar powered air conditioner using BLDC motor, IOP Conference Series: Materials Science and Engineering, vol. 23.

[16] V Venkatesh, A Nazar Ali, R Jaiganesh. V Indiragandhi (2019), Extraction and conversion of exhaust heat from automobile engine in to electrical energy Energy, IOP Conference Series: Materials Science and Engineering, vol. 23.

[17] A Ali Nazar, R Jayabharath, MD Udayakumar (2014) ,An ANFIS Based Advanced MPPT Control of a Wind-Solar Hybrid Power Generation System, International review of modeling and simulations.vol.7, no. 4, pp. 638–643.

[18] A.Nazar Ali and R. Jayabharath (2014), "Performance Enhancement of Hybrid Wind/Photo Voltaic System Using Z Source Inverter with Cuk-sepic Fused Converter." Research Journal of Applied Sciences, Engineering and Technology7.19 pp. 3964-3970.

An Effective Solar Photovoltaic Module Parameter Estimation Technique for Single-Diode Model

M Premkumar[1*], R Sowmya[2], S Umashankar[3], and J Pradeep[4]

[1] Department of EEE, GMR Institute of Technology, Rajam, Andhra Pradesh, India
[2] Department of EEE, National Institute of Technology, Tiruchirapalli, Tamil Nadu, India
[3] Renewable Energy Laboratory, Prince Sulton University, Riyadh, Saudi Arabia
[4] Junior Engineer, Rajasthan Rajya Vidyut Prasaran Nigam, Sikar, Rajasthan, India

***Corresponding Author:** premkumar.m@gmrit.edu.in

Abstract. This paper presents a simplified parameter estimation procedure for a photovoltaic (PV) single diode model (SDM). Based on an iterative method, the module parameters are estimated for the SDM using the information from the datasheet. The five parameters of the SDM are estimated based on the datasheet using two steps. In the first step, the best ideality factor, *n*, value is determined, and in the next step, the value of the shunt resistance, R_{sh}, can be calculated to increase its accuracy. For validating the performance of the estimated parameters, a model that considers the difference in the solar irradiance and the temperature. In the standard testing conditions (STC), the proposed procedure shows the best results relative to other methods. The optimization of the parameters, such as *n*, R_{sh}, and series resistance, R_{se}, allows the minimum error between the values obtained from the proposed technique and datasheet.

1. Introduction

Suitable modelling circuit and estimation of optimum parameters of PV panels are serious subject matter for various applications, such as design, simulation, performance assessment under different environmental conditions, efficiency assessment, and PV system control [1-3]. It is helpful also for the operation of PV systems, the testing and development of the maximum power point tracking (MPPT) techniques [4], the progress of fault detection approaches, the power forecasting, the calculation of losses, and the reproduction of the PV simulator references as a function of temperature and irradiation differences in real-time in commercial PV inverter testing [5]. The curves, such as voltage (V) - current (I) and power (P) - V, generally characterize the PV cells. The manufacturers give the specification of the PV module at STC, such as open-circuit voltage (V_{oc}), short-circuit current (I_{sc}), and maximum power point (MPP), i.e., maximum power (P_{mpp}), maximum current (I_{mpp}), and maximum voltage (V_{mpp}). Few manufacturers provide the values of the short-circuit current (α_{sc}) temperature coefficient and the open-circuit voltage (β_{oc}) temperature coefficient. The parameters in the datasheet cannot represent the PV cell or module behavior, because these are not foreseen in most electric models already in use. Hence, the parameters for the different PV models must be extracted from the datasheet [6].

The electrical equivalent of the PV cell is traditionally represented by an ideal current source and a diode with no hint of shunt resistance (R_{sh}) and the series resistance (R_{se}). Different studies on the development of PV cell models have been presented. These models, however, have different complexity levels. These differences depend on the diode numbers used, the infinite or finite value of R_{sh}, the variable or fixed diode-ideality factor, *n* and various algorithms used to determine the necessary parameters. In the scientific community, the assessment among the PV cell models is an essential topic of research [7]. The electrical equivalent model of the SDM has the parameters, such as diodes saturation current (I_{sd}), photocurrent (I_{ph}), diode ideality-factor (n), and R_{sh} and R_{se} resistances. Nevertheless, the manufacturers of PV modules neither provide these parameters explicitly nor entirely. As a result, electrical PV cell modeling and parameter extraction methods are based on various aspects, for instance, complexity, photovoltaics technology, accuracy, and the speed of estimation [8]. The field of PV cell modeling and parameter estimation is based on numerous research publications. There are several essential investigations on the simulation and estimation of parameters of PV cells are carried out in [9]. The PV cell model, such as SDM, double-diode model, and three diode models, are discussed in the literature. The authors also covered analytical and soft computing approaches and analytical methods for the estimation of parameters. In the review [10], the hybrid method between the analytical and soft computing approaches was recommended. The numerical, analytical, graphical, and heuristic methods are used mainly to extract the photovoltaic cell parameters.

The primary objective of this paper is to present a simplified five parameter assessment method for the SDM of the multi-crystalline PV panel. The presented process uses an iterative method that differs from the optimization-based estimation of the parameters, such as n, $R_{se,}$ and R_{sh}. The presented method has two iterative steps to estimate the model parameters, beginning with the datasheet of the supplier. The results obtained by the presented method are compared with a few literature methods. In addition, electrical models have been used to obtain the V-I characteristics in light of the variation in irradiance and temperature.

The paper is organized as follows: The SDM of the PV module is discussed in Section 2, section 3 presents the simplified five-parameter extraction technique, while section 4 presents results in comparison with other approaches. Lastly, Section 5 concludes the paper.

2. Single-diode PV model of the module

The electrical equivalent circuit of the PV module describes its physical characteristics. The single-diode PV model is being used in many of the fields, specifically to describe the characteristics of the PV module [11]. There are several PV cells connected in parallel (N_p) or series (N_s) in the PV module, and the equivalent circuit structure of the SDM is depicted in Fig. 1, which includes photocurrent (I_{ph}), shunt resistor current (I_{sh}), and the diode current (I_d). Furthermore, the total PV current (I_{pv}) of the cell is given in Eq. (1).

$$I_{pv} = I_{ph} + I_{sh} - I_d \tag{1}$$

Based on the Shockley equation, the diode current, I_d, is written and given in Eq. 2, and the shunt resistance current, I_{sh}, is presented in Eq. 3. The thermal voltage, V_t, of the PV module is given in Eq. 4.

$$I_d = I_{sd}\left[exp\left(\frac{V_{pv} + I_{pv}R_{se})}{V_t}\right) - 1\right] \tag{2}$$

$$I_{sh} = \frac{V_{pv} + I_{pv}R_{se}}{R_{sh}} \tag{3}$$

$$V_t = \frac{knT}{q} \tag{4}$$

Where, q represented the electron charge and is equal to $1.602*10^{-19}$ C, I_{sd} represents the saturation current, n represents the ideality factor, T symbolized the absolute temperature in Kelvin, the output voltage is denoted by V_{pv}, k represents the Boltzmann constant and is equal to $1.3806*10^{-23}$ J/K). Based on the above-said equations, Eq. 1 is rewritten as follows.

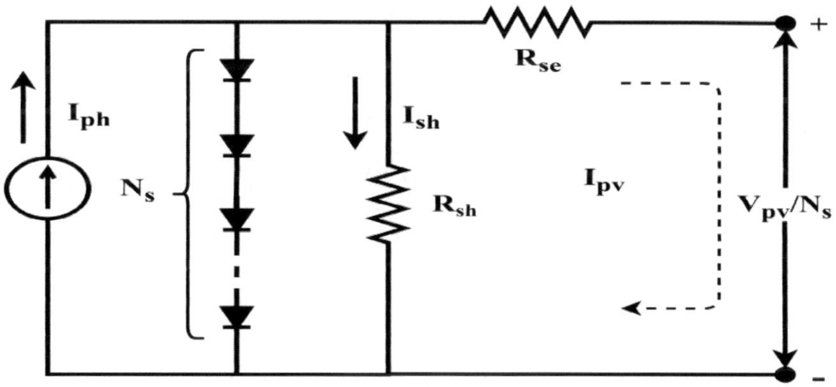

Figure 1. Electrical equivalent circuit of the PV module.

$$I_{pv} = I_{ph} - I_{sd}\left[exp\left(\frac{V_{pv} + I_{pv}R_{se}}{V_t}\right) - 1\right] + \frac{V_{pv} + I_{pv}R_{se}}{R_{sh}} \quad (5)$$

Eq. 5 is expanded to the PV module by considering the series-connected PV cells (N_s) to form the module as follows and assume N_p is equal to 1. The thermal voltage for the module is given in Eq. 7.

$$I_{pv} = I_{ph} - I_{sd}\left[exp\left(\frac{V_{pv} + N_s I_{pv}R_{se}}{V_t}\right) - 1\right] + \frac{V_{pv} + I_{pv}R_{se}N_s}{R_{sh}N_s} \quad (6)$$

$$V_t = \frac{knTN_s}{q} \quad (7)$$

3. Proposed Estimation Technique

The suggested method of estimating the parameters of the SDM involves two iterative phases. The first phase is the calculation of the global value of n. During this iterative process, the effect of the shunt resistor is ignored, and the R_{se} value is also determined. The R_{sh} value is then pulled out in phase 2, using the value from step 1. The presented method is simulated using MATLAB software.

3.1. Phase-I

During this step, Eq. 6 is rewritten by considering the zero effects due to the shunt resistance as Eq. 7. The operating conditions, such as open-circuit and short-circuit, is applied to Eq. 7 and the equations for the four unknown parameters, such as I_{ph}, I_{sd}, n, and R_{se}. During short-circuit, the value of V_{pv} is equal to zero, and the value of I_{pv} is equal to the short-circuit current, I_{sc}. Therefore, Eq. 6 rewritten as follows.

$$I_{sc} = I_{ph} - I_{sd}\left[exp\left(\frac{N_s I_{sc}R_{se}}{V_t}\right) - 1\right] \quad (8)$$

Now, by considering the open-circuit test condition, Eq. 6 is rewritten as follows by assuming the value of V_{pv} is equal to the open-circuit voltage, $V_{oc,}$ and the value of I_{pv} is equal to zero.

$$I_{ph} = I_{sd}\left[exp\left(\frac{V_{oc}}{V_t}\right) - 1\right] \quad (9)$$

From Eq. 8 and Eq. 9, the expression for the diode saturation current, $I_{sd,}$ is as follows.

$$I_{sd} = \frac{I_{sc}}{\left[exp\left(\frac{V_{oc}}{V_t}\right) - 1\right] - \left[exp\left(\frac{N_s I_{sc} R_{se}}{V_t}\right) - 1\right]} \tag{10}$$

The output power during this iterative process is written as follows.

$$P_{pv} = V_{pv} * \left(I_{ph} - I_{sd}\left[exp\left(\frac{V_{pv}+N_s I_{pv} R_{se}}{V_t}\right) - 1\right]\right) \tag{11}$$

Differentiate Eq. 11 with respect to V_{pv} and equate it to zero to derive the expression for R_{se}.

$$\frac{dP_{pv}}{dV_{pv}} = \left(I_{ph} - I_{sd}\left[exp\left(\frac{V_{pv}+N_s I_{pv} R_{se}}{V_t}\right) - 1\right]\right) + V_{pv}\left(-\frac{I_{sd}}{V_t}\left[exp\left(\frac{V_{pv}+N_s I_{pv} R_{se}}{V_t}\right)\right]\right) = 0 \tag{12}$$

By replacing V_{pv} and I_{pv} as V_{mpp} and I_{mpp} in Eq. 12, the expression for the R_{se} is written as follows.

$$R_{se} = \left\{V_t \ln\left[\frac{I_{ph} - I_{sd}}{I_{sd}} \frac{V_t}{V_t + V_{mpp}}\right] - V_{mpp}\right\} \frac{1}{I_{mpp} N_s} \tag{13}$$

Substitute Eq. 9 and Eq. 10 in Eq. 13, and the expression for the R_{se} is rewritten as follows.

$$R_{se} = \frac{V_{oc}}{I_{mpp} N_s} + \frac{V_t}{I_{mpp} N_s} \ln\left[\frac{V_t}{V_t + V_{mpp}}\right] - \frac{V_{mpp}}{I_{mpp} N_s} \tag{14}$$

The value of the diode ideality factor taken from 1 to 2 on an iterative basis to estimate the other parameters by comparing the voltage value given by the manufacturer at the MPP. The variation of the value of n affects the thermal voltage, and hence the value of R_{se}. The maximum power voltage can be obtained by using Eq. 6, and replacing $I_{pv}=I_{mpp}$ and $V_{pv}=V_{mpp}$ and with the highest voltage. The expression for the maximum output voltage is as follows.

$$V_{mpp} = V_t \ln\left(\frac{I_{ph} - I_{sd}}{I_{sd}}\right) + V_t - I_{mpp} N_s R_{se} \tag{15}$$

The initial definition of the parameter, n is provided by the proposed algorithm, which is equal to 1 and the tolerance for the measured voltage of 0.1 V_{max}. The number of iterations is selected as 5000. V_t and R_{se} values are then calculated with the ideality factor, n=1, and started with iterations. When the newly estimated maximum voltage is lower than the theoretical voltage, the value of n is incremented by 0.01; otherwise, it is decremented by 0.01. For the new ideality factor, n, the values, such as I_{ph}, I_{sd}, and V_{mpp}, are recalculated using Eq. 9, Eq. 10, and Eq. 15, respectively. The algorithm continuously runs till the maximum iteration or the error is less than the specified tolerance. Lastly, the R_{se} value is updated with the estimated diode ideality factor.

3.2. Phase-II

The PV model without the effect of R_{sh} is, unfortunately, not accurate, and the MPP with the updated values of R_{se} and n does not correspond with the value stated by the manufacturer. Therefore, the value of R_{sh} is introduced to increase accuracy. A similar procedure is followed in step 2, but the estimated parameter is R_{sh} using the updated values of R_{se} and n. The presented technique provides an initial R_{sh} estimate, which can be obtained using the expression below.

$$P_{mpp} = V_{mpp} * \left(I_{ph} - I_{sd}\left[exp\left(\frac{V_{mpp} + N_s I_{mpp} R_{se}}{V_t}\right) - 1\right] + \frac{V_{mpp} + I_{mpp} R_{se} N_s}{R_{sh} N_s}\right) \tag{16}$$

From Eq. 16, the expression for the R_{sh} is derived as follows.

$$R_{sh} = \frac{V_{mpp}\left(V_{mpp} + I_{mpp}R_{se}N_s\right)}{\left\{V_{mpp}\left(I_{ph} - I_{sd}\left[exp\left(\frac{V_{mpp} + N_s I_{mpp}R_{se}}{V_t}\right) - 1\right]\right) - P_{mpp}\right\}N_s} \tag{17}$$

As similar, the expressions for the updated values of diode saturation current, I_{sd}, and the photocurrent, I_{ph} is derived from Eq. 6 by considering the open-circuit and short-circuit conditions.

$$I_{sd} = \frac{I_{SC}\left(1 + \frac{R_{se}}{R_{sh}}\right) - \frac{V_{oc}}{R_{sh}}}{\left[exp\left(\frac{V_{oc}}{V_t}\right) - 1\right] - \left[exp\left(\frac{N_s I_{sc} R_{se}}{V_t}\right) - 1\right]} \tag{18}$$

$$I_{ph} = I_{sd}\left[exp\left(\frac{V_{oc}}{V_t}\right) - 1\right] + \frac{V_{oc}}{R_{sh}} \tag{19}$$

As discussed earlier, an initial estimate of the R_{sh} is needed in step 2. By considering the values of n, R_{se}, and I_{ph}, the value of Rsh is calculated using Eq. 17. In step 2, the tolerance is 0.001 A, and the number of iterations is equal to 5000. The updated I_{sd} and I_{ph} values are calculated using Eq. 18-19 with the initial estimation of R_{sh}. The method consists of locating the current at the maximum power conditions in Eq. 6 with V_{pv}=V_{mpp}. The calculated power is compared with the maximum power provided by the supplier, and the value of R_{sh} is updated. If the value is below the expected current, the R_{sh} value is incremented by a factor, *0.1 × current iterations*, otherwise decremented by a factor, *0.1 × current iterations*. The error is compared with the acceptable tolerance. This process is repeated until the error exceeds the desired threshold or the maximum number of iterations.
The accuracy of the PV model is evaluated by finding the value of the root mean square error (RMSE). The expression for finding the value of RMSE is given as follows.

$$RMSE = \sqrt{\frac{1}{M}\sum_{j=1}^{M}\left(I_{exp,j} - I_{sim,j}\right)^2} \tag{20}$$

Where, $I_{sim,j}$ denotes the simulated value of the current and $I_{exp,j}$ denotes the measured value of the current. Along with the value of RMSE, the other performance metrics, such as the absolute error (AE) and relative error also calculated as per the following equations.

$$RE_i = \frac{I_{exp} - I_{sim}}{I_{exp}} \tag{21}$$

$$AE_i = \left|I_{exp} - I_{sim}\right| \tag{22}$$

4. Results and Discussions

The presented parameter estimation approach was realized by utilizing the SDM equations in the MATLAB software. The method was selected to test the Kyocera KC200GT module with 54 series-connected PV cells. The specifications of the PV module are shown in Table 1.

The various output data phases of the presented technique for the Kyocera KC200GT PV module is shown in Fig. 2. Fig. 2 illustrates the V-I and P-V characteristics of the PV module by considering the three cases, such as without the effect of R_{sh}, with the effect of the R_{sh}, and the proposed technique by considering the effect of R_{sh}. From Fig. 2, it is noticed that the model by considering the effect of $R_{sh,}$ shows a better result, and matches the datasheet results. Therefore, it is decided to consider the presented equations from Eq. 9 - Eq. 19 to model the proposed technique.

Table 1. Electrical specifications of Kyocera KC200GT PV module.

S. No.	Name of the parameters	Specifications
1	Maximum output voltage, V_{mpp}	26.3 V
2	Maximum output current, I_{mpp}	7.61 A
3	Open circuit voltage, V_{oc}	32.9 V
4	Short circuit current, I_{sc}	8.21 A
5	Temperature coefficient at V_{oc}	0.0138 V/°C
6	Temperature coefficient at I_{sc}	-0.123 A/°C
7	Number of series cells, N_s	54

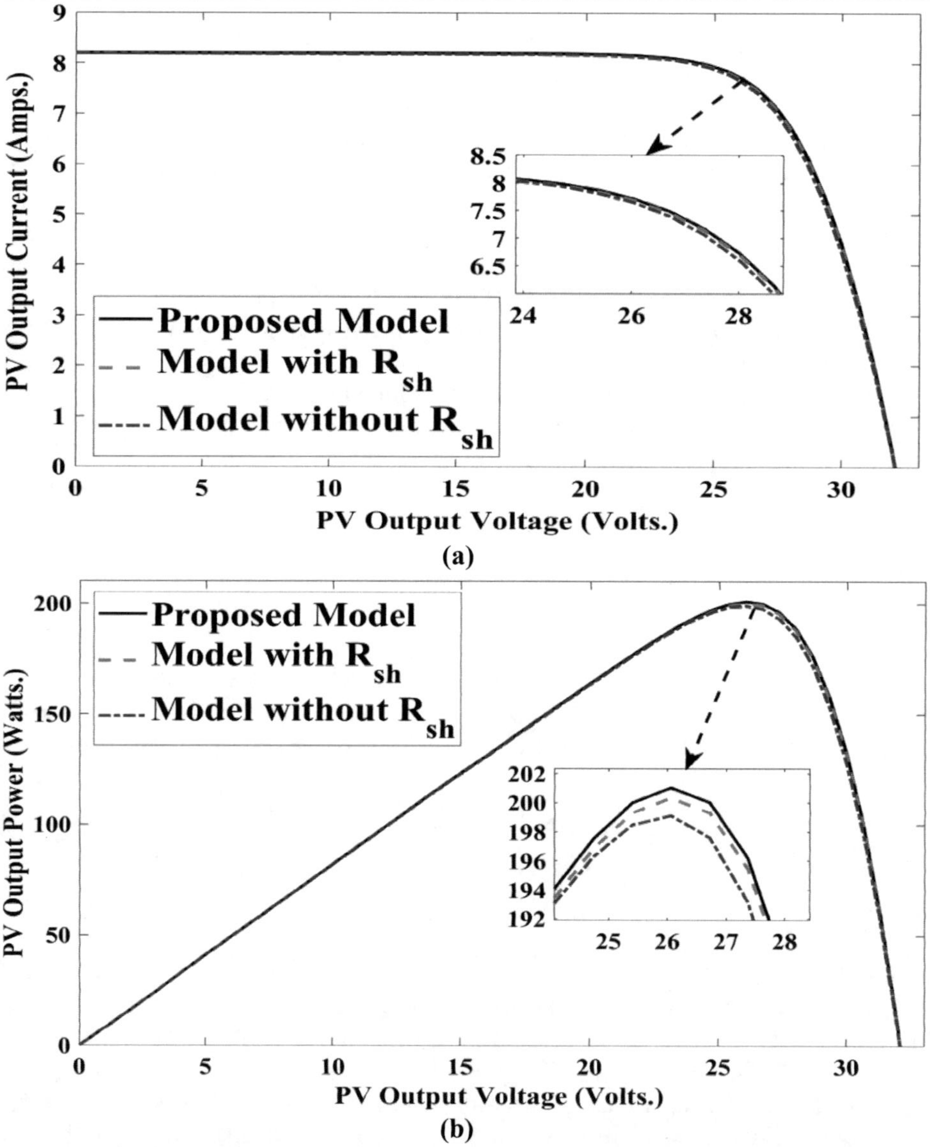

Figure 2. Various output phases of the presented method, (a) V-I characteristics, (b) P-V characteristics.

The PV module behavior is simulated by a model defined by Eq. 6 for the STC, i.e., 1000 W/m^2 solar irradiance and 25 °C temperature. The V-I characteristics of the Kyocera KC200GT PV module with respect to STC is illustrated in Fig. 3.

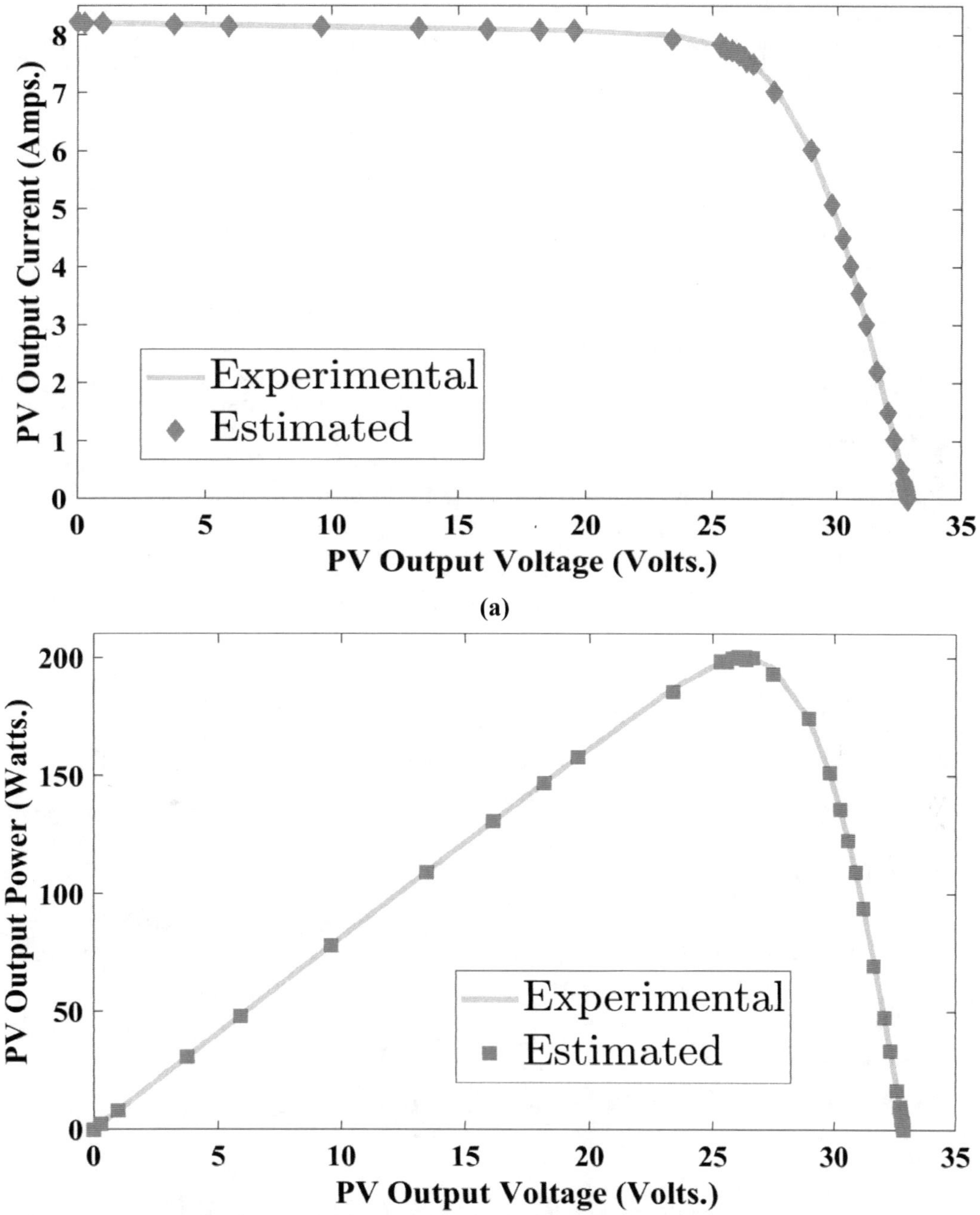

(b)

Figure 3. Characteristics of the Kyocera KC200GT PV module at STC, (a) V-I curve, (b) P-V curve.

The performance comparison of the presented technique and other techniques, such as Nayak [13], Silva [12], and Hejri [14], is presented in Table 2. The accuracy of all methods is assessed by the RMSE, AE, and RE values. The presented technique shows, among comparative approaches in the STC, the best results for RMSE, AE, and RE. The values of RMSE, AE, and RE are 1.985E-03, 5.471E-03, and 3.698E-03, respectively. Table 4 presents the estimated parameters of the Kyocera KC200GT PV module, as a thorough study, via the presented technique at different solar irradiance and temperature values. The assessment metrics, such as RMSE, AE, and RE for all methods, including the proposed method is shown in the bar chart as follows for a better understanding. According to the above discussions, it is concluded that the presented technique which utilizes Eq. 6 gives the excellent performance at STC as compared with the conventional analytical techniques.

Table 2. Performance Comparison for the Kyocera KC200GT PV module at STC.

Technique	Parameters					Metrics		
	I_{ph} (A)	I_{sd} (μA)	n	R_{sh} (Ω)	R_{se} (Ω)	RMSE	AE	RE
[12]	8.193	0.0003	1.000	171.2	0.2710	0.19	9.245E-2	9.988E-2
[13]	8.193	0.0358	1.241	599.9	0.1984	0.15	4.878E-2	6.547E-2
[14]	8.160	0.1501	1.340	951.9	0.1800	0.11	1.326E-2	2.147E-2
Proposed	8.201	0.1781	1.334	951.93	0.2201	1.985E-3	5.471E-3	3.698E-3

The presented technique displays good outcomes in STC when compared to the methods discussed in the literature. Besides, the SDM for the extraction of V-I and P-V curves is normally used in the STC. However, a suitable model must be used to have accurate V-I and P-V curves which approach the expected values.

Table 3. Estimated parameters of the Kyocera KC200GT PV module at different temperature and solar irradiance.

Temperature (°C)	Solar Irradiance (W/m^2)	Parameters				
		I_{ph} (A)	I_{sd} (μA)	n	R_{sh} (Ω)	R_{se} (Ω)
	200	1.592	0.0352	1.239	4389.36	1.0014
	400	3.285	0.0692	1.277	2273.58	0.5179
25 °C	600	4.939	0.1035	1.308	1545.87	0.3478
	800	6.568	0.1388	1.317	1169.55	0.2688
	1000	8.201	0.1781	1.334	951.93	0.2201
50 °C	1000	8.288	0.0558	1.048	1351.22	0.3077

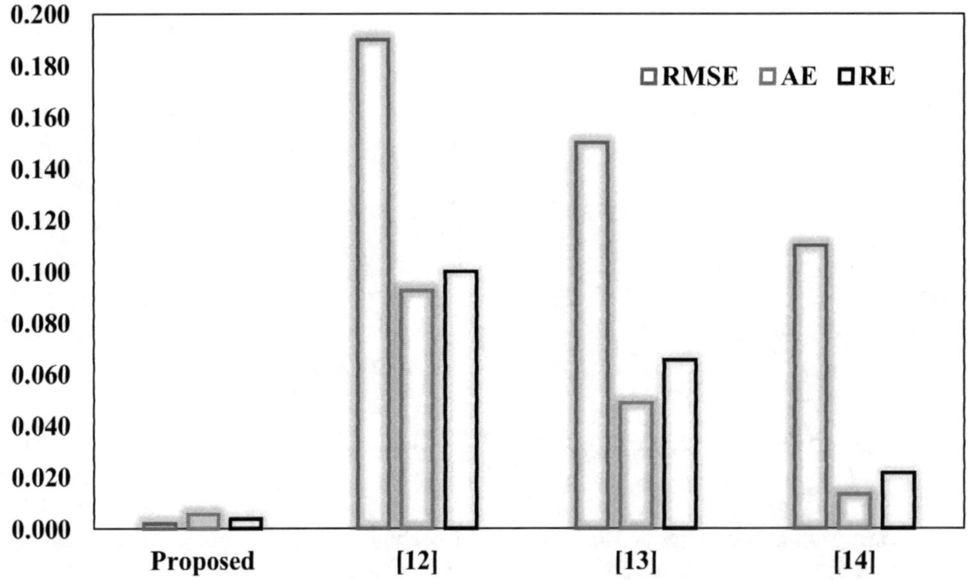

Figure 4. Bar chart of the metrics for all algorithms.

5. Conclusions

This paper presents a simplified parameter estimation technique for the SDM. The presented technique is based on two phases of iteration to estimate the parameters of the SDM. The technique was well-defined with all the necessary equations and simulated through MATLAB software. The commercial Kyocera KC200GT PV module is used for functional testing, and the outcomes of the presented technique are compared with the other methods. The assessment metrics, such as RMSE, AE, and RE, are employed to confirm the accuracy of the proposed technique. The error values were calculated for the PV module under STC. To conclude, when compared to other approaches, decent outcomes were realized. In fact, the RMSE, AE, and RE respectively in STC were equal to 1.985E-3, 5.471E-3, and 3.698E-3, and these values are better than other approaches.

6. References:

[1] Elkholy A and Abou El-Ela AA 2019 Optimal parameters estimation and modelling of photovoltaic modules using analytical method *Heliyon* **5 e02137**
[2] Premkumar M, Subramaniam U, Babu TS Elavarasan RM and Mihet-Popa L 2020 Evaluation of mathematical model to characterize the performance of conventional and hybrid PV array topologies under static and dynamic shading patterns *Energies* **13 3216**
[3] Alaaeddin M, Sapuan S, Zuhri M, Zainudin E and AL-Oqla F 2019 Photovoltaic applications: Status and manufacturing prospects *Renew. Sustain. Ener. Rev.* **102, 318-332**
[4] Premkumar M and Sowmya R 2019 An effective maximum power point tracker for partially shaded solar photovoltaic systems *Ener. Rep.* **5 1445-1462**
[5] Premkumar M and Sowmya R 2019 Certain study on MPPT algorithms to track the global MPP under partial shading on solar PV module/array *Inter. J. Comp. Dig. Sys.* **8 405-416**
[6] Premkumar M Babu TS Subramaniam U and Sowmya R 2020 A new metaphor-less algorithms for the photovoltaic cell parameter estimation *Optik* **208 164559**
[7] Sangram Bana RP and Saini A 2016 A mathematical modeling framework to evaluate the performance of single diode and double diode based SPV systems *Ener. Rep.* **2 171-187**
[8] Kulaksız AA 2011 ANFIS-based parameter estimation of one-diode equivalent circuit model of PV modules *Proc. of the IEEE 12th International Symposium on Computational Intelligence and Informatics, Budapest* pp 415-420
[9] Chin VJ, Salam Z and Ishaque K 2015 Cell modelling and model parameters estimation techniques for photovoltaic simulator application: A review *App. Ener.* **154 500-519**
[10] Rezaee J 2016 Parameter estimation of solar photovoltaic (PV) cells: A review *Renew. Sustain. Ener. Rev.* **61 354-371**
[11] Premkumar M, Kumar C and Sowmya R 2020 Mathematical modelling of solar photovoltaic cell/panel/array based on the physical parameters from the manufacturer's datasheet *Inter. J. Renew. Ener. Dev.* **9 07-22**
[12] Silva E, Bradaschia F Cavalcanti M and Nascimento A 2016 Parameter estimation method to improve the accuracy of photovoltaic electrical model *IEEE J. Photovolt.* **6 278-285**
[13] Nayak D, Mohapatra A and Mohanty K 2013 Parameters estimation of photovoltaic module using nonlinear least square algorithm: A comparative study *Proc. of the Annual IEEE India Conference (INDICON), Mumbai, India* pp 1-6
[14] Hejri M and Mokhtari H 2017 On the comprehensive parametrization of the photovoltaic (PV) cells and modules *IEEE J. Photovolt.* **7 250-258**

Performance Enhancement of Boost Converter Fed Permanent Magnet Synchronous Machine

GN.SachinAmreiss[1], C.Kalavalli[2] and Dr.A.Nazar Ali[3]

[1] Assistant Professor, Department of Electrical and Electronics Engineering, M.Kumarasamy College of Engineering, Karur, Tamilnadu

[2] Assistant Professor, Department of Electrical and Electronics Engineering, K. Ramakrishnan College of Technology, Trichy, Tamilnadu.

[3] Associate Professor, Department of Electrical and Electronics Engineering, Rajalakshmi Engineering College, Rajalakshmi Nagar, Thandalam, Tamilnadu

E.mail: kalavallichn@gmail.com

Abstract: The renewable energy plays a vital role in various fields like electronic and automotive applications. This paper presents the Permanent magnet synchronous machine (PMSM) which is fed by boost converter. A wide range of dc voltage is fed to bridge inverter by boost converter which in turn controls the speed of PMSM. Boost converter is used to boost up the input voltage. An electric vehicle may be powered through a shaft of the motor to run the various electric vehicles like electric aircraft, electric truck, electric space craft, electric locomotive etc. In this paper by using converter topologies increase the performance of the motor speed. The performance of the proposed system is verified using MATLAB/Simulink and presented results demonstrate the performance and analysis of the system to provide high speed by using PID controller.

Key words: PID controller, boost converter, automotive, electric vehicle.

1. Introduction:

PMSM are used for the traction application of electric vehicles. It is normally used for high effective and high performance motors. In PMSM the field excitation is presumed by permanent magnets and it can provoke torque at zero speed [1]. The load is driven by PMSM through a bridge inverter which is fed by boost converter. In order to reduce the PMSM's energy consumption, it can able to improve the forbearance mileage of electromotive vehicles [2]. The bridge inverter is used to convert dc supply to ac supply. The input voltage and output voltage mainly depends on specific device. To solve this problem this inverter can able to control the speed of the synchronous machine [3]. Square wave is suitable for low sensitive applications. Permanent magnet synchronous machine starts to rotate by using inverter ac supply. The proposed method is simulated and hardware is implemented [4].

2. Block Diagram:

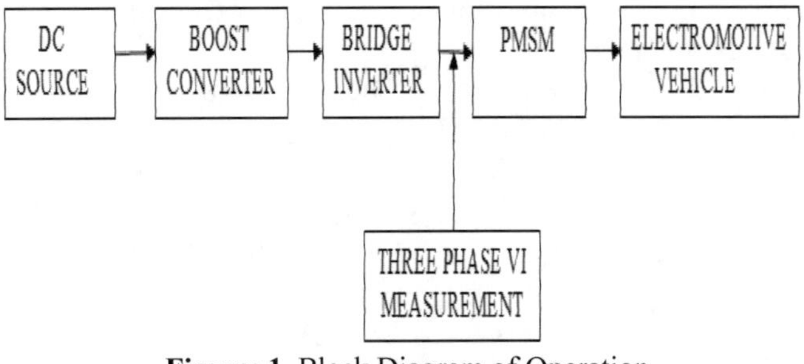

Figure: 1. Block Diagram of Operation

IOP Conf. Series: Materials Science and Engineering **937** (2020) 012015 doi:10.1088/1757-899X/937/1/012015

3. Operation:

A 230V DC voltage is fed to boost converter. 230 V is boosted to 457.6 V by the working of boost converter. Input side of the inductor has the value of 200e-6 H. In that inductor resists the input current Dis-continuous and continuous conduction mode is the two modes of operation on boost converter [5-6].

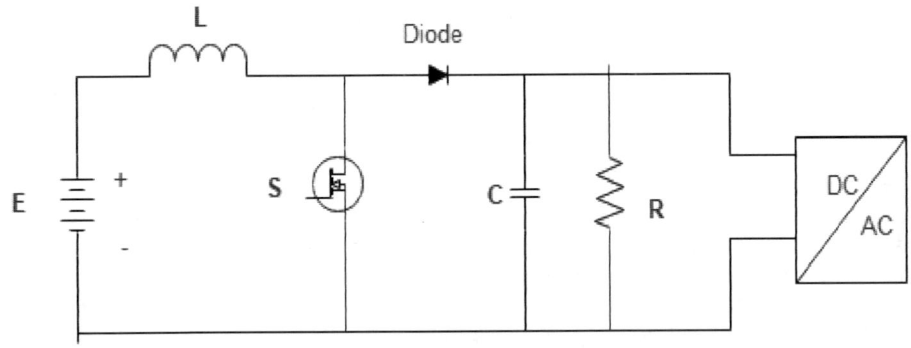

Figure: 2. Circuit Diagram of Boost Converter

3.1. Continuous Conduction Mode:

In this mode, never goes to zero value of inductor current.

Case 1: ON mode

No current passes across the diode while switch is in ON condition
Finally inductor current while switch ON mode equation then,

$$\Delta I_{LOn} = \frac{1}{L}\int_0^{DT} E_i\, dt = \frac{DT}{L} E_i \quad \text{.....(1)}$$

Case 2: OFF mode

Diode will short circuited while switch in OFF condition. Switch is un-energizing and the inductor discharges the energy as well as diode discharges the energy and given to the load and output power [7].

Current through the inductor equation is given as

$$\Delta I_{LOff} = \int_{DT}^{T} \frac{(E_i - E_o)dt}{L} = \frac{(E_i - E_o)(1-D)T}{L} \quad \text{.....(2)}$$

3.2. Dis-Continuous Conduction Mode:

In this mode, the current through inductor goes to zero [5].

$$I_o = \frac{E_i DT}{2L} \cdot \frac{E_i D}{E_o - E_i} = \frac{E_i^2 D^2 T}{2L(E_o - E_i)} \quad \text{.....(3)}$$

$$\frac{E_o}{E_i} = 1 + \frac{E_i D^2 T}{2LI_o} \quad \text{.....(4)}$$

IOP Conf. Series: Materials Science and Engineering **937** (2020) 012015 doi:10.1088/1757-899X/937/1/012015

$$\frac{E_o}{E_i} = \frac{1+\sqrt{1+\frac{4D^2}{K}}}{2} \qquad .....(5)$$

Where,

$K = \frac{2L}{RT}$ [1]

E_o = Output voltage of the converter

E_i = Input voltage of the converter

L=Inductance of the converter

4. Inverter Design:

Thus, the inverter topology used is a bridge inverter energy conserved and utilized efficiently [8]. Gate pulse is given by means of feedback network [7]. Thus, it is necessary to determine the look-up table logic for each and every phase voltage coordinating each other [8]. The gate sequence is produced by using relational operators suggesting the proper logic pattern. The six logic on and off period is controlling the operation the phase and line voltage of the system[9-10].

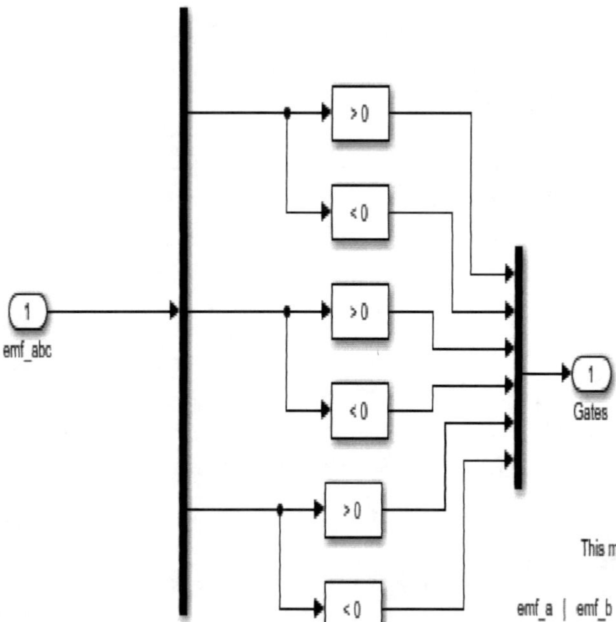

This module implements the following true table

emf_a	emf_b	emf_c	Q1	Q2	Q3	Q4	Q5	Q6
0	0	0	0	0	0	0	0	0
0	-1	+1	0	0	0	1	1	0
-1	+1	0	0	1	1	0	0	0
-1	0	+1	0	1	0	0	1	0
+1	0	-1	1	0	0	0	0	1
+1	-1	0	1	0	0	1	0	0
0	+1	-1	0	0	1	0	0	1
0	0	0	0	0	0	0	0	0

Figure: 3. Lookup Table and Gate Sequence

IOP Conf. Series: Materials Science and Engineering **937** (2020) 012015 doi:10.1088/1757-899X/937/1/012015

4.1. Decoder:

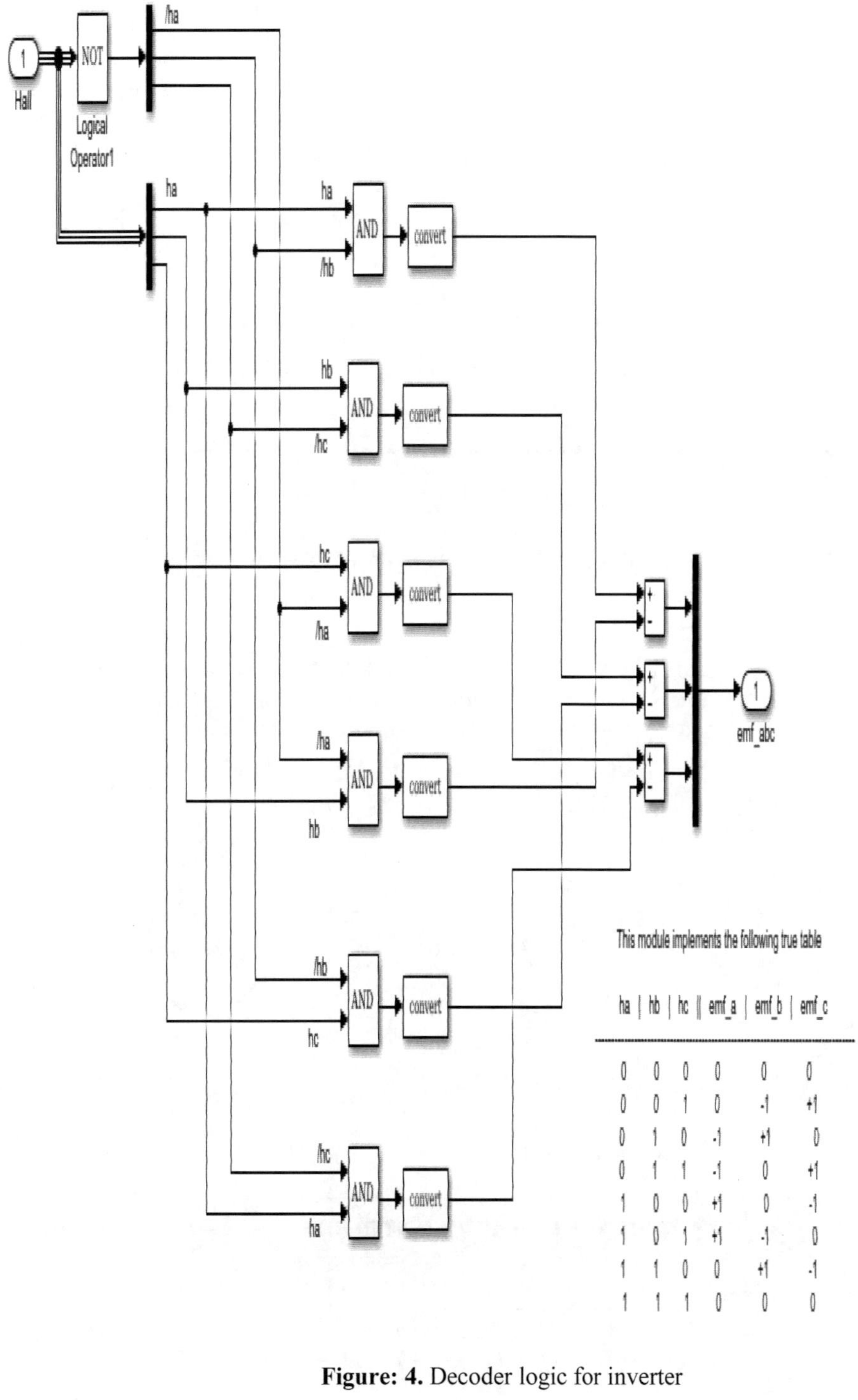

Figure: 4. Decoder logic for inverter

IOP Conf. Series: Materials Science and Engineering **937** (2020) 012015 doi:10.1088/1757-899X/937/1/012015

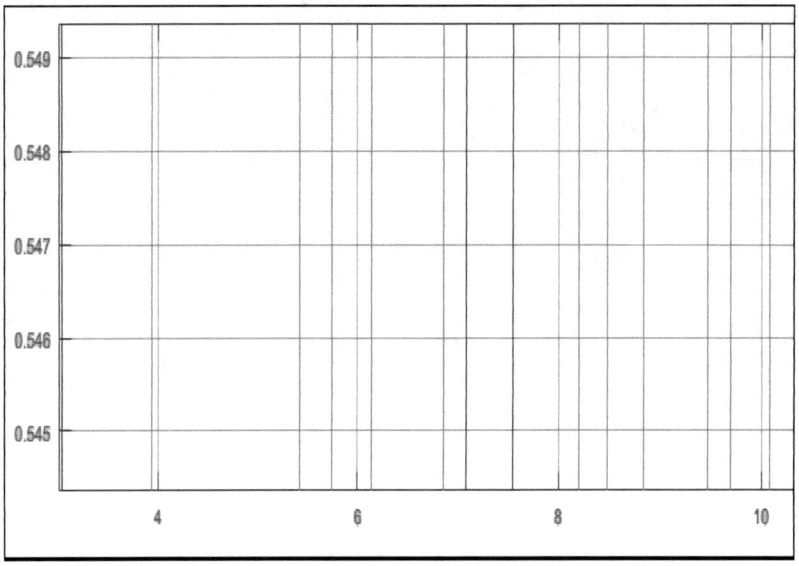

Figure: 5. Waveform of gate pulse for inverter

The above figure 5 represents the gate pulse used for controlling the firing angle of inverter.

5. PID Controller:

Calculated the values of error and compare the measured value and value of error [11-12]. On comparing the set value the measured value is more nominal. The proportional, integral and derivative get summed to determine the output of PID controller. The equation is given as follows,

$$u(t) = K_p e(t) + K_i \int_0^t e(\tau)d\tau + K_d \frac{de(t)}{dt} \quad \text{..... (6)}$$

Here,

K_p=Proportional gain`

K_i= integral gain

K_d= derivative gain

$e(t)$= error

The transfer function of PID controller is

$$L(s) = K_p + \frac{K_i}{s} + K_d s \quad \text{..... (7)}$$

Table: 1 Specifications Of PID controller

GAIN VALUE OF PID CONTROLLER	
K_p	0.0001
K_i	0.8
K_d	0

IOP Conf. Series: Materials Science and Engineering **937** (2020) 012015 doi:10.1088/1757-899X/937/1/012015

6. Specification:

Table: 2 Specification details of converter and PMSM

S.NO	COMPONENTS		RATING
1	DC VOLTAGE	Input voltage	230 V
2	Boost Converter	Inductor	200 mH
		Capacitor	50 μF
		Resistor	20 Ω
		Output Voltage	457.6 V
		Output current	24.75 A
3	PMSM	Output voltage	237.7 V
		Speed of rotor	5000 RPM
		Torque	105 N-m
		Current of stator	120 A

7. Result Analysis:

The model of boost converter fed PMSM is performed in MATLAB/Simulink software [13-14]. The consequence of boost converter gives information of output voltage [15-16]. During simulation output voltage gets doubled when compared to input voltage [17]. After simulation output voltage of the converter is higher than input voltage and it gives higher efficiency of 96%. No controller is used in this converter topology [18-20].

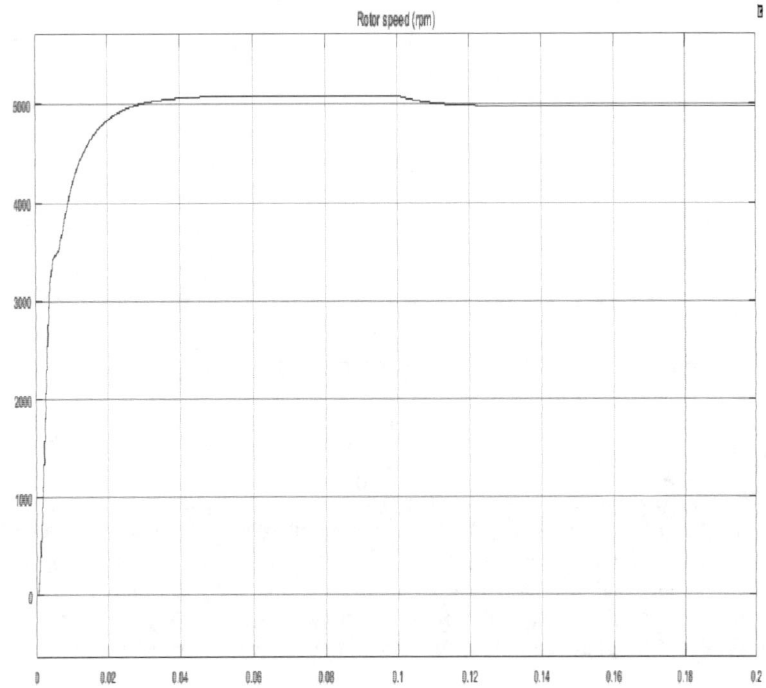

Figure. 6. shows motor speed at 5000 RPM.

IOP Conf. Series: Materials Science and Engineering **937** (2020) 012015 doi:10.1088/1757-899X/937/1/012015

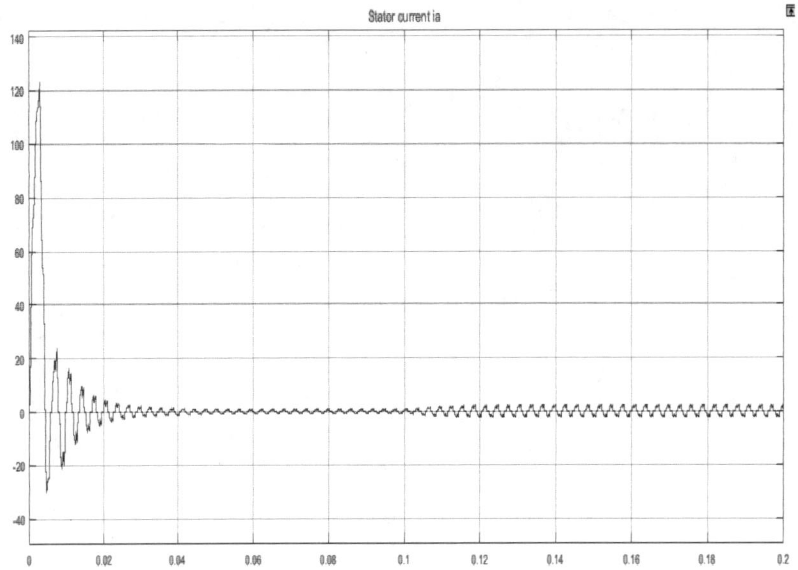

Figure. 7. shows stator current at 120 A

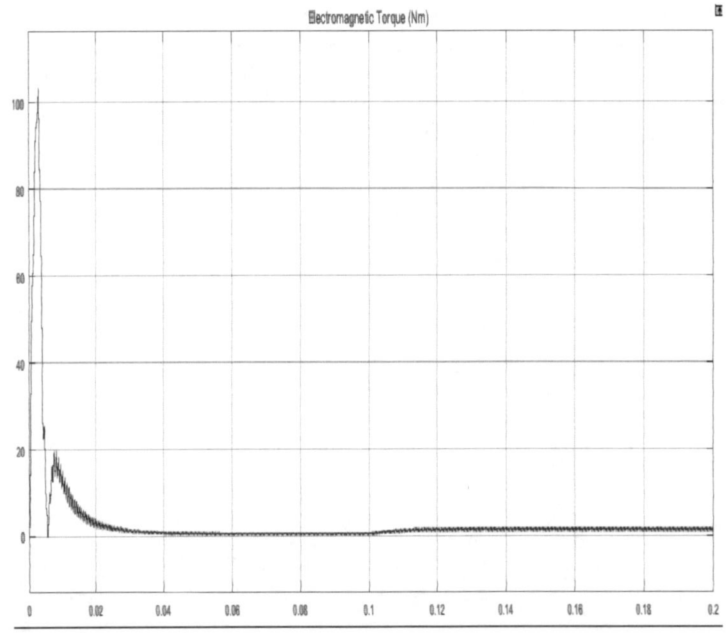

Figure. 8. Shows torque of the motor which is measured at 105 NM.

7. Conclusion:

As Proposed less noise high voltage boost converter was presented in this paper. The required amount of voltage is fed to the bridge inverter by boost converter which controls the speed of the motor. The motor speed can reach about the settle time of 0.2s. The Proposed work mainly used in high voltage applications. Electric vehicles are charged through batteries. The proposed converter design and its performance was observed through simulation and then verified with experimental measurements.

IOP Conf. Series: Materials Science and Engineering **937** (2020) 012015 doi:10.1088/1757-899X/937/1/012015

References:

[1] Erik S. Z. H. Li, and so on. “Design and Control of a Multiple Input DC/DC Converter for Battery/Ultra-capacitor Based Electric Vehicle Power System,” IEEE Applied Power Electronics Conference – APEC 2009, pp. 591-586

[2] Shyam D, Premkumar K, Thamizhselvan T, Nazar Ali A, Vishnu Priya M(2019),Symmetrically Modified Laddered H-Bridge Multilevel Inverter with Reduced Configurationally Parameters, International journal of engineering and advanced technology, Vol 9, issue 1.

[3] A.Nazar Ali and R. Jayabharath (2014), "Performance Enhancement of Hybrid Wind/Photo Voltaic System Using Z Source Inverter with Cuk-sepic Fused Converter." Research Journal of Applied Sciences, Engineering and Technology7.19 pp. 3964-3970.

[4] C.Kalavalli, S.R.Paveethra, S.Murugesan, Dr.A.Nazar Ali, (2020), Design and Implementation Of High Efficiency H6 PV Inverter With Dual Axis Tracking, International Journal of Scientific & Technology Research, Vol 9, issue 02, pp. 4728-31.

[5] S.Dinesh kumar, Mr.J.Praveen Daniel, Dr.N.Senthilnathan, “Three phase shunt active filter using modified algorithm for interfacing renewable energy sources with main grid” International Journal of Applied Engineering Research, Vol 10, No.55, pp 3045-3049, 2015.

[6] A. Saranya; S. Dineshkumar, “Improving voltage stability of power system using facts device” 2017 IEEE International Conference on Electrical, Instrumentation and Communication Engineering (ICEICE) Year: 2017

[7] S. A Nazar Ali, D Sivamani, R Jaiganesh M Pradeep (2019), Solar powered air conditioner using BLDC motor, IOP Conference Series: Materials Science and Engineering, vol. 23.

[8] V Venkatesh, A Nazar Ali,, R Jaiganesh. V Indiragandhi (2019), Extraction and conversion of exhaust heat from automobile engine in to electrical energy Energy‖, IOP Conference Series: Materials Science and Engineering, vol. 23.

[9] A Ali Nazar, R Jayabharath, MD Udayakumar (2014) ,An ANFIS Based Advanced MPPT Control of a Wind-Solar Hybrid Power Generation System, international review of modelling and simulations. vol.7, no. 4, pp. 638–643.

[10] R.Karthikeyan, GN.Sachin Amreiss “PV Based Interleaved Boost Converter for Pumping Applications”, 2018 International Conference on Intelligent and Advanced System (ICIAS).

[11] V. Jayakumar, DC.Kumaresan, R.Karthikeyan “Wind Energy Conversion System and Solar PV Integration” International Journal of Innovative Technology and Exploring Engineering (IJITEE) ISSN: 2278-3075, Volume-8 Issue-7, 1595- 1600, May, 2019.

[12] R Karthikeyan & Dr S Chenthur Pandian “A Novel 3-D Space Vector Modulation Algorithm for Cascaded Multilevel Inverter”, International Review of Electrical Engineering (IREE), November – December 2010 pp: 2860 – 2865, ISSN: 1827- 6679.

[13] A. T. Sankara Subramanian, P. Sabarish, M. D.Udayakumar and T. Vishnu kumar,” Performance Analysis of Various Photovoltaic Configurations under Uniform Shading and Rapid Partial Shading Formations”, Biosc.Biotech.Res. Comm. Special Issue Vol 13 No (3) 2020 Pp-185-192.

[14] P. Sabarish et al 2019 IOP Conf. Ser.: Mater. Sci. Eng. 623 012011.

[15] M D Udayakumar et al 2019 IOP Conf. Ser.: Mater. Sci. Eng. 623 012018.

[16] P. Sabarish, A. T. Sankara Subramanian, S. Murugesan and V. Sureshkumar,”A New Methodology of Arterial Blood Clot Removal Using Bio Molecular Devices for ATPase Nuclear Motors.” Biosc.Biotech.Res.Comm. Special Issue Vol 13 No (3) 2020 Pp-197-201.

[17] Karthick, R and Sundararajan, M. (2018), “A novel 3-D-IC test architecture-a review,” International Journal of Engineering and Technology (UAE), 7.

[18] Karthick, R and Sundararajan, M. (2017), “A Reconfigurable Method for Time Correlated MIMO Channels with a Decision Feedback Receiver,” International Journal of Applied Engineering Research.

[19] AT Sankara Subramanian, P Sabarish, A Nazar Ali(2017),A power factor correction based canonical switching cell converter for VSI fed BLDC motor by using voltage follower technique IEEE International Conference on Electrical, Instrumentation and Communication Engineering (ICEICE), pp.1-8.

[20] S.R.Paveethra, C.Kalavalli, S.Vijayalakshmi, Dr.A.Nazar Ali, D.Shyam (2020), Evaluation Of Voltage Stability Of Transmission Line With Contingency Analysis, International Journal of Scientific & Technology Research, Vol 9, issue 02, pp. 4018-22

IOP Conf. Series: Materials Science and Engineering **937** (2020) 012016 doi:10.1088/1757-899X/937/1/012016

Solar Energy Application in Indian Irrigation System

R Rajesh Kanna[1], M Baranidharan[1], R Raja Singh [12]*, V Indragandhi [12]

[1]School of Electrical Engineering,Vellore Institute of Technology, Vellore, India
[2]Advanced Drives Laboratory, Department of Energy and Power Electronics,Vellore Institute of Technology, Vellore, India
Correspondence: rrajasingh@gmail.com

Abstract. The electricity deficit and high diesel costs influence the pumping needs of urban water supply and irrigation; hence, the use of solar power for water pumping is a viable alternative to traditional water pumping systems dependent on grid power and diesel. In recent decades, attention has been given to clean and renewable energies from the vast majority of countries' priorities in the environmental and economic sectors. Agricultural sector is backbone of Indian economy as population increases demand of water also increases. Extract and transfer of water to agriculture and drinking is equals high-energy consumption. One of the best solutions is to use of an abundant and almost free solar energy to supply energy in all sectors, especially the irrigation sector. In the growth period of crops, when irrigation is often required, there are a high of solar radiation comes from the sun. This study shows that in India, like many other developing countries, the weakness of using this important source, the program of converting solar energy, and photovoltaic pumping systems use. This paper summarizes solar energy development and its application in the irrigation system and comparing them with the conventional diesel pumps. The expected results of this paper are to encourage the politicians, statesmen, and professionals to planning and policy-making the use of the green energy industry in the water sector that could pump water to the villages and agriculture and jump in the energy sector to the economic and Sustainable development.

1. Introduction
It is well known fact that Indian economy is one of the leading economies over the world. Among the various sectors in the Indian economy, agricultural sector provides the largest contribution to develop the nation. In order to achieve extreme manpower utilization and highest return in a given postulated it is required to promote different engineering methods which are recently used in today's global market [1]. Since there is a snowballing in usage of energy, the nonstop decrease in prevailing sources of fossil fuels and the rising issues in relation to the environment pollution, we have been pushed to discover renewable energy sources like solar energy, wind energy, etc. [2]. Solar energy is one of the greatest plentiful resources of energy in this universe. Besides being a solution to today's energy problem, Solar power also acts as a pollution free form of energy. Photovoltaic generation is an effective method for consuming the solar energy [3]. The solar energy projects in most of the developing countries of the world are diverse, such as rural power supply, communication services, water pumping, street lighting, cooling and heating of residential and commercial spaces, heating of water etc. that protected the environment [4]. Handling with the current situation of energy problems in India, irrigation system using solar power can be opted for farmers to manage their agricultural works. It is a finest technique to produce the energy in an eco-friendly manner after a primary investment [5]. One of the most significant usages of solar energy is providing drinking water for small communities which has the around thousand

members [6]. The World Bank estimates the number of people who live in remote rural areas and do not have safe drinking water in developing countries about one billion people [7].

According to the distant villages of the nationwide grid, supplying energy and drinking water to the rural population is very difficult and costly. For this reason, the policy of exploiting local energy sources, such as solar energy, has been recognized for many years as a useful solution. One can mention the benefits of using local energy, improving health and well-being, boosting agriculture through irrigation, access to safe drinking water and providing residential lighting and reducing urban migration. At present, more than 160,000 villages worldwide are based on solar energy [8]. Among the many uses of solar energy, photovoltaic pumping systems have been widely welcomed in the last two decades. According to World Bank reports, up to 1993, ten thousand photovoltaic pumps were installed and operated worldwide, and by the year 1998, there were 60,000 photovoltaic pumping systems [9]. Solar water pumps are very efficient in areas where the expansion of the power grid is not feasible. Even in areas where the connection to the national grid is possible, the use of photovoltaic pumps is preferable for several reasons [10]. Solar pumps can be used in numerous capacities according to the necessity of water [11].Today, the implementation and exploitation of photovoltaic pumps system in developing countries are growing faster than in developed countries. The reasons for this are as follows. restrictions on the distribution of the power grid, money restrictions for the import of oil and The Government's commitment towards rural population, providing electricity, health, and educational programs that have been guaranteed by international organizations [12]. Therefore, Governments in developing countries have decided to invest in the production of photovoltaic pumps system. Most of these studies began in 1979 and continue to this day [13]. The cost-effective economics of the photovoltaic pumps were fixed in comparison with the diesel pumps. In addition, during the survey of the inhabitants of the region, great reliability, and supply of water volume required by photovoltaic pumps were confirmed. Studies of photovoltaic pumps system in India confirm the economic justification and social benefits of these systems [14].

2. Solar pumping system

Photovoltaic is being used for pumping for an extensive kind of processes, especially for water irrigation in agricultural area. It can be also used in refrigeration of fruits and vegetables, fencing, cattle lighting and insects controlling. The techniques which are being used in the PV system is applicable to only those activities that have need of slight power input. They are inappropriate when it comes to large power consuming areas like rice mills and high powered agricultural processing [15]. In current situation, under the hot sun, farmers find difficult to irrigate their plants to maintain their crops green. They suffer because they have no appropriate idea about obtainability of the power. The solar pumping system offers an idea to solve this problem. Generally, the solar pumping system has the following main parts: Electric solar panels pump controller, submersible pump, converter, storage tank and rechargeable battery. The illustration of solar pumping system is shown in figure 1.

2.1. Solar Panel

Solar panels are manufactured with solar cells which comprises of semiconductor materials. The main purpose of solar panels is to change solar power into DC electrical power. The rating of the load determines the number of cells needed and their size. The higher amount of electric power can be produced with the help of collecting solar cells. However, the solar panel should be placed accurately at precise angles towards the sun rays. The size of the pump is being decided in accordance with the photovoltaic cells. The solar submersible pumping system need to be operated along with a PV array of capability with the value ranging between 250 to 550 watt peak, and are being calculated through several typical test. Numerous modules in series & parallel can be utilized to acquire the adequate PV power output. When the PV modules are involved in typical test conditions, it should have a minimum of 74-watt peak.

IOP Conf. Series: Materials Science and Engineering **937** (2020) 012016 doi:10.1088/1757-899X/937/1/012016

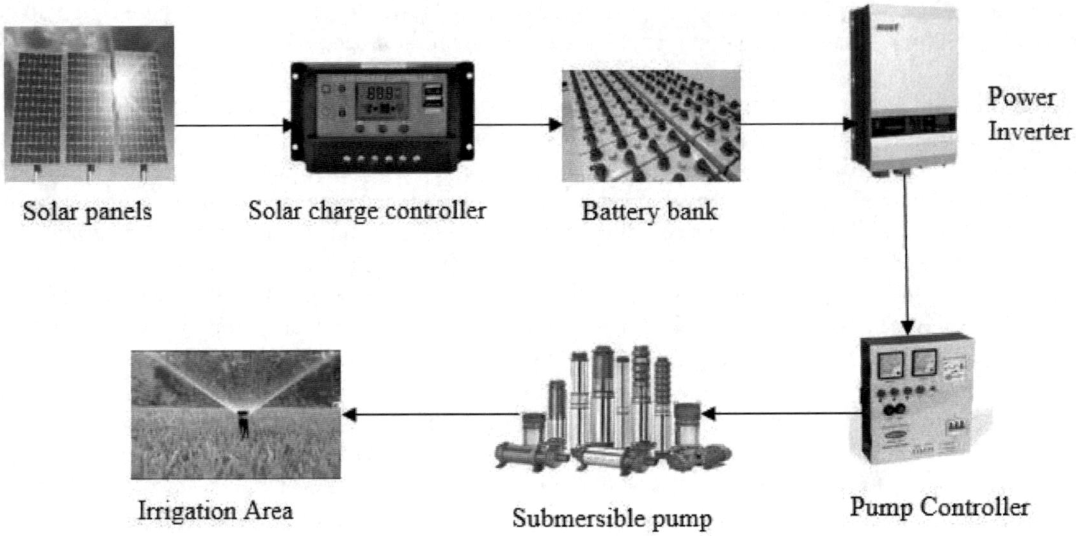

Figure 1. Solar pumping system

2.2. The Submersible Pump
The submersible pump is often placed in a stainless steel case, which is located in a well pit at the joint of the open channel and the natural stream course. The pump controller is used to pump water towards the water tank within a specific time period using the control unit. The size of the solar pumps is determined by the total depth of the well and the longitudinal and local losses of the piping system in pumping from the well and the required water flow.

2.3. Solar charge Controller
A charge controller is very significant electrical device in any solar-power system. It allows correct charging voltages to the batteries. The main purpose of charge controller is to regulate the current as well as voltage from the solar panel. It also used for charging the battery and protect the battery from over and undercharging conditions.

2.4. Battery
The battery is generally used for store current that is formed by the PV panel and delivered to the equivalent loads. The batteries requirement may be varied according to load requirement.

The photovoltaic system being large, needs the solar panel as its component in order to produce and deliver the electricity in rural and urban area's applications. The electric current produced by the photovoltaic panels is direct current (D.C). The voltages of these plates varying according to the variation of temperature and solar radiation. Solar cells are available in three types of monocrystals, polycrystals, and morphs. The pump controller adjusts the power exchanged between the solar panels and the pump. This controller protects the pump against current fluctuations and provides conditions for permanent output. In addition, if the pump is dry, the pump controller can automatically shut down the system, also the pump can be shut down by installing a floating switch connected to the controller when filling the tank. Control circuit permits the battery to charge. A converter circuit transforms DC supply from the battery to AC supply and it gives power to the water pump which is sunken into the well. Before discharging the water into the agricultural arena, the water is drove into an overhead tank for keeping save the water provisionally. The storage mechanism in the solar water pump system can be either water storage method or power storage method. The storage water method most economical way to save water and the most reliable method. Of course, the system can save the power for using the pump during cloudy days and at night with battery embedding. Although this method has high reliability, it also requires higher costs and more maintenance than the first method.

3. Comparison of energy consumption and solar pumps in India with other countries

The addition of solar power capacity into the electrical grids has been adopted by several countries and terrains to deliver a substitute to conventional energy resources. To take advantage of global experiences and compare the situation in India in terms of solar energy projects and consumption activities, especially in irrigation sector, according to a study of broad and wrap existing research and reports created. In this table of solar energy, development activities in various countries, especially in the solar water pumping for irrigation to farms, are summarized. The growth of photovoltaics over the universe is enormously vibrant and differs intensely by country. Nearly 37 countries in the world will generate more than one gigawatt cumulative solar power by 2020. By the end of 2019, a cumulative amount of 629 GW of solar power was installed throughout the world [16]. By early 2020, the leading country for solar power was China with 208 GW, [17] [18] accounting for one-third of global installed solar capacity. The cumulative photovoltaic capacity has been improved as 75 gigawatt (GW) and touched 303 GW, adequate to deliver around 1.8% of the world's entire electricity intake at the end of 2016. The China, the United States, and India were the leading PV panel installers in 2016 through 2019[19] [20]. The Honduras having its PV capacity is presently liable to deliver 12.5 percent of the country's electrical energy, at the same time, Italy, Greece and German is generating from 7 to 8 percent of the domestic electricity intake [21-23].

Table 1. The activities of the different countries in the field of the use of solar energy and pump

Country	Description	Description of activity	Data
China	The world's largest solar power plant construction program with the production capacity of the 200 gigawatts accounting for one-third of global installed capacity.	99% of the area under irrigation pump and photovoltaic systems are most cultures.	2020[24][25]
Japan	The program uses existing solar energy in space and sends to Earth by the year 2030 in Japan. By the end of 2017 cumulative capacity reached 50 gigawatts.	Solar energy production, increasing 1.2 times in one year.	Up to 2017[26]
Unite the States of America	The decision to install 10 million solar panels within a decade and 10% of the electricity needed by the year 2030.at the end of 2017 cumulative capacity touched 50 gigawatts.	Convert the Nevada desert (the location of the nuclear tests) is the world's largest solar laboratory.	Up to 2017[27]

Taiwan	Long-term goal of adding 6500 to 10000 MW for renewable energy sources in the next 20 years. By the end 2019, the capacity of solar energy production touched 4 gigawatts.	Construction of the largest solar power plant of Asia with an area of two hectares with a production capacity of one megawatt of energy.	2019[28]
Jordan	The fortunes of the public to make use of this technology have increased. Hit the number of pumps installed is exact at hand. By the end 2018, the capacity of solar energy production touched 1 gigawatts.	Promote the use of photovoltaic systems are the water pump and the expansion of research in this field.	2018[29]
Turkey	The program of the annual energy production of photovoltaic systems is 500 kW. By the end 2018, the capacity of solar energy production touched 6 gigawatts.	Solar energy collector plates increased from 8 million to 11 million square meters.	2018[30]
India	Installation and operation of photovoltaic water pump 10000 of the goals of the PM-Kusum in this country. . By the end 2019, the capacity of solar energy production touched 37.5 gigawatts.	Installed the 181000-water pump with photovoltaic systems for agricultural areas.	2019[31]

4. A review: usage of solar energy in agricultural area

Water pumps in most essential application in agricultural sector, the use of solar energy for lifting the water is a best option for agricultural development and efficient use of available water resources. There are various stages to implement the solar water pump to the irrigation system, and are as follows, (i) Determine the quantity of water requirement which includes climatic analysis. (ii) Calculate the power requirement for pumping water. (iii) Estimate the rate of energy production and climatic adaptation. The aim of the study was to estimate the rate of energy production and climate adaptation [24-36]. Water requirement for irrigation, which is completed by use CROPWAT8.0 software for two types' crops one is winter wheat it planting time in November and harvest time in July.So during the design stages of the required flow drip irrigation system 33.3m^3/h (25.9lit/s) and 7.5 hours and duration of irrigation in two turns. 27-meter system and the dynamic head static head are estimated to be 5 meters. The pump should be selected as the height of 32 metros of water [15].

The hydraulic energy (HE) requires pumping the 799.2 m^3/day water.

$$HE = \frac{Q \times TDH \times Dw \times Gr}{3.6 \times 10^6} \tag{1}$$

Where:

Q is flow requirement

TDH is a total dynamic head of well.

Dw is water density.

Gr is the gravity of earth.

HE= (799.2 x 32 x 1000 x 9.81)/ (3.6 x10^6)

HE= 69.69 kWh/day

$$\text{Pump capacity} = \frac{HE}{daily\ sun\ hours} \tag{2}$$

=69.69 /6.791=10.26kW=13 hp.

Pump capacity=13hp

4.1. Shaft power

The power which is provided by the motor to the pump shaft is named as the shaft power. The ratio of hydraulic pump to the efficiency of the pump determines the shaft power.

$$\text{Shaft power (Ps)} = \frac{\text{hydraulic power}}{\text{efficiency of pump } (\eta p)} \tag{3}$$

ηp = 80 % = 0.8

Shaft power (Ps) = 13/0.8 = 16 hp

4.2. Motor power

The power used up by the pump motor to change the pump shaft is called as the motor power. We can calculate the motor power by adding the shaft power and power loss that is made due to ineffectiveness in changing electric power into kinetic power [37-48]. It is better defined as the ratio of shaft power to the motor efficiency.

$$\text{Motor power} = \frac{\text{shaft power}}{\text{efficiency of motor } (\eta m)} \tag{4}$$

ηm = 90 % = 0.9

Motor capacity = 16/ 0.9 = 17 = 17 hp

Total energy requirement 12.66 kW capacity available solar submersible in bazar start 0.5 horse power (HP) up to 50 horsepower (HP).

5. Design of solar photovoltaic array

The solar photovoltaic array consists of several photovoltaic solar modules size basically to satisfy the energy needs of pumps has been designed. The connection of the PV modules in series and parallel, depending on the voltage and time required are made. The solar array installed in this field Fixed type can be either track one. The tracking system will considerably improve the efficiency although it adds an additional cost of tracking equipment.40 to 60% of the photovoltaic panels of the overall system costs. The power output of the PV array with respect to the difference in the location of the solar radiation and atmospheric factors, different temperatures, so the appropriate size according to the required page is very important and the process is difficult. The size of the PV array can be done regarding the need to lift the daily required an amount of hydraulic energy of water HE, Average daily radiation on photovoltaic array E and the daily subsystem efficiency. This method makes the system more accurate and reliable.
Peak watt module can be calculated with the equation

$$\text{Solar PV array power(kwp)} = \frac{\text{motor energy requirement}}{\text{F x E}} \quad (5)$$

Where:

F= array mismatch factor (0.85 on average)

E= daily subsystem efficiency (0.25 - 0.5 typically)

Solar array requirement (kW) = 12.66 kW/ (0.5*0.85) =37.23 kW

Table 2. Electrical characteristics of a multicrystalline solar panel.

Electrical characteristics	**Values**
Maximum power (Pmax)	170 W
Voltage at maximum power point (Vmax)	35.4 V
Current at maximum power point (Imax)	4.8 A
Short circuit current (Isc)	5 A
Open circuit voltage (Voc)	44.2 V
Temperature coefficient of short circuit current (Isc)	0.065± 0.015 (%/°C)
Temperature coefficient of open circuit voltage (Vop)	-160± 20mV
Temperature coefficient of power	-0.5± 0.05 (%/°C)
NOCT (°C)	47 ± 2

5.1. Number of series and parallel

Maximum power of the system W_{pa} = 37235W
The maximum voltage of the module V_{mp}=35.4V
The maximum current of the module I_{mp}= 4.8A
Boost converter input voltage it is the maximum voltage of the system V_{ma}=500V
The maximum current of the system I_{ma}= Maximum power of the system/ maximum voltage of the system
Maximum current of the system I_{ma}= W_{pa}/ V_{ma}=37235/500 = 74.47 A
Number of series module N_s = maximum voltage of the system/ Maximum voltage of the module= V_{ma}/ V_{mp} = 500/35.4 = 15
Number of module parallel = Maximum current of the system/ Maximum current of the module = I_{ma}/ I_{mp} =74.47/4.8 = 15
Total number of the PV panel = 15*15= 225

6. Solar water pump comparing with diesel water pump

In this section, a detailed analysis carried out between solar water pump and diesel water pump is the perspective of cost and reliability. Diesel water pumps, are usually cost effective at the initial time, though when it comes to operation and maintenance, they are costly. Unlikely, the solar water pumps are having maximum initial cost and little for operation and maintenance. In consistency, it is more liable to make use of solar water pumps rather than using diesel water pumps. From the environmental point of view, diesel driven water pump is produce various gases which change climate as well as health hazards. This is made possible where diesel water pump gets rusted when it is actually not in use, but solar pumps may survive for years, even if it is not used properly. The preliminary cost of the solar water pump is frequently unapproachable to the project executors to expanse their funds as extreme as possible to grasp the highest number of receivers with the use of a low primary optimum cost.

Table 3. Comparing the solar water pump and diesel water pump

Particulars	Solar water pump	Diesel water pump
Available capacity kW (hp)	0.5(0.75) to 75(100)	3.7(5) to 11(150)
Capital cost	High	Moderate
Running cost	No running cost	Very high
Maintenance cost	Negligible	High
Routine maintenance	Only panel cleaning once a week	Diesel lubrication minor &major servicing required, periodic overhauling must
Part replacement	Moderate	Worn out parts need to be replaced often
Operator	No operator required, auto start / stop possible	Operator required
Other utilization	PV array can be used for electricity generation when the pump is not running	No
Environmental aspect	Silent & pollution free green power	High air pollution and noise pollution
Limitation	Works on solar during day time only	High fuel & maintenance cost

IOP Conf. Series: Materials Science and Engineering **937** (2020) 012016 doi:10.1088/1757-899X/937/1/012016

Table 4: comparison of solar water pump and diesel water pump.

Years	Year 1	year 2	year 3	year 4	Year 5	year 10	year 15	year 20	Total USD
Solar water pump									
Initial cost	46644	0	0	0	0	0	0	0	46644
Servicing cost	0	0	0	0	0	574	0	673	1,247
Maintenance cost	0	0	0	0	0	32163	0	37684	69847
Cumulative total	46644	46644	46644	46644	46644	79381	79381	117738	117738
Diesel water pump									
Servicing cost	500	510	520	531	541	2,873	3,172	3,501	12148
Maintenance cost	0	0	3121	0	1624	8620	9668	6386	29419
Fuel	4380	5037	5793	6661	7661	59399	119472	240299	448702
Fuel delivery costs	2600	2652	2705	2759	2814	14939	16493	18210	63172
Cumulative total	7480	15679	27818	37769	50409	136240	285045	553441	553441

7. Conclusion

The solar power is suitable for agricultural applications such that electrical fencing, threshing, aeration, grinding, irrigation, purification etc. Nowadays the farmers from India are generally using the solar energy in the water sectors especially in irrigation sectors for their agricultural farms. However, farmers are thinking the solar water pump systems initial cost is higher than diesel water pump system but never thought generation and maintenance costs of both systems. One of the most importing things is the solar water pump system PV array can be used for electricity generation when the irrigation is not a requirement. Solar water pumping systems can easily meet the irrigation water requirement for landholding system marginal farmers. Due to facing regularly increasing fuel cost, pump sets is being deployed every year in India. Solar water pump system helps in the reduction of diesel consumption. The solar water pump is no novelty it has existed since three decades also solar water pump are available for different types. And despite being blue, uses up to 30% of the country's productive energy would devour precious resources continue to use most of the electrical energy production. it insists that many costs on society and the environment, based on the frequency in the electric sector and thermal power plants to the environment imposes. From the above, we conclude that, compared with diesel engine pumping system, the solar pumping system is more feasible for irrigating the farms in India. In economic basis, solar water pumps are not more expensive if the solar energy are used continuously, though their capital cost is high.

References

[1] Mr. M. A. Murtaza, Mr. Mragank Sharma, Rohit Yadav "Solar Powered Automatic Irrigation System" International Journal of Engineering Science and Computing, vol.7,issue no.4,April 2017.

[2] Er.Upendra Singh, Mohit Vyas, Gaurav Sharma," Solar Based Smart Irrigation System" International Journal of Recent Research Aspects,vol.3,issue no.1,march 2016.

[3] B. Kalaskar,Prof. Yashoda A. Kale" Solar Powered Automated Irrigation System", International Journal for Scientific Research & Development,vol.5,issue.10,2017.

[4] R. R.Singh, B. A. Kumar, D. Shruthi, R. Panda, and C. T. Raj, "Review and experimental illustrations of electronic load controller used in standalone Micro-Hydro generating plants," Eng. Sci. Technol. an Int. J., vol. 21, no. 5, pp. 886–900, 2018, doi: 10.1016/j.jestch.2018.07.006.

[5] V. Indragandhi, B. K. Vardhan, G. Arunkumar, R. Raja Singh, and S. Banumathi, "Implementation of a DC micro-grid for house hold applications," in 2018 8th International Conference on Power and Energy Systems, ICPES 2018, 2019, pp. 141–145, doi: 10.1109/ICPESYS.2018.8626889.

[6] L. Prisilla, P.S.V. Rooban and L. Arockiam, "A novel method for water irrigation system for paddy fields using ANN," International Journal of Computer Science and Network, Vol.1, No. 2, April 2012.

[7] R. R. Singh, V. Indragandhi, and S. Umashankar, "Impact of Voltage Variation on Hydroelectric Doubly Fed Machines - An Electro-thermomechnaical Investigation," in 2019 IEEE 2nd International Conference on Power and Energy Applications, ICPEA 2019, 2019, pp. 22–27, doi: 10.1109/ICPEA.2019.8818535.

[8] M.A. Salam, A. Ahmed, H. Ziedan, K. Sayed, M. Amery and M. Swify "A Solar-Wind Hybrid Power System for Irrigation in Toshka Area," IEEE Jordan Conference on Applied Electrical Engineering and Computing Technologies, pp. 1-6, Dec. 2011.

[9] Kou Q, Klein SA, Beckman WA. A method for estimating the long-term performance of direct-coupled PV pumping systems. Sol Energy 1998;64:33–40.

[10] Khan MTA, Ahmed MR, Ahmed SI, Khan SI. Design and Performance analysis of water pumping using solar PV. In: Proceedings of the international conference of developments in renewable energy technology (ICDRET) 2012.

[11] Mokeddem A, Midoun A, Kadri D, Said H, Iftikhar RA. Performance of a directly-coupled PV water pumping system. Energy Convers Manag 2011;52:3089–95.

[12] R. R. Singh, H. Mohan, and T. R. Chelliah, "Performance of doubly fed machines influenced to electrical perturbation in pumped storage plant - a comparative electromechanical analysis," in India International Conference on Power Electronics, IICPE, 2017, vol. 2016-November, doi: 10.1109/IICPE.2016.8079516.

[13] Benghanem M, Daffallah KO, Joraid A, Alamri and Jaber A. Performances of solar waterR pumping system using helical pump for a deep well: a case study for Madinah, Saudi Arabia. Energy Convers Manag 2013;6:50–6.

[14] Setiawan A, Purwanto DH, Pamuji DS, Nurul Huda N. Development of a solar water pumping system in Karsts Rural Area Tepus, Gunungkidul through student community services. Energy Procedia 2014;47:7–14.

[15] R. Raja Singh et al., "IoT embedded cloud-based intelligent power quality monitoring system for industrial drive application," Futur. Gener. Comput. Syst., vol. 112, pp. 884–898, 2020, doi: 10.1016/j.future.2020.06.032.

[16] Sitharthan R, Geethanjali M and Pandy TKS 2016 Adaptive protection scheme for smart microgrid with electronically coupled distributed generations *Alexandria Engineering Journal* **55(3)** 2539-2550

[17] Fathima AH, and Palanisamy K 2014 Battery energy storage applications in wind integrated

systems—a review *IEEE International Conference on Smart Electric Grid* 1-8

[18] Prabaharan N and Palanisamy K 2015 Investigation of single-phase reduced switch count asymmetric multilevel inverter using advanced pulse width modulation technique *International Journal of Renewable Energy Research* **5(3)** 879-890.

[19] Jerin ARA, Kaliannan P and Subramaniam U 2017 Improved fault ride through capability of DFIG based wind turbines using synchronous reference frame control based dynamic voltage restorer. *ISA transactions* **70** 465-474

[20] Sitharthan, R, Sundarabalan CK, Devabalaji KR, Nataraj SK and Karthikeyan M 2018 Improved fault ride through capability of DFIG-wind turbines using customized dynamic voltage restorer *Sustainable cities and society* **39** 114-125

[21] Prabaharan N and Palanisamy K 2016 A single-phase grid connected hybrid multilevel inverter for interfacing photo-voltaic system *Energy Procedia* **103** 250-255

[22] Palanisamy K, Mishra JS, Raglend IJ and Kothari DP 2010 Instantaneous power theory based unified power quality conditioner (UPQC) *IEEE Joint International Conference on Power Electronics, Drives and Energy Systems* 1-5

[23] Sitharthan R and Geethanjali M 2017 An adaptive Elman neural network with C-PSO learning algorithm-based pitch angle controller for DFIG based WECS Journal of Vibration and Control **23(5)** 716-730

[24] Sitharthan R and Geethanjali M 2015 Application of the superconducting fault current limiter strategy to improve the fault ride-through capability of a doubly-fed induction generator–based wind energy conversion system *Simulation* **91(12)** 1081-1087

[25] Sitharthan R, Karthikeyan M, Sundar DS and Rajasekaran S 2020 Adaptive hybrid intelligent MPPT controller to approximate effectual wind speed and optimal rotor speed of variable speed wind turbine *ISA transactions* **96** 479-489

[26] Sitharthan R, Devabalaji KR and Jees A 2017 An Levenberg–Marquardt trained feed-forward back-propagation based intelligent pitch angle controller for wind generation system *Renewable Energy Focus* **22** 24-32

[27] Sitharthan R, Sundarabalan CK, Devabalaji KR, Yuvaraj T and Mohamed Imran A 2019 Automated power management strategy for wind power generation system using pitch angle controller *Measurement and Control* **52(3-4)** 169-182

[28] Sundar DS, Umamaheswari C, Sridarshini T, Karthikeyan M, Sitharthan R, Raja AS and Carrasco MF 2019 Compact four-port circulator based on 2D photonic crystals with a 90° rotation of the light wave for photonic integrated circuits applications *Laser Physics* **29(6)** 066201

[29] Sitharthan R, Parthasarathy T, Sheeba Rani S and Ramya KC 2019. An improved radial basis function neural network control strategy-based maximum power point tracking controller for wind power generation system *Transactions of the Institute of Measurement and Control* **41(11)** 3158-3170

[30] Rajesh M and Gnanasekar JM 2017 Path observation based physical routing protocol for wireless ad hoc networks *Wireless Personal Communications* **97(1)** 1267-1289

[31] Palanisamy K, Varghese LJ, Raglend IJ and Kothari DP 2009. Comparison of intelligent techniques to solve economic load dispatch problem with line flow constraints *IEEE International Advance Computing Conference* 446-452

[32] Sitharthan R, Ponnusamy M, Karthikeyan M and Sundar DS 2019 Analysis on smart material suitable for autogenous microelectronic application *Materials Research Express* **6(10)** 105709

[33] Rajaram R, Palanisamy K, Ramasamy S and Ramanathan P 2014 Selective harmonic elimination in PWM inverter using fire fly and fireworks algorithm *International Journal of Innovative Research in Advanced Engineering* **1(8)** 55-62

[34] Sitharthan R, Swaminathan JN and Parthasarathy T 2018 March. Exploration of wind energy in India: A short review *IEEE National Power Engineering Conference* 1-5

[35] Karthikeyan M, Sitharthan R, Ali T and Roy B 2020 Compact multiband CPW fed monopole antenna with square ring and T-shaped strips *Microwave and Optical Technology Letters* **62(2)** 926-932

[36] Sundar D Sridarshini T, Sitharthan R, Madurakavi Karthikeyan, Sivanantha Raja A, and Marcos Flores Carrasco 2019 Performance investigation of 16/32-channel DWDM PON and long-reach PON systems using an ASE noise source *In Advances in Optoelectronic Technology and Industry Development: Proceedings of the 12th International Symposium on Photonics and Optoelectronics* 93

[37] Sitharthan R and Geethanjali M 2014 Wind Energy Utilization in India: A Review *Middle-East J. Sci. Res.* **22** 796–801 doi:10.5829/idosi.mejsr.2014.22.06.21944

[38] Sitharthan R and Geethanjali M 2014 ANFIS based wind speed sensor-less MPPT controller for variable speed wind energy conversion systems *Australian Journal of Basic and Applied Sciences* **8**14-23

[39] Jerin ARA, Kaliannan P, Subramaniam U and El Moursi MS 2018 Review on FRT solutions for improving transient stability in DFIG-WTs *IET Renewable Power Generation* **12(15)** 1786-1799

[40] Prabaharan N, Jerin ARA, Palanisamy K and Umashankar S 2017 Integration of single-phase reduced switch multilevel inverter topology for grid connected photovoltaic system *Energy Procedia* **138** 1177-1183

[41] Rameshkumar K, Indragandhi V, Palanisamy K and Arunkumari T 2017 Model predictive current control of single phase shunt active power filter *Energy Procedia* **117** 658-665

[42] Fathima AH and Palanisamy K 2016 Energy storage systems for energy management of renewables in distributed generation systems *Energy Management of Distributed Generation Systems* 157

[43] Rajesh M 2020 Streamlining Radio Network Organizing Enlargement Towards Microcellular Frameworks *Wireless Personal Communications* 1-13

[44] Subbiah B, Obaidat MS, Sriram S, Manoharn R and Chandrasekaran SK 2020 Selection of intermediate routes for secure data communication systems using graph theory application and grey wolf optimisation algorithm in MANETs IET *Networks* doi:10.1049/iet-net.2020.0051

[45] Singh RR and Chelliah TR 2017 Enforcement of cost-effective energy conservation on single-fed asynchronous machine using a novel switching strategy *Energy* **126** 179-191

[46] Amalorpavaraj RAJ, Palanisamy K, Umashankar S and Thirumoorthy AD 2016 Power quality improvement of grid connected wind farms through voltage restoration using dynamic voltage restorer *International Journal of Renewable Energy Research* **6(1)** 53-60

[47] Singh RR, Chelliah TR, Khare D and Ramesh US 2016 November. Energy saving strategy on electric propulsion system integrated with doubly fed asynchronous motors *IEEE Power India International Conference* 1-6

[48] Singh RR, Mohan H and Chelliah TR 2016 November. Performance of doubly fed machines influenced to electrical perturbation in pumped storage plant-a comparative electromechanical analysis *IEEE 7th India International Conference on Power Electronics* 1-6

Groundwater Monitoring System Using Internet of Things

GN. SachinAmreiss[1], B. Karthikeyan[2] and Dr.K. Premkumar[3]

[1] Assistant Professor, Department of Electrical and Electronics Engineering, M.Kumarasamy College of Engineering, Karur, Tamilnadu

[2] Assistant Professor, Department of Electrical and Electronics Engineering, K. Ramakrishnan College of Technology, Trichy, Tamilnadu.

[3] Associate Professor, Department of Electrical and Electronics Engineering, Rajalakshmi Engineering College, Rajalakshmi Nagar, Thandalam, Tamilnadu

E.mail: karthikeyankrct@gmail.com[2]

Abstract: The major problem among the people is that they are lacking in the awareness of the underground water usage. Thus, they are in the process of designing the "Application" which gives the data about the underground water consumption of the consumers and monitor the ground water usage of the consumers. Also, alarming when the maximum usage of water or wastage of water. This system sends the information of the water usage of that area and automatically recorded in the government water board of that area with quantity of the water usage, If exceeds the limit will have the locking system of the water supply of the usage area with fine. Thus the paper gives the total guidance to the people according the usage of the underground water resources digitally with the application itself

Keywords - Arduino, IoT, Groundwater

1. Introduction

Groundwater is water underneath the outside of the Earth. Groundwater begins as precipitation and the part of the downpour water that penetrates underneath the ground's surface, either normally or falsely, becomes groundwater. The rest of the bit of the precipitation is utilized by plants, dissipates, or becomes surface water spillover which can either add to groundwater levels in different territories or be expanded by groundwater outpourings relying upon the geography the surface water voyages through. The measure of precipitation that gets assimilated and becomes groundwater relies upon the dirt type [1-2]. Highly permeable soils, for example, sandy soils, retain water a lot quicker than soil, for example, mud which has little pores. The immersed soil acts like a wipe and the region where groundwater is available, or soaked, in the dirt are called a spring, which for the most part has a limit characterized as its basin [3-4]. Groundwater bowls are framed normally over a period running from quite a while to over a thousand years in certain topographies. Groundwater is basic to the networks that are based on or approach the spring, or the underground layer of water that fills splits in the stone or sand that makes up the soil. This groundwater layer which makes up around 30 percent of the worlds freshwater gracefully used for certain, reasons including water framework, private drinking water, and metropolitan water supplies. Groundwater is a key bit of the United States ability to flood its farmland. Water system structures speak to the usage of about 53.3 billion [5] gallons of groundwater consistently for agriculture watering. This utilization can be appeared differently in relation to the 1900s when the US simply used 2.2 billion [6] gallons for water system. Furthermore the creatures and aquaculture ventures eat up generally 3.2 billion [7] gallons of groundwater consistently. Information as for groundwater use for drinking water generally speaking is limited, in any case it is surveyed that 33% [8-9] of the all out people depend upon groundwater as the essential wellspring of their drinking water. More than 13 million US nuclear families typically depend upon private ground water wells. Groundwater sources are used by around 33 percent [10] of the vast water structures in the United States. Among open and private water wells around 44 percent of the US people depends upon groundwater for their drinking water. Counting drinking water and developing nearby various jobs of groundwater, for instance, collecting, mining, and thermoelectric ability to give a few models, 79.6 billion gallons each day of new groundwater are used.

2. Motivation

Knowing the groundwater level is noteworthy for a couple of reasons, including understanding spring levels under static conditions and siphoning conditions, choosing how the levels partner with close by surface water sources, and perceiving how surface headway has influenced the spring. Groundwater extraction from siphoning groundwater for surface use has the best effect on the proportion of groundwater [11-12] set aside in a spring and the rate at which it finishes off or energizes. The most serious outcome of over the top groundwater siphoning is that the water table can be lowered. It is critical to screen and comprehend the groundwater levels proceedings boring any wells which will have noteworthy draw down of the spring on the grounds that the water table levels will give a smart thought of the effect of the new well. For water to be pulled back beginning from the most punctual stage, must be directed from a well that compasses underneath the water table [13-14]. The data assembled by checking groundwater can be used to choose the proportion of groundwater that can safely be pulled back before no more water can be siphoned. Close by water managers can shield wells from going dry and hinder the improvement of progressively disastrous quality groundwater into the spring. In the event that groundwater levels decline exorbitantly [15] far, by then the well owner may need to broaden the well, drill another well, or, at any rate, try to cut down the siphon which could end up being expensive for the proprietor. Despite the cost of extending the significance of the well, entering another well, or moving the direct down as the significance to water constructs; the water should now be lifted higher to rise to the top. Sometimes there may be a chance that siphons are used to lift the water (rather than artesian wells), greater essentialness is required to drive the siphon which realizes continuously exorbitant water. In the end a significant well could end up being prohibitively expensive to siphon water from.

Certain regions decreasing the water table level can greatly affect the groundwater's quality. In waterfront freshwater spring's salt water interruption can happen when the various densities of both the saltwater and new water permit the sea water to barge in into the new water aquifer [16-17]. Often beach front groundwater springs bolster huge populaces where the interest for groundwater withdrawals surpasses the new water revive rate permitting salt water to advance into the spring sullying the water.

Worldwide Positioning System (GPS) and nearby situating calculations can be utilized to acquire area and situating data of water. By which the location of the water body with the flow rate is shown in mobile simply. The sensor is used to detect the Moisture content and Temperature of the certain area in the underground water and when the limit exceeds the Notification is sent to the Customer and Government water board automatically.

3 Proposed System

3.1 Water Level Sensor

Water stream sensor comprises of a plastic valve from which water can pass. Water pivots alongside a Hall Effect sensor that sense and measures the water stream [18-19]. The fundamental working guideline behind the working of this sensor is the Hall Effect. As per this standard, in this sensor, a voltage distinction is instigated in the conductor because of the revolution of the rotor. This initiated voltage distinction is transverse to the electric flow.

IOP Conf. Series: Materials Science and Engineering **937** (2020) 012017 doi:10.1088/1757-899X/937/1/012017

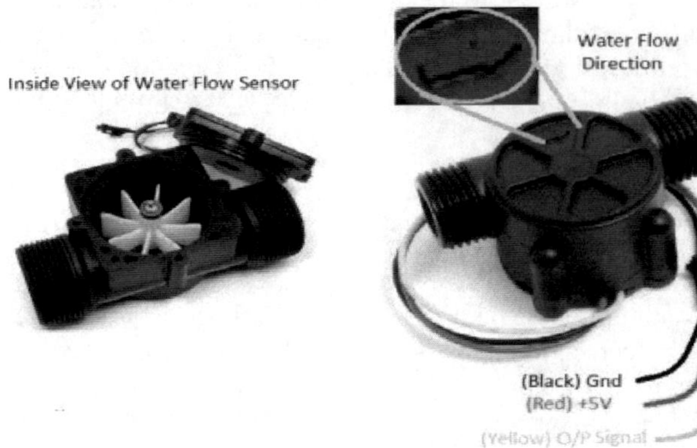

Figure: 1.Water flow sensor

3.2. Arduino

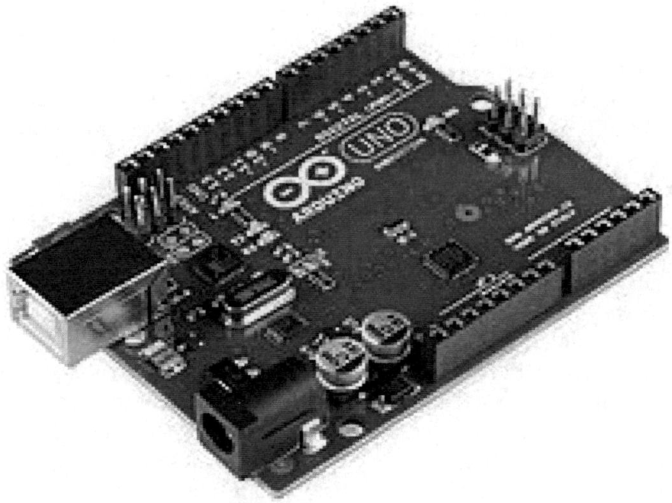

Figure: 2 Arduino-UNO

3.3 Arduino UNO Pinout

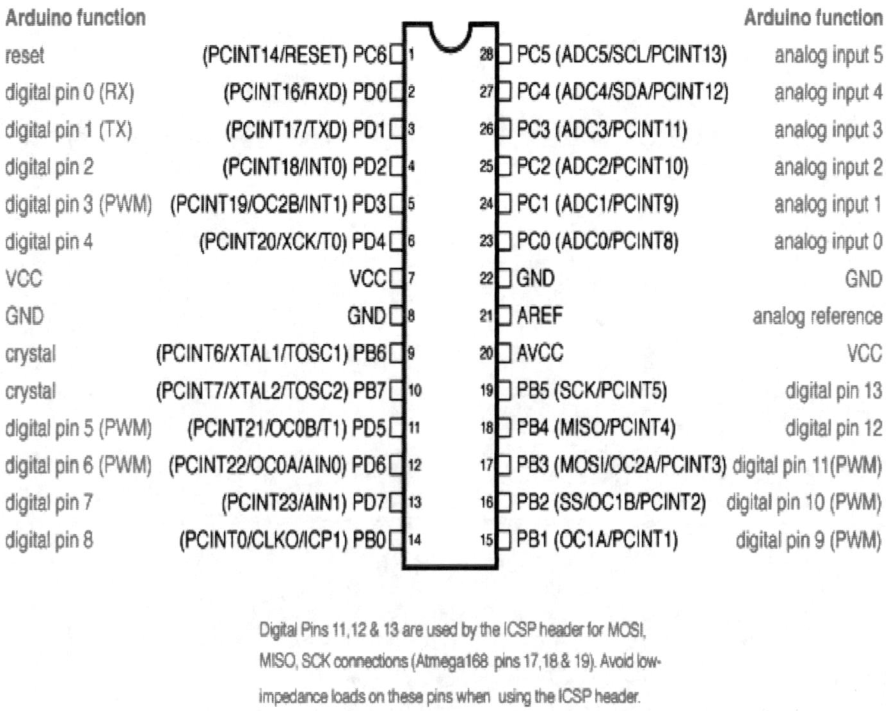

Figure: 3. PIN configuration

4. Implementation

GPS (Global Positioning System) which is used to locate the area of the underground fresh water availability by storing the entire data by this project in the Cloud Computing Technology. The Application provides the Guidance with the Data of water in that area visually to the people in "Google Map"and it help energy conserved and utilized efficiently [20]. The limit of the water is set according to the water availability and quality of water to alarm over usage information to the people and Water Board by measuring the water flow using "Arduino & Sensors". Peoples are always For instance deforestation, depleted wetlands, and urban improvement cause water to overflow a lot quicker than it would normally. This prompts diminished charging of the fundamental aquifer. The surface water collaboration referenced before extra concerns incorporate the expanded expense of water and land subsidence. The expense of water increments as the profundity of the water table increments because of expenses related with the vitality required to lift the water further. Land subsidence, or soaking in, is brought about by the loss of help underground. Groundwater extraction can cause subsidence by leaving a void where the water was and drying the dirt permitting it to shrivel and settle. As groundwater is ceaselessly removed from the dirt the probability builds that the ground will settle to occupy the unfilled spaces left behind. This settling can be the wellspring of significant harm to the nearby networks remembering breaks for establishments, dividers, streets or possibly even sinkholes. It is necessary to determine the Powerful groundwater watching is the best way to deal with secures the local system, ensure a dependable and sensible groundwater gracefully, and guarantee the sum available for at some point later [21]. Help line provision with Water Board Advice also available. The GPS (Global Positioning System) is utilized to recognize the area of the water body with the subtleties is associated with the Google Map secretly. This configuration makes it possible for the Raspberry pi sensor sent the data to the Government when the water is over utilized or squandered [22].

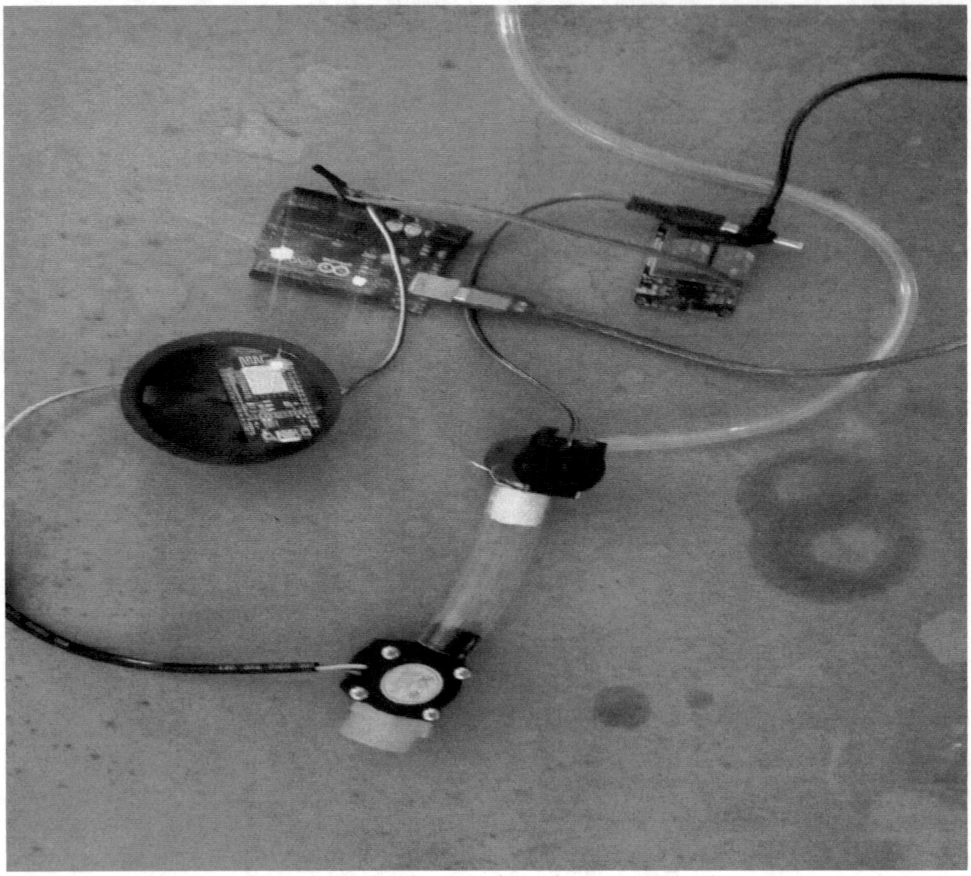

Figure: 4. Hardware implementation of water monitoring system

5. Results

The consumer can view their water consumption through the application shown below and they can communicate to water board through it. It will useful for consumer and immediately they get consumption of water. The flow rate also can able to monitor with help of internet of things. The proposed method is simulated and hardware is implemented [23].

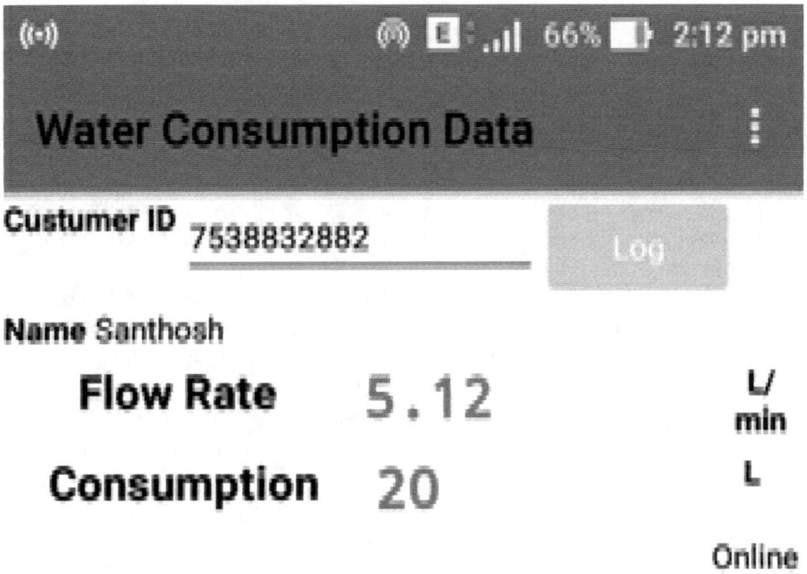

Figure: 5. Results of the Arduino based water monitoring system

6. Conclusion

In this paper, a model brilliant groundwater level observing framework utilizing Arduino is introduced. For this a few sensors are utilized. The gathered information from the all the sensors are utilized for investigation reason for better arrangement of water level just as stream rate checking can be executed. So this application will be the best challenger continuously checking and control framework and use to comprehend all the groundwater level observing related issues.

7. References

[1] Alessio B, Walter D, Valerio P, Antonio P, "Integration of Cloud computing and Internet of Things: A survey", Futur Gener Comput Syst 56: pp. 684–700, 2016.
[2] K. Xu, Y. Qu, K. Yang,"A tutorial on the Internet of Things: From a heterogeneous network integration perspective",IEEE Netw., Vol. 30, No. 2, pp. 102-108, 2016
[3] Prachet Verma, Akshay Kumar, Nihesh Rathod, Pratik Jain, Mallik Arjun, Renu Subramanian,"Towards an IoT based water management system for a campus", IEEE 2015
[4] Al-Fuqaha A et al,"Internet of Things: A Survey on Enabling Technologies, Protocols, and Applications", IEEE Commun Surv Tutorials 17(4), pp. 2347–2376, 2015
[5] Thinagaran Perumal, Md Nasir Sulaiman, Leong.C.Y, "Internet Of Things (IoT) enabled water monitoring system", 2015 IEEE 4th Global Conference on Consumer Electronics (GCCE)..
[6] Prachet Varma, Akshay Kumar, Nihesh Rathod, Pratik Jain, Mallikarjun S, Renu Subramaniyam, Bhardhwaj Amrutur, M.S.Mohan Kumar, Rajesh Sundresan,"IoT based water management System for a Campus IEEE", IEEE First International Smart Cities Conference (ISC2), 2015.
[7] Perumal, T.; Sulaiman, M.N.; Mustapha, N.; Shahi, A.; Thinaharan, R., "Proactive architecture for Internet of Things (IoTs) management in smart homes," Consumer Electronics International Research Journal of Engineering and Technology (IRJET) e-ISSN: 2395-0056 Volume: 05 Issue: 10 | Oct 2018 www.irjet.net p-ISSN: 2395-0072 © 2018, IRJET | Impact Factor value: 7.211 | ISO

9001:2008 Certified Journal | Page 1173 (GCCE), 2014 IEEE 3rd Global Conference on, pp.16,17, 7-10 Oct. 2014.

[8] V. Jayakumar, DC.Kumaresan, R.Karthikeyan "Wind Energy Conversion System and Solar PV Integration" International Journal of Innovative Technology and Exploring Engineering (IJITEE) ISSN: 2278-3075, Volume-8 Issue-7, 1595- 1600, May, 2019.

[9] V. Jayakumar, PL. Somasundaram "Implementation of single phase improved inverter for PV source", International journal of pure and applied mathematics JARIIE-ISSN(O)-2395-4396 Volume 4 No. 3 2018, 118 – 126

[10] S K Saranya & Dr R Karthikeyan "Security for Smart Distribution Grid by Using Wireless Communication", International Journal of Innovative Research in Computer and Communication Engineering (IJIRCCE), Volume 02 Special Issue 01 March 2014, pp: 01 – 09, ISSN: 2320-9801.

[11] Saima Maqbool, Nidhi Chandra,"Real Time Wireless Monitoring and Control of Water Systems using Zigbee 802.15.4", 5th International Conference on Computational Intelligence and Communication Networks, 2013.

[12] Asaad Ahmed Mohammed Ahmed Eltaieb, Zhang Jian Min,"Automatic Water Level Control System", International Journal of Science and Research (IJSR) 2013.

[13] A. T. Sankara Subramanian, P. Sabarish, M. D.Udayakumar and T. Vishnu kumar," Performance Analysis of Various Photovoltaic Configurations under Uniform Shading and Rapid Partial Shading Formations", Biosc.Biotech.Res. Comm. Special Issue Vol 13 No (3) 2020 Pp-185-192.

[14] P. Sabarish et al 2019 IOP Conf. Ser.: Mater. Sci. Eng. 623 012011.

[15] M D Udayakumar et al 2019 IOP Conf. Ser.: Mater. Sci. Eng. 623 012018.

[16] P. Sabarish, A. T. Sankara Subramanian, S. Murugesan and V. Sureshkumar, "A New Methodology of Arterial Blood Clot Removal Using Bio Molecular Devices for ATPase Nuclear Motors." Biosc.Biotech.Res.Comm. Special Issue Vol 13 No (3) 2020 Pp-197-201.

[17] Karthick, R and Sundararajan, M. (2017), "Design and Implementation of Low Power Testing Using Advanced Razor Based Processor," International Journal of Applied Engineering Research, 12.

[18] Karthick, R and Sundararajan, M. (2018), "A novel 3-D-IC test architecture-a review," International Journal of Engineering and Technology (UAE).

[19] A Nazar Ali, D Sivamani, R Jaiganesh M Pradeep (2019), Solar powered air conditioner using BLDC motor, IOP Conference Series: Materials Science and Engineering, vol. 23.

[20] V Venkatesh, A Nazar Ali,, R Jaiganesh. V Indiragandhi (2019), Extraction and conversion of exhaust heat from automobile engine in to electrical energy Energy, IOP Conference Series: Materials Science and Engineering, vol. 23.

[21] A Ali Nazar, R Jayabharath, MD Udayakumar (2014) ,An ANFIS Based Advanced MPPT Control of a Wind-Solar Hybrid Power Generation System,Iinternational review of modelling and Simulations. vol.7, no. 4, pp. 638–643.

[22] A.Nazar Ali and R. Jayabharath (2014), "Performance Enhancement of Hybrid Wind/Photo Voltaic System Using Z Source Inverter with Cuk-sepic Fused Converter." Research Journal of Applied Sciences, Engineering and Technology7.19 pp. 3964-3970.

[23] C.Kalavalli, S.R.Paveethra, S.Murugesan, Dr.A.Nazar Ali, (2020), Design And Implementation Of High Efficiency H6 PV Inverter with Dual Axis Tracking, International Journal of Scientific & Technology Research, Vol 9, issue 02, pp. 4728-31

IOP Conf. Series: Materials Science and Engineering **937** (2020) 012018 doi:10.1088/1757-899X/937/1/012018

Comparison of Emulated Inertia controller with Synchronous Generator

Ratnam Kamala Sarojini[1], K Palanisamy[1]

School of Electrical Engineering, Vellore Institute of Technology, Vellore, Tamil Nadu, India.

kpalanisamy@vit.ac.in

Abstract. The emulated inertia controller (EIC) is designed to emulate the inertia to the power electronic inverter. The EIC helps to integrate the large-scale PV into the grid without affecting the stability. This paper demonstrates the design of EIC control. This paper the highlights the comparison of EIC with the traditional synchronous generator (SG). The frequency response of both EIC and SG are compared for different scenarios.

1. Introduction

The usage of renewable energy sources (RES) in the power generation sector is tremendously increasing due to concern on carbon emissions. The renewable power generation such as PV and wind are integrated to the grid though the power electronic converters. The power electronic based renewable power generators decouple the load from the source. The more penetration of RES in the power generation affects the frequency stability of the system. In general, the SG deals with frequency stability issues in the traditional power system. Whenever there is a change in the system frequency the kinetic energy stored in the rotor of the SG injects to balance the system in the inertial response. As the power generation from the RES is increasing, the inertia constant of the grid is decreasing. The less inertia constant results in more frequency deviation and large rate of change of frequency. To mitigate the inertia related issues, the virtual inertia concept is proposed in the literature. The main idea behind the virtual inertia is to emulate the inertial characteristics of the SG to the power electronic inverter. Several emulated inertia control (EIC) techniques are developed in the literature to address the issue [1].

In [2], the complete mathematical model of the SG is taken to design the inertia emulation algorithm, and it results in numerical instability. In [3], the second-order model of SG is considered to design the control algorithm; here, the calculation of ROCOF leads to complexity in the algorithm. The EIC control scheme is an appropriate solution to the low inertia power system. The EIC mimics the SG characteristics, by regulating the switching pattern of the inverter to offer the virtual inertia and to decrease frequency nadir during disturbances. In [4], the small-signal modelling of EIC with ideal grid-connected inverter is analysed. The implementation of the EIC algorithm in the power system is reported in [5]. The use of inertia emulation in the microgrid is analysed in [6]. Inertia emulation using hybrid energy storage consists of battery and supercapacitor for the single-phase rooftop system is proposed in [7-20]. However, earlier investigation on EIC only emphasis on the controller design, grid-connected applications and hardly compares inverter with EIC and SG. The objective of this paper is to compare the physical structure and performance of the EIC with power electronic inverter with SG [21-30].

The rest of this paper is organized as follows. Section 2 explains the structure of RES based power generation and traditional SG. Section 3 describes the control strategy of EIC. Section 4 presents fundamental comparison of the SG and EIC with inverter. Section 5 deals with the results and discussions. Finally, section 6 gives the conclusion.

2. Structure of Power Generation from RES with Grid-tied Inverters and traditional SG

The general structure of power generation from the renewable and conventional sources is detailed in Figure 1. The Figure 1 consists of the primary energy source, prime mover, energy conversion unit and grid [31]. The primary energy source for the RES based system might be solar, wind, battery, fuel cell etc. whereas in case of traditional energy storage system for the primary energy storage system is Hydraulic, coal etc. The prime mover for the RES based system is DC-DC convertors, for the Conventional system the prime mover is turbine [32-34]. The next stage is the energy conversion unit. The energy conversion unit for the RES based system consists of DC bus capacitor, grid-tied inverter, and a filter. The energy conversion unit for the conventional system consists of the rotor and a synchronous generator.

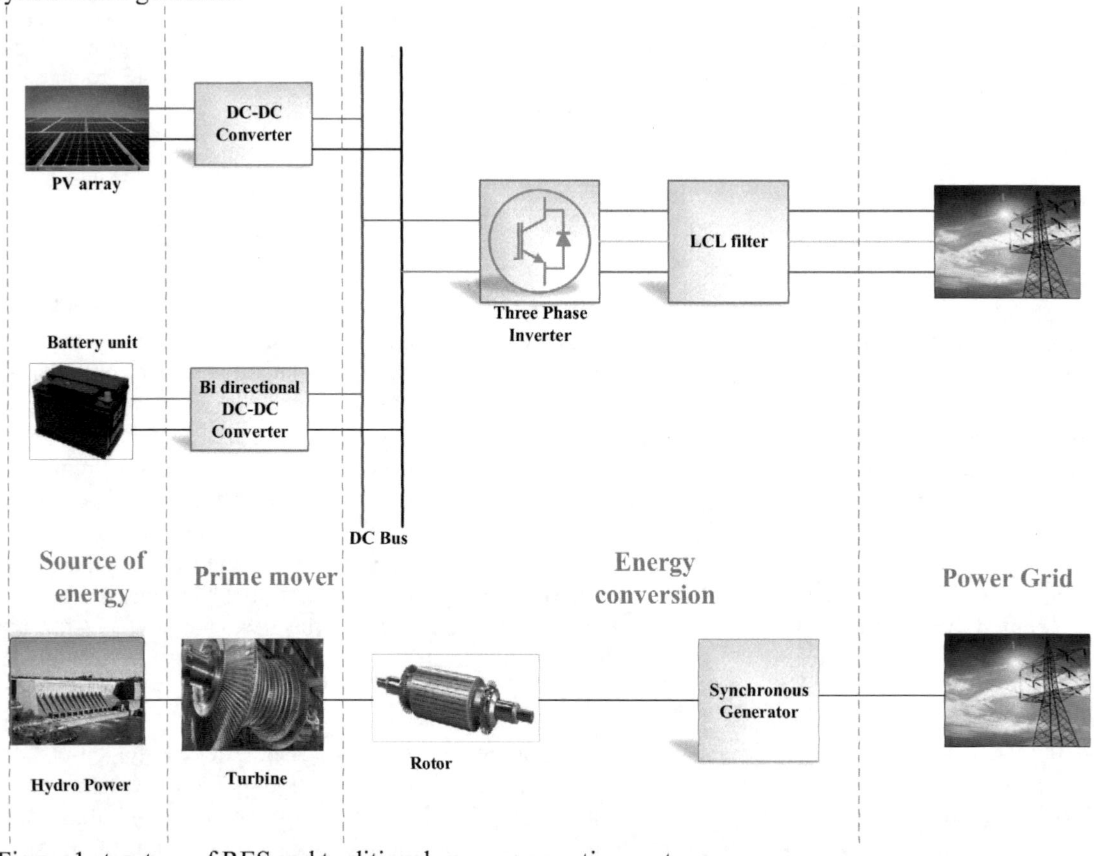

Figure 1 structure of RES and traditional power generation system

3. Control strategy of EIC

EIC enhances frequency stability of the power system and allows more power generation from renewables into the system [8]. EIC uses the swing equation to emulate the inertia. EIC resembles the

IOP Conf. Series: Materials Science and Engineering **937** (2020) 012018 doi:10.1088/1757-899X/937/1/012018

SG characteristics to enhance the frequency stability. Moreover, it emulates inertia to the power electronic inverter like SG to support frequency stability of stand-alone microgrid. The control diagram for the EIC is illustrated in Figure 2. The main concept of EIC is to calculate the extra power required for the frequency deviation and to add this extra power to set point. The EIC responds to frequency variations and alters the load angle. The electrical appearance of both EIC and SG mathematically equivalent. Hence, the inertial response of the EIC is similar to the SG.

The EIC developed based on the swing equation to alter the load angle of the inverter θ. The fundamental concept of EIC is shown in (1)-(3).

$$\left(P_{Set}-P_{Cal}\right)/\,\omega-D_P(\omega-\omega_r)=J\frac{d\omega}{dt} \tag{1}$$

$$\frac{d\theta}{dt}=\omega \tag{2}$$

$$\frac{1}{k}\int(Q_{Set}-Q_{Cal})+D_q(V_m^*-V_m)dt=E \tag{3}$$

The EIC is divided into two control parts. The first part of EIC is used to alters the load angle based on the frequency deviation. The damping term for the given frequency deviation is included to the difference in the set torque (calculated with set power) to the torque output (calculated with output power). The integrator, along with inertia term alters the load angle based on the difference in torque terms. Whereas in the second part of EIC regulates the voltage amplitude. Depends on the reactive power variation, the voltage amplitude can be altered. The voltage amplitude and the load angle are combined to get the reference voltage for the inverter. The reference voltage generated by the EIC is fed to PWM to produce the pulses for the inverter. Whenever there is a change in the demand, the EIC alters the load angle to inject/absorb the power with the help of energy stored in the battery. Implementation of EIC in a stand-alone microgrid reduces the frequency deviations and ROCOF.

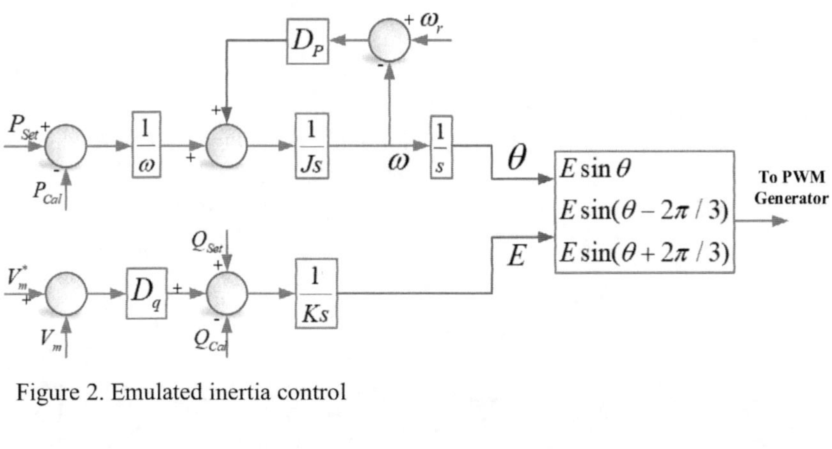

Figure 2. Emulated inertia control

4. Fundamental correlation between EIC with inverter and SG

The inverter with EIC control and SG shows the close resemblances in physical structures, parameter characteristics, mechanism of energy transmission and mathematical models.

4.1. Fundamental comparison in physical structure and parameter

The physical structures for both RES based generation and traditional generation are shown in Figure 1. The detailed physical structure for both the generation is clearly discussed in Section 2. The summary of the relevance of the EIC with inverter and the SG is tabulated in Table 1.

Table I – Relevance of EIC with inverter and SG elements

	Inverter with EIC	**SG**
Source of energy	RES, Battery, SC	Hydraulic, Coal. nuclear
Prime mover	Power electronic converters	Turbines
Energy conversion device	Capacitor	Rotor

The parameters of the EIC with inverter and SG are logically related. The inertia of the rotor of SG is related to the capacitance of the capacitor of the power electronic inverter. The energy stored in the rotor of SG is used in inertial response to compress the frequency deviations in traditional power system. Whereas the energy stored in capacitor/ energy in storage element is used to reduce the frequency deviations. The relevance of the different parameters, EIC with inverter and SG is summarised in Table 2.

Table 2- Relevance of EIC with inverter and SG parameters

Inertia with EIC	**Traditional SG**
Capacitance of capacitor	Inertia of rotor
Filter	Stator of the SG
DC capacitor voltage	Rotor angle
Output voltage of the inverter	SG internal voltage
Output voltage of the filter	Terminal voltage of SG
Energy stored in capacitor	Energy stored in rotor

4.2. Similarity of the mechanism of energy transmission and Its mathematical model

There is a similarity in the mechanism of the energy transmission in the EIC with inverter and in SG is shown in (1). From (1) the dynamic behaviour of rotor of SG and capacitor of the inverter are first order differential equations. Hence, the energy transfer mechanism for the both systems in frequency deviations are similar to maintain the system in equilibrium. The control mechanism of the EIC is to attain power balance and to maintain the stability. The simulation parameters are shown in Table 3.

5. Results and discussions

The simulations are carried out using the Matlab/Simulink environment. The simulations are performed in order to compare the performance of EIC and SG model. The ratings of Power electronic inverter with EIC and SG are considered to be of same rating. To exhibit the similarity of the EIC with inverter and SG several cases were tested. The simulation parameters are shown in Table 3.

Table 3: Simulation parameters

Parameters	Values
Reference frequency (f)	50 Hz
Reference Voltage (V^*)	220 V
Rated DC bus voltage (V_{DC})	500 V
PV cell	$V_{OC} = 37.3$ V, $I_{SC} = 8.2$ A, $V_m = 30.3$, $I_m = 7.5$, $N_S = 15$, $N_P = 2$.
DC bus Capacitor	3300 μF
Inverter side inductance (L_1)	2.1 mH

IOP Conf. Series: Materials Science and Engineering **937** (2020) 012018 doi:10.1088/1757-899X/937/1/012018

Load side inductance (L_2)	0.8 mH
Filter capacitance (C)	12 μF
Frequency Drooping Coefficient (D_P)	50
Inertia coefficient (J)	0.058 Kg.m^2
Voltage drooping coefficient (D_q)	120
Gain (k)	1000 A. s
SG capacity	10 KVA

5.1. Response to sudden load increase

In this case the frequency response of the inverter with EIC control and SG are compared under sudden increase in load. Initially the system is running in steady state and the frequency is maintained at 50 Hz. At t =2 s the sudden 10% increasing of load demand is simulated. The frequency response of inverter is replicating the frequency response of the SG as shown in Figure 3.

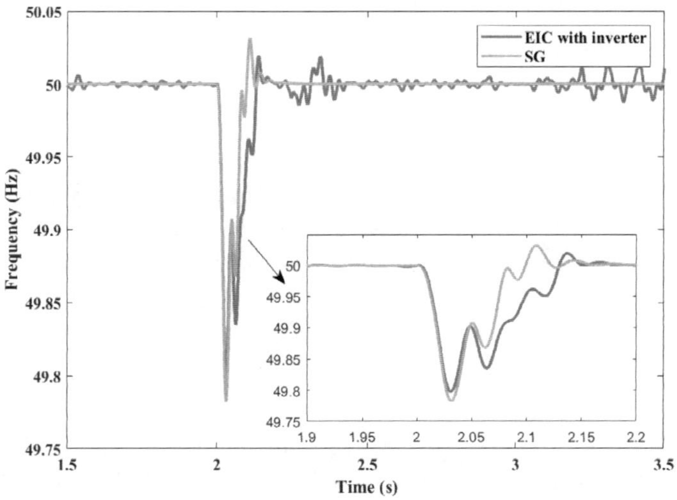

Figure 3. Frequency response under sudden load increase

5.2. Response to sudden load decrease

In this case the frequency response of the inverter with EIC control and SG are compared under sudden decrease in load. The system reaches to steady state after sudden increasing of demand and the frequency is maintained at 50 Hz. At t = 4 s the sudden 10% increasing of load demand is simulated. The frequency response of inverter is replicating the frequency response of the SG as shown in Figure 3.

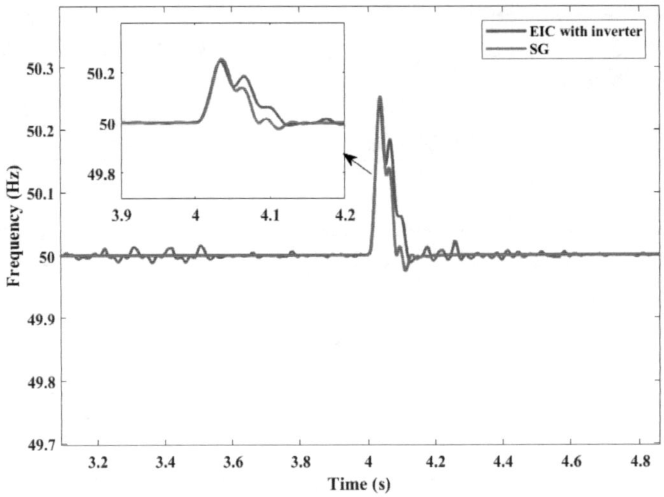

Figure 4. Frequency response under sudden load decrease

From simulation results it was concluded that inverter with EIC control is able to emulate the frequency response like SG. Hence, low inertia in power grids can addressed by using the EIC along with power electronic inverters.

6. Conclusion

In this paper, an EIC was modelled to emulate the inertial response of the SG. The transient behaviour of the grid connected inverter with EIC is compared with SG. It can be concluded that the VSG is showing similar characteristics of SG and can be used in the large penetration of RES system to maintain the stability. On the other hand, power rating of the inverter needs to be increased. This paper summaries the comparison of EIC with inverter and SG in terms of the physical structure and parameters. This paper proved that the EIC controlled inverter exhibit the relevant dynamic response same as of SG.

References

[1]. Ratnam KS, Palanisamy K, and Yang G 2020 Future low-inertia power systems: Requirements, issues, and solutions - A review *Renew. Sustain. Energy Rev* **124** 109773

[2]. Sitharthan R, Geethanjali M and Pandy TKS 2016 Adaptive protection scheme for smart microgrid with electronically coupled distributed generations *Alexandria Engineering Journal* **55(3)** 2539-2550

[3]. Fathima AH, and Palanisamy K 2014 Battery energy storage applications in wind integrated systems—a review *IEEE International Conference on Smart Electric Grid* 1-8

[4]. Prabaharan N and Palanisamy K 2015 Investigation of single-phase reduced switch count asymmetric multilevel inverter using advanced pulse width modulation technique *International Journal of Renewable Energy Research* **5(3)** 879-890.

[5]. Jerin ARA, Kaliannan P and Subramaniam U 2017 Improved fault ride through capability of DFIG based wind turbines using synchronous reference frame control based dynamic voltage restorer. *ISA transactions* **70** 465-474

[6]. Sitharthan, R, Sundarabalan CK, Devabalaji KR, Nataraj SK and Karthikeyan M 2018 Improved fault ride through capability of DFIG-wind turbines using customized dynamic voltage restorer *Sustainable cities and society* **39** 114-125

[7]. Prabaharan N and Palanisamy K 2016 A single-phase grid connected hybrid multilevel inverter for interfacing photo-voltaic system *Energy Procedia* **103** 250-255

[8]. Palanisamy K, Mishra JS, Raglend IJ and Kothari DP 2010 Instantaneous power theory based unified power quality conditioner (UPQC) *IEEE Joint International Conference on Power Electronics, Drives and Energy Systems* 1-5

[9]. Sitharthan R and Geethanjali M 2017 An adaptive Elman neural network with C-PSO learning algorithm-based pitch angle controller for DFIG based WECS Journal of Vibration and Control **23(5)** 716-730

[10]. Sitharthan R and Geethanjali M 2015 Application of the superconducting fault current limiter strategy to improve the fault ride-through capability of a doubly-fed induction generator–based wind energy conversion system *Simulation* **91(12)** 1081-1087

[11]. Sitharthan R, Karthikeyan M, Sundar DS and Rajasekaran S 2020 Adaptive hybrid intelligent MPPT controller to approximate effectual wind speed and optimal rotor speed of variable speed wind turbine *ISA transactions* **96** 479-489

[12]. Sitharthan R, Devabalaji KR and Jees A 2017 An Levenberg–Marquardt trained feed-forward back-propagation based intelligent pitch angle controller for wind generation system *Renewable Energy Focus* **22** 24-32

[13]. Sitharthan R, Sundarabalan CK, Devabalaji KR, Yuvaraj T and Mohamed Imran A 2019 Automated power management strategy for wind power generation system using pitch angle controller *Measurement and Control* **52(3-4)** 169-182

[14]. Sundar DS, Umamaheswari C, Sridarshini T, Karthikeyan M, Sitharthan R, Raja AS and Carrasco MF 2019 Compact four-port circulator based on 2D photonic crystals with a 90° rotation of the light wave for photonic integrated circuits applications *Laser Physics* **29(6)** 066201

[15]. Sitharthan R, Parthasarathy T, Sheeba Rani S and Ramya KC 2019. An improved radial basis function neural network control strategy-based maximum power point tracking controller for wind power generation system *Transactions of the Institute of Measurement and Control* **41(11)** 3158-3170

[16]. Rajesh M and Gnanasekar JM 2017 Path observation based physical routing protocol for wireless ad hoc networks *Wireless Personal Communications* **97(1)** 1267-1289

[17]. Palanisamy K, Varghese LJ, Raglend IJ and Kothari DP 2009. Comparison of intelligent techniques to solve economic load dispatch problem with line flow constraints *IEEE International Advance Computing Conference* 446-452

[18]. Sitharthan R, Ponnusamy M, Karthikeyan M and Sundar DS 2019 Analysis on smart material suitable for autogenous microelectronic application *Materials Research Express* **6(10)** 105709

[19]. Rajaram R, Palanisamy K, Ramasamy S and Ramanathan P 2014 Selective harmonic elimination in PWM inverter using fire fly and fireworks algorithm *International Journal of Innovative Research in Advanced Engineering* **1(8)** 55-62

[20]. Sitharthan R, Swaminathan JN and Parthasarathy T 2018 March. Exploration of wind energy in India: A short review *IEEE National Power Engineering Conference* 1-5

[21]. Karthikeyan M, Sitharthan R, Ali T and Roy B 2020 Compact multiband CPW fed monopole antenna with square ring and T-shaped strips *Microwave and Optical Technology Letters* **62(2)** 926-932

[22]. Sundar D Sridarshini T, Sitharthan R, Madurakavi Karthikeyan, Sivanantha Raja A, and Marcos Flores Carrasco 2019 Performance investigation of 16/32-channel DWDM PON and long-reach PON systems using an ASE noise source *In Advances in Optoelectronic Technology and Industry Development: Proceedings of the 12th International Symposium on Photonics and Optoelectronics* 93

[23]. Sitharthan R and Geethanjali M 2014 Wind Energy Utilization in India: A Review *Middle-East J. Sci. Res.* **22** 796–801 doi:10.5829/idosi.mejsr.2014.22.06.21944

[24]. Sitharthan R and Geethanjali M 2014 ANFIS based wind speed sensor-less MPPT controller for variable speed wind energy conversion systems *Australian Journal of Basic and Applied Sciences* **8**14-23

[25]. Jerin ARA, Kaliannan P, Subramaniam U and El Moursi MS 2018 Review on FRT solutions for improving transient stability in DFIG-WTs *IET Renewable Power Generation* **12(15)** 1786-1799

[26]. Prabaharan N, Jerin ARA, Palanisamy K and Umashankar S 2017 Integration of single-phase reduced switch multilevel inverter topology for grid connected photovoltaic system *Energy Procedia* **138** 1177-1183

[27]. Rameshkumar K, Indragandhi V, Palanisamy K and Arunkumari T 2017 Model predictive current control of single phase shunt active power filter *Energy Procedia* **117** 658-665

[28]. Fathima AH and Palanisamy K 2016 Energy storage systems for energy management of renewables in distributed generation systems *Energy Management of Distributed Generation Systems* 157

[29]. Rajesh M 2020 Streamlining Radio Network Organizing Enlargement Towards Microcellular Frameworks *Wireless Personal Communications* 1-13

[30]. Subbiah B, Obaidat MS, Sriram S, Manoharn R and Chandrasekaran SK 2020 Selection of intermediate routes for secure data communication systems using graph theory application and grey wolf optimisation algorithm in MANETs IET *Networks* doi:10.1049/iet-net.2020.0051

[31]. Singh RR and Chelliah TR 2017 Enforcement of cost-effective energy conservation on single-fed asynchronous machine using a novel switching strategy *Energy* **126** 179-191

[32]. Amalorpavaraj RAJ, Palanisamy K, Umashankar S and Thirumoorthy AD 2016 Power quality improvement of grid connected wind farms through voltage restoration using dynamic voltage restorer *International Journal of Renewable Energy Research* **6(1)** 53-60

[33]. Singh RR, Chelliah TR, Khare D and Ramesh US 2016 November. Energy saving strategy on electric propulsion system integrated with doubly fed asynchronous motors *IEEE Power India International Conference* 1-6

[34]. Singh RR, Mohan H and Chelliah TR 2016 November. Performance of doubly fed machines influenced to electrical perturbation in pumped storage plant-a comparative electromechanical analysis *IEEE 7th India International Conference on Power Electronics* 1-6

IOP Conf. Series: Materials Science and Engineering **937** (2020) 012019 doi:10.1088/1757-899X/937/1/012019

EFFECT OF MOTOR VIBRATION PROBLEMS ON POWER QUALITY OF WATER PUMPING AT RESIDENCY

Nabanita Dutta[1], Palanisamy Kaliannan[1], Umashankar Subramaniam[2]

[1]Department of Energy and Power Electronics, School of Electrical Engineering, VIT Vellore, Vellore 632014, India; nabajhilikbarbi@gmail.com, kpalanisamy@vit.ac.in,
[2]College of Engineering, Prince Sultan University Riyadh, Saudi Arabia, shankarums@gmail.com

Abstract. Water pumping system is an essential machinery for residential usage. Around 150 litres of water are used by a person in India daily for their daily usage. For this reason the water pumping is used regularly. Due to dynamic performance of the pumping unit pump can be operated in both no load and load condition. Though the pumping system was controlled by throttling drive in the early stage but in modern technological advancement time pumps are controlled by variable frequency drives (VFD). VFD provides the pump variable duty requirements with significant energy saving and improved reliability. In every pumping system due to dynamic forces of hydraulic and mechanical component some vibrations are seen but sometimes this vibrations may be excessive which can damage the components of the pump. The excessive vibrations indicate some problems or failures in the system. Power quality issues are one of those problems which are seen during the vibration of the pump motor and induction motor connected with pump. This paper presents the power quality issues in the pump motor in different control method like vvc+ and v/f control method.

1. Introduction

Pumps are of various categories like centrifugal, positive displacement pump etc. [1]. but most of the pumps are used for industrial and residential purposes are centrifugal pump [2]. This centrifugal pump can be controlled by VFD drives in order to save the energy. VFD drives has three controller technology like VVC+, V/f and flus sensor less. In the latest motor control technology, installation of variable frequency drive (VFD) is a cost-effective solution to achieve significant energy savings, and when the drive is applied for pumping system, it can save 10% [13] of the energy used previously. Controlling the motor speed is one way of saving energy, and 68% of the electrical energy is used to operate flow loads like fan, blowers, compressors etc. [3]. by constant speed. The adjustments are needed where output flow fluctuates most of the time. Flow can be controlled by valve control, inlet damper control, inlet valve, outlet valve control etc. VFD is a power electronic device which converts one frequency of alternating current to another frequency for running the AC motor in variable speed to save the energy. Power quality issues are the major issues in the pumping system controlled by VFD drives because of its adjustable speed drives nature and programmable controllers. Continuous operation in resonance condition also creates excessive vibration in the system. The power quality issues are sag, swell, flickering and harmonics problems [4]. Here hardware based experimental research has been done to analysis the effect of THD, flickering, 5th and 7th harmonics problems in the pumping system. Master follower cascaded pump has been used which is controlled by VLT aqua drive and flow rate will be added with same head value when all the pumps are operated in parallal. The block diagram of multistage parallal pumping system has been shown in figure 1.

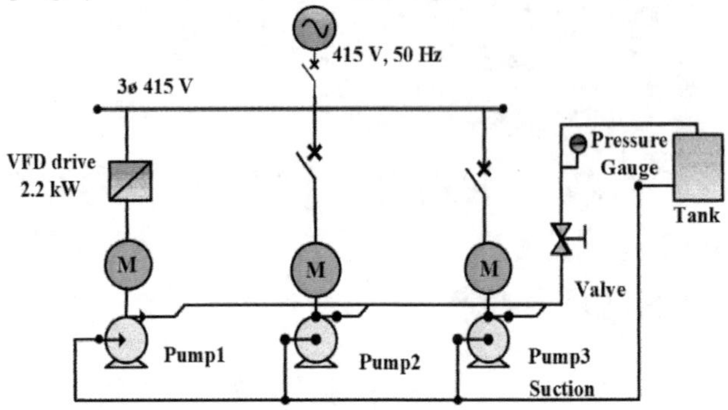

Figure 1: VFD based parallal pumping system

2. Vibration Analysis

Vibration analysis is a common technique now a days used in Industries. The vibration condition occurs due to vibration of pumps, when there is a fault in the installation of the motors or uneven grounds where motors need to be installed and it results in either mechanical or electrical imbalance of the motor. Bad motor vibrations affects the overall equipment efficiency (OEE), which is an important factor that need to be considered while performing motor operations, as it helps to prevent unexpected failures of the equipment [5].

Maximum vibration level measured at no load is less than compared at full load. Pump loading causes increase in vibration level and further causes pump bearing problems [6]. Due to unbalance moving parts within the pumping system, interaction of fluids with the pumps connecting with pipes, movements of pipelines causes vibration [7]. There are various reasons are available for unwanted vibrations like impeller unbalance, hydraulic unbalance, voltage unbalance of pump motor, bearing problems, cavitation and water hammering issues etc. This vibrational problem can be both mechanical and hydraulic problem in the pump. If impeller unbalance, bearing and seal problem creates vibration it will be categorized as mechanical failure and hydraulic failure occurs when cavitation and water hammering issues are seen [8].

Vibration problems can also occur at any time in the installation or operation of a motor. When they occur it is normally critical that one reacts quickly to solve the difficulty. If not solved quickly, one could either expect long term damage of the motor or immediate failure. To solve a vibration problem one must differentiate between cause and effect. Perhaps the support structure is just not stiff enough to minimize the displacement [10].

2.1. Power Quality Issues

The electrical system fails to meet the demand and loads are not operating properly, if any parameters of electrical supplies creates any troubleshoots all these analysis is called power system analysis. The quality of electric power is subjected to set of parameters like changes of voltage and current, continuity of service, transient voltage and current, harmonic contents of waveforms of AC power [9]. The current and voltage at various operating conditions is known as power quality. Because of unhealthy power quality, problems occur such as voltage disturbances for example voltage sags, voltage swells, voltage surges and impulse. Flickering, harmonic invasion and power factor quality. The Harmonics contents and THD can be calculated by measuring the voltage and current signal and their measured values represent the status of power quality. The harmonics have several effects like torque pulsation, acoustic noise and increased power losses. Types of harmonics present in Induction motors that is space harmonics and time harmonics. Due to the different phase windings interaction Space harmonics are generated. The power supply is sinusoidal and that can be reduced by applying a proper machine design [10].

2.2. Total Harmonic Distortion

The measurement of harmonic distortion present in the signal and the distortion is defined by the ratio of sum of the powers of all harmonic components and the power of fundamental frequency. This is called total harmonic distortion (THD) [11]. The series of harmonic components that represent a distorted waveform are often described by the total harmonic distortion (THD). THD can be calculated from the following equations;

$$\text{THD} = \sqrt{\frac{\sum_{n=2} V_n}{V_1}} \qquad (1)$$

Where n is the order of harmonic, V_1 is the rms value of fundamental voltage and V_n is the rms value of voltage harmonic with order n. The voltage distortion limits in IEEE 519 are fairly straight forward, as shown in Table 1. There are only three levels recommended, depending on the voltage level.

Table 1. Voltage distortion limits (IEEE STD 519_2014)

Bus voltage V at PCC	**Individual harmonic (%)**	**Total Harmonic distortion (%)**
$V \leq 1.0$ kV	5.0	8.0
1 kV $\leq V \leq$ 69 kV	3.0	5.0
69 kV $\leq V \leq$161 kV	1.5	2.5
161 kV $< V$	1.0	1.5

2.3. Flicker Problem

The network voltage power supply varies with respect to time due to perturbations. Electrical power quality can be decorated due to interaction of electrical loads with the networks. High power loads can varies and fluctuating in any cases. High power load can draw fluctuating current like large motor drives, arc furnaces, low, variations of voltage with low frequency etc.

The electrical instability can be seen in the system during flickering. Flicker is expressed in terms of two parameters like short term flicker who's PST for severity is 10 min and short term severity is possible to assess and for long term severity time requires 2 h in which time long term severity PLT is assessed. The equations will show the process of calculation of flicker [12].

$$P_{ST} = \sqrt{0.0314P_{0.1} + 0.0525P_{1Ss} + 0.0657P_{3s} + 0.28P_{10s+0.08}P_{50s}} \qquad (2)$$

$$P_{LT} = \sqrt[3]{\frac{\sum_{i=1}^{N} P_{ST_i}^3}{N}} \qquad (3)$$

P_{ST} period number within P_{LT} observations is N. Measurements would be required for calculation of P_{LT}.

3. Hardware setup:

Using VLT AQUA Drive FC 202, the experiment has been carried out and the readings have taken at PCC (Point of Common Coupling)[Figure 2]. The pressure is set in the drive and the mode of control is selected and the valve position is changed starting with fully open, semi open or semi closed and fully closed [16] [18] [20]. The motor chosen here is Induction motor. VLT Aqua drive is connected with the system for the higher performance of AC motor for wastewater applications [14] [15]. The device has a wide range of standard features, which help to improve the performance of the system [21-33]. Freely programmable warnings and alerts help the adaptation to the application of perfect system integration. Hardware results also shown in figure 3.

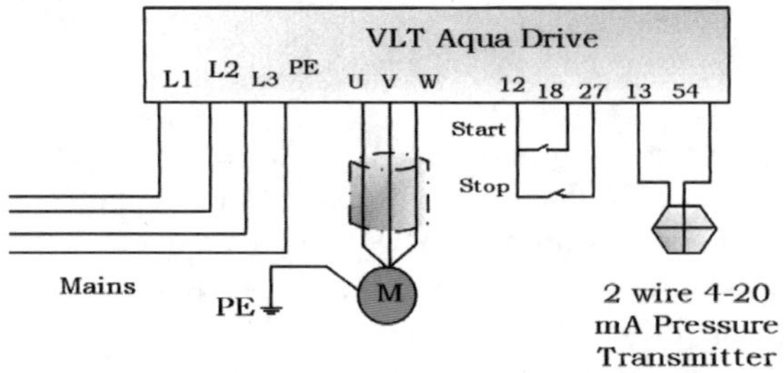

Figure 2: VLT Aqua drive schematic diagram

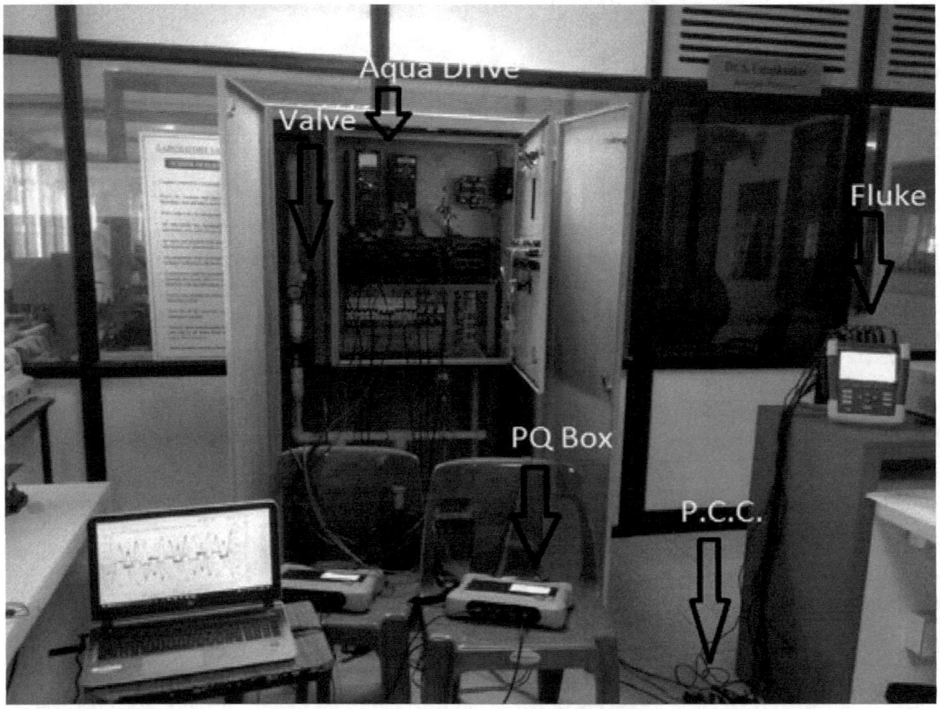

Figure 3: Hardware Test Bed setup

4. Vibration analysis in case of bad motor vibration

For the vibration analysis purpose two condition taken i.e., motor running under tightly and loosely connected condition with its base plate. To control the motor speed Danfoss aqua drive is used [17]. Motor is run at different pressure bar (0.3, 0.6, 1, 1.6 bar) under different value conditions (fully open, semi-closed, fully closed) at different control scheme such as VVC+ and V/F control scheme. For more analysis vibration level is increase by again loosening base plate of motor also more reading taken as taken previously taken for low vibration level. The experiment has been done in three case studies like valve fully open, semi closed and fully closed condition. In three conditions 3rd, 5th, 7th and 11th harmonics, voltage and current THD, flickering, frequency values are measured [19].

4.1. Electrical power quality analysis in case of VVC+ control scheme used when motor running Vibration condition

The measurements were analysed and represented in results and observation shows the THD, 5th, 7th,11th and instantaneous flicker values under VVC+ control scheme in case of bad motor vibration operations. It is shown that the THD value is changed depending on the number of units operation its values change between 3 and 4.5%, however these values are nearest to the permissible limits which are shown in Table 1 but there values are not exceed these limits. Also the 5th and 7th harmonic contents

IOP Conf. Series: Materials Science and Engineering **937** (2020) 012019 doi:10.1088/1757-899X/937/1/012019

are higher. The instantaneous flicker values are high which means variable fluctuation of the supply voltage due to unbalance of torque operation which results from bad motor vibration.
In the first case valve is fully closed then the changes of voltage THD, current THD, flickering and frequency will be shown by the figures [Figure 4-7].

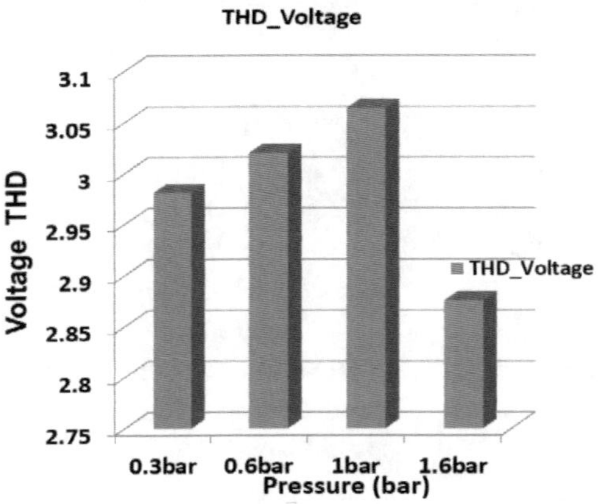

Figure 4: Voltage THD at fully closed condition

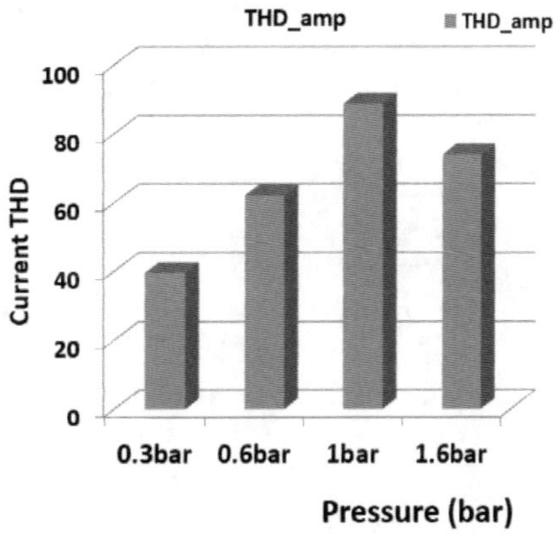

Figure 5: Current THD at fully closed condition

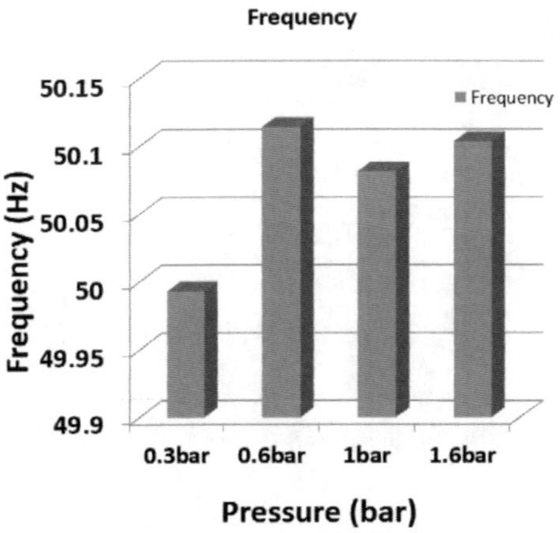

Figure 6: Frequency at fully closed condition

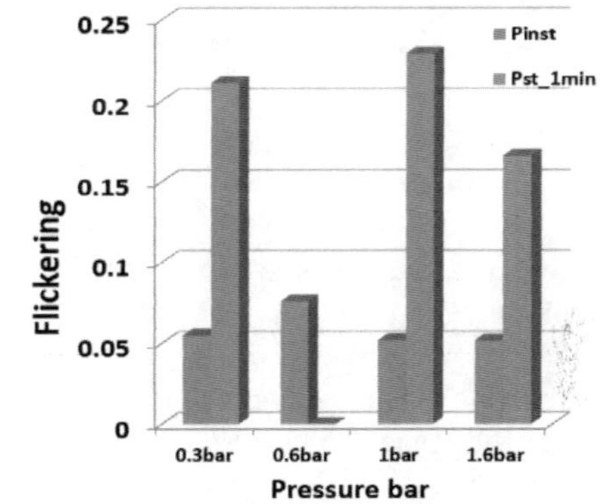

Figure 7: Flickering at fully closed condition

The second case is when valves are semi closed values are collected for voltage and current THD, frequency and flickering [Figure 8-11].

IOP Conf. Series: Materials Science and Engineering **937** (2020) 012019 doi:10.1088/1757-899X/937/1/012019

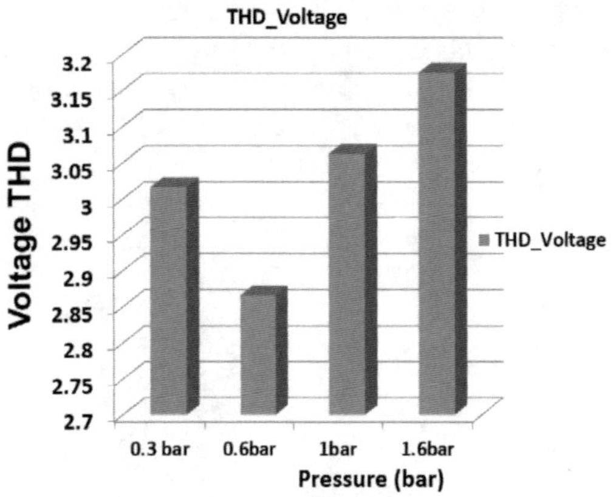

Figure 8: Voltage THD at semi closed condition

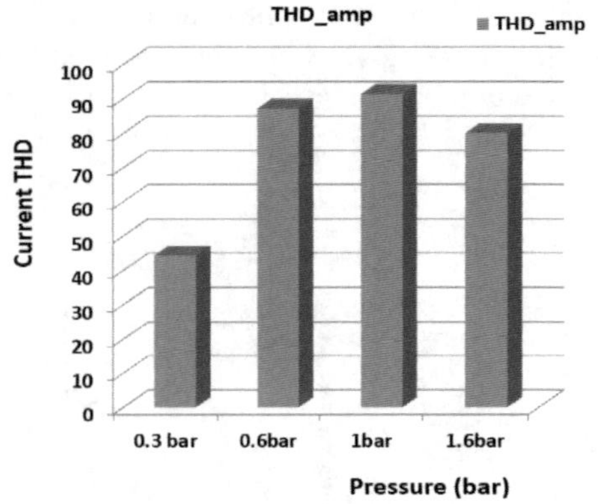

Figure 9: Current THD at semi closed condition

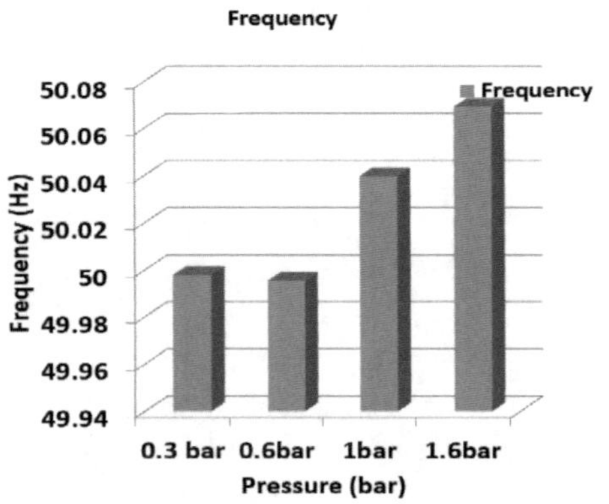

Figure 10: Frequency at semi closed option condition

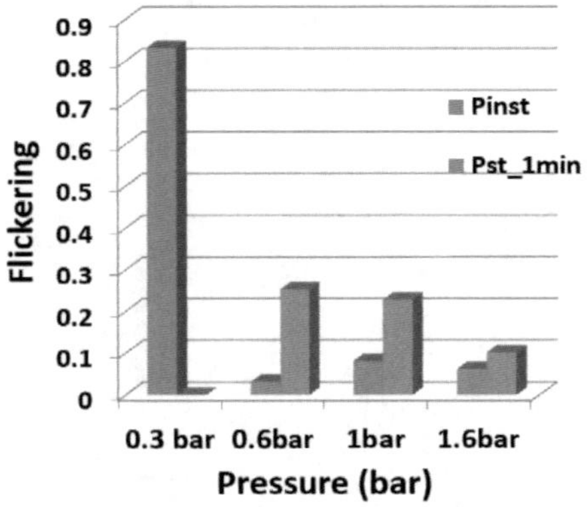

Figure 11: Flickering at semi closed

Now in case 3 the value of voltage and current THD, frequency and flickering values are collected when valves are fully open [Figure 12-15].

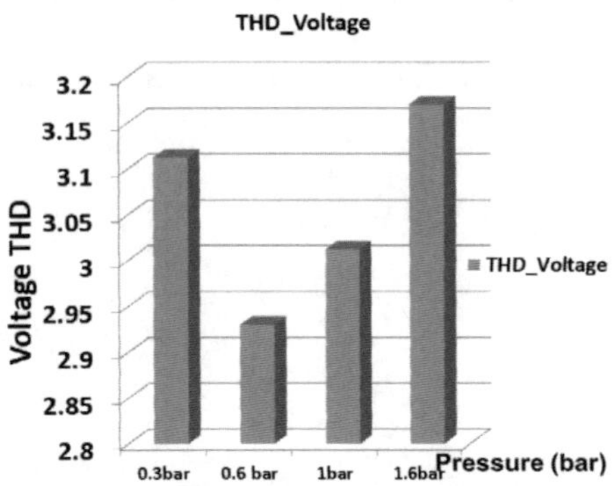

Figure 12: Voltage THD at fully open condition

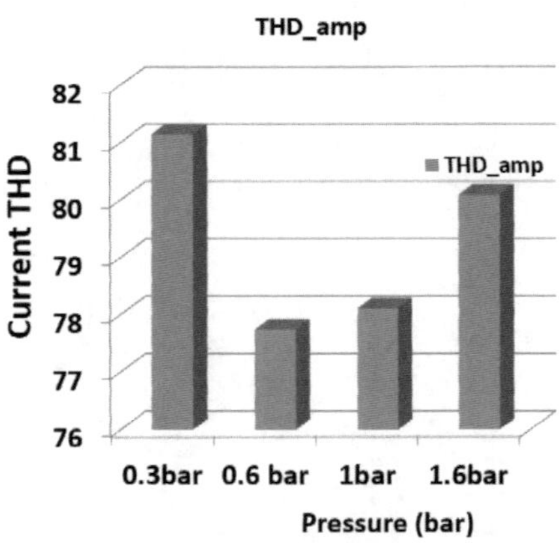

Figure 13: Current THD at fully open condition

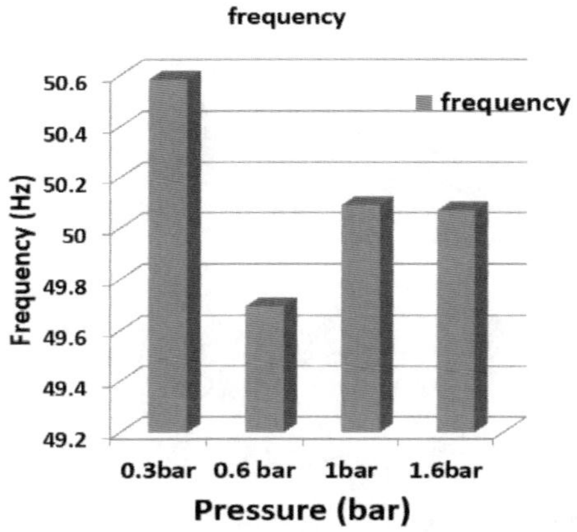

Figure 14: Frequency at fully open condition

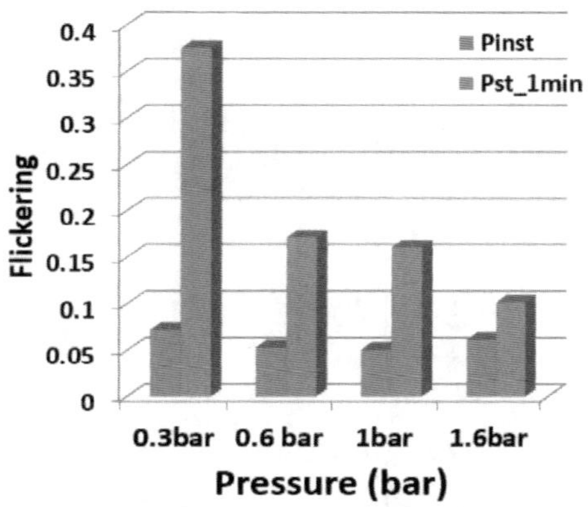

Figure 15: Flickering at fully open condition

4.2. Electrical power quality analysis in case of V/F control scheme used when motor running Vibration condition

The measurements were analysed and represented in results and observation shows the THD, 3rd, 5th, 7th, 11th and instantaneous flicker values under V/F control scheme in case of bad motor vibration operations. It is shown that the THD value is more than the THD value when motor run during VVC+ control scheme. Also the 5th and 7th harmonic contents are higher. The instantaneous flicker values are also high which means variable fluctuation of the supply voltage due to unbalance of torque operation which results from bad motor vibration. The VVC+ and V/F control are the two major controller technique for VFD based pump. Here the characteristics of VVC + and V/F are compared for power quality issues. Here also three cases have been investigated like fully closed, semi closed and fully open condition.

Here case 1 is when vales are fully closed that times the nature of voltage, current THD, flickering and frequency values are collected for V/F control [Figure 16-19].

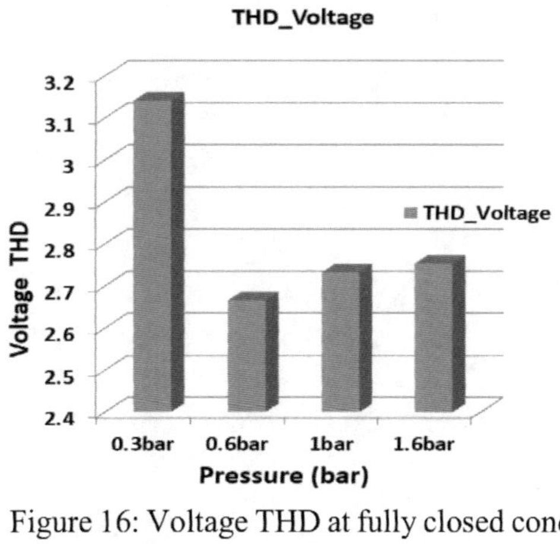

Figure 16: Voltage THD at fully closed condition

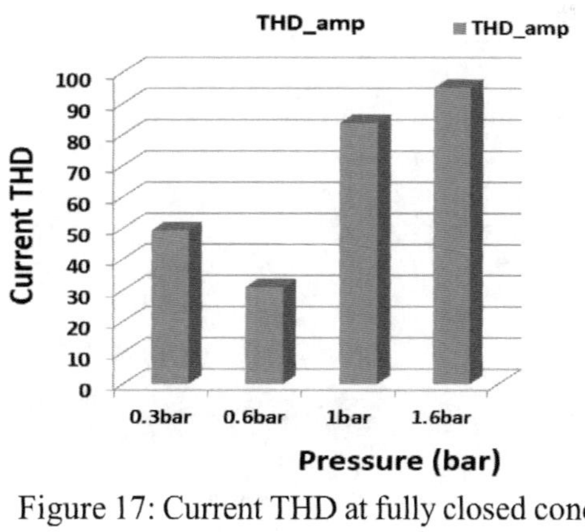

Figure 17: Current THD at fully closed condition

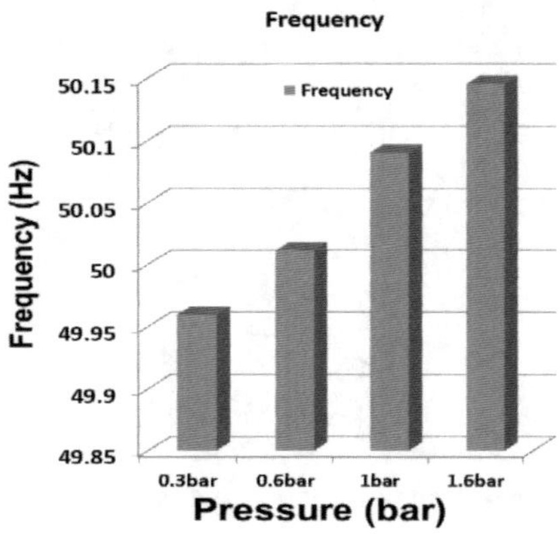

Figure 18: Frequency at fully closed condition

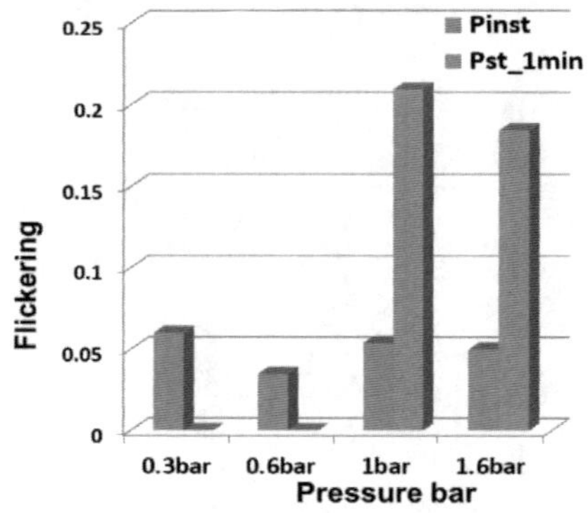

Figure 19: Flickering at fully closed condition

In case 2 when valves are semi closed the voltage, current THD, flickering, frequency values are collected [Figure 20-23].

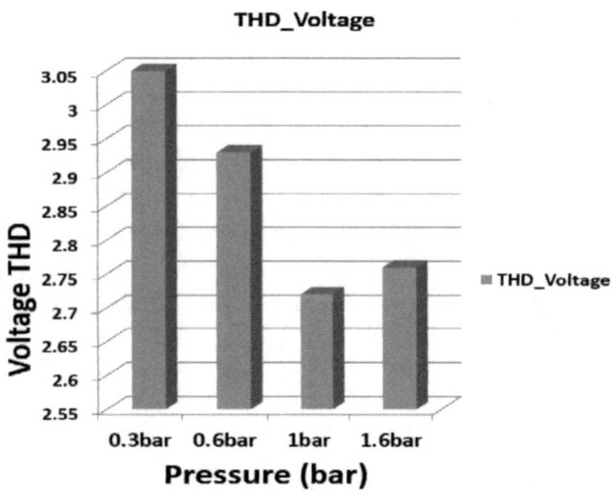

Figure 20: Voltage THD at semi closed condition

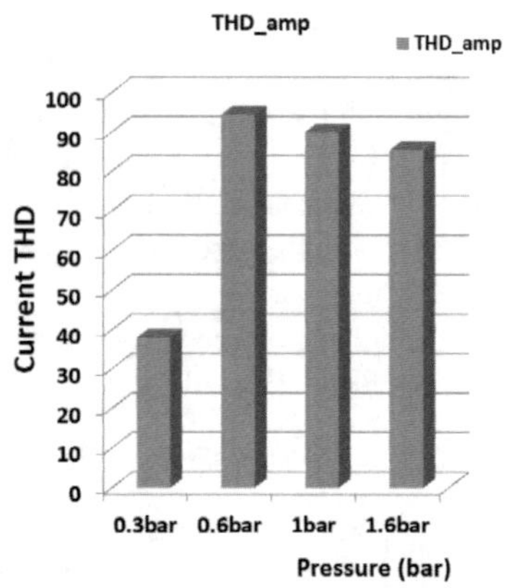

Figure 21: Current THD at semi closed condition

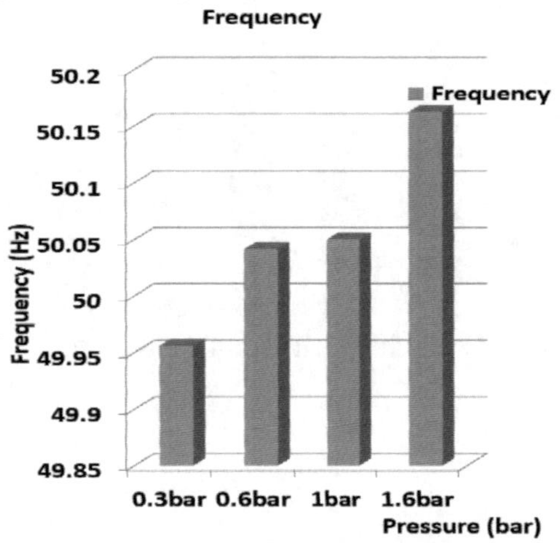

Figure 22: Frequency at semi closed condition

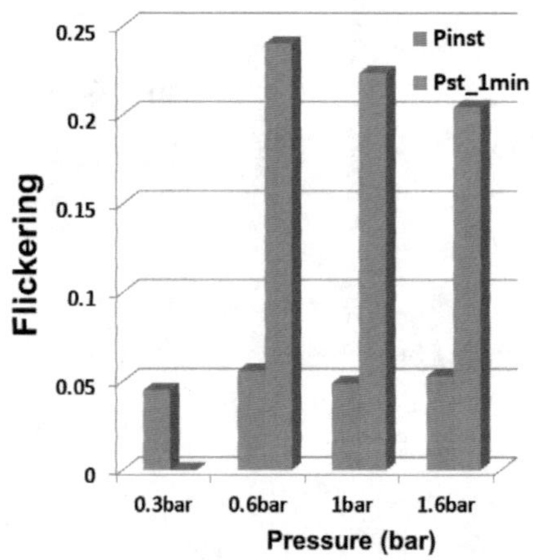

Figure 23: Flickering at semi closed condition

In the 3rd case when all the valves are open then voltage, current THD, frequency, flickering values are collected [Figure 24- 27].

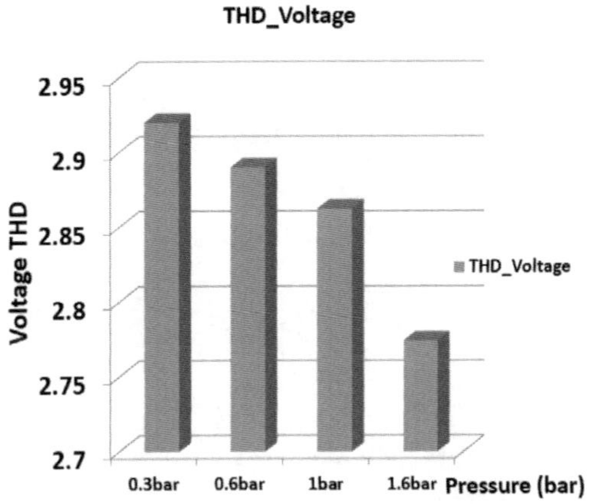

Figure 24: Voltage THD at fully open condition

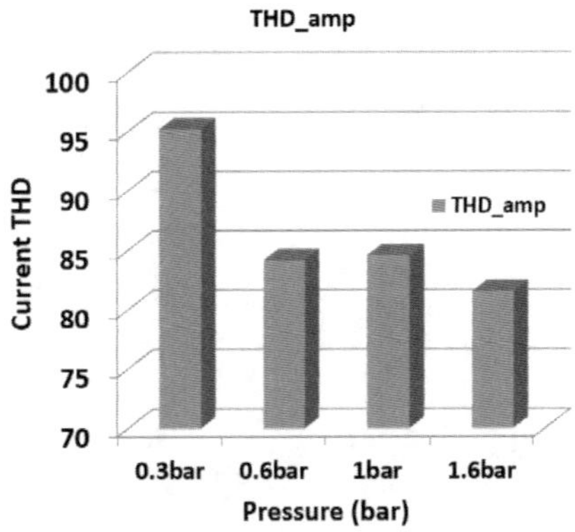

Figure 25: Current THD at fully open condition

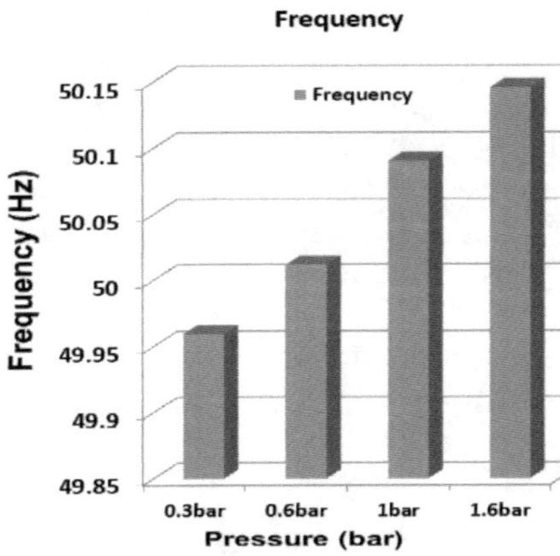

Figure 26: Frequency at fully open condition

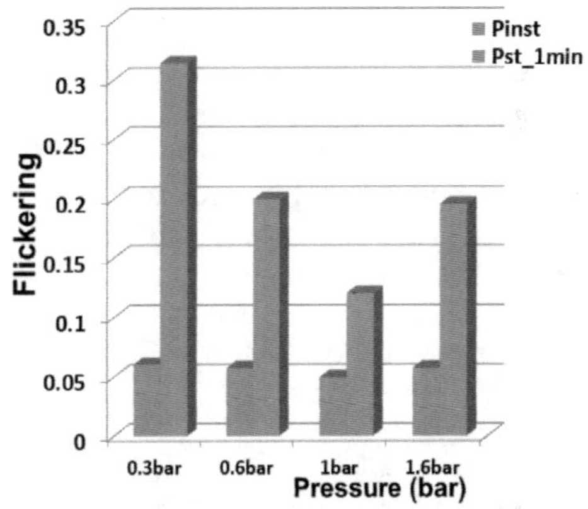

Figure 27: Flickering at fully open condition

Table 2: Valve operations condition for VVC+ control

	VVC + Control											
	Valve fully closed				Valve semi closed				Valve fully open			
	0.3 bar	0.6 bar	1 bar	1.6 bar	0.3 bar	0.6 bar	1 bar	1.6 bar	0.3 bar	0.6 bar	1 bar	1.6 bar
THD voltage	2.9811 volt	3.02 volt	3.064 volt	2.8755 volt	3.016 volt	2.866 volt	3.0625 volt	3.1746 Volt	3.11 volt	2.93 volt	3.01 volt	3.17 volt
THD Current	39.57 amp	62.2 amp	88.6782 amp	74.02 amp	44.29 amp	87.222 amp	91.4113 amp	80.094 amp	81.15 amp	77.74 amp	78.11 amp	80.1 amp
Frequency	49.9931 Hz	50.114 Hz	50.0814 Hz	50.1035 Hz	49.998 Hz	49.9955 Hz	50.0398 Hz	50.0693 Hz	50.58 Hz	49.7 Hz	50.09 Hz	50.1 Hz
Flickering												
Pinst	0.0551	0.076	0.052	0.518	0.833	0.032	0.0814	0.062	0.07	0.053	0.05	0.06
Pst_1 min	0.211	0	0.229	0.166	0	0.2525	0.2277	0.102	0.38	0.17	0.16	0.11

Table 3: Valve operations condition for V/F control

	V/F Control											
	Valve fully closed				Valve semi closed				Valve fully open			
	0.3 bar	0.6 bar	1 bar	1.6 bar	0.3 bar	0.6 bar	1 bar	1.6 bar	0.3 bar	0.6 bar	1 bar	1.6 bar
THD voltage	3.14 volt	2.665 volt	2.73 volt	2.75 volt	3.05 volt	2.93 volt	2.72 volt	2.76 volt	2.92 volt	2.89 volt	2.86 volt	2.77 volt
THD Current	49.38 amp	31.03 amp	87.11 amp	94.92 amp	37.98 amp	94.37 amp	89.89 amp	85.36 amp	95.2 amp	84.2 amp	84.6 amp	81.54 amp
Frequency	49.93 Hz	49.96 Hz	49.98 Hz	50.09 Hz	49.96 Hz	50.04 Hz	50.05 Hz	50.16 Hz	49.9 Hz	50.0 Hz	50.1 Hz	50.2 Hz
Flickering												
Pinst	0.06	0.035	0.054	0.05	0.045	0.056	0.049	0.053	0.06	0.06	0.05	0.05
Pst_1 min	0	0	0.2094	0.1845	0	0.2401	0.2233	0.240	0.31	0.19	0.12	0.19

From Table 2 and Table 3 it has been concluded that in the case of valve fully closed condition except in 1.6 bar the voltage THD is lower when the motor is running in VVC+ control rather than V/F control. Current THD also lower most of the cases in VVC+ control expect in 0.6 bar pressure condition. Average Flicker are more dominant when motor run under V/F method. Similarly when valve is fully open condition voltage and current THD is more in V/F control. Average flicker is more dominant when the motor run under V/F control. In valve semi closed condition voltage THD is higher in VVC+ control in 1 bar pressure and current THD is higher in VVC+ control most of the cases. Average Flicker are more dominant when motor run under VVC+ method.

The harmonics values of the system like voltage harmonics and current harmonics are collected to analysis the power quality issues in water pumping system. Here also in both VVC+ and V/F control method the process have been done [Figure 28-39].

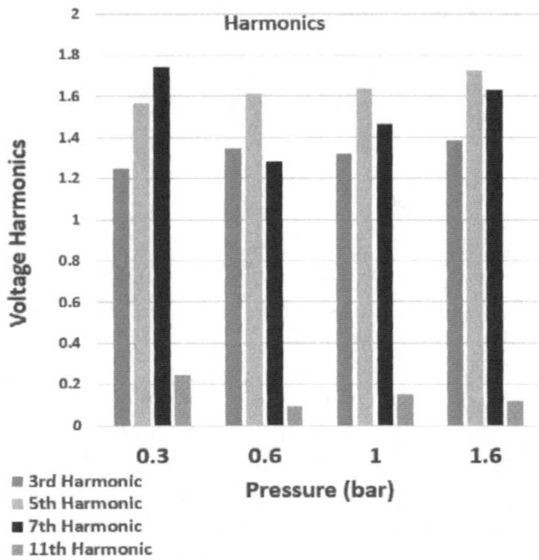

Figure 28: Voltage harmonics for fully open condition (VVC+)

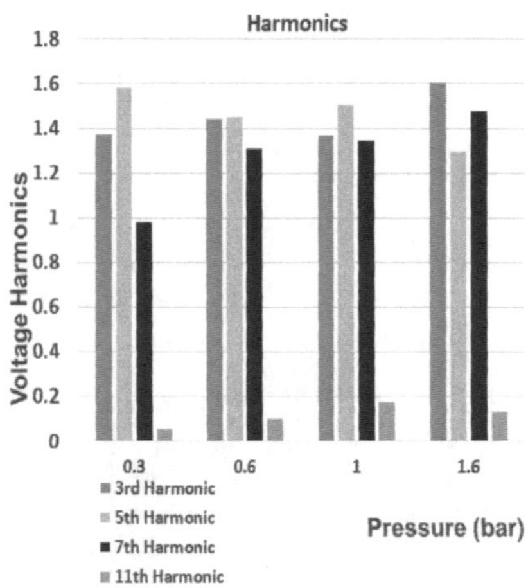

Figure 29: Voltage harmonics for fully open condition (V/F)

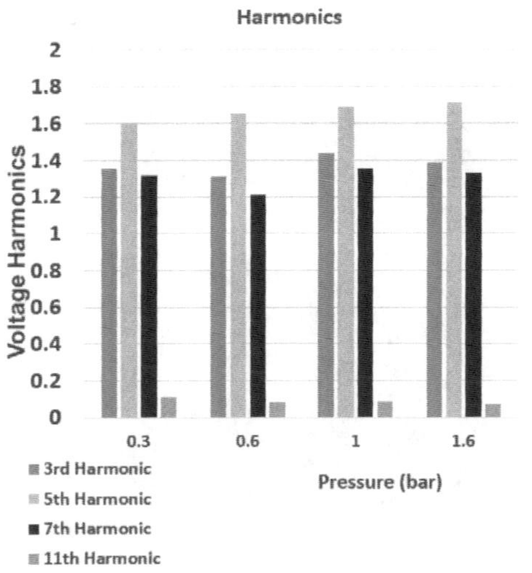

Figure 30: Voltage harmonics for semi closed condition (VVC+)

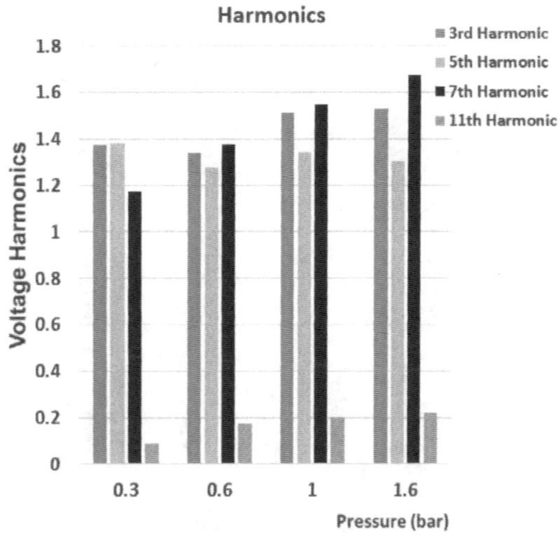

Figure 31: Voltage harmonics for semi closed condition (V/F)

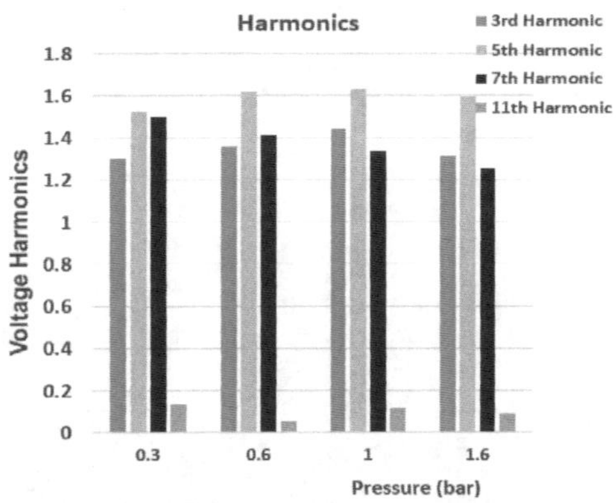

Figure 32: Voltage harmonics for fully closed condition (VVC+)

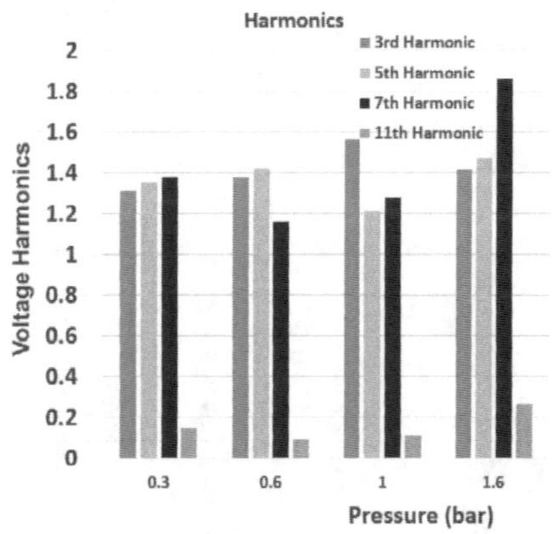

Figure 33: Voltage harmonics for fully closed condition (V/F)

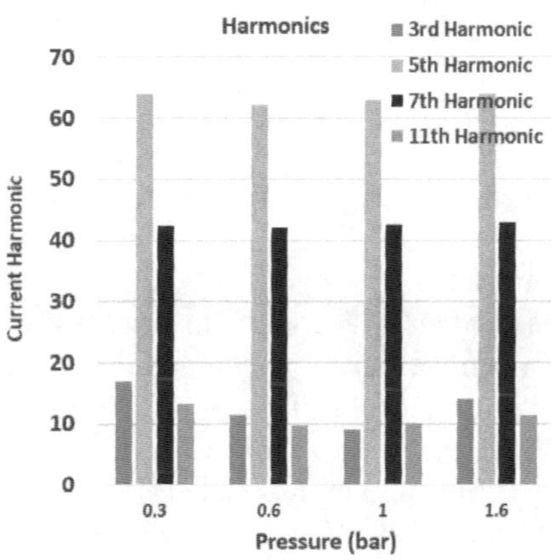

Figure 34: Current harmonics for fully open condition (VVC+)

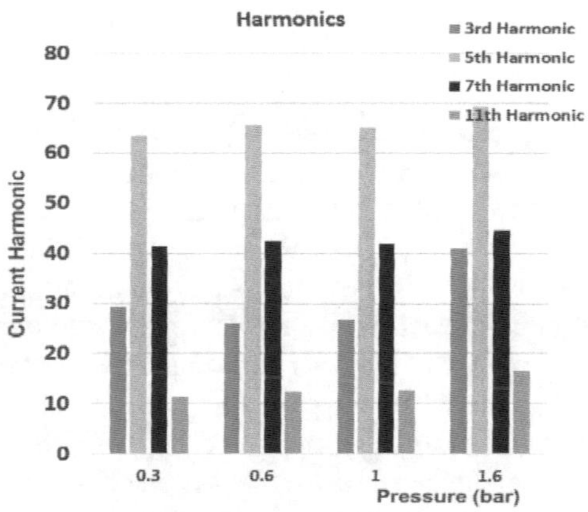

Figure 35: Current harmonics for fully open condition (V/F)

IOP Conf. Series: Materials Science and Engineering **937** (2020) 012019 doi:10.1088/1757-899X/937/1/012019

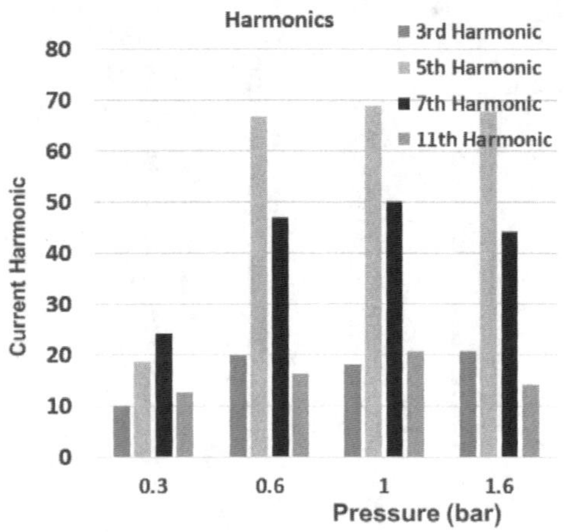

Figure 36: Current harmonics for semi closed condition (VVC+)

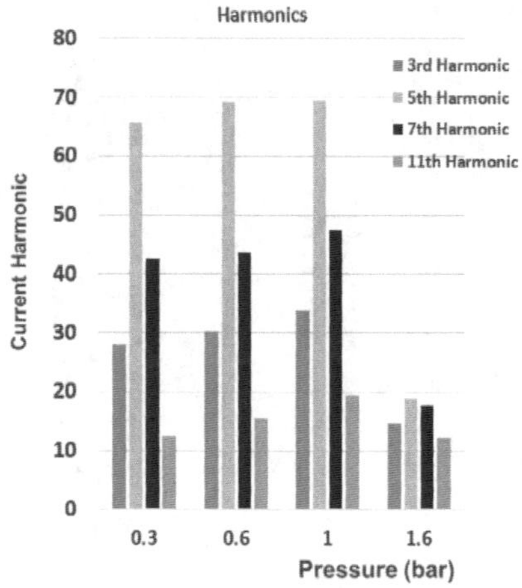

Figure 37: Current harmonics for semi closed condition (V/F)

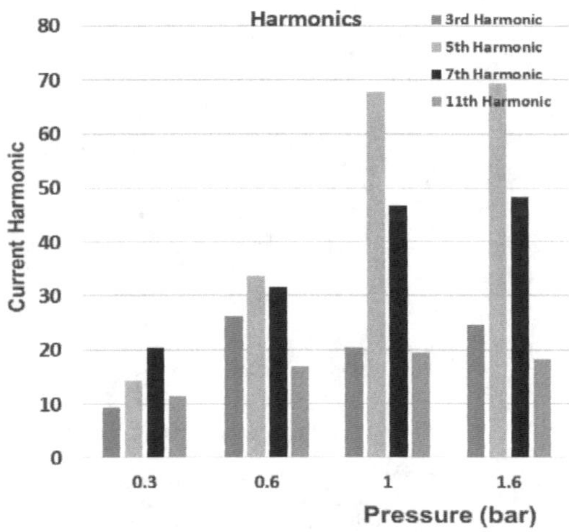

Figure 38: Current harmonics for fully closed condition (VVC+)

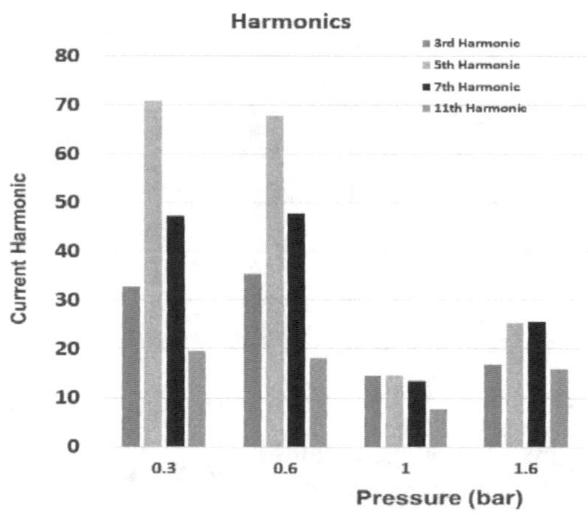

Figure 39: Current harmonics for fully closed condition (V/F)

5. Comparison among VVC+ and V/F method

The overall experimental results shows that the voltage THD and current THD is almost high in the case of V/F control in three cases than VVC+ control in different pressure bar. The flicker is dominant in most of the time for V/F control method. In the case of harmonics also it is seen from the figures and table 3 that in percentage of harmonic in VVC+ control is lesser than V/F control in three cases in different pressure bar for both voltage and current harmonics. When value is at open condition overall performance of motor is good under V/F method even though vibration level is high. So VVC+ method is this case when power quality issues taken in consideration. When value is at fully -Closed condition overall performance of motor is good under VVC+ method (Except 0.6pressure bar) even though vibration level is high. While V/F method is suitable only for 0.6 pressure bar. So overall VVC+ method is preferable when value is at Fully-Closed condition under vibrating condition (Table 4).

Table 4: Comparison between Normal condition and under vibrating condition

Case	Normal	VVC+	V/F
THD	About 4.5%	Less than7%	About 9%
3th harmonics	3.2%	Less than 5.4%	More than 6.8%
5th harmonics	2-2.5%	Less than 3.2%	More than4.3%
7th harmonics	About1.3-1.4%	Less 2.1%	More than 3.2%
11th harmonics	About 1.2%	Less 1.8%	More than2.1%
Flicker (at 1min)	Less	Small flicker value	Large flicker value

Conclusion

The analysis of the dynamic measurements and power quality showed that; as motor vibration level is increase it affects not only dynamic performance but also electrical power quality. Here a comparison between electrical power quality analysis and dynamic performance of motor under vibration condition has been done and result shows that under poor fixing or weakness of baseplate or foundation leads to high levels of vibration. The THD level is more in V/f method than VVC+ method also the harmonics content of 3rd,5th,7th,11th is more in V/F control than VVC+ method .The results also show that VVC+ control method is best suitable under vibration condition than V/F control method.

Acknowledgement

Authors are like to thank the Danfoss Advance Drives laboratory of VIT Vellore for their modern setup and allow authors to complete the experiment and Danfoss Pvt Ltd, Chennai for their support in technical solution and Prince Sultan University, Riyadh, Saudi Arabia for technical guidance.

Reference

[1]. Sitharthan R, Geethanjali M and Pandy TKS 2016 Adaptive protection scheme for smart microgrid with electronically coupled distributed generations *Alexandria Engineering Journal* **55(3)** 2539-2550

[2]. Fathima AH, and Palanisamy K 2014 Battery energy storage applications in wind integrated systems—a review *IEEE International Conference on Smart Electric Grid* 1-8

[3]. Prabaharan N and Palanisamy K 2015 Investigation of single-phase reduced switch count asymmetric multilevel inverter using advanced pulse width modulation technique *International Journal of Renewable Energy Research* **5(3)** 879-890.

[4]. Jerin ARA, Kaliannan P and Subramaniam U 2017 Improved fault ride through capability of DFIG based wind turbines using synchronous reference frame control based dynamic voltage restorer. *ISA transactions* **70** 465-474

[5]. Sitharthan, R, Sundarabalan CK, Devabalaji KR, Nataraj SK and Karthikeyan M 2018 Improved fault ride through capability of DFIG-wind turbines using customized dynamic voltage restorer *Sustainable cities and society* **39** 114-125

[6]. Prabaharan N and Palanisamy K 2016 A single-phase grid connected hybrid multilevel inverter for interfacing photo-voltaic system *Energy Procedia* **103** 250-255

[7]. Palanisamy K, Mishra JS, Raglend IJ and Kothari DP 2010 Instantaneous power theory based unified power quality conditioner (UPQC) *IEEE Joint International Conference on Power Electronics, Drives and Energy Systems* 1-5

[8]. Sitharthan R and Geethanjali M 2017 An adaptive Elman neural network with C-PSO learning algorithm-based pitch angle controller for DFIG based WECS Journal of Vibration and Control **23(5)** 716-730

[9]. Sitharthan R and Geethanjali M 2015 Application of the superconducting fault current limiter strategy to improve the fault ride-through capability of a doubly-fed induction generator–based wind energy conversion system *Simulation* **91(12)** 1081-1087

[10]. Sitharthan R, Karthikeyan M, Sundar DS and Rajasekaran S 2020 Adaptive hybrid intelligent MPPT controller to approximate effectual wind speed and optimal rotor speed of variable speed wind turbine *ISA transactions* **96** 479-489

[11]. Sitharthan R, Devabalaji KR and Jees A 2017 An Levenberg–Marquardt trained feed-forward back-propagation based intelligent pitch angle controller for wind generation system *Renewable Energy Focus* **22** 24-32

[12]. Sitharthan R, Sundarabalan CK, Devabalaji KR, Yuvaraj T and Mohamed Imran A 2019 Automated power management strategy for wind power generation system using pitch angle controller *Measurement and Control* **52(3-4)** 169-182

[13]. Sundar DS, Umamaheswari C, Sridarshini T, Karthikeyan M, Sitharthan R, Raja AS and Carrasco MF 2019 Compact four-port circulator based on 2D photonic crystals with a 90° rotation of the light wave for photonic integrated circuits applications *Laser Physics* **29(6)** 066201

[14]. Sitharthan R, Parthasarathy T, Sheeba Rani S and Ramya KC 2019. An improved radial basis function neural network control strategy-based maximum power point tracking controller for wind power generation system *Transactions of the Institute of Measurement and Control* **41(11)** 3158-3170

[15]. Rajesh M and Gnanasekar JM 2017 Path observation based physical routing protocol for wireless ad hoc networks *Wireless Personal Communications* **97(1)** 1267-1289

[16]. Palanisamy K, Varghese LJ, Raglend IJ and Kothari DP 2009. Comparison of intelligent techniques to solve economic load dispatch problem with line flow constraints *IEEE International Advance Computing Conference* 446-452

[17]. Sitharthan R, Ponnusamy M, Karthikeyan M and Sundar DS 2019 Analysis on smart material suitable for autogenous microelectronic application *Materials Research Express* **6(10)** 105709

[18]. Rajaram R, Palanisamy K, Ramasamy S and Ramanathan P 2014 Selective harmonic elimination in PWM inverter using fire fly and fireworks algorithm *International Journal of Innovative Research in Advanced Engineering* **1(8)** 55-62

[19]. Sitharthan R, Swaminathan JN and Parthasarathy T 2018 March. Exploration of wind energy in India: A short review *IEEE National Power Engineering Conference* 1-5

[20]. Karthikeyan M, Sitharthan R, Ali T and Roy B 2020 Compact multiband CPW fed monopole antenna with square ring and T-shaped strips *Microwave and Optical Technology Letters* **62(2)** 926-932

[21]. Sundar D Sridarshini T, Sitharthan R, Madurakavi Karthikeyan, Sivanantha Raja A, and Marcos Flores Carrasco 2019 Performance investigation of 16/32-channel DWDM PON and long-reach PON systems using an ASE noise source *In Advances in Optoelectronic Technology and Industry Development: Proceedings of the 12th International Symposium on Photonics and Optoelectronics* 93

[22]. Sitharthan R and Geethanjali M 2014 Wind Energy Utilization in India: A Review *Middle-East J. Sci. Res.* **22** 796–801 doi:10.5829/idosi.mejsr.2014.22.06.21944

[23]. Sitharthan R and Geethanjali M 2014 ANFIS based wind speed sensor-less MPPT controller for variable speed wind energy conversion systems *Australian Journal of Basic and Applied Sciences* **8**14-23

[24]. Jerin ARA, Kaliannan P, Subramaniam U and El Moursi MS 2018 Review on FRT solutions for improving transient stability in DFIG-WTs *IET Renewable Power Generation* **12(15)** 1786-1799

[25]. Prabaharan N, Jerin ARA, Palanisamy K and Umashankar S 2017 Integration of single-phase reduced switch multilevel inverter topology for grid connected photovoltaic system *Energy Procedia* **138** 1177-1183

[26]. Rameshkumar K, Indragandhi V, Palanisamy K and Arunkumari T 2017 Model predictive current control of single phase shunt active power filter *Energy Procedia* **117** 658-665

[27]. Fathima AH and Palanisamy K 2016 Energy storage systems for energy management of renewables in distributed generation systems *Energy Management of Distributed Generation Systems* 157

[28]. Rajesh M 2020 Streamlining Radio Network Organizing Enlargement Towards Microcellular Frameworks *Wireless Personal Communications* 1-13

[29]. Subbiah B, Obaidat MS, Sriram S, Manoharn R and Chandrasekaran SK 2020 Selection of intermediate routes for secure data communication systems using graph theory application and grey wolf optimisation algorithm in MANETs IET *Networks* doi:10.1049/iet-net.2020.0051

[30]. Singh RR and Chelliah TR 2017 Enforcement of cost-effective energy conservation on single-fed asynchronous machine using a novel switching strategy *Energy* **126** 179-191

[31]. Amalorpavaraj RAJ, Palanisamy K, Umashankar S and Thirumoorthy AD 2016 Power quality improvement of grid connected wind farms through voltage restoration using dynamic voltage restorer *International Journal of Renewable Energy Research* **6(1)** 53-60

[32]. Singh RR, Chelliah TR, Khare D and Ramesh US 2016 November. Energy saving strategy on electric propulsion system integrated with doubly fed asynchronous motors *IEEE Power India International Conference* 1-6

[33]. Singh RR, Mohan H and Chelliah TR 2016 November. Performance of doubly fed machines influenced to electrical perturbation in pumped storage plant-a comparative electromechanical analysis *IEEE 7th India International Conference on Power Electronics* 1-6

IOP Conf. Series: Materials Science and Engineering **937** (2020) 012020 doi:10.1088/1757-899X/937/1/012020

Resource assessment and energy yield estimation for 160 MW solar-wind hybrid project using system advisory model

G Anilkumar[1], A Chakradhar[2], G V Brahmendra Kumar[3], K Palanisamy[4*],

[1,2]School of Renewable Energy, JNT University, Kakinada, 533003, India.
[3,4]Centre for Smart Grid Technologies, Vellore Institute of Technology, Vellore, 632014, India.
Email: gorantlaanilkumar@gmail.com, brahmendrakumar.g@gmail.com,
*Corresponding author: kpalanisamy@vit.ac.in

Abstract. A techno-economic and environmental cost benefit analysis and assessment for the proposed 160 MW (Solar-120 MW and Wind-40 MW) hybrid renewable energy project with grid integration is carried out for electricity generation at Ramagiri & Muthavakuntla village locations of Anathapuram district in Andhra Pradesh state of India. Manual estimations are carried out based on recent research publications in similar line of studies and simulation studies are carried out using System Advisory Model (SAM) software under individual and hybrid operating modes with authenticated data resources to determine annual energy yield as well as sizing along with their economic and environmental concerns for solar and wind energy resources. Finally, comprehensive annual energy benefits are presented that shows techno-economic-environmental viability of this proposed project at the aforementioned location.

1. Introduction

Allocation of available, affordable and accessible power supply to different societal sectors remains as a challenging task for developing countries especially due to prevalence of high costs in addition to transmission losses and power shortages. As per the recent reports, still there is a dire need to enhance the accessibility of electrical power with emerging demands for power consumption around the world [1][2]. In this regard, renewable energy-based power plants can offer a potentially viable economic solution to meet growing energy demands [3]. Intermittent nature of wind and solar radiation over day-night and seasonal cycles impose power supply interruption leading to risk generation to rely on single renewable energy supply source. Integration of solar and wind energy under hybrid mode to grid remains as a plausible solution to overcome these issues [4-14]. Several probabilistic and deterministic approaches are available to develop hybrid system models with optimum resource combinations and assess overall system performance [15-25]. This work evaluates the technical, economic and environmental feasibility of proposed 160 MW (Solar-120 MW and Wind-40 MW) hybrid project located at Ramagiri and Muthavakuntla village, Ananthapuram district, Andhra Pradesh state of India whose geographical coordinates were N 14°20'32.76" and E 77°30'29.02" using System Advisory Model (SAM) based on ground, satellite and integrated data bases in Indian context and annual comprehensive benefit analysis based on recent similar line of research studies [25-33].

2. Site Climatology

The proposed project land encompasses 928 acres of land out of which 800 acres for Solar Park and 90 acres for wind farm. Initially, appropriate selection of location with adequate energy resources can be attained by considering parametric constraints of the requisite hybrid system. Here in this case, energy yield estimation pertaining to availability of solar radiation and wind speed at the proposed site location which in turn depends on weather and climate factors over proposed site location, energy demand, seasonal variations in load and intensity of energy resources and so on.

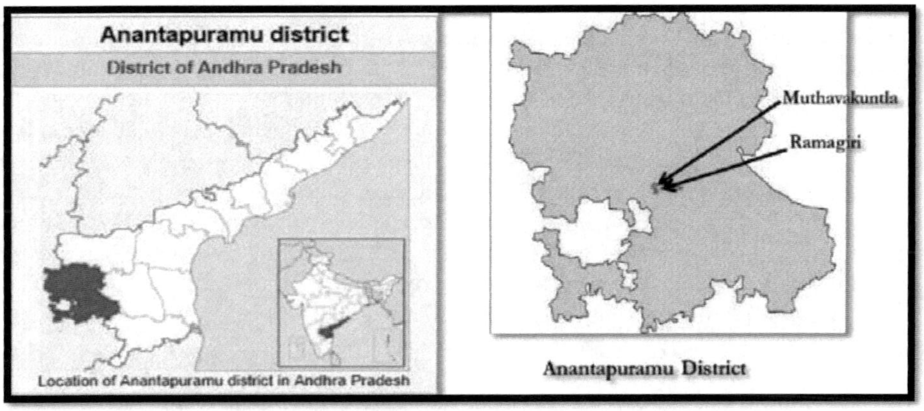

Figure 1. Proposed 160 MW Solar-Wind hybrid project location.

Figure 1 shows the proposed 160 MW Solar-Wind hybrid project location. The area that falls under this site location comes under semi-arid climatic zone of India with almost hot and dry conditions throughout a year. Summers usually start in the month of February with recorded average high temperatures at 37°C and peak temperatures in the month of May whereas monsoon starts in the month of September that lasts until November with 250 mm of observed precipitation followed by dry and mild winter in late November that lasts until early February with little humidity that exhibits temperatures averaging in between 22-23°C. Overall annual average rainfall within the district is observed to be 535 mm with mean seasonal rainfall distribution of 316 mm during southwest monsoon (June-September) while September and October months are noticed to be the wettest months of the year. The water requirement can be met through deep tube well while the maximum and minimum level of the ground water table is around 300 feet to 900 feet.

3. Technical Evaluation

The Proposed 160 MW wind-solar hybrid study was analyzed allowing for 40 MW generation facilities from wind farm and 120 MW of Solar PV capacity. Site characterization along with overall resource assessments were carried out for energy yield estimations. Wind resource assessments have been carried out by using Ramgiri wind mast, such as contour maps and wind characteristics observed at project site location and then mapping it with solar installations. The solar-wind resource assessment and overall energy yield estimation as well as photovoltaic (PV) system sizing have been carried out by using PVSyst-6.8.1 and SAM-2018 software versions. Studies pertaining to resource assessment for solar as well as wind energy generation were carried out by collecting required climatic resource data base from standard Indian Meteorological Data (IMD) and established National Solar Resource Data Base (NSRDB) data bases.

3.1 Solar output system evaluation

The solar radiation parametric data such as GHI, DNI and DHI are collected serially at periodic intervals of half-hourly and hourly along with meteorological time series data from 2004 to 2014. This data is downloaded from SAM software for Ramagiri location with temporal resolution of 1 km at geological coordinates mentioned above. Using this data, potential availability of usable solar energy for power generation can be predicted in and around proposed site at Ramagiri location. Data collected from individual sources for both solar and wind energy resources on daily, weekly, monthly and yearly basis were compared with real-time data collected by the project developer Solar Energy Corporation of India (SECI) for reference which showed GHI of 2152 $kWh/m^2/year$ with an average GHI and DHI of 179.3 kWh/m^2 and 69.5 kWh/m^2 respectively and an average temperature of 26.3°C.

As on today, there are various solar PV and wind turbine generator technologies are available for adoption throughout the world [4]. Different available technologies were taken into consideration for the sake of analysis from both commercial and technological perceptions for the proposed site location and identified the most suitable technologies for yielding hybrid energy using Siemens Gamesa G97-2.0 MW wind turbine for wind power generation and Canadian solar CS6U-325wp solar PV module for solar power generation. PV system sizing for was carried for both DC and AC systems by considering solar PV module ratings and balance of system components with appropriate DC-AC ratios using SAM software. Table 1 shows the simulation results obtained for solar system, while figure 2 and figure 3 represents monthly energy production and monthly system power output over a year respectively. It is clearly evident that there is a deviation from desired output from table 1, figure 2 and figure 3 which can be mainly attributed to generic loss factors associated with solar PV system and environmental concerns.

Table 1. Simulation (SAM) results of the proposed 120 MW PV system.

Metric	Value
Annual energy (year 1)	241,380,112 kWh
Capacity factor (year 1)	17.7 %
Energy yield (year 1)	1,548 kWh/kW
Performance ratio (year 1)	0.77

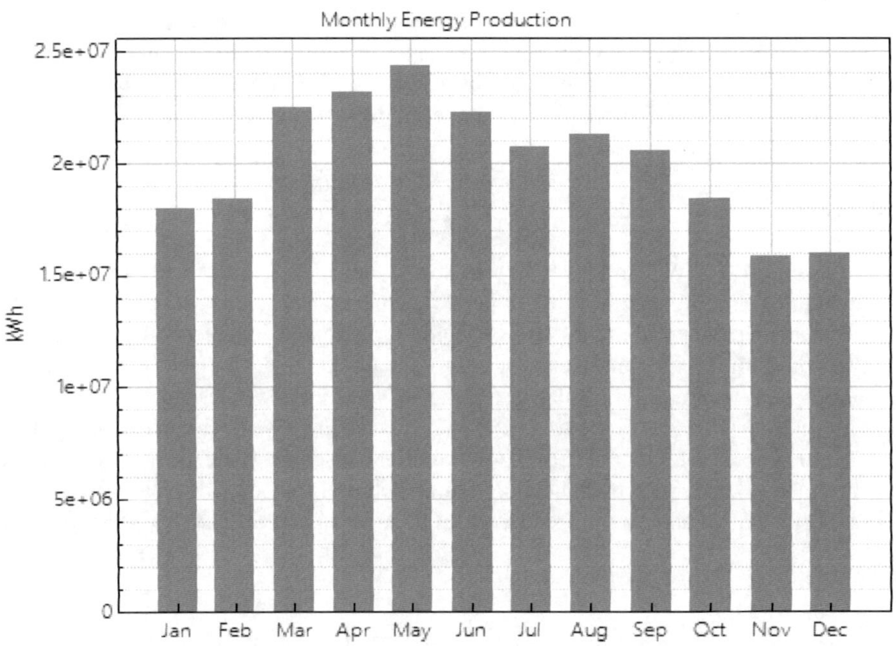

Figure 2. Simulated monthly PV energy production over a year.

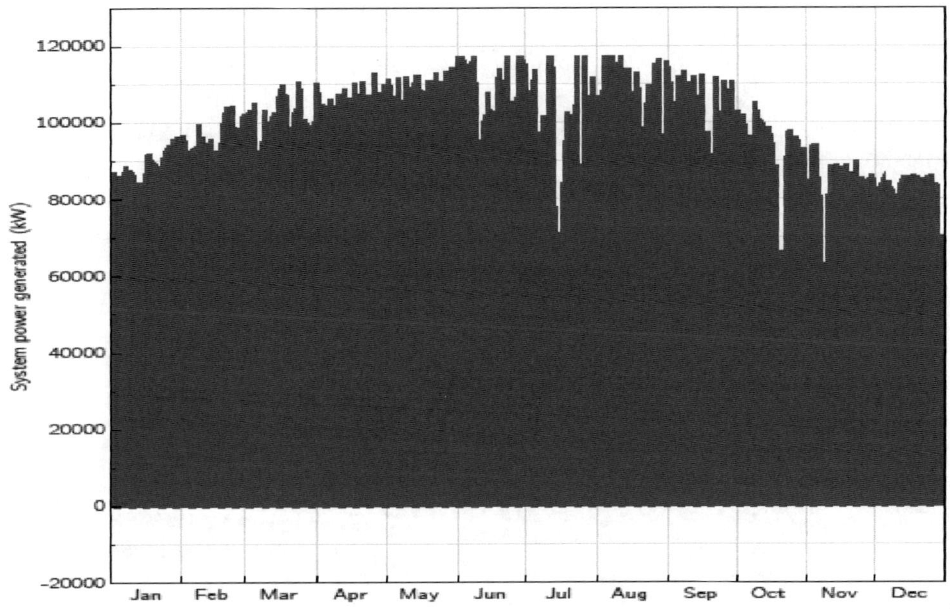

Figure 3. Simulated monthly PV system power output over a year.

3.2 Wind energy system output evaluation

Wind resource assessment was carried out by considering wind data sources from two masts of National Institute of Wind Energy (NIWE) available nearer to proposed project area, wind data from the 78 m height wind mast installed by a private developer, MERRA-2 re-analysis data from the node point nearest to the site, wind data from the 100 m height wind mast available at Ramagiri site.

Table 2. Wind data at proposed site over a year.

Month	Temperature (^{0}C)	Pressure (mbr)	Air Density (Kg/m^3)	100 m north WS (m/s)	100 m south WS (m/s)	80m WS (m/s)	50m WS (m/s)
Jan	22.1	958.9	1.12	5.1	5.1	4.9	4.5
Feb	26.24	957.5	1.11	5.5	5.5	5.2	4.8
Mar	29.98	955.4	1.09	5.4	5.4	5.2	4.8
Apr	31.43	953.5	1.08	5.1	5.1	4.9	4.5
May	28.88	950.3	1.09	5.1	5	4.9	4.5
Jun	26.65	950.1	1.10	8.5	8.5	8.2	7.8
Jul	25.73	950.4	1.10	9.7	9.7	9.5	9.1
Aug	25.06	951.3	1.10	10.7	10.7	10.5	10.1
Sep	26.01	953.6	1.10	5.9	5.9	5.7	5.4
Oct	26.15	955.9	1.11	5	5	4.8	4.4
Nov	24.41	957.2	1.11	5.1	5.1	4.9	4.5
Dec	23.25	957.5	1.12	5	5	4.8	4.4
Avg	**26.3**	**954.3**	**1.1**	**6.34**	**6.32**	**6.11**	**5.73**

Wind resource data is collected by using wind mast incorporated by SECI at proposed site and collected data regarding wind speeds at different reference hub heights such as 50 m, 80 m and 100 m in north and south direction as shown in table 2 to estimate actual energy production and the corresponding monthly mean wind speed over a year at proposed site location is shown in figure 4. From table 2 mean values of 26.3°C temperature, 954.3 mbar pressure, 1.1 kgm^{-3}, air density can be observed with mean wind speed variability ranging from 5.73 ms^{-1} to 6.34 ms^{-1} with respect to variation in height from 50 m to 100 m indicating good wind resource for wind power generation.

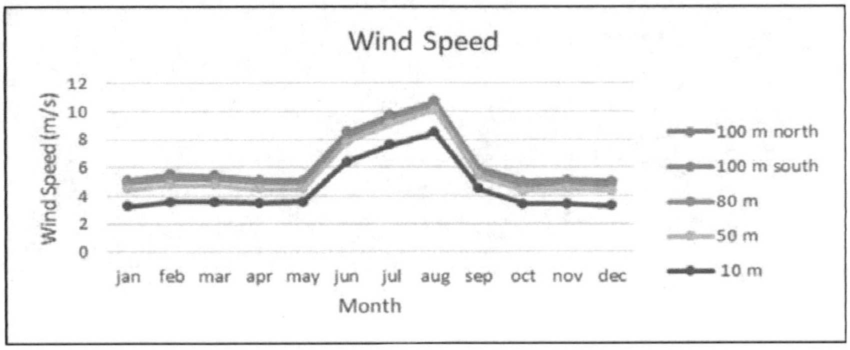

Figure 4. Monthly mean wind speed over a year at proposed site.

Annual energy Production was estimated for the 40 MW wind farm of this proposed project by considering GamesaG97 wind turbine model with each turbine rated at power generation capacity of 2 MW for 104 m hub height level and 97 m rotor diameter along with a shear coefficient of 0.14 and presumed total losses around 14% based on site characteristics represented in table 2. Using SAM software, it has been estimated that 126464.32MW of annual energy can be generated with 36.1% capacity factor. A data of the annual wind energy production is presented table 3. Further, the monthly energy production and power production from wind farm distribution were displayed in following figure 5 and figure 6 respectively. A maximum monthly generation of 1.3 x 10^{7} kWh was noticed in the month of January while almost constant system power production around 34.5 MW was noticed with partial no power production for some days in the months of June, July and August over a year.

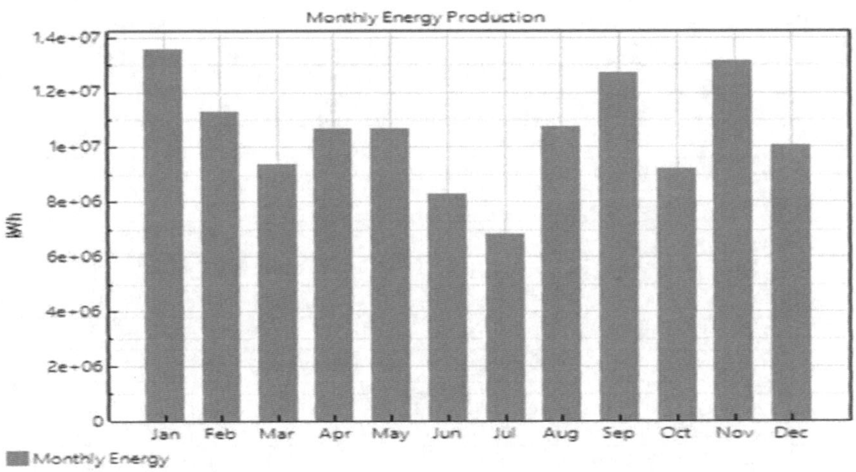

Figure 5. Monthly energy production using wind farm at the proposed site over a year.

Table 3. Annual wind energy production.

Metric	Value
Annual energy (year 1)	126,464,320 kWh
Capacity factor (year 1)	36.1%

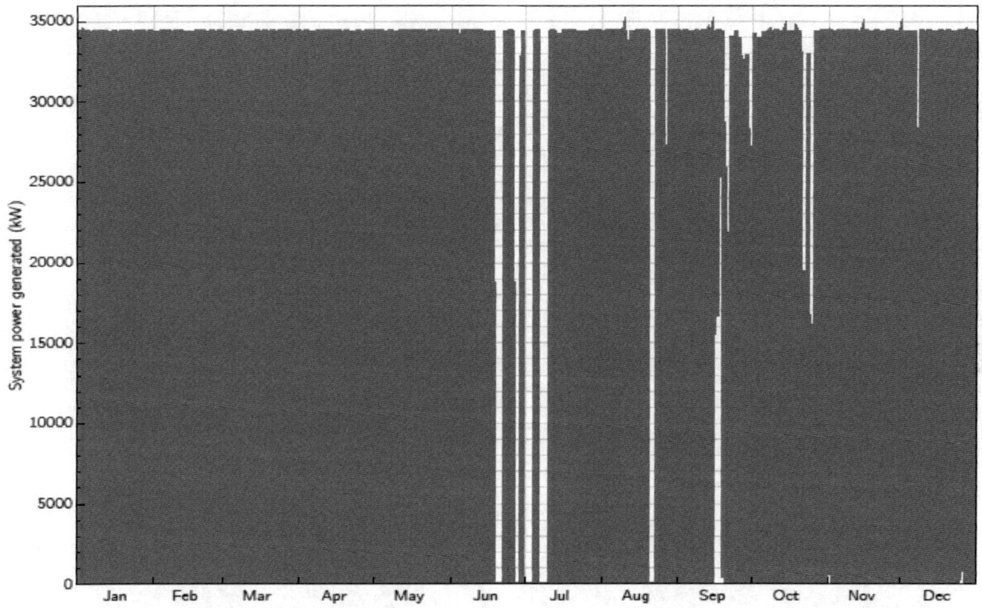

Figure 6. Power production from wind farm over a year at proposed site.

3.3 Hybrid system evaluation

By combining both solar and wind input parameters with Marco simulated parameter using generic system at the proposed project site location, it has been observed that nearly around 367844.44 MW of power can be generated with a capacity factor of 21.4%. Table 4 shows the simulation results for proposed hybrid system, while figure 7 and figure 8 shows monthly distributions of energy and power for the base case year 2018. A maximum monthly energy generation of 3.5 x 10^7 kWh in the month of May and a maximum system power production in the months of May, June and July around 150 MW can be clearly noticed.

Table 4. Simulation results of proposed 160 MW hybrid system.

Metric	Value
Annual energy (year 1)	367,844,448 kWh
Capacity factor (year 1)	21.4%

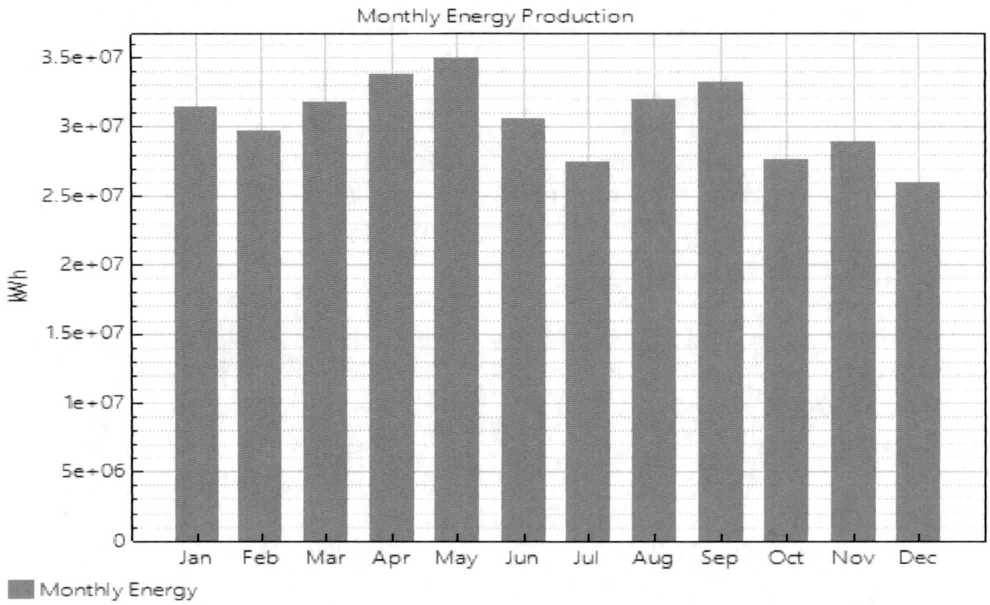

Figure 7. Simulated monthly energy generation of proposed hybrid system.

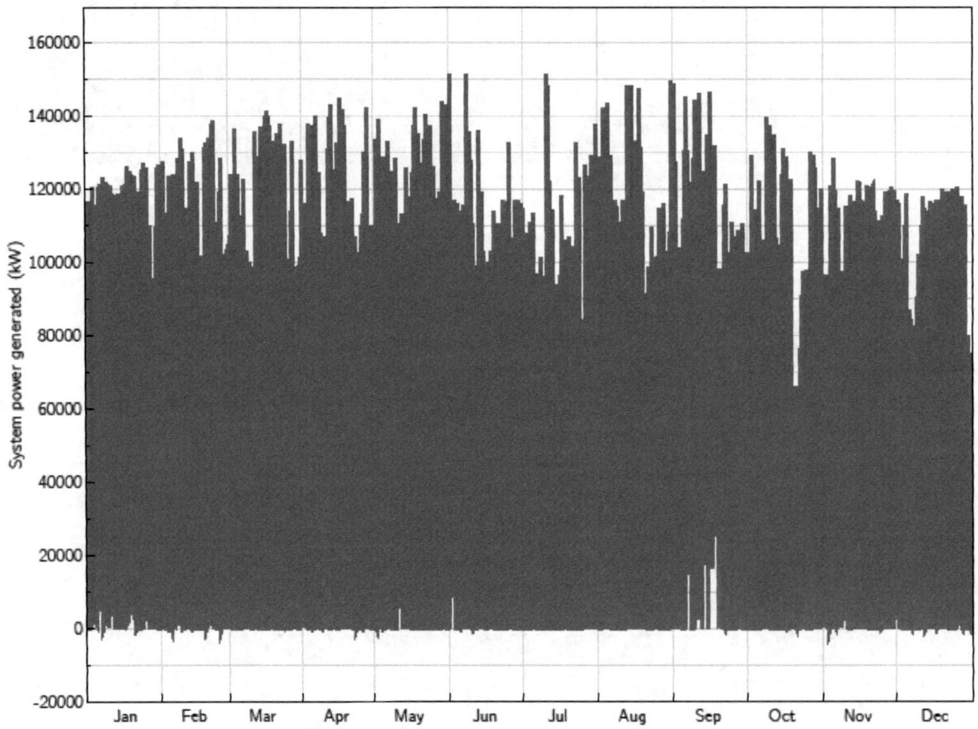

Figure 8. Simulated monthly power generation of proposed hybrid system.

4. Environment-Economic Evaluation

Hybrid power generation systems based on Solar PV and wind energy resources were recognized to be environmentally benign and sustainable with lower carbon footprint. Here environmental and economic evaluation of this hybrid system was carried out by considering technical characteristics and meteorological data of the proposed site in addition to other influencing aspects such as transportation, installation, operation and maintenance (O&M) costs using life cycle cost (LCC) assessment methods wherein costs incurred at each stage of system development, variation in O&M costs, tariff rates, discount rates and tax rates are taken into account in Indian context [7][8]. Further, this method considers all the costs involved, process flow and structure of the hybrid system by identifying energy supply, materials used, left out wastes to the environment before grid integration. The methodological analysis incorporated in this evaluation is based on the methodology outlined in [9] and the results obtained are presented in ensuing section.

5. Results and Discussion

The climatic condition at the proposed project site is favorable for the development of hybrid power generation based on solar as well as wind energy resources due to the abundance of sunshine and air flow throughout the year. The overall performance benefits and techno economic feasibility of the proposed 160 MW hybrid system are presented in the following subsections.

5.1 Energy yield estimation

In summary, the solar energy yield estimation is observed to be 241380.1 MWh per year using simulated software SAM whereas the theoretical estimation shows 2390174.5 MWh per year based on the methodology outlined [7]. Further, wind energy yield estimation is observed to be 126464.3 MWh per year using simulated software SAM whereas the theoretical estimations shows that wind energy is 128076 MWh per year. Finally, by integrating both solar and wind energy sources at the proposed location to form a hybrid energy source, estimates showed that 367844.4 MWh per year can be generated using SAM simulation software whereas theoretical estimations showed 367094.3 MWh indicating an efficiency of 15.9% and 15.8% respectively. The following table 5 shows a glimpse of energy yield estimation results that are obtained using manual and simulation approaches.

Table 5. Energy yield estimation results of the proposed hybrid system.

Source Mode	Manual energy yield outputs MWh/year	Simulated energy yield outputs MWh/year
Solar	239017.5	241380.1
Wind	128076.8	126464.3
Hybrid	367094.3	367844.4

5.2 Environmental benefits

Based on the methodological calculations outlined, and manual estimations were made and presented in table 6 which shows the resulting environmental benefits of the proposed 160 MW hybrid (solar-wind) system wherein one can clearly notice that the proposed project reduces the carbon footprint by saving 334281499 $kgCo_2$ while estimated saving based on SAM software simulations shows 334865340 $kgCo_2$. From economic point of view, the carbon benefit of the hybrid system is estimated to be nearly around 2.5 Cr/year.

Table 6. Environmental Benefit of system.

Estimation Methodology	Carbon Benefit (kgCo₂)	Carbon Benefit (Cr/year)
Using manual energy output data	334281499	2.5
Using simulated energy output data	334865340	2.5

5.3 Economic benefits

The Annual economic benefit of the proposed hybrid solar-wind energy system is calculated to be around 77.3 crores using methodological procedure outlined in [7] whereas it shows 77.5 crores based on SAM simulations. Further, the annual comprehensive benefit of the hybrid system is found to be around 80 crores based on the aforementioned methodological evaluation [7]. The following table 7 shows a glimpse of results that are obtained using manual and simulation approaches for assessment of economic benefit of the proposed system.

Table 7. Economic Benefit of the proposed hybrid system.

Estimation Methodology	Annual Economic Benefit of the system (Cr/year)	Annual Comprehensive Benefit of the system (Cr/year)
Using Manual energy Output	77.3	80
Using Simulated energy data	77.5	80

6. Conclusion

A well-designed hybrid power generation system based on solar and wind energy resources can be more consistent and economical which can significantly reduce not only fossil fuel consumption and associated carbon emissions but also offers fruitful economic benefits. The above results are shown by careful consideration of weather data, characteristics and losses of solar and wind energy resources over the proposed project site location using which one can predict the energy generation through hybrid system of similar kind along with procedures adopted for environmental and economic analysis. The study findings showed the techno economic and environmental feasibility of the proposed 160 MW hybrid power generation system with negative GHG emission. The present study can be extended by considering Battery Energy Storage System (BESS) and its sizing not only to meet Jawaharlal Nehru National Solar Mission (JNNSM) targets set by Indian government but also to meet prevailing demands in view of curtailment avoidance, deviation settlement avoidance, and ramp rate control towards grid stability for reliable power supply to consumers.

***Acknowledgments**:* The authors express their gratitude to the National Renewable Energy Laboratory (NREL) for permitting to utilize SAM software, Solar Energy Corporation of India (SECI); JNT University, Kakinada, and Centre for Smart Grid Technologies, Vellore Institute of Technology, Vellore, India for their support towards financial, instrumental and informative material support.

References

[1]. Sitharthan R, Geethanjali M and Pandy TKS 2016 Adaptive protection scheme for smart microgrid with electronically coupled distributed generations *Alexandria Engineering Journal* **55(3)** 2539-2550

[2]. Fathima AH, and Palanisamy K 2014 Battery energy storage applications in wind integrated systems—a review *IEEE International Conference on Smart Electric Grid* 1-8

[3]. Prabaharan N and Palanisamy K 2015 Investigation of single-phase reduced switch count asymmetric multilevel inverter using advanced pulse width modulation technique *International Journal of Renewable Energy Research* **5(3)** 879-890.

[4]. Jerin ARA, Kaliannan P and Subramaniam U 2017 Improved fault ride through capability of DFIG based wind turbines using synchronous reference frame control based dynamic voltage restorer. *ISA transactions* **70** 465-474

[5]. Sitharthan, R, Sundarabalan CK, Devabalaji KR, Nataraj SK and Karthikeyan M 2018 Improved fault ride through capability of DFIG-wind turbines using customized dynamic voltage restorer *Sustainable cities and society* **39** 114-125

[6]. Prabaharan N and Palanisamy K 2016 A single-phase grid connected hybrid multilevel inverter for interfacing photo-voltaic system *Energy Procedia* **103** 250-255

[7]. Palanisamy K, Mishra JS, Raglend IJ and Kothari DP 2010 Instantaneous power theory based unified power quality conditioner (UPQC) *IEEE Joint International Conference on Power Electronics, Drives and Energy Systems* 1-5

[8]. Sitharthan R and Geethanjali M 2017 An adaptive Elman neural network with C-PSO learning algorithm-based pitch angle controller for DFIG based WECS Journal of Vibration and Control **23(5)** 716-730

[9]. Sitharthan R and Geethanjali M 2015 Application of the superconducting fault current limiter strategy to improve the fault ride-through capability of a doubly-fed induction generator–based wind energy conversion system *Simulation* **91(12)** 1081-1087

[10]. Sitharthan R, Karthikeyan M, Sundar DS and Rajasekaran S 2020 Adaptive hybrid intelligent MPPT controller to approximate effectual wind speed and optimal rotor speed of variable speed wind turbine *ISA transactions* **96** 479-489

[11]. Sitharthan R, Devabalaji KR and Jees A 2017 An Levenberg–Marquardt trained feed-forward back-propagation based intelligent pitch angle controller for wind generation system *Renewable Energy Focus* **22** 24-32

[12]. Sitharthan R, Sundarabalan CK, Devabalaji KR, Yuvaraj T and Mohamed Imran A 2019 Automated power management strategy for wind power generation system using pitch angle controller *Measurement and Control* **52(3-4)** 169-182

[13]. Sundar DS, Umamaheswari C, Sridarshini T, Karthikeyan M, Sitharthan R, Raja AS and Carrasco MF 2019 Compact four-port circulator based on 2D photonic crystals with a 90° rotation of the light wave for photonic integrated circuits applications *Laser Physics* **29(6)** 066201

[14]. Sitharthan R, Parthasarathy T, Sheeba Rani S and Ramya KC 2019. An improved radial basis function neural network control strategy-based maximum power point tracking controller for wind power generation system *Transactions of the Institute of Measurement and Control* **41(11)** 3158-3170

[15]. Rajesh M and Gnanasekar JM 2017 Path observation based physical routing protocol for wireless ad hoc networks *Wireless Personal Communications* **97(1)** 1267-1289

[16]. Palanisamy K, Varghese LJ, Raglend IJ and Kothari DP 2009. Comparison of intelligent techniques to solve economic load dispatch problem with line flow constraints *IEEE International Advance Computing Conference* 446-452

[17]. Sitharthan R, Ponnusamy M, Karthikeyan M and Sundar DS 2019 Analysis on smart material suitable for autogenous microelectronic application *Materials Research Express* **6(10)** 105709

[18]. Rajaram R, Palanisamy K, Ramasamy S and Ramanathan P 2014 Selective harmonic elimination in PWM inverter using fire fly and fireworks algorithm *International Journal of Innovative Research in Advanced Engineering* **1(8)** 55-62

[19]. Sitharthan R, Swaminathan JN and Parthasarathy T 2018 March. Exploration of wind energy in India: A short review *IEEE National Power Engineering Conference* 1-5

[20]. Karthikeyan M, Sitharthan R, Ali T and Roy B 2020 Compact multiband CPW fed monopole antenna with square ring and T-shaped strips *Microwave and Optical Technology Letters* **62(2)** 926-932

[21]. Sundar D Sridarshini T, Sitharthan R, Madurakavi Karthikeyan, Sivanantha Raja A, and Marcos Flores Carrasco 2019 Performance investigation of 16/32-channel DWDM PON and long-reach PON systems using an ASE noise source *In Advances in Optoelectronic Technology and Industry Development: Proceedings of the 12th International Symposium on Photonics and Optoelectronics* 93

[22]. Sitharthan R and Geethanjali M 2014 Wind Energy Utilization in India: A Review *Middle-East J. Sci. Res.* **22** 796–801 doi:10.5829/idosi.mejsr.2014.22.06.21944

[23]. Sitharthan R and Geethanjali M 2014 ANFIS based wind speed sensor-less MPPT controller for variable speed wind energy conversion systems *Australian Journal of Basic and Applied Sciences* **8**14-23

[24]. Jerin ARA, Kaliannan P, Subramaniam U and El Moursi MS 2018 Review on FRT solutions for improving transient stability in DFIG-WTs *IET Renewable Power Generation* **12(15)** 1786-1799

[25]. Prabaharan N, Jerin ARA, Palanisamy K and Umashankar S 2017 Integration of single-phase reduced switch multilevel inverter topology for grid connected photovoltaic system *Energy Procedia* **138** 1177-1183

[26]. Rameshkumar K, Indragandhi V, Palanisamy K and Arunkumari T 2017 Model predictive current control of single phase shunt active power filter *Energy Procedia* **117** 658-665

[27]. Fathima AH and Palanisamy K 2016 Energy storage systems for energy management of renewables in distributed generation systems *Energy Management of Distributed Generation Systems* 157

[28]. Rajesh M 2020 Streamlining Radio Network Organizing Enlargement Towards Microcellular Frameworks *Wireless Personal Communications* 1-13

[29]. Subbiah B, Obaidat MS, Sriram S, Manoharn R and Chandrasekaran SK 2020 Selection of intermediate routes for secure data communication systems using graph theory application and grey wolf optimisation algorithm in MANETs IET *Networks* doi:10.1049/iet-net.2020.0051

[30]. Singh RR and Chelliah TR 2017 Enforcement of cost-effective energy conservation on single-fed asynchronous machine using a novel switching strategy *Energy* **126** 179-191

[31]. Amalorpavaraj RAJ, Palanisamy K, Umashankar S and Thirumoorthy AD 2016 Power quality improvement of grid connected wind farms through voltage restoration using dynamic voltage restorer *International Journal of Renewable Energy Research* **6(1)** 53-60

[32]. Singh RR, Chelliah TR, Khare D and Ramesh US 2016 November. Energy saving strategy on electric propulsion system integrated with doubly fed asynchronous motors *IEEE Power India International Conference* 1-6

[33]. Singh RR, Mohan H and Chelliah TR 2016 November. Performance of doubly fed machines influenced to electrical perturbation in pumped storage plant-a comparative electromechanical analysis *IEEE 7th India International Conference on Power Electronics* 1-6

IOP Conf. Series: Materials Science and Engineering **937** (2020) 012021 doi:10.1088/1757-899X/937/1/012021

On line monitoring and control of grid-interfaced PV system using zigbee

R Ramkumar[1], D Sivamani[2], K Preetha[3], D Shyam[4]
[1]Assistant Professor, Department of Electrical and Electronics Engineering, K.Ramakrishnan College of Technology, Trichy, Tamilnadu.
[2]Assistant Professor, Department of Electrical and Electronics Engineering, Rajalakshmi Engineering College, Rajalakshmi Nagar, Thandalam, Tamilnadu
[3]Assistant Professor, Department of Electrical and Electronics Engineering, Dhanalakshmi Srinivasan Engineering College, Perambalur, Tamilnadu.
[4]Assistant Professor, Department of Electrical and Electronics Engineering, Rajalakshmi Engineering College, Rajalakshmi Nagar, Thandalam, Tamilnadu

E.mail: ramkumarr.eee@krct.ac.in

Abstract. Grid connected photovoltaic system is used to watch the soundness and execution of the framework. The easiest strategy is to gather and transmit the information across information links. Customary wired checking framework gives dependable arrangement in information transmission yet experiences a few restrictions. The utilization of information links likewise builds establishment and upkeep cost. Information links are influenced by the sun pillar and downpour. Checking station should be found sensibly near the observing plant. To conquer these issue Zigbee based remote checking framework is created for web based observing of a lattice associated photovoltaic framework. A PC-based application coordinated with online capacity is planned and executed so as to permit controller of the framework just as simple access of the information over the web. Boundaries like temperature, illumination, PV power yield and framework inverter power yields are checked. To approve the presentation, the framework has been executed on 100Wp network associated photovoltaic framework.

1. Introduction

As the worldwide vitality request increments with the developing total populace, nations everywhere throughout the world are putting increasingly more accentuation on the improvement of sustainable power source. Among numerous wellsprings of sustainable power source, sun based vitality is viewed as the most encouraging and solid vitality source [1]. In the light of this, legislatures in numerous nations have given different motivations to arrangement sunlight based vitality based force plants, to supplement the current force plants which are running on petroleum product. So as to guarantee dependability and unwavering quality of a PV framework, observing framework is frequently liked. As matter of actuality, numerous ongoing sun based vitality transformation frameworks have included observing capacity as a fundamental piece of the framework to guarantee information can be gathered and broke down in precise way [2].

Ordinary wired checking framework gives dependable arrangement in information transmission yet experiences a few restrictions. Aside from the physical imperatives during laying of the information links, the utilization of these links likewise expands establishment and support cost. In addition, for open air application, for example, PV frameworks, ceaseless introduction to sun shaft and rains may diminish the life expectancy of the framework. To defeat these issues, remote observing framework is preferred over its link based partner. In this paper, a Zig Bee-based remote observing framework is structured and worked as a substitution to the customary link based checking framework for a matrix tied PV framework. This configuration makes it possible for the different parts of the framework, from plan to development and testing, is point by point here. Other than that, a PC-based application coordinated with electronic capacity is structured and executed so as to permit controller of the framework just as simple access of the information over the web [3].

2. Existing system

2.1 Wired system

So as to guarantee dependability and unwavering quality of a PV framework, observing framework is frequently liked. As far as information move instrument, both wired and remote frameworks have been presented before For wired frameworks, information transmissions are generally done utilizing RS232 link or RS485 link with the checking frameworks being PC-based. Customary wired observing framework gives relilable arrangement in information transmission however experiences a few restrictions, for example, physical imperatives during laying of the information links, the utilization of these links additionally builds establishment and support cost. Plus, for open air application, for example, PV frameworks, consistent introduction to sun pillar and downpours may lessen the life expectancy of the framework. In this paper, the MATLAB result also shown [4].

2.2 Wireless sytem

To conquer these issues, remote checking framework is preferred over its link based partner. For framework utilizing remote information transmission, a more extensive assortment of information transmission innovation has been accounted for, for example, the utilization of Satellite, GSM, Bluetooth, WiFi, Zigbee and other vague RF gadgets.

2.3. *Satellite*

A communication satellite is an artifical satellite that relays and amplifiers radio telecommunication signals via a transponder. Wireless communication uses electromagnetic waves to carry signals. Information transmission by means of satellite was accounted for to be moderate, taking around 8 to 12min and require significant expense of establishment.

2.4. *GSM*

The Global System of Mobile correspondence is progressively solid with exactness of information transmission by means of SMS upto 100%. It additionally show low retransmission rate and low all out information misfortune pace of roughly 2.73% and 0.66% individually. The principle downside of GSM is its high working expense, as the client needs to pay for the information transmission administration.

2.5. *RF Device*

RF information transmission was a serious mainstream methods for remote information move framework. Radio correspondence has the chance of sending and getting a tremendous measure of data requiring little to no effort of transmission, and it is additionally a decent option in far off region which doesn't have phone lines. Its primary burden is the trouble in acquiring authorization for the transmission recurrence and the significant expense of its establishment.

2.6. *Bluetooth*

Bluetooth is one of the straightforward remote systems administration framework however it just spread the short separations contrasted with different remote systems administration framework.

2.7. *Wi-Fi*

Wi-Fi is the another elective innovation in remote correspondence where it is large information move geography. In any case, the main burden is the expense of Wi-Fi gadget is generally higher contrasted with Wi-Fi.

3. Proposed method

To improve the performance and reduces complexity restriction that is brought about by the current framework, another framework proposed here [5]. ZigBee is that it apportions exceptional schedule opening to stay away from information impact. At the point when the past information is still in support, ZigBee will spare the approaching information to a unique schedule opening and hold up until the cradle is cleared before recovering the information from the time allotment. In this manner, it is necessary to determine the information crash can be maintained a strategic distance from[6]. Moreover, the ZigBee work geography permit the joining of different remote hub which makes it upgradable to help enormous system limit up to 65,536 hubs contrasted with 7 and 32 hubs for ZigBee and Wi-Fi individually, While working in work geography, ZigBee is open minded to disappointment, for example, gadget drop, where it empowers the gadget to build up another connection between accessible hubs. These highlights make ZigBee a well known alternative for information transmission [7-8].

3.1. Design Specification

3.1.1. Solar panel

This block consists of solar panel to give electrical power to charge battery. It is a 22V and 51W Solar panel. The power output is varied with sun intensity, so at day time power is stored in battery and used at night (or) low intensity time.

3.1.2. Battery charger control

This block consists of a full wave bridge rectifier circuit which converts the generated AC voltage to DC.

3.1.3. Battery

This block consists of 12V/7AH for battery to store the power.

3.1.4. Temperature sensor

For this application, the heat has for fever sensors must be some place in the scope of 0°C and 100°C. Two units of LM35 Precision Centigrade Temperature Sensor are used to check the vibe and board temperature. The sensor has an evaluating range from +2°C to +150° C.Thusly, it has a touch of room over direct temperature sensors balanced in Kelvin as it encourage the arrangement technique using progressed multimeter and thermocouple. Since temperature is a moderate advancing variable, only a solitary data is obtained for each assessing stretch.[17-19]

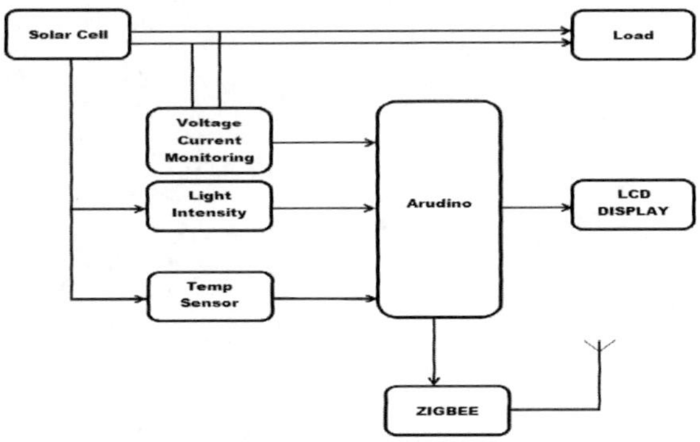

Figure 1 Square Diagram of transmitter

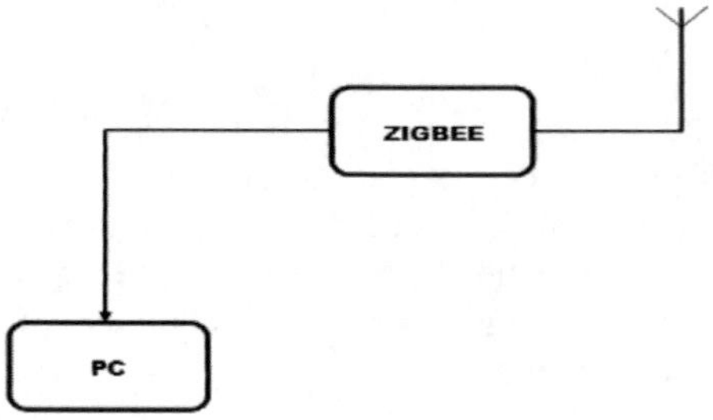

Figure 2 Square Diagram of receiver

3.1.5. *Solar current and inverter current detecting*
Current sensor for DC and AC current must have the option to gauge input current scope of 0-80A LA25-NP is utilized as current transducer and it can quantify up to 36A DC and 25A AC current.
The gadget bundle takes into account simple usage by the client. Ordinary applications incorporate engine control, load recognition and the executives, exchanged mode power supplies, and over current deficiency security. The gadget comprises of an exact, low-balance, straight lobby sensor circuit with a copper conduction way situated close to the outside of the kick the bucket. [16]

3.1.6. Light intensity measurement
Light Dependent Resistance (LDR) connected with 3.3K ohm resistor. As light falls on LDR, opposition of LDR diminishes, because of which simple voltage is produced, later apply this voltage to Arduino.

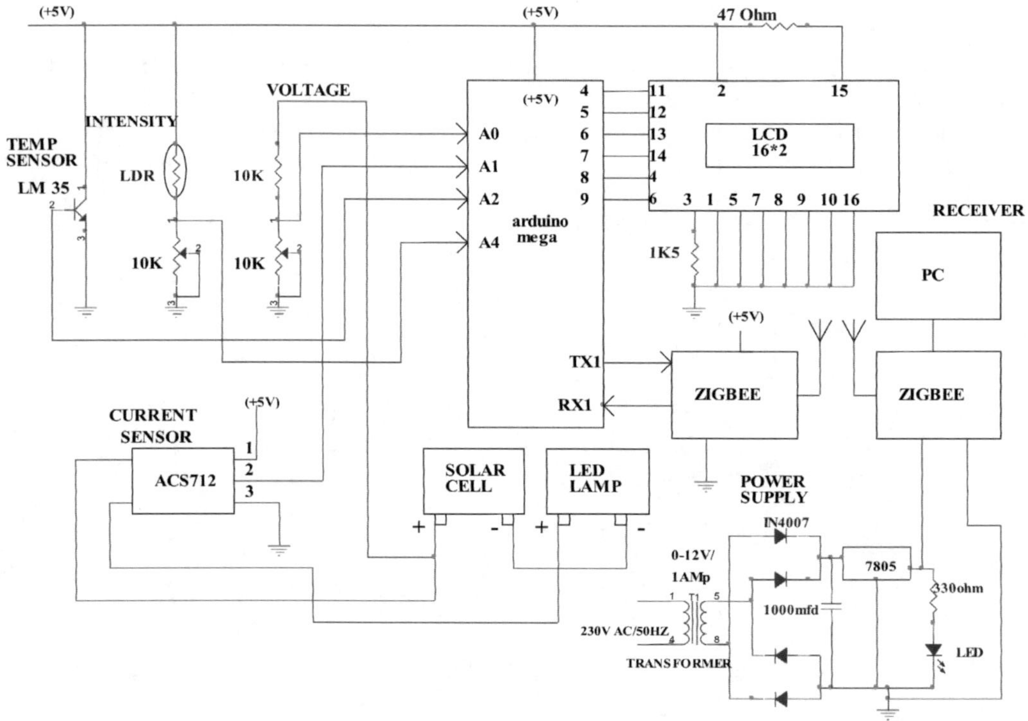

Figure 3 Circuit diagram

IOP Conf. Series: Materials Science and Engineering **937** (2020) 012021 doi:10.1088/1757-899X/937/1/012021

4. Circuit specification

In this project, four sensors are used. The temperature sensor type LM35 which is connected to the Arduino analog input pin A0. For the measurement of light intensity, LDR is used which is connected to the Arduino analog pin A1. ACS712 is used for current measurement that is connected to the Arduino analog input pin A2. Voltage from the solar panel is connected to the voltage divider, whose output is feeded to the Arduino analog input pin A3. Supply for the sensor units is given by the power supply circuit. The Arduino here is used for the data logging purpose. The stored data is viewed at that place itself by using 16*2 LCD display and then it is transmitted via the ZigBee transmitter which is interfaced with the Arduino. The transmitted data is received by using ZigBee receiver which is located at the far end from the transmitter section [9-10].

Transmission of the data is done by the serial port USB cable. The ZigBee receiver is connected to the personal computer using serial port to USB convertor cable and it is continuously monitor by personal computer [11-12].

4.1. ARDUINO

Arduino is a device for seeming well and good and control a greater amount of the physical world than your PC [13]. It's an open-source physical figuring stage dependent on a basic microcontroller board, and an improvement domain for composing programming for the board. Arduino can be utilized to create intuitive items, taking contributions from an assortment of switches or sensors, and controlling an assortment of lights, engines, and other physical yields. Arduino undertakings can be independent, or they can be spoken with programming running on PC (e.g. Flash, Processing, MaxMSP.) The sheets can be collected by hand or bought preassembled; the open-source IDE can be downloaded for nothing [14-15].

Arduino can detect the earth by accepting contribution from an assortment of sensors and can influence its environmental factors by controlling lights, engines, and different actuators. The microcontroller on the board is customized utilizing the Arduino programming language (in light of Wiring) and the Arduino improvement condition (in view of Processing). Arduino tasks can be independent or they can speak with programming on running on a PC (for example Streak, Processing, MaxMSP). Arduino is a cross-stage program. Arduino is utilized in following OS.

- Windows
- Mac OS X
- Linux

4.1.1. Description

This is the new MEGA2560 R3. Notwithstanding all the highlights of the past board, the MEGA currently utilizes an ATMega16U2 rather than the ATMega8U2 chip. This mulls over speedier trade rates and more memory. The capacity to have the Uno appears as a console, mouse, joystick, and so on [16].

4.1.2. Characteristics

- Schematic plan of the open source advancement interface free download, and furthermore as indicated by the requirements of their own changes.
- Download the program is basic and helpful.
- Just with the sensor, a wide scope of electronic segments association, (for example, LED light, bell, keypad, photo resistor, and so on.), make a wide range of fascinating things
- Utilizing the rapid small scale preparing controller (ATMEGA328).
- The improvement of language and advancement condition is exceptionally basic, straightforward, and entirely reasonable for fledglings to learn [17-18].

IOP Conf. Series: Materials Science and Engineering **937** (2020) 012021 doi:10.1088/1757-899X/937/1/012021

Figure 4 Arduino Mega 2560 Kit

5. Results and discussion

5.1. Transmission side

Sun based radiation; a pyranometer from Li-Cor is set on the PV board. The chose pyranometer has an estimating range from 0 W/m2 to 1276 W/m2, and creates a corresponding voltage yield of 0–5 V. This estimating range is satisfactory for this venture, since the irradiance level of the site area is relied upon to be under 1000 W/m2, The sun powered irradiance is estimated once every examining span

Figure 5 Hardware transmission kit

5.2. Receiver side

The force created and vitality from PV module, vitality yield from converter and combined force produced from PV module, are shown. Utilizing the "View" button, clients can likewise observe the constant estimations of the checked boundaries as show, as illustrated in Fig. 5 & 6

Figure 6 Hardware receiver kit

5.3. LCD display

Figure 6 LCD display

5.4. System output

Since the information estimations are done at an inspecting time span min, the invigorate pace of the website page is set to be 1 min too., the data indicated will reliably have a 1 min delay from the constant information gathered by the observing framework.

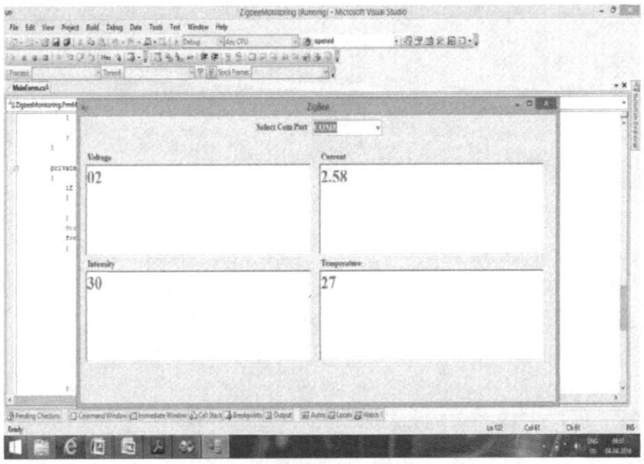

Figure 7 System output

By and by, considering the way that the framework is proposed to screen all the boundaries 24 h daily, this postponement can be viewed as decent.

6. Conclusion

In this task, the usage of an online remote checking framework for matrix associated PV converter is introduced. Subtleties in general procedure of the usage, beginning from framework structure plan, equipment execution, control framework programming to the electronic checking interface have been secured. Other than having control work which can screen the framework distantly, the framework is likewise outfitted with online capacity which made it available at wherever and whenever through the web. Usage results show that the framework can work with great execution.

7. References

[1] Alena, R., Gilstrap, R., Baldwin, J., Stone, T., & Wilson, P. (2011). Fault tolerance in ZigBee Wireless sensor networks. Aerospace Conference, 2011, 1–15.

[2] A.Nazar Ali and R. Jayabharath (2014), "Performance Enhancement of Hybrid Wind/Photo Voltaic System Using Z Source Inverter with Cuk-sepic Fused Converter." Research Journal of Applied Sciences, Engineering and Technology7.19 pp. 3964-3970.

[3] C.Kalavalli, S.R.Paveethra, S.Murugesan, Dr.A.Nazar Ali, (2020), Design And Implementation Of High Efficiency H6 PV Inverter With Dual Axis Tracking, International Journal of Scientific & Technology Research, Vol 9, issue 02, pp. 4728-31.
[4] S.R.Paveethra, C.Kalavalli, S.Vijayalakshmi, Dr.A.Nazar Ali, D.Shyam(2020), Evaluation Of Voltage Stability Of Transmission Line With Contingency Analysis, International Journal of Scientific & Technology Research, Vol 9,issue 02, pp. 4018-22
[5] Shyam D, Premkumar K, Thamizhselvan T, Nazar Ali A, Vishnu Priya M(2019),Symmetrically Modified Laddered H-Bridge Multilevel Inverter with Reduced Configurationally Parameters ,International journal of engineering and advanced technology ,Vol 9, issue 1
[6] A Ali Nazar, R Jayabharath, MD Udayakumar(2014) ,An ANFIS Based Advanced MPPT Control of a Wind-Solar Hybrid Power Generation System, international review of modeling and simulations,vol.7, no. 4, pp. 638–643.
[7] Ramkumar, R., and N. Tejaswini. "A novel low cost three arm Ac automatic voltage regulator." *Advances in Natural and Applied Sciences*, vol. 10, no. 3, 2016
[8] Ramkumar, R., et al. "PV based SEPIC converter fed four switch BLDC motor drive." *Advances in Natural and Applied Sciences*, vol. 10, no. 3, 2016
[9] R. Ramkumar, B. Karthikeyan, Dr. A. Rajkumar, V. Venkatesh and A. Anton Amala Praveen," Design and Implementation of IOT Based Smart Library Using Android Application," Biosc.Biotech.Res.Comm.Special Issue Vol 13 No (3) 2020 Pp-56-62.
[10] B.Karthikeyan, R. Ramkumar, L. Nagarajan, A. Anton Amala Praveen and S. Vijayalakshmi," ANN Based MPPT Controller for PEMFC System with Interleaved Resonant PWM High Step up Converter" Biosc.Biotech.Res.Comm. Special Issue Vol 13 No (3) 2020 Pp-44-47.
[11] Dr. P. R. Sivaraman, R. Jaiganesh, P. Ragupathy and R. Ramkumar," HI-TECH Electoral Machine for Election Commission of India," Biosc.Biotech.Res.Comm. Special Issue Vol 13 No (3) 2020 Pp-13-17.
[12] D. Sivamani, P. Thirusenthil Kumaran, C. Gnanavel and L. Nagarajan," Powering Smart Street Lighting Cities Using Visible Light Communication and Light Fidelity," Biosc.Biotech.Res.Comm.Special Issue Vol 13 No (3) 2020 Pp-180-184.
[13] A Nazar Ali, D Sivamani, R Jaiganesh M Pradeep (2019), Solar powered air conditioner using BLDC motor, IOP Conference Series: Materials Science and Engineering, vol. 23
[14] Ramkumar, R., et al2019. Intelligent system to monitor and diagnose performance deviation Industrial Equipment.IOPconf.Series: Materials science and Engineering, Vol 623(2019)012012.
[15] A. T. Sankara Subramanian, P. Sabarish, M. D.Udayakumar and T. Vishnu kumar," Performance Analysis of Various Photovoltaic Configurations Under Uniform Shading and Rapid Partial Shading Formations", Biosc.Biotech.Res. Comm. Special Issue Vol 13 No (3) 2020 Pp-185-192.
[16] P. Sabarish et al 2019 IOP Conf. Ser.: Mater. Sci. Eng. 623 012011.
[17] M D Udayakumar et al 2019 IOP Conf. Ser.: Mater. Sci. Eng. 623 012018.
[18] P. Sabarish, A. T. Sankara Subramanian, S. Murugesan and V. Sureshkumar,"A New Methodology of Arterial Blood Clot Removal Using Bio Molecular Devices for ATPase Nuclear Motors." Biosc.Biotech.Res.Comm. Special Issue Vol 13 No (3) 2020 Pp-197-201.

IOP Conf. Series: Materials Science and Engineering **937** (2020) 012022 doi:10.1088/1757-899X/937/1/012022

Performance analysis of solar updraft tower using Ansys fluent

G Anilkumar[1*], G R Krishna[2], G V Brahmendra Kumar[3] and K Palanisamy[4]

[1,2] School of Renewable Energy, JNT University, Kakinada, 533003, India
[3,4]Centre for Smart Grid Technologies, Vellore Institute of Technology, Vellore, 632014, India.
Email: brahmendrakumar.g@gmail.com, kpalanisamy@vit.ac.in
*Corresponding author: gorantlaanilkumar@gmail.com

Abstract. The solar updraft tower (SUT) is a renewable energy power plant design concept for generating electricity by utilizing convective air flow due to low temperature heat absorption from incoming solar radiation. A parametric study on the performance power output of the SUT at various canopy angles and different insolation levels were carried out to investigate the relationship between them with respect to key parameters i.e. in terms of updraft tower velocity, temperature, pressure and height to choose turbine position for maximum air velocity absorption and energy conversion using Ansys-fluent software version 15 for analysis. The results obtained were analyzed comparatively to draw conclusions. The results identified using simulation software mentioned above were validated with earlier results and identified significant improvement in the power generated at various collector angles, when compared with the earlier works on SUT.

1. Introduction

Allocation Solar Updraft Tower (SUT) utilizes three basic governing principles i.e. greenhouse effect, convection and wind power generation to convert incoming solar radiation into electricity [1]. As direct and diffuse solar radiation strikes the surface of glass roof or canopy, reflection, refraction, transmission and absorption take place depending on the incidence angle and optical characteristics of the canopy which can be measured in terms of physical parameters like refractive index, thickness and extinction coefficient [2]. Apart of the transmitted radiation through canopy was absorbed while another part gets reflected back towards the canopy and undergoes further reflection[3-6]. This leads to multiple reflections in-between canopy and surface of the ground so that enhanced energy absorption takes place at the surface of the ground due to transmittance-absorptance product of the ground resulting in production of hot air beneath the canopy. The buoyant air rises up in plant chimney by natural convection and initiates forced convection by allowing more amount of air to flow with in the vicinity of canopy [7-12]. Through mixed convection, the lower part of canopy gets heated up due to presence of warm air while a part of energy is absorbed by the cooler part of the ground surface in addition to radiation exchange between warm ground surface and canopy [13-20]. Further, due to natural and forced convection, heat exchange occurs

IOP Conf. Series: Materials Science and Engineering **937** (2020) 012022 doi:10.1088/1757-899X/937/1/012022

over canopy surface with adjacent ambient air. While the air parcels flow upward from canopy to chimney, its temperature increases and velocity remains constant due to increase in height of the canopy [21,22]. Finally, air parcels get cooled at the terminal walls of the chimney where heat energy is converted into kinetic energy [23-30]. The difference in pressure across the chimney base and its outlet can be known by measuring density difference of respective air parcels which in turn depend on associated inlet and outlet temperatures across the chimney. The requisite pressure difference required to drive the turbine can be regulated by minimizing chimney's entrance, exit and friction losses. As the air passes through the turbine, the kinetic energy associated with it drives the turbine blades which in turn drive the generator [30-35].

2. Methodology

The objective of this work is to develop a simplified design model of solar updraft tower power plant to investigate the effects of the canopy angle engulfment and turbine location on fluid flow. Decreasing the overall tower height would decrease the overall construction cost and improve the appearance of the chimney which leads to decreased power output and reduced efficiency of the system. This indicates that there is a scope for optimization. Thus, in search of optimizing the same, the goal of this work is to conduct parametric study and identify maximum power output by varying canopy orientation with respect to ground at various insolation levels using simplified two-dimensional simulation model based on Ansys fluent version 15 software [2].

3. Solar updraft tower design

The experimental prototype of SUT design can be carried out in four different phases such as geometry phase, meshing phase, solution setup phase and modelling phase which are presented here as follows.

3.1 Phase-I: Geometry Selection

Ansys fluent 15 software pack with full version was used for modelling SUT design with prototype dimensions as shown in table 1. The corresponding geometry can be modelled by opening fluent in Ansys 15 workbench wherein by selecting, geometry → 2D analysis model → XY plane → optional selection of angles and parameters can be further selected for modification. Using this procedure, SUT design based on the opted parametric dimensions displayed in table 1 can be simulated. After attaining the desired tower structure, one can save the same by procedural clicking of options in menu bar → concept → sketches → surface → generate → save the SUT design in a folder.

Table 1. Dimensions of SUT prototype.

Parameters	Dimensions
Height of the tower	2 m
Length of the base	3.2 m
Diameter of the tower	0.4 m
Angle of canopy	0,5,10 degrees
Height of the inlet	0.127,0.15,0.17 m

3.2. Phase-II: Meshing

After completing the geometry, select the mesh option in the Ansys workbench. Meshing is a process that involves dimension setup, refinement, face mapping of the desired structure (SUT) usually achieved by fragmenting the desired component into similar individual panel components to study the uniform distribution effect of applied load on the entire structural composition. Since each element has its own stiffness while loading, the stress developed on the entire structure can be calculated by summing up all individual stiffness's of the components to attain global stiffness matrix.

The mesh of each part in SUT is generated separately as its components can be split into three different zones i.e. canopy zone, transition section between canopy and tower zone. A secondary refinement of the mesh can be considered when panel size is too wide or sensitive to obtain better quality of the mesh. As an example, a strong pressure gradient will exist at the transition zone i.e. between canopy and tower zone where secondary refinement is recommended due to its sensitiveness in the computational domain. Finally, skewness ratio is examined to check desired mesh quality. Here the system geometry considered for examination consists of tower with height 2 meters (m) and 0.4 m diameter having a base length of 3.2 m and provision to vary not only the orientation of canopy angle between 0°-10° but also the height of the inlet between 0.127 m to 0.17 m.

3.3 Phase-III: Solution Setup

Once the geometry and meshing were done for desired SUT structure, general settings for the solution setup has to be applied by opting Ansys workbench 15 → click on Solution setup → launch fluent with double precision. The general setting applied for examination is showed in table 2.

Table 2. Solution setup general settings.

Type	**Pressure based**
Velocity formulation	Absolute
Type	Steady
2D space	Planar
Gravitational acceleration	-9.8 m/s^2

3.4 Phase-IV: Modeling

After completing the above three phases successfully, the presumptions required for the SUT modeling has to be carried out by choosing requisite options and setting desired options such as Energy equation → ON, K-Epsilon model with standard mode with buoyancy effect [6], select air as material (boussinesq) [7], Operating conditions → 300K, insertion of boundary conditions like insolation (heat flux) values at different parts, setting solution initialization as standard initialization. Finally, clicking on 'Run calculations' by insertion of desired value for number of iterations to be carried out.

4. Results and Discussion

By following the above-mentioned procedure, simulation results can be obtained where contour maps and graphs pertaining to different parameters under consideration can be observed for desired SUT power plant structure. Here it is intended to analyze the variations in basic physical parameters such as temperature, pressure and velocity by varying the canopy angles and insolation levels to explore their interdependence for extracting maximum power output. The following observations were noticed [8] while carrying out parametric study on the original design of the simulation prototype.

4.1 Contour maps at 0° canopy angle and 600 Wm^{-2}

Figure 1a, 1b and 1c represents contour maps of variation in pressure, temperature and air velocity at 0° canopy angle and 600 Wm^{-2} while table 3 displays variations in pressure, temperature and velocity with respect to variation in height at 0° canopy angle at 600 Wm^{-2} insolation. From figure 1a, it can be observed that the pressure is least at center of the tower and increases with increase in the height of the tower. Pressure varies from inlet to the outlet of the tower which creates an updraft and leads to mass flow in the tower wherever pressure difference occurs. From figure 1b, it can be noticed that, as the heat flux emanate from the base, the temperature below the canopy is identified to be more than the outside temperature and the temperature at inlet of the tower is identified to be maximum whereas with the increase in height of the tower, there is a decrease in the value of temperature. From figure 1c, it is clearly evident that air velocity remains constant throughout the tower but there is a change in velocity at middle and bottom of the tower with maximum velocity at a height of 0.3 m, indicating suitability for turbine placement position to attain maximum turbine efficiency.

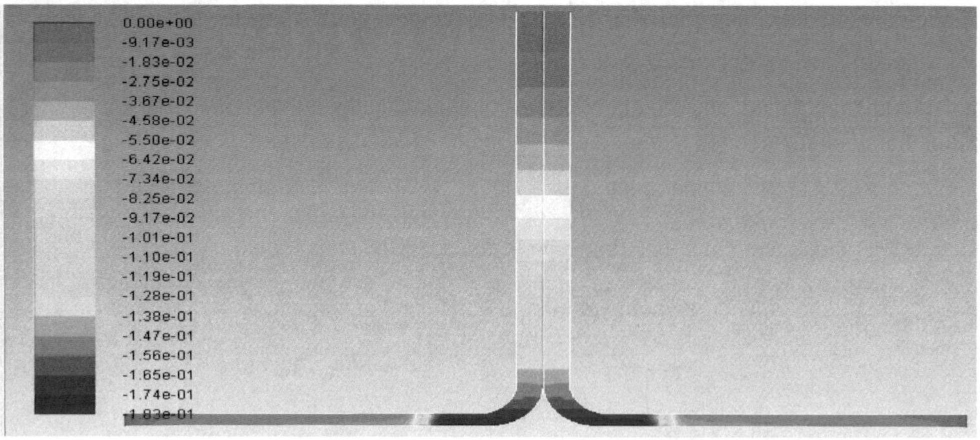

Figure 1a. Pressure contour maps of SUT at 0° canopy angle and 600 Wm^{-2}.

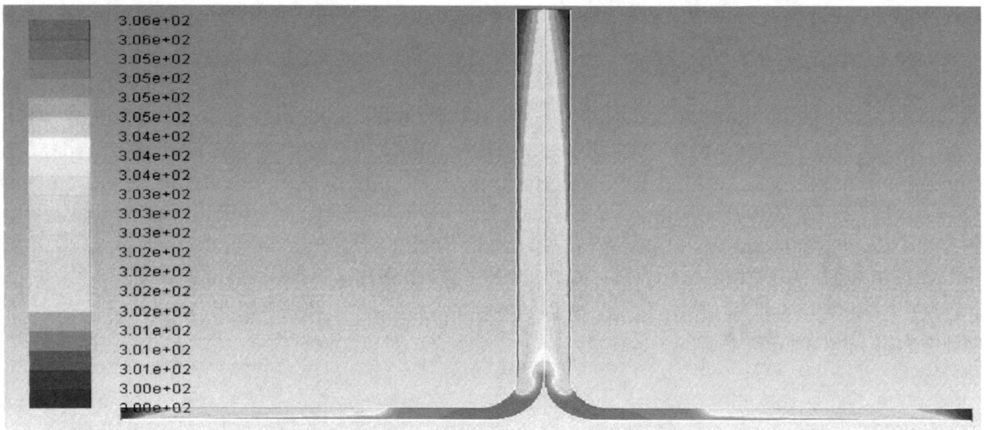

Figure 1b. Temperature contour maps of SUT at 0° canopy angle and 600 Wm^{-2}.

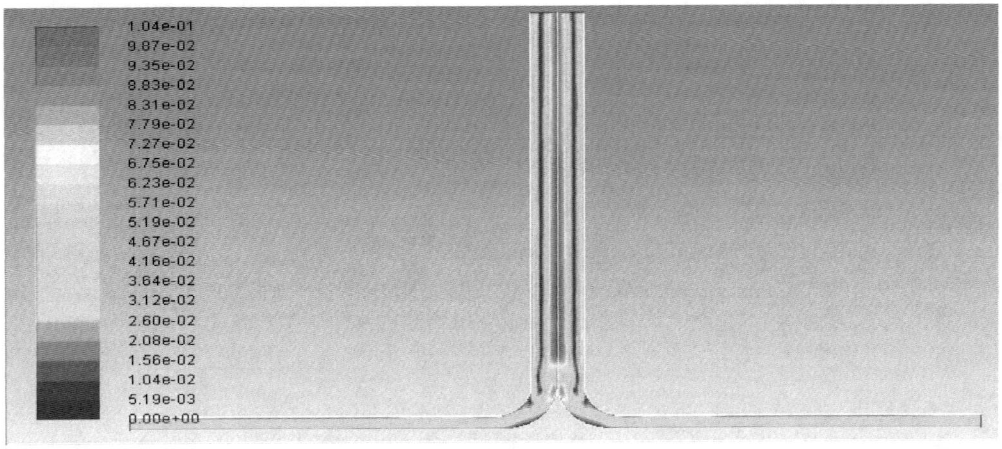

Figure 1c. Air velocity contour maps of SUT 0° canopy angle and 600 Wm^{-2}.

Table 3. Parametric values for constant insolation at 0° canopy value (without slope).

Height (m)	Pressure (Pa)	Temperature (K)	Velocity (ms^{-1})
0.5	-0.08	303.3	0.105
1	-0.045	302.5	0.101
1.5	-0.02	302	0.098
2	0	301.7	0.096

4.2 Pressure contour maps obtained at 5 ° canopy angle with varying Insolation

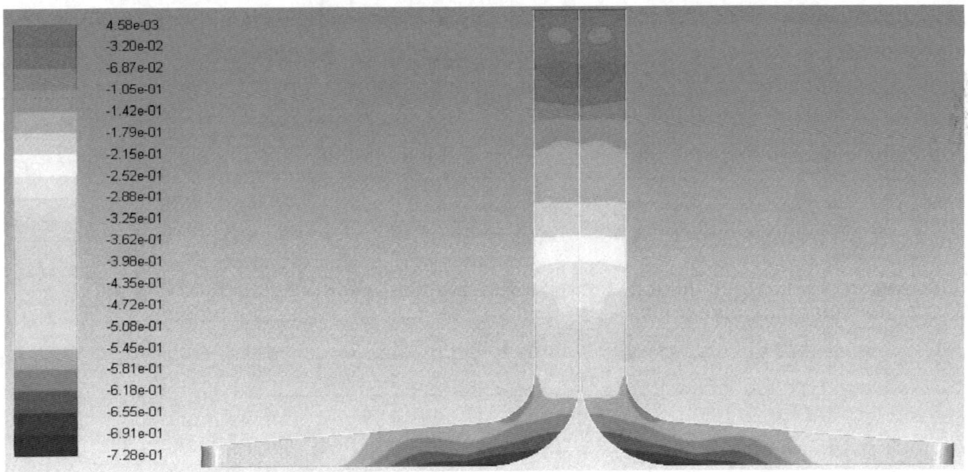

Figure 2a. Pressure contour maps of SUT plant at 600 Wm^{-2} at 5° canopy value.

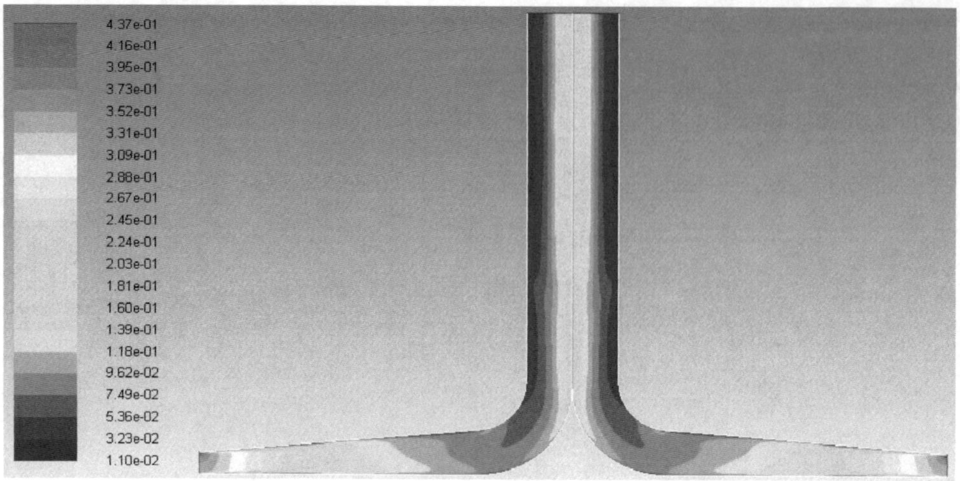

Figure 2b. Pressure contour maps of SUT plant at 800 Wm^{-2} at 5° canopy value.

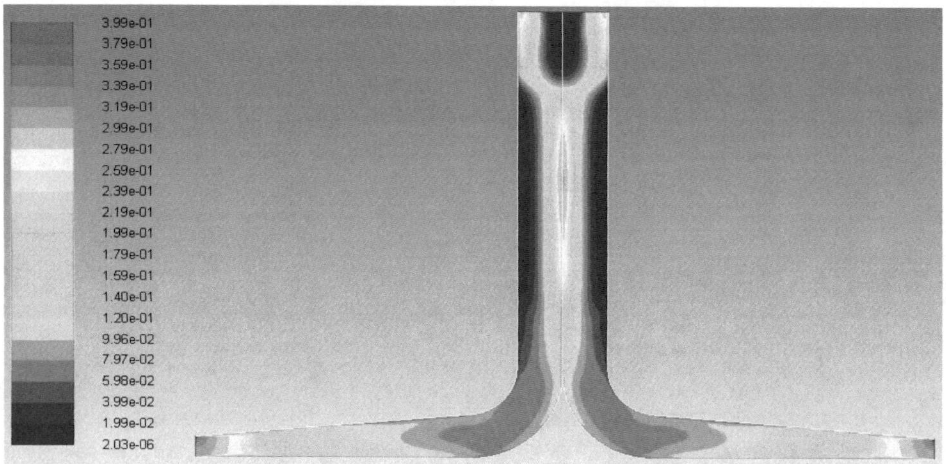

Figure 2c. Pressure contour maps of SUT plant at 1000 Wm^{-2} at 5° canopy value.

Table 4. Pressure values for different insolation at 5° canopy angle.

Height (m)	600 Wm^{-2}	800 Wm^{-2}	1000 Wm^{-2}
0.5	-0.52	-0.26	-0.22
1	-0.28	-0.19	-0.18
1.5	-0.14	-0.13	-0.11
2	-0.06	-0.04	0.2

Figure 2a, 2b and 2c represents contour maps of variation in pressure at 5° canopy angle with respect to varying insolation at 600 Wm^{-2}, 800 Wm^{-2} and 1000 Wm^{-2} while table 4 displays variations in pressure with respect to variation in height of the tower at 5° canopy angle at different insolation levels of

600 Wm^{-2}, 800 Wm^{-2} and 1000 Wm^{-2}. From figure 2a, it can be observed that the pressure at the inlet of the tower is low and gradually increase at the outlet of the tower. Thus, pressure difference is created in the tower. From the figure 2b it can be noticed that at 800 Wm^{-2}, pressure at the inlet of the tower is low and gradually increase at the outlet of the tower there by creating pressure difference in the tower. From figure 2c, it is clearly evident that the pressure at the inlet of the tower is low and gradually increases towards the outlet of the tower. Thus, pressure difference is created in the tower.

4.3. Temperature contours obtained for different insolation values

Figure 3a, 3b and 3c represents contour maps of variation in temperature at 5° canopy angle with respect to varying insolation at 600 Wm^{-2}, 800 Wm^{-2} and 1000 Wm^{-2} while table 5 displays variations in temperature with respect to variation in height of the tower at 5° canopy angle at different insolation levels of 600 Wm^{-2}, 800 Wm^{-2} and 1000 Wm^{-2}. From figure 3a, it can be observed that at 600 Wm^{-2} insolation, change in temperature is low with maximum temperature at the base due to absorption and reflection of solar radiation beneath the canopy. From figure 3b i.e. at 800 Wm^{-2} insolation, no significant change with respect to figure 3a except that tower temperature slightly increases when compared to temperature beneath the canopy. From figure 3c i.e. at 1000 Wm^{-2}, it is clearly evident that temperature inside the tower is more than the canopy temperature.

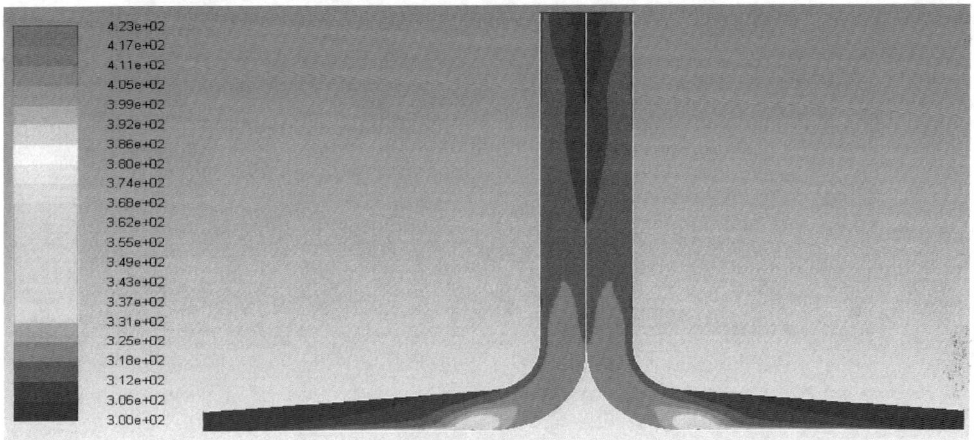

Figure 3a. Temperature contour maps of SUT plant at 600 Wm^{-2} at 5° canopy value angle.

IOP Conf. Series: Materials Science and Engineering **937** (2020) 012022 doi:10.1088/1757-899X/937/1/012022

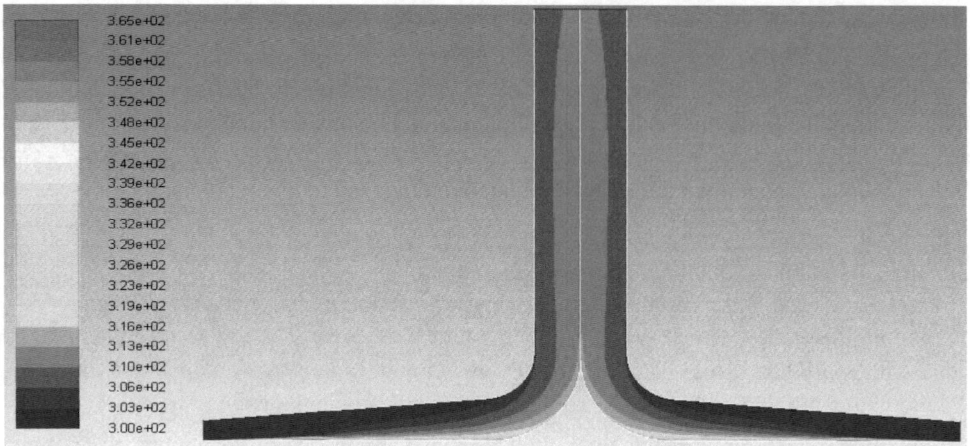

Figure 3b. Temperature contour maps of SUT plant at 800 Wm^{-2} at 5° canopy angle.

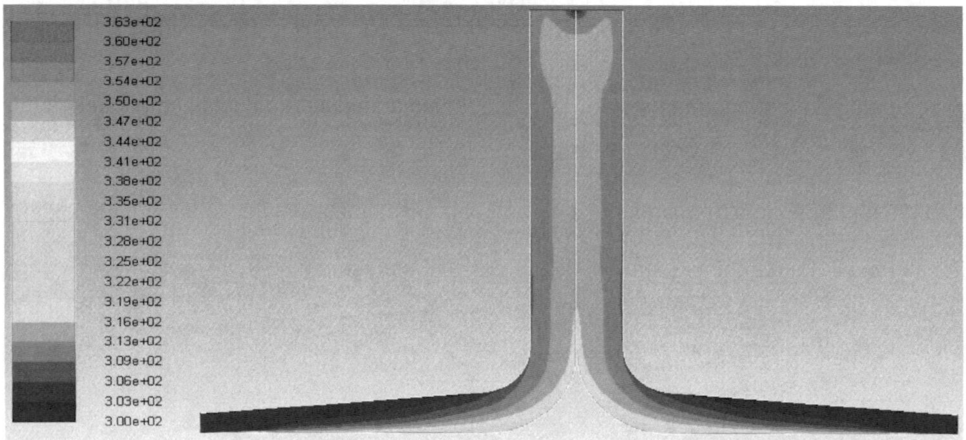

Figure 3c. Temperature contour maps of SUT plant at 1000Wm^{-2} at 5° canopy angle.

Table 5. Temperature values for different insolation levels at 5° canopy value.

Height (m)	600 Wm^{-2}	800 Wm^{-2}	1000 Wm^{-2}
0.5	309.5	313	315.8
1	307	311.5	314.4
1.5	304	310.5	313.6
2	303.5	310	313.2

4.4. Velocity contours obtained for different insolation values

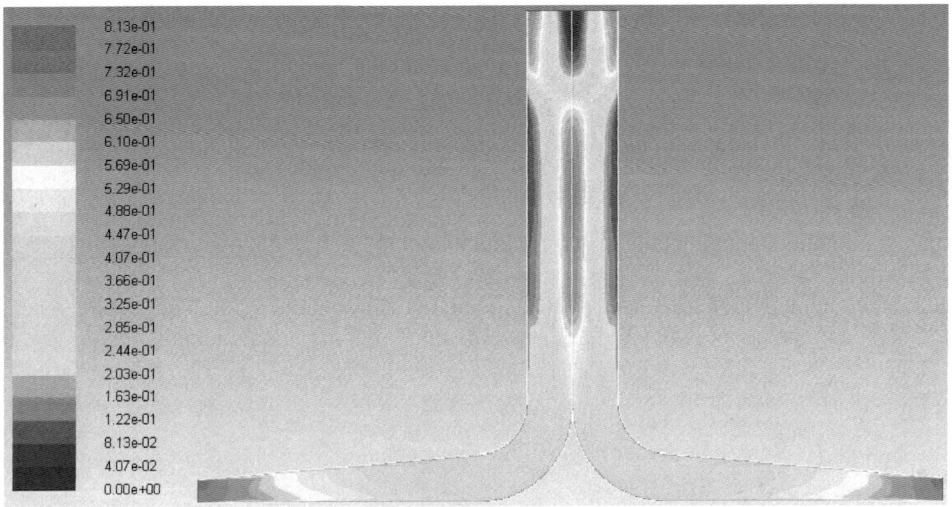

Figure 4a. Velocity contour maps of SUT plant at 600 Wm^{-2} at 5° canopy angle.

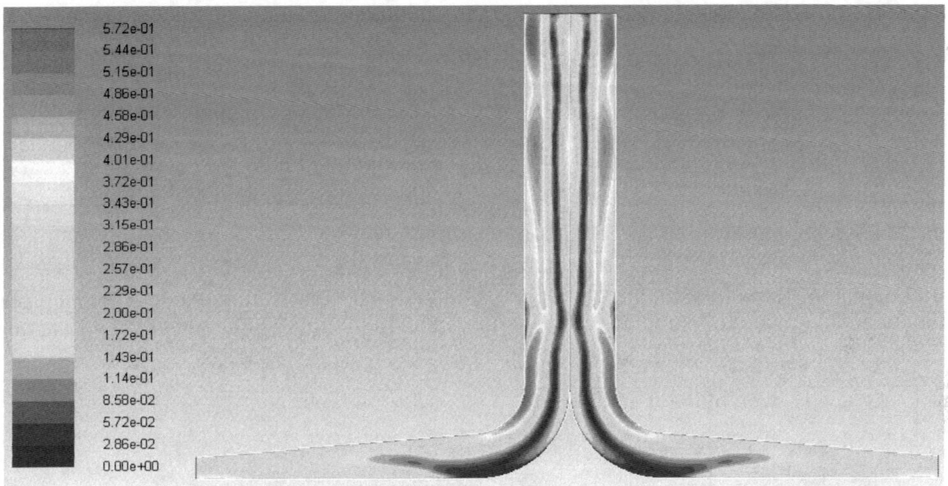

Figure 4b. Velocity contour maps of SUT plant at 800Wm^{-2} at 5° canopy angle

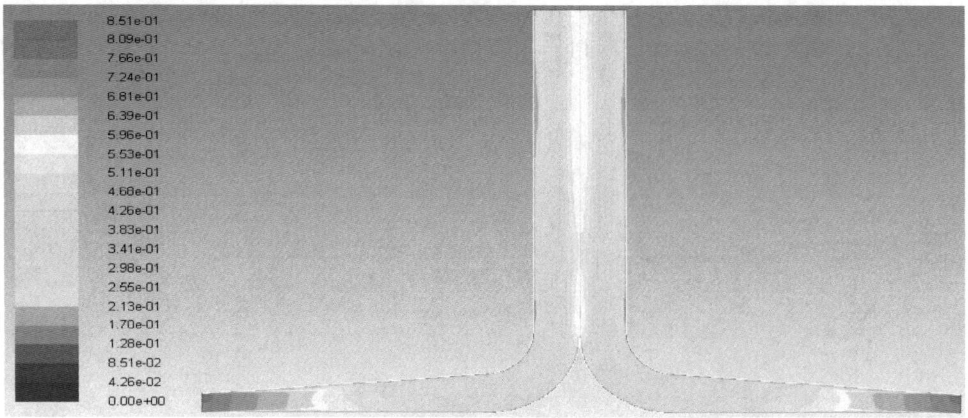

Figure 4c. Velocity contour maps of SUT plant at 1000 Wm^{-2} at 5° canopy angle.

Table 6. Velocity values for different insolation levels at 5° canopy value.

Height (m)	600 Wm^{-2}	800 Wm^{-2}	1000 Wm^{-2}
0.5	0.19	0.5	0.58
1	0.16	0.51	0.62
1.5	0.25	0.56	0.65
2	0.21	0.55	0.68

Figure 4a, 4b and 4c represents contour maps of variation in velocity profile at 5° canopy angle with respect to varying insolation at 600Wm^{-2}, 800Wm^{-2} and 1000Wm^{-2} while table 6 displays variations in velocity with respect to variation in height of the tower at 5° canopy angle at different insolation levels of 600 Wm^{-2}, 800 Wm^{-2} and 1000 Wm^{-2}. From figure 4a, a maximum velocity at the center of the tower can be observed that at 600 Wm^{-2} and 800 Wm^{-2} insolation levels, indicating suitability of turbine placement for attaining maximum efficiency whereas at 1000 Wm^{-2}, maximum velocity at the inlet periphery of the collector can be clearly noticed from figure 4c, indicating desirable placement of turbine to extract maximum efficiency with an option to place more than one turbine.

4.5 Contour maps obtained at 10° canopy angle

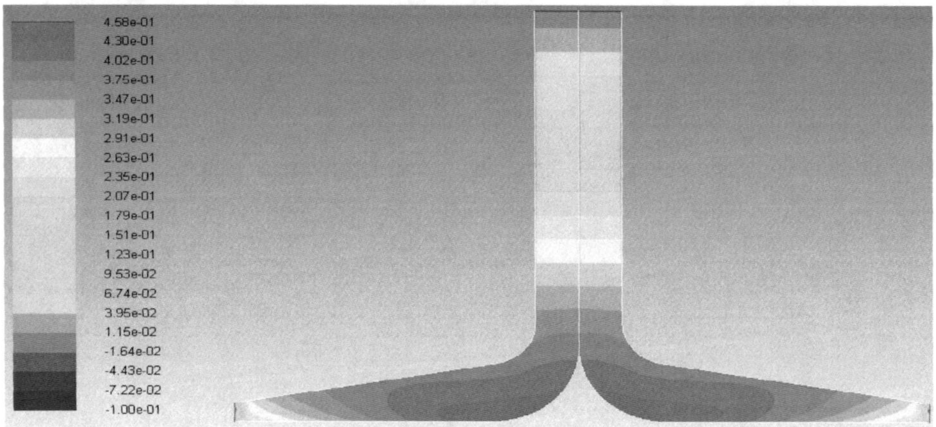

Figure 5a. Pressure contour maps of SUT plant at 600Wm^{-2} at 10° canopy angle.

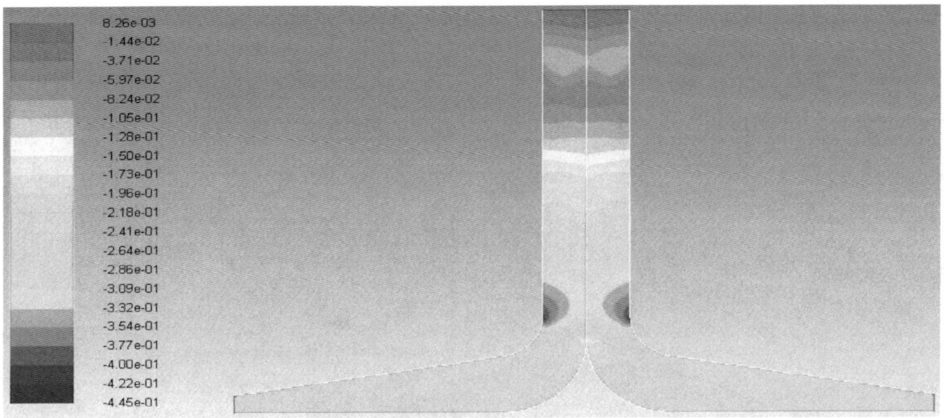

Figure 5b. Pressure contour maps of SUT plant at 800 Wm^{-2} at 10° canopy angle.

Figure 5c. Pressure contour maps of SUT plant at 1000 Wm^{-2} at 10° canopy angle.

Table 7. Pressure values for different insolation levels at 10° canopy value.

Height (m)	600 Wm^{-2}	800 Wm^{-2}	1000 Wm^{-2}
0.5	-0.41	-0.34	-0.28
1	-0.26	-0.24	-0.21
1.5	-0.15	-0.12	-0.1
2	-0.04	-0.03	0.02

Figure 5a, 5b and 5c represents contour maps of variation in pressure profile at 10° canopy angle with respect to variation in insolation at 600 Wm^{-2}, 800 Wm^{-2} and 1000 Wm^{-2} while table 7 displays variations in pressure with respect to variation in height of the tower at 10° canopy angle at different insolation levels of 600 Wm^{-2}, 800 Wm^{-2} and 1000 Wm^{-2}. At 600 Wm^{-2}, it can be observed that the pressure at inlet of the tower is high and gradually decreases at the outlet of the tower from figure 5a thereby allowing air flow within the tower due to pressure difference. By examining the figure 5b and figure 5c i.e. at 800 Wm^{-2} and 1000 Wm^{-2} insolation levels, pressure at the inlet of the tower is observed to be low and increases gradually at the outlet of the tower leading to occurrence of pressure difference within the tower.

4.6 Temperature contours obtained for different insolation values

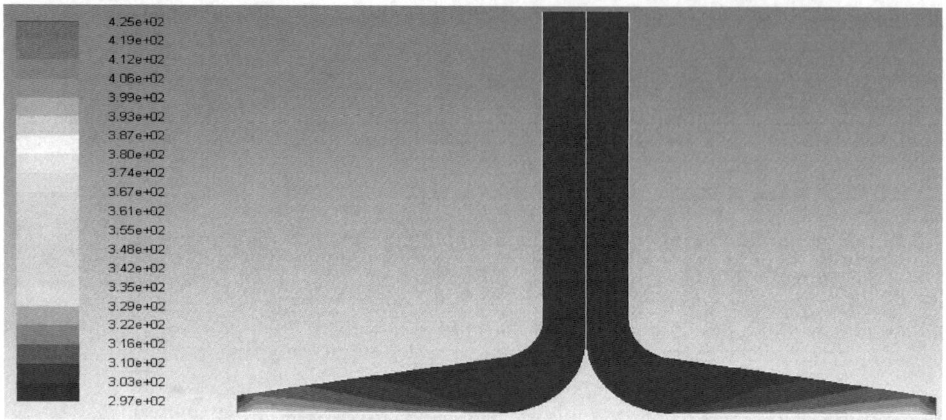

Figure 6a. Temperature Contour maps of SUT plant at 600 Wm^{-2} at 10° canopy angle.

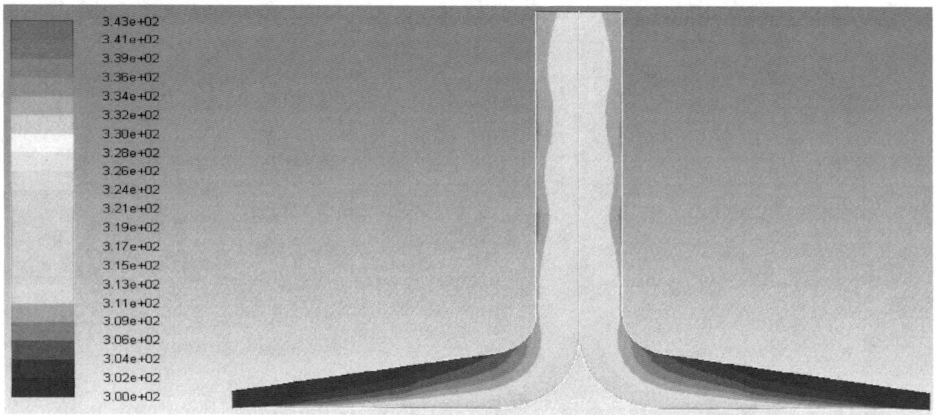

Figure 6b. Temperature Contour maps of SUT plant at 800 Wm^{-2} at 10° canopy angle.

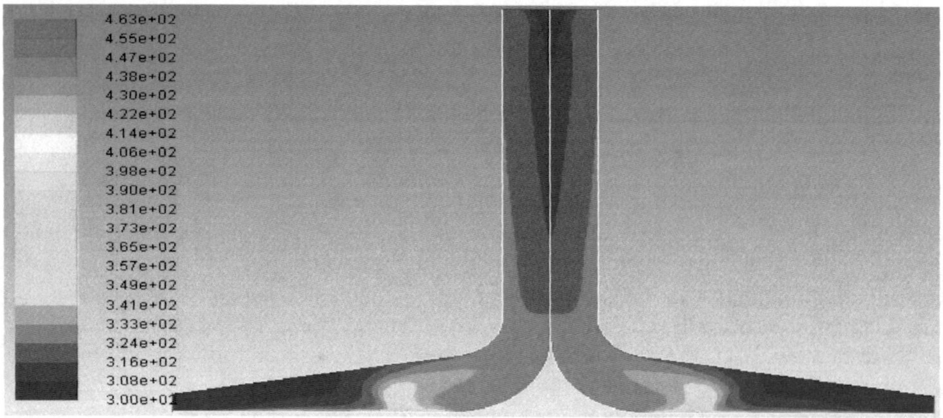

Figure 6c. Temperature Contour maps of SUT plant at 1000 Wm^{-2} at 10° canopy angle.

Table 8. Temperature values for different insolation levels at 10° canopy value.

Height (m)	**600 Wm^{-2}**	**800 Wm^{-2}**	**1000 Wm^{-2}**
0.5	309	317.1	327.5
1	307	311.8	317
1.5	306	311.6	313.3
2	305	311.2	312.1

Figure 6a, 6b and 6c represents contour maps of variation in temperature profile of SUT at 10° canopy angle with respect to variation in insolation at 600 Wm^{-2}, 800 Wm^{-2} and 1000 Wm^{-2} while table 8 displays variations in temperature with respect to variation in height of the tower at 10° canopy angle at different insolation levels of 600 Wm^{-2}, 800 Wm^{-2} and 1000 Wm^{-2}. At 600 Wm^{-2} insolation i.e. from figure 6a, it can be observed that the temperature at the base is high due absorption of incoming radiation and reflection from base beneath the canopy whereas at 800 Wm^{-2} and 1000 Wm^{-2} insolation, no significant change with respect to figure 6a, except that the inside temperature of tower is slightly more than temperature corresponding to area underneath canopy.

4.7 Velocity contours obtained for different insolation values

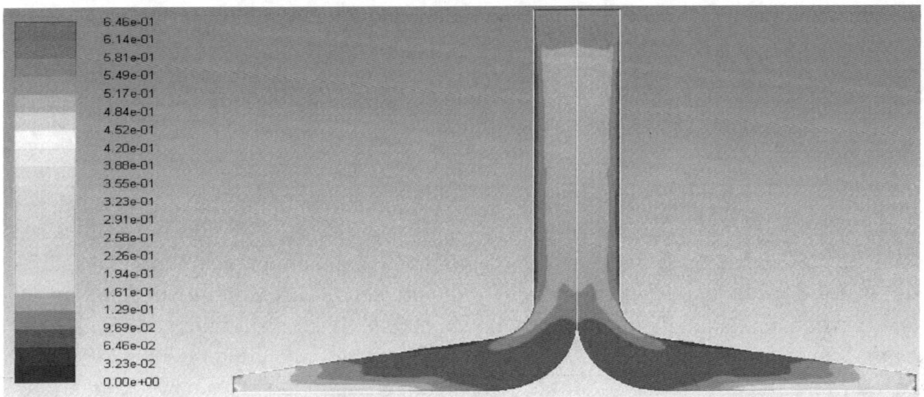

Figure 7a. Velocity contour maps of SUT plant at 600 Wm^{-2} at 10° canopy angle.

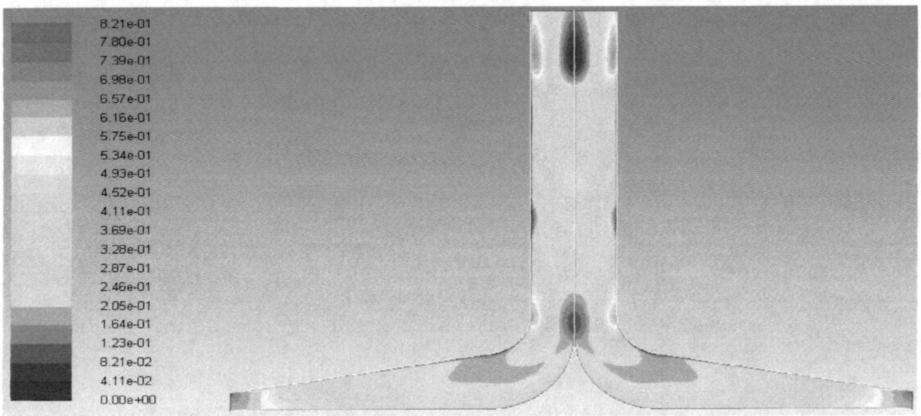

Figure 7b. Velocity contour maps of SUT plant at 800 Wm^{-2} at 10° canopy angle.

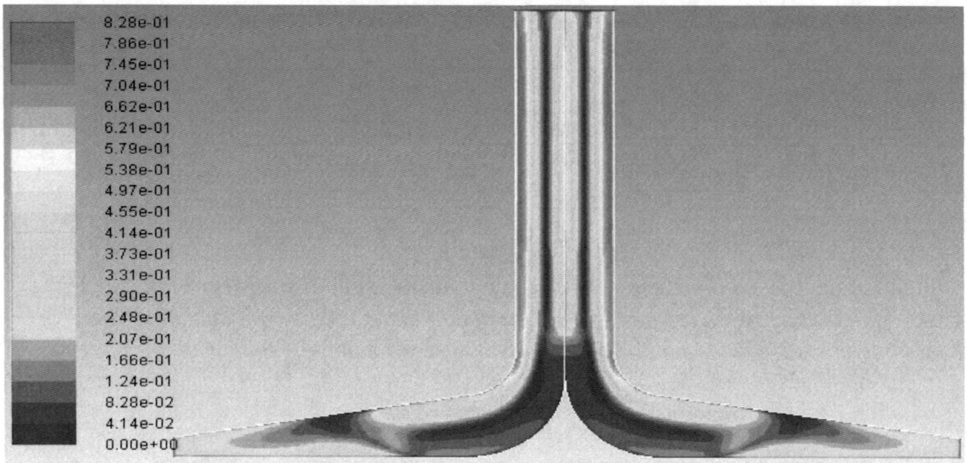

Figure 7c. Velocity contour maps of SUT plant at 1000 Wm^{-2} at 10° canopy angle.

Table 9. Velocity values for different insolation levels at 10° canopy value.

Height (m)	600 Wm^{-2}	800 Wm^{-2}	1000 Wm^{-2}
0.5	0.102	0.12	0.14
1	0.140	0.26	0.34
1.5	0.136	0.28	0.38
2	0.130	0.08	0.36

Figure 7a, 7b and 7c represents contour maps of variation in velocity profile (blue color in the legend) of SUT at 10° canopy angle with respect to variation in insolation at 600 Wm^{-2}, 800 Wm^{-2} and 1000 Wm^{-2} while table 9 displays variations in velocity with respect to variation in height of the tower at 10° canopy angle at different insolation levels of 600 Wm^{-2}, 800 Wm^{-2} and 1000 Wm^{-2}.

At 600 Wm^{-2} insolation level i.e. from figure 7a maximum velocity at the center of the tower can be observed, indicating the suitability of turbine placement for maximum efficiency. From figure 7b, at 800 Wm^{-2} maximum velocity inside the tower can be examined in addition to distribution of velocity profile across the tower due to enhanced recirculation of air inside the tower. Although velocity remains constant, velocity maximum can be noticed at the inlet of the collector thereby indicating suitability of turbine placement beneath the canopy for attaining maximum efficiency of SUT while a constant velocity profile with a maximum velocity at the inlet of the collector. Here it should be noticed that power generated by the turbine may decrease due to recirculation of air.

5. Conclusion

Simulation was conducted using the Ansys fluent software, which solved the energy equation using the Boussinesq approximation. The standard k–epsilon turbulence model was used together incorporating the solar ray tracing model for a more accurate result.

The inlet and outlet ambient temperature was assumed at 300 K. All walls were considered as adiabatic except for the solar collector and the soil cover beneath the solar collector. The solar collector modeling parameters were defined as non-slip, convective heat transfer and radiation. And as for the soil layer, it was assumed as a thermal storage with a heat absorption coefficient of 0.8. The numerical model was also validated using previous numerical results obtained from IIT Dhanbad and Marian College of engineering Thiruvananthapuram.

The parametric study was conducted for the angle of inclination of the slope collector, with fixed height of the solar tower configuration with either a fixed entrance or a fixed exit and different solar radiation for the optimized geometry. The objective of varying these angles and insolation is to determine the most suitable position for installation of the power generation turbine. It was important to have a high velocity at the upstream of the turbine. The power generation turbine utilized the updraft velocity, which means the greater the velocity, the faster the turbine can turn to generate more power.

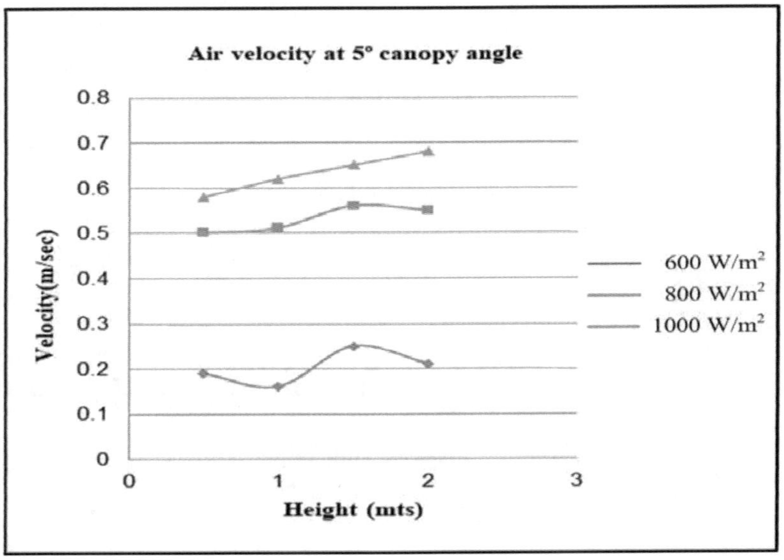

Figure 8a. Air velocity vs Height at 5° canopy angle.

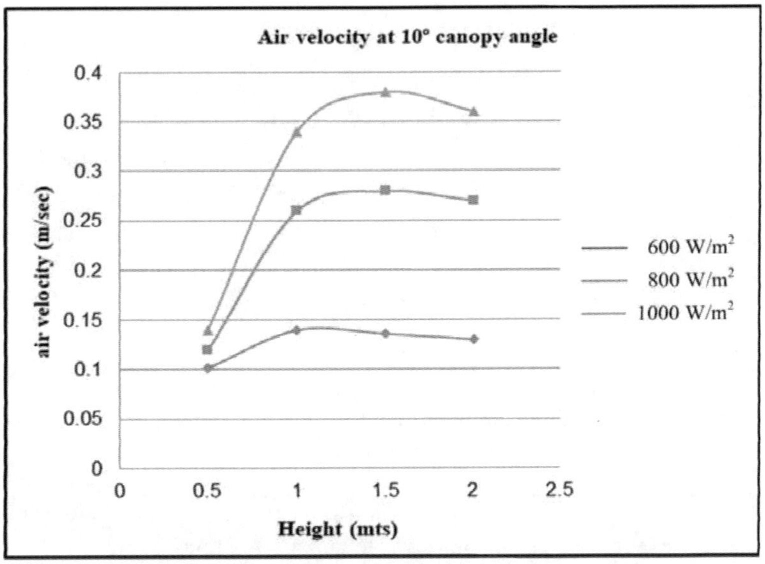

Figure 8b. Air velocity vs Height at 10° canopy angle.

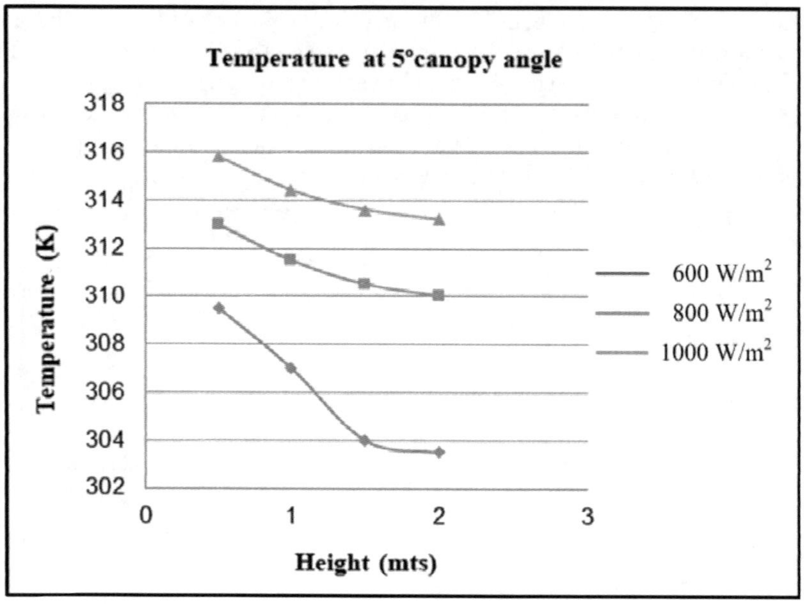

Figure 9a. Temperature vs Height at 5° canopy angle.

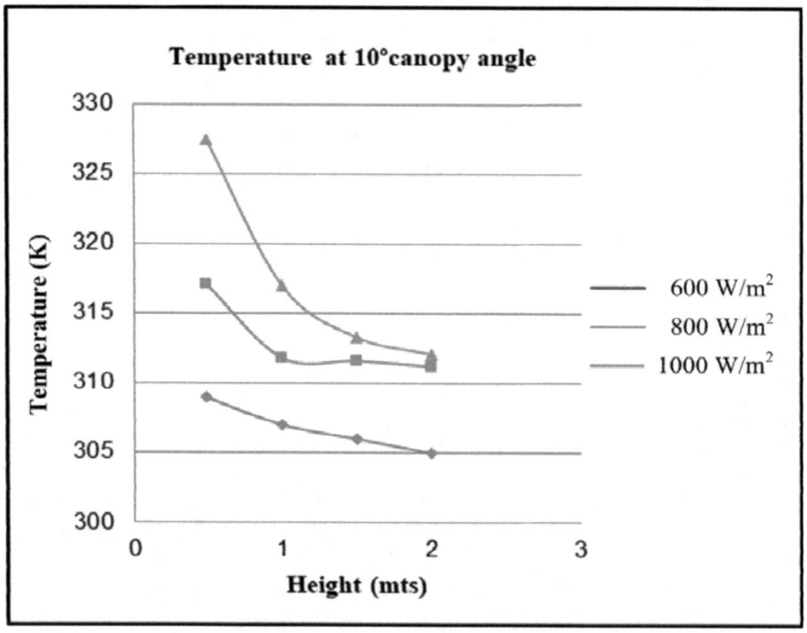

Figure 9b. Temperature vs Height at 10° canopy angle.

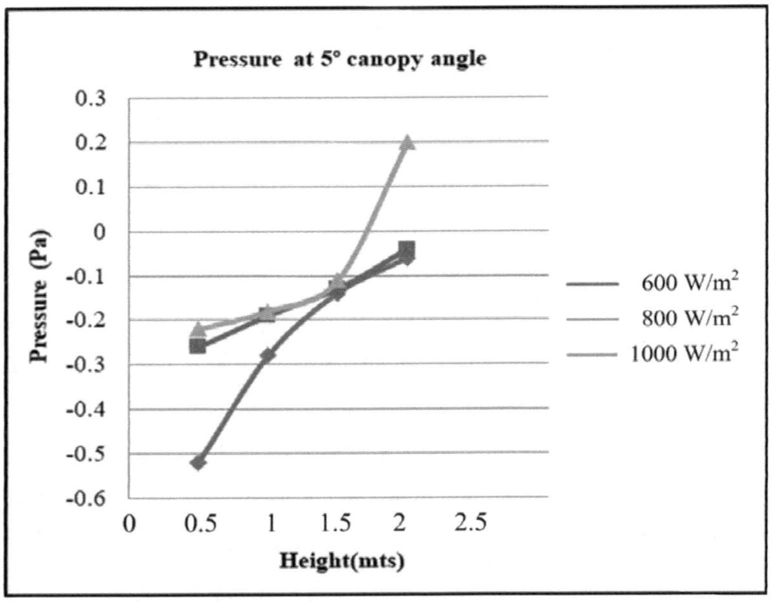

Figure 10a. Pressure vs Height at 5° canopy angle.

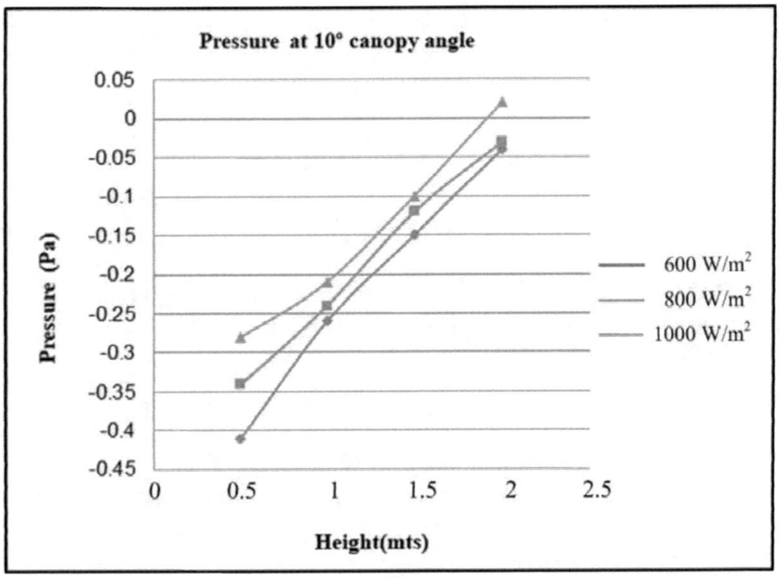

Figure 10b. Pressure vs Height at 10° canopy angle.

Figures 8a, 9a and 10a represents the parametric relationships of air velocity, temperature and pressure with respect to variation in height at 5° canopy angle while figures 8b, 9b and 10b represents the parametric relationships of air velocity, temperature and pressure with respect to variation in height at 10° canopy angle. From these figures, one can clearly notice that, it is possible to determine the most suitable position for installation of the power generation turbine by deploying site specific design consideration values in Ansys fluent software to extract maximum efficiency from SUT. This simulation was carried out to analyze the performance of the SUT using a two-dimensional steady state and energy equation. The air flow inside the updraft tower was assumed as to be planar where it was simulated with k-epsilon model using Ansys Fluent software. The above results are validated and selection of canopy angle depending on the latitude and longitude of a specific location can yield desired results in successful implementation of SUT power plant.

Acknowledgments: The authors express their gratitude to the JNT University, Kakinada, and Smart Grid Research Lab, Vellore Institute of Technology, Vellore, India for their entire support towards this research work.

References

[1]. Sitharthan R, Geethanjali M and Pandy TKS 2016 Adaptive protection scheme for smart microgrid with electronically coupled distributed generations *Alexandria Engineering Journal* **55(3)** 2539-2550
[2]. Fathima AH, and Palanisamy K 2014 Battery energy storage applications in wind integrated systems—a review *IEEE International Conference on Smart Electric Grid* 1-8
[3]. Prabaharan N and Palanisamy K 2015 Investigation of single-phase reduced switch count asymmetric multilevel inverter using advanced pulse width modulation technique *International Journal of Renewable Energy Research* **5(3)** 879-890.
[4]. Jerin ARA, Kaliannan P and Subramaniam U 2017 Improved fault ride through capability of DFIG based wind turbines using synchronous reference frame control based dynamic voltage restorer. *ISA transactions* **70** 465-474
[5]. Sitharthan, R, Sundarabalan CK, Devabalaji KR, Nataraj SK and Karthikeyan M 2018 Improved fault ride through capability of DFIG-wind turbines using customized dynamic voltage restorer *Sustainable cities and society* **39** 114-125
[6]. Prabaharan N and Palanisamy K 2016 A single-phase grid connected hybrid multilevel inverter for interfacing photo-voltaic system *Energy Procedia* **103** 250-255
[7]. Palanisamy K, Mishra JS, Raglend IJ and Kothari DP 2010 Instantaneous power theory based unified power quality conditioner (UPQC) *IEEE Joint International Conference on Power Electronics, Drives and Energy Systems* 1-5
[8]. Sitharthan R and Geethanjali M 2017 An adaptive Elman neural network with C-PSO learning algorithm-based pitch angle controller for DFIG based WECS Journal of Vibration and Control **23(5)** 716-730
[9]. Sitharthan R and Geethanjali M 2015 Application of the superconducting fault current limiter strategy to improve the fault ride-through capability of a doubly-fed induction generator–based wind energy conversion system *Simulation* **91(12)** 1081-1087
[10]. Sitharthan R, Karthikeyan M, Sundar DS and Rajasekaran S 2020 Adaptive hybrid intelligent MPPT controller to approximate effectual wind speed and optimal rotor speed of variable speed wind turbine *ISA transactions* **96** 479-489
[11]. Sitharthan R, Devabalaji KR and Jees A 2017 An Levenberg–Marquardt trained feed-forward back-propagation based intelligent pitch angle controller for wind generation system *Renewable Energy Focus* **22** 24-32
[12]. Sitharthan R, Sundarabalan CK, Devabalaji KR, Yuvaraj T and Mohamed Imran A 2019 Automated power management strategy for wind power generation system using pitch angle controller *Measurement and Control* **52(3-4)** 169-182
[13]. Sundar DS, Umamaheswari C, Sridarshini T, Karthikeyan M, Sitharthan R, Raja AS and Carrasco MF 2019 Compact four-port circulator based on 2D photonic crystals with a 90° rotation of the light wave for photonic integrated circuits applications *Laser Physics* **29(6)** 066201

[14]. Sitharthan R, Parthasarathy T, Sheeba Rani S and Ramya KC 2019. An improved radial basis function neural network control strategy-based maximum power point tracking controller for wind power generation system *Transactions of the Institute of Measurement and Control* **41(11)** 3158-3170

[15]. Rajesh M and Gnanasekar JM 2017 Path observation based physical routing protocol for wireless ad hoc networks *Wireless Personal Communications* **97(1)** 1267-1289

[16]. Palanisamy K, Varghese LJ, Raglend IJ and Kothari DP 2009. Comparison of intelligent techniques to solve economic load dispatch problem with line flow constraints *IEEE International Advance Computing Conference* 446-452

[17]. Sitharthan R, Ponnusamy M, Karthikeyan M and Sundar DS 2019 Analysis on smart material suitable for autogenous microelectronic application *Materials Research Express* **6(10)** 105709

[18]. Rajaram R, Palanisamy K, Ramasamy S and Ramanathan P 2014 Selective harmonic elimination in PWM inverter using fire fly and fireworks algorithm *International Journal of Innovative Research in Advanced Engineering* **1(8)** 55-62

[19]. Sitharthan R, Swaminathan JN and Parthasarathy T 2018 March. Exploration of wind energy in India: A short review *IEEE National Power Engineering Conference* 1-5

[20]. Karthikeyan M, Sitharthan R, Ali T and Roy B 2020 Compact multiband CPW fed monopole antenna with square ring and T-shaped strips *Microwave and Optical Technology Letters* **62(2)** 926-932

[21]. Sundar D Sridarshini T, Sitharthan R, Madurakavi Karthikeyan, Sivanantha Raja A, and Marcos Flores Carrasco 2019 Performance investigation of 16/32-channel DWDM PON and long-reach PON systems using an ASE noise source *In Advances in Optoelectronic Technology and Industry Development: Proceedings of the 12th International Symposium on Photonics and Optoelectronics* 93

[22]. Sitharthan R and Geethanjali M 2014 Wind Energy Utilization in India: A Review *Middle-East J. Sci. Res.* **22** 796–801 doi:10.5829/idosi.mejsr.2014.22.06.21944

[23]. Sitharthan R and Geethanjali M 2014 ANFIS based wind speed sensor-less MPPT controller for variable speed wind energy conversion systems *Australian Journal of Basic and Applied Sciences* **8**14-23

[24]. Jerin ARA, Kaliannan P, Subramaniam U and El Moursi MS 2018 Review on FRT solutions for improving transient stability in DFIG-WTs *IET Renewable Power Generation* **12(15)** 1786-1799

[25]. Prabaharan N, Jerin ARA, Palanisamy K and Umashankar S 2017 Integration of single-phase reduced switch multilevel inverter topology for grid connected photovoltaic system *Energy Procedia* **138** 1177-1183

[26]. Rameshkumar K, Indragandhi V, Palanisamy K and Arunkumari T 2017 Model predictive current control of single phase shunt active power filter *Energy Procedia* **117** 658-665

[27]. Fathima AH and Palanisamy K 2016 Energy storage systems for energy management of renewables in distributed generation systems *Energy Management of Distributed Generation Systems* 157

[28]. Rajesh M 2020 Streamlining Radio Network Organizing Enlargement Towards Microcellular Frameworks *Wireless Personal Communications* 1-13

[29]. Subbiah B, Obaidat MS, Sriram S, Manoharn R and Chandrasekaran SK 2020 Selection of intermediate routes for secure data communication systems using graph theory application and grey wolf optimisation algorithm in MANETs IET *Networks* doi:10.1049/iet-net.2020.0051

[30]. Singh RR and Chelliah TR 2017 Enforcement of cost-effective energy conservation on single-fed asynchronous machine using a novel switching strategy *Energy* **126** 179-191

IOP Conf. Series: Materials Science and Engineering **937** (2020) 012022 doi:10.1088/1757-899X/937/1/012022

[31]. Amalorpavaraj RAJ, Palanisamy K, Umashankar S and Thirumoorthy AD 2016 Power quality improvement of grid connected wind farms through voltage restoration using dynamic voltage restorer *International Journal of Renewable Energy Research* **6(1)** 53-60

[32]. Singh RR, Chelliah TR, Khare D and Ramesh US 2016 November. Energy saving strategy on electric propulsion system integrated with doubly fed asynchronous motors *IEEE Power India International Conference* 1-6

[33]. Sujatha K and Punithavathani DS 2018 Optimized ensemble decision-based multi-focus imagefusion using binary genetic Grey-Wolf optimizer in camera sensor networks *Multimedia Tools and Applications* **77(2)** 1735-1759

[34]. Krishnamoorthy S, Punithavathani S and Priya JK 2017 Extraction of well-exposed pixels for image fusion with a sub-banding technique for high dynamic range images *International Journal of Image and Data Fusion* **8(1)** 54-72

[35]. Singh RR, Mohan H and Chelliah TR 2016 November. Performance of doubly fed machines influenced to electrical perturbation in pumped storage plant-a comparative electromechanical analysis *IEEE 7th India International Conference on Power Electronics* 1-6

Non-invasive optical technique for urodynamic testing using a wireless near infrared spectroscopy device

R Ramkumar[1], G Jagan[2], D Sivamani[3], P Ragupathy[4]
[1]Assistant Professor, Department of Electrical and Electronics Engineering, K.Ramakrishnan College of Technology, Trichy, Tamilnadu.
[2] Associate Professor, Department of Biomedical Engineering, RVS Educational Trust Group of Institution, Dindigul, Tamilnadu.
[3]Assistant Professor, Department of Electrical and Electronics Engineering, Rajalakshmi Engineering College, Rajalakshmi Nagar, Thandalam, Tamilnadu
[4]Assistant Professor, Department of Electrical and Electronics Engineering, Rajalakshmi Engineering College, Rajalakshmi Nagar, Thandalam, Tamilnadu

E-mail: sivamani.d@rajalakshmi.edu.in

Abstract. A real time wearable medical device for monitoring of urological parameters like bladder pressure, urine flow rate is measured by using cyst metric test and uroflow metric test. It is a invasive method, Patients needs local anesthesia, and patient may feel discomfort for few days. Such problems will be overcome by using a Near Infrared Spectroscopy Device. Near Infrared Spectroscopy is an optical technique, utilizes light in the NIR Spectrum ranges between 700nm – 1000nm has a high penetration depth in the living tissue. In urology, NIRS has been utilized to screen bladder muscle work and to analyze bladder brokenness. Since, it is a Noninvasive method; it is compatible for patients like Multiple Sclerosis, Stroke, Spinal Card injury, elderly patients with incontinence and neurological disorders. With such a gadget utilized as a sensor with a caution, it is thus achievable to caution the subject when the volume of pee in his/her bladder arrives at a pre-decided edge of bladder limit. This technique would possibly empower patients in danger for urinary maintenance to shield themselves from renal harm.

1. Introduction

The expression "urologic ailments" portrays a wide assortment of conditions, all identified with the preparing and completing of pee of the body [1]. They can influence men, ladies, and offspring everything being equal. These maladies influence quite certain pieces of the body. In females, they include the urinary lot. In guys, they influence the urinary plot or potentially the regenerative organs. Urologic infection can include innate or procured brokenness of the urinary system. The reason for illnesses of the body is normal to the urinary parcel. Colonization by microscopic organisms, protozoa or parasites can cause disease.

Uncontrolled cell development can cause neoplasia. There are numerous urologic issues and illnesses. Coming unext are a determination of a portion of the maladies recognized by the American Urological Association Foundation (AUAF) as normal. The proposed method is simulated and hardware is implemented [2].

1.1 Benign Prostatic Hyperplasia (BPH)

Kind prostatic hyperplasia (BPH) is a developed prostate. It is an expansion in the size of the prostate organ. BPH is normal in more seasoned men. It isn't brought about by prostate disease. Men with BPH may encounter a regular inclination to pee. They may likewise have a frail stream of pee when they do go and an inclination that the bladder isn't unfilled after pee. In this paper, your primary care physician may decide to simply screen this condition or endorse prescriptions like alpha-blockers for treatment [3]. It might likewise be tended to through medical procedure on the off chance that it is serious.

1.1.1 Incontinence

Incontinence is lost bladder control. It brings about the undesirable arrival of pee. This condition can be awkward and humiliating for victims. In any case, it is a long way from extraordinary. As indicated by the AUAF, in excess of 15 million individuals in the U.S. have incontinence (AUAF, 2011). There are various things that can cause incontinence. A few of the most common causes include:

- Diabetes
- Labor
- Hyperactive Bladder
- Prostate enlargement
- Spinal Cord trauma
- Hemorrhoids

1.1.2. Urinary Tract Infections (UTIs)

UTIs are the aftereffect of bothersome yeast, microorganisms, or infections that attack the urinary plot and cause disease. They are significantly more typical in ladies, in spite of the fact that men can get them as well. As per the AUAF, approximately 40 percent of females and 12 percent of guys will have an UTI that causes recognizable indications sooner or later in their lives (AUAF, 2011). A consuming sensation during pee is one of the manifestations of an UTI. Others incorporate an incessant inclination to pee and the inclination that the bladder isn't totally vacant subsequent to peeing. Performance and reduces complexity problem can normally clear up most UTIs inside a couple of days [4].

1.1.3. Kidney and Urethral Stones

Stones create in the kidneys when there are precious stones in the pee and little particles encompass and gather on these gems. Urethral stones are ones that move from the kidney into the ureter (the cylinders that convey pee from the kidneys to the bladder. These stones can square pee stream and cause a lot of agony. Numerous individuals wind up ousting these stones from the body without clinical assistance however this isn't generally the situation [5-6].

1.1.4. Urological disorders

Some other common urological conditions include:

- Prostatic Cancer
- Urinary Bladder Cancer
- Urinary Bladder Prolapsed

2. Structure of urinary tract

The urinary tract is the body's seepage framework for expelling squanders and additional water. The urinary tract incorporates two kidneys, two ureters, a bladder, and a urethra. Blood courses through the kidneys and the kidneys sift through squanders and additional water, making pee. The pee goes down two thin cylinders called the ureters.[4] The pee is then put away in a solid, expand like organ called the bladder. The bladder swells into a round shape when it is full and gets littler as it purges. At the point when the bladder discharges, pee streams out of the body through the urethra Kidneys are a couple of bean-formed organs found along the back mass of the stomach cavity. The left kidney is found somewhat higher than the correct kidney in light of the fact that the correct side of the liver is a lot bigger than the left side. The kidneys, in contrast to different organs of the stomach cavity, are found back to the peritoneum and contact the muscles of the back. [11]

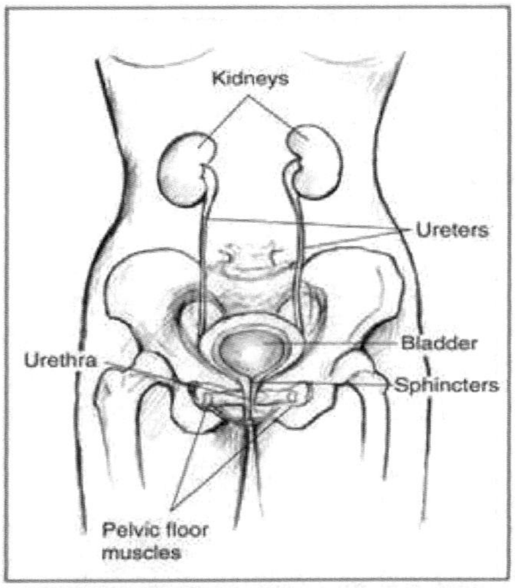

Figure1. Structure of Urinary Tract

The kidneys are encircled by a layer of fat that holds them set up and shields them from physical harm. The kidneys channel metabolic squanders, overabundance particles, and synthetic compounds from the blood to shape pee. The ureters are a couple of cylinders that convey pee from the kidneys to the urinary bladder. The ureters are around 10 to 12 inches in length and run on the left and right sides of the body corresponding to the vertebral section [7-8]. Gravity and peristalsis of smooth muscle tissue in the dividers of the ureters push pee toward the urinary bladder. The closures of the ureters expand marginally into the urinary bladder and are fixed at the purpose of passage to the bladder by the ureterovesical valves. These valves keep pee from streaming back towards the kidneys [9].

The urinary bladder is a sac-like empty organ utilized for the capacity of pee. The urinary bladder is situated along the body's midline at the mediocre finish of the pelvis. Pee entering the urinary bladder from the ureters gradually occupies the empty space of the bladder and stretches its flexible dividers. The dividers of the bladder permit it to stretch to hold somewhere in the range of 600 to 800 milliliters of pee [10-11]The urethra is the cylinder through which pee goes from the bladder to the outside of the body. The female urethra is around 2 inches in length and closures second rate compared to the clitoris and better than the vaginal opening. In guys, the urethra is around 8 to 10 inches in length and finishes at the tip of the penis. The urethra is additionally an organ of the male regenerative framework as it helps sperm out of the body through the penis. The progression of pee is through the urethra. Urinary framework cross-area is constrained by the inner and outer urethral sphincter muscles. The inside urethral sphincter is made of smooth muscle and opens automatically when the bladder arrives at a specific set degree of enlargement. The opening of the inner sphincter brings about the impression of expecting to pee [12]

The outer urethral sphincter is made of skeletal muscle and might be opened to permit pee to go through the urethra or might be held shut to postpone pee [13]. Pee is the way toward discharging pee from the urinary bladder through the urethra and out of the body. The procedure of pee starts when the muscles of the urethral sphincters unwind, permitting pee to go through the urethra. While the sphincters unwind, the smooth muscle in the dividers of the urinary bladder agreement to oust pee from the bladder.

2.1. Block diagram

The pressure information from the abdominal skin is collected using the optical sensor. The emitter (LED) delivers IR signals which are reflected by the tissue surface and is collected by the detector (phototransistor) in the sensor. [17-19]

The sensor output is given to the PIC 16F887. Microcontroller performs the analog to digital conversion of the obtained bio signals serially. The output from the PIC is given to the level converter which transfers the digital signal from TTL to RS232 series. This signal can be used for communication with the

Personal Computer. This data is connected to the computer using a serial port connector (DB9) and the obtained pressure signals are then analyzed using LABVIEW.
It is necessary to determine the the urinary flow rate, the signals are processed. Thus the intra bladder pressure and urinary flow rate is determined [14].

2.2. components

- Sensor
- PIC16F887 Microcontroller
- Level Converter (MAX 232)
- DB-9 Connector
- LABVIEW

NIR Sensors use condition of craftsmanship optical parts including elite thin band pass NIR Channels over a wide scope of the infrared range to make precise, steady, hearty estimation. Chemo metric methods are utilized to recognize the territory of the NIR range required for a given estimation; these particular close to infrared frequencies are amplified in the NIR sensor to make an estimation which is heartless toward process changes, for example, encompassing lighting, relative moistness, item stature vacillations and inside item varieties, for example, crop-source changes, occasional contrasts or shading variety subsequently, steady item quality can be accomplished. A NIR investigation conveys quick precise estimations of tests taken from the procedure then [15].

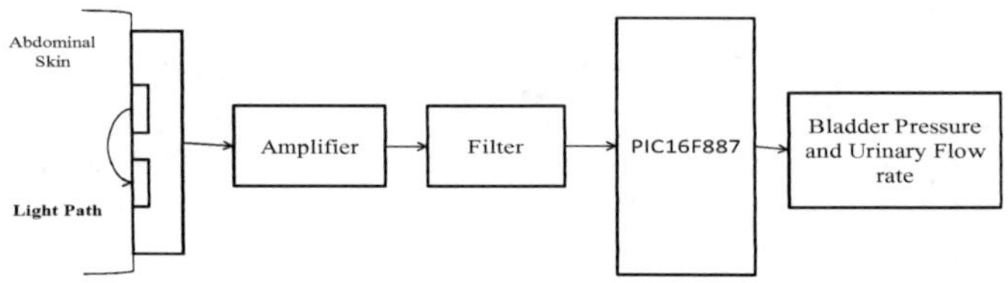

Figure2. Block diagram

3. Software description

The software simulation of this project is done using LABVIEW software.

3.1. Lab view

Lab VIEW (Laboratory Virtual Instrument Engineering Workbench) is a framework plan stage and improvement condition for a visual programming language from National Instruments. It is graphical programming language based on structured data flow which consists of graphical objects or G-programming elements. [16]
It has 2 windows namely,

1) Block diagram- which is meant for programming (shows the graphical source code).
2) Front panel- which gives the graphical view of the output to be displayed (shows the graphical user interface items connected to the source code).

The front panel consists of various controls and indicators for both input and output whereas the block diagram has the terminals corresponding to the front panel controls and indicators as well as constants, functions, sub VI s, structures(program control elements like for loop, while loop etc), wires etc.

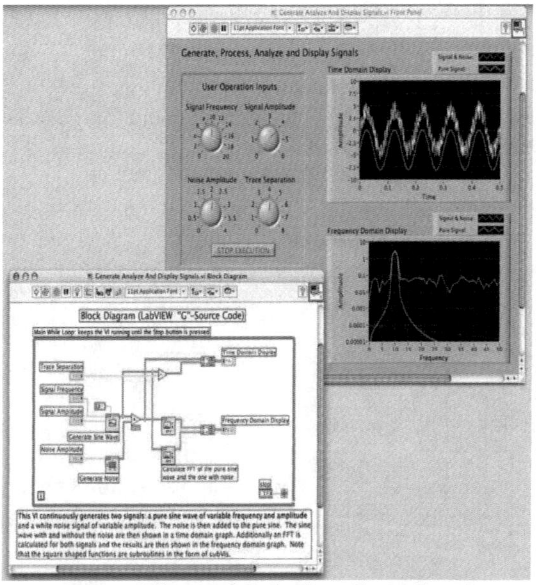

Figure 3 Lab view Front Panel

.lvproj - extension for projects in labview.
.vi - extension for files in labview.

Hardware's integrated with labview workbench makes it a powerful tool. It provides a complete development environment with project explorer, debugging tools, source code control and more.

3.2. Graphical Programming

Lab VIEW ties the making of UIs (called front sheets) into the movement cycle. Lab VIEW programs/subroutines are called virtual instruments (VIs). Each VI has three sections: a square diagram, a front board and a connector board. The last is used to address the VI in the square charts of other, calling VIs. The front board is created using controls and markers. Controls are inputs – they license a customer to smoothly information to the VI. Pointers are yields – they show, or show, the results subject to the data sources given to the VI. The back board, which is a square outline, contains the graphical source code. The aggregate of the things set on the front block will show on the back board as terminals. The back board in like manner contains structures and limits which perform technique on controls and deftly data to markers. The structures and limits are found on the Capacities palette and can be determined to the back board. In general controls, markers, structures and limits will be insinuated as center points. Center points are related with one another using wires – for instance two controls and a marker can be wired to the extension work with the objective that the pointer shows the all out of the two controls. Suitably a virtual instrument can either be run as a program, with the front board filling in as a UI, or, when dropped as a middle onto the square diagram, the front board depicts the wellsprings of information and yields for the given place through the connector board. This proposes every VI can be handily endeavored before being brought as a subroutine into a more prominent program. The graphical way of thinking in addition permits non-programming modelers to accumulate programs by moving virtual portrayals of lab hardware with which they are beginning at now ordinary. The Lab VIEW programming condition, with the included models and documentation, makes it easy to make little applications. This is a good situation on one side, yet there is moreover a specific peril of scarcely thinking about the capacity required for top notch G programming. For complex figuring's or enormous degree code, it is immense that the originator has a far reaching information on the exceptional Lab VIEW sentence structure and the geology of its memory the board. The most powerful Lab VIEW improvement structures offer the chance of building

IOP Conf. Series: Materials Science and Engineering **937** (2020) 012023 doi:10.1088/1757-899X/937/1/012023

self-ruling applications [17-18]. Moreover, it is conceivable to make spread applications, which award by a customer/server plot, and are in this way less hard to execute considering the normally equal nature.
The image above is a layout of an essential Lab VIEW program demonstrating the dataflow source code as the square outline in the lower left packaging and the data and yield factors as graphical things in the upper right packaging. The two are the fundamental sections of a Lab VIEW program implied as a Virtual Instrument..The programming reproduction of the bladder filling recognition unit is finished utilizing LABVIEW. The intra bladder pressure signal is reenacted utilizing a nonstop sine wave signal and is intensified, separated and afterward this simple sign is carefully changed over and shown. The urinary stream rate which is determined utilizing the intra bladder pressure and the bladder volume is shown and when the bladder pressure surpasses the ordinary level (13-15mmHg), the LED alert sparkles advising that the patient is prepared for maturation [19].

4. Results and discussion

In this work, NIR (Near infrared) sensor is used to detect the bladder pressure and urinary flow rate by noninvasive method. Patients with disorders such as, spinal cord injury, neurological disorders and elderly patients with incontinence. This method will be highly useful. The device indicates alarm when it reaches the abnormal level. The below figure shows, the simulation output with normal and abnormal condition.

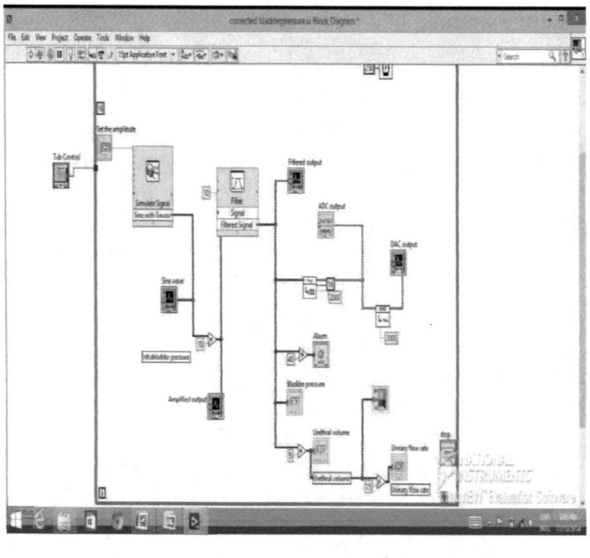

Figure 4. Block Diagram in Lab view

IOP Conf. Series: Materials Science and Engineering **937** (2020) 012023 doi:10.1088/1757-899X/937/1/012023

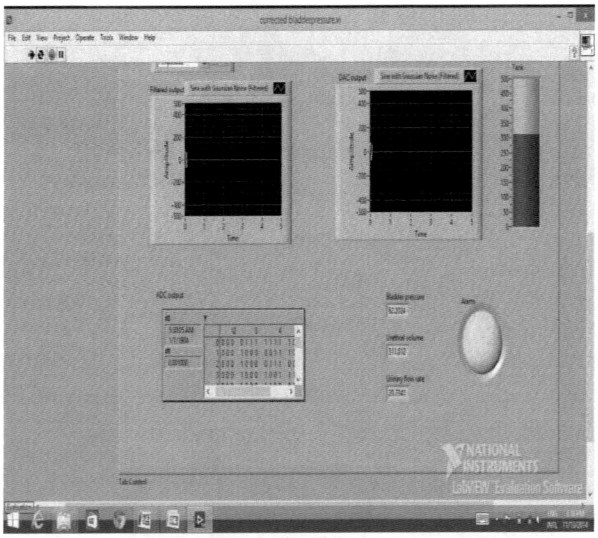

Figure 5 Indication of alarm during abnormal condition

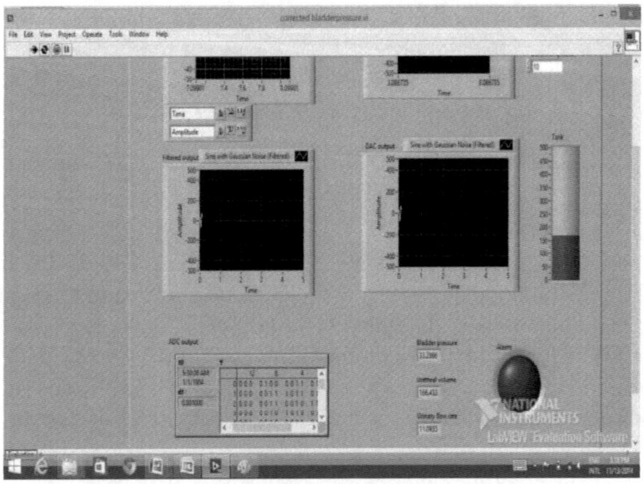

Figure 6 Indication for normal condition

5. Conclusion

This Project presents a simple but effective technique for monitoring bladder pressure and urine flow rate by noninvasive method. It is a wearable medical device for monitoring the mentioned parameter. These parameters are used to screen the obstruction in the flow of urine and can screen some diseases like BPH, BOO and Prostate cancer. The test is noninvasive and may be used to assess bladder and sphincter function. The contraption is prepared for isolating between when the bladder is unfilled or contains a little volume of pee and when it ends up being full by using the osmosis properties of water at a recurrence of 950nm. Urological checking with this NIRS instrument is plausible and creates information of likely indicative worth. The simulation output is successfully executed using Lab view software and the hardware is ready to be implemented.

6. References

[1] AlexandreGoguin, Frederic Lesage, HuguesLeblond, Melanie Pelegrini-Issac, Serge Rossignol, And HabibBenali, Member, IEEE, "A Low-Cost Implantable Near –Infrared Imaging System of Spinal Cord Activity in the Cat", VOL, 4, NO, 5,October 2010.

[2] C.Kalavalli, S.R.Paveethra, S.Murugesan, Dr.A.Nazar Ali, (2020), Design and Implementation of High Efficiency H6 PV Inverter with Dual Axis Tracking, International Journal of Scientific & Technology Research, Vol 9, issue 02, pp. 4728-31.

[3] S.R.Paveethra, C.Kalavalli, S.Vijayalakshmi, Dr.A.Nazar Ali, D.Shyam(2020), Evaluation Of Voltage Stability Of Transmission Line With Contingency Analysis, International Journal of Scientific & Technology Research, Vol 9,issue 02, pp. 4018-22

[4] Shyam D, Premkumar K, Thamizhselvan T, Nazar Ali A, Vishnu Priya M(2019),Symmetrically Modified Laddered H-Bridge Multilevel Inverter with Reduced Configurationally Parameters ,International journal of engineering and advanced technology ,Vol 9, issue 1

[5] Ramkumar, R., et al2019. Intelligent system to monitor and diagnose performance deviation in Industrial Equipment.IOPconf.Series:Materials science and Engineering,Vol 623(2019)012012.

[6] N. K. Kristiansen, S. Ringgaard, H. Nygaard, and J. C. Djurhuus, "MRIassessment of the influence of body position on the shape and position of the urinary bladder," Scandia. J. Urol. Nephrol., vol. 38, no. 1, pp.53–61, Jan. 2014.

[7] C. H. Van der Vaart, J. R. J. de Leeuw, J. P.W. R. Roovers, and A. P. M. Heintz, "The effect of urinary incontinence and overactive bladder symptoms on quality of life in young women," *BJU Int.*, vol. 90, no. 6, pp. 544–549, Oct. 2017.

[8] Ramkumar, R., and N. Tejaswini. "A novel low cost three arm Ac automatic voltage regulator." *Advances in Natural and Applied Sciences*, vol. 10, no. 3, 2016

[9] Ramkumar, R., et al. "PV based SEPIC converter fed four switch BLDC motor drive." *Advances in Natural and Applied Sciences*, vol. 10, no. 3, 2016

[10] R. Ramkumar, B. Karthikeyan, Dr. A. Rajkumar, V. Venkatesh and A. Anton Amala Praveen," DesignandImplementation of IOT Based Smart Library Using Android Application," Biosc.Biotech.Res.Comm.Special Issue Vol 13 No (3) 2020 Pp-56-62.

[11] B.Karthikeyan, R. Ramkumar, L. Nagarajan, A. Anton Amala Praveen and S. Vijayalakshmi," ANN Based MPPT Controller for PEMFC System with Interleaved Resonant PWM High Step up Converter" Biosc.Biotech.Res.Comm. Special Issue Vol 13 No (3) 2020 Pp-44-47.

[12] Dr. P. R. Sivaraman, R. Jaiganesh, P. Ragupathy and R. Ramkumar," HI-TECH Electoral Machine for Election Commission of India," Biosc.Biotech.Res.Comm. Special Issue Vol 13 No (3) 2020 Pp-13-17.

[13] D. Sivamani, P. Thirusenthil Kumaran , C. Gnanavel and L. Nagarajan," Powering Smart Street Lighting Cities Using Visible Light Communication and Light Fidelity," Biosc.Biotech.Res.Comm.Special Issue Vol 13 No (3) 2020 Pp-180-184.

[14] A Ali Nazar, R Jayabharath, MD Udayakumar(2014) ,An ANFIS Based Advanced MPPT Control of a Wind-Solar Hybrid Power Generation System, international review of modeling and simulations,vol.7, no. 4, pp. 638–643.

[15] A. T. Sankara Subramanian, P. Sabarish, M. D.Udayakumar and T. Vishnu kumar," Performance Analysis of Various Photovoltaic Configurations under Uniform Shading and Rapid Partial Shading Formations", Biosc.Biotech.Res. Comm. Special Issue Vol 13 No (3) 2020 Pp-185-192.

[16] P. Sabarish et al 2019 IOP Conf. Ser.: Mater. Sci. Eng. 623 012011.

[17] M D Udayakumar et al 2019 IOP Conf. Ser.: Mater. Sci. Eng. 623 012018.

[18] P.Sabarish, A. T. Sankara Subramanian, S. Murugesan and V. Sureshkumar,"A New Methodology of Arterial Blood Clot Removal Using Bio Molecular Devices for ATPase Nuclear Motors." Biosc.Biotech.Res.Comm. Special Issue Vol 13 No (3) 2020 Pp-197-201.

[19] A Nazar Ali, D Sivamani, R Jaiganesh, M Pradeep (2019), Solar powered air conditioner using BLDC motor, IOP Conference Series: Materials Science and Engineering, vol. 23

IOP Conf. Series: Materials Science and Engineering **937** (2020) 012024 doi:10.1088/1757-899X/937/1/012024

Design and development of wireless networking for solar PV panel cleaning robots

G Anilkumar[1*], K Naveen[2], M T V S H K Pavan Kumar[3], G V Brahmendra Kumar[4], K Palanisamy[5], Avinash Patil[6], Deepak Sachan[7]

[1,2,3]School of Renewable Energy, JNT University, Kakinada, 533003, India.
[4,5]Centre for Smart Grid Technologies, Vellore Institute of Technology, Vellore, 632014, India.
[6,7]Technology Development Lab, BHEL R&D Division, Hyderabad, 500093 India.
Email: koreddinaveen@gmail.com, mahantpavankumar@gmail.com, brahmendrakumar.g@gmail.com, kpalanisamy@vit.ac.in, avinashp@bhel.in, deepaksachan@bhel.in
*Corresponding author: gorantlaanilkumar@gmail.com

Abstract. This paper discusses about one of the possible solutions to overcome the influence of dust on the surface of solar PV panel. A decrease in power output with the increase in particle deposition has been identified in large scale solar PV power generation where there were many numbers of solar panels connected in the form of arrays and each array requires a robot to carry out efficient cleaning with in stipulated time. Autonomous automatic cleaning operation along with self-control and monitoring of accumulated dust with proper coordination through networking between robots is prerequisite option to attain efficient cleaning system in large scale solar power plants. To accomplish the same, selection of appropriate communication technology for real time wireless networking of solar cleaning robots with low power consumption and data delay to operate over wide area is required. Here an attempt had been made to bring in an automated robot cleaning system using low power wide area network (LPWAN) based on a network of ESP 8266 Node MCU's associated with a set of sensors under different configurations which showed successful results for implementation in industrial scale large solar PV power plants.

1. Introduction

Allocation Operation and Maintenance of solar photovoltaic (PV) panels should be carried out frequently not only to improve its efficiency but also to prevent the damage caused by the accumulated dust due to environmental pollution. The main influence that exhibit impact on operating parameters is PV panel efficiency. There is a dire need to explore technical solution to this problem in an efficient manner although there were some existing methods. A study on several dust removal techniques has been carried out for example [1,2] which suggested that removal of dust particles from the surface of solar PV panel can be done by using nylon brush with automated robotic system where as [3] mentioned about the energy losses that arise due to deposition of dust particles on solar panel. A significant amount of power loss was noticed by the research work carried out by earlier researchers [1-4]. In case of PV bases large scale power generation, the intensity of power loss further increases. Classification of solar PV cleaning mechanisms are presented in figure 1 which shows a glimpse of several methodologies and techniques such as natural cleaning, mechanical cleaning, self-cleaning and electrostatic cleaning that are available in the market to remove the accumulated dust on the solar PV panel. Here, an attempt had been made to develop an automated cleaning system using low power wide area network (LPWAN) based on a network of ESP 8266 Node MCU's associated with a set of sensors under different configurations which showed successful results for implementation in industrial scale large solar PV power plants.

The efficiency of the solar PV panel has reduced to 40% after one year of solar system installation [5]. The reasons that are attributed were due to high temperature, panel pitch and orientation [6], accumulation of dust, snow, sea salt and bird droppings [7]. Among those, the accumulation of dust and other particles on the solar PV panel reduces the system efficiency of the panel about 50% [8]. Moreover, these particles can be removed from the surface of the panel by providing proper cleaning mechanism. Few research studies were carried out to deploy different software-based prototype cleaning mechanisms. By implementing arm controller and gear motor-based cleaning method, the output increased about 35% [7]. A microcontroller based automatic dust cleaning system was developed to increase the panel efficiency of about 1.6% to 2.2% by cleaning the panel every two hours [9-15]. By providing a self-cleaning mechanism with software and hardware, the operating cost reduces and the overall efficiency can be increased [16-25]. From the simulation, the accumulated dust on the panel can be detected and provide the cleaning mechanism which increases the performance of the cleaning mechanism [26-30]. A dust cleaning mechanism with a panel cooling system showed an improvement in the performance of the PV panel [31-34]. An automatic dust cleaning system operated during the day time and turned off during night time for solar street light panels showed an increase in the overall system efficiency [35]. Among the above discussed research studies, it has been identified that, most of the developed or proposed automated cleaning system were implemented with dry cleaning mechanism to avoid the panels from short circuits and allowed to operate when dust was accumulated on the panel.

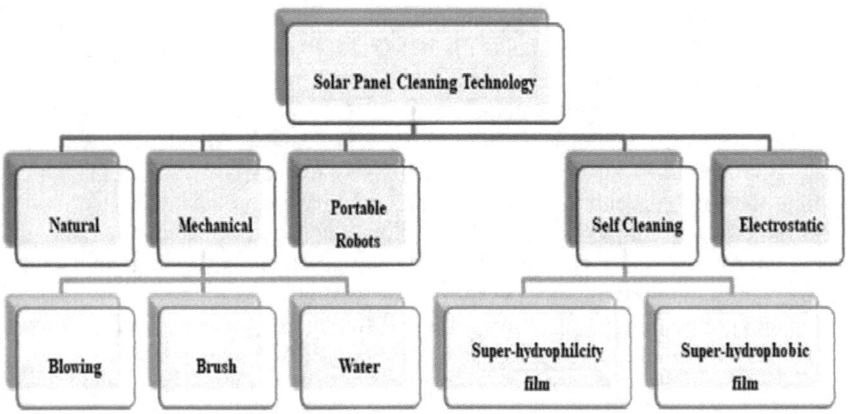

Figure 1. Solar PV panel cleaning methods.

2. Cleaning robot for industrial solar panels (CRISP)

2.1 Construction

Figure 2 represents the recently developed CRISP robot at BHEL R&D, Hyderabad. The solar PV panel cleaning robot is designed in a modular fashion with two-axis movement in X and Y direction on the edges of the panel. It has a long aluminium guide rail along the modular frame which allows the robot to move along the horizontal direction with a set of motorized dummy wheels that can travel on top and bottom edge of the solar panel eliminating the capital investment of additional rails in addition to a cleaning trolley with a pair of rotating microfiber brushes that can move up and down in vertical direction by making use of motorized winch and a wire rope to clean the surface of the solar PV panel. An IR sensor and DC motor is used to adjust the height of cleaning brushes. Proximity sensors are used to detect the top and bottom position of the cleaning trolley for automatic up and down movement. Limit switches are used to detect the start, end, and position of the robot. Distance between the microfiber brush and solar panel surface is adjustable. The robot is parked in a parking station where its batteries are automatically charged by an independent solar panel in the day time and during the night time, the robot cleans the solar panels. The robot is controlled by a PLC controller. An android based application or web-based application is used as HMI for controlling the robot.

Figure 2. CRISP.

2.2 Mechanical Design

The construction of the CRISP can be categorized in two ways those are mechanical design and electrical control design. The mechanical design of the CRISP comprises of top drive, bottom drive, rope winch, trolley, parking station. The top drive of the CRISP consists of two motorized wheels which are fixed at the right and left side of the top extrusion bar. The right-side motorized wheel can be coupled to the rotary encoder which is used for monitoring the accurate distance traveled by the robot. The top extrusion bar also consists of four dummy wheels which are used to provide support to the robot. The bottom drive also consists of two motorized wheels which are fixed at the left and right side of the bottom extrusion bar. It consists of a rotary encoder which is used to control the speed of the robot. The winch is one of the major components involved in the cleaning cycle by which the trolley with brushes is moved up and down for cleaning the solar panels. Two motors are coupled to the winch pulley which rotates by winding and unwinding the wire rope of the trolley around the pulley. The trolley is fixed with two brushes to carry out the cleaning operation of the solar panels. These brushes are coupled to the motors which are fixed on one side of the trolley. For efficient cleaning of the panels, the height of the brushes can be adjusted according to the solar panel positioning. The trolley is based on the main frame of the robot on free running wheels and movement of the trolley is made possible through the wire rope of which one end is tied to the trolley and the other end to the winch pulley at the top. The parking station consists of a solar panel with charge controller and hard stops for mechanical stop and electrical contact. This station provides the parking facility to the robot. It also consists of parking cover which is made with nylon used to protect the robot trolley from rain and UV rays. It provides the charging facility to the robot batteries during the presence of solar radiation through the electrical contacts which are mounted on the mechanical hard stops.

2.3 Electrical Design

Electrical control design of CRISP comprises of PLC and extension module, solar panel, charge controller, rechargeable battery, 12 V high torque dc motors, Wi-Fi router, 12/24 V DC-DC converter, 7805 voltage regulator, dual DC motor driver, digital DC volt & ammeter, rotary encoder, limit switch, proximity sensor, current sensor, rain sensor, DHT11 sensor, panel lamp indicators, emergency push button switch. Using preprogrammed simulation program, PLC controls the speed and direction of the motors (such as top drive, bottom drive, winch, and trolley brush motors) with proximity sensor, photo eye sensor, limit switch, encoder, rain sensor, current sensor and potential divider as an input. It can transfer the signal to the Wi-Fi router through the category 5 Ethernet cable for control and monitoring of the robot with Wi-Fi connected devices (such as mobile, laptop, tab, etc.). A transistor I/O module is connected to this PLC to extend the number of inputs and outputs. While the above mentioned sensors and electronic devices will carry out their usual functions in the system design, CRISP robot is equipped with a dedicated self-cleaning solar PV panel with 120 Wp maximum peak power, 19.95 V maximum output voltage, 23.26 V of open circuit voltage, 6.02 A maximum current output, 6.36 A of short circuit current that can be wired to attain a maximum system voltage of 600 V had been selected to charge two parallel connected 12V-24 Ah lead-acid batteries during day time through charge controller.

2.4 Working

Initially, the robot would be at home position in the parking station provided, where the batteries of the robot get charged from its own solar panel through a charge controller. By considering the shadow effect of the robot on the solar panels, the robot is made to be idle during the sun hours of the day and made to start the cleaning cycle during the non-operative hours of the solar panels. The movement of the robot can be considered as a two axes movement, where the cleaning of the panels is done in the vertical movement of the trolley, while the robot moves in forward and backward direction in the horizontal movement covering the entire area of the array of the solar panels. The cleaning cycle can be considered in four phases viz., Upward movement, Downward movement, Forward movement, and Backward movement.

During the Upward movement, the trolley is pulled up by the winch with the brushes rotating resulting in the dust being wiped off from the panel surface. This upward movement will take place until the top proximity sensor gets activated. Once the sensor senses the trolley at the top, the winch changes the direction unwinding the wire rope which results in the downward direction of the trolley.

During the Downward movement, the dust wiped off in the upward direction is completely pushed downward with the brushes and is pushed out of the panel surface making the panel surface completely clean. The downward movement takes place until the bottom proximity sensor is activated. Once the sensor is activated, the winch and the brushes stop rotating and the robot is ready to move in the forward direction.

During the Forward movement, the robot is supposed to move forward as the cleaning is done in that particular position. Now the robot has to move a distance which is equal to the length of the brush since the cleaning done is for one length of the brush. This is achieved by programming a set number of counts for the encoder. As there are separate encoders for top and bottom drive, the robot moves until both drives are moved for set length. This eliminates any tilt in the movement of the robot due to an imbalance in the load. Once both the drives move for the set number of counts, the cleaning starts in the upward direction and downward direction. Once cleaning is done again, the robot moves in the forward direction. This takes place till the top and bottom forward limit switches are on. Once the top and bottom forward limit switches are on, the last upward and downward movements are done for the last time and the top drive and bottom drive change their direction.

During the Backward movement, the robot comes to the home position at the parking station as the top drive and bottom drive move in a backward direction until the top and bottom backward limit switches are on. Once these limit switches are on, the robot contacts the charge controller which is connected to the solar panel and the batteries get recharged until the next cleaning cycle starts. It can be operated in manual mode also by pressing respective switches to move left, right, up, down and trolley brushes.

3. Wireless networking of CRISP

Wireless network resembles a set of devices that can be interconnected based on certain communication methodologies and techniques without using wires. Here in this case, devices refer to robots that are used for solar panel cleaning mechanism which can be grouped together wirelessly and form a wireless local area or wireless wide area network (WLAN or WWAN). Coordinated control and monitoring can be attained in-between robots and the HMI from a central control station to enhance the system performance. A PLC along with Wi-Fi router can be used in CRISP to accomplish the same. But a PLC with Wi-Fi option is not economically viable in addition to excess power consumption. Instead, an ESP 8266 NodeMCU microcontroller unit with inbuilt Wi-Fi antenna can be considered that allows access point as well as device control capability to HMI to attain desired wireless networking of CRISP robots. It is possible to network these individual NodeMCU's wirelessly by integrating them with IoT in a smart manner. The data corresponding to the status of each robot along with its associated PV system parameters in CRISP can be accessed via NodeMCU by the user to store in cloud data base and can achieve desired control and monitoring operations using existing communication methodologies and techniques.

The NodeMCU module can be configured as a station, so as to connect it to the Wi-Fi network. It can also be configured as a soft access point (soft-AP), to establish its own Wi-Fi network. Therefore, it is possible to connect other stations to such NodeMCU module so that it exhibits the ability to operate both in station as well as soft access point mode. This provides the possibility of integrating a large network such as a star or mesh network by using the same network devices. Here the developed prototype assumes each NodeMCU as an individual CRISP and collects the data from requisite sensors discussed in previous section for monitoring and control. Different wireless networking topologies (star, radial and long range) are taken into design considerations based on NodeMCU and are presented as follows.

3.1 Star topology

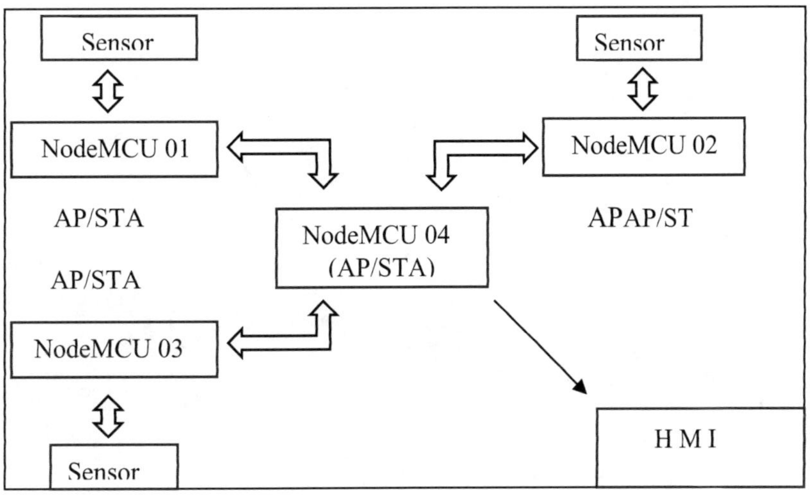

IOP Conf. Series: Materials Science and Engineering **937** (2020) 012024 doi:10.1088/1757-899X/937/1/012024

Figure 3. Basic structure of star topology.

The Digital Humidity and Temperature (DHT11) sensor is used in this topology to measure and transfer temperature and humidity data values, and in this network, the sensor can be powered ON and OFF from the monitor which is displayed in the HMI. Figure 3 above shows the basic structure of the wireless networking star topology. The network topology for controlling the robot and monitoring of temperature and humidity values of the DHT11 sensor can be explained with the help of following flow chart shown in figure 4.

Initially, all the NodeMCU's are configured in AP mode and the HMI is connected to the NodeMCU 04 with proper service set identifier (SSID) and Password. Once the connection is established, access the NodeMCU by calling its internet protocol (IP) address in the web page, then the user interface (UI) containing the home page will be shown on the HMI. Then select any one of the NodeMCU's to read the temperature and humidity values of the respective DHT by pressing the request button. Then the connected AP (NodeMCU 04) is configured as a station mode and it will connect to the selected NodeMCU and sends a request to it. Then the respective DHT is powered ON and will read and store the temperature and humidity values. Next, the selected NodeMCU is configured as a station mode and connects to the NodeMCU 04 and posts the temperature and humidity data. Then the DHT is powered OFF and the NodeMCU is configured as an access point (AP).

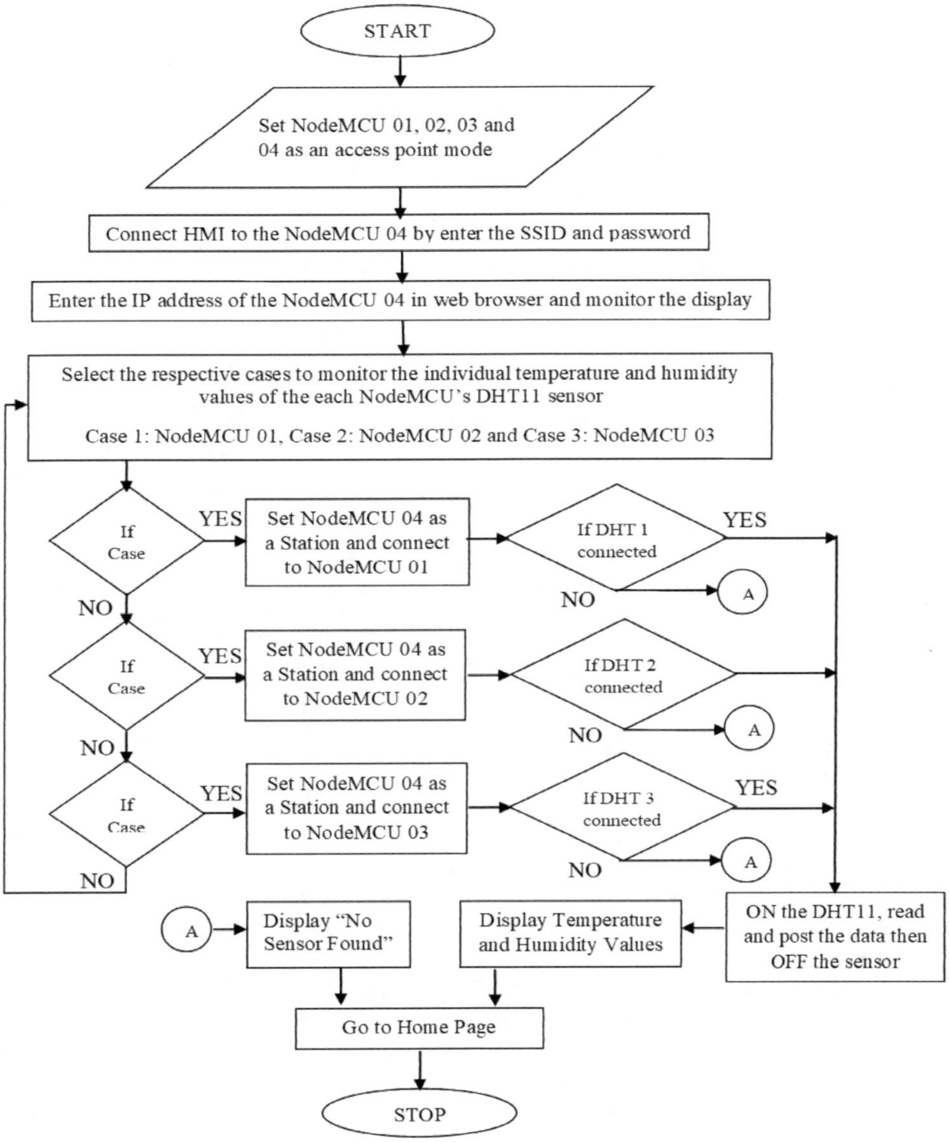

Figure 4. flow chart of star topology.

In the HMI monitor, whenever the read button is pressed, the temperature and humidity values will be appeared on a new page. If no DHT sensor is connected to the respective NodeMCU, "No Sensor Found" will appear on the screen. In this topology, the DHT sensor is OFF unless the request button of the respective NodeMCU on the home page is pressed. So, the power can be saved. This topology is used not only to read the temperature and humidity values on the monitor but also it will control the DHT sensor ON and OFF. With the reference of this topology, it can control the motors and LED to switch ON and OFF for emergency purpose. The next step in this topology is including the LED indicators and motors which are to be controlled either ON or OFF and display on the monitor to know the status. This topology's range is less than 100 meters only and this topology support the users to control the robots individually. So this wireless network topology can be used for local area purpose.

3.2 Radial topology

In this topology, the NodeMCU's can be networked wirelessly in the form of a serial connection, so it is called radial topology. Where the NodeMCU 01 is not directly communicated with the HMI, it will communicate to the HMI through other NodeMCU's which means that the NodeMCU 01 will communicate with NodeMCU 02, and then it will communicate with NodeMCU 03 later it will communicate with NodeMCU 04. The process is reversed while the communication from HMI to NodeMCU 01. This topology is mainly suitable for long-range wireless networking. The one better thing in this topology is the medium NodeMCU also serves its own data along with the data of the adjacent NodeMCU to the HMI. And the NodeMCU 04 gathers the data from all the NodeMCU's and sends it to the HMI, at the same time it will also send the data of its own sensors.

In this wireless networking topology, data of all the NodeMCUs can be displayed on the same web page. So, this topology can be used to monitor all the NodeMCU's data on a single monitor at the same time. In this topology, the DHT11 sensor is used for the monitor and control from the HMI.

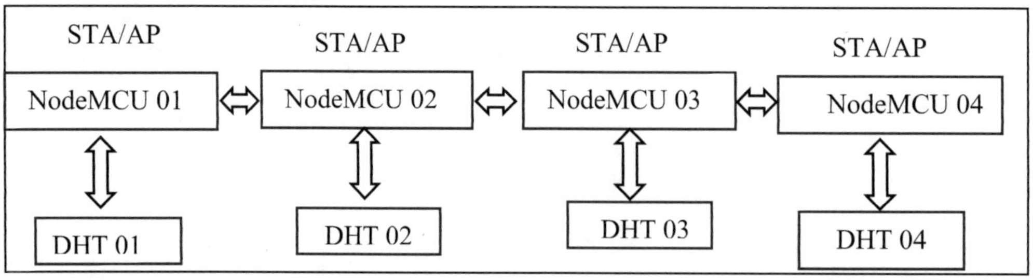

Figure 5. basic Structure of Radial topology.

Figure 5 shows a complete architecture of wireless networking with radial topology. In this topology, all the four NodeMCU's can be used to transfer the data to HMI. And if anyone sensor fails to prepare the weather report, then it will indicate on the monitor that the sensor is not found. Further, the DHT will not be powered ON until the Request is sent from the HMI. The complete network performance of radial wireless networking topology can be explained with the use of flow chart which is shown in figure 6. Initially, all the NodeMCU's are set up as an access point and the HMI will connect to the NodeMCU 04 by entering the proper SSID and Password. Then enter the IP address of the NodeMCU 04 in the web browser to display the home page on the monitor.

Whenever request button is pressed which is displayed on the monitor, the DHT4 sensor is Powered ON and measures the Temperature and Humidity values and stores it. Then the NodeMCU 04 sets up in STA mode and requests the NodeMCU 03 and is again set up as an AP. Then the NodeMCU 03 setup as an STA mode then connected to the NodeMCU 02 and the DHT3 sensor is Powered ON and it measures the Temperature values and stores it and again set up as an AP. Next, the NodeMCU 02 setup as an STA mode then connected to the NodeMCU 01 and the DHT2 sensor is Powered ON and it measures the Temperature values and stores it and again set up as an AP. Up next, the NodeMCU 01 DHT sensor is Powered ON and it measures the Temperature and humidity values and setup as an STA mode then connected to the NodeMCU 02 and transfers the data, DHT1 sensor Powered OFF and again set up as an AP.

The NodeMCU 02 receives the DHT01 data and adds it to the DHT02 data of temperature and humidity values and setup as an STA mode then connected to the NodeMCU 03 and transfers the data, DHT02 sensor powered OFF and again set up as an AP. Next, the NodeMCU 03 receives the DHT01 and DHT02 data and add it to the data of DHT03 and setup as an STA mode then connected to the NodeMCU 04 and transfer the data, DHT03 sensor powered OFF and again set up as an AP.

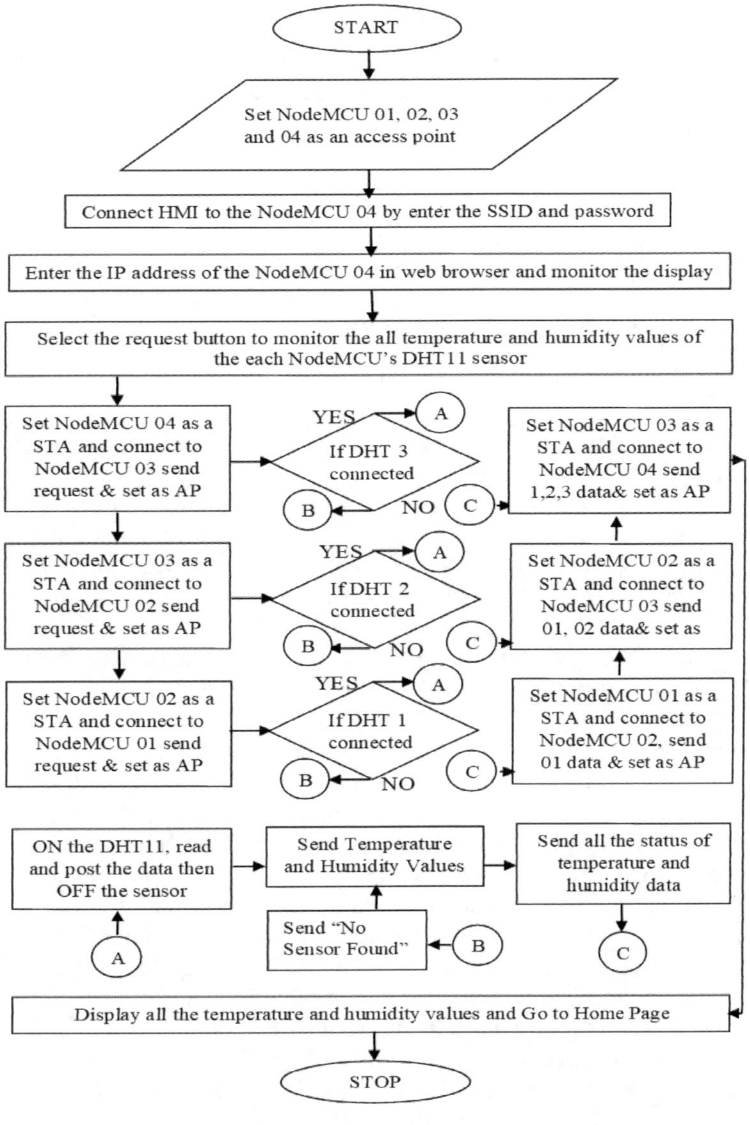

Figure 6. Flow chart of Radial topology

Finally, the NodeMCU 04 receives the DHT01, DHT02, and DHT03 data and adds it to the DHT04 data of Temperature and humidity values and stores it. Then, DHT04 powered OFF. So whenever Read button on the monitor is pressed, each DHT's measured value will be displayed on the new web page. Moreover, if any DHT is not connected to the NodeMCU, then it will show "No Sensor Found". This topology is suitable for long-range data transfer and control of application. In the next step, the motor and LED indicator can be included in this radial topology. To transfer the data in this topology, it will take the time delay to setup the AP or STA modes. So, this problem can be eliminated by the use of next topology with repeaters for long range.

3.3 long-range topology

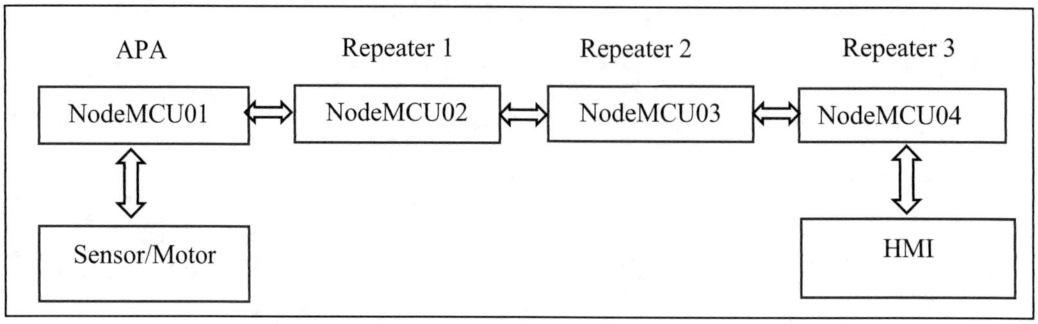

Figure 7. Basic structure of Long-range topology.

This topology is an advanced one which is being used for long-range applications with instant control and monitoring of the devices. Before starting this topology the NodeMCU is configured as a Repeater. While setup, it will ask the station mode and access point mode settings. In the AP settings, the SSID and Password can be set up for repeater. Whereas in STA settings, enter the SSID and Password of the other AP device which we want to connect to it. The repeater can be set up as an auto mesh mode to connect automatically to the nearest AP devices based on their signal strength. Figure 7 gives the basic structure of wireless networking with long-range topology by repeaters. Where the NodeMCU 01 as an access point and the DHT sensor, Motor and LED indicator are connected to it. In Repeater 1, the station (STA) SSID and Password belong to NodeMCU 01 and in Repeater 2, the STA SSID and Password belong to Repeater 1 and in Repeater 3, the STA SSID and Password belong to Repeater 2. Now the Repeater3 will give access to the HMI. Then the HMI can connect to any one of the Repeater to control and monitor the devices which are connected to the NodeMCU 01 based on the required range.

In this topology, power, speed and direction of the motor can be controlled, LED indicator power and brightness can be controlled and the weather report from the DHT sensor can be displayed. The complete working performance can be explained based on the flow chart which is shown in figure 8. Initially, the HMI is connected to the Repeater 3 or 2 or 1 based on the range we required with proper SSID and Password. After opening a web page and entering the IP address of NodeMCU 01, the web page will appear which include the weather report, motor control and LED control. There will be some buttons which are used to control the Motor and LED indicator. The weather report is updated for every refresh of the page. To control the Motor, ON and OFF buttons are provided to power it on and off. In addition, the direction of the motor can be controlled by pressing the 'change direction button' on the display. The speed control is also possible by pressing the increase and decrease buttons. For control of the LED indicator, ON and OFF buttons are provided to power the LED and the brightness of the LED can be increased or decreased by pressing respective buttons which are displayed on the HMI.

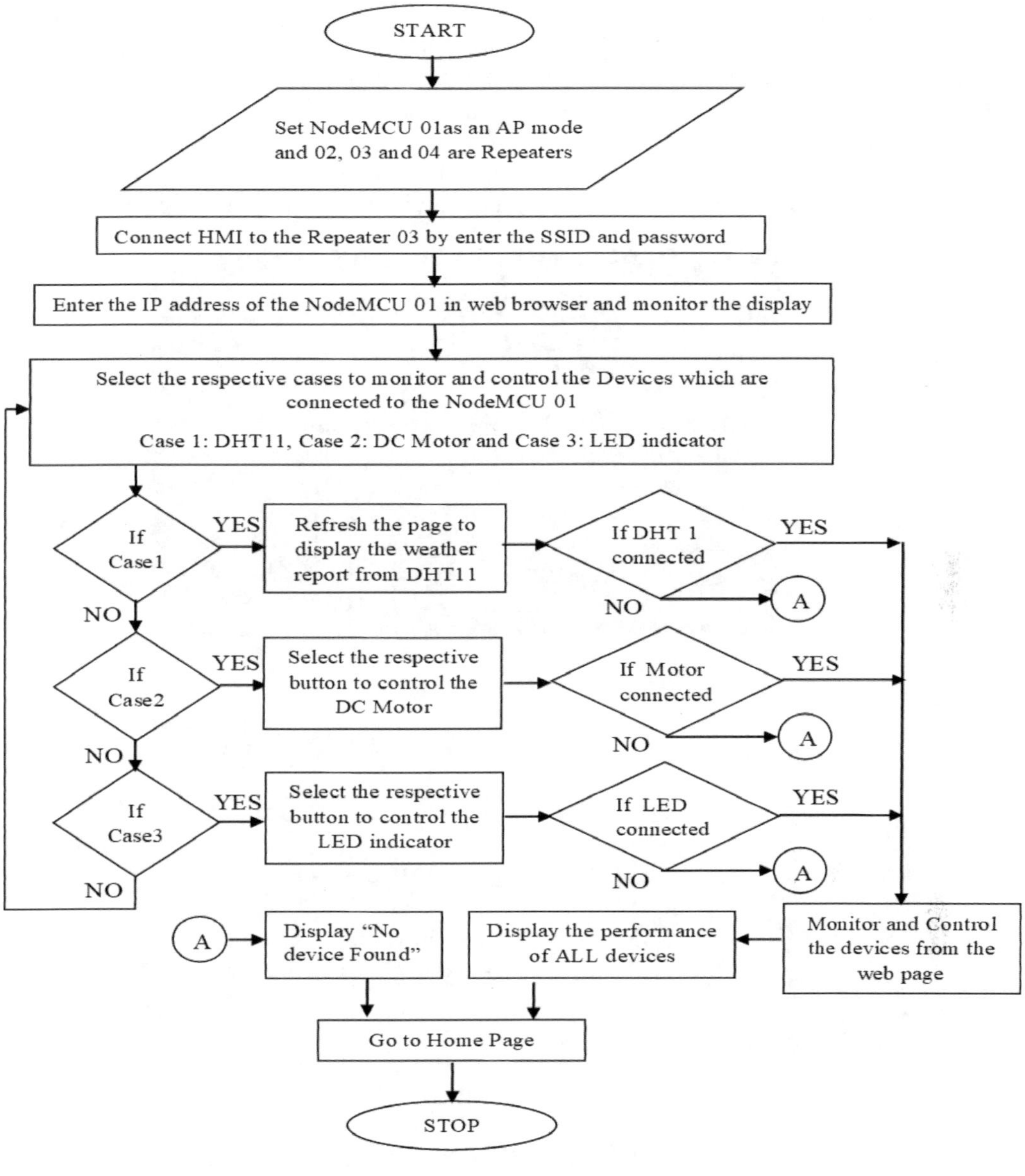

Figure 8. Flow chart of Long-range topology.

4. Results and Discussion

Different types of wireless networking topologies are designed by considering star, radial and long-range topologies for development of CRISP at BHEL. The development procedure followed in successful development and implementation under Arduino IDE environment with ESP 8266 NodeMCU for star topology is presented in figure 9. Before starting the wireless networking of these devices, open the Arduino IDE software in PC and prepare the individual code to each NodeMCU. Then all the NodeMCU's are fixed to the breadboard and connected to the PC through the USB cable. Also, connect the DHT11 sensors to the

NodeMCU's with respective pins as per the code. In this topology, the DHT11 is not connected to the NodeMCU 03 to check the variation with NodeMCU's connected with DHT11. After connections, select the respective ports and upload the code which is written in the software. In this topology, the range from each NodeMCU to Access Point is 50 meters but those all are placed near to each other to understand how the data is being transferred and controlled through the serial monitor. After uploading the code into the respective NodeMCU's, open the serial monitor. Now connect the HMI to the NodeMCU 04 by entering the proper SSID and Password. Then open the web browser and type the IP address of NodeMCU 04 (192.168.4.4), then the web page will appear which is shown in figure 10. The results of star topology can be studied under two scenarios which are based on whether the DHT11 is connected to NodeMCU or not.

Figure 9. Wireless networking with a star topology.

Figure 10. Home page for star topology.

In case 1, the DHT11 is not connected to the NodeMCU 03, so it should display "No Sensor Found". So, whenever we press the REQUEST3 button on the home page it will send the request to the NodeMCU 03 but here no sensor is connected to NodeMCU, so it will show "No Sensor Found" which is shown in figure 11b. This web page can be obtained by pressing the READ3 button on the home page. From figure 11a, clearly shows the data transfer and control mechanism of DHT11 sensor between Access Point and NodeMCU 03.

In case 2, the DHT11 is connected to the NodeMCU 01 to display the temperature and humidity values. By clicking on REQUEST1 button on the home page, request is initiated at NodeMCU 01 and turns ON the DHT11 which sends the requisite data pertaining to temperature and humidity on LCD display by

IOP Conf. Series: Materials Science and Engineering **937** (2020) 012024 doi:10.1088/1757-899X/937/1/012024

pressing the READ1 button on the home page as shown in figure 12b while figure 12a clearly shows the data transfer between Access Point and NodeMCU 01 to control DHT11 sensor.

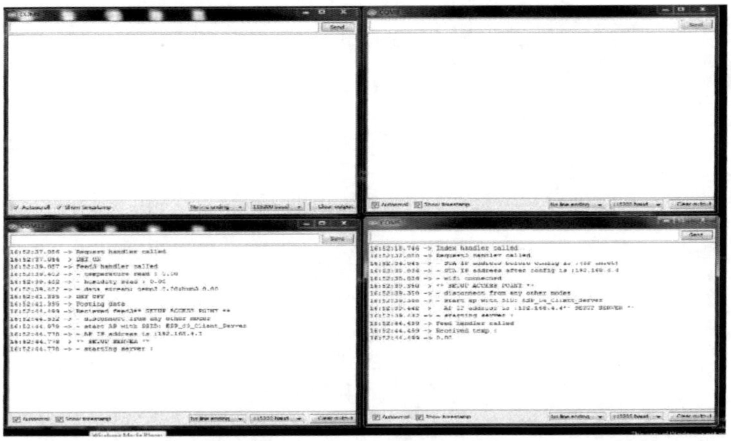

Figure 11. Star topology NodeMCU 03 results (a) Serial monitor.

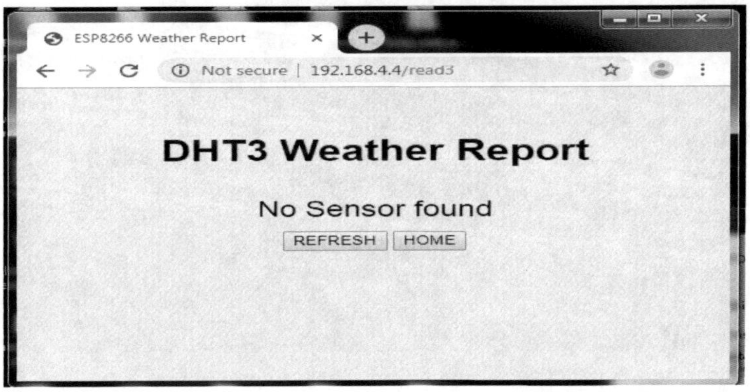

(b) Web page.

Figure 12. Star topology NodeMCU 01 results (a) Serial monitor.

(b) Web page.

Figure 13. Wireless networking with radial topology.

Figure 14. Home page for radial topology.

Radial topology is an extension to star topology to increase the range of wireless networking. In radial topology, four NodeMCU's and two DHT11 sensors are used for wireless networking and those are connected in a radial topology as shown in figure 13. Here in this topology all the four NodeMCU's were considered as a robot and the fourth NodeMCU is used as an access point to establish communication between individual NodeMCU's and central control station.

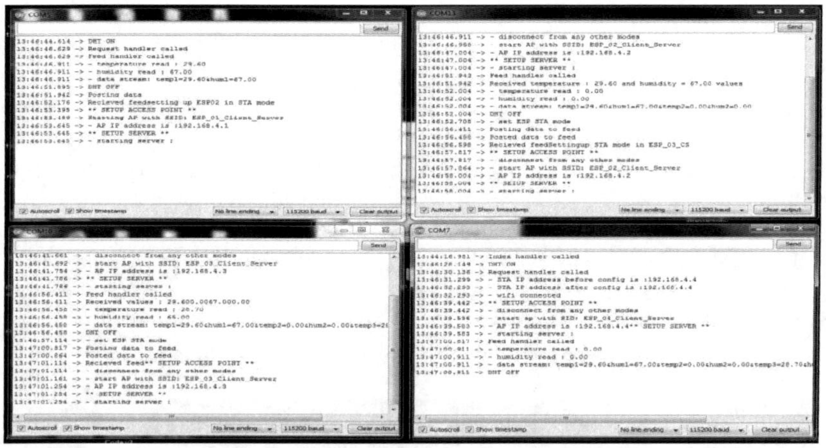

Figure 15. Radial topology results (a) Serial monitor.

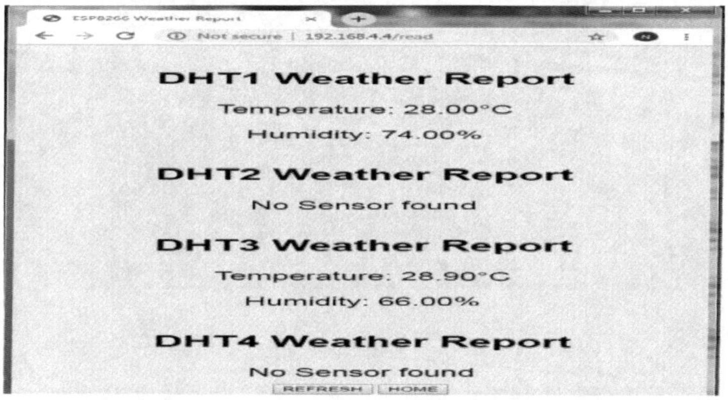

(b) Web page.

The development procedure followed in successful development and implementation under Arduino IDE environment with ESP 8266 NodeMCU for radial topology initially starts by opening the Arduino IDE software in PC and prepare the individual code to each NodeMCU before starting the wireless networking of these devices. By using the USB cable, connect all the NodeMCUs which are fixed on the breadboard to the PC. Then connect the DHT11 sensors to the NodeMCU's with respective pins as mentioned in the code. In this topology also, the DHT11 is not connected to the NodeMCU02 and NodeMCU04 to check the variation with connected NodeMCU's. After connections, select the respective ports and upload the code which is written in the software. The range from each NodeMCU to AP is selected as 50 meters but for the sake of convenience all the NodeMCU's were placed nearer to each other to understand how the data is being transferred and DHT11 is controlled through the serial monitor. After uploading the code into the respective NodeMCUs, open the serial monitor and select Wi-Fi mode on devices and connect to the NodeMCU 04 by providing access to connect it by entering the proper SSID and

Password. Then open the web browser and type the IP address of NodeMCU 04 (192.168.4.4) which displays the web page as shown in figure 14. The results of this radial wireless networking topology can be monitored on the web page and with the help of a serial monitor, it can be observed that how the data can be transferred from NodeMCU 01 to NodeMCU 04 or vice versa.

By pressing the REQUEST button on the home page, the request can be sent from NodeMCU 04 to NodeMCU 01. The DHT11 sensors will be powered ON and start to send the data to the NodeMCU 04 through the NodeMCU 02 and NodeMCU 03. The NodeMCU 03 transfers the data of its DHT11 sensor along with NodeMCU01 data. Similarly, the NodeMCU02 transfers the data of its DHT11sensor data along with NodeMCU 01 and NodeMCU 02 sensors data. Finally, the NodeMCU 04 gets all the sensors data and adds its sensor data to them and stores it in a string. Then the DHT11 sensors will be powered OFF which is shown in figure 15a. And by pressing the READ button on the home page, the weather report will be displayed on the new web page which is shown in figure 15b. In this topology, the DHT11 sensors are not connected to the NodeMCU02 and NodeMCU 04. So only the temperature and humidity values of NodeMCU 01 and NodeMCU 03 are displayed. For NodeMCU 02 and NodeMCU 04, "No Sensor Found" is displayed. Although the range can be extended in radial network topology, the data transfer delay is a serious concern of about 8 to 10 seconds. So, in order to overcome this problem, the repeaters can be used as a medium to transfer the data from NodeMCU 01 to HMI. In this topology, the NodeMCU 01 consists of DHT11 sensor, LED indicator and one DC motor driver which are shown in figure 16.

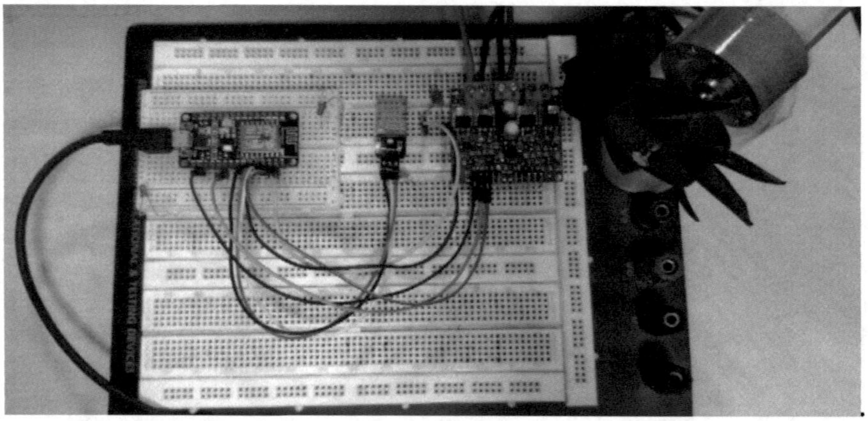

Figure 16. Wireless networking with long-range topology.

Figure 17. Repeaters 01, 02 and 03 at different locations.

IOP Conf. Series: Materials Science and Engineering **937** (2020) 012024 doi:10.1088/1757-899X/937/1/012024

Long range topology consists of three repeaters which are used to increase the range of the wireless network and operate the devices which are connected to the NodeMCU instantly. The developmental procedure can be initiated by preparing the requisite code for NodeMCU 01 in Arduino IDE software environment and then connecting NodeMCU 01 to the PC through USB cable followed by allowing all the devices to get connected to NodeMCU 01 pins as per the code configuration of the software thereby port selection and code launching can be carried out. Now configure the other three NodeMCU's as repeaters using Flash Downloader tool provided by espressif which are shown in figure 17. While configuring as a repeater, it should be reminded that STA has to be setup followed by AP settings in which the SSID and Password in STA settings are similar to that of SSID and Password of other NodeMCU's which provides access to it. For this topology, the three repeaters are placed at different locations in transmission and distribution laboratory (TDL) at BHEL by maintaining a distance of about 50 to 70 meters between every two repeaters. Now the power is allowed to each repeater with a USB cable to attain access to other stations and connect the HMI to the repeater 03 by entering proper SSID and Password. Next, go to a web browser and type the IP address of NodeMCU 01 (192.168.4.4) to display relevant web page as shown in figure 18.

The results of long-range repeaters can be obtained instantly on web page wherein the weather report from the DHT11 sensor can be updated and displayed for every refresh of the web page. The motor can be controlled in three ways i.e. switching the power supply in ON and OFF conditions for motor operation, by changing the direction of shaft rotation, and controlling the speed of the motor either by increasing or decreasing. The LED can be controlled in two ways i.e. by switching the power supply in ON and OFF conditions and by increasing or decreasing the brightness of the LED. The devices can be controlled by pressing the respective buttons on the web page.

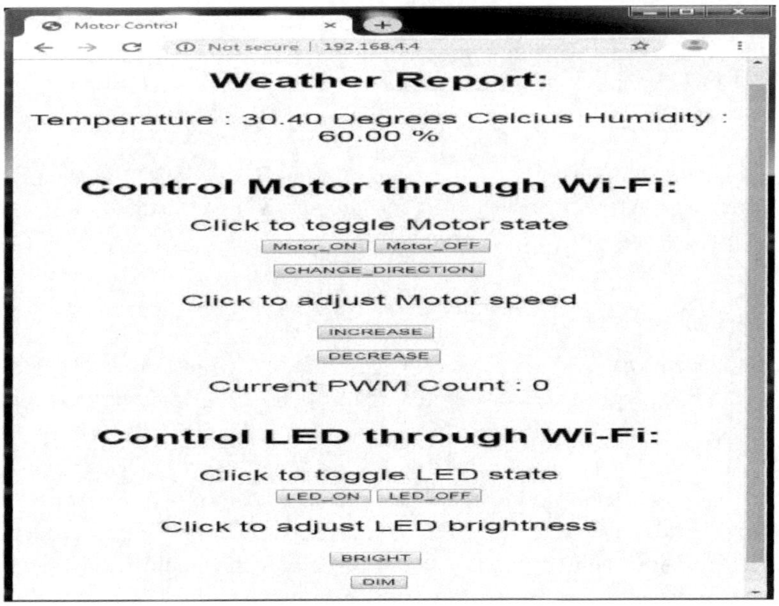

Figure 18. Home page for long-range topology.

5. Conclusion

The main objective of this work is to provide wireless networking for the solar panel cleaning robots which are used to clean the dust accumulated on the solar panel. This work emphasizes on the development of CRISP robot to control and monitor robots from a long distance by establishing a network of many robots connected to a central control station. The CRISP is developed to overcome the problem of accumulation of dust on the solar panel with advanced technology and better performance at BHEL Hyderabad. It is controlled wirelessly using Wi-Fi from the HMI. Different wireless networking topologies are designed and developed by utilizing NodeMCU as microcontroller and Wi-Fi device for CRISP. Replacement of PLC and Wi-Fi router in CRISP with NodeMCU showed significant reduction on incurred capital.

In star topology, all the robots are wireless networked to one NodeMCU which is an access point interconnected with HMI to control each robot individually. Using this topology, it has been identified that the robots were controlled within a limited range i.e. about 50 to 70 meters from the access point. To increase the range of the wireless network the radial topology was developed where the range of wireless networking was increased from 50 meters to 200 meters with the help of three more NodeMCU's. Those three NodeMCU's are used not only to increase the range of the network but also made to control an individual robot. But a delay of around 4 seconds was observed between each NodeMCU of the network. As these are networked in serial format, to control and monitor the NodeMCU 01, it takes around 15 seconds. To overcome this problem, long-range wireless networking topology was taken into consideration wherein the repeaters (additional NodeMCU's) are introduced for wireless networking of the robots. By setting up the STA and AP settings in repeaters, wireless network coverage can be enhanced to the desired level over a long distance so that demerits of previous two topologies can be minimized.

The work can be further improvised by utilizing wireless networking with a mesh topology for CRISP robots which are used in large solar power plants and supplemental wireless bypass networking topology in which, if one NodeMCU fails to connect to the next NodeMCU, it connects to the next available NodeMCU to maintain the network. The NodeMCU has only one analog read pin which can be extended with the multiplexer. Extension of these wireless networking topologies to other similar products like street light controlling can also be considered towards networking with smart grid operations. Hence, it can be concluded that utilization of CRISP with appropriate wireless networking topology for monitoring and control of solar PV panel cleaning in an efficient manner will remain as a viable option for industrial scale large solar power plants.

***Acknowledgments**:* The authors express their gratitude to the BHEL R&D Division, Hyderabad; JNT University, Kakinada and Centre for Smart Grid Technologies, Vellore Institute of Technology, Vellore, India for their entire support towards this research work.

References

[1]. Sitharthan R, Geethanjali M and Pandy TKS 2016 Adaptive protection scheme for smart microgrid with electronically coupled distributed generations *Alexandria Engineering Journal* **55(3)** 2539-2550

[2]. Fathima AH, and Palanisamy K 2014 Battery energy storage applications in wind integrated systems—a review *IEEE International Conference on Smart Electric Grid* 1-8

[3]. Prabaharan N and Palanisamy K 2015 Investigation of single-phase reduced switch count asymmetric multilevel inverter using advanced pulse width modulation technique *International Journal of Renewable Energy Research* **5(3)** 879-890.

[4]. Jerin ARA, Kaliannan P and Subramaniam U 2017 Improved fault ride through capability of DFIG based wind turbines using synchronous reference frame control based dynamic voltage restorer. *ISA transactions* **70** 465-474

[5]. Sitharthan, R, Sundarabalan CK, Devabalaji KR, Nataraj SK and Karthikeyan M 2018 Improved fault ride through capability of DFIG-wind turbines using customized dynamic voltage restorer *Sustainable cities and society* **39** 114-125

[6]. Prabaharan N and Palanisamy K 2016 A single-phase grid connected hybrid multilevel inverter for interfacing photo-voltaic system *Energy Procedia* **103** 250-255

[7]. Palanisamy K, Mishra JS, Raglend IJ and Kothari DP 2010 Instantaneous power theory based unified power quality conditioner (UPQC) *IEEE Joint International Conference on Power Electronics, Drives and Energy Systems* 1-5

[8]. Sitharthan R and Geethanjali M 2017 An adaptive Elman neural network with C-PSO learning algorithm-based pitch angle controller for DFIG based WECS Journal of Vibration and Control **23(5)** 716-730

[9]. Sitharthan R and Geethanjali M 2015 Application of the superconducting fault current limiter strategy to improve the fault ride-through capability of a doubly-fed induction generator–based wind energy conversion system *Simulation* **91(12)** 1081-1087

[10]. Sitharthan R, Karthikeyan M, Sundar DS and Rajasekaran S 2020 Adaptive hybrid intelligent MPPT controller to approximate effectual wind speed and optimal rotor speed of variable speed wind turbine *ISA transactions* **96** 479-489

[11]. Sitharthan R, Devabalaji KR and Jees A 2017 An Levenberg–Marquardt trained feed-forward back-propagation based intelligent pitch angle controller for wind generation system *Renewable Energy Focus* **22** 24-32

[12]. Sitharthan R, Sundarabalan CK, Devabalaji KR, Yuvaraj T and Mohamed Imran A 2019 Automated power management strategy for wind power generation system using pitch angle controller *Measurement and Control* **52(3-4)** 169-182

[13]. Sundar DS, Umamaheswari C, Sridarshini T, Karthikeyan M, Sitharthan R, Raja AS and Carrasco MF 2019 Compact four-port circulator based on 2D photonic crystals with a 90° rotation of the light wave for photonic integrated circuits applications *Laser Physics* **29(6)** 066201

[14]. Sitharthan R, Parthasarathy T, Sheeba Rani S and Ramya KC 2019. An improved radial basis function neural network control strategy-based maximum power point tracking controller for wind power generation system *Transactions of the Institute of Measurement and Control* **41(11)** 3158-3170

[15]. Rajesh M and Gnanasekar JM 2017 Path observation based physical routing protocol for wireless ad hoc networks *Wireless Personal Communications* **97(1)** 1267-1289

[16]. Palanisamy K, Varghese LJ, Raglend IJ and Kothari DP 2009. Comparison of intelligent techniques to solve economic load dispatch problem with line flow constraints *IEEE International Advance Computing Conference* 446-452

[17]. Sitharthan R, Ponnusamy M, Karthikeyan M and Sundar DS 2019 Analysis on smart material suitable for autogenous microelectronic application *Materials Research Express* **6(10)** 105709

[18]. Rajaram R, Palanisamy K, Ramasamy S and Ramanathan P 2014 Selective harmonic elimination in PWM inverter using fire fly and fireworks algorithm *International Journal of Innovative Research in Advanced Engineering* **1(8)** 55-62

[19]. Sitharthan R, Swaminathan JN and Parthasarathy T 2018 March. Exploration of wind energy in India: A short review *IEEE National Power Engineering Conference* 1-5

[20]. Karthikeyan M, Sitharthan R, Ali T and Roy B 2020 Compact multiband CPW fed monopole antenna with square ring and T-shaped strips *Microwave and Optical Technology Letters* **62(2)** 926-932

[21]. Sundar D Sridarshini T, Sitharthan R, Madurakavi Karthikeyan, Sivanantha Raja A, and Marcos Flores Carrasco 2019 Performance investigation of 16/32-channel DWDM PON and long-reach PON systems using an ASE noise source *In Advances in Optoelectronic Technology and Industry Development: Proceedings of the 12th International Symposium on Photonics and Optoelectronics* 93

[22]. Sitharthan R and Geethanjali M 2014 Wind Energy Utilization in India: A Review *Middle-East J. Sci. Res.* **22** 796–801 doi:10.5829/idosi.mejsr.2014.22.06.21944

[23]. Sitharthan R and Geethanjali M 2014 ANFIS based wind speed sensor-less MPPT controller for variable speed wind energy conversion systems *Australian Journal of Basic and Applied Sciences* **8**14-23

[24]. Jerin ARA, Kaliannan P, Subramaniam U and El Moursi MS 2018 Review on FRT solutions for improving transient stability in DFIG-WTs *IET Renewable Power Generation* **12(15)** 1786-1799

[25]. Prabaharan N, Jerin ARA, Palanisamy K and Umashankar S 2017 Integration of single-phase reduced switch multilevel inverter topology for grid connected photovoltaic system *Energy Procedia* **138** 1177-1183

[26]. Rameshkumar K, Indragandhi V, Palanisamy K and Arunkumari T 2017 Model predictive current control of single phase shunt active power filter *Energy Procedia* **117** 658-665

[27]. Fathima AH and Palanisamy K 2016 Energy storage systems for energy management of renewables in distributed generation systems *Energy Management of Distributed Generation Systems* 157

[28]. Rajesh M 2020 Streamlining Radio Network Organizing Enlargement Towards Microcellular Frameworks *Wireless Personal Communications* 1-13

[29]. Subbiah B, Obaidat MS, Sriram S, Manoharn R and Chandrasekaran SK 2020 Selection of intermediate routes for secure data communication systems using graph theory application and grey wolf optimisation algorithm in MANETs IET *Networks* doi:10.1049/iet-net.2020.0051

[30]. Singh RR and Chelliah TR 2017 Enforcement of cost-effective energy conservation on single-fed asynchronous machine using a novel switching strategy *Energy* **126** 179-191

[31]. Amalorpavaraj RAJ, Palanisamy K, Umashankar S and Thirumoorthy AD 2016 Power quality improvement of grid connected wind farms through voltage restoration using dynamic voltage restorer *International Journal of Renewable Energy Research* **6(1)** 53-60

[32]. Singh RR, Chelliah TR, Khare D and Ramesh US 2016 November. Energy saving strategy on electric propulsion system integrated with doubly fed asynchronous motors *IEEE Power India International Conference* 1-6

[33]. Sujatha K and Punithavathani DS 2018 Optimized ensemble decision-based multi-focus imagefusion using binary genetic Grey-Wolf optimizer in camera sensor networks *Multimedia Tools and Applications* **77(2)** 1735-1759

[34]. Krishnamoorthy S, Punithavathani S and Priya JK 2017 Extraction of well-exposed pixels for image fusion with a sub-banding technique for high dynamic range images *International Journal of Image and Data Fusion* **8(1)** 54-72

[35]. Singh RR, Mohan H and Chelliah TR 2016 November. Performance of doubly fed machines influenced to electrical perturbation in pumped storage plant-a comparative electromechanical analysis *IEEE 7th India International Conference on Power Electronics* 1-6

Optical modeling and Performance Design Of A Fresnel Lens For CPV Units

Sarah El Himer[1], Salima Al Ayane[2], Ali Ahaitouf[2]
[2]Laboratory of Intelligent Systems, Georessources and Renewable , Electrical Engineering department, Faculty of Sciences and Technologies, Sidi Mohamed Ben Abdellah University -Fez PO. Box 2202 Fez Morocco

[1]E-mail: sarah.elhimer@usmba.ac.ma

Abstract. In CPV technology, a focus on keeping low costs has led most recent designs to use concentration ratios of 1000X or higher. For that, many high-concentration photovoltaic systems use refractive Fresnel lenses because it is one of the main transmission-type concentrators. In this work, a solar concentrator system composed from a typical Fresnel lens made of PMMA, with an effective diameter of 350mm and a squared solar cell of 10x10mm is proposed. The design of a Fresnel lens is introduced; we studied and analysed a squared design configurations with multiple tuning of focal lengths, facet spacing and an extended range of wavelengths. The results show that the Fresnel lens is performant in the visible spectrum with facet spacing of 1mm and the focal length between 265 to 350mm. The acceptance angle of the whole system was between 0.6 to 0.8° and the optical efficiency obtained by simulations was between 80 to 90%.

1. Introduction

CPV technology is becoming a trend to create a cost-effective system with a high level of concentration compared to other photovoltaic technologies, [1][2]. This technology increases the efficiency of cell conversion and reduces cell size, which also reduces the contribution of photovoltaic costs to the entire system [3][4]. The CPV system is based on the use of optical elements that can be refractive (Fresnel lenses), reflective (Mirror), or both.

CPV technology is becoming a trend to create a cost-effective system with a high level of concentration compared to other photovoltaic technologies, [1][2]. This technology increases the efficiency of cell conversion and reduces cell size, which also reduces the contribution of photovoltaic costs to the entire system [3][4]. The CPV system is based on the use of optical elements that can be refractive (Fresnel lenses), reflective (Mirror), or both.

Each CPV system is characterized by its concentration ratio which given by:

$$C = C_{geo} \times \eta \tag{1}$$

Where C_{geo} denotes the geometrical concentration and η the optical efficiency. Those parameters are respectively given by:

$$C_{geo} = \frac{A_{in}}{A_{out}} \quad (2)$$

$$\eta = \frac{P_{in}}{P_{out}} \quad (3)$$

Where Ain and Aout are the input and the output aperture of CPV system as indicated at fig.1

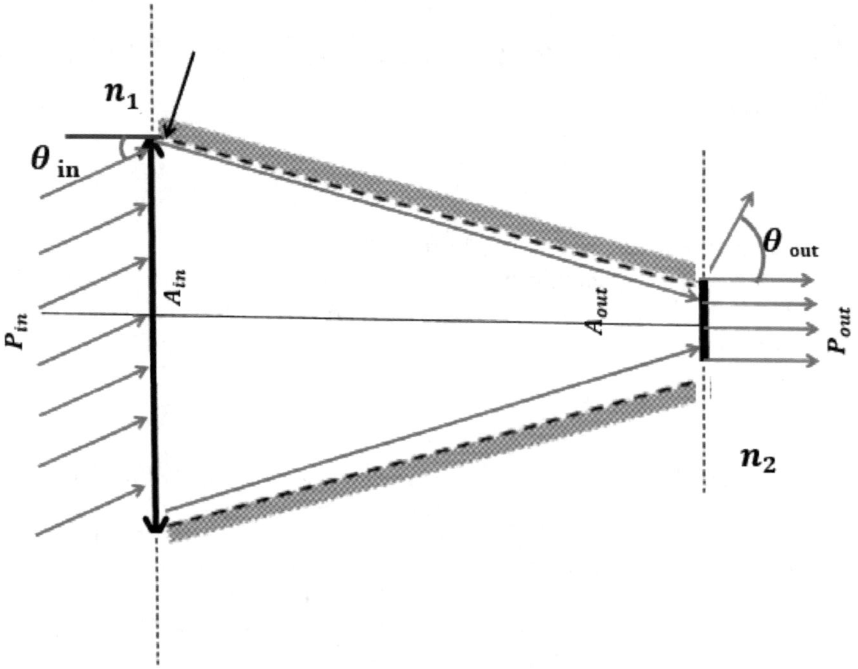

Figure 1. Schematic view of a solar concentrator

The Fresnel lens [5][6] has been widely used as a concentrator in the CPV. It is considered as one of the main transmission-type concentrators which has several advantages on engineering such as lighter weight, larger allowable error of track and easier to receive light etc…[7]. Moreover this optical element has two disadvantages. The first, the optical concentrator, has a long focal length that increases the volume and weight of the system. The second, the distribution of the irradiance on the solar cell, is non-uniform and creates a hot spot, which decreases the efficiency of the conversion and decreases the reliability, the lifetime of the photovoltaic system [8]. Tver'yanovich et al. [8] has early studied how profile effected of the performance of solar-engineer Fresnel lens. Khalil et al. [9] designed a linear focusing flat Fresnel lens and found that the edge-prism has bad transmittance. Leutz et al. [10] designed a convex-shaped Fresnel lens by using edge ray principle. This Fresnel lens has a short focal length and high concentration ratio. The tested optical efficiency reaches to 72%. Xinglong et al. [11] analysed the relation between transmittance and incident angle quantitatively, which were based on Fresnel lens and the reversibility of light.

In this work, we propose a solar concentrator system composed from a typical Fresnel lens made of PMMA, with an effective diameter of 350mm and a squared solar cell with 10x10mm. The design of a Fresnel lens is introduced; we studied and analysed a squared design configurations including several focal lengths, facet spacing and an extended range of wavelengths.

The rest of the paper is organized as follows: section 2 introduces the design and theoretical background of Fresnel lens; section 3 is dedicated to analyse serval studies and cases of Fresnel lenses.

2. Design of Fresnel lens

IOP Conf. Series: Materials Science and Engineering **937** (2020) 012025 doi:10.1088/1757-899X/937/1/012025

The CPV optics constitute a typical design problem that contains both the bundle coupling problem for obtaining maximum acceptance-concentration product and the prescribed irradiance to get uniform irradiance distribution on the solar cell area [12]. The design method includes the inner saw teeth of Fresnel lens that work by total internal reflection and refraction. The Fresnel lens is composed of several prisms as shown in Fig.2,a. A typical Fresnel lens with grooves facing inwards will be presented here. The design has an analytical solution and follows Tver'yanovich [8]. In accordance with Fig. 2, several parameters can be set up to describe the prism which are: $\theta_a$is an angle between the normal of the receiver and the refractive ray, α_p is an angle between the normal of the Fresnel lens's facet and the incident ray, β_p is an angle between the normal direction of the Fresnel lens's facet and the refractive ray, f is the focal length and d is the Fresnel lens diameter as illustrated in Fig.2,b.

$$n \sin \alpha_p = \sin \beta_p \quad (4)$$

$$\sin \theta = \frac{D}{2f} \quad (5)$$

$$\beta_p = \theta + \alpha_p \quad (6)$$

$$F_{\#} = \frac{f}{d} \quad (7)$$

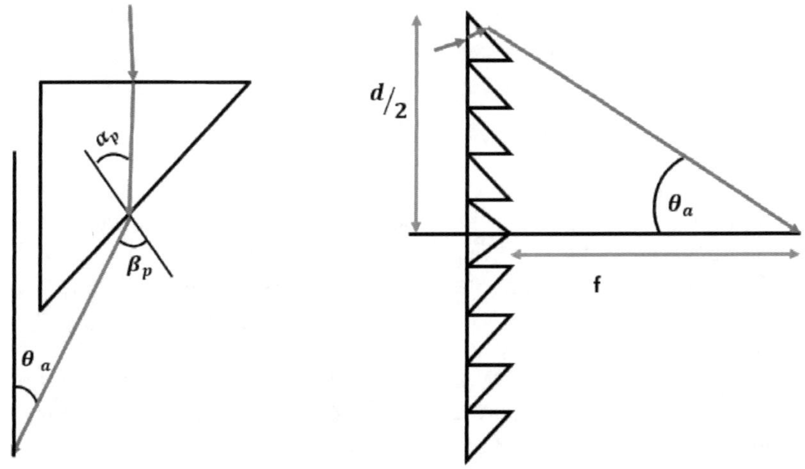

(a). Fresnel lens prism **(b).** Fresnel lens

Figure 2. Fresnel lens design

Using the above equation, we have designed a square Fresnel lens with 350mm as diameter and 265mm as focal length as sketched in Fig.3. The material used in this design is polymethyl methacrylate (PMMA), which is a typical optical plastic with a refractive index of 1.49

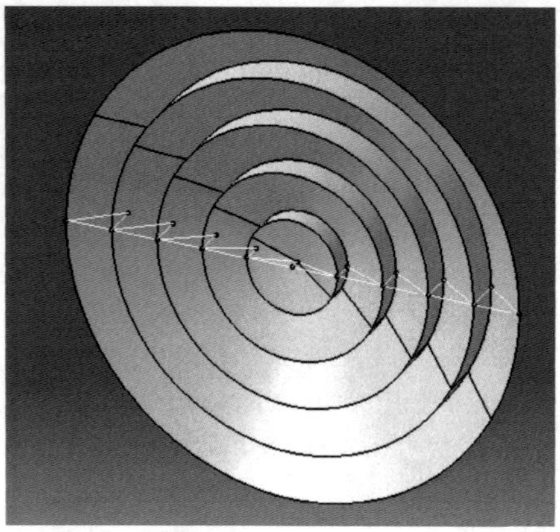

(a). Fresnel lens in 3D

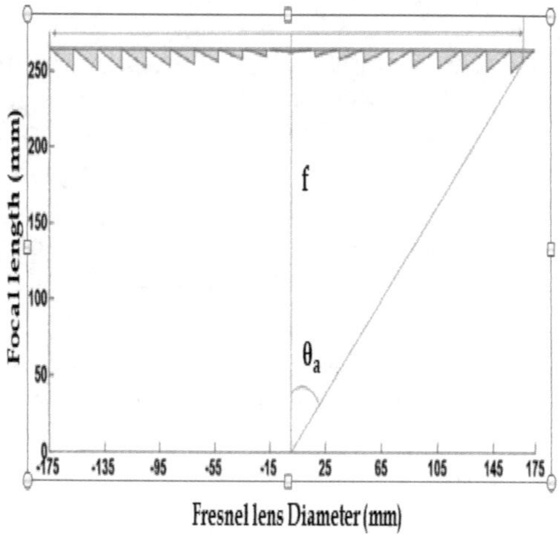

(b). Fresnel lens in 2D

Figure 3. Design of Fresnel lens

3. Simulation Results

3.1. Effect of facet spacing on Fresnel lens Performance

In order to evaluate our design. The optical performance are analysed by using the software, TRACEPRO. Firstly, we will discuss the influence of the facet spacing on the intensity distribution, optical efficiency and the acceptance angle. For this test, the solar spectrum is the visible range and the focal length is fixed to 265mm. Figures 4 and 5 show the evolution of the intensity and the optical efficiency according to facet spacing.

The intensity decreases from 1.2 108 W/m2 to 9.5 107. It should be noted that this figure illustrates the central intensity of the receiver. Concerning the optical efficiency, it increases slightly for facet spacing less than 1 mm and begins to decrease with increasing the facet spacing. In fact, for a fixed diameter lens, and by increasing the facet spacing (the number of prism decreases), the angle of the prism will be increased, and consequently, the size of the prism will be larger, this will produce losses resulting from the total internal reflection and geometrical losses (the losses that occur due to the divergence of the optical beam) for the rays that coincide with the ends of the prisms.

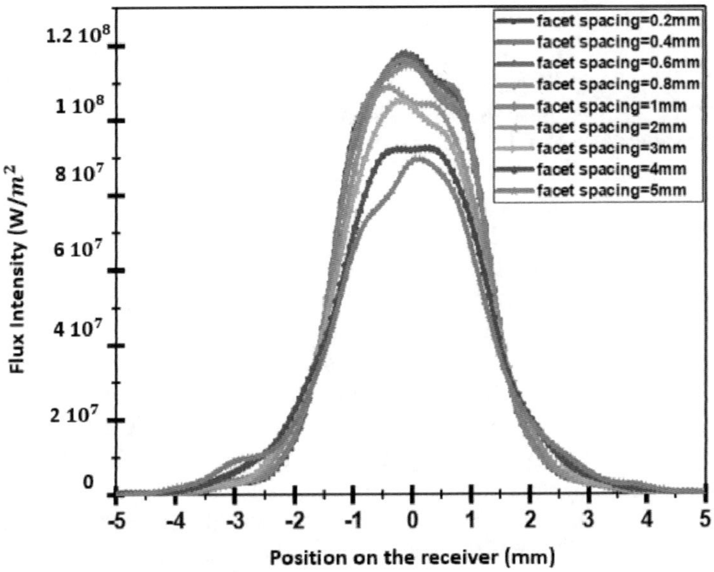

Figure 4. Intensity in function of facet spacing

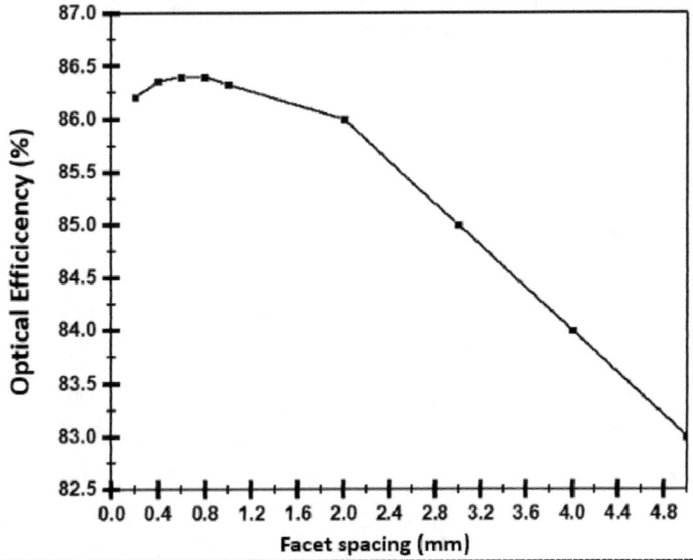

Figure 5. Optical Efficiency in function of facet spacing

Table 1. Acceptance angle Vs facet spacing .

Facet Spacing (mm)	0.2	0.4	0.6	0.8	1	2	3	4
Acceptance Angle (°)	0.6	0.7	0.7	0.8	0.7	0.6	0.6	0.6

IOP Conf. Series: Materials Science and Engineering **937** (2020) 012025 doi:10.1088/1757-899X/937/1/012025

3.2. Effect of focal length on Fresnel lens Performance

In this part, we have varied the focal length of our Fresnel lens from 175 to 700mm and evaluate its optical performances. At this moment, the solar spectrum used is extended from 300 to 700nm (Visible spectrum) and the facet spacing is fixed to 1mm.

Fig.6 shows the simulation of our design. We can observe that for the short focal length some of light light are reflected by TIR and don't reach the receiver, however, for the long focal length, all rays reach the receiver.

Fig.7 illustrates the evolution of intensity distribution on the receiver for several focal lengths. The intensity start to increase reach a maximum from 265 mm to 350 mm (f-number from 0.75 to 1) as focal length, and decrease with increasing focal length in the case of intensity, but the observed peak becomes larger indicating some homogenization of the intensity repartition on the receiver. In fact, one can easily notice the high and narrow peak for the focal lengths between 175 and 350mm.

Fig.8 shows the optical efficiency versus the focal length. The optical efficiency start to increase also reach a maximum in the focal range 265 mm to 350 mm (f-number from 0.75 to 1) as focal length, and start to decrease, beyond these values, but more slowly.

Table1 resumes the acceptance angle of the whole concentrator system. We observe that the acceptance angle is stable from 175mm to 350mm and starts to decrease with the larger focal lengths.

Finally, we can say that the focal length cannot be short because some of light are reflected by TIR causing an unbearable optical efficiency as shown in fig.6 and fig.8. For the long focal lengths, they would result in a too deep, thus expensive module

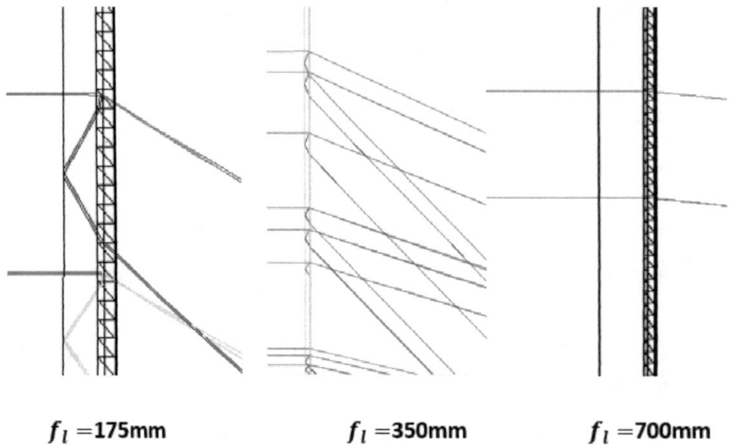

Figure 6. Simulation of Fresnel lens

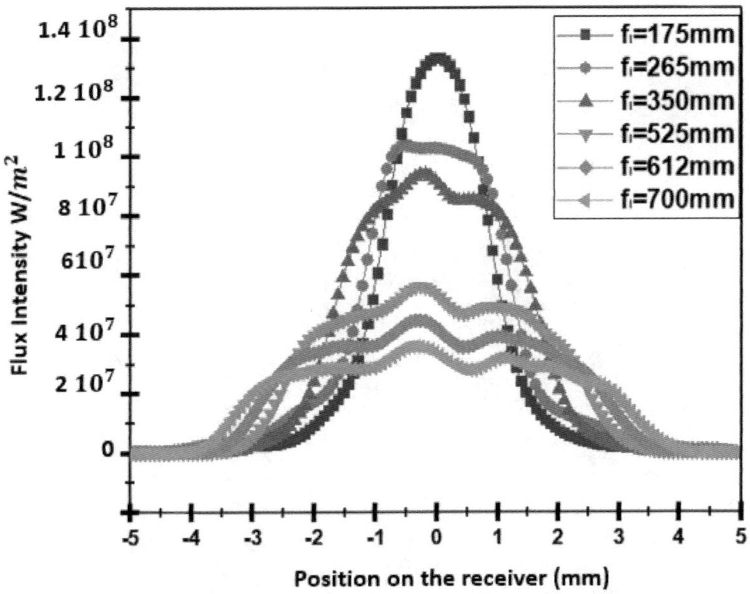

Figure 7. Flux intensity on the receiver for several focal length

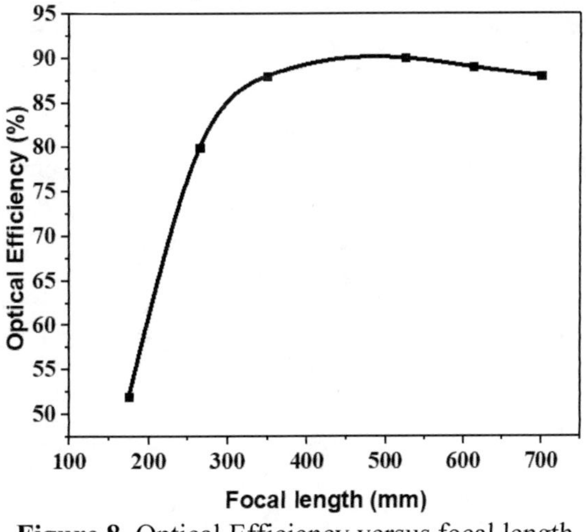

Figure 8. Optical Efficiency versus focal length

Table 2. Acceptance angle Vs focal length

Focal length (mm)	175	265	350	525	700	175	265	350
Acceptance Angle (°)	0.7	0.8	0.8	0.7	0.6	0.6	0.6	0.6

3.3. Effect of wavelength on Fresnel lens Performance

For the current study, we set the focal length to 265mm, facet spacing to 1mm and we vary the wavelength from 0.1 to 1.4µm. We obtained the evolution of intensity and the optical efficiency according to the wavelength variation. As fig.9 and Fig.10 show the effect of wavelength variation on intensity and optical efficiency. It is noted that the intensity and the optical efficiency are high in the ultra violet and the visible and decrease with the increase of the wavelength. These results are due for two reasons; the first is the transmission of the PMMA material as a function of the wavelength. PMMA transmission is higher in the UV and visible (92 %), however, it is low in the infra-red, the second reason is the evolution of the irradiance depending on the length of wave, since, one finds a strong irradiance in the UV and the visible one and weak for the infrared. The acceptance angle in this case is still stable whatever the wavelength as sketched in table 3.

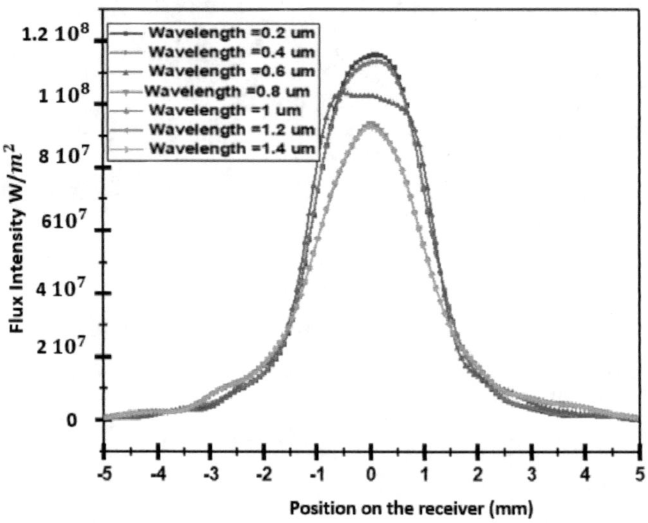

Figure 9. Flux intensity in function of wavelength

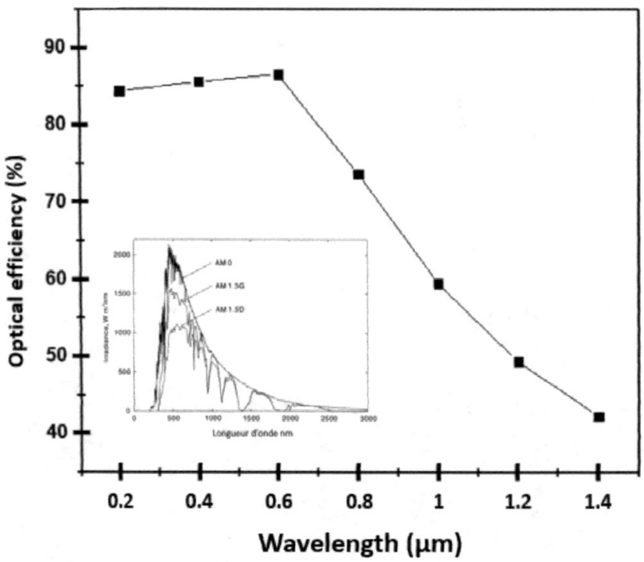

Figure 10. Optical Efficiency in function of wavelength

Table 3. Acceptance angle Vs focal length

wavelength(µm)	0.2	0.6	1	1.4	0.2	0.6	1	1.4	ALL

Acceptance Angle (°)	0.8	0.8	0.8	0.8	0.8	0.8	0.8	0.8	0.8

In order to valid and to generalize the study results. An additional test configuration is performed. The test concerned the diameter of the lens. We have varied the diameter of Fresnel lens from 50 mm to 400mm.

We found at each time that the optical efficiency is high when the f-number was at 0.75 and at 1. We found also that it high at the visible spectrum

4. Conclusion

We have presented in this paper a performance design of Fresnel lens and detailed configurations studies with multiple tuning of focal lengths, facet spacing and an extended range of wavelengths.

Results show that the intensity and the optical efficiency start to increase reaching a maximum from 265 mm to 350 mm (f-number from 0.75 to 1) as focal length, and decrease with increasing focal length in the case of intensity, but the optical efficiency is still stable. We observed that the focal length could not be shorter because some of light are reflected by TIR causing an unbearable optical efficiency. For the long focal lengths, they would result in a too deep, thus expensive module. Concerning the acceptance angle of the whole concentrator system. We observe that the acceptance angle is stable from 175mm to 350mm and starts to decrease with the increment of the focal length.

We have also seen that the intensity and the optical efficiency is higher in the case of the ultra violet and it decreases with the increment of the wavelength. The acceptance angle in this case is still stable whatever the wavelength. Concerning the facet spacing, the intensity distribution and the optical efficiency increase with the increment of facet spacing. Concerning the CIC77ance angle, we observed that is stable and higher from 0.6mm to =1mm.

From this study, we can say that the Fresnel lens is performant in the visible spectrum with facet spacing of 1mm and the focal length between 265 to 350mm. The acceptance angle of the whole system was between 0.6 to 0.8° and the optical efficiency was between 80 to 90%.

References

[1] Cristóbal, A., Martí Vega, A., Luque López, A. (Eds.), Next Generation of Photovoltaics, *Springer Series in Optical Sciences*. Springer-Verlag, Berlin Heidelberg, Berlin, Heidelberg, 2012.

[2] Cristóbal, A., Martí Vega, A., Luque López, A. (Eds.), Next Generation of Photovoltaics, *Springer Series in Optical Sciences*. Springer-Verlag, Berlin Heidelberg, Berlin, Heidelberg, 2012

[3] Vossier, A., Chemisana, D., Flamant, G., Dollet, A., Very high fluxes for concentrating photovoltaics: considerations from simple experiments and modeling. *Renew. Energy,* **38**, 31–39, 2012

[4] Gordon, J.M., Katz, E. a., Feuermann, D., Huleihil, M., Toward ultrahigh-flux photovoltaic concentration. *Appl. Phys. Lett.* **84**, 3642, 2004.

[5] Vallerotto, G., Victoria, Marta., Askins, S., Herrero, R., Dominguez, C., Anton, I., Sala, G., Design and modeling of a cost-effective achromatic Fresnel lens for concentrating photovoltaics. *Opt. Exp.* **24**, 89–92, 2016..

[6] R. Leutz, A. Suzuki, Nonimaging Fresnel lenses: design and performance of solar concentrators, *Springer*, 2012

[7] W. T. Xie, Y. J. Dai, R. Z. Wan, K. Sumathy. Concentrated solar energy applications using Fresnel lenses: A review. *Renewable and Sustainable Energy Reviews*,15: 2588-2606, 2011

[8] Tver'yanovich EV. Profiles of solar-engineering Fresnel lenses. *Applied Solar Energy*, **19(6)**, 36-39, 1983.

[9] Khalil E.J. Al-Jumaily, Munadhila.K.A. Al-Kaysi. The study of the performance and efficiency

of flat linear Fresnel lens collector with sun tracking system in Iraq. *Renewable Energy*, **14(1-4),** 41-48, 1998,.

[10] Leutz R, Suzuki A, Akisawa A, Kashiwagi. Design of a non-imaging Fresnel lens for solar concentrators. **Solar Energy**, **65(6),** 379-387, 1999.

[11] Xinglong Ma, Hongfei Zheng, Meng Tian. Optimize the shape of curved-Fresnel lens to maximize its transmittance. *Solar Energy*, **127(2016)**, 285-293, 2016.

[12] L. W. James, “Use of refractive secondaries in photovoltaic concentrators,” *Tech. Rep. SAND,* 89-7029, Sandia National Laboratories, Alburquerque, NM, USA, 1989.

Implementation of a new Bi-Directional Switch multilevel Inverter for the reduction of harmonics

S.Vijayalakshmi[1], Dr P R Sivaraman[2] Dr R Karthick[3] Dr.A.Nazar Ali [4]
[1] Assistant Professor, Department of Electrical and Electronics Engineering, K. Ramakrishnan College of Technology, Trichy, Tamilnadu.
[2] Assistant Professor, Department of Electrical and Electronics Engineering, Rajalakshmi Engineering College, Rajalakshmi Nagar, Thandalam, Tamilnadu.
[3] Assistant Professor, Department of Electronics and Communication Engineering, Sethu Institute of Technology, Karriyapatti, Tamilnadu.
[4] Associate Professor, Department of Electrical and Electronics Engineering, Rajalakshmi Engineering College, Rajalakshmi Nagar, Thandalam, Tamilnadu

E-mail: nazarali.a@rajalashmi.edu.in

Abstract. In the current scenario, multilevel inverter plays a major role in heavy duty drive systems. MLI is one of the most advanced power converters for highly efficient AC output. But there is more to the requirement of number of control switches and total number of input sources which often generates more losses. This paper presents a new multilevel inverter design with limited number of control switches and with just one DC source to reduce these demerits. It uses a smaller number of Bi-directional switches. So that complexity of the circuit is reduced. Finally, an eleven-level inverter is designed and simulated using MATLAB/SIMULINK and THD reduction has been analyzed. This paper also provides a comparative statement of required components for various conventional methods.

1. Introduction

Staggered inverters are a truly appropriate setup to arrive at high power evaluations and excellent yield waveforms. In the group of staggered inverter, topologies dependent on arrangement associated H-spans are especially appealing a result of their measured quality and straight forwardness of control structure for the uncompromising electric drives would require electronic forced control inverters to meet the needs.Staggered converters do not cursively generate low mutilation voltages, but also minimize dv / dt-tension, thus reducing problems with electromagnetic compatibility. Shockingly, staggered converters have a few detriments. One inconvenience is the large number of semi-conductor switch requirement. the possibility of using low voltage control switches, each switch requires a driver circuit. This may make the general framework gradually very expensive and complex.

The Harmonics can be reduced by increasing the levels. It is compulsory that an output voltage with low THD is required, but greater the number of levels requires more elements; also the control will be very complex.

For eleven level yield voltages, five helper switches, four principle switches and five capacitors are required. To demonstrate the decrease in segment numbers accomplished by this rearranged H-connect staggered inverter arrangement, Table I present the quantity of part required to actualize an eleven level inverter utilizing disentangled H-connect staggered inverter and three recently characterized ones:
The two which were considered the usual staggered stages, the clipped diode and the tapped condenser, and another Fig. shows the block diagram of simplified H-bridge inverter. the symmetric course setup. The new topology accomplishes an around 40%.

The new arrangement decreases the quantity of diodes and capacitors, when contrasted and diode clamped setup. The new arrangement lessens the quantity of capacitors, when contrasted and the

IOP Conf. Series: Materials Science and Engineering **937** (2020) 012026 doi:10.1088/1757-899X/937/1/012026

capacitor clamped setup. Furthermore, since five capacitors are connected in parallel with the basic dc control supply, no critical condenser voltage swing is delivered during ordinary activity, keeping away from a problem that in some other staggered design can restrain working extent.

2. Block Diagram:

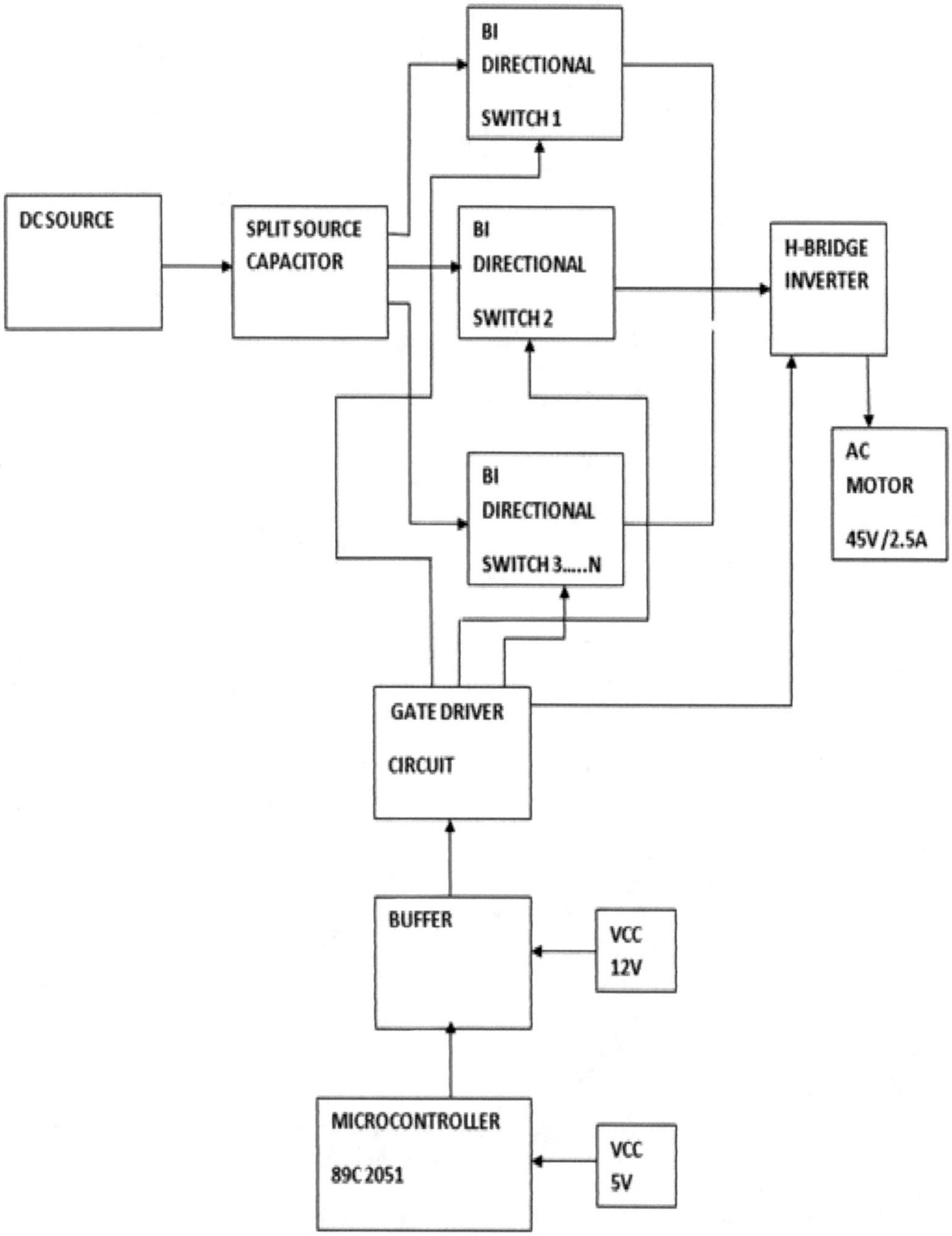

Figure 1. Block Diagram

IOP Conf. Series: Materials Science and Engineering **937** (2020) 012026 doi:10.1088/1757-899X/937/1/012026

Circuit has single DC source with some number of split source capacitor with some number of bi-directional bridge structure. Each bridge is responsible for producing two levels, one from positive and the other from negative. Only one H-Bridge is connected, and this produces three level output (i.e.) one positive level, one negative level and one zero level. Gate driver circuit is used to trigger the switches.

Table 1 : Components of Eleven level Inverter for various types of MLI

S. no.	Types	Number of switches needed	Number of Diodes Needed	Number of Capacitors Needed
1	Diode Clamped Type Inverter	20	90	10
2	Flying Capacitor Type Inverter	20	-	55
3	Cascaded H-Bridge Type Inverter	20	-	-
4	Modified H-Bridge Type Inverter	8	16	5

Reduction of the quantity of simple circuit needed just eight operated power switches instead of t wenty necessary ineach of the other three arrangements. The helper switch voltage and current ratings are lower than the once required by the principle-controlled switches.

3. Circuit Diagram

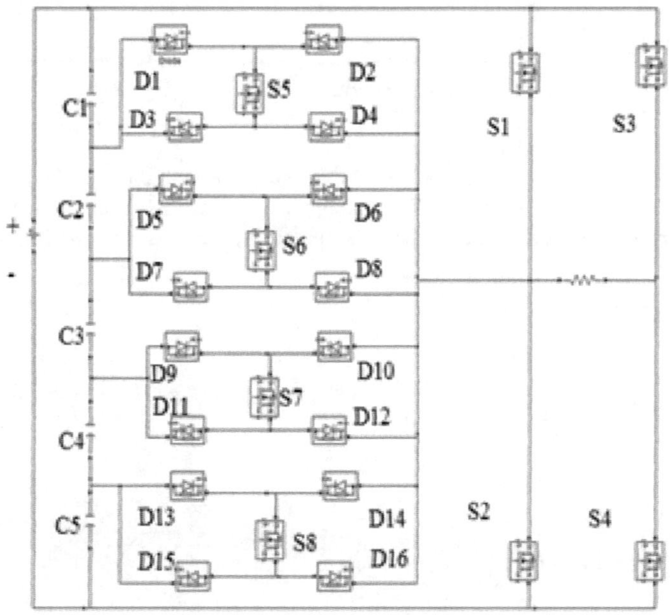

Figure 2. Circuit Diagram

The circuit diagram has five splitter Capacitors, these capacitors are used to divide the source voltage equally. In order to equal the voltage level, the value of all the capacitors should be same. Then there are four switches surrounded by 4 diodes each, these are responsible for producing two level output. And the H-bridge is used to produce three level ouput.so that totally 2+2+2+2+3=11 levels can be achieved.

IOP Conf. Series: Materials Science and Engineering **937** (2020) 012026 doi:10.1088/1757-899X/937/1/012026

4. Modes of Operation

4.1. Zero Output Level

The two primary switches S2 and S4 are ON, short-circuiting the load. All remaining control switches turned down; the voltage at the load terminal becomes zero thus which produces circulating current.

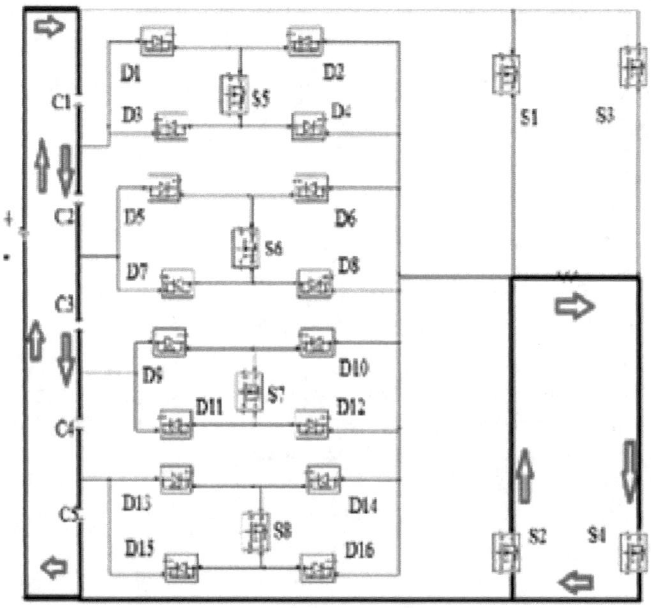

Figure 3. Mode-1 Diagram

4.2. Higher Output Level (Vdc/5)

The assistant adjustment S8 is closed, associating output +ve terminal, through diodes D13 and D16 also the adjustment switch S4 is going closed, interfacing the output Lower terminal point to ground. All remaining control switches are down; the potential at the output terminal is Vdc/5.

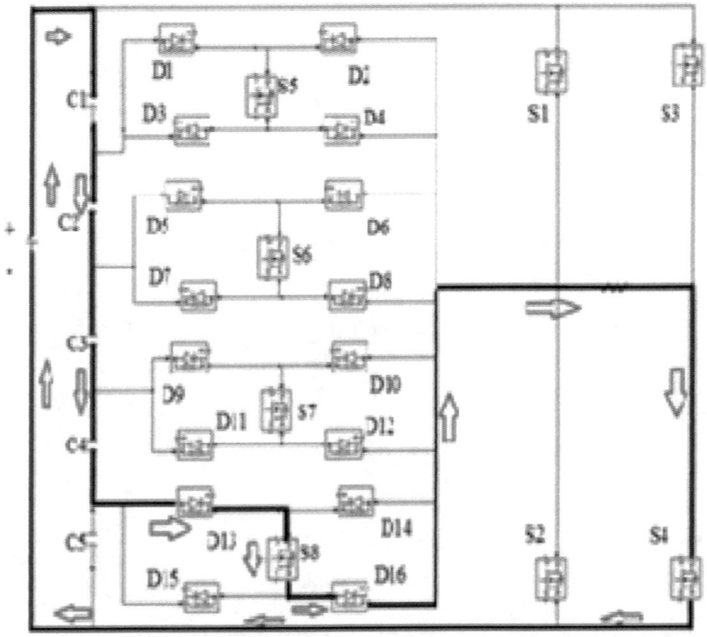

Figure 4. Mode-2 Diagram

4.3. Higher Output Level (2Vdc/5)

The helper switch S7 is closed, Connecting to the output +ve terminal point, through diodes D9 and D12, also S4 is closed, associating the lower terminal point to ground. Remaining control switches are Down, the potential at the terminal is 2Vdc/5.

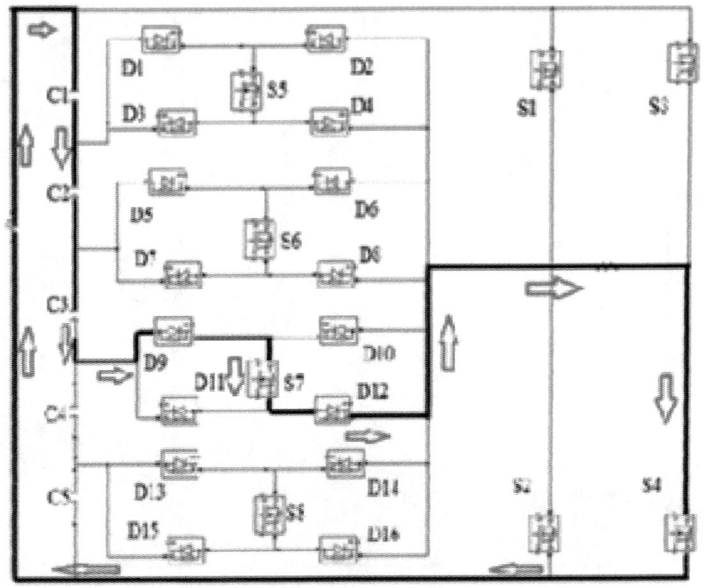

Figure 5. Mode-3 Diagram

4.4. Higher Output Level (3Vdc/5)

The helper switch S6 is closed, connecting to the output +ve terminal, through diodes D5 and D8, and S4 is closed, interfacing the lower terminal point to ground. All remaining control switches are down, the potential at stack terminal is 3Vdc/5.

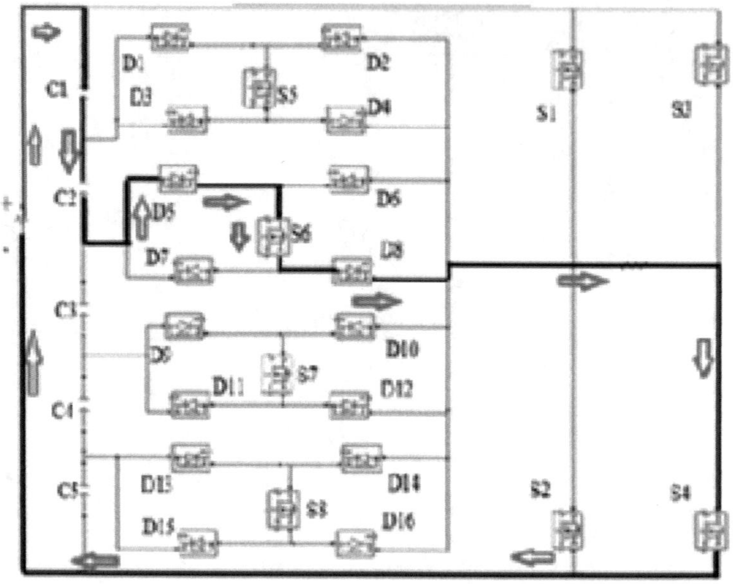

Figure 6. Mode-4 Diagram

4.5. Higher Output Level (4Vdc/5)

The assistant switch S5 is closed Connecting to the output +ve terminal point, through diodes D1 and D4, also S4 is closed, associating the output lower terminal point to ground. All remaining control switches are down, the potential at stack terminal is 4Vdc/5.

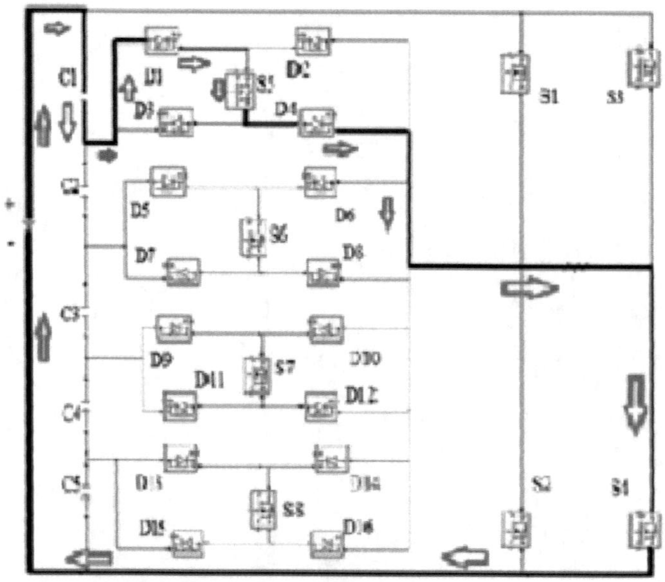

Figure 7. Mode-5 Diagram

4.6. Higher Output Level (5Vdc/5=Vdc)

S1 is closed, interfacing output +ve terminal point to Vdc and contact S4 is closed, associating the output lower terminal point to ground. All remaining control switches are down. The potential at stack terminal is Vdc.

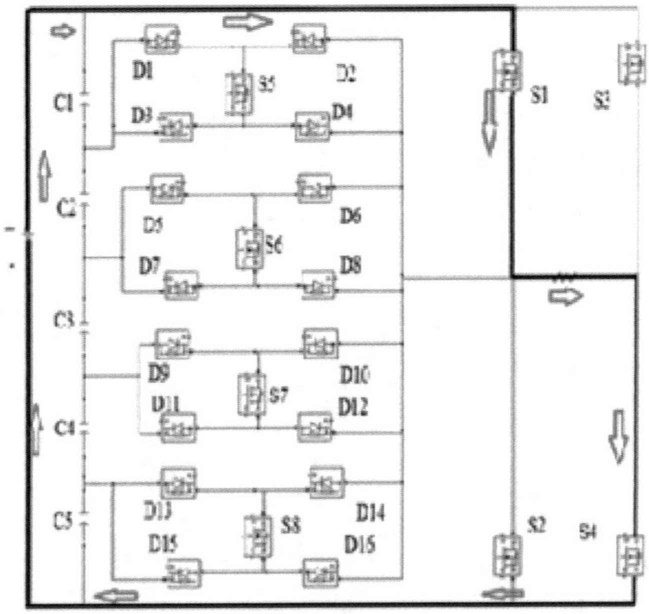

Figure 8. Mode-6 Diagram

4.7. Lower Output Level (-Vdc/5)

The assistant switch S5 is closed Connecting to the output +ve terminal point through diodes D3 and D2, also S3 is closed, associating the output lower terminal. All remaining control switches are down, the potential at stack terminal is - Vdc/5.

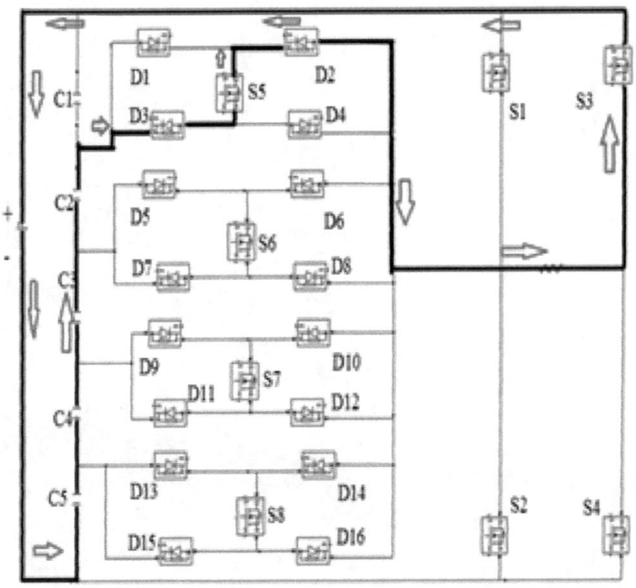

Figure 9. Mode-7 Diagram

4.8. Lower Output Level (-2Vdc/5)

The assistant switch S6 is closed Connecting to the output +ve terminal, through diodes D7 and D6, also S3 is closed, associating the output lower terminal. All remaining control switches are down, the potential at stack terminal is - 2Vdc/5.

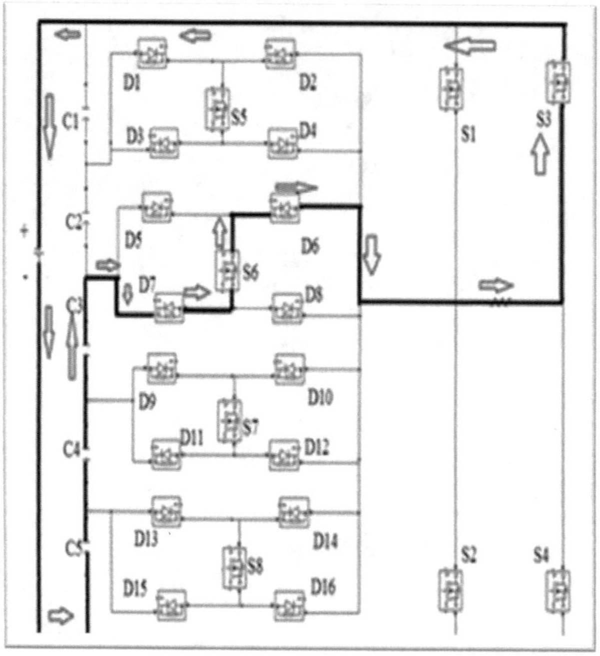

Figure 10. Mode-8 Diagram

4.9. Lower Output Level (-3Vdc/5)

The helper switch S7 is getting closed Connecting to the output +ve terminal, through diodes D11 and D10, also S3 is closed, interfacing the output lower terminal. All remaining control switches are down, the potential at stack terminal is - 3Vdc/5.

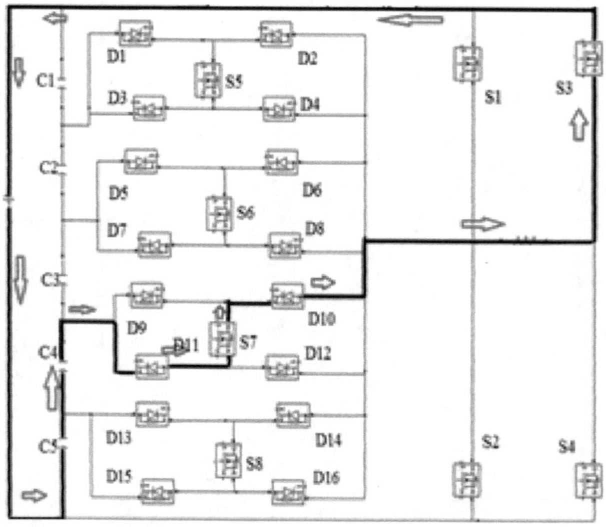

Figure 11. Mode-9 Diagram

4.10. Lower Output Level (-4Vdc/5)

The assistant switch S8 is closed Connecting to the output +ve terminal, through diodes D15 and D14, also S3 is closed, associating the output lower terminal. All remaining control switches are down, the potential at stack terminal is - 4Vdc/5.

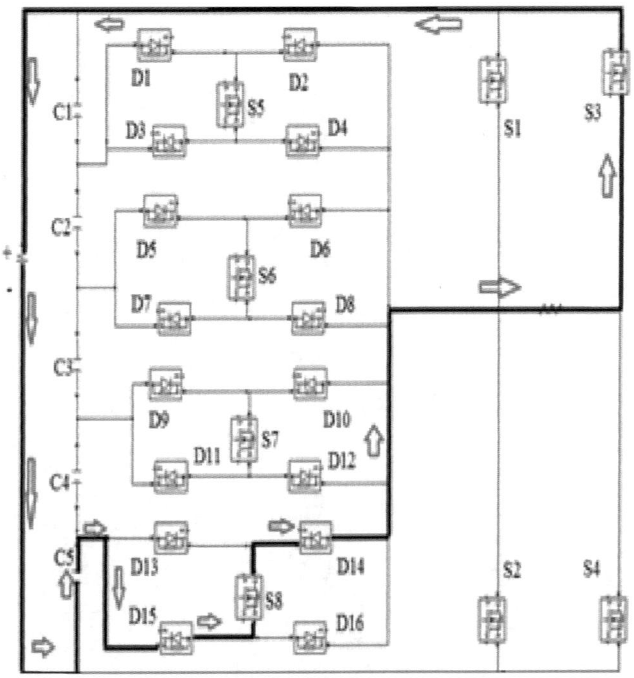

Figure 12. Mode-10 Diagram

4.11. Lower Output Level (-5Vdc/5=-Vdc)

S2 is Triggered, associating the output +ve terminal also S3 is closed, interfacing the load lower terminal. All remaining control switches are reduced, output potential at the output terminal is – Vdc.

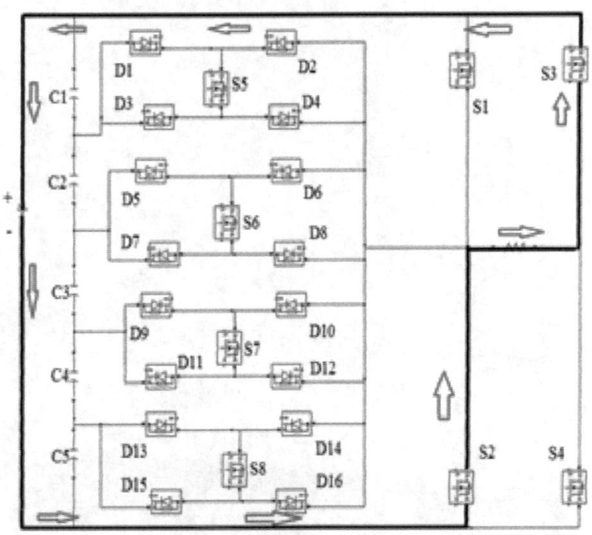

Figure 13. Mode-11 Diagram

5. Output voltage according to switch condition

Table 2 : Switch Condition

Volt	S1	S2	S3	S4	S5	S6	S7	S8

Vdc	**ON**	OFF	OFF	**ON**	OFF	OFF	OFF	OFF
4Vdc/5	OFF	OFF	OFF	**ON**	**ON**	OFF	OFF	OFF
3Vdc/5	OFF	OFF	OFF	**ON**	OFF	**ON**	OFF	OFF
2Vdc/5	OFF	OFF	OFF	**ON**	OFF	OFF	**ON**	OFF
Vdc/5	OFF	OFF	OFF	**ON**	OFF	OFF	OFF	**ON**
0	OFF	**ON**	OFF	**ON**	OFF	OFF	OFF	OFF
-Vdc/5	OFF	OFF	**ON**	OFF	**ON**	OFF	OFF	OFF
-2Vdc/5	OFF	OFF	**ON**	OFF	OFF	**ON**	OFF	OFF
-3Vdc/5	OFF	OFF	**ON**	OFF	OFF	OFF	**ON**	OFF
-4Vdc/5	OFF	OFF	**ON**	OFF	OFF	OFF	OFF	**ON**
-Vdc	OFF	**ON**	**ON**	OFF	OFF	OFF	OFF	OFF

6. Simulation Result

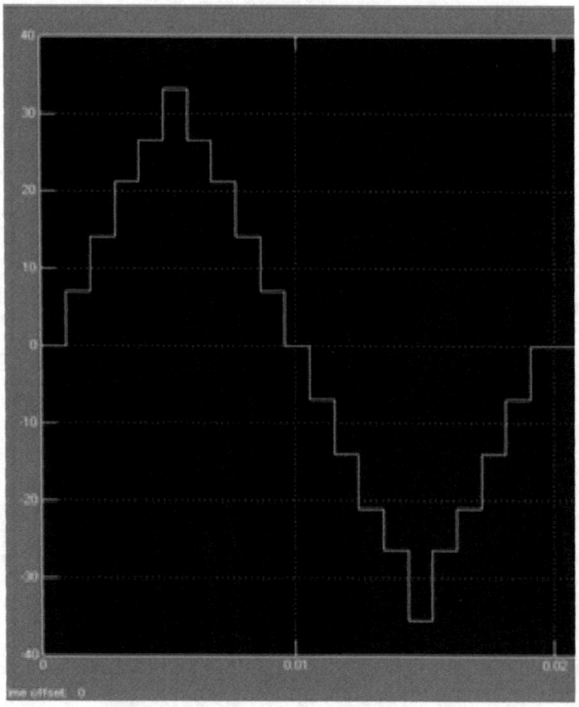

Figure 14. Simulation Output

7. Total Harmonic Distortion

IOP Conf. Series: Materials Science and Engineering **937** (2020) 012026 doi:10.1088/1757-899X/937/1/012026

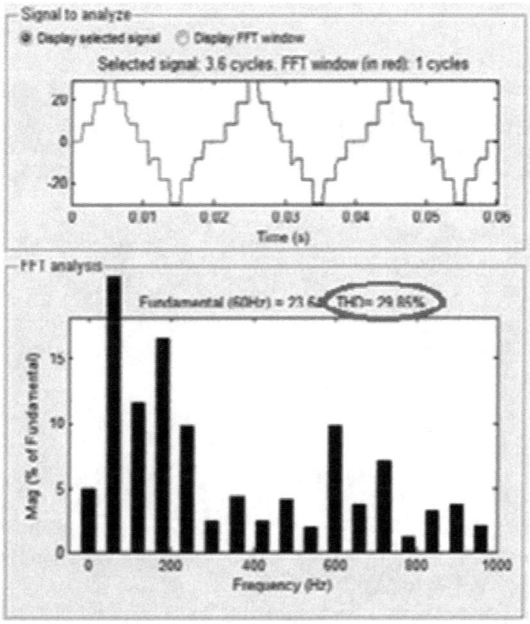

Figure 15. Simulation results of Seven Level Inverter Total Harmonic Distortion

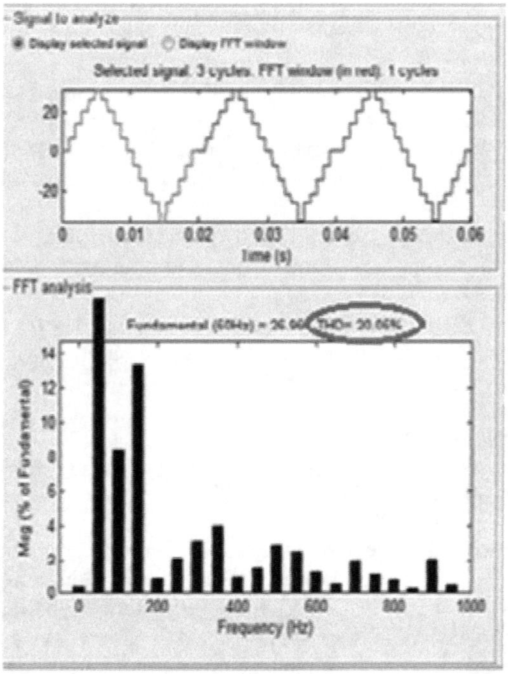

Figure 16. Simulation results of Eleven Level Inverter Total Harmonic Distortion

8. Conclusion

For huge electrical drives a staggered inverter with single dc source has been proposed. Reproductio n and test results have shown that the switches are controlled by a control device at the critical recur rence; these converters have low yield voltage,low THD and high productivity.

References

[1] A Nazar Ali, D Sivamani, R Jaiganesh M Pradeep, Solar powered air conditioner using BLDC motor, IOP Conference Series: Materials Science and Engineering, vol. 23,2019
[2] V Venkatesh, A Nazar Ali,, R Jaiganesh. V Indiragandhi , Extraction and conversion of exhaust heat from automobile engine in to electrical Energy‖, IOP Conference Series: Materials Science and Engineering, vol. 23,2019.
[3] A Ali Nazar, R Jayabharath, MD , Udayakumar, An ANFIS Based Advanced MPPT Control of a Wind-Solar Hybrid Power Generation System, international review of modelling and simulations. vol...7, no. 4, pp. 638– 643, 2014.
[4] A.Nazar Ali and R. Jayabharath, "Performance Enhancement of Hybrid Wind/Photo Voltaic System Using Z Source Inverter with Cuk-sepic Fused Converter." Research Journal of Applied Sciences, Engineering and Technology7.19 pp. 3964-3970,2014
[5] AT SankaraSubramanian,PSabarish, A Nazar Ali , A power factor correction based canonical switching cell converter for VSI fed BLDC motor by using voltage follower technique‖IEEE International Conference on Electrical, Instrumentation and Communication Engineering (ICEICE), pp.1-8.2017
[6] C.Kalavalli, S.R.Paveethra, S. Murugesan, Dr.A.Nazar Ali, Design and Implementation Of High Efficiency H6 PV Inverter With Dual Axis Tracking, International Journal of Scientific & Technology Research, Vol 9, issue 02, pp. 4728-31, 2020
[7] S.R.Paveethra, C.Kalavalli, S.Vijayalakshmi, Dr.A.Nazar Ali, D.Shyam, Evaluation Of Voltage Stability of Transmission Line With Contingency Analysis, International Journal of Scientific & Technology Research, Vol 9, issue 02, pp. 4018-22,2020
[8] Premkumar, Kamaraj et al. (2018), Antlion Algorithm Optimized Fuzzy PID Supervised On-line Recurrent Fuzzy Neural Network Based Controller for Brushless DC Motor, Electric Power Components and Systems, 45, 20, pages.2304-2317.
[9] K Premkumar et al. (2018), Novel bacterial foraging-based ANFIS for speed control of matrix converter-fed industrial BLDC motors operated under low speed and high torque, Neural Computing and Applications, 29, 12, pages.1411–1434.
[10]Premkumar et al. (2015), GA-PSO optimized online ANFIS based speed controller for Brushless DC motor, Journal of Intelligent & Fuzzy Systems, 28, 6, pages.2839-2850.
[11]A Nazar Ali, R Jeyabharath, MD Udayakumar‖ Cascaded Multilevel Inverters for Reduce Harmonic Distortions in Solar PV Applications‖ Asian Journal of Research in Social Sciences and Humanities, Vol 6,pp.703-715.
[12]A.Nazar Ali and R. Jayabharath. "Performance Enhancement of Hybrid Wind/Photo Voltaic System Using Z Source Inverter with Cuk-sepic Fused Converter." Research Journal of Applied Sciences, Engineering and Technology7.19 (2014): 3964-3970.
[13]M D Udayakumar et al 2019 IOP Conf. Ser.: Mater. Sci. Eng. 623 012018
[14]L.M. Tolbert and F.Z. Peng," multilevel inverters for large automotive drives," in conference, Rec.all electric compact vehicle second int.conf. Dearborn, June 8-12,1997.
[15]F.Z. Peng and J.S. Lai," Dynamic performance and control of a static var generator using cascaded multilevel inverters", IEEE Trans.Ind. Applica., vol 33, may/june-1997.
[16]Y. Cheng, C. Qian, M. L. Crow, S. Pekarek, and S. Atcitty, "A comparison of diode-clamped and cascaded multilevel converters for a STATCOM with energy storage," IEEE Trans. Ind. Electron., vol. 53, no. 5, pp. 1512– 1521, Oct. 2006.
[17]S.Vijayalakshmi, P.Sabarish, S.R.Paveethra, Dr.P.R.Sivaraman, Dr.V.Venkatesh," Exploration and applications of electronic ballast for high power discharge lamps at high frequency through power factor modification", IJSTR,volume-9,issue02,feb-2020,ISSN 2277-8616.
[18]R. Stala, S. Pirog, M. Baszynski, A. Mondzik, A. Penczek, J. Czekonski, and S. Gasiorek, "Results of investigation of multi cell converters with balancing circuit—Part II," IEEE Trans. Ind. Electron., vol. 56, no. 7, pp. 2620–2628, Jul. 2009.
[19]S.R.Paveethra, S. Vijayalakshmi. C.Kalavalli, S. Murugesan and Dr.S.Jeyasudha," MPPT based Harmonic reduction in PV module for grid connect ed system" Biosc.Biotech.Res.Comm.Special issue vol 13 No (4) 2020Pp-43-48.
[20]M. Klabunde, Y. Zhao, and T.A. Lipo," Current control of a 3-level rectifier /inverter drive system", in conf. Rec.IEEE-IASAnnu. Meeting, oct-1994.
[21]Karthick, R and Sundararajan, M.(2017), "PSO based out-of-order (ooo) execution scheme for HT-

MPSOC" Journal of Advanced Research in Dynamical and Control Systems, 9.
[22]Karthick, R and Sundararajan, M. (2017), "A Reconfigurable Method for Time Correlated MIMO Channels with a Decision Feedback Receiver," International Journal of Applied Engineering Research,
[23]Karthick, R and Sundararajan, M. (2017), "Design and Implementation of Low Power Testing Using Advanced Razor Based Processor," International Journal of Applied Engineering Research,12.
[24]Karthick, R and Sundararajan, M. (2018), "A novel 3-D-IC test architecture-a review," International Journal of Engineering and Technology (UAE),7.

Liquid metal level measurement techniques

R Reshma[1], Uppu Ramachandraiah[1], K R Devabalaji[1], and R. Sitharthan[2]

[1]Department of Electrical and Electronics Engineering
Hindustan Institute of technology and science – Chennai
[2]School of Electrical Engineering, Vellore Institute of Technology, Vellore.

reshmarajendran444@gmail.com

Abstract. Various types of liquid metal level measurement are used in various applications. Liquid metals have excellent electrical properties which can also be utilized for level measurement. Many types of level measurement exist such as pressure transducers, floats, inductive devices, resistance tubes, visual etc. Many liquid metals have corrosive or toxic nature. This mandates indirect or non-intrusive means of measurement. So techniques are being experimented even now for the easy attainment of accurate measurements. This paper gives a brief description on various level measurement techniques used to measure the level of liquid metals. Detailed description about a mutual inductance based level probe is also presented in this paper.

1. Introduction

The Liquid level measurement is important in various industrial processes. Different techniques are used to satisfy the requirements of level measurement. Direct and indirect methods of measurement can be used. Level measurement can be in a continuous interval or at discrete points [1-5]. The continuous level measurement provides the exact level of liquid present in certain location. In case of discrete or point level measurement a reference point is given and measurement indicates whether the liquid is above or below the particular reference point [5-10]. Many parameters affect the measurement of liquid metal level liquid metal. Some of these parameters are temperature, pressure, density, corrosive nature, electrical and acoustic properties etc. Normally the level measurement can be done visually by using metre scales [11-15]. The points will be marked in the scale and by placing this in the container the quantity of liquid present is identified. Many a times the liquid metals are stored in closed vessels where visual measurement is not possible [15-20]. Another method consists of placement of a sensor inside the container and producing signal when the liquid comes in contact with the sensor. Each sensor will have different principles and according to this the measurement is done [21-25]. Pressure at the bottom of a container is directly related to the height of the liquid at a constant temperature and thus by measuring static pressure also, the liquid level in a tank or container can be measured [26-30]. The level measurement for liquid metal is challenging because most of the liquid metals have high melting point which requires use of special materials to fabricate sensors [31-35].

IOP Conf. Series: Materials Science and Engineering **937** (2020) 012027 doi:10.1088/1757-899X/937/1/012027

2. Methods of level measurement

For non-corrosive liquids like water, level can be measured simply by observing the level of the liquid through naked eyes or we can just insert a scale and measure the level. But this method is not very accurate because of errors in measurement due to manual errors.

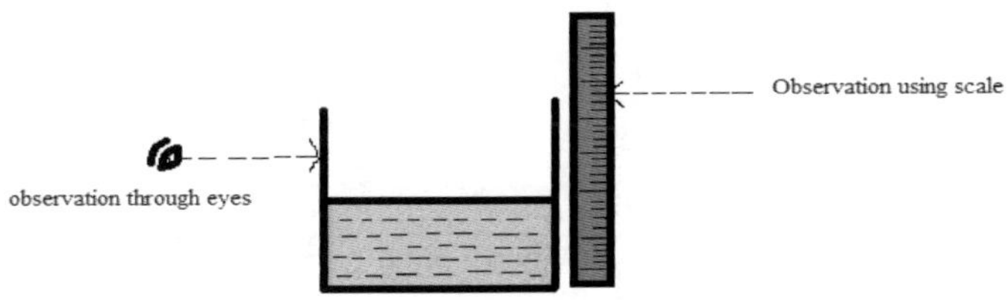

Figure 1: schematic of visual level measurement.

2.1 Direct and indirect methods of level measurement.

Generally there are two types of level measurement i.e. the direct method and the indirect method. In the direct method of level measurement, the varying level of the liquid is used as a means for obtaining the measurement. This is also the simplest method of measuring the liquid level. The level indicators used here are sight glass, Float type and Hook-type level indicator. The sight glass continuously indicates the liquid level within the tank or a vessel. But here the accuracy depends upon the cleanliness of the fluid and the glass. In the float type, a float will be present on the surface of the liquid and as the level changes the float indicates the variation.

2.1.1 Hook-type level measurement. In the hook-type of level indicator, the hook is pushed below the liquid surface and gradually is raised until the point is just about to break through the surface of the liquid. The reading is very accurate. In the indirect methods the level is measured in terms of variables which changes according to the change in the liquid level.

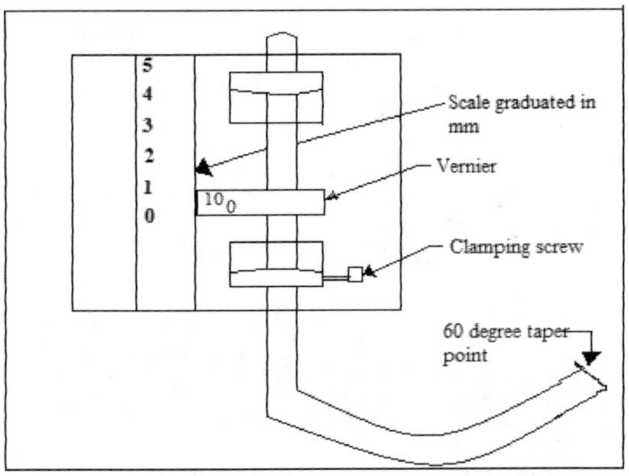

Figure 2: Hook-type level measurement.

The indirect method is based on Hydrostatic head (pressure transmitter), Electrical capacitance method and Radiation method.

IOP Conf. Series: Materials Science and Engineering **937** (2020) 012027 doi:10.1088/1757-899X/937/1/012027

2.1.2 *Pressure transmitter for level measurement.* The pressure transmitter can be used to measure the liquid level inside a tank, in the river and other body of liquid. The liquid level can be easily measured through this way. The static pressure at the bottom of a liquid is proportional to the height of the liquid. A transmitter can also be used to measure and record the hydrostatic pressure and thus liquid level can be measured and recorded. The pressure transmitter gives a signal which indicates the level of the liquid. The disadvantage of this method is that, there will be temperature variations and thus the liquid density changes. Thus temperature corrections are required.

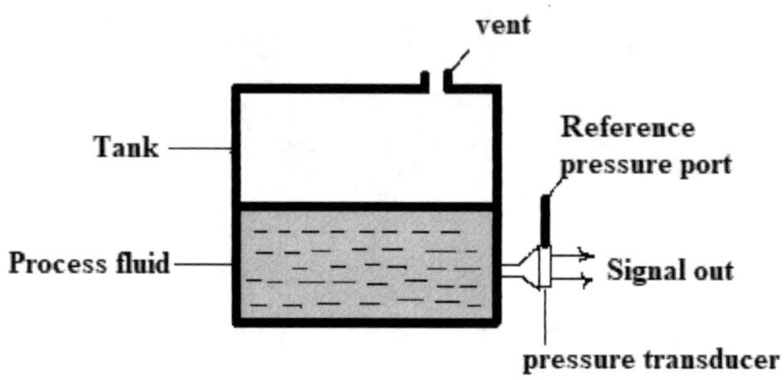

Figure 3: Level measurement using pressure transmitter.

2.1.3 *Electrical capacitance method.* The device consist of an insulated capacitance probe firmly attached parallel to the metal wall of the container. The liquid is non-inductive. Here the probe and the wall acts as two parallel plates and liquid act as dielectric. If we use a conductive liquid, the probe and the liquid acts as capacitance plates and insulation acts as dielectric. The capacitance value is calibrated in terms of level and the measurement can be performed. When the liquid inside the container increases, the capacitance value increases. Selection of correct electrode is important. This is temperature sensitive. The capacitance level indicators find applications in measurement of granular solid levels.

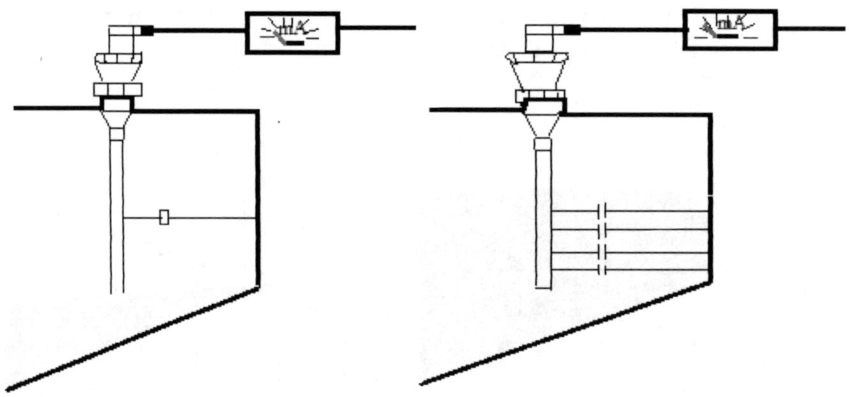

Figure 4: Electrical capacitance method.

2.1.4 *Radiation method of level measurement.* A gamma rays source and a detector is present on either side of the tank whose level is to be measured. This is a non-contact method. A thin layer of gamma ray is released and this will penetrate to the tank and loses its intensity. At the detector end, the reception is inversely proportional to the tank wall thickness and the intermediate medium. The amount of radiation received is inversely proportional to the

IOP Conf. Series: Materials Science and Engineering **937** (2020) 012027 doi:10.1088/1757-899X/937/1/012027

amount of fluid in the tank. The radiation signal is calibrated in terms of level provides the required measurement. The change in density will show effects in output. This can be operated even at high temperature and provides good accuracy and response.

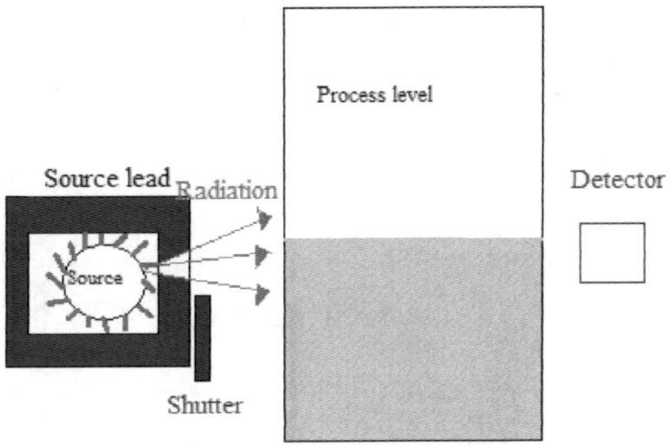

Figure 5: Radiation level measurement.

3. Dipstick level measurement

Dipsticks are easy and quick level measurement of liquid. The design of dipstick level measurement is simple. In this type of measurement a metal strip is used. The metal strip is connected to a handle. The handle is usually made up of plastic or metal. The metal strip is having markings and these markings indicate the level of the liquid. Normally the dipstick is inserted into the container in which the level is to be measured. And after a couple of minutes it is removed and the readings are noted.

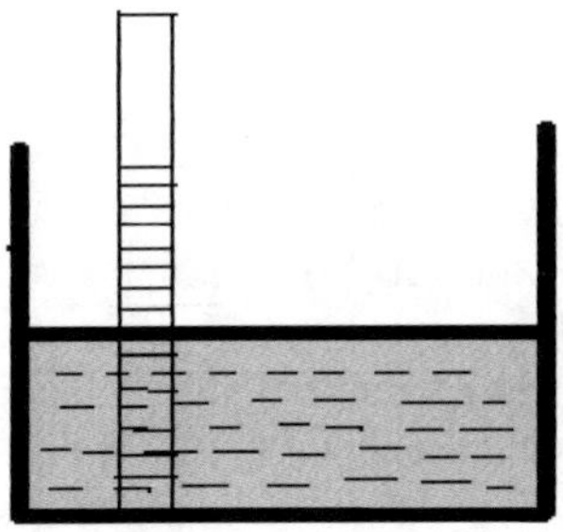

Figure 6: Dipstick level measurement.

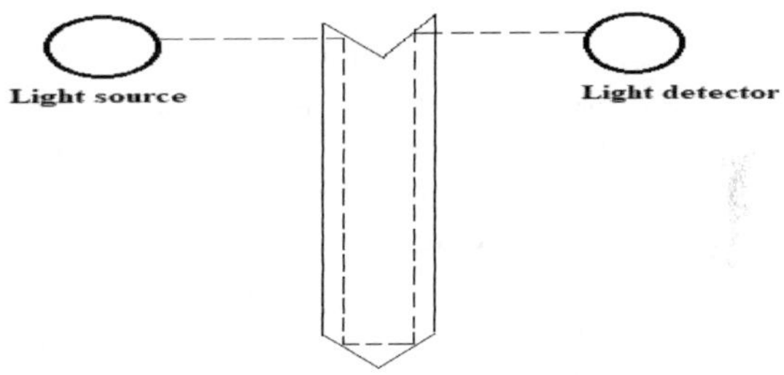

Figure 7: Optical Dipstick.

Apart from the simple dipstick, Optical dipstick type of level measurement is also used for liquid level measurement. In this type of measurement a light source is present and the source is reflected from the mirror and passes through the oblique end of the dipstick, and enters the detector after a reflection from the second mirror. When the oblique end of the dipstick comes in contact with liquid, the internal reflection properties are altered and the light no longer enters the detector. To move the instrument up and down and to measure its position a suitable mechanical drive is used. Thus the liquid level is monitored. This is applicable in larger or deeper tanks.

4. Float displacement type level measurement

The float displacement method works on Archimedes Principle. When a body is placed in a liquid, it is buoyed (bounced) up by a force which is equal to the weight of the displaced liquid, and the evident change in the weight of the body is directly proportional to the level of the liquid in which the setup is placed. A torque tube is the most commonly used for this setup. The displacer is attached to the torque tube and the rotary motion is used for the control. Displacers can be used for both point level and

IOP Conf. Series: Materials Science and Engineering **937** (2020) 012027 doi:10.1088/1757-899X/937/1/012027

continuous level indication. Floats and displacers are present in many shapes like cylindrical, spherical etc. Displacer float switches can be used to detect the high and low levels in the tanks and vessels.

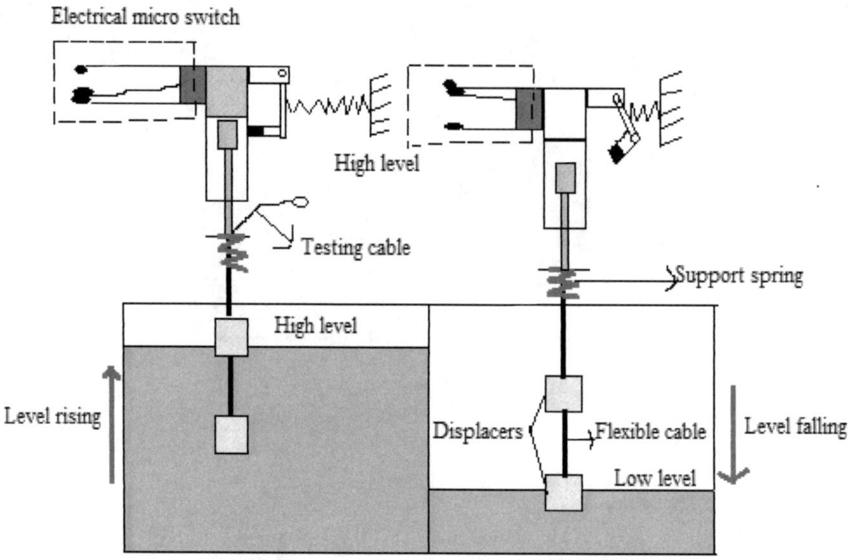

Figure 8: Float displacement type level measurement.

This gives high accuracy. Can be mounted internally and externally. This is highly reliable in clean liquids. For externally mounted apparatus the cost increases. Only limited range can be used. More length is difficult to balance and is bulky.

5. Spark plug level measurement

The spark plug level sensor is basically a stainless steel sensor probe. The principle is based on grounding the sensor in contact with the sodium. This consists of a central electrode which contain resistors. As the liquid sodium level rises and touches the end segment of the level probe, the electronic circuit gets grounded through the sodium. This gives indication liquid metal contact with probe. The main advantages of this type of level measurement is that the maintenance is easy. The electronics and the construction is simple. However, above 250°C the sodium vapor deposition short circuits the probe and this gives spurious signals.

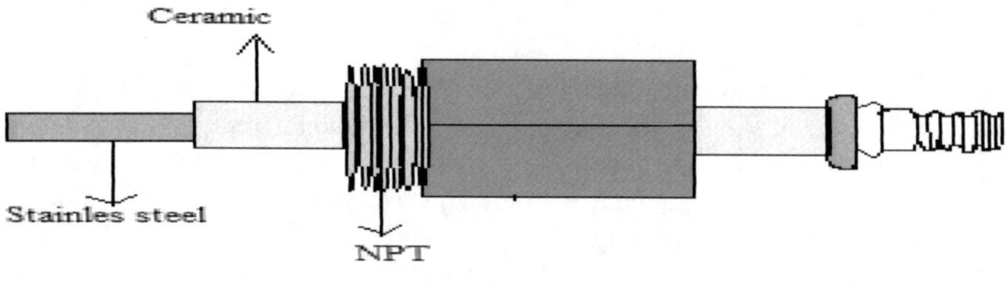

Figure 9: Spark plug level measurement.

The operating temperature of the spark plug will be the actual physical temperature at the tip of the spark plug.

6. Optical type level measurement

The main property used here for level measurement is the reflection property. If we used is an optical fibre level detector then refraction will be the property. This type of measurement can be done for both point and continuous level measurement. The infrared or visible light sources can be used for the detection of the level. Light is passed to the liquid whose level is to be detected, the corresponding reflection is received by the light sensors which are mainly quartz crystals or prisms. This measurement can be calibrated in terms of level measurement units. Due to the refractive index of liquid prism when there is no liquid present the light intensity will remain the same and in presence of the liquid the intensity gets diminished. These sensors can be also used in corrosive, organic and aqueous liquid according to the proper choice of material of construction and placement location. The alternative application for optical sensors is by using laser light. But laser lights are limited use for industrial purposes because it is difficult to maintain and more costly.

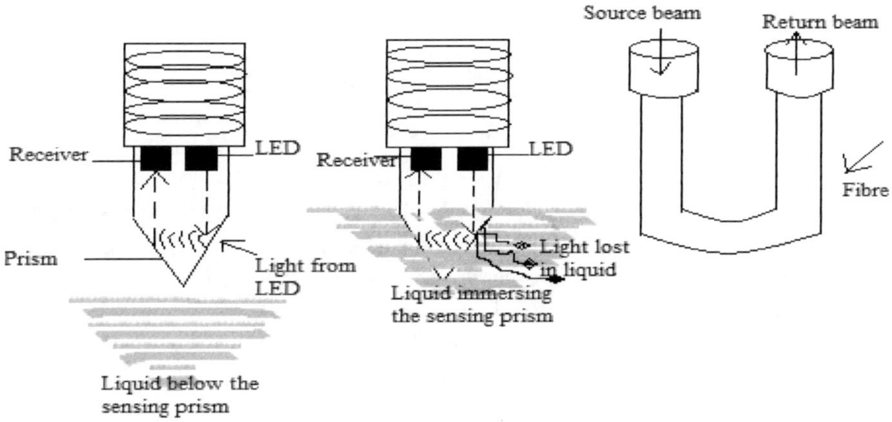

Figure 10: Optical type level measurement.

The cost is low and it is small device set up. This measurement gives high precision and accuracy. The set up should be cleaned or else it will affect the performance. If there is any change in the reflectivity of the medium the process is adversely affected.

7. Ultrasonic type of level measurement

In ultrasonic type of level measurement, the ultrasonic waves are used. This method is used for both point and continuous monitoring. The propagation time of the sound waves in the ultrasonic region is the principle used for level measurement here. Attenuation of the vibrating diaphragm face is one of the method in which it operates. The other one is absorbing the acoustic energy when it is travelling towards the receiver from the source. The time required for the echo produced to return is measured. There are two different orientations for the ultrasonic sensor. One is when we place it at the top of the tank/container, it is the region where the speed of the sound will be around 330 meters per second. Here the vapor space depth above the liquid is measured. Another orientation is when we keep it at the bottom of the tank/container, the depth of the liquid in container is measured. As the sound velocity changes proportionally with the square root of the temperature, the temperature compensation can be provided. When we apply an electrical signal, the required mechanical wave is generated. For this purpose piezoelectric crystal is used. Smooth surface are used for better reflection because rough surfaces produces diffused reflection. Turbulence and foam have a major role in this type of measurement as they prevent the sound wave from being properly reflected. The ultrasonic transducer is used both for transmitting and receiving the acoustic energy. So it is exposed to mechanical vibrations called 'ringing'. The vibration should fade before the echo signal is processed. There is a small range near the face of the transducer, where it is not able to detect the object. This is called blanking zone.

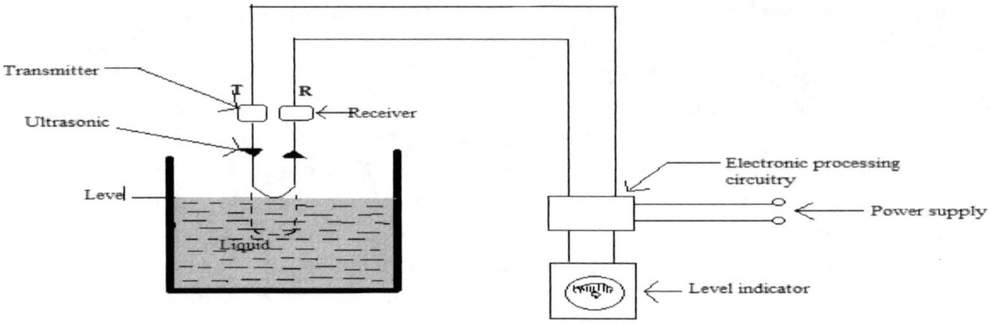

Figure 11: Ultrasonic type level measurement.

The advantage of this setup is that only low maintenance is required. Self-cleaning can be done. Some disadvantages are that temperature compensation is required, these are subjected to multiple interferences. Dirt and rough reflecting surface may affect the accuracy.

8. Float actuated type resistive level measurement

The relationship between the resistance and the current flow is been used here for accurately measuring the level. The commonly used design is a probe which consists of two conductive strips. The gold-plated steel will be at the base of one strip and the other strip will be of an elongated wire resistor. They form a complete electric circuit by connecting it together at the bottom. A low voltage supply is given to the upper end of the strips. A process material is present and a flexible plastic sheath is used for isolating the strips from the process material. When the level of the process material rises, the pressure forces/pushes the resistance strips together up to the interface. This shorts the circuit below the interface level and the resistance gets reduced proportionally. Here the liquid level position is converted to an electrical signal. The float is used as the primary transducer. And this transforms the variation in the liquid level to suitable displacement. The displacement is sensed by the secondary transducer which is a resistive potentiometric device. This device converts the displacement to electrical signal. When there is a liquid level change the float starts to move according to this change in liquid. The variation in float position triggers the arm. The slider then moves over the resistive element of the potentiometer. The resistance of the circuit changes and this change in resistance is directly proportional to the liquid level in the tank.

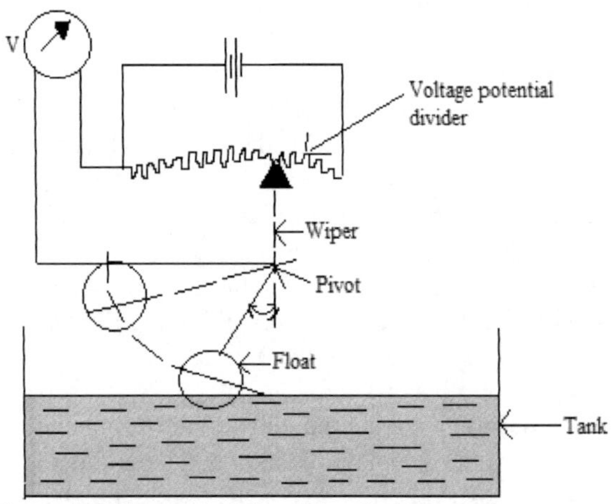

Figure 12: Float actuated resistive level measurement.

The level can be continuously recorded with the addition of contact probes. The danger can be eliminated by using low voltage. To any desired point signals can be transmitted. This setup is simple to calibrate. The disadvantage is that it is unsafe to use this in explosive atmosphere. This require large number of contact rods and can be corroded by corrosive liquids. If there is any variation in the conductivity of the liquid serious errors may occur.

9. Radar type of level measurement

In the RADAR type of level measurement, a radar sensor is used. The microwaves are transmitted by the antenna system of the radar sensor to the surface of the liquid. The waves get reflected by the surface of the product and is received back by the antenna system. The time interval between the emissions to the reception of the signals is proportional to the level of the liquid in the container. Both radar type of level measurement and ultrasonic type of level measurement use time of flight concept but the only difference is that in ultrasonic measurement sound waves are used whereas in radar type of level measurement radio waves are used. Radio waves have very high frequency and are electromagnetic in nature. The radar level instruments works at its best when there is a large difference in permittivity between the two substances. The power reflection factor (R) is defined by the ratio of reflected power to transmitted power at any interface of the material in the tank. Probes are used in guided-wave radar instruments. The probes can be single metal rod, parallel pairs of metal rod or coaxial type of rod can be used. Non-contact radar instruments depend mainly on antennas. The antennas should be kept cleaned and dry so they are separated from the interior of the vessel by placing a dielectric window which is made of substances like plastic which are relatively transparent to electromagnetic waves.

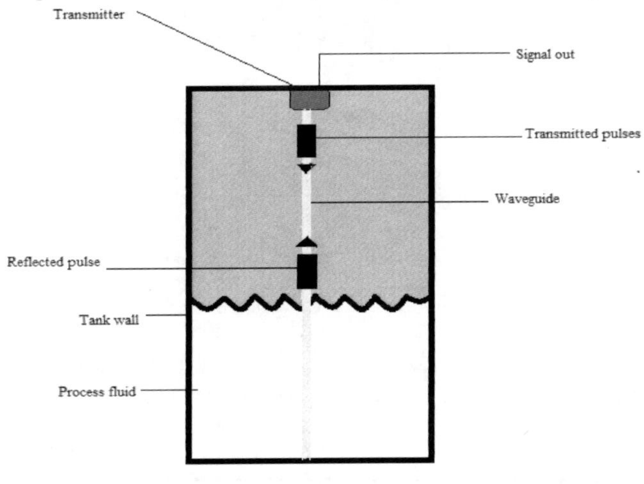

Figure 13: Radar type of level measurement.

The accuracy of the radar level instruments changes as the temperature or pressure varies. These are the factors influencing the dielectric constant. When the radar level instrument transmit the energy some gets reflected back to the source and the remaining propagates further. The guided-wave radar instruments receives a greater percentage of their transmitted power when compared to the non-contact type radar level instruments. There is no moving part present in these type of level instruments. The setup has low maintenance and easy installation. The disadvantages are they are not applicable for interface-interface measurement. If vapor is present, it weakens the radar signals and this reduces the accuracy of level measurement.

10. Mutual inductance level probe type of level measurement

The mutual inductance level probe (MILP) type of level measurement works on the principle of change in mutual inductance with the change occurring in level of liquid metal like sodium which have good electrical conductivity. This setup consists of a primary winding and a secondary winding. The windings

IOP Conf. Series: Materials Science and Engineering **937** (2020) 012027 doi:10.1088/1757-899X/937/1/012027

are wounded on a non-magnetic former in a bifilar fashion. The primary winding is excited with an AC constant current source. This produces an alternating flux which induces an EMF in the secondary. As the sodium level rises, an alternating flux induces an EMF in the liquid metal. Since the liquid metal is conductive and continuous, the induced EMF produces alternating current in liquid metal. These circulating currents are called eddy currents and produces flux which opposes the main flux and the net flux is reduced. Thus the secondary EMF also gets reduced. The decrease in the secondary EMF is linearly proportional to increase in liquid metal level. When the temperature of the sodium is varied, the resistance of liquid metal also changes causing a variation in eddy currents. Thus the secondary voltage changes. Thus, the secondary voltage of the probe is a function of liquid metal level and also its temperature. Temperature compensation is required and is carried out by various methods. The level probe assembly consists of a stainless steel pipe called "pocket" and a probe. The probe consists of a sensing element, a bobbin supporting tube probe flange, and terminal box. The probe is placed inside the pocket. The sensitivity of the level probe can be increased by increasing the number of turns of the windings.

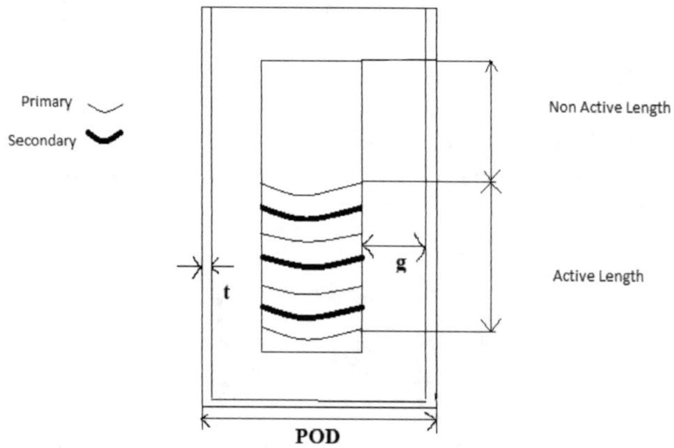

Figure 14: Schematic of Mutual inductance level probe (MILP).

POD – Pocket outer diameter, T – Thickness of the pocket, AL- Active length, NAL – Non-active length. The active length is the bobbin length and the non-active length is the area where there is no winding.

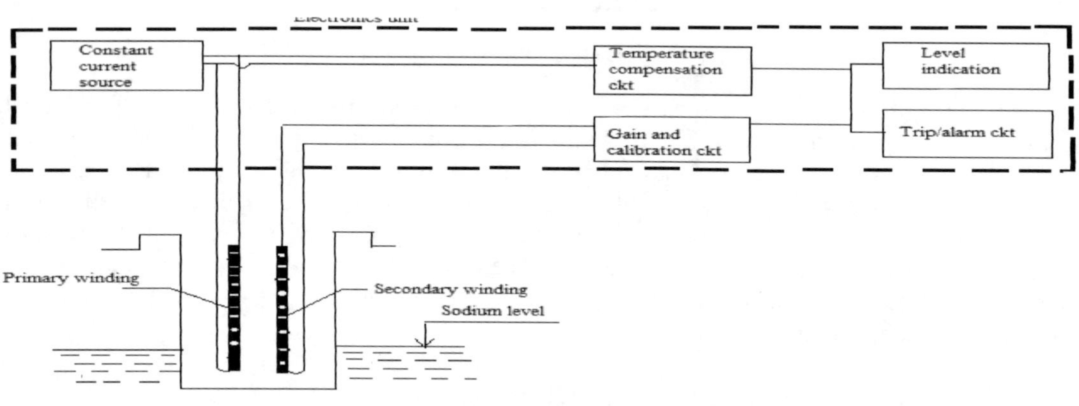

Figure 15: Mutual inductance type continuous level probe.

The advantage of the Mutual inductance level probe over resistance type level probe is the former works on non-contact principle and the latter works based on its contact with sodium. Since the mutual inductance level probe is kept inside the pocket its maintenance is easy. The non-wetting of sodium with the sensor does not affect the operation of the mutual inductance sensor.

11. Conclusion

Various methods of liquid level measurement are presented in this paper. Each level measurement technique has different working principles. Every level measurement system includes the interaction between the sensing device, the element or the system, and the type of substance inside the container. In this paper we thoroughly reviewed various techniques of level measurement, the kind of sensors used, their working principle and the advantages and disadvantages of the level measurement techniques. A method of measuring liquid metal level by making use of mutual inductance principle is also depicted in the form of a mutual inductance level probe. This probe is a non-contact probe and is used quite frequently in liquid metal systems like fast breeder reactors which have liquid sodium as the liquid metal coolant. Research and development is still in progress in these sensing techniques to make them more compact and maintenance free.

Acknowledgments

I sincerely thank Dr.Uppu Ramachandraiah, Head of Department of Electrical and Electronics Engineering, Dr. K.R Devabalaji, Assistant Professor, Department of Electrical and Electronics Engineering, for the continuous encouragement and guidance.

References

[1]. Sitharthan R, Geethanjali M and Pandy TKS 2016 Adaptive protection scheme for smart microgrid with electronically coupled distributed generations *Alexandria Engineering Journal* **55(3)** 2539-2550

[2]. Fathima AH, and Palanisamy K 2014 Battery energy storage applications in wind integrated systems—a review *IEEE International Conference on Smart Electric Grid* 1-8

[3]. Prabaharan N and Palanisamy K 2015 Investigation of single-phase reduced switch count asymmetric multilevel inverter using advanced pulse width modulation technique *International Journal of Renewable Energy Research* **5(3)** 879-890.

[4]. Jerin ARA, Kaliannan P and Subramaniam U 2017 Improved fault ride through capability of DFIG based wind turbines using synchronous reference frame control based dynamic voltage restorer. *ISA transactions* **70** 465-474

[5]. Sitharthan, R, Sundarabalan CK, Devabalaji KR, Nataraj SK and Karthikeyan M 2018 Improved fault ride through capability of DFIG-wind turbines using customized dynamic voltage restorer *Sustainable cities and society* **39** 114-125

[6]. Prabaharan N and Palanisamy K 2016 A single-phase grid connected hybrid multilevel inverter for interfacing photo-voltaic system *Energy Procedia* **103** 250-255

[7]. Palanisamy K, Mishra JS, Raglend IJ and Kothari DP 2010 Instantaneous power theory based unified power quality conditioner (UPQC) *IEEE Joint International Conference on Power Electronics, Drives and Energy Systems* 1-5

[8]. Sitharthan R and Geethanjali M 2017 An adaptive Elman neural network with C-PSO learning algorithm-based pitch angle controller for DFIG based WECS Journal of Vibration and Control **23(5)** 716-730

[9]. Sitharthan R and Geethanjali M 2015 Application of the superconducting fault current limiter strategy to improve the fault ride-through capability of a doubly-fed induction generator–based wind energy conversion system *Simulation* **91(12)** 1081-1087

[10]. Sitharthan R, Karthikeyan M, Sundar DS and Rajasekaran S 2020 Adaptive hybrid intelligent MPPT controller to approximate effectual wind speed and optimal rotor speed of variable speed wind turbine *ISA transactions* **96** 479-489

[11]. Sitharthan R, Devabalaji KR and Jees A 2017 An Levenberg–Marquardt trained feed-forward back-propagation based intelligent pitch angle controller for wind generation system *Renewable Energy Focus* **22** 24-32

[12]. Sitharthan R, Sundarabalan CK, Devabalaji KR, Yuvaraj T and Mohamed Imran A 2019 Automated power management strategy for wind power generation system using pitch angle controller *Measurement and Control* **52(3-4)** 169-182

[13]. Sundar DS, Umamaheswari C, Sridarshini T, Karthikeyan M, Sitharthan R, Raja AS and Carrasco MF 2019 Compact four-port circulator based on 2D photonic crystals with a 90° rotation of the light wave for photonic integrated circuits applications *Laser Physics* **29(6)** 066201

[14]. Sitharthan R, Parthasarathy T, Sheeba Rani S and Ramya KC 2019. An improved radial basis function neural network control strategy-based maximum power point tracking controller for wind power generation system *Transactions of the Institute of Measurement and Control* **41(11)** 3158-3170

[15]. Rajesh M and Gnanasekar JM 2017 Path observation based physical routing protocol for wireless ad hoc networks *Wireless Personal Communications* **97(1)** 1267-1289

[16]. Palanisamy K, Varghese LJ, Raglend IJ and Kothari DP 2009. Comparison of intelligent techniques to solve economic load dispatch problem with line flow constraints *IEEE International Advance Computing Conference* 446-452

[17]. Sitharthan R, Ponnusamy M, Karthikeyan M and Sundar DS 2019 Analysis on smart material suitable for autogenous microelectronic application *Materials Research Express* **6(10)** 105709

[18]. Rajaram R, Palanisamy K, Ramasamy S and Ramanathan P 2014 Selective harmonic elimination in PWM inverter using fire fly and fireworks algorithm *International Journal of Innovative Research in Advanced Engineering* **1(8)** 55-62

[19]. Sitharthan R, Swaminathan JN and Parthasarathy T 2018 March. Exploration of wind energy in India: A short review *IEEE National Power Engineering Conference* 1-5

[20]. Karthikeyan M, Sitharthan R, Ali T and Roy B 2020 Compact multiband CPW fed monopole antenna with square ring and T-shaped strips *Microwave and Optical Technology Letters* **62(2)** 926-932

[21]. Sundar D Sridarshini T, Sitharthan R, Madurakavi Karthikeyan, Sivanantha Raja A, and Marcos Flores Carrasco 2019 Performance investigation of 16/32-channel DWDM PON and long-reach PON systems using an ASE noise source *In Advances in Optoelectronic Technology and Industry Development: Proceedings of the 12th International Symposium on Photonics and Optoelectronics* 93

[22]. Sitharthan R and Geethanjali M 2014 Wind Energy Utilization in India: A Review *Middle-East J. Sci. Res.* **22** 796–801 doi:10.5829/idosi.mejsr.2014.22.06.21944

[23]. Sitharthan R and Geethanjali M 2014 ANFIS based wind speed sensor-less MPPT controller for variable speed wind energy conversion systems *Australian Journal of Basic and Applied Sciences* **8**14-23

[24]. Jerin ARA, Kaliannan P, Subramaniam U and El Moursi MS 2018 Review on FRT solutions for improving transient stability in DFIG-WTs *IET Renewable Power Generation* **12(15)** 1786-1799

[25]. Prabaharan N, Jerin ARA, Palanisamy K and Umashankar S 2017 Integration of single-phase reduced switch multilevel inverter topology for grid connected photovoltaic system *Energy Procedia* **138** 1177-1183

[26]. Rameshkumar K, Indragandhi V, Palanisamy K and Arunkumari T 2017 Model predictive current control of single phase shunt active power filter *Energy Procedia* **117** 658-665

[27]. Fathima AH and Palanisamy K 2016 Energy storage systems for energy management of renewables in distributed generation systems *Energy Management of Distributed Generation Systems* 157

[28]. Rajesh M 2020 Streamlining Radio Network Organizing Enlargement Towards Microcellular Frameworks *Wireless Personal Communications* 1-13

[29]. Subbiah B, Obaidat MS, Sriram S, Manoharn R and Chandrasekaran SK 2020 Selection of intermediate routes for secure data communication systems using graph theory application and grey wolf optimisation algorithm in MANETs IET *Networks* doi:10.1049/iet-net.2020.0051

[30]. Singh RR and Chelliah TR 2017 Enforcement of cost-effective energy conservation on single-fed asynchronous machine using a novel switching strategy *Energy* **126** 179-191

[31]. Amalorpavaraj RAJ, Palanisamy K, Umashankar S and Thirumoorthy AD 2016 Power quality improvement of grid connected wind farms through voltage restoration using dynamic voltage restorer *International Journal of Renewable Energy Research* **6(1)** 53-60

[32]. Singh RR, Chelliah TR, Khare D and Ramesh US 2016 November. Energy saving strategy on electric propulsion system integrated with doubly fed asynchronous motors *IEEE Power India International Conference* 1-6

[33]. Sujatha K and Punithavathani DS 2018 Optimized ensemble decision-based multi-focus imagefusion using binary genetic Grey-Wolf optimizer in camera sensor networks *Multimedia Tools and Applications* **77(2)** 1735-1759

[34]. Krishnamoorthy S, Punithavathani S and Priya JK 2017 Extraction of well-exposed pixels for image fusion with a sub-banding technique for high dynamic range images *International Journal of Image and Data Fusion* **8(1)** 54-72

[35]. Singh RR, Mohan H and Chelliah TR 2016 November. Performance of doubly fed machines influenced to electrical perturbation in pumped storage plant-a comparative electromechanical analysis *IEEE 7th India International Conference on Power Electronics* 1-6

IOP Conf. Series: Materials Science and Engineering **937** (2020) 012028 doi:10.1088/1757-899X/937/1/012028

A novel IOT based domestic automation system for load monitoring and efficient control

A T Sankara Subramanian[1], S Priyadharsini[2], S Palaniyappan[3] and Dr A Nazar Ali[4]

[1, 2, 3&4] Department of Electrical and Electronics Engineering,
[1]K.Ramakrishnan College of Technology, Trichy, Tamilnadu, India.
[2]Government College of Engineering, Tiruneveli, Tamilnadu, India.
[3]SRM TRP Engineering College, Trichy, Tamilnadu, India
[4]Rajalakshmi Engineering College, Chennai, Tamilnadu, India

Email: nazarali.a@rajalakshmi.edu.in

Abstract. Home and industrial automation are really relevant as it offers flexibility and allows things easy to use. Wireless home automation control is built and deployed using Wi-Fi technologies to monitor specific home appliances with essential protection and also to secure the network. Devices installed in each room have their own work boards attached to the personal computer or mobile devices through the control panel and the user may be aware every day of their energy use by sophisticated tracking methods. The suggested method is also very beneficial for users in managing their electricity consumption and effective load management may also be carried out.

1. Introduction

The Internet of Things (IoT) is an Internet communication that allows the interaction and exchanging of data with the actual items, vehicles , home appliances and those in hardware applications, sensors, activators and networking. Each object is special, yet can fit into the network infrastructure, via its integrated computing unit [1].

The IoT allows artifacts to be detected or remotely controlled through existing internet infrastructure, goals and to make incorporation of the physical world with desktop-based systems [2]. It resulting in enhanced efficiency , accuracy and financial advantages in addition to the reduced human involvement. IoT with Mobile applications are going to gather useful data utilizing numerous current technologies and enabling themselves for data transfer among other devices.

Now days there are variety of services and the system that can aimed to reduce the energy consumption, so we devolp IOT based load monitoring and efficent control [3]. In this system we use a separate display unit for monitoring and knowing the energy consumption. Our system can be installed in industries or in apartments, lodges, hotels and highly populated residential buildings.

Automatic monitoring and controlling is the way to reduce the human efforts by using the mobile application and web pages [4-5]. We can control many loads as required by using mobile application, the main purpose of mobile application is to control the loads by customer themselves. So the proposed system is user friendly and the customer can control the load through a simle mobile application and also do monitoring process.

2. Existing System

The new network comprises of stat device Wi-Fi router[6], local computer, rectifier, transmitter, panel, relay drivers and relays, which are included in the latest tracking system by way of an energy meters and a locally-based database[7]. This is not managed over long distances with the present PIC controller as only a load can be attached. We would have to write the coding for further charge and then upload the coding in a pic controller if we were going to attach the additional load reset [8].

However, we use atmega328 to link to this charge anytime as per requirement. That is what the atmega328 has to offer. For converting DC supply to AC supply, the existing system rectifier is used. Nonetheless, we can regulate ac supply directly. Compared to our method, the current framework is quite complex [9]. The Wi-Fi router is found in the new network. We are using the implementation of Wi-Fi modem in our network. Similarly, on the internet, the ios app is really simple for end users to manage load.

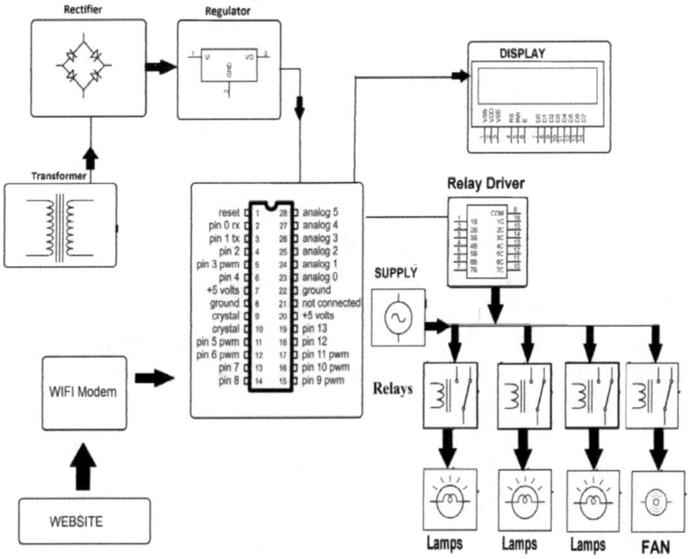

Figure 1. Block diagram of existing system

Figure 1 shows the block diagram of existing system which is typically used to learn our current Process Block diagrams, with less comprehensive explanations at a higher degree, and are meant to explain basic principles without reference to implementation specifics. It provides information on the design and physical layout of electrical components [8].

Typically an AC power supply transfers the voltage from a wall outlet (main supply) to the desired voltage. There may also be some filtering [10]. AC power supplies can be broken into one-phase and three-phase devices for standard applications. "The key disparity between single phase and three phase AC capacity is the transmission constancy at 230v, 50 Hz.

An electronic part is a voltage regulator device that helps to preserve a constant degree of voltage control. A Potential regulator can provide constructive feedback and can use a clear feedback system. Feedback voltage regulation feature contrasts the present output voltage with a certain specified reference voltage. The controlling factor is usually regulated if the output potential is too high to achieve lower output potential [9].

Using PIC16F57, the memory of the system and data is isolated [11]. Data is 8-bit, 16-bit memory, and 32-bit width in new versions. The software directions may be 12, 14, 16 or 24 bits in bit counting per PIC unit. The device's hardware features vary from SMD 6-pin, 8-pin Drop, up to 142-pin IC with dedicated O/I connectors [10]. The guidance can even be extended for more efficient chips that include instructions for automated signal processing functions. The developer offers machine tools with many forms of applications. It offers reduced-cost or costless software resources and the re-programming of series programed machines; low-cost production programmers and high-processing programmers are supported [12]. A relay is a circuit powered electrically. Many relays are fitted with an electromagnet, although certain practical concepts such as solid phase relays are often included. Relaxation is used when a specific low-power signal may be operated on a circuit or when several circuits can be controlled by a single signal. The first reliefs were used as amplifiers in long range telegraph circuits: they reproduced and conveyed the signal through one circuit again. Relays have been commonly used for logical operations of phone lines and early computers. ATMEGA328, MCU (Wi-Fi module), ACS712 (30A, Existing sensor) etc. are the framework that is suggested. With the aid of block diagram and circuit diagrams we are now learning about it in depth.

Both load management and effective regulation are seen in the device block diagram suggested. However, only load control is required in the current framework. And for the control and tracking, we use ATMEGA328 and Node MCU primarily. We use the current sensor to display how often current is actually being used. And also for the same operation, the voltage sensor is used. We may track the watt-hour with this sensor. It implies that in a hour or one day we know what more power we used. This control program would allow end users to realize how much electricity they used for an hour. This control device allows the end consumer rising energy use. They use a pic device in this current framework. It is very challenging to attach new load to the pic unit [13]. However, we use ATMEGA328 in the proposed framework rather than in the pic unit. We can load this ATMEGA328 anytime we want. In contrast with the pic unit, it is easy to install load in ATMEGA328. In our method, Ac supply is included in our planned single step 230v. And to transform the stock, we use a Bridge Corrector. Throughout our plan we should incorporate protection load regulation as the voltage level rises or the load should immediately decrease OFF. Figure 2 shows the Proposed System Block diagram for higher level and less comprehensive definitions are usually used to explain basic principles without concern for execution specifics. It displays the output power and information regarding the design of the electrical and physical elements [14].

The AC power supply usually takes the voltage from the Electrical outlet (main source) and reduces it to the required voltage some filtering may often take place. In current implementations, AC power supplies can be split into single phase and three phase networks. The main distinction between single phase and three phase AC control is the constancy of transmission. We use 230v, 50 Hz in the current model. A relay is a circuit powered electrically. While some design ideas like stable-state relays frequently are used, other relays use an electromagnet to physically control a switch. Relays come into action when a new low-power signal is needed for the communication, or where a single signal is sufficient to monitor multiple circuits. For long distance telegraph systems, the whole first devices were used as transducers: they repeated the signal from one trunk and retrieved the signal to another. Switching devices are used extensively for rational processes of telephone exchanges and energy conserved and utilized efficiently [15].

A relay device that can manage the power required to monitor an electric motor or other freight directly is called a contactor. Rather of utilizing the semiconductor devices for shifting, solid-state relays monitor power circuits with no moving parts. Electric circuits defend against over challenge or loss by utilizing relays with optimized electrical characteristics and often several working coils; such tasks are done by automated instruments often regarded as "protective relays" in current electric power systems. Magnetic relays allow one pulse of inductive power in one direction and a diverted pulse in another to reverse the contacts in one direction. There is no response of repetitive pulses from the same signal. For systems where lost control is not to pass contacts magnetic latching relays are helpful. Both single and dual spindles are necessary for magnetic locks relays. The switch runs in one direction on a single rotating system while the control is applied with a particular polarity and resets when the polarity is inverted. It is necessary to determine the If polarized voltage is added to a reset spin on a dual spinning unit, the contacts are modified [16].

Magnetic lock relays operated by AC have single coils that are used to vary between running and restarting instructions by steering diodes. This has been used in long-distance telegraph systems, storing and sending a signal from one circuit to another. It has A / D (8-can in TQFP and QFN / MLF packages), internal oscillator programmable control signal and five power saving device modes. The system operates from 1.8 to 5.5 volts.. The system will reach an output of 1 MIPS per MHz, manage power usage, and work speed by executing high-performance instructions in a single clock cycle. At first, though, the device and the instructions that it approved were almost without English-language records.

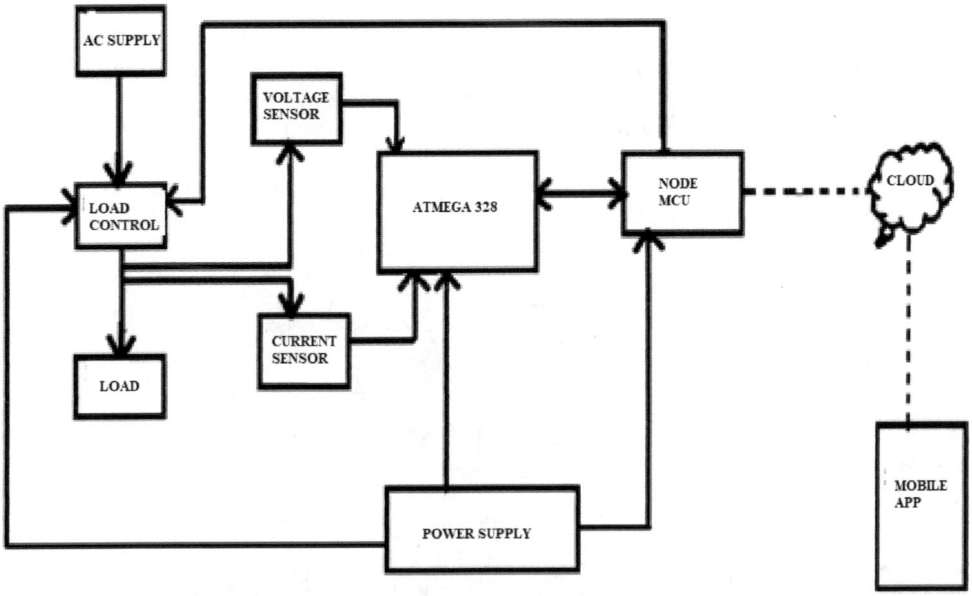

Figure 2. Block diagram of proposed system

The extremely low price and the very small number of external components on the board, which suggested a very cheap amount, attracted many hackers to investigate and convert the processor, chip and software on this project.

3. Software calculation

A cross-platform code written in the computer language Java is Arduino's advanced production framework. The GNU General Public Authorization, version 2 is the root code of the IDE. Use special application layout laws, the Arduino IDE will support the languages C and C++. The Arduino IDE provides the Wiring project program library with various can inlet and outlet procedures. The user-written code contains only two simple functions that are compiled and connected to the software stub main) (for beginning the sketch and the program code loop [17]. The GNU series of software used with the IDE release in an interactive cyclic executive system. The Arduino IDE uses the avrdude program to transform program file into a hexadecimal text format, which has been fed onto the controller board by a feeding function in the relevant software of the component.

3.1. Voltage Measurement

VolrVoltage_Value1 = Voltage_Value * (5.0/1023.0) *67

3.2. Current Measurement

DoubleRMSCurrent = (((maxCurrent - 516)*0.707)/11.8337)

3.3. Power Measurement

Power= voltage * current

3.4. Energy Measurement

Kilo = kilo + (RMS Power * (2.05/60/60/1000))

3.5. Consumption bill amount

Bill amount = kilowatt*3.5(Tariff can be changes based on requirement)

IOP Conf. Series: Materials Science and Engineering **937** (2020) 012028 doi:10.1088/1757-899X/937/1/012028

4. Hardware arrangement

This scheme is based strictly on load handling in our program. We use various devices including a resistor bridge and a voltage control device. The performance will be shown on the LCD monitor utilizing Arduino IDE tools. The entire system is connected to a rough board. Both load management and effective regulation are seen in the device block diagram suggested. However, only load control is required in the current framework. And for the control and tracking, we use ATMEGA328 and Node MCU primarily. We use the current sensor to display how much current is actually being used. And also for the same operation, the voltage sensor is used [18-19].

Figure 3. Hardware setup of proposed system

Figure 3 shows the hardware setup of the proposed system with a motor load. A relay is a circuit powered electrically. Many relays are fitted with an electromagnet, although certain practical concepts such as solid state relays are often included. Relaxation is used when a specific low-power signal may be operated on a circuit or when several circuits can be controlled by a single signal. A current sensor is a system which senses the electric coil of wire and produces a corresponding signal. This signal is used to show the calculated current, or to store and use it in a device for future study. The voltage sensor may calculate the existence of a voltage, i.e. on insulating wires, without having a metal touch. A voltage monitor is made up of a voltage resistive divider. Included in a casted resin (for voltages from 1-72kV), the combined resistors display a weak inductance rating [20].

5. Simulations of the Proposed System

The circuits meant for simulating the proposed system are as follows,

5.1. Simulation circuit for Voltage Measurement

The potential measurement device consists of AC-DC Converter IC, Zener diode, condenser, transformer and variable resistance. The conversation device goes down to 12V ac, and moves into the correction unit and receives 12V dc. 230V AC. This is pushed via the filter device and three volts dc supply is received. The programming is now in micro C. The code could also be written using the Java language. The system generates and creates the code. The main benefit of using Arduino is that the software can be modified without flickering the controllers except in the runtime. This makes writing code simpler and more versatile. The simulation circuit for calculating voltage is shown in Figure 4. Drawing and simulating in Proteus is even simpler.

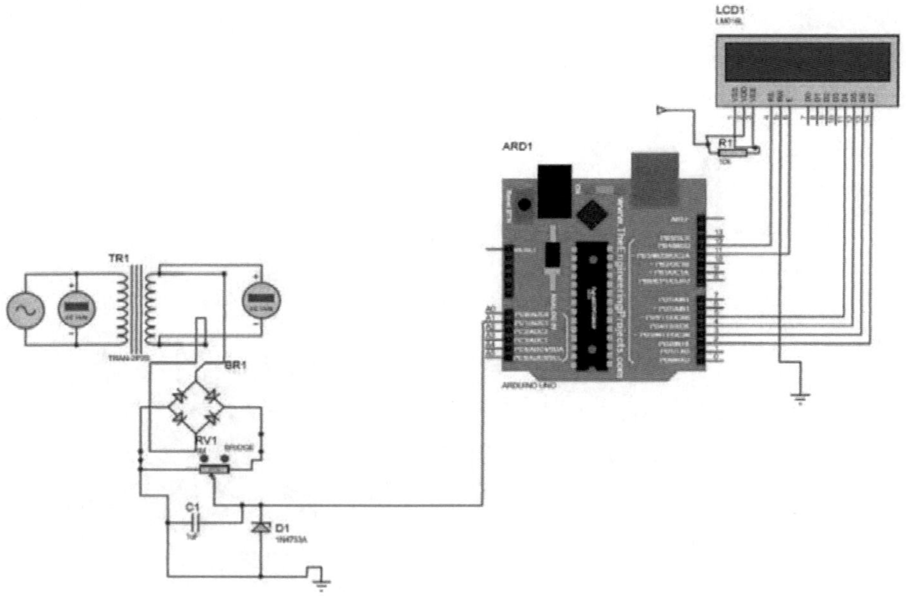

Figure 4. Measuring Voltage in Proteus

5.2. Simulation circuit for Measuring current

The sensor used to measure current ACS712 is being used in the segment. The current calculation approximation as seen in Figure 5.

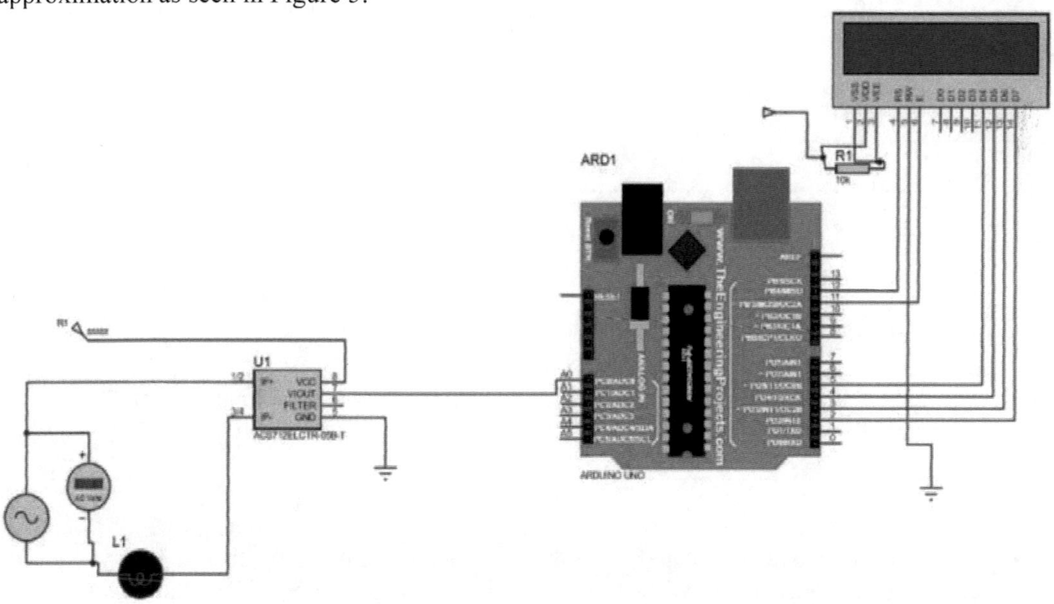

Figure 5. Simulation circuit of current measurement.

5.3. Complete Proteus Simulation

Figure 6 shows the complete simulation diagram of the proposed system with a load which appears to be a simple lamp load for simulation purpose.

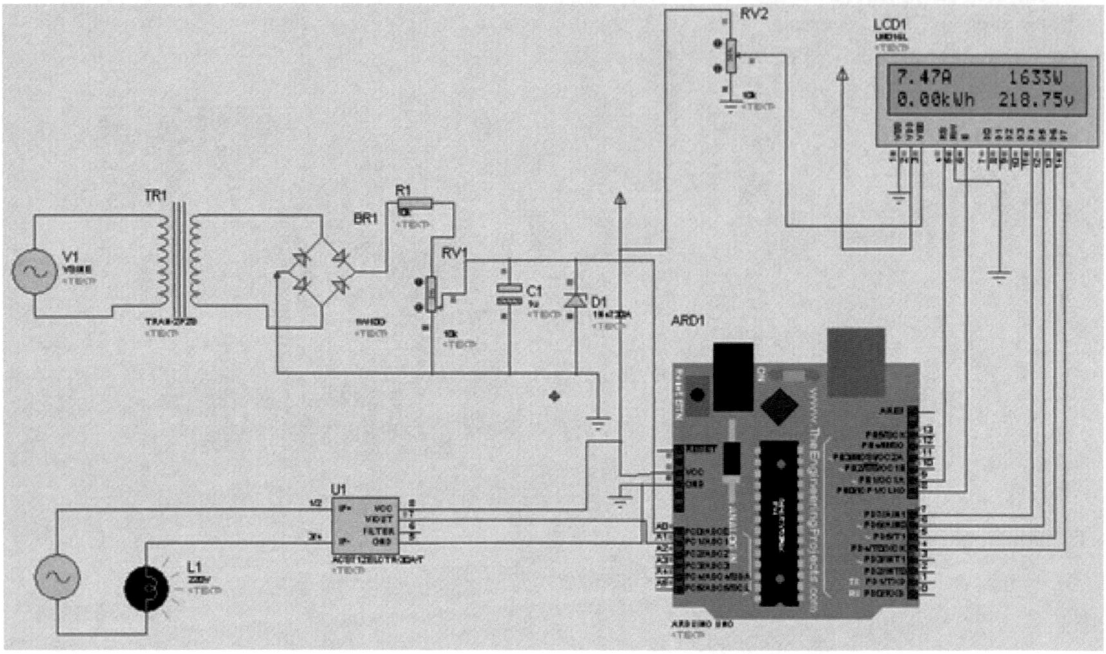

Figure 6. Complete Simulation Diagram with load

6. Hardware Results

Hardware result for under voltage, normal voltage and over voltage diagram given below. Table 1 shows the reference voltage parameters taken for handling loads. The reference voltages are based on the nature of the load selected and as per IEEE norms.

Table 1. Voltage reference parameters

Description	Range	Condition
Under voltage	Below 180v	Relay is ON , Load is OFF
Normal voltage	180v to 230v	Relay is OFF Load is ON
Over voltage	Above 250v	Relay is ON Load is OFF

6.1. During Under Voltage condition

If the measured voltage value is less than that of the reference value of 180V, then the relay is turned ON to protect the load and the Load will get OFF. The LCD display under this condition is shown in the figure 7. In the figure the LCD display interfaced to the Arduino controller shows the active or inactive behavior of the Relay and the load.

Figure 7. Output display for Under Voltage

Figure 8. Output display for Over Voltage

Figure 9. Output display for Normal Voltage

6.2. During Over Voltage Condition

If the measured voltage value is greater than that of the reference value of 250V, then the relay is turned ON to protect the load from over voltage and the Load will get OFF. The LCD display under this condition is shown in the figure 8. In the figure the LCD display interfaced to the Arduino controller shows the active or inactive behavior of the Relay and the load. This overvoltage if not prevented then it will definitely damage the load.

6.3 During Normal Operating Condition

If the measured voltage value is well within the normal values, i.e. 180V to 230V, then the relay is turned OFF and the Load will be turned ON. The LCD display under this condition is shown in the figure 9. In the figure the LCD display interfaced to the Arduino controller shows the active or inactive behavior of the Relay and the load.

7. Conclusion

This work implies a good plant efficiency and demand management principles includes an awareness of service levels and consumption calculation added to clear knowledge of the operation and the load which is managed. The paper also proposed a modified smart monitoring and management system with the function of manipulating data at high speed by profibus technology and digital meter current era. The customized system proposed is designed as an automatic system for demand tracking and load control with exact management of different load during peak period based on system requirements and actual time control of system output. Power quality could also be evaluated and recorded in a control module by multiplying the voltage and current.

8. References

[1] M D Udayakumar et al 2019 IOP Conf. Ser.: Mater. Sci. Eng. 623 012018.

[2] A. T. S. Subramanian, P. Sabarish and A. N. Ali, "A power factor correction based canonical switching cell converter for VSI fed BLDC motor by using voltage follower technique," 2017 IEEE International Conference on Electrical, Instrumentation and Communication Engineering (ICEICE), Karur, 2017, pp. 1-8, doi: 10.1109/ICEICE.2017.8191932.

[3] A. T. Sankara Subramanian, P. Sabarish, M. D.Udayakumar and T. Vishnu kumar," Performance Analysis of Various Photovoltaic Configurations Under Uniform Shading and Rapid Partial Shading Formations", Biosc.Biotech.Res. Comm. Special Issue Vol 13 No (3) 2020 Pp-185-192.

[4] B.Karthikeyan, R. Ramkumar, L. Nagarajan, A. Anton Amala Praveen and S. Vijayalakshmi," ANN Based MPPT Controller for PEMFC System with Interleaved Resonant PWM High Step up Converter" Biosc.Biotech.Res.Comm. Special Issue Vol 13 No (3) 2020 Pp-44-47.

[5] S.R.Paveethra, S.Vijayalakshmi, C. Kalavalli, S.Murugesan and Dr.S.Jeyasudha," MPPT based Harmonic Reduction in PV Module for Grid Connected System," Biosc.Biotech.Res.Comm. Special Issue Vol 13 No (4) 2020 Pp-43-48.

[6] P. Sabarish et al 2019 IOP Conf. Ser.: Mater. Sci. Eng. 623 012011

[7] A T Sankara Subramanian et al 2019 IOP Conf. Ser.: Mater. Sci. Eng. 623 012012

[8] Dr.A. Rajkumar, R. Jai Ganesh, V. Suresh Kumar, T. Vishnu Kumar and T. Ram Kumar," Implementation of High Efficient Single Input Triple output DC-DC Converter," Biosc.Biotech.Res.Comm. Special Issue Vol 13 No (3) 2020 Pp-48-55.

[9] R. Ramkumar, B. Karthikeyan, Dr. A. Rajkumar, V. Venkatesh and A. Anton Amala Praveen," Design and Implementation of IOT Based Smart Library Using Android Application," Biosc.Biotech.Res.Comm. Special Issue Vol 13 No (3) 2020 Pp-56-62.

[10] S.Murugesan, C.Kalavalli, S.Vijayalakshmi, S.R.Paveethra and M.V.Suganyadevi," Development of IoT Architecture for Physically Challenged People Using BLYNK Server," Biosc.Biotech.Res.Comm. Special Issue Vol 13 No (4) 2020 Pp-49-53.

[11] S.Jeyasudha, A. Anton Amala Praveen and B. Geethalakshmi," Fractional Order Controlled Multilevel Boost Hybrid Converter," Biosc.Biotech.Res.Comm. Special Issue Vol 13 No (3) 2020 Pp-202-210.

[12] R. Jaiganesh, L. Nagarajan, A. Arthi and V. Venkatesh," Anti Theft Control of Automatic Teller Machine Using Wireless Sensors," Biosc.Biotech.Res.Comm. Special Issue Vol 13 No (3) 2020 Pp-18-22.

[13] P. Sabarish, A. T. Sankara Subramanian, S. Murugesan and V. Sureshkumar,"A New Methodology of Arterial Blood Clot Removal Using Bio Molecular Devices for ATPase Nuclear Motors." Biosc.Biotech.Res.Comm. Special Issue Vol 13 No (3) 2020 Pp-197-201.

[14] A Nazar Ali, D Sivamani, R Jaiganesh M Pradeep (2019), Solar powered air conditioner using BLDC motor, IOP Conference Series: Materials Science and Engineering, vol. 23.

[15] V Venkatesh, A Nazar Ali,, R Jaiganesh. V Indiragandhi (2019), Extraction and conversion of exhaust heat from automobile engine in to electrical energy Energy, IOP Conference Series: Materials Science and Engineering, vol. 23.

[16] A Ali Nazar, R Jayabharath, MD Udayakumar (2014) ,An ANFIS Based Advanced MPPT Control of a Wind-Solar Hybrid Power Generation System, international review of modeling and simulations. vol.7, no. 4, pp. 638–643.

[17] A.Nazar Ali and R. Jayabharath (2014), "Performance Enhancement of Hybrid Wind/Photo Voltaic System Using Z Source Inverter with Cuk-sepic Fused Converter." Research Journal of Applied Sciences, Engineering and Technology 7.19 pp. 3964-3970.

[18] K Premkumar et al. (2018), Novel bacterial foraging-based ANFIS for speed control of matrix converter-fed industrial BLDC motors operated under low speed and high torque, Neural Computing and Applications, 29, 12, pp.1411–1434.

[19] Premkumar et al. (2015), GA-PSO optimized online ANFIS based speed controller for Brushless DC motor, Journal of Intelligent & Fuzzy Systems, 28, 6, pp.2839-2850.

[20] R. Jai Ganesh,A.Nazar Ali, S.Kodeeswaran, B.Karthikeyan, L.Nagarajan (2020), Iot based Water Management System for Highly Populated Residential Buildings, International Journal Of Disaster Recovery And Business Continuity, Vol 11,issue 01,pp.4018-22.

A review on issues, challenges and various optimization techniques in microgrid

B Goutham[1] , H A Anoop[2]
Assistant Professor[1 2]
Department of Electrical and Electronics Engineering[1 2]
Vidyavardhaka College of Engineering Mysuru India[1 2]

Email-[1]goutham.b@vvce.ac.in
[2]anoopha@vvce.ac.in

Abstract. A drastic change in the increase of carbon emission and green-houses gases has increased the demand of the use of Renewable Energy Sources (RESs). The integration of renewable energy in a power system network is due to the fact that the available natural resources are depleting at a faster rate. Microgrid with different configurations along with the development of smart grids ensures a paradigm shift from centralised to decentralized control. The installations done at the demand side of distributed generation marks change in the forthcoming years in the electricity markets both at the generating and distribution end. With the introduction of Microgrid the hedging of investment is done in different sectors of the electric industry such as generation operations, etc., easily. This review paper gives an idea about fundamental definition of a Microgrid (MGs); classification of Microgrid, issues, and challenges associated and optimization techniques for the overall stability and security of the grid.

1. Introduction

Many people today are facing the problem of access to electricity. The access problems approximates to one in seven [1-10]. Mainly people living in developing countries of Africa and South Asia. Poverty in energy sector reduces income, growth of nation and retards all round development of Humans. Increase in global climatic issues resulted in effective usage of distributed energy resources like wave tidal, wind, solar, combined with different energy storage Technologies [11-15]. The problem shortage of power leads to the load shedding. This led to the development of Microgrids [16-20]. Microgrid is an individual group of electrical sources and the load, the concept of Microgrid was first put by Consortium of Electric Reliability Technology Solutions (CERTS). The power supply control and operation of Microgrids is done by power electronic devices [16-28]. Even Microgrids are found to be efficient compared to that of traditional methods of electrical systems of involving transformation transmission losses blackouts and lost productivity [29-35]. Fig 1 shows the typical line diagram of MG.

IOP Conf. Series: Materials Science and Engineering **937** (2020) 012029 doi:10.1088/1757-899X/937/1/012029

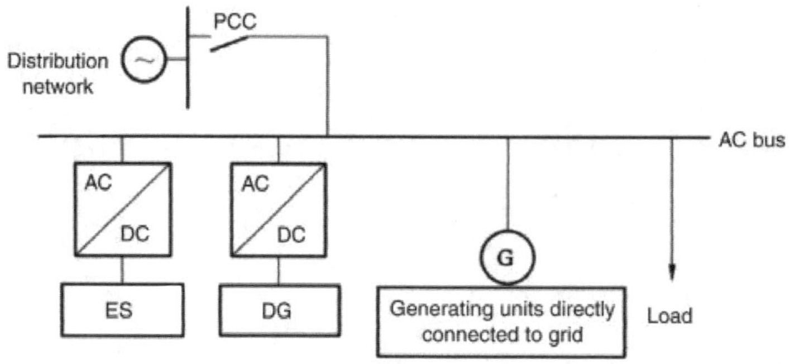

Figure 1.Typical Line diagram of Microgrid System

MGs are generally operated in two modes i.e., on grid mode and off grid mode (Islanded mode).In grid connected mode it operates as normal DG and during off-grid modes it operates as independently providing reliable power supply. During the condition on grid mode MGs operate normally under the fault condition the breaker isolates at a point of common coupling. When MGs operated off-grid many researchers have shown that fault level of Microgrid varies abruptly in the different modes of conditions.

During the on grid condition even the fall current is very high the protections can be easily done by conventional Protection Scheme but during off grid condition the drop in the current is very sharp and even zero crossing does not occur the problem will be serious if the Microgrids are inverter interfaced.

2. Types of microgrid

Microgrid choice depends on its stability, connectivity, available source and available infrastructure. MGs which is serving thousands of customers in a single local community and restricted only to that community is called a local community MGs they may be AC MGs or DC MGs and even they may be centralized or decentralized. MGs that operate completely in a decentralized independent islanded mode i.e., off grid mode are known as remote off grid MGs. These types of MGs came to existence because of the economic conditions or due to geographical limitations. Remote off grid MGs many depend on the distributed energy resources. Microgrid can be broadly classified as AC Microgrid, DC Microgrid and Hybrid Microgrid.

In MGs connectivity plays an important role. Usage of AC equipment's results in AC MGs .The grid operates with the AC power. Fig 2 shows the block diagram of A MGs. Most of the equipment's used installed and loads connected will work with AC power. The sources in the grid and the load are connected through common point of coupling (PCC). Exchange of power between loads and source takes from for Common Point of Coupling. Regulation of active and reactive power and the stability of the grid are maintained by using advanced power electronic converters. In AC MGs convection sources like DC generator is used to maintain stability of grid under the disturbance condition. Advanced power electronic converters convert DC power from distributed energy resource to AC for effective usage of dc power.AC MGs can be classified into Coupled AC MGs and decoupled AC MGs.AC MGs directly connected to utility through transformer is known as AC Coupled MGs. AC MGs connected to utility loads through power electronic converters are called as decoupled AC MGs.

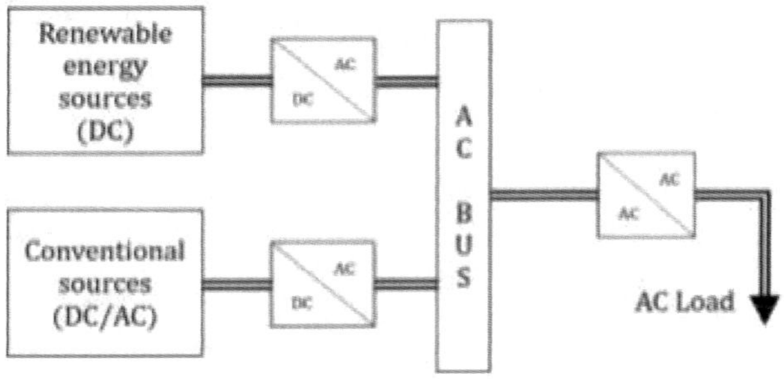

Figure 2. AC Microgrid System

Island operation of electrical systems prefers DC Microgrids. DC Microgrid usually preferred due to low voltage DC operations of equipment and because of its high stability and low losses. The block diagram of DC MGs shown in the figure 3.The major generation of power is from the renewable source like photovoltaic cells which generator DC power. The islanded mode of operation in electrical systems usually prefers DC MGs .The DC load will be usually low voltage power rating equipment's. The connection of load in DC MGs can be done directly from the DC sources with the fewer interfaces of converters. There are many advantages of DC MGs like low power loss reduction ,less voltage drop and increase in the electric line capacity .DC Microgrid are broadly classified as monopolar DC MGs, bipolar DC MGs and homopolar DC MGs configuration.

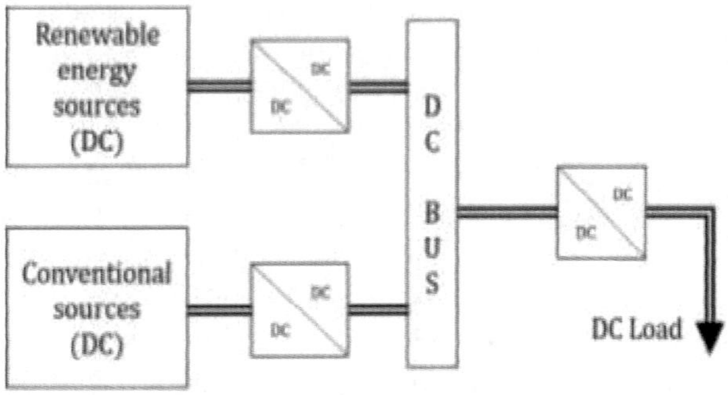

Figure 3.DC Microgrid system

AC MGs are suitable for Grids with dynamic modes having voltage standards. DC MGs are suitable for optimised cost effective DERs. Hybrid Microgrid is has combined advantages over Convention AC MGs and recent developing DC MGs .The loads connected to these types of grids are of both ac and dc. The bus considered in grid will be usually AC bus. The power generation in the grid is from both renewable and conventional sources. Fig 4 shows the block diagram of Hybrid MGs.

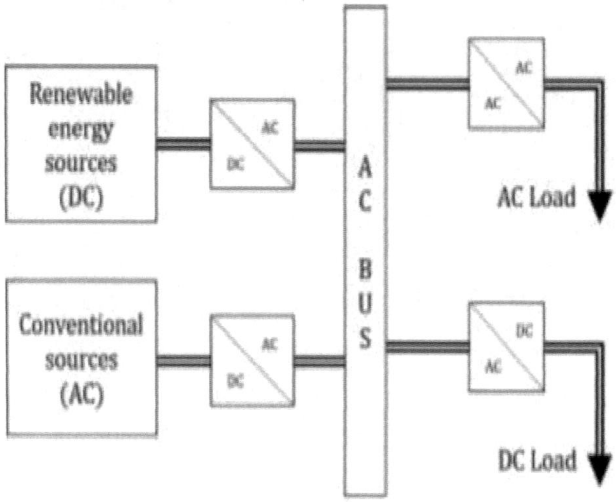

Figure 4. Hybrid Microgrid system

3. Issues and Challenges in Microgrid

There are many challenges and issues faced by Microgrids. MGs should operate in both on- grid and off-grid mode. During these types of conditions switching over operations is performed, there is a chance of loss of stability of the MGs .And this may result in reliability of power supply. The controlling process of MGs has become a difficult task because of the connection of more and more distributed energy resources and leading to load supply fluctuations. The issue is overcome by proper design of high efficient power electronic converters [9].

The adoption of renewable energy sources results in necessity of usage of energy storage devices. There are many types of energy storage devices with each facing different problems. The mechanical energy storage devices face the problem of low energy density and high self-recharging for a short time. Electrochemical Energy Storage devices face the problem of low energy density, low life expectancy, environmental pollution, safety issues due to flammability. If we move towards the chemical energy storages using synthetic gases it will have low energy conversion efficiency of 40% to 50% and high initial cost with low security.

Because of the usage of power electronic components issue power quality is faced like increase of harmonic content, which causes the compromise of power quality. Another issue of power quality is due to voltage transient occurs in the renewable energy sources because of me power deviation occurs that is dc link used in between the converters causes electromagnetic interference in the system electromagnetic interference causes voltage fluctuation in the system also affect the functioning of other devices.

3.1. Protection Issues in Microgrid

The sources and energy storage will contribute to the total fault current and the difficult part here is in this system the absence of natural occurrence of zero crossing correct in DC MGs [12]. When MGs are protected using voltage based protection that generation should be converted to DC quantity using DC frames. Here they will use a communication link which will be deployed for identification of inner zone and Outer zone. And problems will be faced in the communication link and during the differentiation of zones, which in turn leads to the operation of protective devices.[2]Have identified the problems and discussed the Protection Scheme of solid circuit breakers so no natural zero current lower impedance system. But the coordination of the connected equipment's in parallel and the direction for converters in a grid connected to DC MGs is a challenge. The problem of the fasting voltage Restoration for improving the better quality of DC voltage in a DC Microgrid has been discussed [2].

3.2. Problems Encountered Due to Bidirectional Power Flow

The flow of power in MGs will take place in both the directions. Conventional methods of protection for these systems fail [8]. Protection schemes here fail to identify backward and forward faults results in undesired tripping of the system [3].The usage of bipolar converter will result in causing the circulating current. Interruptions and isolation of short circuit faults .The problem of lacking common ground in a DC Microgrid are found due to the usage of bipolar converters. In bidirectional power converters Interruptions and isolation of short circuit faults and the problem of lacking common ground in a DC Microgrid is studied [3].

The Control method for bidirectional power controller has been suggested [9]. The inner loop control method suppresses the inter-circulating current. Outer of control loop method to manage the DC voltage [4]. They are using voided of conventional d-q axes strategy of control and used the three axis d-q-0 control strategy. Hence the supress of circulating current in generated ac side is done by parallel multiple bipolar converts [10].

4. Optimization Techniques in Microgrid

The RESs connected are said to be intermittent in nature, which man's the power output is entirely dependent on the distributed energy sources such as solar, wind etc., DERs play a vital role in the integration of the Microgrid. Distributed generation has few advantages such as reduction in the generation costs compared with generation done at large scale which tend to improve the power reliability, reduces blocking of grid, help in reducing transmission line losses and provide the support in case of reactive power. Most of the micro grids under operation contemplate partial load demand by preserving fossil fuels for the next generation and aiming at achieving zero carbon emission.
To overcome the limitations such as intermittent nature of the RESs, this causes unreliable operation to meet the load demand. The feasible solution would be employing a power electronics based Energy Storage System (ESS). Voltage and frequency of the system can be controlled to meet the generation and load demands. The benefits of using the ESS is to meet the excess load demands in certain cases and provide improved power quality and back-up power during power outages/shortages.

Suggesting a control technique for the established grid architecture tend to provide a stable and efficient operation. Some objective functions such as area of installation, Preference of the user, Power capacity of the Microgrid, equipment installed in the Microgrid, regulations and policies and types of tariffs.

Various mathematical optimization techniques can be employed for the control of the Microgrid through numerical simulations. Few are listed below:

4.1. Linear Programming

Two-dispatch optimizers [5] can be implemented for the management of energy systems. A Mixed integer linear programming [MILP] and improved genetic algorithm code can be used for the unit commitment and economic dispatch of such Microgrid units. To counter the limitations of the above said method, for non-linear loading constraints an improved algorithm with a novel technique can also be used.

A novel approach is used for the double layer coordinated control for management of energy in MGs. The schedule layer obtains the scheme of economic operation based on the load forecasting data available and the dispatch layer provides a real time data of the power controllable units [6].

In [7] a design approach which is based on formulation of Integer Linear Programming (ILP) has an acronym (CoSMNET) .Cost aware Smart Microgrid communication network can allow power transactions to be economic among smart micro grids.

4.2. Nonlinear Programming

For Microgrid operating in Islanded mode an EMS algorithm based on MINLP is used for different scenarios. The performance evaluation of a hybrid renewable energy system can be measured by a framework for energy consumption and also technical challenges which arise in the system frequently. The capital and the operating cost annually can be minimized by considering the Non-linear mixed integer problem.

4.3. Dynamic Programming

An algorithm is employing dynamic programming for energy optimization analysis and power management in Microgrid with diesel generators and renewable energy sources. Framework is suggested for control of Microgrid and performing operational planning for long term. Optimization of the unit commitment is done by a dynamic programming model.

4.4. Stochastic Programming

The energy and reserve scheduling method stochastic in nature is to meet demand responses which are different and facilitate the customers. To minimize the cost and maximize comfort preferences a multi objective dynamic stochastic Programming is preferred. Coordinated scheduling and day ahead cooling problems can be solved by improved Particle Swarm optimization (PSO) technique.

5. Conclusion

An ever-increasing load demand results in a need to generate more power. The thermal generation which requires fossil fuels that deplete day by day to meets most of the demand. To present fossil fuels for future generation and rising carbon emissions, alternate generation approaches are required to meet the demand for loads. Globally, in view of energy scarcity, Microgrid acts as the effective alternative to the conventional grid. Microgrid will greatly reduce the loss compared to the conventional centralized power supply network and save investment in transmission and distribution construction. This paper attempted to provide a review of Microgrid where it can efficiently use the energy in a cascade to higher energy efficiency, thus enhancing the protection and reliability of power supply. Microgrid will significantly minimize or remove the direct gird impact caused by access to distributed power supply, and offers a new path for the wide-ranging deployment of new and renewable energy in connected grid. Nearly 95% of the MG's are fitted with renewable energy sources. Due to the erratic nature of renewable integration various issues, consequences, and resolutions arrived have been discussed, off-grid and on-grid operating modes, and their interaction with appliances, and effective security schemes are provided. Different mathematical optimization techniques were discussed.

References

[1]. Sitharthan R, Geethanjali M and Pandy TKS 2016 Adaptive protection scheme for smart microgrid with electronically coupled distributed generations *Alexandria Engineering Journal* **55(3)** 2539-2550

[2]. Fathima AH, and Palanisamy K 2014 Battery energy storage applications in wind integrated systems—a review *IEEE International Conference on Smart Electric Grid* 1-8

[3]. Prabaharan N and Palanisamy K 2015 Investigation of single-phase reduced switch count asymmetric multilevel inverter using advanced pulse width modulation technique *International Journal of Renewable Energy Research* **5(3)** 879-890.

[4]. Jerin ARA, Kaliannan P and Subramaniam U 2017 Improved fault ride through capability of DFIG based wind turbines using synchronous reference frame control based dynamic voltage restorer. *ISA transactions* **70** 465-474

[5]. Sitharthan, R, Sundarabalan CK, Devabalaji KR, Nataraj SK and Karthikeyan M 2018 Improved fault ride through capability of DFIG-wind turbines using customized dynamic voltage restorer *Sustainable cities and society* **39** 114-125

[6]. Prabaharan N and Palanisamy K 2016 A single-phase grid connected hybrid multilevel inverter for interfacing photo-voltaic system *Energy Procedia* **103** 250-255

[7]. Palanisamy K, Mishra JS, Raglend IJ and Kothari DP 2010 Instantaneous power theory based unified power quality conditioner (UPQC) *IEEE Joint International Conference on Power Electronics, Drives and Energy Systems* 1-5

[8]. Sitharthan R and Geethanjali M 2017 An adaptive Elman neural network with C-PSO learning algorithm-based pitch angle controller for DFIG based WECS Journal of Vibration and Control **23(5)** 716-730

[9]. Sitharthan R and Geethanjali M 2015 Application of the superconducting fault current limiter strategy to improve the fault ride-through capability of a doubly-fed induction generator–based wind energy conversion system *Simulation* **91(12)** 1081-1087

[10]. Sitharthan R, Karthikeyan M, Sundar DS and Rajasekaran S 2020 Adaptive hybrid intelligent MPPT controller to approximate effectual wind speed and optimal rotor speed of variable speed wind turbine *ISA transactions* **96** 479-489

[11]. Sitharthan R, Devabalaji KR and Jees A 2017 An Levenberg–Marquardt trained feed-forward back-propagation based intelligent pitch angle controller for wind generation system *Renewable Energy Focus* **22** 24-32

[12]. Sitharthan R, Sundarabalan CK, Devabalaji KR, Yuvaraj T and Mohamed Imran A 2019 Automated power management strategy for wind power generation system using pitch angle controller *Measurement and Control* **52(3-4)** 169-182

[13]. Sundar DS, Umamaheswari C, Sridarshini T, Karthikeyan M, Sitharthan R, Raja AS and Carrasco MF 2019 Compact four-port circulator based on 2D photonic crystals with a 90° rotation of the light wave for photonic integrated circuits applications *Laser Physics* **29(6)** 066201

[14]. Sitharthan R, Parthasarathy T, Sheeba Rani S and Ramya KC 2019. An improved radial basis function neural network control strategy-based maximum power point tracking controller for wind power generation system *Transactions of the Institute of Measurement and Control* **41(11)** 3158-3170

[15]. Rajesh M and Gnanasekar JM 2017 Path observation based physical routing protocol for wireless ad hoc networks *Wireless Personal Communications* **97(1)** 1267-1289

[16]. Palanisamy K, Varghese LJ, Raglend IJ and Kothari DP 2009. Comparison of intelligent techniques to solve economic load dispatch problem with line flow constraints *IEEE International Advance Computing Conference* 446-452

[17]. Sitharthan R, Ponnusamy M, Karthikeyan M and Sundar DS 2019 Analysis on smart material suitable for autogenous microelectronic application *Materials Research Express* **6(10)** 105709

[18]. Rajaram R, Palanisamy K, Ramasamy S and Ramanathan P 2014 Selective harmonic elimination in PWM inverter using fire fly and fireworks algorithm *International Journal of Innovative Research in Advanced Engineering* **1(8)** 55-62

[19]. Sitharthan R, Swaminathan JN and Parthasarathy T 2018 March. Exploration of wind energy in India: A short review *IEEE National Power Engineering Conference* 1-5

[20]. Karthikeyan M, Sitharthan R, Ali T and Roy B 2020 Compact multiband CPW fed monopole antenna with square ring and T-shaped strips *Microwave and Optical Technology Letters* **62(2)** 926-932

[21]. Sundar D Sridarshini T, Sitharthan R, Madurakavi Karthikeyan, Sivanantha Raja A, and Marcos Flores Carrasco 2019 Performance investigation of 16/32-channel DWDM PON and long-reach PON systems using an ASE noise source *In Advances in Optoelectronic Technology and Industry Development: Proceedings of the 12th International Symposium on Photonics and Optoelectronics* 93

[22]. Sitharthan R and Geethanjali M 2014 Wind Energy Utilization in India: A Review *Middle-East J. Sci. Res.* **22** 796–801 doi:10.5829/idosi.mejsr.2014.22.06.21944

[23]. Sitharthan R and Geethanjali M 2014 ANFIS based wind speed sensor-less MPPT controller for variable speed wind energy conversion systems *Australian Journal of Basic and Applied Sciences* **8**14-23

[24]. Jerin ARA, Kaliannan P, Subramaniam U and El Moursi MS 2018 Review on FRT solutions for improving transient stability in DFIG-WTs *IET Renewable Power Generation* **12(15)** 1786-1799

[25]. Prabaharan N, Jerin ARA, Palanisamy K and Umashankar S 2017 Integration of single-phase reduced switch multilevel inverter topology for grid connected photovoltaic system *Energy Procedia* **138** 1177-1183

[26]. Rameshkumar K, Indragandhi V, Palanisamy K and Arunkumari T 2017 Model predictive current control of single phase shunt active power filter *Energy Procedia* **117** 658-665

[27]. Fathima AH and Palanisamy K 2016 Energy storage systems for energy management of renewables in distributed generation systems *Energy Management of Distributed Generation Systems* 157

[28]. Rajesh M 2020 Streamlining Radio Network Organizing Enlargement Towards Microcellular Frameworks *Wireless Personal Communications* 1-13

[29]. Subbiah B, Obaidat MS, Sriram S, Manoharn R and Chandrasekaran SK 2020 Selection of intermediate routes for secure data communication systems using graph theory application and grey wolf optimisation algorithm in MANETs IET *Networks* doi:10.1049/iet-net.2020.0051

[30]. Singh RR and Chelliah TR 2017 Enforcement of cost-effective energy conservation on single-fed asynchronous machine using a novel switching strategy *Energy* **126** 179-191

[31]. Amalorpavaraj RAJ, Palanisamy K, Umashankar S and Thirumoorthy AD 2016 Power quality improvement of grid connected wind farms through voltage restoration using dynamic voltage restorer *International Journal of Renewable Energy Research* **6(1)** 53-60

[32]. Singh RR, Chelliah TR, Khare D and Ramesh US 2016 November. Energy saving strategy on electric propulsion system integrated with doubly fed asynchronous motors *IEEE Power India International Conference* 1-6

[33]. Sujatha K and Punithavathani DS 2018 Optimized ensemble decision-based multi-focus imagefusion using binary genetic Grey-Wolf optimizer in camera sensor networks *Multimedia Tools and Applications* **77(2)** 1735-1759

[34]. Krishnamoorthy S, Punithavathani S and Priya JK 2017 Extraction of well-exposed pixels for image fusion with a sub-banding technique for high dynamic range images *International Journal of Image and Data Fusion* **8(1)** 54-72

[35]. Singh RR, Mohan H and Chelliah TR 2016 November. Performance of doubly fed machines influenced to electrical perturbation in pumped storage plant-a comparative electromechanical analysis *IEEE 7th India International Conference on Power Electronics* 1-6

Analysis and Scope of Hybrid Renewable Energy System

Prajna Jandial
Vellore Institute of Technology, Vellore, India

prajna.jandial2019@vitstudent.ac.in

Abstract. Shifting to Renewable Energy Sources has become the need of the hour, but the dependency of these sources on natural, uncontrollable factors, poses a great challenge. A stable power supply is necessary for a hundred percent shift to renewable sources of energy. This reliability can be increased in many ways, but here in focus, are Hybrid Renewable Energy Systems (HRES). HRES is an attempt to make a stable energy source that is clean and sustainable. The paper in detail presents the scope of HRESs keeping India in focus.

1. Introduction

Non-renewable sources of energy by definition, are sources of energy that are limited. Meaning, these sources will not be replenished in many lifetimes. Non-renewables mainly refer to coal, oil, petroleum, natural gas, and also include nuclear and biomass. Most power plants that are used to generate electricity on a large-scale burn fossil fuel [1-10]. Being limited is not the only drawback of these fossil fuels, as they also cause a lot of pollution and add to the ever-increasing climate problem. Climate change is real, and its effects, shifting weather patterns and rising sea levels have never been more evident. A fundamental change in the way we generate electricity has become necessary [11-20].

Renewable energy, as the name suggests, replenishes quickly. These sources are more sustainable and the key to solving the climate problem. The need for power in our day to day lives will only increase with technological advancement. Hence, renewables are an obvious solution. To name a few, solar and wind are widely used sources of renewable energy. Though they do come with their own set of problems. As these sources of energy are heavily dependent on natural factors, that humans cannot control and often cannot predict, these systems become very unreliable [21-30].

A hybrid energy system has two or more renewable energy sources working together increasing efficiency [31-35]. The objective is to make an overall system that can be relied upon to generate clean electricity. About 5,000 trillion kWh per year solar energy is incident over India's land area with most parts receiving 4-7 kWh per sq. m per day, and wind power potential of 302 GW at 100 meter above ground level is present [6]. With HRES this potential could utilized better. This paper will discuss the potential and feasibility of Hybrid Renewable Energy Systems.

IOP Conf. Series: Materials Science and Engineering **937** (2020) 012030 doi:10.1088/1757-899X/937/1/012030

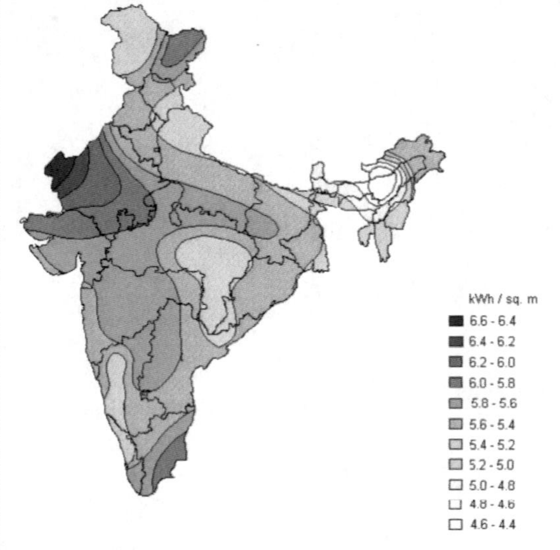

Figure 1. Solar radiations on India [1].

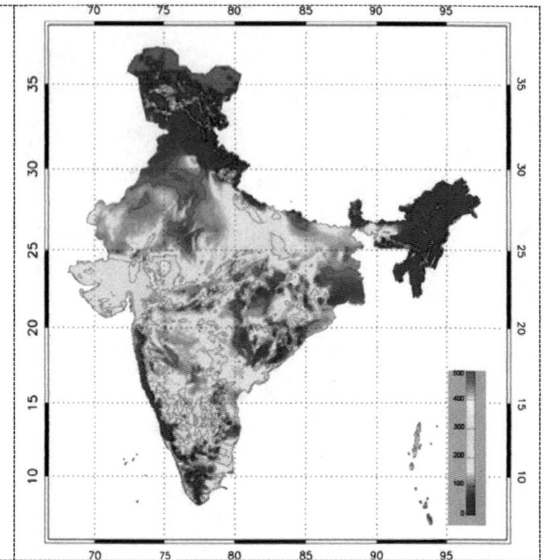

Figure 2. Wind power density map at 80 m (W/m2)[2].

2. Present Status of energy in India:

India has an estimated renewable energy potential of about 900 GW from sources like Wind – 102 GW, Bio-energy – 25 GW, Small Hydro – 20 GW, and Solar power – 750 GW. Here is a table showing Indian power generation as of April 2020.

Table 1. Power generation in India as of April 2020 [3]

Fuel	Megawatt (MW)	% of total
Coal	1,98,525	54.2
Lignite	6,610	1.7
Gas	24,955	6.7
Diesel	510	0.1
Hydro	45,699	12.4
RES	87,269	23.6
Nuclear	6780	1.9

From the given table it is clear that 62.8% of the energy produced in India comes from thermal generation, that is 2,30,600 MW. There is a lot of renewable potential in India that is currently not being used. For a developing country, electricity plays a key role in enabling the progress of technology and infrastructure. Shifting to renewables will not only accelerate the progress of India, but also facilitate in a more sustainable nation in the long run.

India's electricity sector is very active when it comes to utilization of RES. As of December 2011, India produced 22.4 GW electricity which was beyond the total installed electricity in Austria by all technologies. India was also the first country in the world to set up a Ministry of non-conventional energy resources in early 1980 [4], [5].

3. Problems with RES:

Power generation becomes very much dependant on natural conditions in the case of RES. For example, in the case of solar panels, if it rains or if it is cloudy, the power output will be low. As humans cannot control these conditions, the integration of these sources into the main grid has become a challenge. Other challenges with RES are the dependence on optimum location and large areas of land required for set up. The cost of installation is also high [7].

4. Hybrid Renewable Energy System (HRES):

Hybrid Renewable Energy Systems are setups that integrate two or more renewable energy sources to get a more reliable system. These systems are location dependant and accordingly try to maximize output.

4.1. Different combination of HRES

Here are some examples of HRES combinations.

4.1.1. Solar-wind Solar-Wind HRES is an integrated system that uses solar panels and wind turbines together to generate electricity. Here is the basic setup of a solar-wind HRES.

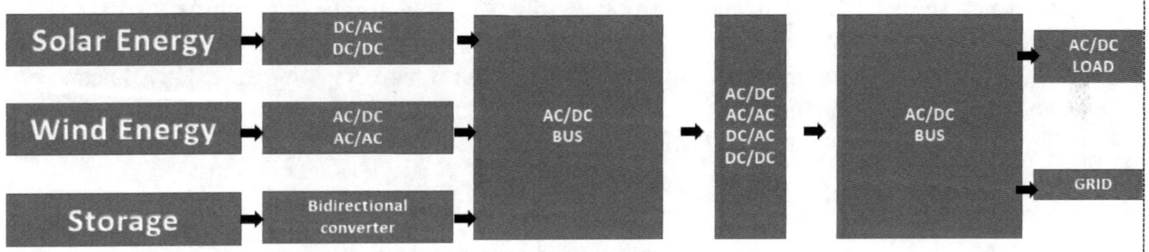

Figure 3. Solar-wind basic component [8].

4.1.2. Solar-Biomass A combination of solar energy and biomass energy HRES. This system can have further applications where it is used to in desalination and other applications. This hybrid system has a lot of potential in electrifying rural areas. With its added applications it can accelerate development immensely [8].

4.1.3. Hydro-Wind A wind-hydro system generates electric energy combining wind turbines and pumped storage. These systems can also be used to meet water supply needs, by using a wind turbine to provide energy for the pumping system, national grid need not be used [9]. Hydro-wind pumped storage power stations are also being developed which could increase wind penetration in islands are also being developed, and discussed in many research papers [10], [11].

4.1.4. Solar-Wind-Biomass Any type of combination can be developed according to needs. A combination of all three according to the location can also be feasible. For any HRES this basic flowchart is applicable.

IOP Conf. Series: Materials Science and Engineering **937** (2020) 012030 doi:10.1088/1757-899X/937/1/012030

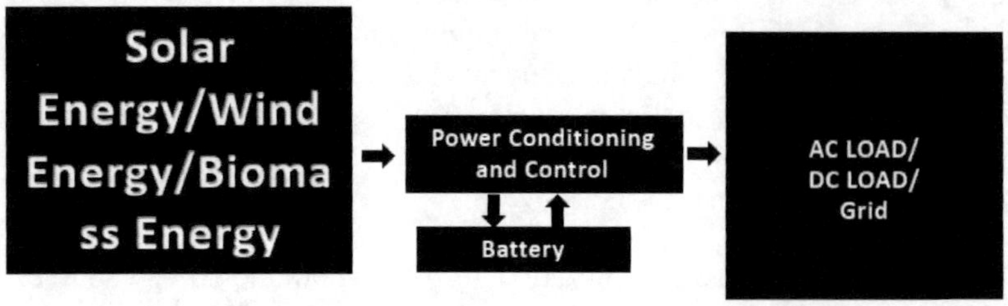

Figure 4. Flowchart [12].

This kind of system, where three sources are used can be installed in remote places where it is not feasible to install power grid. It can give independent and constant electricity to that region.

4.2. Scope of HRES in India

India is a country where 75% of the population lives in rural areas. These areas do not have adequate electricity, water supply, education, and transport. With HRES, electricity can become easily accessible in rural India. Using systems that integrate rain-water harvesting can also help overcome the water problem. Developing villages has been a challenge for India for a very long time. HRESs can be the solution to many of the problems.

Here is the location-wise suitability of HRES.

Table 2. location-wise HRES recommendation [13]

Geographical feature	Type of HRES	Recommendation
High Altitude	Biomass-wind-fuel cell, photovoltaic-wind, photovoltaic-biomass	Photovoltaic-biomass
Mountain	Biomass-wind-fuel cell, photovoltaic-wind, photovoltaic-biomass	Photovoltaic-biomass
Plain	Photovoltaic-biomass, hydro wind, solar flower, combined HRES plant, biomass wind, photovoltaic-wind	Combined HRES plant
Semi Desert	Wind-fuel cell, wind-photovoltaic, wind biomass, photovoltaic-biomass, photovoltaic wind-biomass	Photovoltaic-Wind-Biomass
Desert	Wind-fuel cell, wind-photovoltaic, wind biomass, photovoltaic-biomass, photovoltaic wind-biomass	Photovoltaic-Wind-Biomass

The given table is just a recommendation and based on the location a more specific HRES can be applicable.

There are limitations to HRESs as well. There will be converter loss associated, and also a cost of establishment. As HRESs are still upcoming there is a lot of research that can be done to improve these systems.

India with its diversity in weather conditions could benefit a lot with such systems, specially in remote areas where installation and the reliability of the national grid is often not feasible. These systems can independently function and because of their increased reliability can power regions effectively. HRES also exploit India RES potential more effectively, and utilisation of such systems could benefit in long term sustainability of the country.

5. Conclusion:

This paper established what Hybrid Renewable Energy Systems are, and analysed their potential. There is still a lot optimisation required in hybrid systems. Overall, these systems could make Renewable Energy Sources more reliable and help in their integration to the main grid. Rural India could benefit a lot with HRESs and see immense growth and development. HRESs could be very beneficial for remote regions, and can have further applications in water supply, desalination and rain-water harvesting. Systems that could integrate not just power generation, but also have more functionality (as mentioned above: water supply, desalination and rain-water harvesting), can accelerate rural development in India.

6. Reference:

[1]. Sitharthan R, Geethanjali M and Pandy TKS 2016 Adaptive protection scheme for smart microgrid with electronically coupled distributed generations *Alexandria Engineering Journal* **55(3)** 2539-2550

[2]. Fathima AH, and Palanisamy K 2014 Battery energy storage applications in wind integrated systems—a review *IEEE International Conference on Smart Electric Grid* 1-8

[3]. Prabaharan N and Palanisamy K 2015 Investigation of single-phase reduced switch count asymmetric multilevel inverter using advanced pulse width modulation technique *International Journal of Renewable Energy Research* **5(3)** 879-890.

[4]. Jerin ARA, Kaliannan P and Subramaniam U 2017 Improved fault ride through capability of DFIG based wind turbines using synchronous reference frame control based dynamic voltage restorer. *ISA transactions* **70** 465-474

[5]. Sitharthan, R, Sundarabalan CK, Devabalaji KR, Nataraj SK and Karthikeyan M 2018 Improved fault ride through capability of DFIG-wind turbines using customized dynamic voltage restorer *Sustainable cities and society* **39** 114-125

[6]. Prabaharan N and Palanisamy K 2016 A single-phase grid connected hybrid multilevel inverter for interfacing photo-voltaic system *Energy Procedia* **103** 250-255

[7]. Palanisamy K, Mishra JS, Raglend IJ and Kothari DP 2010 Instantaneous power theory based unified power quality conditioner (UPQC) *IEEE Joint International Conference on Power Electronics, Drives and Energy Systems* 1-5

[8]. Sitharthan R and Geethanjali M 2017 An adaptive Elman neural network with C-PSO learning algorithm-based pitch angle controller for DFIG based WECS Journal of Vibration and Control **23(5)** 716-730

[9]. Sitharthan R and Geethanjali M 2015 Application of the superconducting fault current limiter strategy to improve the fault ride-through capability of a doubly-fed induction generator–based wind energy conversion system *Simulation* **91(12)** 1081-1087

[10]. Sitharthan R, Karthikeyan M, Sundar DS and Rajasekaran S 2020 Adaptive hybrid intelligent MPPT controller to approximate effectual wind speed and optimal rotor speed of variable speed wind turbine *ISA transactions* **96** 479-489

[11]. Sitharthan R, Devabalaji KR and Jees A 2017 An Levenberg–Marquardt trained feed-forward back-propagation based intelligent pitch angle controller for wind generation system *Renewable Energy Focus* **22** 24-32
[12]. Sitharthan R, Sundarabalan CK, Devabalaji KR, Yuvaraj T and Mohamed Imran A 2019 Automated power management strategy for wind power generation system using pitch angle controller *Measurement and Control* **52(3-4)** 169-182
[13]. Sundar DS, Umamaheswari C, Sridarshini T, Karthikeyan M, Sitharthan R, Raja AS and Carrasco MF 2019 Compact four-port circulator based on 2D photonic crystals with a 90° rotation of the light wave for photonic integrated circuits applications *Laser Physics* **29(6)** 066201
[14]. Sitharthan R, Parthasarathy T, Sheeba Rani S and Ramya KC 2019. An improved radial basis function neural network control strategy-based maximum power point tracking controller for wind power generation system *Transactions of the Institute of Measurement and Control* **41(11)** 3158-3170
[15]. Rajesh M and Gnanasekar JM 2017 Path observation based physical routing protocol for wireless ad hoc networks *Wireless Personal Communications* **97(1)** 1267-1289
[16]. Palanisamy K, Varghese LJ, Raglend IJ and Kothari DP 2009. Comparison of intelligent techniques to solve economic load dispatch problem with line flow constraints *IEEE International Advance Computing Conference* 446-452
[17]. Sitharthan R, Ponnusamy M, Karthikeyan M and Sundar DS 2019 Analysis on smart material suitable for autogenous microelectronic application *Materials Research Express* **6(10)** 105709
[18]. Rajaram R, Palanisamy K, Ramasamy S and Ramanathan P 2014 Selective harmonic elimination in PWM inverter using fire fly and fireworks algorithm *International Journal of Innovative Research in Advanced Engineering* **1(8)** 55-62
[19]. Sitharthan R, Swaminathan JN and Parthasarathy T 2018 March. Exploration of wind energy in India: A short review *IEEE National Power Engineering Conference* 1-5
[20]. Karthikeyan M, Sitharthan R, Ali T and Roy B 2020 Compact multiband CPW fed monopole antenna with square ring and T-shaped strips *Microwave and Optical Technology Letters* **62(2)** 926-932
[21]. Sundar D Sridarshini T, Sitharthan R, Madurakavi Karthikeyan, Sivanantha Raja A, and Marcos Flores Carrasco 2019 Performance investigation of 16/32-channel DWDM PON and long-reach PON systems using an ASE noise source *In Advances in Optoelectronic Technology and Industry Development: Proceedings of the 12th International Symposium on Photonics and Optoelectronics* 93
[22]. Sitharthan R and Geethanjali M 2014 Wind Energy Utilization in India: A Review *Middle-East J. Sci. Res.* **22** 796–801 doi:10.5829/idosi.mejsr.2014.22.06.21944
[23]. Sitharthan R and Geethanjali M 2014 ANFIS based wind speed sensor-less MPPT controller for variable speed wind energy conversion systems *Australian Journal of Basic and Applied Sciences* **8**14-23
[24]. Jerin ARA, Kaliannan P, Subramaniam U and El Moursi MS 2018 Review on FRT solutions for improving transient stability in DFIG-WTs *IET Renewable Power Generation* **12(15)** 1786-1799
[25]. Prabaharan N, Jerin ARA, Palanisamy K and Umashankar S 2017 Integration of single-phase reduced switch multilevel inverter topology for grid connected photovoltaic system *Energy Procedia* **138** 1177-1183
[26]. Rameshkumar K, Indragandhi V, Palanisamy K and Arunkumari T 2017 Model predictive current control of single phase shunt active power filter *Energy Procedia* **117** 658-665

[27]. Fathima AH and Palanisamy K 2016 Energy storage systems for energy management of renewables in distributed generation systems *Energy Management of Distributed Generation Systems* 157
[28]. Rajesh M 2020 Streamlining Radio Network Organizing Enlargement Towards Microcellular Frameworks *Wireless Personal Communications* 1-13
[29]. Subbiah B, Obaidat MS, Sriram S, Manoharn R and Chandrasekaran SK 2020 Selection of intermediate routes for secure data communication systems using graph theory application and grey wolf optimisation algorithm in MANETs IET *Networks* doi:10.1049/iet-net.2020.0051
[30]. Singh RR and Chelliah TR 2017 Enforcement of cost-effective energy conservation on single-fed asynchronous machine using a novel switching strategy *Energy* **126** 179-191
[31]. Amalorpavaraj RAJ, Palanisamy K, Umashankar S and Thirumoorthy AD 2016 Power quality improvement of grid connected wind farms through voltage restoration using dynamic voltage restorer *International Journal of Renewable Energy Research* **6(1)** 53-60
[32]. Singh RR, Chelliah TR, Khare D and Ramesh US 2016 November. Energy saving strategy on electric propulsion system integrated with doubly fed asynchronous motors *IEEE Power India International Conference* 1-6
[33]. Sujatha K and Punithavathani DS 2018 Optimized ensemble decision-based multi-focus imagefusion using binary genetic Grey-Wolf optimizer in camera sensor networks *Multimedia Tools and Applications* **77(2)** 1735-1759
[34]. Krishnamoorthy S, Punithavathani S and Priya JK 2017 Extraction of well-exposed pixels for image fusion with a sub-banding technique for high dynamic range images *International Journal of Image and Data Fusion* **8(1)** 54-72
[35]. Singh RR, Mohan H and Chelliah TR 2016 November. Performance of doubly fed machines influenced to electrical perturbation in pumped storage plant-a comparative electromechanical analysis *IEEE 7th India International Conference on Power Electronics* 1-6

HERIC Inverter- A SEPIC based transformer-less converter design and simulation for isolated standalone PV system

M.D.Udayakumar[1], K.S.Gowthaman[2], Prabhu.A[3] and Dr.A.Nazar Ali[4]

[1,2,3&4] Department of Electrical and Electronics Engineering,
[1]Assistant Professor, K.Ramakrishnan College of Technology, Trichy, Tamilnadu.
[2]Assistant Professor, Government College of Engineering, Thanjavur, Tamilnadu
[3]Assistant Professor, Sardar raja College of Engineering, Alangulam, Tamilnadu.
[4]Associate Professor Rajalakshmi Engineering College, Chennai, Tamilnadu.

E.mail: uday.krct@gmail.com

Abstract. Photovoltaic transformer-less inverter due to its high efficiency, low power loss and cost plays an important role in energy market. In this paper solar PV based 1φ transformer-less HERIC converter for standalone isolated PV system has been designed and analysed. To control the solar PV output, a dc-dc SEPIC converter is modelled and designed with the feature of Maximum Power Point Tracking. Contrary to other inverter topologies H5, H6, NPC (neutral point clamping), HERIC inverter comprises low leakage current. The entire proposed system is designed and simulation is carried out in MATLAB environment.

1. Introduction

Due to increasing demand in electricity, energy has been obtained from various sources such as sun, wind, fossil fuels etc., On account of global warming considerations pollution free energy source usage has increased [1]. The availability of solar energy has utilized to draw electricity. In the energy conversion process power electronic components play a vital role [2].

Photovoltaic inverter contributes a major role in the solar Photovoltaic system overall performance. As the number and size of electronic component expand, the system overall cost increases. So, efforts have taken to overcome the above problem. In order to reduce the size of photovoltaic inverter, transformer-less prototype has emerged recently. Various inverter topology such as full bridge, half bridge, H-bridge, H5, H6, H7 topology, HERIC has designed [3-4]. Transformer based inverter topology produce common mode leakage current when integrated with grid[5]. The leakage current must be within the limits otherwise it led to electromagnetic interference and safety problems. Hence transformer-less inverter topology is developing nowadays to eliminate the leakage current problems. A high performance low leakage current HERIC inverter is used [6-7]. To control the solar PV system output, SEPIC converter is chosen among various dc-dc converter topologies such as cuk converter, buck converter, boost and buck-boost converters. Contrary to boost converter, the non-inverting mode of operation is offered by SEPIC converter. In addition, it is common that both input and output current waveforms are fluctuating in the boost converter but the proposed converter has the advantage of drawing continuous input current with lower pulsation [8-10]. As above mentioned, merits make SEPIC converter more suitable for Solar PV system. The well-known P&O Maximum Power Point Tracking method is used to withdraw the possible maximum power from solar PV panel.

The PV module is designed and simulated in MATLAB/Simulink environment. Figure 1 gives the block diagram of the proposed standalone Photovoltaic system [11-12]. A double frequency SPWM technique is made use for modulation of the inverter with different power converter switching combinations.

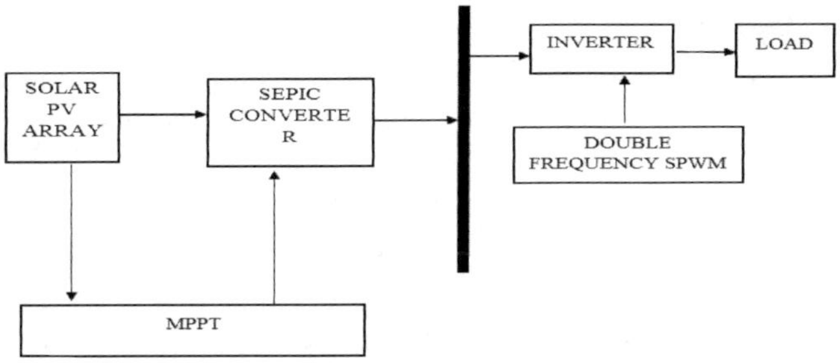

Figure 1: Schematic block of proposed system

2. Proposed system design

The test system is modeled in such a way that the solar PV array with at most power P_{mpp} =1.5 kW for feeding a load under STC condition (25^0 C) of irradiation 900 W/m^2. The module specifications are shown in Table.1.It is necessary to determine [13] the number of PV modules is needed for connecting either in parallel combination or series combination is calculated as indicated by equations (1), (2) and (3).

$$I_{MPP} = \frac{P_{MPP}}{V_{MPP}} = \frac{1500}{120} = 12.5A \tag{1}$$

Equation (1) represents the input current at MPP.

$$N_P = \frac{I_{MPP}}{I_M} = \frac{12.5}{8.21} = 1.2 \approx 2 \tag{2}$$

Equation (2) shows the required number of parallel modules

$$N_S = \frac{V_{MPP}}{V_M} = \frac{120}{26.4} = 4.54 \approx 5 \tag{3}$$

Equation (3) shows the required number of series modules

Thus, a five module series connected and two module parallel connected solar PV array is designed successfully and energy conserved and utilized efficiently [14]

Table 1: Specifications of 200W PV panel

Key Parameters	Numerical Value
Peak power	200 W
V_{MPP}	26.4 V
I_{SC}	8.21A
I_{MPP}	8.215 A

IOP Conf. Series: Materials Science and Engineering **937** (2020) 012031 doi:10.1088/1757-899X/937/1/012031

Number of cells	54
Series resistance	0.39385
Series resistance	412.404

3. SEPIC Converter Design

The SEPIC converter is modeled using MATLAB Simulink software as shown in Figure 2 [15]. For designing of SEPIC converter, the parameters value has been calculated using equations (4), (5), (6) and (7).

The duty cycle can be determined using the equation (4)

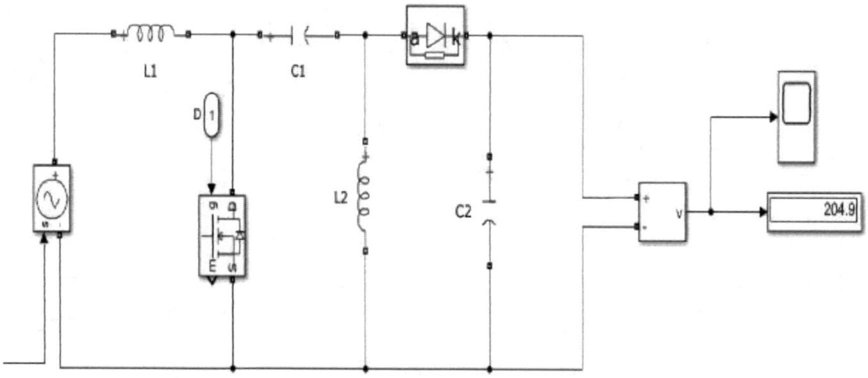

Figure 2: Simulation of SEPIC converter

$$D = \frac{V_{OUT}}{V_{OUT} + V_{IN}} = \frac{250}{250+120} = 0.676 \quad (4)$$

$$I_{OUT} = \frac{P_{MPP}}{V_{OUT}} = \frac{1500}{250} = 6A \quad (5)$$

$$L_{1(MIN)} = L_{2(MIN)} = \frac{1 * V_{IN(MIN)} * D}{2 * \Delta I_L * f_{sw}} = \frac{1*120*0.676}{2*2*8000}$$

$$L_{1(MIN)} = L_{2(MIN)} = 2.028mH \quad (6)$$

If equivalent series resistance is low, then C_1 & C_2 can be determined by equation (7).

$$C_1 = C_2 = \frac{I_{OUT} * D}{\Delta V_{RPL} * f_{sw}} = \frac{6*0.676}{0.01*250*8000}$$

$$C_1 = C_2 = 202.8\mu F \quad (7)$$

IOP Conf. Series: Materials Science and Engineering **937** (2020) 012031 doi:10.1088/1757-899X/937/1/012031

4. Design of MPPT Algorithm

In this article, the well-known MPPT technique Perturb and observe method is employed. Fig. 3 shows the flow chart of operation of Perturb and Observe method. Power variance and voltage variation shall decide the action needed to increase or decrease the reference voltage. It can be observed by varying the duty ratio of PWM signal given for the purpose of switching the dc-dc converter. The proposed method is simulated and hardware is implemented [16].

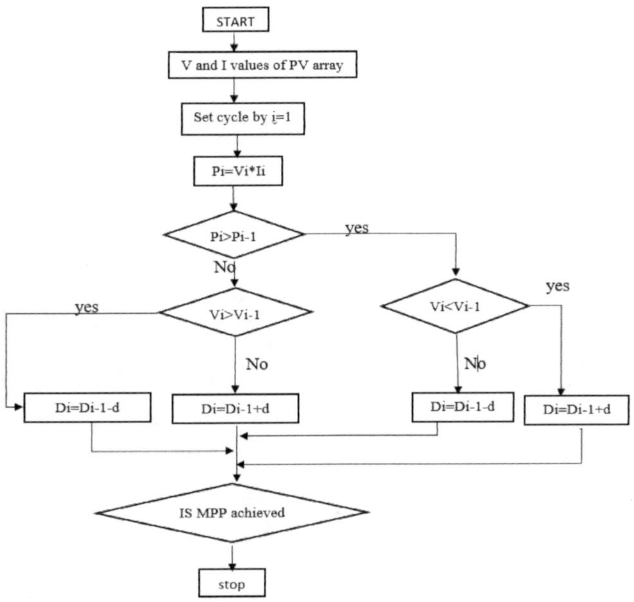

Figure 3: Flowchart for Perturb and Observe MPPT technique

5. HERIC Inverter Topology

The simulation diagram of HERIC topology is shown in figure 4. HERIC inverter with additional switches provides isolation between Solar PV array and grid. The reactive power exchange is prevented which increases the efficiency of inverter and performance and reduces complexity problem in inverter [17].

Table 2: Parameters for designing SPWM

Key Parameters	Numerical Value
Carrier wave amplitude	1V
Carrier frequency	8kHz
Reference wave amplitude	0.8V
Reference wave frequency	50Hz

Table 2 gives the parameters value for designing the sinusoidal pulse width modulation technique which generates pulses required for triggering the inverter.

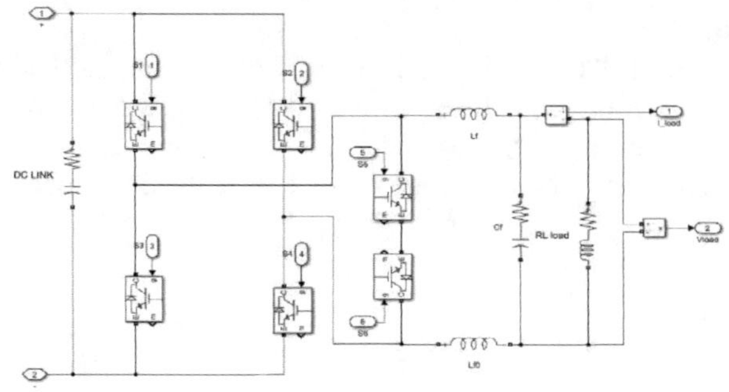

Figure 4: Simulation model of HERIC inverter

6. Simulation Results

The PV system with battery storage based single-phase transformer-less HERIC inverter with R Load is designed and simulated by MATLAB Simulink [18]. Fig.5 gives the output values of solar PV voltage as well as current of 120 V and 12.5 A respectively as designed for 1.5KW.

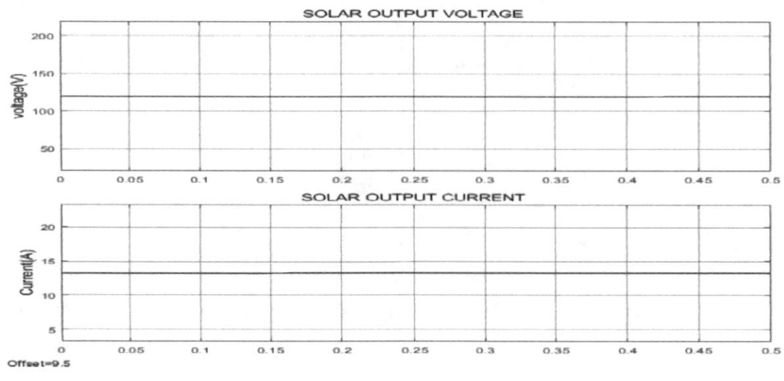

Figure 5: Output waveforms

As shown in fig.3, the simulation is done at 67% duty cycle for SEPIC converter and the shape of output waveforms reveal that for the input values from the designed solar panel with MPPT. The output voltage is constant at 250V expected voltage level and does not differ with input variations. In addition, there has been a substantial decrease of overshoot has been noted [19-20].

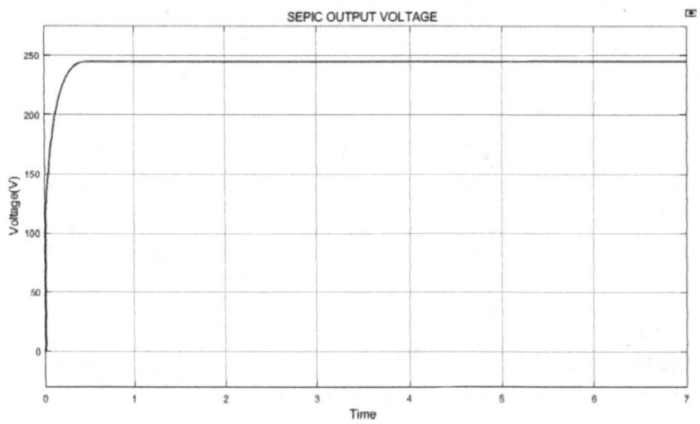

Figure 6: Output voltage from SEPIC converter

Fig.7 shows the output waveforms for the proposed inverter with R load. It shows due to the effects of output voltage drop of SEPIC converter, there is a change in inverter output. The SEPIC converter mainly focuses on extracting maximum power from PV system.

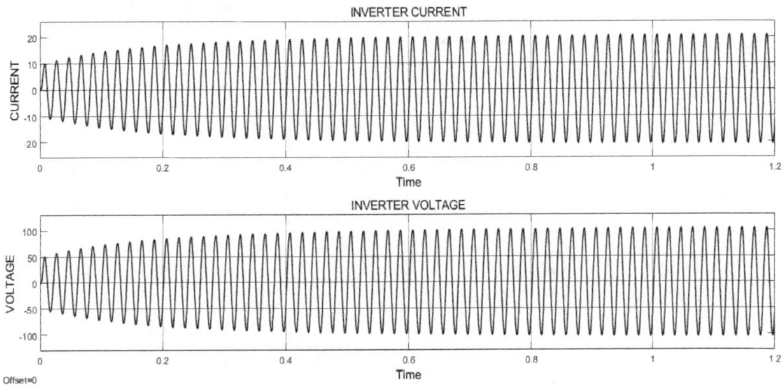

Figure 7: Output waveforms For R Load

7. Conclusion

The proposed model of solar PV system with HERIC inverter has been designed and simulation result has been obtained. The solar PV array output is regulated using MPPT based SEPIC converter for variation in temperature and irradiation. It is observed that the variation in solar radiation and temperature has influence on output of HERIC inverter.

8. References

[1] S. Lee and H. Do, "Zero-Ripple Input-Current High-Step-Up Boost–SEPIC DC–DC Converter With Reduced Switch-Voltage Stress," in IEEE Transactions on Power Electronics, vol. 32, no. 8, pp. 6170-6177, Aug. 2017, doi: 10.1109/TPEL.2016.2615303.

[2] S. Padmanabhan and K. Kaliyappan, "High Performance MPPT Based on Variable Speed Generator Driven by Wind Power Generation in Battery Applications," Journal of Electrical Engineering and Technology, vol. 9, no. 1, pp. 205–213, Jan. 2014.

[3] S. V. Araujo, P. Zacharias and R. Mallwitz, "Highly Efficient Single-Phase Transformerless Inverters for Grid-Connected Photovoltaic Systems," in IEEE Transactions on Industrial Electronics, vol. 57, no. 9, pp. 3118-3128, Sept. 2010.

[4] K. Gupta, M. S. Joshi and V. Agarwal, "On the control and design issues of single phase transformerless inverters for photovoltaic applications," 2014 IEEE 6th India International Conference on Power Electronics (IICPE), Kurukshetra, 2014, pp. 1-6, doi: 10.1109/IICPE.2014.7115805.

[5] T. Esram and P. L. Chapman, "Comparison of Photovoltaic Array Maximum Power Point Tracking Techniques," in IEEE Transactions on Energy Conversion, vol. 22, no. 2, pp. 439-449, June 2007.

[6] A. T. Sankara Subramanian, P. Sabarish, M. D.Udayakumar and T. Vishnu kumar," Performance Analysis of Various Photovoltaic Configurations Under Uniform Shading and Rapid Partial Shading Formations", Biosc.Biotech.Res. Comm. Special Issue Vol 13 No (3) 2020 Pp-185-192.

[7] B.Karthikeyan, R. Ramkumar, L. Nagarajan, A. Anton Amala Praveen and S. Vijayalakshmi," ANN Based MPPT Controller for PEMFC System with Interleaved Resonant PWM High Step up Converter" Biosc.Biotech.Res.Comm. Special Issue Vol 13 No (3) 2020 Pp-44-47.

[8] R.Jai Ganesh ,S.Kodeeswaran, M.Kavitha, T.Ramkumar, Performance analysis of piezoelectric energy harvesting system employing bridgeless power factor correction boost rectifier", https://doi.org/10.1016/j.matpr.2020.02.085.

[9] Dr.A. Rajkumar, R. Jai Ganesh, V. Suresh Kumar, T. Vishnu Kumar and T. Ram Kumar," Implementation of High Efficient Single Input Triple output DC-DC Converter," Biosc.Biotech.Res.Comm. Special Issue Vol 13 No (3) 2020 Pp-48-55.

[10] S.R.Paveethra, S.Vijayalakshmi, C. Kalavalli, S.Murugesan and Dr.S.Jeyasudha," MPPT based Harmonic Reduction in PV Module for Grid Connected System," Biosc. Biotech.Res.Comm. Special Issue Vol 13 No (4) 2020 Pp-43-48.

[11] A Nazar Ali, D Sivamani, R Jaiganesh M Pradeep (2019), Solar powered air conditioner using BLDC motor, IOP Conference Series: Materials Science and Engineering, vol. 23.

[12] V Venkatesh, A Nazar Ali,, R Jaiganesh. V Indiragandhi (2019), Extraction and conversion of exhaust heat from automobile engine in to electrical energy Energy, IOP Conference Series: Materials Science and Engineering, vol. 23.

[13] A Ali Nazar, R Jayabharath, MD Udayakumar (2014) ,An ANFIS Based Advanced MPPT Control of a Wind-Solar Hybrid Power Generation System, international review of modeling and Simulations. vol.7, no. 4, pp. 638–643.

[14] A.Nazar Ali and R. Jayabharath (2014), "Performance Enhancement of Hybrid Wind/Photo Voltaic System Using Z Source Inverter with Cuk-sepic Fused Converter." Research Journal of Applied Sciences, Engineering and Technology7.19 pp. 3964-3970.

[15] Subramanian, A.T sankar, P. sabarish and A. Nazar Ali , "*A Power factor correction based Canonical switching cell converter for VSI fed BLDC motor by using voltage follower Technique.*"Electrical, Instrumentation and communication Engineering (ICEICE),2017 IEEE International conference on IEEE, 2017

[16] C.Kalavalli, S.R.Paveethra, S.Murugesan, Dr.A.Nazar Ali, (2020), Design And Implementation Of High Efficiency H6 PV Inverter with Dual Axis Tracking, International Journal of Scientific & Technology Research, Vol 9,issue 02, pp. 4728-31.

[17] Shyam D, Premkumar K, Thamizhselvan T, Nazar Ali A, Vishnu Priya M(2019),Symmetrically Modified Laddered H-Bridge Multilevel Inverter with Reduced Configurationally Parameters , International journal of engineering and advanced technology, Vol 9, issue 1.

[18] S.R.Paveethra, C.Kalavalli, S.Vijayalakshmi, Dr.A.Nazar Ali, D.Shyam(2020), Evaluation Of Voltage Stability Of Transmission Line With Contingency Analysis, International Journal of Scientific & Technology Research, Vol 9,issue 02, pp. 4018-22

[19] Premkumar, Kamaraj et al. (2018), Antlion Algorithm Optimized Fuzzy PID Supervised On-line Recurrent Fuzzy Neural Network Based Controller for Brushless DC Motor, Electric Power Components and Systems, 45, 20, pp.2304-2317.

[20] K Premkumar et al. (2018), Novel bacterial foraging-based ANFIS for speed control of matrix Converter-fed industrial BLDC motors operated under low speed and high torque, Neural Computing and Applications, 29, 12, pp.1411–1434.

Vitality Burglary Recognition and Issue Investigation Dependent on IOT

R Saranraj[1], Aruljothi Kuppannan[2], V Sunil Kumar[3], M Kishore[4], S Lalith[5]

[1] Assistant Professor/ EEE, K. Ramakrishnan college of Engineering, Trichy, India.
[2] Assistant Professor/ EEE, J.K.K.N college of Engineering & Technology, Erode, India.
[3] IV Year/EEE Department, K. Ramakrishnan college of Engineering, Trichy, India.
[4,5] III Year/EEE Department, K. Ramakrishnan college of Engineering, Trichy, India.

Abstract. The proposed idea on this paper pursuits on the format and usage rendition of electrical vitality theft recognition and fault assessment. The unlawful use of intensity must be fathomed through advanced way, with no human cooperation. The reason for this work is to give a usage approach to control burglary discovery and controlling which licenses violators to be recognized at a faraway region. This design integrates various approach for problems faced by way of energy distribution device inclusive of power theft and transmission line fault. It consists of IOT era to discover the electrical theft and transmission line fault. In the proposed approach we offer alert to the authorized electricity issuer via an alert message which eliminates the diverse problems associated with the meter studying and theft detection and it blocks the energy deliver to the specific vicinity.

Keywords: **IOT, Potentiometer, Arduino Uno R3.**

1. Introduction

Generally electrical power involve many operational losses in various stages from supplier to consumer. We can described the losses in technically however distribution and transmission losses can't be actually evaluated with the end realities. This represents the inclusion of nontechnical parameter in transmission and conveyance of power. In addition specialized misfortunes emerge clearly and are hastened because of influence scattering in transmission lines, transformers. The gadget forestalls the unlawful use of power. At this factor of mechanical improvement the issue of unlawful use of vitality can be illuminated with no human control utilizing IOT. It consists of IOT approach to discover the electrical theft and transmission line fault. Moreover gathering the meter readings for billing approaches from all clients is a difficult and time consuming mission which requires a tremendous variety of labors. By implementing our challenge we will locate the exact region of the fault and theft. In all likelihood it saves time to discover the vicinity and reduces the labor value.

2.Design Components

The components used are

- Arduino Uno
- LCD display 2x16

IOP Conf. Series: Materials Science and Engineering **937** (2020) 012032 doi:10.1088/1757-899X/937/1/012032

- Step down transformer
- LED ,bulb
- IOT module
- Potentiometer
- Relay

Arduino Uno R3

Arduino is an open-supply device with programming layout, client arrange the plans and makes microcontrollers and microcontroller packs for building advanced gadgets. Arduino board plans utilize a choice of microchips and controller.

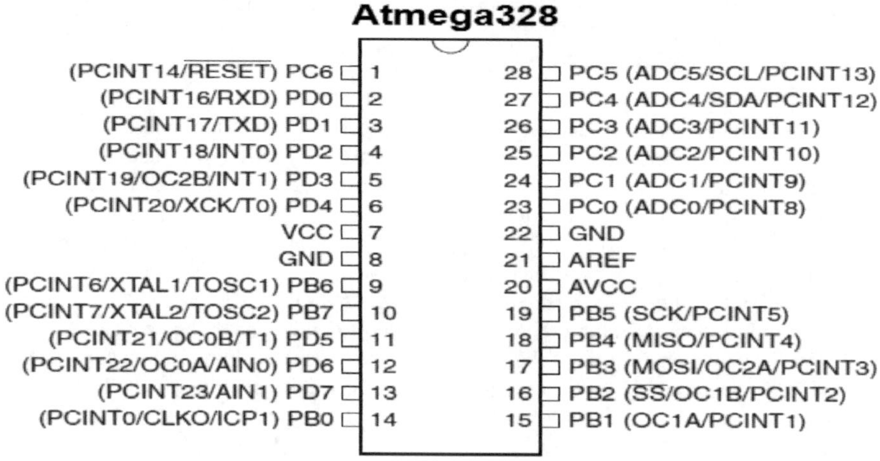

Figure 1. Arduino Uno Pin Diagram

Internet of things (IOT)

The web of things or IOT, is a framework that is connected between the gadgets, simple, mechanical and computerized machines, contraptions which are furnished with specific identifiers (UIDs) and the possibility to sending insights over a system without interactions. The IOT defines "Network of Internet related item able to gather and transfer records". IOT is the concept of connecting any device with an ON and OFF transfer to the net after which provide an appropriate output. .

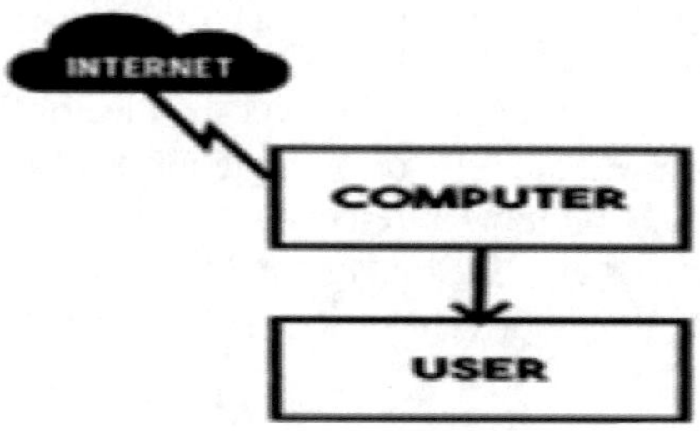

Figure 2. IOT Interface

3.Block Diagram

The below diagram illustrates the general block diagram to analyze theft and fault analysis.

IOP Conf. Series: Materials Science and Engineering **937** (2020) 012032 doi:10.1088/1757-899X/937/1/012032

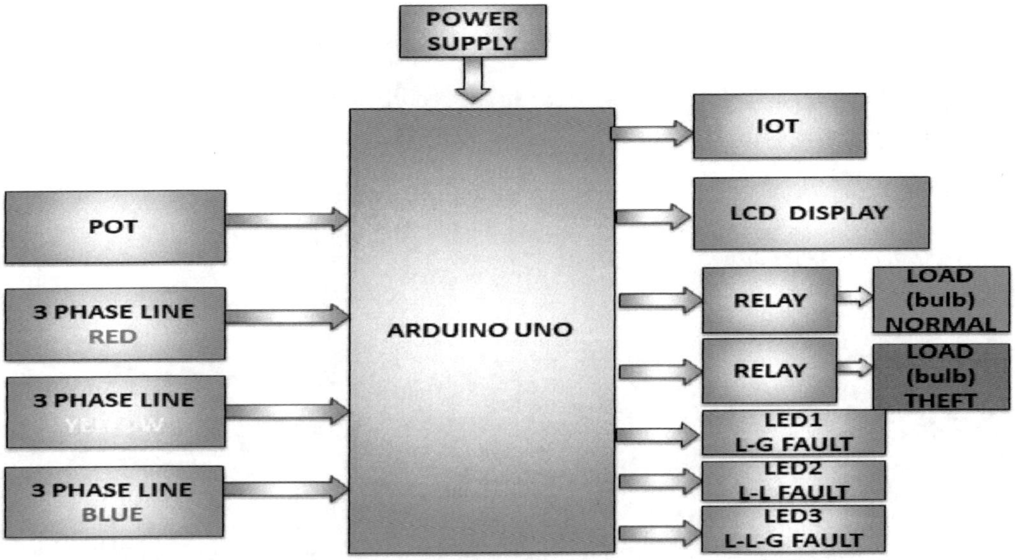

Figure 3. Block Diagram

Energy theft detection and fault analysis based on IOT. This measures the faults in transformers LED is used for indication of faults in transformer. In this system energy theft detection implemented using POT. The condition about this system is updated to the web server using IOT.

The working process is initiated by supplying an AC voltage. The applied AC voltage is given to the step down transformer which coverts 220v to 12 v AC supply. Thus the working process is categorized into two different cases.

Case 1- Energy theft

By taking the AC supply it is given to the Power supply unit. The power supply unit contains the bridge rectifier .the output from the bridge rectifier is given to the relay. As power supply to load is normal it displays **(P-N) POWER –NORMAL.** When an another external load is acted upon the same supply (it means that theft) then it detect us that energy theft has been happened and it displays **(P-T-D) POWER –THEFT DETECTED.** And suddenly the power supply is disconnected form the particular location. And the location data is provided by IOT.

Case 2-Fault analysis

There are mainly three faults occurs in receiving end [load side] namely L-G, L-L & L-L-G faults [L refers line and G refers Ground]. Around 70 percent of deficiencies are of L-G Fault. By using Arduino we can detect what is the nature of the fault and location of the fault. The installation of the device can be installed in home, distribution units.

4.Mathematical Analysis

On each occasion the input energy passes from provider to the consumer and the entire amount of electricity gets by means of the receiver are not equal indicates a possibility of robbery of energy. Following mathematical evaluation happens in the comparison unit of the theft detection device.

Total sent power = Total consumed power + Loss (No Energy Theft)
Total sent power ≠ Total consumed power + Loss (Energy Theft Occurs)
Where,
Sent power refers Meter facts of the strength introduced to the patron ,
Consumed power refers Meter records of the electricity consumed by using a patron

5.Software Results

The below image shows the simulation setup for the project .this is done by using EMBEDDED C in PROTEUS

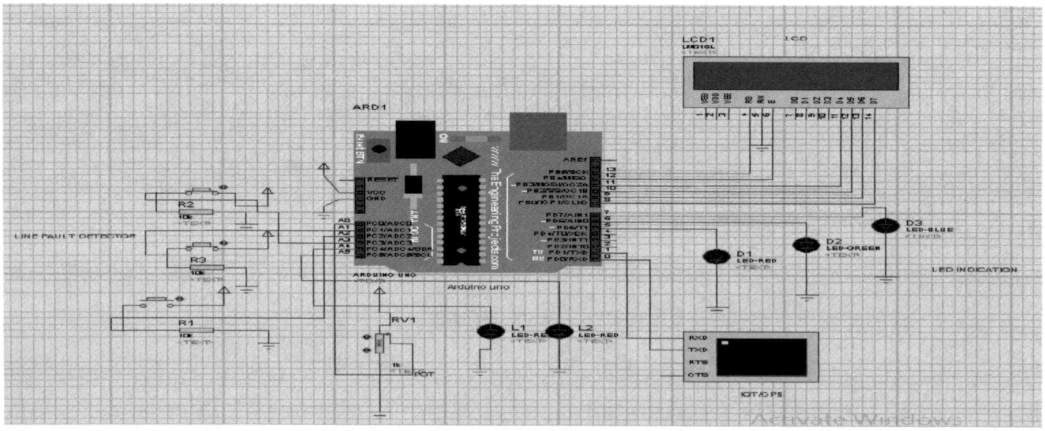

Figure 4. Simulation Diagram-1

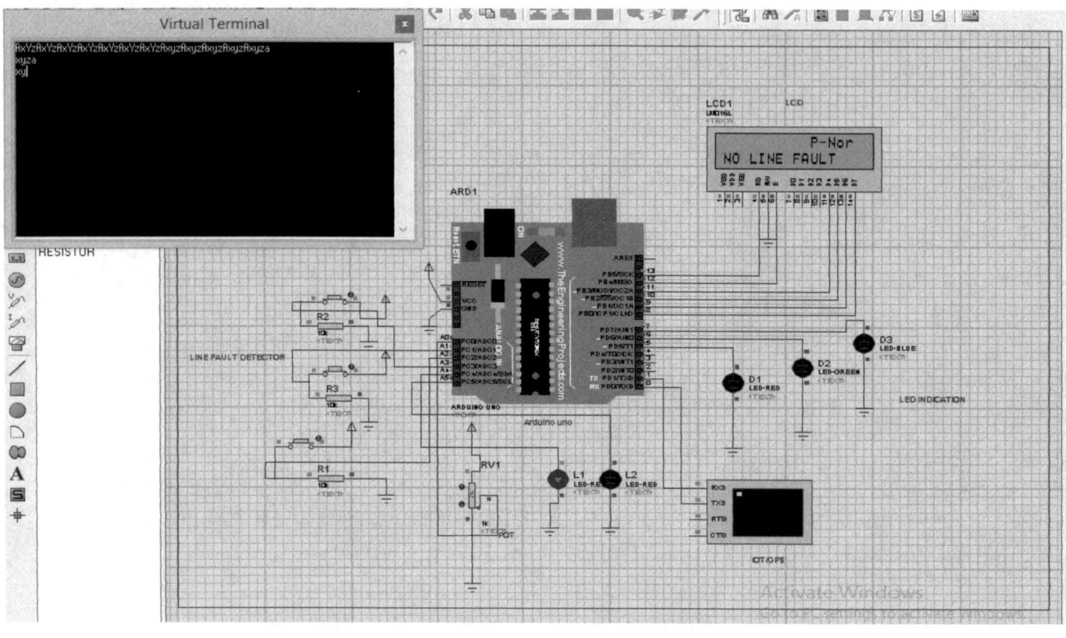

Figure 5. Simulation Diagram with Output

IOP Conf. Series: Materials Science and Engineering **937** (2020) 012032 doi:10.1088/1757-899X/937/1/012032

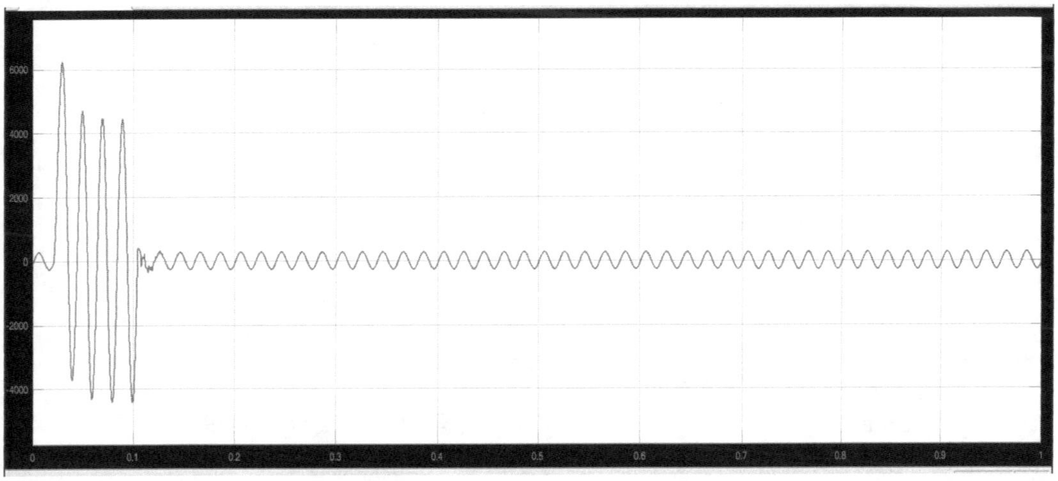

Figure 6. Current Waveform during Normal Condition

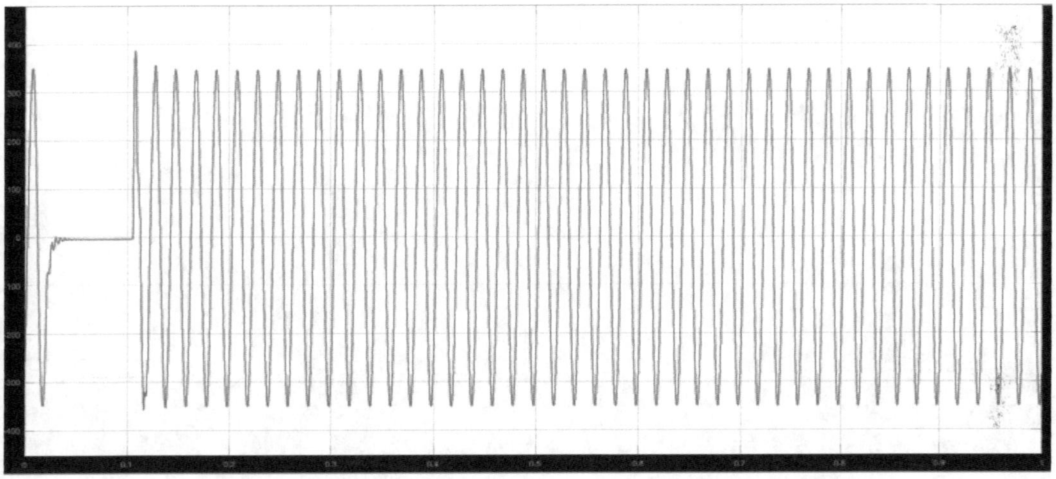

Figure 7. Current Output Waveform on Fault Checking Condition

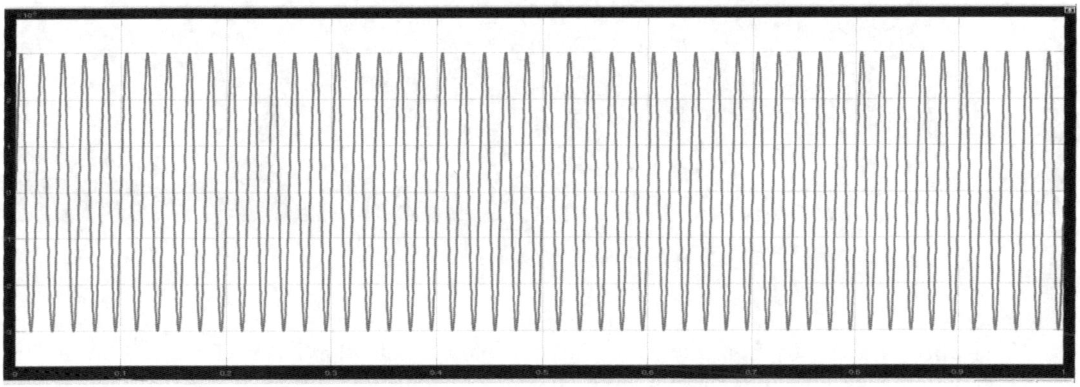

Figure 8. Voltage Waveform during Normal Condition

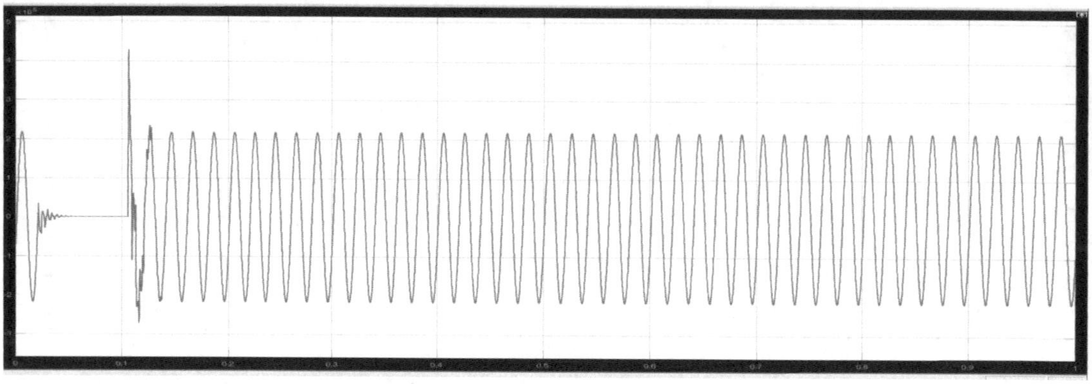

Figure 9. Voltage Output Waveform on Fault Checking Condition

6.Conclusion

With help of robbery detection gadget, energy Distribution Company will conquer the unlawful power theft and will deliver profit to Distribution Company. As well as we are able to locate the fault vicinity without difficulty between the transmission line and the nature of the fault occurred

Figure 10. Hardware Setup

References

[1] Dhayalini. K, 2016, Optimization Of Wind-Thermal Coordination Dispatch Using Flower Pollination Algorithm, International Journal of Advancement in Chemistry, vol. 12, no.16, pp. 4963 – 4970.

IOP Conf. Series: Materials Science and Engineering **937** (2020) 012032 doi:10.1088/1757-899X/937/1/012032

[2] R. Manivasagam and V. Dharmalingam, "Power quality problem mitigation by unified power quality conditioner: An adaptive hysteresis control technique," International Journal of Power Electronics, vol. 6, pp. 403-425, 2014.
[3] A.Subramaniya siva ,M.Bhavani " Mitgation of harmonics by shunt active power filter using synchronous detection method " IJETT volume 4 issue 6 2013
[4] A.Subramaniya siva, S.Sathiesh kumar & T.Santhosh Kumar " Investigation of Harmonics & optimal Power flow in IEEE 14 bus system using ETAP software" IJSTR vol 9, issue 4 ,2020.
[5] Ronald Armstron-su-gao- Lei Lei —A zero-inventory production and distribution problem with a fixed customer sequence‖, Springer science business Media,LLC 2008.
[6] A.Subramaniya siva, N.Vinothini, S.Sathiesh kumar "Piezoelectric based electric energy generation from Sound energy" IJSTR vol 9, issue 4 ,2020.
[7] R.Illango, T.Santhosh Kumar, S.Sathiesh kumar "Load flow & transient stability analysis for IEEE5 bus system using UPFC" IJSTR vol 9, issue 4 ,2020.
[8] R. Manivasagam and D. Aarthi, "Design of UPFC using ten switch converter with switch reduction," International Journal for Scientific Research & Evelopment, vol. 5, pp. 490-493, 2017.
[9] R.Manivasagam & D Aarthi —Design and Implementation of UPFC Using Ten Switch Converter With Switch Reduction‖, International Journal Of Engineering And Computer Science, vol.6,Issue.6, 2017 ,Page no. 21861- 21869.
[10] Titus Sigamani, Ram Prakash Ponraj " Modified Multi level boost inverter for hybrid energy systems" IJSTR vol 9, issue 4 ,2020.
[11] Manivasagam Rajendran ,Saranraj rajendran "Wind turbine manufacturing supply chain "IJSTR vol 9, issue 3 ,2020.
[12] Mendalka, M.; Gadaj, M.; Kulas, L.; Nyka, K., WSN for intelligent street lighting system, Information Technology (ICIT), 2010 2nd International Conference on , vol., no., p.99,100, 28-30 June 2010.
[13] Sunil Chopra and Peter Meindi, —Supply Chain Management –Strategy Planning and Operation, Third Indian Reprint , Pearson Educaion, New jersey, 2004.
[14] R.Saranraj, P.Vigneshwaran ,C.Dinesh Kumar "Domestic LPG monitoring and regulating Systems using IOT"IJREAM vol 5,issue 12,2020
[15] Michael H.Veatch and Francis De Vericourt "ZeroInventory Conditions for a Two-Part-Type Make-to –stock production system" Kluwer Academic Publishers, Queueing systems Vol 43,2003,pp.251266. in IEEE Xplore Digital Library, pp: 96-99.

IOP Conf. Series: Materials Science and Engineering **937** (2020) 012033 doi:10.1088/1757-899X/937/1/012033

Case study: Solar electric vehicles in India.

Mamta L Prajapati[1], Prof. (Dr.) P J Gundaliya[2], Dr. Anal Sheth[3]

[1]T.F.G.Polytechnic Adipur Kutch, 370205, India
[2]L.D.Engg, Ahmadabad, 380015, India
[3]CEPT University, Ahmadabad, 380009 India
[1] Email id: mistrymamta23@gmail.com
[2]Email id: pjgundaliya@ldce.ac.in
[3]Email id: anal.sheth@gmail.com

Abstract. Over exploitation of fossils fuel to meet humans luxurious in present day life has created imbalance in the ecosystem so to have a balance "Green revolution" is required. It means that there is need of using renewable energy resources instead of non-renewable energy which do not harm the environment and also does not contribute in increasing carbon footprint. In context to this each and every sector in the world is adopting alternative resources (renewable energy) to reduce somehow the degradation of environment so that the ecosystem can be balanced and healthy livelihood is achieved. Seeing the condition of the global environment many researchers are attempting to find technological advances which can efficiently transform this renewable energy into mechanical energy in that they have succeeded India has abundance of resource of renewable energy; amongst them solar energy is available with minimum limitation and in abundance. If transportation sector shifts to renewable energy utilization, a good amount of green house gases(GHG) can be minimized. This paper exemplifies the cases of solar electric vehicles including solar cars, solar e–rickshaws, solar tricycles, solar buses and solar vans, that have been successfully implemented in India. These examples demonstrate that the solar electric vehicles can transform a future scenario on a larger scale for India and also other countries.

Keywords: Renewable Energy, Solar Energy, Eco-System, Electric vehicles

1. Introduction

Our existence depends on energy; without which our sustainability might not exist. Energy is an essential view to individuals' life and is also a key requirement for various sectors, including agribusiness, transportation as well as for industrial exercises that are imperative to the social and financial development of each nation (F.M.Uhammad-Sukkiet al., 2012). A country's development is directly related to its transportation infrastructure. A majority share of the world's oil request is for transit (Richardson, 2013).With ever increasing population, transportation demand in India has also increased. The demand for energy is growing day by day world-over. After the oil crisis in 1973, there is a global movement to resort to alternative sources of renewable resource apart from conventional resources such as coal, gas and petroleum (Salunke et al., 2017). Due to good earning, comfort, door to door facility and user friendliness, many people choose private transport mode which has resulted into increase in numbers of vehicles that are driven by fossil fuels. The move towards privatization of transport mode has resulted in increase in air and noise pollution in India. Even when considering public transport, present day in India it utilizes petrol, diesel, etc which cause a worry for unclean natural and energy security (Sheth and Sarkar, 2017).

2. Renewable energy and Environment

The government bodies' panel on climate change has predicted that the GHG emission could ascend by 25–90% by 2030 in contrast with the year 2000, and the Earth could get hotter by 3°C this century. Indeed, even with a little ascent in temperature of 1–2.5°C, it could actuate genuine impacts, incorporating diminished harvest yields in tropical territories prompting expanded danger of craving and spread of different diseases (Adheesh, 2016). Solar based Electric vehicles (EVs) are one of the most encouraging innovations in the transportation division to decrease GHG outflows in the post-2020 time period (IPCC 2012; IEA 2013).

Further, according to the IPCC the transportation division is blameworthy for roughly 23% of CO_2 emission on the planet. Every single area has utilized electricity as the principle wellspring of energy. As indicated by the focal electrical position 2014, Service of intensity, GOI, India lattice is profoundly dirtied with 950 Gco2/Kwh contrasted with different nations on the planet (Sheth and Sarkar, 2017).Thus it is being realized that the consumption of renewable energy is vital and necessary. Researchers have found out various new technologies to deal with it and many positive findings have been revealed.

India's GHG emission from transportation has ascend from roughly 80 million tons for every year in 1994 to around 119 million tons for every year in 2000. In 2004, the transportation sector in India contributed around 8 percent of the nation's energy based GHG emission. Around 90 percent of these were from street transport, contrasted with a worldwide normal of 72 percent. As compared to other countries, India is far away in utilizing renewable energy. To reduce the pollution level and to balance our ecosystem, there is need to prioritize renewable energy, especially in public transportation sector. India's promise under the Paris Understanding is to decrease the carbon impression of its economy by 33-35% by 2030, when contrasted with the 2005 levels (www.investindia.gov.in). Different positive effects can be expected from the presentation of EVs, including proprietor vehicles working costs, decreased CO_2 discharges.

India was the primary nation on the world to set up a Service of renewable power source asset in mid 1980. Various renewable energy resources available are solar energy, wind energy, hydropower energy, geothermal energy, biomass energy, etc. From point view of renewable energy resources, India has vast energy resource potential especially that of solar energy. Conversely, hydropower, biomass and geothermal energy have been developed for a considerable length of time, and five year development plan for these inexhaustible renewable solar energy were equivalent with traditional energy (fossil fuels). Smart grid technologies have proved a cleaner energy flexibly that is progressively effective, affordable and increasingly feasible (Ellaben et al, 2014). Conventional energy is popular because of ease of utilization and availability. Exploitation of conventional energy resources has come to that extent that these are getting scarce and on verge of extinction. But recent developments in replacing conventional energy by renewable energy are gaining importance. For example, in a small scale to investigate maintainability through utilization of renewable energy source in the vehicle division demonstrates that solar energy based electric transports can give as equivalent to diesel based transport with help from GOI in type of subsidy. (Sheth and Sarkar, 2017)

3. Innovative developments in solar transport

Alternative vehicles advances, for example, electric vehicles (EVs), are being acquainted with diminish the world's reliance on oil for transportation and breaking point transportation related CO_2 emanations. Richardson (2013) has likewise noticed that the non conventional energy sources are being created and to

IOP Conf. Series: Materials Science and Engineering **937** (2020) 012033 doi:10.1088/1757-899X/937/1/012033

replace petroleum based vehicles, reducing ozone depleting substance emission just as the outflow of different toxins, for example, nitrous oxides (NOx) and sulfur dioxide (SO2). In future, up gradation of technology may reveal more efficient and economical technique in the market which may be more appropriate in terms of techno-commercial feasibility.

Rand (2009) observed that innovation in solar car have done a lot and now it is no longer a curiosity. As encounters and results of the World Sun powered Test, the vehicles are assuming a significant job in spread out the advancement of battery-powered battery innovation, and in giving both a feature and a proving ground for the most recent advances. **Rose (2011)** observed that E-bike innovation is developing and there is proof that an expanding scope of light electric vehicles will challenge the obstructions which are utilized to define vehicles that are classified as 'bikes'. While the heft of the current market for these vehicles is ruled by China, there are indications of developing deals in different nations. Improving battery innovation, alongside creative item configuration, is probably going to bring about a progressively different scope of vehicles, with lower weight and improved execution. A study by **Devan et al (2020)** dwells upon a tricycle used by physically challenged persons. A tricycle is a three-wheeled vehicle driven by human-powered. Discussing of this work, an eco-friendly motorized tricycle operated by solar cells was developed for disabled people to enable them to drive without much effort. As a result of this, the GHG radiation is also get rid of and the operational cost is minimized. Thus, the solar-powered tricycle is both eco-friendly and advisable. Across the world countries such as Malaysia, America, UK and Malta are giving huge subsidy to individuals when they register for electric vehicles (EVs). Many cities in India like Ahmadabad and Hyderabad have also successfully implemented EVs.

4. Developments in solar transport in India

Similarly, there have been many innovative developments in the sector of solar transport in India as well. This section illustrates some case studies of such developments in India.

(a) Solar buses:

According to study by **Naiket al (2019)**, the capa-buses (super capacitor), the electric vehicles have been worked alongside the electric transports to investigate the electrified transportation structure in the Guwahati city. The proposed framework is reasonable for different circumstances in all the periods of a year. The benefits that the Guwahati city gains, by the usage of the proposed framework are (a) the diminished activity of inside burning motor vehicles in urban areas and therefore the improved air quality in city district and (b) the advantage to the basic resident to participate in decreasing carbon impression based transportation framework in the city.

Students at Lovely Professional University have developed India's first driverless, solar-powered bus. The vehicle was showcased at the 106th edition of the Indian Science Congress (INC). The driverless bus uses renewable energy that is delivered with an electric motor and solar power. With this set up, the bus can manage to reach a top speed of 30kmph and can transit up to 70 km per single full energy. The bus has a capacity to seat 10 to 30 people. (Source: https://www.financialexpress.com)

(b) Solar train:

Indian Railroads on July 14, 2017 started first solar based fuelled DEMU (diesel electrical numerous unit) train from the Safdarjung rail line station in Delhi. The train runs from SaraiRohilla in Delhi to Farukh Nagar in Haryana. A total of 16 sunlight based boards, each creating 300 Wp, are fitted in six mentors. The sunlight based boards achieve out 17 units of intensity in a day which empowers the lighting framework in the mentor. As of now solar oriented boards are introduced on non-air conditioning mentors as it were. By sparing an approximated 1.2 lakh kilo liter of diesel consistently, the railroads spare a use of INR 672 crore for each year. The sunlight based force likewise helps in limiting 2.7 lakh huge amounts of carbon dioxide outflow every year. (Source: https://economictimes.indiatimes.com)

(c) Solar e-rickshaw:

"ElecRic", an electric rickshaw model launched in Bangalore in 2016, runs on solar power, has zero-tailpipe emissions and is available at a nominal price of INR 2 lakh. The model has a range of 110 km with one full charge that takes about five hours for charging. The maintenance cost of the solar electric rickshaw is estimated to be INR 1 to INR 5 per km. (source :https://economictimes.indiatimes.com/)

(d) Solar car:

In 2013, the "e20" model of electric car has been launched by Mahindra group in India in order to meet the company's "5cs" framework: Clean, convenient, connective, clever, and cost effective. The car can be charged via a common 15 ampere outlet, or using solar panels. Many other electric car models from various manufacturers have entered the market to boost sustainability benefits.(source: https://www.businessinsider.in)

(e) Solar van/taxi:

Individual small scale efforts have also been recorded such as attempts by Chitre (year) from Nagpur to convert his regular car to a solar-powered one. With 25 years efforts to perfect the innovation, the e-vehicle has so far travelled a distance of 3,500 kilometres hassle-free. (Source: https://www.thebetterindia.com)

In context to promote, the government of India is launching many schemes to promote use of non-conventional energy. Seeing the vulnerable condition MHRD has announced extensive use of solar energy (to an extent of 1OO GW) and has declared various subsidy in residential, agricultural and industrial sectors; as a response of which good initiatives are being taken by various private companies to use solar energy resulting into good saving in sense of economy as well as environment. Jawaharlal Nehru national solar mission (JNNSM) launched in 2010 is one of the promotional policies, under nationwide action plan on climate change (NAPCC-2008). Until today, under the spotless advancement system, nine tasks were started in India's transportation area concentrating on the development of new patterns in transport including the electrical vehicles, mode shift transportation, and clean petroleum derivative based, and so on. (Nallapaneni et al, 2007)

5. Conclusion

India has abundant potential for solar energy due to its tropical region that is why the concept of solar based electric vehicles can be optimistically taken forward for wider implementation. This paper exemplifies the cases of solar electric vehicles including solar cars, solar e–rickshaws, solar tricycles, solar buses and solar vans, that have been successfully implemented in India. These examples demonstrate that the solar electric vehicles can be a future scenario on a larger scale for India and also other countries. The innovative use of solar energy in transportation can lead to socially, eco-friendly and many other benefits thus leading to a sustainable system.

References

[1] Adepetu, A., & Keshav, S. (2017). The relative importance of price and driving range on electric vehicle adoption: Los Angeles case study. Transportation, 44(2), 353–373. https://doi.org/10.1007/s11116-015-9641-y

[2] Adheesh, S. R., Shravanth Vasisht, M., & Ramasesha, S. K. (2016). Air-pollution and economics: Diesel bus versus electric bus. Current Science, 110(5), 858–862. https://doi.org/10.18520/cs/v110/i5/858-862

[3] Allwood, J. M., & Tekkaya, A. E. (2012). Writing a review paper. Journal of Materials Processing Technology, 212(1), 1–2. https://doi.org/10.1016/j.jmatprotec.2011.09.003

[4] Anonimo. (2014). Format for a review paper. 1. Retrieved from http://www.ugresearchsci.ucla.edu/pdfs/USJ Submission Guidelines for Review Articles 2013-2014.pdf

[5] Aydin, D., Casey, S. P., & Riffat, S. (2015). The latest advancements on thermochemical heat storage systems. Renewable and Sustainable Energy Reviews, 41, 356–367. https://doi.org/10.1016/j.rser.2014.08.054
[6] Bhaskar Naik, M., Kumar, P., & Majhi, S. (2019). Smart public transportation network expansion and its interaction with the grid. International Journal of Electrical Power and Energy Systems, 105(December 2017), 365–380. https://doi.org/10.1016/j.ijepes.2018.08.009
[7] Brozynski, M. T., & Leibowicz, B. D. (2018). Decarbonizing power and transportation at the urban scale: An analysis of the Austin, Texas Community Climate Plan. Sustainable Cities and Society, 43, 41–54. https://doi.org/10.1016/j.scs.2018.08.005
[8] Buchholz, O. S., Van Der Ham, A. G. J., Veneman, R., Brilman, D. W. F., & Kersten, S. R. A. (2014). Power-to-Gas: Storing surplus electrical energy a design study. Energy Procedia, 63, 7993–8009. https://doi.org/10.1016/j.egypro.2014.11.836
[9] Channi, H. K. (2019). Solar Pavement : Smart Means of Transportation Solar Pavement : Smart Means of Transportation. (February).
[10] Connolly, D., Mathiesen, B. V., & Ridjan, I. (2014). A comparison between renewable transport fuels that can supplement or replace biofuels in a 100% renewable energy system. Energy, 73, 110–125. https://doi.org/10.1016/j.energy.2014.05.104
[11] Derby, G. S. (1907). Ligation of the common carotid artery for malignant recurrent hemorrhage of the vitreous. Journal of the American Medical Association, XLIX(2), 107–110. https://doi.org/10.1001/jama.1907.25320020011001c
[12] Devan, P. K., Bibin, C., Krrishana, A. G., Ashwin, M., Bala Kumar, K., & Dinesh, K. B. (2020). A comprehensive review on solar tricycles. Materials Today: Proceedings, (xxxx), 10–12. https://doi.org/10.1016/j.matpr.2020.03.826
[13] Ellabban, O., Abu-Rub, H., & Blaabjerg, F. (2014). Renewable energy resources: Current status, future prospects and their enabling technology. Renewable and Sustainable Energy Reviews, 39, 748–764. https://doi.org/10.1016/j.rser.2014.07.113
[14] Götz, M., Lefebvre, J., Mörs, F., McDaniel Koch, A., Graf, F., Bajohr, S., … Kolb, T. (2016). Renewable Power-to-Gas: A technological and economic review. Renewable Energy, 85, 1371–1390. https://doi.org/10.1016/j.renene.2015.07.066
[15] Goss, W. P., & Mcgowan, J. G. (n.d.). United States and its energy reserves . (For example , see Perry , 1972 or Vogely , 1969). Howell, E., Neal, D., & Kieffer, D. (2018). Changing the paradigm of transportation: Lightweight composites used in solar car in intercollegiate competition. Reinforced Plastics, 62(4), 190–193. https://doi.org/10.1016/j.repl.2017.06.087
[16] Gül, T., Kypreos, S., Turton, H., & Barreto, L. (2009). An energy-economic scenario analysis of alternative fuels for personal transport using the Global Multi-regional MARKAL model (GMM). Energy, 34(10), 1423–1437. https://doi.org/10.1016/j.energy.2009.04.010
[17] Ifaei, P., Khiabani, H., Piran, M. J., & Yoo, C. K. (2020). Techno-econo-environmental feasibility of retrofitting urban transportation system with optimal solar panels for climate change mitigation – A case study. Journal of Cleaner Production, 251, 119639. https://doi.org/10.1016/j.jclepro.2019.119639
[18] Koopmans, T. (2016). Optimum Utilization of the Transportation System Author (s): Tjalling C . Koopmans Source : Econometrica , Vol . 17 , Supplement : Report of the Washington Meeting (Jul ., 1949), Published by : The Econometric Society Stable URL : http://www.jstor.org. 17, 136–146.
[19] Lewandowska-Bernat, A., & Desideri, U. (2017). Opportunities of Power-to-Gas technology. Energy Procedia, 105, 4569–4574. https://doi.org/10.1016/j.egypro.2017.03.982
[20] Lund, H., Werner, S., Wiltshire, R., Svendsen, S., Thorsen, J. E., Hvelplund, F., & Mathiesen, B. V. (2014). 4th Generation District Heating (4GDH). Integrating smart thermal grids into future sustainable energy systems. Energy, 68, 1–11. https://doi.org/10.1016/j.energy.2014.02.089
[21] Mahian, O., Kianifar, A., Kalogirou, S. A., Pop, I., & Wongwises, S. (2013). International Journal of Heat and Mass Transfer A review of the applications of nanofluids in solar energy. Heat and Mass Transfer, 57(2), 582–594. https://doi.org/10.1016/j.ijheatmasstransfer.2012.10.037
[22] Majumder, S., De, K., Kumar, P., & Rayudu, R. (2019). A green public transportation system using E-buses: A technical and commercial feasibility study. Sustainable Cities and Society, 51(August 2019), 101789. https://doi.org/10.1016/j.scs.2019.101789
[23] Mathiesen, B. V., Lund, H., Connolly, D., Wenzel, H., Ostergaard, P. A., Möller, B., … Hvelplund, F. K. (2015). Smart Energy Systems for coherent 100% renewable energy and transport solutions. Applied Energy, 145, 139–154. https://doi.org/10.1016/j.apenergy.2015.01.075

[24] Muhammad-Sukki, F., Munir, A. B., Ramirez-Iniguez, R., Abu-Bakar, S. H., Mohd Yasin, S. H., McMeekin, S. G., & Stewart, B. G. (2012). Solar photovoltaic in Malaysia: The way forward. Renewable and Sustainable Energy Reviews, 16(7), 5232–5244. https://doi.org/10.1016/j.rser.2012.05.002

[25] Nallapaneni M., Reddy G. R. and Miriyam A.(2007), "Potential emission reductions from India's transport sector: a view from the green transportation projects under CDM", Energy Procedia, Vol. 147, pp. 438-444

[26] Olabi, A. G. (2013). State of the art on renewable and sustainable energy. Energy, 61, 2–5. https://doi.org/10.1016/j.energy.2013.10.013

[27] Rand, D. A. J. (2009). Encyclopedia of Electrochemical Power Sources. Encyclopedia of Electrochemical Power Sources, 1(December 1982), 359–366. https://doi.org/10.1016/B978-044452745-5.00380-4

[28] Richardson, D. B. (2013). Electric vehicles and the electric grid: A review of modeling approaches, Impacts, and renewable energy integration. Renewable and Sustainable Energy Reviews, 19, 247–254. https://doi.org/10.1016/j.rser.2012.11.042

[29] Rose, G. (2012). E-bikes and urban transportation: Emerging issues and unresolved questions. Transportation, 39(1), 81–96. https://doi.org/10.1007/s11116-011-9328-y

[30] Salunke, P. M., Kulkarni, H. B., Waghchavare, A. M., & Kurle, S. S. (2017). Solar Energy and its Applications-Need , Overview. 5(Xii), 2102–2111.

[31] Settou, B., Settou, N., Gouareh, A., Negrou, B., Mokhtara, C., & Messaoudi, D. (2019). GIS-based method for future prospect of energy supply in Algerian road transport sector using solar roads technology. Energy Procedia, 162, 221–230. https://doi.org/10.1016/j.egypro.2019.04.024

[32] Sheth, A. (2017). Financial Analysis of Solar Electric Bus in India. 5(7), 145–153.

[33] Tsongidis, N. I., Asimakopoulou, A. G., Pantoleontos, G., & Konstandopoulos, A. G. (2019). Transportation and solar-aided utilization of CO 2 : Technoeconomic analysis of spanning routes of CO 2 conversion to solar fuels. Journal of CO2 Utilization, 30(June 2018), 142–157. https://doi.org/10.1016/j.jcou.2019.01.010

[34] Uyar, T. S., & Beşikci, D. (2017). Integration of hydrogen energy systems into renewable energy systems for better design of 100% renewable energy communities. International Journal of Hydrogen Energy, 42(4), 2453–2456. https://doi.org/10.1016/j.ijhydene.2016.09.086

[35] Verclas, K. (2018). Electric vehicle and solar energy pilot: Opportunity to address suburban energy challenges. Electricity Journal, 31(6), 48–56. https://doi.org/10.1016/j.tej.2018.06.007

[36] Winebrake, J. J., & Creswick, B. P. (2003). The future of hydrogen fueling systems for transportation: An application of perspective-based scenario analysis using the analytic hierarchy process. Technological Forecasting and Social Change, 70(4), 359–384. https://doi.org/10.1016/S0040-1625(01)00189-5

[37] Yu, H., Wei, Y. M., Tang, B. J., Mi, Z., & Pan, S. Y. (2016). Assessment on the research trend of low-carbon energy technology investment: A bibliometric analysis. Applied Energy, 184, 960–970. https://doi.org/10.1016/j.apenergy.2016.07.129

[38] Zahabi, S. A. H., Miranda-Moreno, L., Patterson, Z., & Barla, P. (2017). Impacts of built environment and emerging green technologies on daily transportation greenhouse gas emissions in Quebec cities: a disaggregate modeling approach. Transportation, 44(1), 159–180. https://doi.org/10.1007/s11116-015-9631-0

Design, analysis and optimal sizing of standalone PV/diesel/battery hybrid energy system using HOMER

M Thirunavukkarasu[1], **Yashwant Sawle**[2]

Vellore Institute of Technology, Vellore-632014, Tamil Nadu, India[1, 2]

thirubabu18@gmail.com, yashsawle@gmail.com[2]

Abstract: Now that the world's most important problems are the lack of fossil fuel reserves, we need to find a solution to preserve non-renewable resources and replace them with other renewable resources such as solar power. The next focus is on the rising level of undesired gases, such as carbon dioxide, etc. There is a need to regulate the impact of such greenhouse gases. This paper mainly intentional to situate optimal scheduling and to meet the power demand profile by the utilization of available solar energy resources. Considering the above data we have put forth an attempt to scrutinize the leverage of distributed Generation considering solar energy, alone can vary, when associated together they give a dependable wellspring of vitality as a DG is utilized as reinforcement unit for the crisis.So consolidating a solar powered energy source with a diesel generator provides reliable generation, utilizing independent sunlight based renewable power system for a chose area in Vellore region, Tamil Nadu, India.The principle goal of this paper is to decide the optimal sizing dependent on different designs, moreover reduce the cost parameters such as, total net preset Cost (TNPC), system cost of Energy (COE), unmet electric load, CO_2 emissions by utilizing HOMER Software. It is found that the PV/Diesel/converter combination provides optimal results which providing vitality with 0% unmet load at the minimum electricity cost, which is diminished from$ 0.672 to $0.319/Kwh.

Keywords: Solar energy, Backup system, HOMER, Cost of energy, Pollutant Emission

1. Introduction

Due to rising communities, and because of fast industrialization a sustainable power source with better proficiency is emphatically rely on socio economic advancement for any countries. As developing nations like India has relies upon fossil derivative to satisfy daily energy need, which confronting an intense vitality emergency which influences the lives of the individuals as well as a significant hindrance in the turn of events and progress of the nation. Despite the unpredictability of solar or wind power, complete hybrid power plants are now operating in many parts of the world [1]. The single sources of activity of these creating units are not skilful in imperatives of cost, productivity and dependability. Combining these diverse renewable resources is an effective alternative to building a hybrid energy system. By joining these divergent renewable sources with fossil derivatives to frame a composite imperativeness system is a viable substitute choice [2].

A new research has found that solar battery storage devices are more affordable to meet social load requirements than diesel and wind power batteries [3]. A detailed analysis in the middle of solar PV-Diesel generator and solar PV-fuel cell hybrid systems, as utilized for a 5kW operation, is given in [4] which is more effectiveness in producing the optimum outcomes on the basis of cost-effectiveness.

In [5] execution of the stand-alone hybrid solar photovoltaic /diesel engine with an electric generator/battery system has been marked by conscious design for an inaccessible society in

Bangladesh. The effect of energy storage has been assessed regarding the cost of vitality and the net present expense of the framework. Most appropriate sizing of the hybrid power system has been inspected by HOMER programming.Author [6] suggested techno-economic feasibility investigation for electric power siring through hybrid system which includes biomass, wind and solar photovoltaic. They pointed out that the designed hybrid structure is less expensive contrasted with typical production of electricity.

The author reviewed the economic feasibility of the earliest hybrid system with a solar photovoltaic and coal fired power stations in the Inner Mongolia locale of China. They utilized their conclusion to make proposals for the improvement of this plan kind of hybrid innovation [7].Authors [8] discusses the difficulties in improving the utilization of secluded Small solar photovoltaic in distant zones. In consideration of the intrinsic complexity of the generation of solar power and the load criteria, the author has developed a method for estimating the existence of the different PV technologies. The aim of this proposed work is the simulation and optimization of hybrid solar power generation / diesel engine / power storage for residential load power supply. Systems implementations were shown in Figure 1.

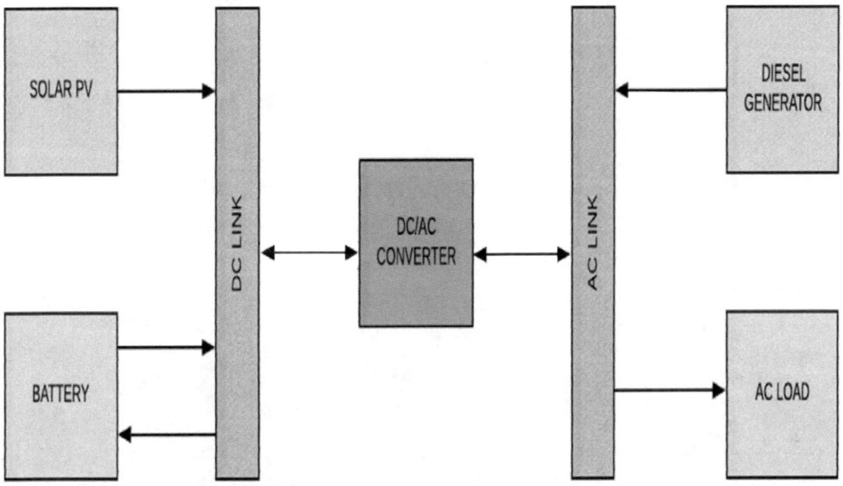

Figure 1.Schematic diagram comprises of solar PV, diesel generator and Battery system

2. Hybrid Optimization Model for Electric Renewables (HOMER) software program

Homer Technology was established by the American National Renewable Energy Laboratory in 1993 [9]. It is a system model that can be used to access combinations of different components of a grid connected system and the performance requirements of an off-grid system. HOMER executes the following three basic functions, begins with simulation then performs optimization and finally a sensitive examination of the energy systems being planned. Throughout the simulation process, the HOMER simulations has been carried out for each hour of the year and assess its technological viability along with the relevant terms of the life cycle cost, such as acquisition, repair, service and maintenance.

In the optimization procedure, HOMER models numerous setups to determine the most suitable option that meets the requirements of technical feasibility in terms of lowest life cycle cost. Insensitive analysis, HOMER plays out multiple optimizations with considerations of various data parameters to decide the impact of vulnerability and inconsistency on the model info [10], [17].

IOP Conf. Series: Materials Science and Engineering **937** (2020) 012034 doi:10.1088/1757-899X/937/1/012034

3. Zone of analysis and description of load

The chosen region is the district of Vellore, Tamil Nadu, nation of India, and the position of this research region on the chart at latitudes 12°,55.0'N and longitude 79°,7.9'E, where the usability of the renewable vitality supply were immense in this analysis area, insofar as 'solar, biomass and biogas,' were each of these are used in stand-alone format.

Standard measurement of normal energy usage is 11.26 KWh/day; average system load is 0.47 Kilowatt, peak demand output is 2.09 KW, because the position selected has so little requirement. Load data is estimated and averaged over 24 hours to create an energy demand profile as seen in Figure. 2(a) and Figure. 2(b) displays the annual load profile info.

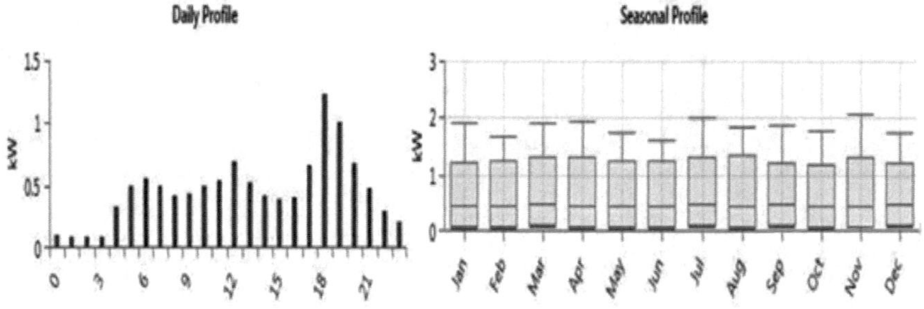

Figure 2(a). Shows the demand profile hourly and monthly

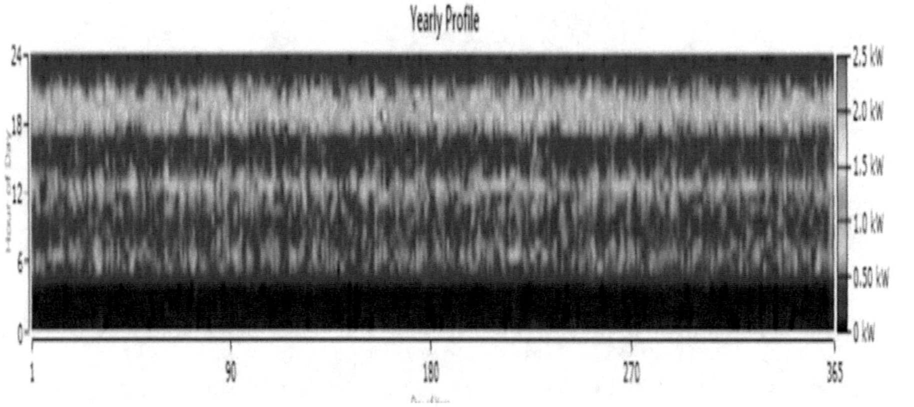

(b)

Figure 2(b). Shows load profile annually

3.1. Solar potential

The better understanding of solar radiation and the clearness index is important for a chosen location to simulate and design of a solar system at HOMER.Average solar information was acquired from NASA via HOMER software, the solar based information comprising of the clearness index and solar radiation ranges from 0.476 to 0.664 and 4.694 kWh/m^2/day to 6.923 kWh/m^2/day respectively as indicated in Figure 3. and Table 1.December month has the lowest solar radiation of 4.694 kWh/m^2/ day and the maximum solar irradiation point of 6.923 kWh /m^2 /day in the month of April.Be that as it may, a month with base sun oriented radiation esteem is adequate to utilize sun based energy for power generation.

The sun-dependent PV device converts irradiance dependent on sunshine into sun-powered vitality to satisfy electric demand. It functions simply when the sun is shining, and the abundance vitality generated by the solar photovoltaic is utilized to recharging the backup batteries, which can be used to fulfill the required load particularly at night when no sun-oriented vitality is present. The solar photovoltaic type utilized for this examination is a generic flat plate with its cost of capital, replacement cost and operational and maintenance cost given in Table 2.

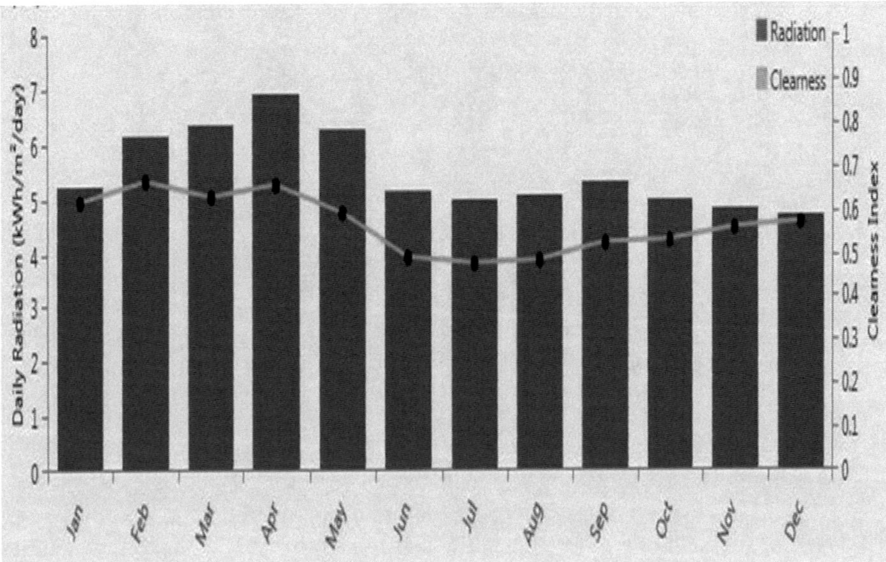

Figure 3. Monthly average solar data for the selected location

Table 1. Meteorological reports collected from NASA shows average values.

S.no	Months	Clearness Index	Daily solar Radiation $Kwh/m^2/day$
1	January	0.616	5.232
2	February	0.664	6.166
3	March	0.628	6.348
4	April	0.656	6.923
5	May	0.592	6.271
6	June	0.491	5.153
7	July	0.476	5.005
8	August	0.483	5.072
9	September	0.532	5.325
10	October	0.529	4.998
11	November	0.558	4.815
12	December	0.571	4.694
13	Annual average	6.225	5.50

4. System components

The following components are considered to model a system which includes solar PV, rechargeable battery storage systems, and diesel generator for back-up. Figure 4. Defines the proposed configurations in which the demand for electrical load is AC load, the Solar system is has been integrated to a DC connection, the diesel engine source is connected to an AC connection, and the battery is linked with DC-link.

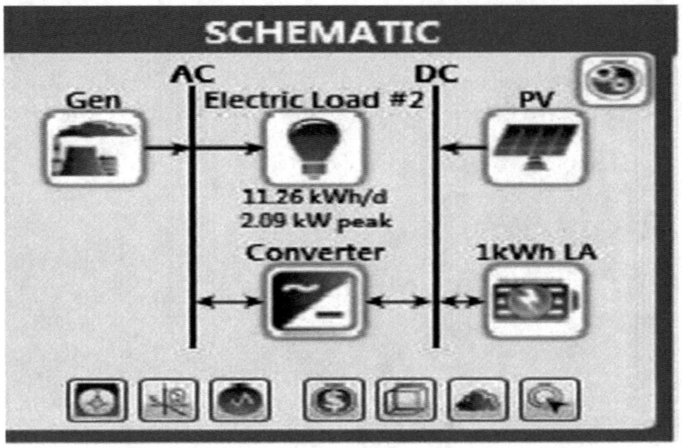

Figure 4. Design of a hybrid device in HOMER

4.1. Solar PV

The sun based PV system changes over the sunlight based irradiance into sun powered vitality to meet the electrical demand. It essentially works when the sun sparkles, and the abundance vitality produced by the PV is utilized to charge the backup batteries, which can be utilized to fulfill the desired load particularly during the night when there is no sun oriented vitality. The solar photovoltaic type utilized for this examination is a generic flat plate with its cost of capital, replacement cost and operational and maintenance cost given in Table 2.

4.2. Converter

In this analysis a generic converter is used with an efficiency of 95 % and 100 % rectifier relative capacity. The cost specifics of the converter, which include capital costs, costs of substitution, running and repair expenses and other basic criteria, are shown in Table 2. In this scenario, HOMER optimizer selects the converter size to match the needs.

4.3. Diesel generator

The diesel engine is integrated into hybrid systems to supplement for power needs where the availability of renewable resources is insufficient to reach the real load requirement. The cost of fuel is varying for each area owing to shipping costs. Generally, the fuel rate is greater in the regional areas, where extra freight expenses to be charge for the distribution of gasoline to certain towns. Diesel cost is varying from $0.7 to $1.0 depending on the venue. The Auto size Genset is the diesel generator which is chosen for this analysis. The specifics of the expenses are provided in Table 2.

4.4. Battery

A generic lead-acid rechargeable battery with a nominal capacity of 1 kWh and 12 V nominal voltage is taken into consideration for this feature. It also has a roundtrip efficiency of 80%. For a minimum SOC of 40% and a maximum of 100%, the battery has a life of up to 10 years. The battery bank stores unused electrical power from the RES network and is be used to service the load when the RES fails to support the load.

Backup diesel generator provides the energy when the load is not serviced by the RES and in the state of battery capacity is completely discharged. The excess energy of the solar system is transferred to the battery reserve, which can be used to assist the load when the renewable network falls. Diesel generator serves as a contingency supply that supply to the load when all renewable sources and the batteries struggle to reach the necessary load specifications. The expense information of the battery is reported in Table 2. And HOMER optimizer has choosing the required size of battery capacity.

Table 2. List of systems cost components used in hybrid systems

S.No	Component	Size (KW)	Capital cost ($)	Replacement cost ($)	O & M cost($/year)	Lifetime (Years)
1	Solar PV panel	1	700	700	10	25
2	Diesel generator	1	300	300	0.030/op.hour	15000 Hours
3	Battery	1	230	230	10	10
4	Converter	1	250	250	0	15

5. Economics

The viability of the designed hybrid system can be examined by distinguishing the financial parameters, for example, levelized cost of vitality, system net present cost, system capital expense, and salvage cost. HOMER simulation and renders the optimal outcome based on these economic parameters.

5.1. Net present cost (NPC)

The NPC is the crucial expense parameter in HOMER programming [11]. The NPC is defined by the calculation of all the expenses borne by the program over the life span of the system, less the value of the incomes earned during the venture's life expectancy. Net present cost comprising Initial capital cost, Replacement cost, expenses for operational and maintenance, fuel cost, etc. The NPC is mathematically expressed by equation (1) as [12].

$$\text{Net present cost} = \frac{\text{Cannual.Tot}}{\text{CRF(i,Lproject)}} \tag{1}$$

Where,

$C_{annual.Tot}$ is absolute annual cost
CRF is a capital recovery factor
i is the interest rate
$L_{project}$ is designed project life time

5.2. Levelized cost of energy (LCOE)

Levelized cost of energy is the amount of useful power generated by the system. As determined by HOMER the LCOE is the proportion of the complete yearly expense of power created to the total useful power generated by the system [13]. The LCOE equation is presented in Eq. (2) as [14].

$$\text{Levelized cost of energy (LCOE)} = \frac{\text{Cannual.Tot}}{(Eprimary.AC + Eprimary.DC + Egridsales)} \tag{2}$$

Where,

$C_{annual.Tot}$ is absolute annual cost
$E_{primaryAC}$is Primary AC load served
$E_{primary\ DC}$is Primary DC load served
$E_{grid\ sales}$ is absolute grid sales

IOP Conf. Series: Materials Science and Engineering **937** (2020) 012034 doi:10.1088/1757-899X/937/1/012034

5.3. Renewable fraction (RF)

Total renewable power produced by the sustainable power sources contrasted with the total power produced by the whole system is known as the renewable fraction [15]. For this investigation, the RF is required to be as high as conceivable to diminish the impact of greenhouse emissions which is ozone-depleting substance discharge by the traditional diesel generator even though it will have an impact on the NPC because of cost of the renewable power sources. It can be mathematically expressed as equation (3) [16-18].

$$\text{Renewable fraction (\%)} = 1 - \frac{\Sigma \text{Pdiesel}}{\Sigma\ \text{Prenewable}} \tag{3}$$

Where,

P_{diesel} - Output provided by the diesel generator

$P_{renewable}$ - Output from renewable sources

6. HOMER simulation results

HOMER simulates and gives optimization results, it is discovered that to satisfy the need profile 3.45 kW solar photovoltaic panel, 2.30 kW generators, 10 KWh lead- acid Battery and 1.01 kW converters are the ideal estimating for over a framework. In this framework, introductory expense is 5656 $ and working expense is 872.61 $/year, cost of power is 0.319 $, in the interim, there is a renewable fraction with the ideal setup is 75.1%. In addition, the total power generation of the hybrid system is provided as 6693 kWh/year; the total generation from PV array and the diesel generator is almost 84.7%, and 15.3%, respectively.

Besides, the amount of the fuel required to run generator is 385 litres per year and the generator runs for 1117 hours per year, while for the base case, the measure of diesel expended is 2457 litres for every year and its operating time is 8760 hours per year. This implies around 2000 litres of fuel and practically over 70% decrease in working time of diesel generator is achieved in optimal configuration.Optimization results of HOMER programming for the various configurations has shown in figure 5.

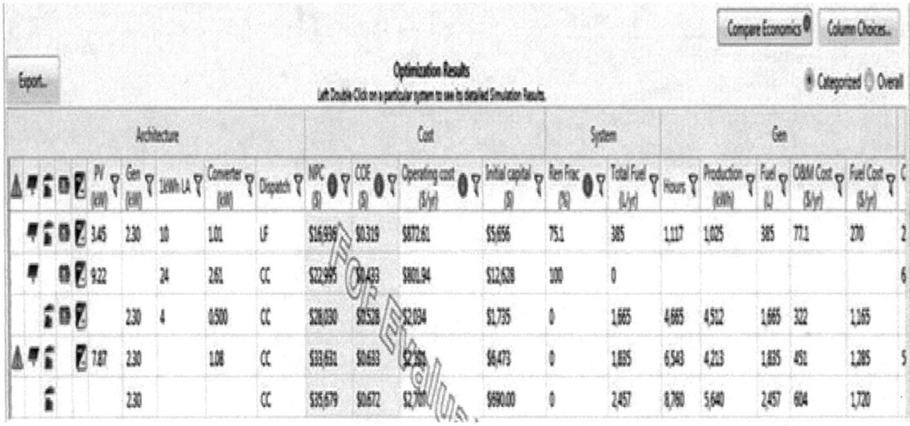

Figure 5. Optimization results from HOMER

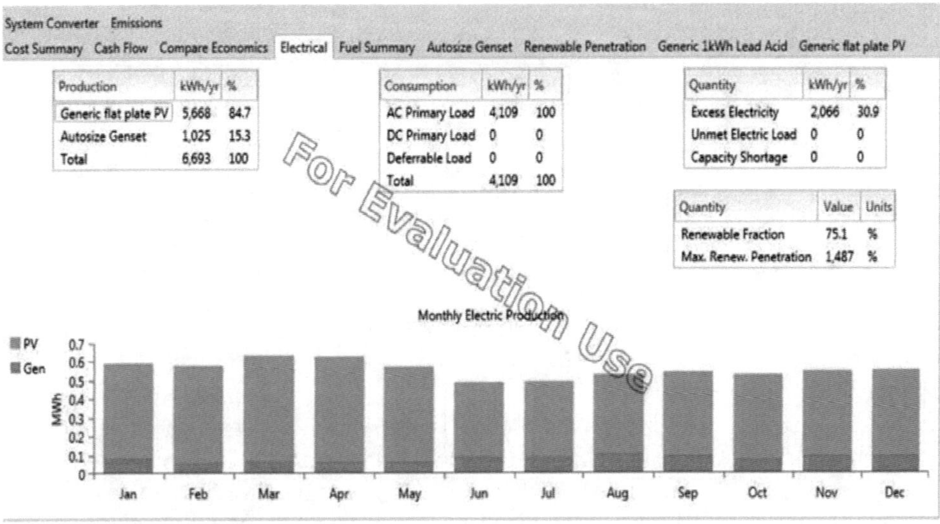

Figure 6. Power generation of Hybrid energy systems

The Power delivered by the system is utilized to take care of AC essential burden. The normal utilization of the AC load is 4109 KWh/year. It very well may be seen that the average electricity generated by Solar PV system is 5668 KWh/year and diesel engine generator is 1025 KWh/ Year. The excess electricity available is 2066 KWh/year with zero percent unmet loads. Figure 6. Shows monthly average power produced throughout hybrid energy systems.It is noticed from Figure 7. shows that the proposed integrated green energy system saves expenses over the project life time with operating cost as $2707 for base case as diesel generator while it is for optimal lowest cost system as $872.61 and Figure 8. Shows Total Net present expenses for optimal configuration and is estimated to be $16936, while it is $35679 for the base case of diesel system. By concerning NPC, there is an extraordinary decrease in their expense by practically half in ideal crossover system.

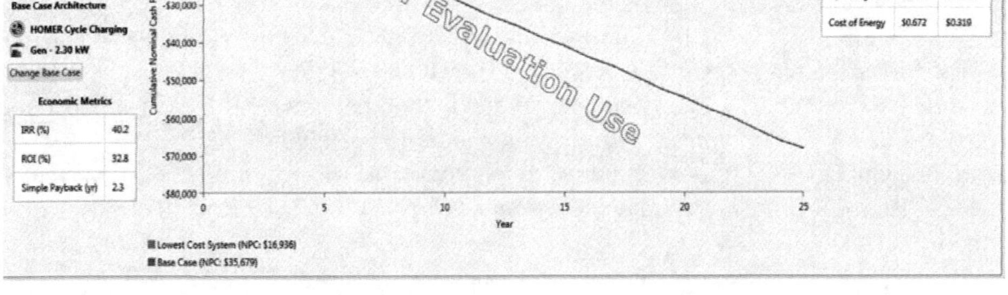

Figure 7 . Saves the money over the project lifetime

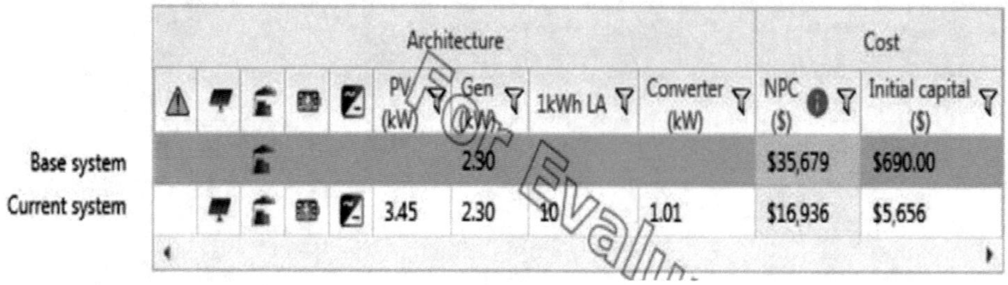

	Architecture					Cost	
	PV (kW)	Gen (kW)	1kWh LA	Converter (kW)		NPC ($)	Initial capital ($)
Base system		2.30				$35,679	$690.00
Current system	3.45	2.30	10	1.01		$16,936	$5,656

Figure 8. Net present cost flow summary of the base case and current system

7. Conclusion

In this paper, we have analyzed the results obtained from simulation and optimization of a hybrid system which consists of Solar photovoltaic, diesel engine generator, and battery for the generations of electrical power have been carried out using HOMER. The combination of a 3.45 kW PV, a 2.30 kW capacity diesel generator, 10 kW batteries and a 1.01 kW power converter with load following strategy is the best solution to this scenario analysis at NPC $ 16936, COE 0.319$ / kWh, and the cost of work $872.61 is more affordable than the base case structure with COE $ 0.672. So the solar photovoltaic-diesel-battery hybrid framework has prompted financially increasingly advantageous structure, by considering the load needs of these networks and the sustainable resources accessible in their individual locales.

References

[1] Bagen, B. R., (2005). "Evaluation of different operating strategies in small standalone power systems". *IEEE Transactions on energy conversion*, 20(3), 654–60.

[2] Deshmukh, M.K., Deshmukh, S.S., (2008) "Modeling of hybrid renewable energy systems". *Renewable Sustainable Energy Reveiw;*12(1), 235–49

[3] Thiam DR. Renewable decentralized in developing countries: Appraisal from microgrids project in Senegal. *Renewable Energy*, vol. 35; 2010. p. 1615-1623

[4] Raj AS, Ghosh PC. Standalone PV-diesel system vs. PV-H2 system: An economic analysis. *Energy* 2012;42:270–80.

[5] B.K. Das, F. Zaman, Performance analysis of a PV/Diesel hybrid system for a remote area in Bangladesh: effects of dispatch strategies, batteries, and generator selection, *Energy* 169(2019)
263–276.

[6] Sawle, Y., Gupta, S.C. and Bohre, A.K., Techno-Economic Scrutiny of HRES through GA and PSO Technique, Inderscience, *International Journal of Renewable Energy Technology*, Vol.9 No.1/2 pp.84 – 107, 2018.

[7] Zhao Y, Hong H, Jin H. Appropriate feed-in tariff of solar–coal hybrid power plant for China's Inner Mongolia region. *ApplTherm Eng.* 2016;108:378–387.

[8] Shrestha, G. B., and L. Goel. 1998. A study on optimal sizing of stand-alone photovoltaic stations. *IEEE Transactions on Energy Conversion* 13(4):373–8.

[9] Sawle, Y., Gupta, S.C. "Optimal sizing of photovoltaic/wind hybrid energy system for rural electrification." *In Power India International Conference (PIICON),* 2014 6th IEEE, pp. 1-4. IEEE, 2014.

[10] HOMER, The Micro -Power Optimization Model, ver.2.68 Beta, *NREL*, 2009.

[11] T. Givler, P. Lilienthal, Using HOMER Software, NREL's Micropower Optimization Model, to explore the Role of Gensets in Small Solar Power Systems; Case Study: Sri Lanka, *National*

Renewable Energy Lab., Golden, CO (US), 2005

[12] Sawle, Y., Gupta, S.C. and Bohre, A.K., A Novel Methodology for Scrutiny of Off-Grid Hybrid Renewable System, John Wiley & Sons, *International Journal of Energy Research*.2018; vol. 42 pp.570-586.

[13] B.K. Das, N. Hoque, S. Mandal, T.K. Pal, M.A. Raihan, A techno-economic feasibility of a stand-alone hybrid power generation for remote area application in Bangladesh, *Energy 134* (2017) 775–788.

[14] A. Hiendro, R. Kurnianto, M. Rajagukguk, Y.M. Simanjuntak, Techno-economic analysis of photovoltaic/wind hybrid system for onshore/remote area in Indonesia, Energy 59 (2013) 652–657.

[15] Sawle, Y., Gupta, S.C. and Bohre, A.K. "Socio-techno-economic design of hybrid renewable energy system using optimization techniques." *Renewable Energy*119 (2018): 459-472.

[16] Sawle, Y., Gupta, S.C. and Bohre, A.K., Review of hybrid renewable energy systems with comparative analysis of off-grid hybrid system, *Renewable and Sustainable Energy Reviews,*vol. 81, June 2017, pp. 2217-2235.

[17] Sawle, Y., Gupta, S.C. and Bohre, A.K., PV-wind hybrid system: A review with case study. *Cogent Engineering Journal*, (2016). 3(1), 1189305.

[18] Sawle, Y., Gupta, S.C. and Bohre, A.K., "Optimal sizing of standalone PV/Wind/Biomass hybrid energy system using GA and PSO optimization technique" *1st International Conference on Power Engineering, Computing and Control, PECCON-2017, 2-4 March 2017, VIT University, Chennai Campus, Elsevier, Energy Procedia* 00 (2017), Vol. 117, 690–698, 2017.

UAV based Mapping System for Precision Agriculture

S.Sheeba Rani[1], J.Janet[2], K.C.Ramya[3], R. Sitharthan[4],T.Kesavan[5], S Shrivastava[4]

1 Associate Professor, Department of EEE, Sri Krishna College of Engineering and Technology, Coimbatore
2 Professor, Department of CSE, Sri Krishna College of Engineering and Technology, Coimbatore
3 Associate Professor, Department of EEE, Sri Krishna College of Engineering and Technology, Coimbatore
4 School of Electrical Engineering, Vellore Institute of Technology and Science, Vellore
5 Assistant Professor, Department of EEE, Sri Krishna College of Engineering and Technology, Coimbatore

E-mail:sheebaranis@skcet.ac.in

Abstract. Agriculture the backbone of our nation as well many nations is essential to feed the people and smart farming is the much-needed disruptive technology to help the farmers increase their yield. Airborne vehicles and relays supplement the automation system to monitor the growth of the plants, individually and record the observations to estimate the level of pesticides and herbicides to be applied. However, there are no simulation platforms designed to help these farmers to collect and process the data gathered. The proposed work consists of a sensor fusion framework that will give the user a complete farming scenario which is focused on data gathering with airborne UAV or relay mechanism, line of sight deployment, coverage area, identify sensor placement, and usage of fog computing paradigm as a backend computing support. The paper focuses on providing an evaluation benchmark for energy consumption, system resource usage, packet delivery ratio and transmission delay.

1. Introduction

As technology advances, many digital ways for enhancing agriculture have been used over the past decades [1]. These inventions are generally focused on enhancing the productivity of crops by collecting and analyzing agricultural land. This process is accomplished with the use deployed sensors. Internet of Things is adopted for precision agriculture and gathering of information which is further used for decision-making based on the data. The presence of fog has paved way to many new challenges in smart agriculture.

It is therefore crucial that we device a system with sensor deployment to obtain the maximum yield in precision farming. But, the performance of the system will be affected due to many issues faced by the sensors. When the sensors are used to gather data, it is found that they consume more energy due to continuous periodic usage which reduces the network lifetime, overall [2]. Hence, it is essential to employ an informed deployment system that can survey the complete land for smart farming applications, in order to reach the desired goal.

In today's development in smart farming, there are many works which make use of actual deployment that doesn't use prior knowledge leading to improper utilization of the resources. Based on surveys, it has also been observed that there are very limited toolkits developed which exists designed solely for the development of smart farming coupled with network configuration species. Some studies show that actual sensors [3] are used in emulated models for decision making with data obtained. However in these network properties, there are couple of things that are ignored such as energy consumption of underlying communication model, overlapping regions of deployed sensors leading to network congestion, and packet rate error.

Moreover, there are very limited framework which enables data processing in fog locations. To overcome these issues, there is need for developing a smart-farming simulation stack that also supports sharing and backend computing environments. Data driven modern sensor deployments are being employed in the recent years. This method is focused on not just monitoring the land, but also for intervention of the user depending on the situation analysis made through observation of the situation [5-15]. The development in high-speed networks and quick accessibility of the edges using cloud has increased the efficiency of the system, thereby increasing the productivity. The following are the important features of the toolkit:

- Use of different mobility models to support mobile and static sensor deployment. The sensors are specifically designed in terms of storage, data rate, transmission range and network parameters.
- Offers support for sensor energy models to give an overview of the realistic scenarios based on the simulation results.
- To support gathering of data using UAVs, a complete data of the UAV configuration has been inbuilt.
- Facility to use non-relay-based and relay based data communication mechanisms.
- Uses a broker who is used to collect UAV data which is further communicated with the other brokers to execute the scheduled policies based on analysis.
- The fog computing concept is used for delay-sensitive applications
- Using the fog computing model in the backend, it provides early warning system and data-driven monitoring [16-20].

2. System Modeling

Consider 'F' to be a farm with 'L' as length and 'W' as width. The dimensions of the farm is represented by L.W and 'S' represents sensors. The system developed is built such that there are N heterogeneous sensors with varying transmission times Ts and energies Es. At the location Sxy the sensors are either mobile deployed or are static [21-30]. They also used to communicate information, apart from sensing, to the central node. The UAVs are used to obtain data from the sensors and then use this data to be sent to the main server where it is analyzed and used in real-time. Every UAV used is configured with an energy Eu
and communication range Ru. These two sensors are designed in such a way that they operate in two modes: active mode and sleep mode [31-35].

Total Energy consumed Et is represented as

$$Et = Eactive + Esleep \tag{2.1}$$

where Eactive and Esleep are the energies consumed in active and sleep modes, respectively.

Here,

$$Eactive = Ewup + Emsr + Epcs + Etmt + Ercp \tag{2.2}$$

$$Esleep = Psleep \,.\, Tsleep \tag{2.3}$$

Where,

Tsleep= Time sleeping

Psleep=Power consumption

Ercp=Energy consumption for data reception

Etmt= Energy consumption for data transmission

Epcs= Energy consumption for data processing

Emsr= Energy consumption for sensing data measurements
Ewup= Energy consumption for wakeup

These energies are configured depending on the active duration and the frequency of the sensor processing unit. Data is gathered from M-UAVs and any interaction within the UAVs and the deployed sensors is done through uplink with OFDMA. 'K' simultaneous connections are supported by the UAVs for gathering information. Moreover, the system designed by us is programmed to share data between nearby fog/edge locations with UAVs [6]. There is possibility for high packet rate error occurrence as the deployed sensors are mobile. Hence Line of Sight (LoS) has to be fixed to enable effective communication.

The probability of Line of Sight is computed as follows:

$$pLoS= 1/(1+\rho*\exp(\sigma[\Phi-\mu])) \quad (2.4)$$

Where σ and ρ are the constants depending on the type of area and communication frequency, Φ is the elevation angle between the UAV Uxy and the sensors Sxy.

As the altitude of UAV increases, so does the probability which indicates that if and only if the LoS probability is close to 1 and the deployed sensors are assigned to UAV jth such that Φ is ≥ pLoS. Hence the this condition can be defined using the formula

$$dij \leq hj/(\sin(pLoS)) \quad (2.5)$$

Where,
hj is the altitude of the UAV
dij is the distance between jth UAV and ith sensor node

If it is assumed that there are m x n sensors deployed, then tc is the coverage time wherein UAVs is calculated as follows:

$$t_c = \sum_m \sum_n T_{mn} + \frac{2}{M} \sum_k \sum_j S_{kj} \quad (2.6)$$

Where,
Skj is the number of sensors
M is the count of UAVs utilized for gathering data
Tmn represents the time required to move the Unmanned Aerial Vehicle from one sensor to the other

B brokers are used to keep track of the fog locations and are used for proper use of the available fog information near the end users. The resources for the neighboring locations are allocated by the broker. This will aid in decreasing the communication delay, thereby enabling IoT based delay sensitive application. The distance between the nodes is represented as the network delay td such that

$$td=da+db \quad (2.7)$$

Where,
db represents the delay between the assigned fog location and the UAV
da is the network delay between UAV and the sensor

The network cost is represented as C which is a linear function represented as follows:

$$C = \varphi . (di,j + \sum\_(k\in B \ \& \ j\neq k)\ d\_(j,k)) \quad (2.8)$$

Where,

di,j is the distance between the local broker j and the ith UAV
φ is a constant and j
d_(j,k) is the distance between the brokers leasing compute resources and the local broker j ∈ B

3. Proposed Methodology

The method proposed is based on INET5 and OMNet++4, which is the extention of FogNetSim++ [4]. An outline of the execution model is represented in the Figure X . The core modules are as described below:

3.1. Communication Module:

This module is made up of many devices and different features. The sensors used in this module are built to sustain short-range communication and hence their communication mechanism also differs. These sensors communicate with the nearby sensor nodes to exchange data which is in turn conveyed to the gateway node. This means of communication requires efficient planning. Surveys show many significant contributions in route planning for sensor nodes which employ overhead of data routing and leader election which will lead to the consumption of energy at a significant scale. To overcome this, we propose gathering of data using airborne vehicles. When the time to gather data comes, the UAV is triggered to alert the ground sensors and a connection is established between the two points. Data is shared by the sensors with the UAV and after sharing the data, the sensors discard the information [10]. Similarly, the data gathered using UAVs will impose new challenges. These conflicts are resolved using a callout mechanism deviced instead of the traditional approach of leader-based grouping model. However, the proposed method is not economical in terms of energy. In the proposed work, the callout mechanism used will assume that prior information is saved in the UAV regarding unique identification number and deployed sensors. When the sensors are called out, the UAV sends request to the sensors which responds. This method can be further used in real time collaboration of UAV to UAV to gather data.

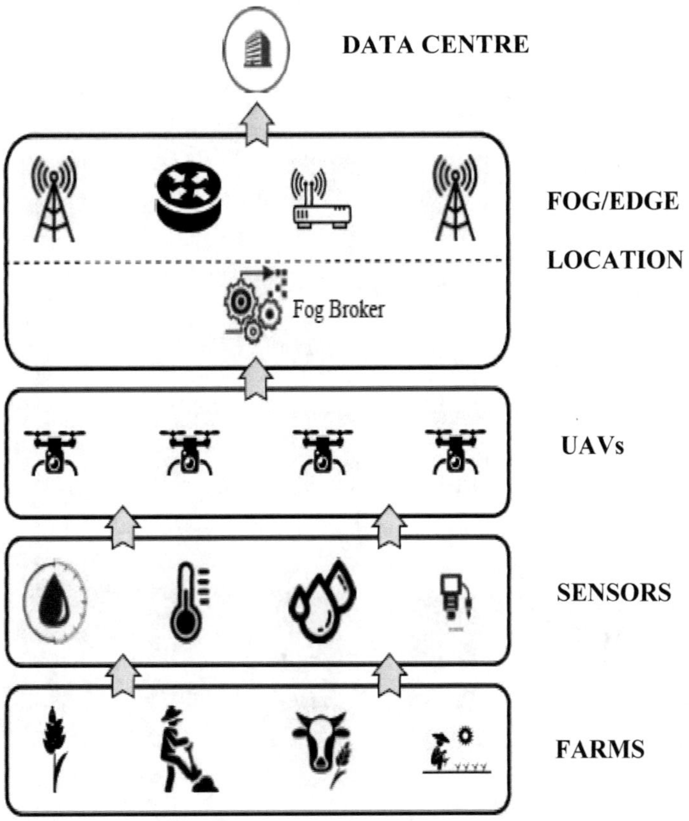

Figure 1. Layered Architecture of Toolkit for Smart Farming

3.2. Fog Nodes

These nodes are resources that are used to store, process and receive the information temporarily from the UAV streams. There are q ∈ 1, 2,….m fog resources at the fog location .Hence, depending on the real-time scenario, specification of every location can be configured accordingly [8].

3.3. Sensors

The proposed work makes use of mobile and static deployment to collect data real-time and increase farm productivity. The sensors are built with different battery life, storage and transmission ranges. These sensors are capable of forwarding the data to the neighbouring nodes using ad hoc ways. The figure X.2 gives an illustration of a sensor node. The UAVs are used too gather information from the mobile sensors and static sensors resulting in a mobile gateway node. In general, data is gathered by the sensors and forwarded to the gateway [7]. But, this will result in unnecessary wastage of energy during route finding and leader selection. To overcome this defect, a mobile gateway concept is employed by us which will increase the lifetime of the sensors and also optimize it for accurate data gathering. The limited storage availability and energy is also modeled accordingly. Figure X.2 represents the architecture of UAV.

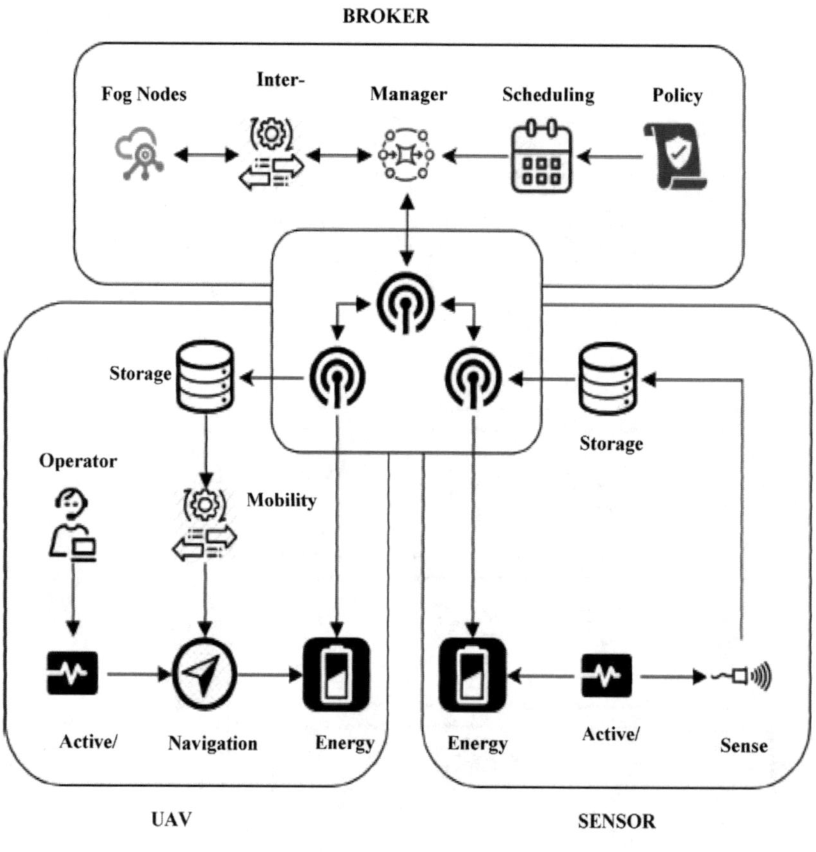

Figure 2. Internal Architecture

3.4. Broker

The fog resources are managed by the brokers and a broker is assigned for every fog location. There are M UAVs assigned to a fog location and broker nodes are used to administer these locations. These nodes further communicate with each other to find an apt resource for the UAV. At the fog location, the broker node is placed as a fixed entity as shown in the architecture, thereby managing scheduling and communicating with the fog nodes. Fault tolerance is enabled using a secondary which will kick in on failure of broker nodes do not reproduce well when printed out—your diagrams may lose vital information when downloaded and printed by other researchers. Try to ensure that lines are no thinner than 0.25 pt. Note that some illustrations may reduce line thickness when the graphic is imported and reduced in size (scaled down) inside Microsoft Word

3.5. Mobility Models

These modules enable movement of the mobile sensors in specific patterns. Though there are several static sensor deployed, the system is designed such that it supports the use of mobile sensors for various purposes. One such use is the utilization of UAVs to obtain information from the sensors on-ground. Vehicle mobility, linear mobility, circle mobility, mass mobility, nomadic community, column mobility, Gauss Markov and random way-points are some examples of mobility models that are support by the proposed framework [9]. There is also an option to extend the mobility models to suit the user requirements.

4. Results and Discussion

Evaluation of the proposed framed is carried out to analyse its performance in terms of UAVs and sensor energy consumption, transmission and processing delay, and system resource usage. UAVs are employed to gather information from the ground sensors. This data is then sent to the broker nodes which will allocate fog resources to perform the task completion. The tasks are generated as MIPs with the help of remote or local resources. Meanwhile, UAVs are also used to evaluate the transmission delay and packet delivery ratio which is important to measure the performance of the wireless sensors at a large-scale [11]. The proposed work also supports the use of fog nodes depending on the number needed by the user. Since the sensors are fixated to a particular location, it might fail due to loss of energy, which is the biggest drawback of the system. The track of fog devices that are available is maintained by the broker with the help of data gathered from the fog nodes.

5. Conclusion

It is observed that the simulation tools play a large part in supporting research in many fields. There are no such tools designed specifically for smart farming purpose and almost all the work is dependent on sensors. Moreover, users are also unaware of sensors congestion phenomena, transmission range and placement of the devices. Hence the proposed framework gives a complete farming ecosystem which is inclusive of fog location, UAVs and sensors. It will give the user access to gather data via sensor relays or UAVs, deploy sensors and also perform fog computation. The work carried out is also equipped with facility to measure transmission delay, energy of the sensor and packet delivery ratio. Future work can be carried out as an extension of this work by incorporating WCN using ANN for intelligent systems and big data analytics for smart farming.

6. References

[1]. Sitharthan R, Geethanjali M and Pandy TKS 2016 Adaptive protection scheme for smart microgrid with electronically coupled distributed generations *Alexandria Engineering Journal* **55(3)** 2539-2550

[2]. Fathima AH, and Palanisamy K 2014 Battery energy storage applications in wind integrated systems—a review *IEEE International Conference on Smart Electric Grid* 1-8

[3]. Prabaharan N and Palanisamy K 2015 Investigation of single-phase reduced switch count asymmetric multilevel inverter using advanced pulse width modulation technique *International Journal of Renewable Energy Research* **5(3)** 879-890.

[4]. Jerin ARA, Kaliannan P and Subramaniam U 2017 Improved fault ride through capability of DFIG based wind turbines using synchronous reference frame control based dynamic voltage restorer. *ISA transactions* **70** 465-474

[5]. Sitharthan, R, Sundarabalan CK, Devabalaji KR, Nataraj SK and Karthikeyan M 2018 Improved fault ride through capability of DFIG-wind turbines using customized dynamic voltage restorer *Sustainable cities and society* **39** 114-125

[6]. Prabaharan N and Palanisamy K 2016 A single-phase grid connected hybrid multilevel inverter for interfacing photo-voltaic system *Energy Procedia* **103** 250-255

[7]. Palanisamy K, Mishra JS, Raglend IJ and Kothari DP 2010 Instantaneous power theory based unified power quality conditioner (UPQC) *IEEE Joint International Conference on Power Electronics, Drives and Energy Systems* 1-5

[8]. Sitharthan R and Geethanjali M 2017 An adaptive Elman neural network with C-PSO learning algorithm-based pitch angle controller for DFIG based WECS Journal of Vibration and Control **23(5)** 716-730

[9]. Sitharthan R and Geethanjali M 2015 Application of the superconducting fault current limiter strategy to improve the fault ride-through capability of a doubly-fed induction generator–based wind energy conversion system *Simulation* **91(12)** 1081-1087

[10]. Sitharthan R, Karthikeyan M, Sundar DS and Rajasekaran S 2020 Adaptive hybrid intelligent MPPT controller to approximate effectual wind speed and optimal rotor speed of variable speed wind turbine *ISA transactions* **96** 479-489

[11]. Sitharthan R, Devabalaji KR and Jees A 2017 An Levenberg–Marquardt trained feed-forward back-propagation based intelligent pitch angle controller for wind generation system *Renewable Energy Focus* **22** 24-32

[12]. Sitharthan R, Sundarabalan CK, Devabalaji KR, Yuvaraj T and Mohamed Imran A 2019 Automated power management strategy for wind power generation system using pitch angle controller *Measurement and Control* **52(3-4)** 169-182

[13]. Sundar DS, Umamaheswari C, Sridarshini T, Karthikeyan M, Sitharthan R, Raja AS and Carrasco MF 2019 Compact four-port circulator based on 2D photonic crystals with a 90° rotation of the light wave for photonic integrated circuits applications *Laser Physics* **29(6)** 066201

[14]. Sitharthan R, Parthasarathy T, Sheeba Rani S and Ramya KC 2019. An improved radial basis function neural network control strategy-based maximum power point tracking controller for wind power generation system *Transactions of the Institute of Measurement and Control* **41(11)** 3158-3170

[15]. Rajesh M and Gnanasekar JM 2017 Path observation based physical routing protocol for wireless ad hoc networks *Wireless Personal Communications* **97(1)** 1267-1289

[16]. Palanisamy K, Varghese LJ, Raglend IJ and Kothari DP 2009. Comparison of intelligent techniques to solve economic load dispatch problem with line flow constraints *IEEE International Advance Computing Conference* 446-452

[17]. Sitharthan R, Ponnusamy M, Karthikeyan M and Sundar DS 2019 Analysis on smart material suitable for autogenous microelectronic application *Materials Research Express* **6(10)** 105709

[18]. Rajaram R, Palanisamy K, Ramasamy S and Ramanathan P 2014 Selective harmonic elimination in PWM inverter using fire fly and fireworks algorithm *International Journal of Innovative Research in Advanced Engineering* **1(8)** 55-62

[19]. Sitharthan R, Swaminathan JN and Parthasarathy T 2018 March. Exploration of wind energy in India: A short review *IEEE National Power Engineering Conference* 1-5

[20]. Karthikeyan M, Sitharthan R, Ali T and Roy B 2020 Compact multiband CPW fed monopole antenna with square ring and T-shaped strips *Microwave and Optical Technology Letters* **62(2)** 926-932

[21]. Sundar D Sridarshini T, Sitharthan R, Madurakavi Karthikeyan, Sivanantha Raja A, and Marcos Flores Carrasco 2019 Performance investigation of 16/32-channel DWDM PON and long-reach PON systems using an ASE noise source *In Advances in Optoelectronic Technology and Industry Development: Proceedings of the 12th International Symposium on Photonics and Optoelectronics* 93

[22]. Sitharthan R and Geethanjali M 2014 Wind Energy Utilization in India: A Review *Middle-East J. Sci. Res.* **22** 796–801 doi:10.5829/idosi.mejsr.2014.22.06.21944

[23]. Sitharthan R and Geethanjali M 2014 ANFIS based wind speed sensor-less MPPT controller for variable speed wind energy conversion systems *Australian Journal of Basic and Applied Sciences* **8**14-23

[24]. Jerin ARA, Kaliannan P, Subramaniam U and El Moursi MS 2018 Review on FRT solutions for improving transient stability in DFIG-WTs *IET Renewable Power Generation* **12(15)** 1786-1799

[25]. Prabaharan N, Jerin ARA, Palanisamy K and Umashankar S 2017 Integration of single-phase reduced switch multilevel inverter topology for grid connected photovoltaic system *Energy Procedia* **138** 1177-1183

[26]. Rameshkumar K, Indragandhi V, Palanisamy K and Arunkumari T 2017 Model predictive current control of single phase shunt active power filter *Energy Procedia* **117** 658-665

[27]. Fathima AH and Palanisamy K 2016 Energy storage systems for energy management of renewables in distributed generation systems *Energy Management of Distributed Generation Systems* 157

[28]. Rajesh M 2020 Streamlining Radio Network Organizing Enlargement Towards Microcellular Frameworks *Wireless Personal Communications* 1-13

[29]. Subbiah B, Obaidat MS, Sriram S, Manoharn R and Chandrasekaran SK 2020 Selection of intermediate routes for secure data communication systems using graph theory application and grey wolf optimisation algorithm in MANETs IET *Networks* doi:10.1049/iet-net.2020.0051

[30]. Singh RR and Chelliah TR 2017 Enforcement of cost-effective energy conservation on single-fed asynchronous machine using a novel switching strategy *Energy* **126** 179-191

[31]. Amalorpavaraj RAJ, Palanisamy K, Umashankar S and Thirumoorthy AD 2016 Power quality improvement of grid connected wind farms through voltage restoration using dynamic voltage restorer *International Journal of Renewable Energy Research* **6(1)** 53-60

[32]. Singh RR, Chelliah TR, Khare D and Ramesh US 2016 November. Energy saving strategy on electric propulsion system integrated with doubly fed asynchronous motors *IEEE Power India International Conference* 1-6

[33]. Singh RR, Mohan H and Chelliah TR 2016 November. Performance of doubly fed machines influenced to electrical perturbation in pumped storage plant-a comparative electromechanical analysis *IEEE 7th India International Conference on Power Electronics* 1-6

[34]. Sumithra M, Abraham A, Rajesh M and Nithiyanandam N 2019 Mining social media content to predict peer trust level in social networks *International Journal of Engineering and Advanced Technology* **6S3** 1959-1964

[35]. Venkatesh K, Ali MJS, Nithiyanandam N and Rajesh M 2019 Challenges and Research Disputes and Tools in Big Data Analytics *International Journal of Engineering and Advanced Technology* **6S3** 1949-1952

IOP Conf. Series: Materials Science and Engineering **937** (2020) 012036 doi:10.1088/1757-899X/937/1/012036

Noise Resilient distributed Voltage and Frequency Control for Low Voltage Islanded Microgrid

S Shrivastava[1], J Sahu[2] and R Sitharthan[1]

[1]School of Electrical Engineering, Vellore Institute of Technology, Vellore, TN 632014, India

[2]School of Mechanical Engineering, Vellore Institute of Technology, Vellore, TN 632014, India

mailto: sonam24390@gmail.com

Abstract. Motivated by advancement in distributed control schemes of microgrid (MG) network, this paper presents a noise resilient secondary control scheme for low voltage (LV) islanded MG network. For a LV MG the grid lines are highly resistive, hence conventional droop control results in poor stability and power sharing. A combination of droop-boost primary control with a noise resilient secondary distributed control is presented in this work. The presented control scheme achieves accurate power sharing and voltage and frequency restoration under noisy communication links for LV MG network. The communication links used for information sharing in secondary control are subjected to noisy environment; therefore the data received is corrupted with noise. The presented control scheme counteracts the noise effect for LV MG network. The performance of presented control scheme is evaluated by simulating a 4-bus MG test system in MATLAB/SimPowerSystem environment under different test scenarios. The obtained results envisage that the presented control scheme restores the voltage/frequency and provide accurate active power sharing in short time under load perturbation, varying noise parameters, and plug and play operating conditions.

1. Introduction

In the last decade electrical power system has undergone plenty of changes that make it more decentralised to distributed power system. The small scale generators geographically located to the user end at low voltage (LV) level is becoming new trend, which involve more and more renewable energy sources (RESs). These low voltage feeders comprises of several distributed generators (DGs), storage devices and loads appears as a distributed network known as microgrid (MG). The MG can operate either in grid tied mode and support bidirectional power flow or can operate bidirectional power flow or can operate autonomously as a self-regulatory islanded system [1]. Despite having small structure the MG have different operating characteristics than the conventional grid, hence different control structure is required for control and operation of inertia-less, low voltage MG network. The MG control structure is evolved from centralised to decentralized then further aggregated as distributed control structure. A hierarchical three level control is best suited approach for the control of MG network that comprises of primary control level, secondary control level and tertiary control level [2]. The primary control depends on the type of MG network. In this paper a droop-boost control scheme is adapted for achieving accurate power sharing and stable operating voltage and frequency. The secondary control level synchronizes the MG voltage and frequency to

IOP Conf. Series: Materials Science and Engineering **937** (2020) 012036 doi:10.1088/1757-899X/937/1/012036

nominal values and compensates for the deviations caused by primary control level [3]. The tertiary control level is a supervisory control layer responsible for optimization of power distribution cost and bidirectional power flow optimization [4].

Inspired from the networked control theory of multi agent system (MAS) several distributed control methods are presented in literature which utilizes neighbours information [5, 6]. The accuracy of these control schemes depends upon the grid parameter such as line impedance. A secondary control methodology addressed in [7] works on the presumption of dominant inductive line impedance. For LV MG network these schemes are subjected to power transients and instability [8]. In [7, 9] voltage control is modelled as a tracking synchronization problem while utilizing feedback linearization to convert nonlinear system dynamics of MG to a linear system. The performance of the controller depends on the topology of information sharing, noisy channel of communication, link failure, packet dropout, and communication link delays. Such problems are solved by many researchers using different secondary control approaches. The work in [10] presents a distributed restoration approach for frequency control of MG network. The unmodeled dynamics and effect of communication link delays for islanded MG are addressed in [11, 12] respectively. The information sharing communication topology is of vital importance in operation of distributed secondary control and it affects the controller performance. In [10, 13] address these issues to reduce the MG instability and high oscillations under large signal disturbances for highly inductive MG. In the contrary this paper introduces a noise resilient distributed secondary synchronization scheme for voltage and frequency both in a low voltage MG network. The main contributions of the paper are as follows.

- Noise resilient distributed secondary synchronization scheme is presented for voltage and frequency restoration in a LV MG network.
- The presented controller only uses sparse communication network and is robust against noise parameter uncertainty and high signal disturbance such as loss of DG unit.

The rest of the paper is in the following order. Section II discusses the preliminaries on modelling of LV MG, primary control, and graph theory and communication noise. Section III and IV discusses the presented secondary restoration scheme for output DG voltage and frequency respectively. Case studies for performance evaluation are presented in section V. Conclusion is presented in Section VI.

2. Preliminaries

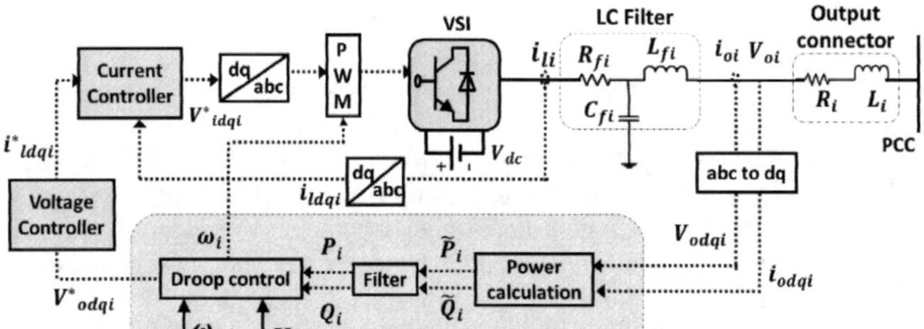

Figure 1. Single line design of VSI based $\boldsymbol{i^{th}}$DG unit.

2.1. Primary control of low voltage microgrid

The microgrid is a group of interconnected DG units, storage devices and loads. It can be seen as a three layer structure that includes electrical, control and communication layers. The electrical layer consists of a DG unit, transmission lines, energy storage element and loads. Figure 1 shows the one line diagram of a DG unit. The DG unit consist of DC output voltage of renewable energy source unit, inverter, filter (LC) and output RL connector. The control scheme is of DG is a hierarchical one which includes three layers primary, secondary and tertiary control. The primary layer has three control loops i.e., voltage control loop, output of which is fed to current control loop and power control loops to implement the droop control. The primary droop control dynamics for DGs units can be given as,

$$\omega_i = \omega_{ni} - m_{Pi}^{\omega} P_i \tag{1}$$

$$V_i = V_{ni} - n_{Qi}^{V} Q_i \tag{2}$$

where V_{ni} and ω_{ni} are reference voltage magnitude and angular frequency; P_i and Q_i are real and reactive power; m_{Pi}^{ω} and n_{Qi}^{V} are droop coefficients of ith DG unit respectively.

However, in case of low voltage MG network, the line impedance is highly resistive in nature where operating frequency and DG output voltages are closely associated to real and reactive power [15]. Hence, the modified droop-boost characteristics for primary control is given as,

$$\omega_i = \omega_{ni} + k_Q^{\omega} Q_i \tag{3}$$

$$V_i = V_{ni} - k_P^{V} P_i \tag{4}$$

where k_Q^{ω} and k_P^{V} are boost and droop coefficients for frequency and output voltage control respectively. The model for DG unit with its inner control loop can be given as [14]

$$\frac{dx_i}{dt} = \psi_i(x_i) + g_i^V(x_i)\, u_i^V + g_i^{\omega}(x_i)\, u_i^{\omega} + k_i(x_i) D_i \tag{5}$$

$$y_{iV} = h_{iV}(x_i) = V_i \tag{6}$$

$$y_{i\omega} = h_{i\omega}(x_i) + d_i u_i^{\omega} = \omega_i = \omega_{ni} + k_Q^{\omega} Q_i \tag{7}$$

where $x_i = [\alpha_i, P_i, Q_i, i_{ldi}, i_{lqi}, V_{odi}, V_{oqi}, i_{odi}, i_{oqi}]^T$; α_i is angle of DG reference frame; $x_i = [\omega_{ref}, V_{bdi}, V_{qdi}]^T$ is known disturbance. The secondary control inputs for primary droop-boost control level will be discussed in next section. This control layer synchronizes the output DG voltage and operating frequency to their nominal values.

2.2. Preface on graph theory

The adaptation of graph theory is taken from the work in [6]. A weighted, non-directional and well-connected graph $\mathcal{G} = (\mathcal{V}_{\mathcal{G}}, \mathrm{E}_{\mathcal{G}}, \mathrm{A}_{\mathcal{G}})$, with $\mathcal{V}_{\mathcal{G}}$ as set of nodes, i.e., $\{1, \ldots, \mathrm{N}\}$, $\mathrm{E}_{\mathcal{G}}$ is set of edges, i.e., $\mathrm{E}_{\mathcal{G}} \subseteq (\mathcal{V}_{\mathcal{G}} \times \mathcal{V}_{\mathcal{G}})$, $\mathrm{A}_{\mathcal{G}}$ is weighted adjacency matrix. For an edge $(\mathrm{i}, \mathrm{j}) \in \mathrm{E}_{\mathcal{G}}$, $\mathrm{a_{ij}} = \mathrm{a_{ij}} = 1$, else $\mathrm{a_{ij}} = \mathrm{a_{ij}} = 0$. The degree matrix $\mathcal{D} = \mathrm{Diag}\{\mathrm{d_i}\}$ of graph , where $\mathrm{d_i} = \sum_{\mathrm{j} \in \mathcal{N}_{\mathrm{i}}} \mathrm{a_{ij}}$. $\mathcal{N}_{\mathrm{i}}$ is the set of neighbouring DG units defined as $\mathcal{N}_{\mathrm{i}} = \{\mathrm{j} \in \mathcal{V}_{\mathcal{G}} : (\mathrm{i}, \mathrm{j}) \in \mathrm{E}_{\mathcal{G}}, \mathrm{j} \notin \mathrm{i}\}$. The Laplacian matrix $\mathcal{L}_{\mathcal{G}} = \mathcal{D} - \mathrm{A}_{\mathcal{G}}$.

2.3. Preface on communication noise

In secondary control layer, the communication between the agents is compulsory requirement. The communication link irrespective of its type is impaired with the noise due to cyber-attack, sensor uncertainty and climate conditions. In this work, generalized Gaussian noise is considered to simulate the noisy communication links between DGs. The probability distribution function (PDF) for noise considered in this work is given as [14]

$$p(n(t)) = \frac{1}{\sqrt{2\pi\sigma^2}} exp\left(\frac{-1}{2\sigma^2}\left(n(t) - mean(n(t))\right)^2\right) \tag{8}$$

where σ^2 is the noise signal variance of $n(t)$. Various PDFs of communication noise with different noise variances are shown in figure 2, where mean is zero.

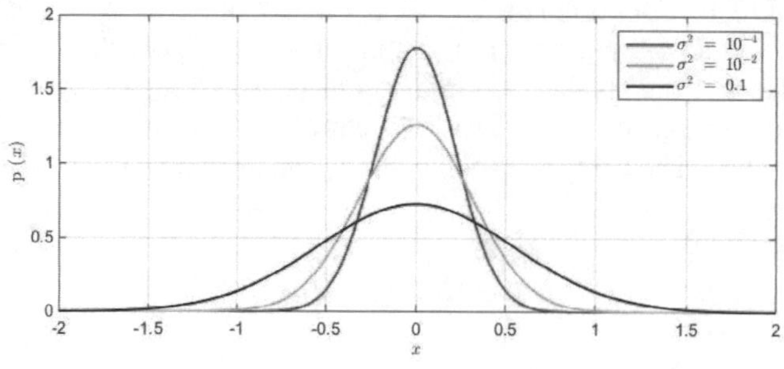

Figure 2. Probability distribution function of AWGN with different variance.

3. Distributed noise resilient voltage control

Feedback linearization is employed to convert the nonlinear DG system dynamics of ith DG into linear system. The state space model of voltage control for MG network in matrix form is given as follows [10, 14]

$$\dot{\boldsymbol{y}}_{iV} = \boldsymbol{A}\boldsymbol{y}_{iV} + \boldsymbol{B}\boldsymbol{u}_i^V, \forall\, \boldsymbol{i} \tag{9}$$

where, $\boldsymbol{y}_{iV} = \begin{bmatrix} y_{iV} & y_{iV,1} \end{bmatrix}^T$; $\boldsymbol{B} = [\boldsymbol{0} \;\; \boldsymbol{1}]^T$; and $\boldsymbol{A} = \begin{bmatrix} \boldsymbol{0} & \boldsymbol{1} \\ \boldsymbol{0} & \boldsymbol{0} \end{bmatrix}$. Also, the leader dynamics $\dot{\boldsymbol{y}}_0 = \boldsymbol{A}_0\boldsymbol{y}_0(\boldsymbol{t})$, with $\boldsymbol{y}_0 = [\boldsymbol{V}_{nom} \;\; \boldsymbol{0}]^T$; $\boldsymbol{u}_i^V = \boldsymbol{\psi}_i(\boldsymbol{x}_i) + \boldsymbol{g}_i^V(\boldsymbol{x}_i)\,\boldsymbol{u}_i$; with $\boldsymbol{u}_i$ denoted the voltage control input.

Considering a low voltage MG being a networked multi-agent system (MAS) with $\boldsymbol{N}$ DG units modelled into agents. The noisy communication channel links existing between the agents causes inaccurate output voltage and operating frequency regulation. Assuming the measured voltage information states depraved with noise can be given as.

$$\widetilde{\boldsymbol{y}}_{iV}(\boldsymbol{t}) = \boldsymbol{y}_{iV}(\boldsymbol{t}) + \boldsymbol{\rho}_i(\boldsymbol{t}), \boldsymbol{i} = \boldsymbol{1}, \dots, \boldsymbol{N} \tag{10}$$

where $\boldsymbol{\rho}_i(\boldsymbol{t})$ is the noise uncertainty. In this paper, we have assumed Additive White Gaussian Noise (AWGN) with $\boldsymbol{mean} = \boldsymbol{0}$ and $\boldsymbol{\sigma}^2$ as variance.

The noise resilient secondary synchronization control law can be designed as [14]

$$\boldsymbol{u}_i^V = -\boldsymbol{C}_{0V}\widetilde{\boldsymbol{y}}_i(\boldsymbol{t}) - \boldsymbol{\alpha}_V\boldsymbol{C}_V\left[\sum_{j\in\mathcal{N}_i}\left(\widetilde{\boldsymbol{y}}_i(\boldsymbol{t}) - \widetilde{\boldsymbol{y}}_j(\boldsymbol{t})\right) + \boldsymbol{q}_i\big(\widetilde{\boldsymbol{y}}_i(\boldsymbol{t}) - \boldsymbol{y}_0(\boldsymbol{t})\big)\right] + \boldsymbol{\delta}_i^V(\boldsymbol{t}) \tag{11}$$

where $\boldsymbol{\delta}_i^V(\boldsymbol{t})$ is the corrective auxiliary control signal to attenuate the noise effect.

The noise auxiliary control signal is designed as [14]

$$\boldsymbol{\delta}_i^V(\boldsymbol{t}) = \boldsymbol{C}_{0V}\widehat{\boldsymbol{\rho}}_i(\boldsymbol{t}) + \boldsymbol{\alpha}_V\boldsymbol{C}_V\left[\sum_{j\in\mathcal{N}_i}\left(\widehat{\boldsymbol{\rho}}_i(\boldsymbol{t}) - \widehat{\boldsymbol{\rho}}_j(\boldsymbol{t})\right) + \boldsymbol{q}_i\big(\widehat{\boldsymbol{\rho}}_i(\boldsymbol{t})\big)\right] \tag{12}$$

where

$$\dot{\widehat{\boldsymbol{\rho}}}_i = -\boldsymbol{\beta}_V\boldsymbol{A}^T\boldsymbol{P}\big(\widetilde{\boldsymbol{y}}_i(\boldsymbol{t}) - \widehat{\boldsymbol{y}}_i(\boldsymbol{t}) - \widehat{\boldsymbol{\rho}}_i(\boldsymbol{t})\big), \boldsymbol{t} \geq \boldsymbol{0} \tag{13}$$

and

$$\widehat{\boldsymbol{y}}_i(\boldsymbol{t}) = \boldsymbol{A}_0\widehat{\boldsymbol{y}}_i(\boldsymbol{t}) - \boldsymbol{\alpha}_V\boldsymbol{B}\boldsymbol{C}_V\left[\sum_{j\in\mathcal{N}_i}\left(\widehat{\boldsymbol{y}}_i(\boldsymbol{t}) - \widehat{\boldsymbol{y}}_j(\boldsymbol{t})\right) + \boldsymbol{q}_i\big(\widehat{\boldsymbol{y}}_i(\boldsymbol{t}) - \boldsymbol{y}_0(\boldsymbol{t})\big)\right] + \big(\boldsymbol{\beta}_V\boldsymbol{A}^T\boldsymbol{P} + \boldsymbol{\gamma}_V\boldsymbol{I}_N\big)\big(\widetilde{\boldsymbol{y}}_i(\boldsymbol{t}) - \widehat{\boldsymbol{y}}_i(\boldsymbol{t}) - \widehat{\boldsymbol{\rho}}_i(\boldsymbol{t})\big), \boldsymbol{t} \geq \boldsymbol{0} \tag{14}$$

with β_V and γ_V are design coefficients; $\hat{\rho}_i(t)$ and $\hat{y}_i(t)$ are the estimated values for noise and corrupted state for i^{th} DG unit; P is solution of linear matrix inequality and given as

$$I_N \otimes (A_0^T P + PA_0 - 2\gamma_V P) - \alpha_V \mathcal{F}_\mathcal{G} \otimes (C_V^T B^T P + PBC_V) < 0 \quad (15)$$

3.1. Theorem 1

A N DG unit networked MAS with connected undirected graph $\mathcal{G}$ is considered with dynamics given in equation (5) to equation (7). The control protocol designed in equation (12) and (13) provide a Lyapunov stable system, while restoring the voltage magnitude to nominal value i.e., $V_{ref} = 230\,V$ under noisy communication channel for LV MG.

3.2. Proof

The compact for of controller designed in equation (12) can be obtained using equation (9) as follows

$$\dot{y}(t) = \left[I_N \otimes A_0 - \alpha_V \mathcal{F}_\mathcal{G} \otimes BC_V\right] y(t) + \left(\alpha_V G_\mathcal{G} \otimes BC_V\right) y_0(t) - (I_N \otimes A)\rho + (I_N \otimes B)\delta^V(t), t \geq 0 \quad (16)$$

where $y(t) = [y_1^T(t), \dots, y_N^T(t)]^T$, $\rho = [\rho_1^T, \dots, \rho_N^T]^T$, and $\delta^V(t) = [\delta_1^T(t), \dots, \delta_N^T(t)]^T$. The compact form of $\hat{y}(t)$ can be given as

$$\dot{\hat{y}}(t) = \left[I_N \otimes A_0 - \alpha_V \mathcal{F}_\mathcal{G} \otimes BC_V\right] \hat{y}(t) + \left(\alpha_V G_\mathcal{G} \otimes BC_V\right) y_0(t) - \left(\alpha_V \mathcal{F}_\mathcal{G} \otimes BC_V + I_N \otimes BC_0\right)\hat{\rho} + (I_N \otimes B)\delta^V(t) + \psi(t), t \geq 0 \quad (17)$$

where $\hat{y}(t) = [\hat{y}_1^T(t), \dots, \hat{y}_N^T(t)]^T$ and $\psi(t) = [\psi_1^T(t), \dots, \psi_N^T(t)]^T$ with $\psi_i(t) = -\dot{\hat{\rho}}_i(t) + \gamma_V e_i(t)$.

Defining global error vector as $e_i(t) = \tilde{y}_i(t) - \hat{y}_i(t) - \hat{\rho}_i(t)$ and $\tilde{\rho}_i(t) = \rho_i - \hat{\rho}_i(t)$, and their compact form can be given as follows

$$\tilde{\rho}(t) = (I_N \otimes \beta_V A^T P) e(t), t \geq 0 \quad (18)$$

$$\dot{e}(t) = (A_e - \gamma_V I_N) e(t) - (I_N \otimes A)\tilde{\rho}(t), t \geq 0 \quad (19)$$

where $A_e = I_N \otimes A_0 - \alpha_V \mathcal{F}_\mathcal{G} \otimes BC_V$ and $e(t) = [e_1^T(t), \dots, e_N^T(t)]^T$ and $\tilde{\rho}(t) = [\tilde{\rho}_1^T(t), \dots, \tilde{\rho}_N^T(t)]^T$.

Based on the findings in [14], the Lyapunov stability of equation (18) and equation (19) is comparable with the control scheme's Lyapunov stability presented in equation (11) in terms of Lyapunov criterion. Considering the Lyapunov stability equation as

$$V(e, \tilde{\rho}) = \sum_{i=1}^{N} \left(e_i^T P e_i + \beta_V^{-1} \tilde{\rho}_i^T \tilde{\rho}_i\right) = e^T (I_N \otimes P) e + \beta_V^{-1} \tilde{\rho}^T \tilde{\rho} \quad (20)$$

Now finding the time derivative of equation (20) along the trajectories of (18) and (19) gives

$$\dot{V}(e, \tilde{\sigma}) = e^T \left(I_N \otimes \left(A_0^T P + PA_0 - 2\gamma_V P\right) - \alpha_V \mathcal{F}_\mathcal{G} \otimes \left(C_V^T B^T P + PBC_V\right)\right) e \leq 0, t \geq 0 \quad (21)$$

Now following the approach similar to [14], it can be shown that $\left(I_N \otimes \left(A_0^T P + PA_0 - 2\gamma_V P\right) - \alpha_V \mathcal{F}_\mathcal{G} \otimes \left(C_V^T B^T P + PBC_V\right)\right)$ is Hurwitz. Hence, equation (18) and equation (19) are Lyapunov stable for all $(e_0, \tilde{\rho}_0) \in \mathbb{R}_N \times \mathbb{R}_N$. This concludes the proof.

3.3. Remark 1

Using Theorem 1, $\lim_{t\to\infty} \tilde{\rho}(t) = 0$, and $\lim_{t\to\infty} e(t) = 0$, implies that $\lim_{t\to\infty}\left(y(t) - \hat{y}(t)\right) = 0$, i.e., state estimate $\hat{y}(t)$ coincide to the correct information state $y(t)$ for $t \geq 0$. Hence, presented controller in equation (11) guarantees voltage restoration for all DGs to nominal value for LV MG with noisy communication link.

4. Distributed noise-resilient frequency control

In this section a distributed secondary noise resilient operating frequency control input is presented using a analogous approach adapted in section 3. Defining the auxiliary control input after feedback linearization as follows

$$\boldsymbol{y_{i\omega} = \dot{\omega}_i = \dot{\omega}_{ni} + k_Q^{\omega} Q_i = u_i^{\omega}} \tag{22}$$

The corrupted measured frequency state is considered as follows

$$\tilde{y}_{i\omega}(t) = y_{i\omega}(t) + \rho_i(t), i = 1, \dots, \mathrm{N} \tag{23}$$

The secondary noise resilient frequency control signal with the auxiliary corrective control signal can be devised as follows [14]

$$\boldsymbol{u_i^{\omega} = -C_{0\omega}\tilde{y}_{i\omega}(t) - \alpha_{\omega} C_{\omega} \left[\sum_{j\in\mathcal{N}_i}\left(\tilde{y}_{i\omega}(t) - \tilde{y}_{j\omega}(t)\right) + q_i\left(\tilde{y}_{i\omega}(t) - y_{0\omega}(t)\right)\right] + \delta_i^{\omega}(t)} \tag{24}$$

and

$$\boldsymbol{\delta_i^{\omega}(t) = C_{0\omega}\hat{\rho}_i(t) + \alpha_{\omega} C_{\omega} \left[\sum_{j\in\mathcal{N}_i}\left(\hat{\rho}_i(t) - \hat{\rho}_j(t)\right) + q_i\left(\hat{\rho}_i(t)\right)\right]} \tag{25}$$

4.1. Theorem 2
The controller presented in equation (24) and equation (25) restores the operating frequency of all DG units to the nominal value under noisy communication link for a LV MG network, while guaranteeing accurate active power sharing among DGs.

4.2. Proof
This theorem can be proved similar to the Theorem 1.

5. Performance evaluation
The time domain case studies using MATLAB/Simulink framework in this section envisage the validity of the presented noise resilient secondary layer control scheme for islanded LV MG system (230 Volts, 50 Hz) as shown in figure. 3. The presented scheme is suitable for any communication topology with arbitrary *N* DG units. The test system and proposed control parameters are given in table 1 and table 2 respectively. The one line MG test system and communication network topology considered for information sharing is given in figure 3 and figure 4 respectively. The parameters of noise resilient distributed secondary control are selected as follows. $\alpha_V = 200$, $\beta_V = 40$, $\gamma_V = 80$, $\alpha_\omega = 200$, $\beta_V = 80$, and $\gamma_\omega = 100$. The reference operating voltage magnitude and frequency are $V_{ref} = 230\, V$ and $f_{ref} = 50\, Hz$ respectively.

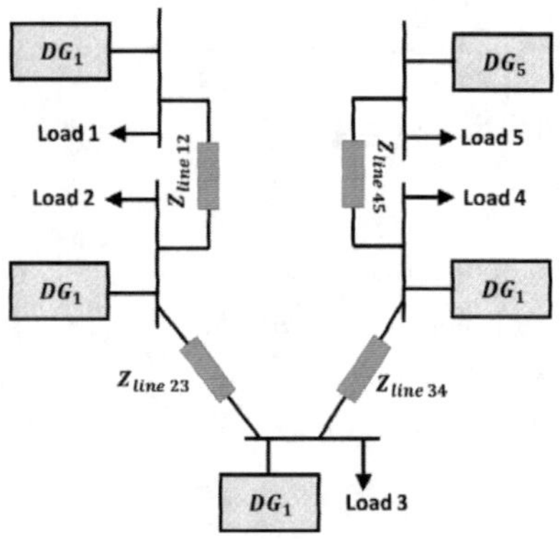

Figure 3. One line diagram test MG system.

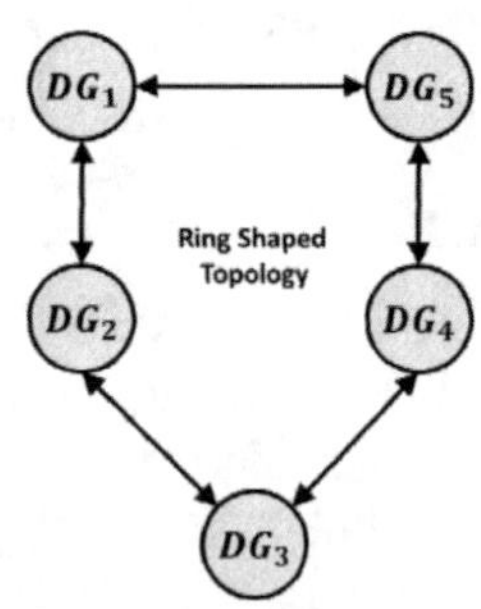

Figure 4. Communication graph.

IOP Conf. Series: Materials Science and Engineering **937** (2020) 012036 doi:10.1088/1757-899X/937/1/012036

5.1. Robustness to load perturbation

This case study shows the effectiveness of presented control scheme with load changing environment. The droop-boost primary control layer active with the start of simulation and the secondary control scheme is activated at

Table 1. Parameters of MG test system.

	DG #1 & DG #2 & DG #3				**DG #4 & DG #5**			
DGs	k_P^{ω}	9.4×10^{-5}			k_P^{ω}	12.5×10^{-5}		
	k_Q^{v}	1.3×10^{-3}			k_Q^{v}	1.5×10^{-3}		
	R_i	0.03Ω			R_i	0.03Ω		
	L_i	0.35mH			L_i	0.35mH		
	R_f	0.1Ω			R_f	0.5Ω		
	L_f	1.35mH			L_f	0.27mH		
	C_f	50μF			C_f	50μF		
	K_{PV}	0.1			K_{PV}	0.05		
	K_{IV}	420			K_{IV}	390		
	K_{PC}	15			K_{PC}	10.5		
	K_{IC}	20000			K_{IC}	16000		
Lines	$Z_{Line\ 12}$		$Z_{Line\ 23}$		$Z_{Line\ 34}$		$Z_{Line\ 45}$	
	$R_{Line\ 1}$	0.35Ω	$R_{Line\ 2}$	0.35Ω	$R_{Line\ 3}$	0.35Ω	$R_{Line\ 4}$	0.35Ω
	$L_{Line\ 1}$	0.07mH	$L_{Line\ 2}$	0.07mH	$L_{Line\ 3}$	0.07mH	$L_{Line\ 4}$	0.07mH
Loads	**Load 1**	**Load 2**	**Load 3**	**Load 4**	**Load 5**			
	$R = 300\Omega$	$R = 40\ \Omega$	$R = 50\ \Omega$	$R = 50\ \Omega$	$R = 50\ \Omega$			
	$L = 477\text{mH}$	$L = 64\ \text{mH}$	$L = 64\ \text{mH}$	$L = 64\ \text{mH}$	$L = 95\ \text{mH}$			

t= 2s. It restores the operating output terminal voltage and frequency in short time duration. Further, at t=3s, load of $R = 300\ \Omega$ and $L = 47\text{mH}$ is added to load 1, and at t=4s, load connected to DG 3 is switched off. The results in figure 5 and figure 6 show that the voltage and frequency restored to the nominal values very quickly after controller activation and maintained within permissible limits after experiencing slight deviations at small signal disturbances i.e., load variation. For simulation all the communication links are subjected to AWGN with $\sigma^2 = 10^{-2}$ and link delay of 100ms.

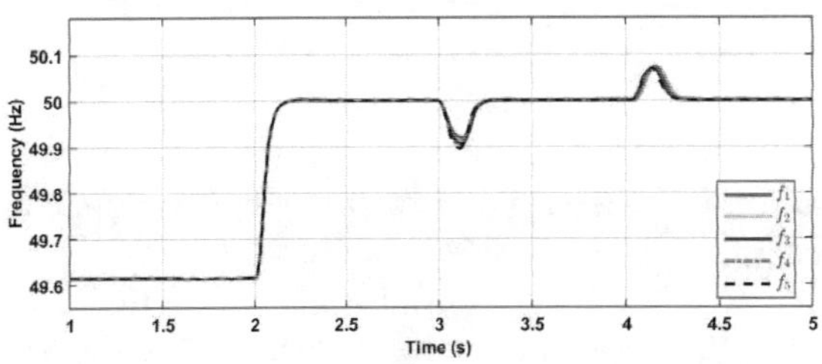

Figure 5. Frequency restoration Performance

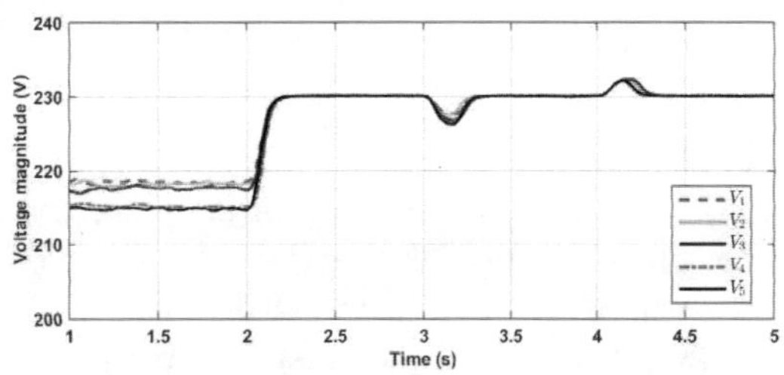

Figure 6. Voltage restoration Performance

5.2. Plug and Play (PNP) operation

The presented control methodology is evaluated for PNP case study in this subsection. The DG 5 is switched off and switched on again intentionally at t=3s and t=4s respectively. The communication topology for PNP operation is shown in figure 7. The obtained simulation results given in figure 8 and figure 9 depict the efficacy of presented control method for PNP operation and the voltage and frequency are maintain within permissible limit during high signal disturbance i.e., DG disconnection. The result in figure 10 and figure 11 show that, the total active and reactive power requirements are apportioned among DGs 1-4, while DG 5 is not associated to MG system. Also, operating output voltage and frequency are maintained within the nominal values i.e., 230 volts, 50 Hz with small transients at point of disturbance. Hence, presented scheme supports the PNP operation of LV MG network under noisy environment.

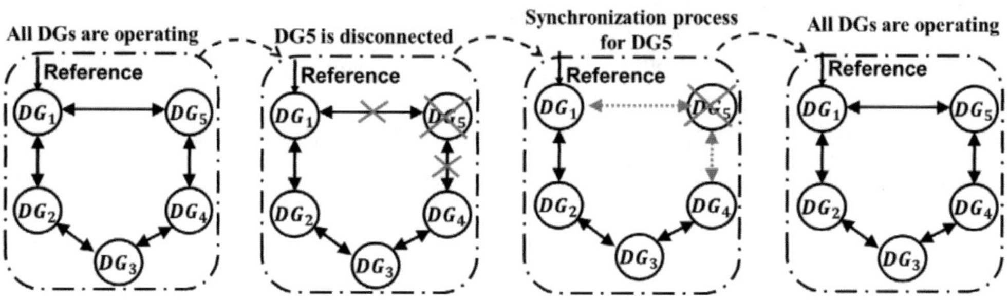

Figure 7. Communication graph for PNP operation

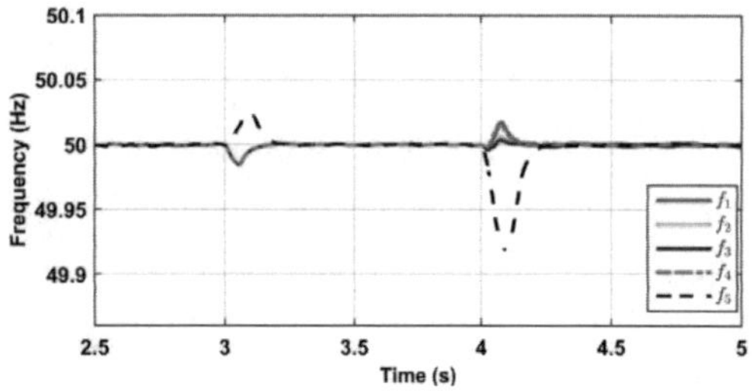

Figure 8. Frequency under PNP operation

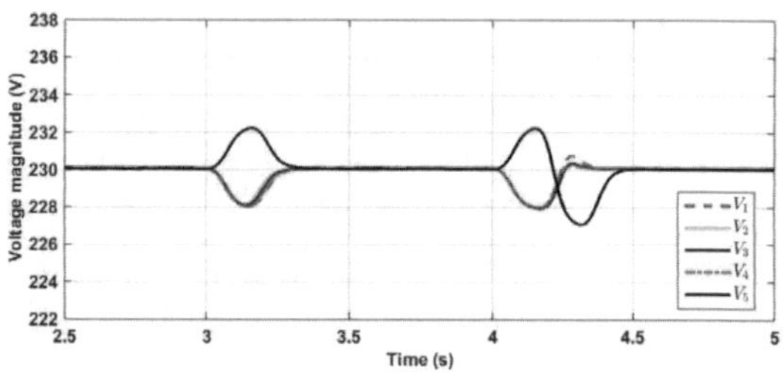

Figure 9. Voltage under PNP operation

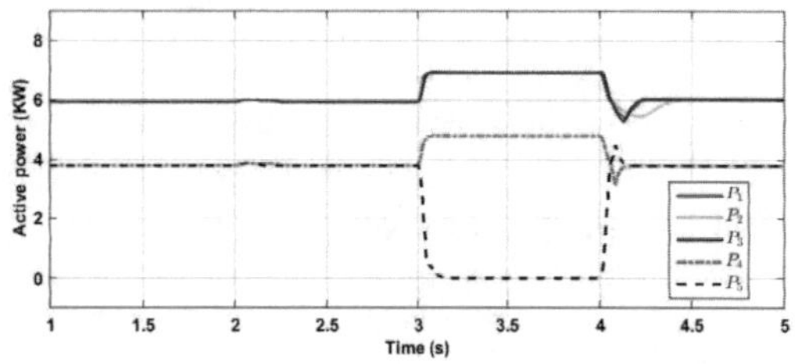

Figure 10. Active power under PNP operation

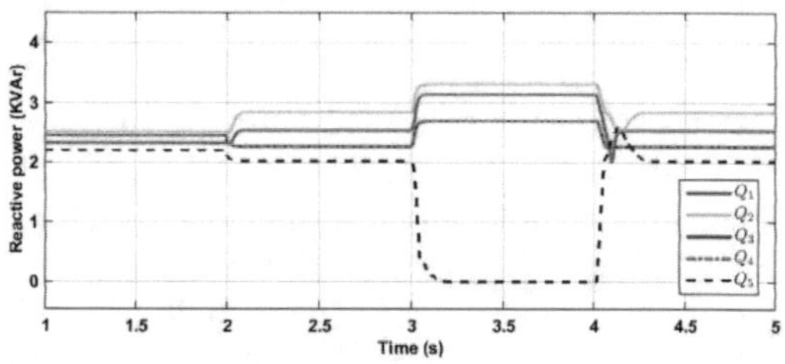

Figure 11. Reactive power under PNP operation

5.3. Robustness under variation in noise parameter

In this case study, the efficacy of the presented noise resilient secondary control method is evaluated under time- varying noise variance σ^2. The noise variance σ^2 varies in three stages: (a) $t \leq 2.5, \sigma^2 = 10^{-2}$; (b) $2.5 < t \leq 3.5, \sigma^2 = 10^{-1}$; and (c) $3.5 < t \leq 5, \sigma^2 = 0.3$. The results obtained in figure 12 and figure 13 clearly envisages the operating voltage magnitude and frequency of DGs are synchronized under permitted limit with different noise level respectively.

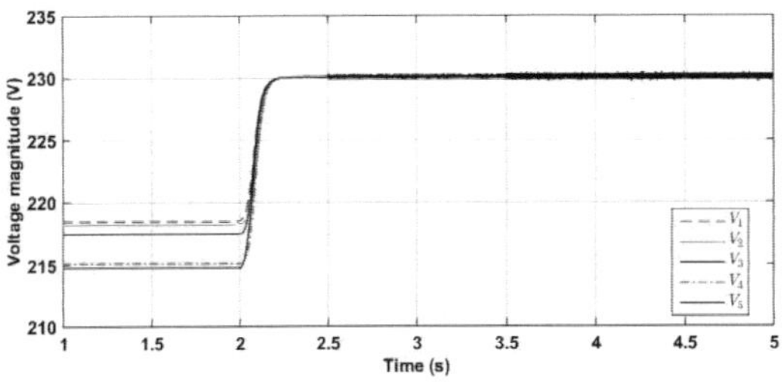

Figure 12. Voltage synchronizatio with time varying σ^2

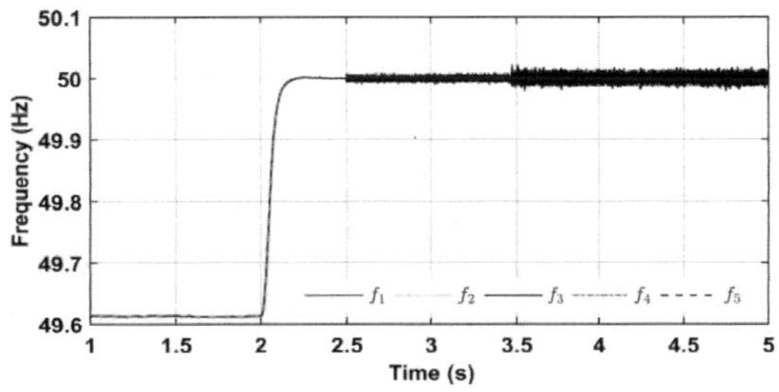

Figure 13. Frequency synchronization with time varying σ^2

6. Conclusion

The work presented in this paper deals with noisy communication links in secondary control layer. A noise resilient distributed secondary control methodology for a low voltage islanded MG network. This work considers a low voltage grid where the lines are dominantly resistive in nature and the information sharing communication channels among the DGs are corrupted with AWGN. The presented a primary droop-boost control augmented with noise resilient distributed secondary control scheme to restore the DG operating output terminal voltage and frequency while providing definitive real power sharing in finite time. Further, the control presented is self-reliant and does not depend on the noise type and required only neighbours information for control and corrective signal, hence can be implemented over a minimal communication network with connected graph. The stability proof is derived using rigorous Lyapunov analysis while considering complete nonlinear dynamics of LV MG network. The simulation results shows that the presented method envisage robust and resilience performance even during load perturbation, loss of DG unit and time varying noise parameter variation. The accurate reactive power sharing is still a problem for future consideration for authors along with extension of presented work for hybrid AC/DC MG and DC microgrid.

References

[1] Guo F, Wen C, Mao J, and Song Y-D 2015 Distributed secondary voltage and frequency restoration control of droop-controlled inverter-based microgrids *IEEE Transactions on Industrial Electronics* **62**(7) pp 4355–4364

[2] Guerrero J M, Vasquez J C, Matas J, Vicuna L G De, and Castilla M 2011 Hierarchical control of droop-controlled ac and dc microgridsa general approach toward standardization *IEEE Transactions on Industrial Electronics* **58**(1) pp 158–172

[3] Bidram A and Davoudi A 2012 Hierarchical structure of microgrids control system *IEEE Transactions on Smart Grid* **3**(4) pp 1963–1976

[4] Shafiee Q, Guerrero J M, and Vasquez J C 2014 Distributed secondary control for islanded microgridsa novel approach *IEEE Transactions on Power Electronics* **29**(2) pp 1018–1031

[5] Li Z, Zang C, Zeng P, Yu H, and Li H 2016 Mas based distributed automatic generation control for cyber-physical microgrid system *IEEE/CAA Journal of Automatica Sinica* **3**(1) pp 78–89

[6] Shrivastava S, Subudhi B and Das S 2018 Distributed voltage and frequency synchronisation control scheme for islanded inverter-based microgrid *IET Smart Grid* **1**(2) pp 48–56

[7] Bidram A, Lewis FL and Davoudi A 2014 Distributed control systems for small scale power networks: Using multiagent cooperative control theory *IEEE Control Systems* **34**(6) pp 56–77

[8] N L Soultanis, S A Papathanasiou and N D Hatziargyriou 2007 A Stability Algorithm for the Dynamic Analysis of Inverter Dominated Unbalanced LV Microgrids *IEEE Transactions on Power Systems* **22**(1) pp 294-304

[9] Bidram A, Davoudi A, Lewis FL and Guerrero JM 2013 Distributed cooperative secondary control of microgrids using feedback linearization *IEEE Transactions on Power Systems* **28**(3) pp 3462–3470

[10] Dehkordi NM, Baghaee HR, Sadati N and Guerrero J M 2019 Distributed noise-resilient secondary voltage and frequency control for islanded microgrids *IEEE Transactions Smart Grid* **10**(4) pp 3780–3790

[11] Romero ME and Seron MM 2019 Ultimate boundedness of voltage droop control with distributed secondary control loops *IEEE Transactions Smart Grid* **10**(4) 4107–4115

[12] Liu S, Wang X and Liu PX 2015 Impact of communication delays on secondary frequency control in an islanded microgrid *IEEE Transactions Industrial Electronics* **62**(4) pp 2021–2031

[13] Hu J and Feng G 2010 Distributed tracking control of leader–follower multi-agent systems under noisy measurement *Automatica* **46**(8) pp 1382–1387

[14] Shrivastava S, Subudhi B and Das S 2018 Noise-resilient voltage and frequency synchronisation of an autonomous microgrid *IET Generation, Transmission & Distribution* **13**(2) pp 189-200

[15] Shrivastava S, Subudhi B and Das S 2017 Voltage and frequency synchronization of a low voltage inverter based microgrid *4th International Conference on Power, Control & Embedded Systems (ICPCES)* IEEE pp 1-6

IOP Conf. Series: Materials Science and Engineering **937** (2020) 012037 doi:10.1088/1757-899X/937/1/012037

Generation of energy from Solar and its protections

A. Rajkumar [1] J. Nagendran[2] J. Subramaniyan[3] R. Jai Ganesh[4]

[1] Professor, Department of Electrical and Electronics Engineering,
K. Ramakrishnan College of Technology, Trichy, Tamilnadu.
[2] Assistant Professor, Department of Electrical and Electronics Engineering,
Dhanalakshmi Srinivasan Engineering College Perambalur.
[3] Assistant Professor, Department of Electrical and Electronics Engineering,
SRM TRP Engineering College. Samayapuram, Trichy
[4] Assistant Professor, Department of Electrical and Electronics Engineering,
K. Ramakrishnan College of Technology, Trichy, Tamilnadu.

E-mail: arkumar77@gmail.com

Abstract. In this article, a seven-level inverter powered by solar has been proposed to achieve a sinusoidal output voltage with high efficiency and enhanced power quality. This system consists of active inverter and flipped condenser clamping. It gives output voltage level of 2/3. By connecting the switched condenser branch in the front or back end series to the inverter output a seven-level output voltage wave shape can be obtained with ample non-work switching states to match the voltage of switching condensers. The consequence of the proposed method is validated using MATLAB SIMULINK for simulation. Comparing with conventional seven-level inverter topology, the proposed design used a minimal number of switches using a basic modulation process. The demand for electricity that is so high in India might be a key driver for the use of renewable power as an option for energy requirements. To satisfy this need, this article introduces a solar smart flower layout and then modeled using the resources provided in the solid function system. And use a set Photovoltaic panels harvest additional power mostly during 12 pm to 2:00 Pm in Nigeria which resulted in less renewable energy. Consequently, there is a need to increase the energy consumption of PV solar panel by developing a solar tracker cannot be overlooked. Photovoltaic cells must be aligned with the sun in order to just get optimum energy.

The microcontroller is being used to power the motor dependent on information gathered from the LDRs. The findings of this research have clearly demonstrated that the monitoring solar panel generates more electricity compared to a conventional panel. The location of solar panel including voltage was going to monitor by using IoT .Light Dependent Resistors LDRs are being used to detect the strength of lighting and thus the PV solar panel is appropriately addressed to control maximum energy. The system uses motors to regulate the amount of the solar panel. The microcontroller is being used to power the motor centered on signals obtained from the LDRs. The outcome of this work has demonstrated clearly that the monitoring solar panel generates more power compared to a conventional panel. The location of solar panel and volts was going to monitor through using IoT.

1. Introduction

As the energy consumption and the international pollution rises with economic and population growth, the based on energy issues and global reflects light are today a matter of growing concern. It is the whole of the hour to leverage traditional renewable resources that are also abundant that is obtained from assets which are obviously replenished naturally such as sunlight , wind , rain, geothermal and solar, etc. Solar energy was among the most significant renewable energy potential which is available in nature especially in tropical areas. In particular India has fairly long warm days for even more than ten months and partially cloudy weather for much of the days of the remaining 2 months. Solar cell is an electrical signal that transforms energy from sunlight heat into electricity through photoelectric effect. Solar cells are turning points of photovoltaic (pv) systems, better known as solar panels. Solar panel is a group of photo voltaic panels which is directly attached and positioned on structural support to consume sun rays as sources of energy is a form or heating. The last several years are intelligently advertised via modern electronic display advertising of solar 'intelligent flower and many aren't engaged. It may well be due to the really high quality and lack of access of service. This chapter gives some studies that attempt to create a layout for solar smart flower but instead examine it all through simulator using great work before completing the experience of its fabricated. The smart flower has always been a-in-one ground installed solar panel with a monitor that tracks the sun. It nickname came from its architecture the solar cells are mounted on individual "pink flowers"that open at the start of the day. Once the sun sets, the smart flower petals lay flat and a self cleaning cycle steps in. In comparison to solar cells, the intelligent flower device includes a double-axis tracker which makes it easy for its "flowers" to face the sun via the sky during day. Due to this monitoring feature, the smart flower will generate substantially more energy than a similarly sized roof solar array device – up to 51 % more, as per smart flower homepage. The 12-petal, 194 square foot installation of a new with 2.5 kilowatts (kW) of energy production, which is equal to a 4 kW static roof array. There will be two models including its smart flower available in the United States, each of which is produced in Austria: a smart flower and the smart flower-plus. The smart flower plus provides energy storage capabilities through embedded batteries in addition to regular solar energy production that the smart flower provides.

2. Existing System

Existing machine built while using Arduino Uno Microcontroller, four LDRs and three stepper motors. The machine is powered by set of technologies and software coding. In device production, four light based resistors LDRs would be used for collecting peak light intensity. Three drivers are being used to shift the solar array as per the maximum incoming light regulated by LDRs. This same program controls this same vertical tilt and longitudinal motion of the solar panel. And this can obey the position of the sun, not just the longitudinal revolution as well as the horizontal motion as per the incidence natural sunlight mostly on solar panel.

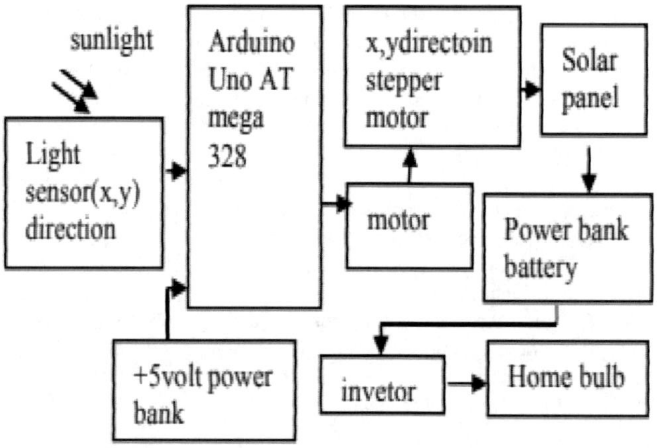

Figure 1. Block diagram of the existing system

2.1 Drawbacks of Existing Systems

- Solar trackers are usually designed for climatic conditions with next to no winter wanting to make someone to a much more effective analysis in hotter areas. Set racking incorporates harsh climatic conditions quite effectively than tracking devices.

- Fixed monitoring systems offer the most ground adaptability unlike single-axis tracking systems.

- For existing solar panels, that were without control systems typical level of efficiency ranges from 10 percent to 4 percent-a level it should increase markedly if the current concern persists.
- For hydraulic monitoring system we had also chosen a single - axis solar mechanical tracking system that used an Electric motors for our layout.

3. Proposed system

The components consists of a solar panel, physical system for installing panel, Motor controller for spinning the physical set up, analog to digital, Micro control,, communication and transport, Level connector and a IOT unit. The solar monitoring system consists of a solar panel, microcontroller and indicators. For this device to work there has to be energy released through the sun. The LDRs act as the detectors to measure the strength of light reaching the photovoltaic power. The LDR instead sends details to the microcontroller. The motor, relay circuit is then formed. The rain sensor will detect the rain so if rainfall arrive that perhaps the time instantly solar cell shuttered and switch to the way down. Voltage divider is used to screen the reference voltage with battery. The location of solar array and volt was tracking by using IoT .Light Dependent Resistors are used to feel the strength of daylight and therefore the PV solar cell is modified accordingly to control additional power.

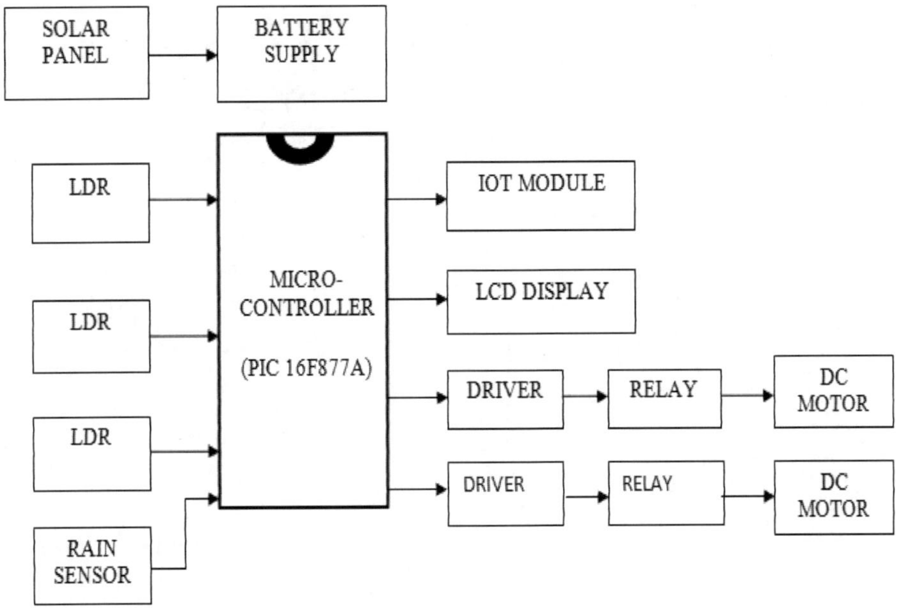

Figure 2. Block Diagram of Proposed System

In this article is developed mainly in to the 3 categories, building of hardware, monitoring system and device control module. This solar tracker utilizes the pic controller panel, dc motor, 3 LDR and, 3 resistors as well as adjust the solar array towards to the heat or a sky light. The solar panels have the benefit of improving the utilization of collecting solar energy while the planet begins its noble revolution. The triple LDRs are acting as light sensors. They are positioned for each sides of solar array and three dc motors are being used to adjust the solar panel. Since the same intensity of daylight falls on LDRs, then electric motors do not turn. Solar panel will still follow a direct sunlight and therefore will look toward this sun to even get charging all the period and provide the output maximum capabilities. The power density can be did receive from either a static array of solar panels at a particular time. The figure 3 shows the schematic diagram of the proposed system.

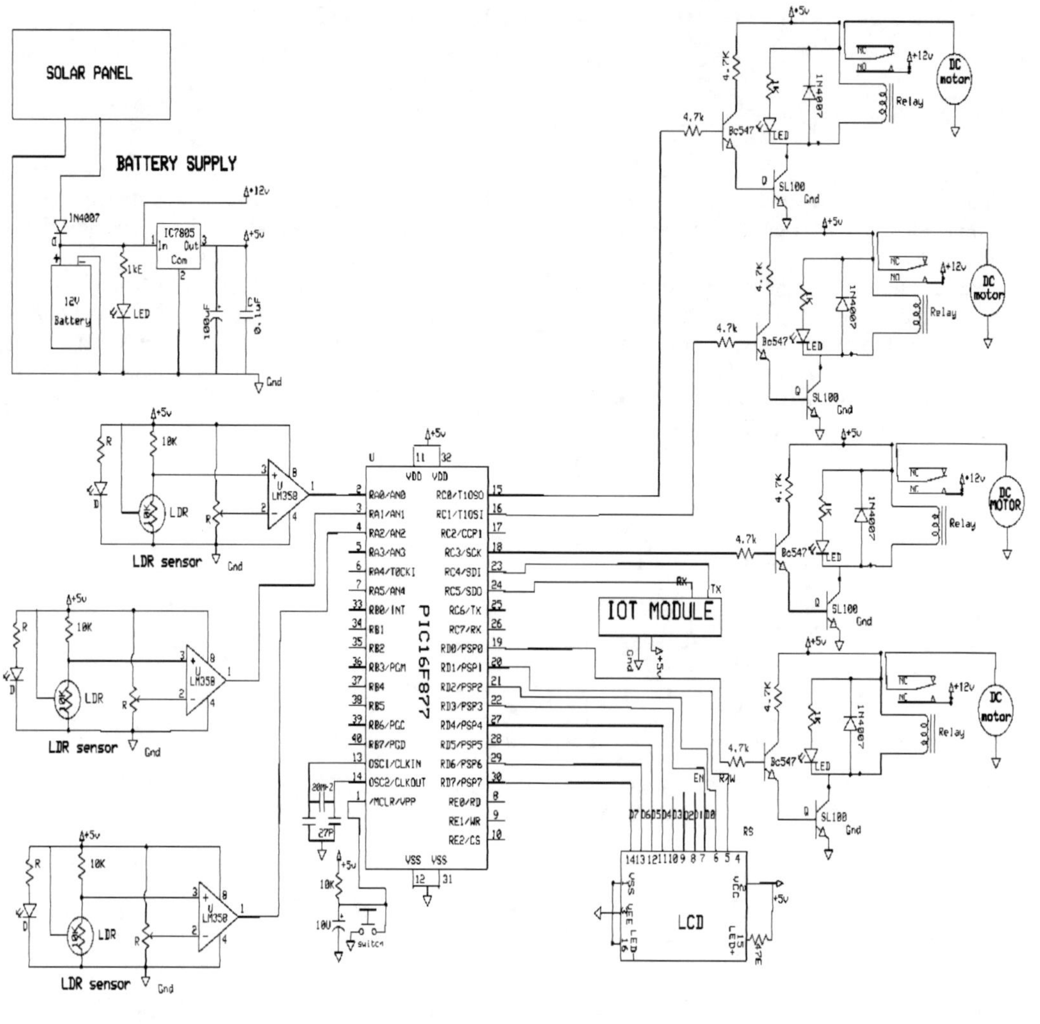

Figure 3. Schematic Diagram of Proposed System

4. Results and Discussion

Based on the above discussion the solar PV panel has been protected from the many environmental changes such as rain, humidity and other natural disturbances. Based on the LDR and rain sensor rating the sensor will senses the temperature and irradiance or humidity level from the environment.

A DC motor was proposed to control the PV panel to fold and rotate to prevent from water other environmental factors. The hardware of the proposed work is shown in the figure 4 and figure 5. The figure 4 will represent the controller parts and sensors arrangements. In the figure 5 represent the physical look of PV panel arrangements.

IOP Conf. Series: Materials Science and Engineering **937** (2020) 012037 doi:10.1088/1757-899X/937/1/012037

Figure 4. Controller parts and sensors arrangements

Figure 5. Physical look of PV panel arrangements

5. Conclusion

Using this approach, the solar tracker can be victorious in maintaining a solar panel at a perpendicular angle to the sun. In addition, precipitation sensors can be navigated to maintain the device working during raining. Other than, in hereafter, because once standard energy may not have been adequate for us, there will also be an actually constitute of trying to manipulate alternative energy source. Through using this immediate management, it is possible for a single individual to construct the device directly.

References

[1] Sustainable Energy Development Authority (SEDA) Malaysia. Available: http://seda.gov.my. Accesed:[05 June,2016].
[2] G. K. Singh, "Solar power generation by PV (photovoltaic) technology: A review," Energy, vol. 53, pp. 1-13, 2013.
[3] S. Kucuksari, A. M. Khaleghi, M. Hamidi, Y. Zhang, F. Szidarovszky, G. Bayraksan, and Y.-J. Son, "An Integrated GIS, optimization and simulation framework for optimal PV size and location in campus area environments," Applied Energy, vol. 113, pp. 1601-1613, 2014
[4] R. Al Abri, E. F. El-Saadany, and Y. M. Atwa, "Optimal placement and sizing method to improve the voltage stability margin in a distribution system using distributed generation," IEEE Transactions on power systems, vol. 28, pp. 326-334, 2013.
[5] B. Vonk, M. Gibescu, E. Veldman and J. Sloothweg, “Automatic PV productions profile generation using geographic and historical weather data”, in 19th International Universities Power Engineering Conference (UPEC), 2014, pp. 1-6
[6] H. Rowlands, B. P. Kemery, and I. Beausoleil-Morrison, "Managing solar-PV variability with geographical dispersion: an Ontario (Canada) case-study," Renewable Energy, vol. 68, pp. 171-180, 2014.
[7] S. Sharma, K. Kumar Jain, A. Sharma, “Solar cells: In research and Applications – A review”, Materials Science and Applications, vol. 6, pp. 1145-1155, 2015
[8] O. Ellabban, H. Abu-Rub, and F. Blaabjerg, "Renewable energy resources: Current status, future prospects and their enabling technology," Renewable and Sustainable Energy Reviews, vol. 39, pp. 748-764, 2014.
[9] E. Romero-Cadaval, G. Spagnuolo, L. G. Franquelo, C. A. RamosPaja, T. Suntio, and W. M. Xiao, "Grid-connected photovoltaic generation plants: Components and operation," IEEE Industrial Electronics Magazine, vol. 7, pp. 6-20, 2013.
[10]R. G. Vieira, F. K. O. M. V. Guerra, M. R. B. G. Vale, M. M. Araujo, “ Comparative performance analysis between static solar panels and single-axis tracking system on a hot climate region near to the equator”, Renewable and Sustainable Energy Rerviews, vol. 64(2016), pp. 672- 681, 2016
[11] R. Kansal, "PIC based automatic solar radiation tracker," Asian Journal of Chemistry, pp. 21-28, 2009. [12]. R. A. Ferdaus, M. A. Mohammed, S. Rahman, S. Salehin, and M. A. Mannan, "Energy efficient hybrid dual axis solar tracking system," Journal of Renewable Energy, vol. , 2014.
[12]Dr.A. Rajkumar, R. Jai Ganesh, V. Suresh Kumar, T. Vishnu Kumar and T. Ram Kumar,” Implementation of High Efficient Single Input Triple output DC-DC Converter,” Biosc.Biotech.Res.Comm. Special Issue Vol 13 No (3) 2020 Pp-48-55.
[13]A. T. Sankara Subramanian, P. Sabarish, M. D.Udayakumar and T. Vishnu kumar,” Performance Analysis of Various Photovoltaic Configurations Under Uniform Shading and Rapid Partial Shading Formations”, Biosc.Biotech.Res. Comm. Special Issue Vol 13 No (3) 2020 Pp-185-192.
[14]P. Sabarish et al 2019 IOP Conf. Ser.: Mater. Sci. Eng. 623 012011
[15]A T Sankara Subramanian et al 2019 IOP Conf. Ser.: Mater. Sci. Eng. 623 012012
[16]M. Pradeep et al 2019 IOP Conf. Ser.: Mater. Sci. Eng. 623 012017
[17]M D Udayakumar et al 2019 IOP Conf. Ser.: Mater. Sci. Eng. 623 012018
[18]P. Thirusenthil Kumaran, P. Pushpakarthick, G. V. Chidambarathanu, M. Venkatachalam, Xavier Raja Durai, R.Jaiganesh, “Power quality in distribution grids ,” International Journal of Innovative Technology and Exploring Engineering (IJITEE) ISSN: 2278-3075, Volume-9, Issue-1, November 2019.
[19]S.Vijayalakshmi, P.Sabarish, S.R.Paveethra, Dr.P.R.Sivaraman, Dr.V.Venkatesh, “Exploration And

Applications Of Electronic Balance For High Power Discharge Lamps At High Frequency Through Power Factor Modification", INTERNATIONAL JOURNAL OF SCIENTIFIC & TECHNOLOGY RESEARCH VOLUME 9, ISSUE 02, FEBRUARY 2020 ISSN 2277-8616.

[20] C.Kalavalli, S.R.Paveethra, S.Murugesan Dr.A.Nazar Ali Dr.V.Venkatesh, "Design And Implementation Of High Efficiency H6 PV Inverter With Dual Axis Tracking", INTERNATIONAL JOURNAL OF SCIENTIFIC & TECHNOLOGY RESEARCH VOLUME 9, ISSUE 02, FEBRUARY 2020 ISSN 2277-8616.

[21] S.R.Paveethra, C.Kalavalli, S.Vijayalakshmi, Dr.A.Nazar Ali, D.Shyam," Evaluation Of Voltage Stability Of Transmission Line With Contingency Analysis", INTERNATIONAL JOURNAL OF SCIENTIFIC & TECHNOLOGY RESEARCH VOLUME 9, ISSUE 02, FEBRUARY 2020 ISSN 2277-8616.

[22] R.Jai Ganesh ,S.Kodeeswaran, M.Kavitha, T.Ramkumar, Performance analysis of piezoelectric energy harvesting system employing bridgeless power factor correction boost rectifier", https://doi.org/10.1016/j.matpr.2020.02.085.

[23] Murugesan. S., and M. V. Suganyadevi. "Hybrid Renewable Energy Parameter Monitoring and Control of Smart Street Light Using IoT." International Journal of Scientific and Technology Research(IJSTR), ISSN: 2277-8616, Volume 8, Issue 10, pp.645–651, October 2019.

[24] Praveen, A. Anton Amala, and S. Murugesan. "Integration of Hybrid Energy Sources for Grid Connected AC Distribution System." Proceedings of the 4th IEEE International Conference on Advances in Electrical and Electronics, Information, Communication and Bio-Informatics, AEEICB 2018. Institute of Electrical and Electronics Engineers Inc., 2018. Proceedings of the 4th IEEE International Conference on Advances in Electrical and Electronics, Information, Communication and Bio-Informatics, AEEICB 2018. Web.

Analyzing the performance of Cascaded H-Bridge Multilevel Inverter with three phase multiwinding transformer and single phase supply

P.Kathirvel[1], V.Karpagam[2], K.Vijayakumar[3], A.Venkatesh[4]

[1,4] Assistant Professor, Department of EIE, Dr.Mahalingam College of Engineering & Technology, Pollachi, Tamilnadu.
[2] Assistant Professor(SS), Department of EIE, Dr.Mahalingam College of Engineering & Technology, Pollachi, Tamilnadu.
[3] Associate Professor, Department of EIE, Dr.Mahalingam College of Engineering & Technology, Pollachi, Tamilnadu.

*Email: kathir178@drmcet.ac.in

Abstract. A multi-level inverter type cascaded H-bridge design with a different harmonic reduction technique is provided. The proposed design includes the DC source, hybrid cascaded multilevel inverter, transformer and load circuit. For a hybrid, multi-level inverter, the 7-level stepped form of the wave is achieved similar to the sinus wave. The transformer supplies the seven-tier hybrid inverter with cascading into the load. The 5th and 7th harmonics were efficiently extracted with the algorithm from the line current. The simulation and research was carried out to show how the system proposed works. The observed harmonic detail results, the THD are approximately 12.27% only. The 3rd level harmonics magnitude is 5.2% and 5th level harmonics is only 8.2%. Simulation results from SIMULINK/MATLAB software and hardware experimental results are tabulated. The results are explained with graphically.

Index Terms - Multilevel inverter, Single phase DC link, Total Harmonic distortion (THD), Flying capacitors, Linear load.

1. Introduction

A broad rule has been developed in several applications for the multilevel converter. This multi-grade converter can be used on both sides of energy supply, as an electrical connection between wind, solar power, biomaterial, waste & bagasse, high power photovoltaic modules. Cascaded H-Bridge multi-level inverter is one of the commonest multi-level topologies. In both 1 and 3 stages, the transformation is probable. You are using the H-Bridge button and diode. A multi-level inverter needs at least three voltage levels. The converters are supplied with a number of voltage sources for switches and condensers. The unit produces steady output voltages with fewer harmonic fluctuations when the MOSFET switch is triggered and deactivated. Multi-level inverters are used as one of the latest power converter zones in production plant and public electricity boards because the disadvantages of standard pulse width modulators [6] can be managed [11]. Three different H-bridge cell types, such as diode clamped or floating condensers, with a separate dc source can be described [7].

1.1 Inverter Type Clamped Diode

The diode-clamped inverter is used for the diodes and supplies condensers in series connected by several voltage levels through different phase values. Send diodes to lower the load on other power supplies with the set voltage frequency. The input DC's average stress is 50%. The switches, diodes and circuit condensers solve this problem. The dc-Link tension of the condenser terminals is regulated on three levels because of condenser safety problems [1, 8].

1.2 Flying type Capacitor Inverter

A number of condenser-connected switching diodes exist. The condensers relay the small voltage to the water. The conditions are similar to those of the diode-clamped inverter. In this type of multi-level inverter, clamping diodes are not required. The voltage is half the voltage of the DC signal.

1.3 H-Bridge Cascaded Cells with different DC Sources

Each H-Bridge module includes a separate dc-link voltage [1, 7] and a new multicolored H-bridge cascade inverter form. — The inverter form is different and calls for fewer components – steps, including condensers and mouse switches. The same dc relation voltage is to be used separately [4] in that H-bridge network. There's some pain. The new goods have been equipped with reduced reactors and condensers, substituted with separate data sources which need at least three independent sources. The two-way switch approach was taken to identify the difference in input and output; while the source of input is adequate, it contributes to the complex configuration of the circuit with the implementation of two-way switches.

2. Multilevel Inverter with Cascaded H-Bridge

A one dc power supply and three separate 3 phase Low Frequency transformers are available on the Cascaded H-bridge multilateral inverter. Compared to conventional three-phase multi-level transformers, the latest design will that a number of transformers. This makes the installation of a simple, powerful inverter. The Newto Raphson process, based on each switch's operation, calculates every relay corner for each switch. The linearization technique can be used in each area for all relay angles. Through this process, low-harmonic tension components are removed [3].

The Cascaded multi-level inverter promises high-powered electrical energy topology [2] due to its low EMI and low frequency control methods [6]. The current stage includes a 2n + 1 level of n-DC sources for a cascading multi-level inverter. The use of this inverter will avoid the use of too many different DC sources in many applications. In the Cascading H-bridges inverter known as the HC-bridge multilevel inverter (H-bridge) [4], a first source may be used with other (n-1)-DC sources to eliminate the need for DC sources to be linked to a separate engine drive. HCMLI can be used for a specific motor powertrain.

Fig. 1 demonstrates a design circuit with four separate H-bridges for inverter devices H-bridge [9]. This has a nine-level output voltage. The voltage is low, with H-bridge cells linked by series.

2.1 Working principle of Hybrid Cascaded Multilevel Inverter (HCMLI)

Condensers can be used to operate a Cascaded Multi-Level Inverter with one DC source as a DC source for all but the first source [3]. Build a two H-bridge inverter as shown in the Fig.1. A battery or fuel cell with a Vdc voltage is DC source for start of the H Bridge while the dc source is a Vc condenser for the next H bridge. The ultimate voltage of the H start is V1 and the corresponding H-bridge is V2.

$$v(tf) = v1\ (ts) + v2\ (tn) \tag{1}$$

IOP Conf. Series: Materials Science and Engineering **937** (2020) 012038 doi:10.1088/1757-899X/937/1/012038

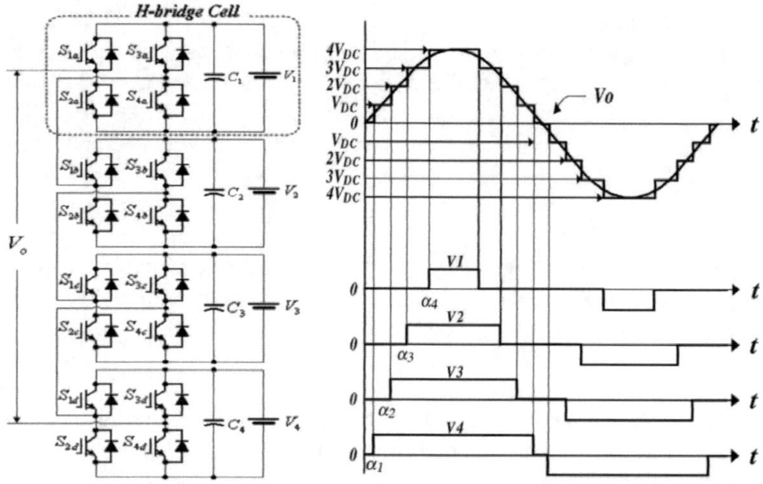

(a) (b)

Figure 1. Traditional 9 –level cascaded type H-bridge cell multilevel inverter (a) Circuit diagram (b) output waveforms

The Next Bridge output voltage can be set to 0, -Vdc, Vdc and next output voltage from −Vc, 0, or Vc is adjusted by switching off and on by activating and disabling the starting bridge switches. The reverser voltage also can be set to nine end-rate values – Vdc+Vc, Vdc, −Vc, Vdc and Vc. In other situations the reverser voltage can also be set at 9 final-rate values. The inverter tension is then available at the end of the device. All possible voltage levels should be used in the condenser voltage control scheme. A Regular Vdc/2, -3Vdc/2, Vdc, −vdc, – vdc/2, Vdc/2, 3Vdc/2 can be used. If the VC with a Vdc/2 voltage capacitor is made, Fig. 4 displays the final values as shown in Table 1. and Fig. 2.

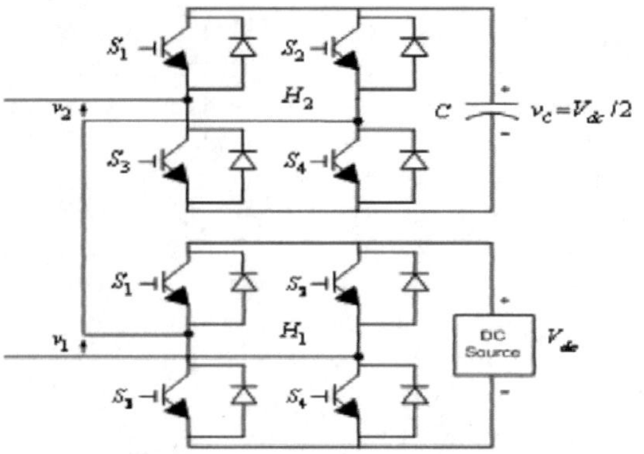

Figure 2. 1ø Multilevel Inverter

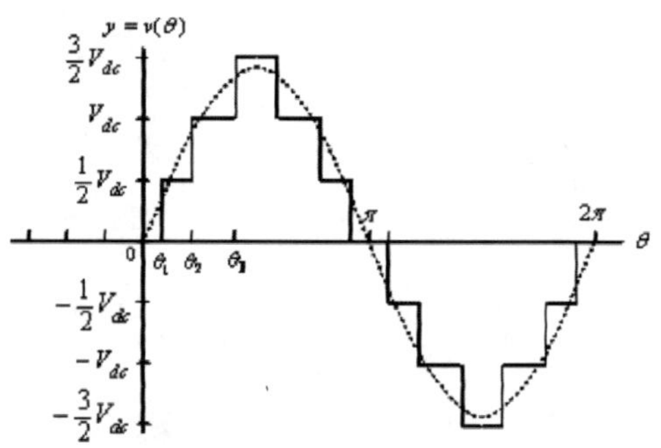

Figure 3. 7- Different level 1- Step same Voltage Waveform

The Fig. 4 displays θ1 ≤ θ < θ2 power for the v1 = Vdc and v2 = − Vdc/2 and waveform voltage of H-bridge voltages. When θ1 ≤ θ < θ2 is used, Fig. 3 will be produced when V1 = Vdc and V2 −Vdc/2 are selected.

IOP Conf. Series: Materials Science and Engineering **937** (2020) 012038 doi:10.1088/1757-899X/937/1/012038

Table 1. Seven-level inverter output voltages

θ	V_1	V_2	$V = V_1 + V_2$
$0 \le \theta < \theta_1$	0	0	0
$\theta_1 \le \theta < \theta_2$	0	$V_{dc}/2$	$V_{dc}/2$
$\theta_1 \le \theta < \theta_2$	V_{dc}	$-V_{dc}/2$	$V_{dc}/2$
$\theta_2 \le \theta < \theta_3$	V_{dc}	0	V_{dc}
$\theta_1 \le \theta < \pi/2$	V_{dc}	$V_{dc}/2$	$3V_{dc}/2$

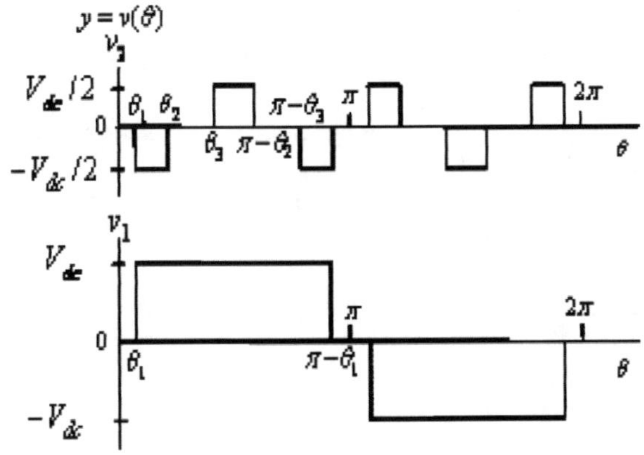

Figure 4. H-Bridge Voltages V1 And V2 Control

3. Modulation Control

For multi-level inverter modulation control, PWM control and space vector methods are used. The more losses are caused by the high-frequency procedures [8]. This makes it possible to monitor HCMLI using methods like selective harmonic removal procedure, simple frequency switches or active harmonic removal methods. We use the basic frequency approach from the proposed work [5].

3.1 Voltage switching control with 7 level equal step output

The expansion of the 7-stage corresponding voltage waveform from the Fourier series shown in the Fig. 3

$$V(wt) = \sum_{n=1,3,5}^{\infty} \frac{2v_{dc}}{n\pi} \left((cosn\theta1 + cosn\theta2 + cosn\theta3)\right)(\sin(nwt) \qquad (2)$$

Where n is the multi-level inverter's harmonic output voltage number [6]. Likewise, with V1 as the ideal fundamental voltage, the angles of switch θ1, θ2, and θ3 will be set in such a way as to remove the V(ωt) = V1. sin(ωt) and higher V(nωt) harmonics. The goal of this project is to accomplish the fundamental and remove the 5th and 7th harmonics. The solution for the following equations can be formulated by (2):

$$\cos(\theta 1) + \cos(\theta 2) + \cos(\theta 3) = 3 \qquad (3)$$

$$\cos(5\theta 1) + \cos 5(\theta 2) + \cos(5\theta 3) = 0 \qquad (4)$$

$$\cos 7(\theta 1) + \cos(\theta 71) + \cos(7\theta 3) = 0 \qquad (5)$$

The model suggested by three unknown inspired equations θ_1, θ_2, and θ_3.

3.2 Hybrid Cascaded Multilevel Inverter Features
The core features are as follows:

• Cascaded inverters need to be converted into DC Power contact sources to one dc of electrical AC. PWM (SHEPWM) technologies will be the only choice for selective harmonic elimination. New dc sources, including fuel cells and biomass, must be produced for various renewable energy sources.

• In the consequent mode it is not possible to link the DC connection sources and converters as a short circuit between the synchronization can be installed.

3.3 Hybrid Cascaded Multilevel Inverter- Advantages
Cascaded inverter's key advantages are:

• Less parts are needed to achieve the same output voltage than diode clamped and flying power inverters.
• There are no other clamping diodes or voltage balance condensers needed for each stage.
• The architecture and functionality of all the circuits are perfect.
• In order to reduce system voltage decreases and stress, different soft switching strategies are necessary.

4. Proposed Cascaded H-Bridge Multi - Level Inverter with Multi - Winding Transformer

The topology of multi-wind transformers can be called a multi-source topology type. A four-cell multi-coiling inverter is shown in the Fig. 5

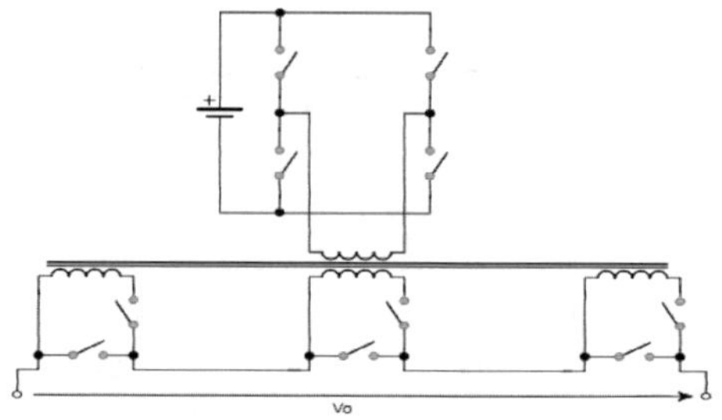

Figure 5. Symbol of multi-winding transformer

4.1 Circuit configuration
Fig. 6 displays the new multi-level network inverter supply configuration [2]. A rising source dc and many low frequency transformers are available. For three-stage transformers, the number of necessary parts and system volume can be reduced. The main side is linked to the VDC, zero and VDC supply bridges. The transformer's secondary side is linked to the production stage of the show. The three-phase terminal is connected to a delta to limit the third harmonic portion of the device. The Figs appears in 6 and 7 display transformers in three stages. The VAS voltage is connected serially with three terminal outputs. The multilevel inverter can be transmitted as a multi-level cascading inverter for H-bridge in this particular configuration. In the Fig. 6 The ultimate H-bridge inverter voltages associated with the Kth transformer are Vak, Vbk, and Vck voltages. The final voltages for each of the three phase values therefore belong to VAk, VBk and VCk[9].

IOP Conf. Series: Materials Science and Engineering **937** (2020) 012038 doi:10.1088/1757-899X/937/1/012038

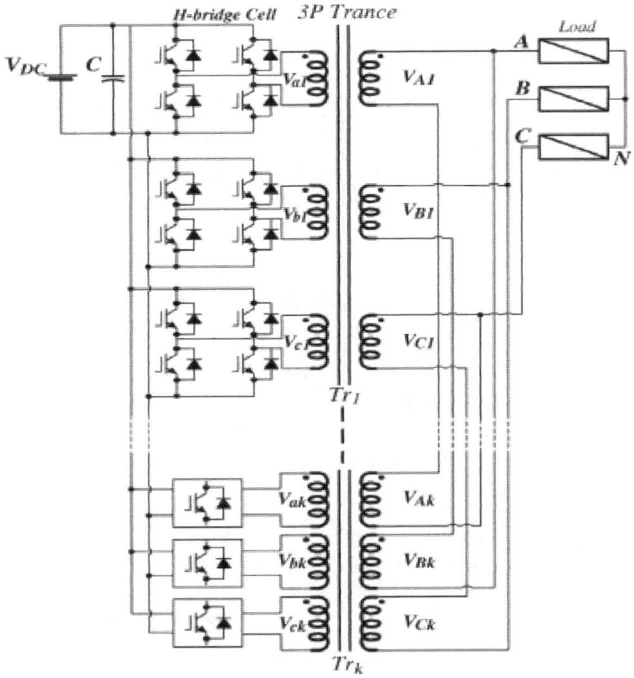

Figure 6. The given design of multilevel inverter

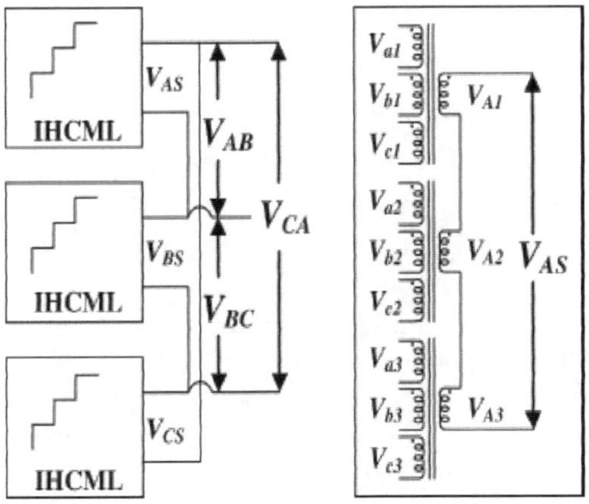

Figure 7. The minimized design of the given multilevel inverter

5. Harmonic Reduction Technique

The LOH angles are used by four different equations, especially the fifth and seventh or 11th voltages. Equation 2.2 adjusts the required RMS voltage simultaneously. For fifth, seventh and eleventh grade harmonics, the remaining equations are employed. The THD is absolutely popular.

IOP Conf. Series: Materials Science and Engineering **937** (2020) 012038 doi:10.1088/1757-899X/937/1/012038

The four equations are

$$\cos \propto_1 + \cos \propto_2 + \cos \propto_3 + \cos \propto_4 = 3\pi m \tag{6}$$

$$\cos 5 \propto_1 + \cos 5 \propto_2 + \cos 5 \propto_3 + \cos 5 \propto_4 = 0 \tag{7}$$

$$\cos 7 \propto_1 + \cos 7 \propto_2 + \cos 7 \propto_3 + \cos 7 \propto_4 = 0 \tag{8}$$

$$\cos 11 \propto_1 + \cos 11 \propto_2 + \cos 11 \propto_3 + \cos 11 \propto_4 = 0 \tag{9}$$

The MathCAD software has found the angles of control and complies with the above
$\alpha_1 = 12.844^o$; $\alpha_2 = 29.918^o$;
$\alpha_3 = 50.983^o$; $\alpha_4 = 64.239^o$;
The total harmonic distortion is defined as

$$\mathrm{THD} = \sqrt{(V_3^2 + V_5^2 + V_7^2 + \ldots\ldots\ldots V_{31}^2)} \tag{10}$$

The transcended equations are converted into polynomials first. The four equations are then rendered by the variables. Consider Cos $\theta_i = x_i$;

$$\cos(5\theta) = 5cos\theta - 20cos^3\theta + 16cos^5\theta \tag{11}$$

$$\cos(7\theta) = -7cos\theta + 56\, cos^3\theta - 112\, cos^5\theta + 64\, cos^7\theta \tag{12}$$

$$\cos(11\theta) = -11cos\theta + 220\, cos^3\theta - 1232\, cos^5\theta + 2816\, cos^7\theta - 1024\, cos^{11}\theta \tag{13}$$

The polynomials are formed by

$$P_1(X) = X_1 + X_2 + X_3 + X_4 - m \tag{14}$$

$$P_2(X) = \textstyle\sum_{i=1,3,5,7}(5\, X_i + 20^3_{xi} + 16^5_{xi} \tag{15}$$

$$P_3(X) = \textstyle\sum_{i=1,3,5,7}(-11\, X_i + 220^3_{xi} - 1232^5_{xi} + 2816^7_{xi} + 1024^9_{xi} \tag{16}$$

There are four unknown equations. Consequently, all measurements must be made. It led to a series of polynomial equations rather than trigonometric. There is a well-known theory that many polynomial equations can be solved, but the grade is very high. The resulting theory is used to solve this sequence of equations. α_1, α_2, α_3, α_4 which increases the overall harmonic voltage following the MathCAD phase, is the final solution.

6. Simulink model of CMLI

Figs 8 and 9 displays simulation proposed the 3ø CMLI specification. Three separate subsystems make up this architecture. A subsystem frame including it is specified in each subsystem block. Each subsystem builds into a group to reduce the complexity of design. Growing subsystem has a CMLI level of 7. To measure the increased voltage values of the circuit, a voltage measurement unit is used.

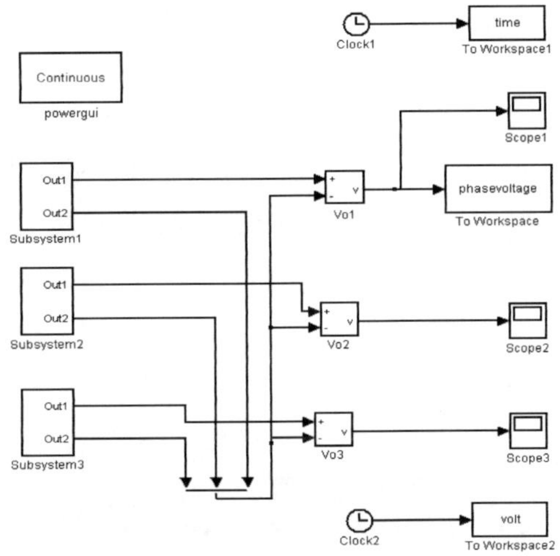

Figure 8. Proposed simulation design of 3ø- CMLI

6.1 Subsystem Model of CMLI

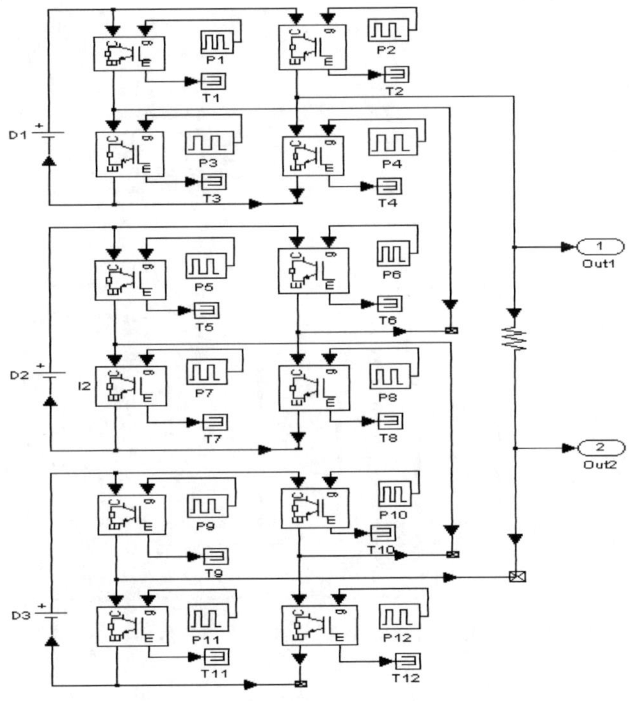

Figure 9. Subsystem model of CMLI

6.2 Output Voltage

Fig. 10 displays the final voltage of the 110V input voltage with the CMLI tri-phase simulation. The final tension of every block is 110 V and the total tension is 330 V.

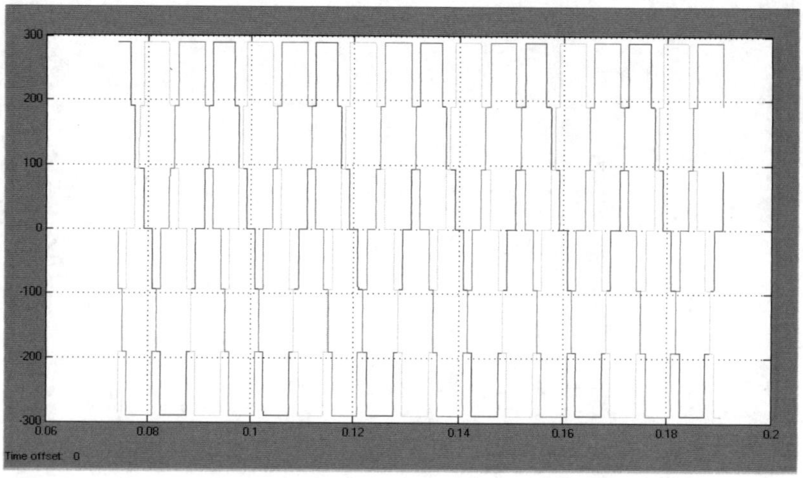

Figure 10. Final voltage of three phase CMLI

6.3 Harmonic Profile

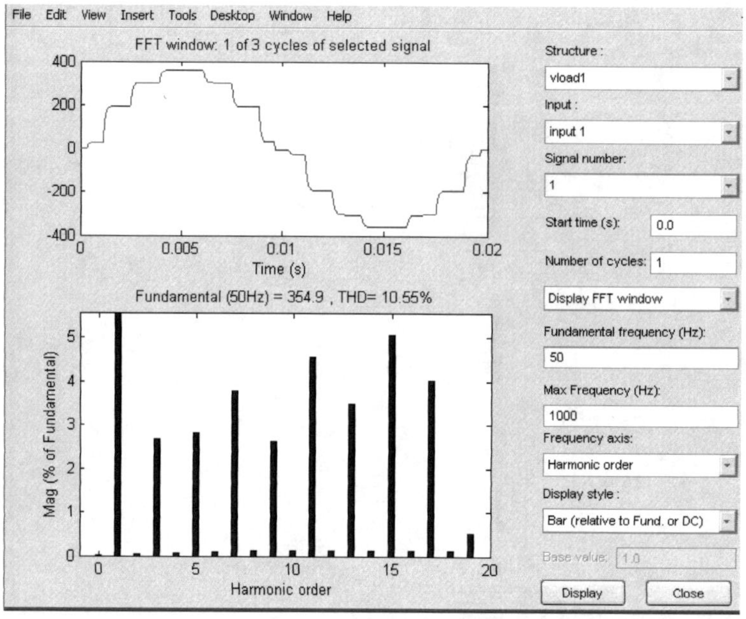

Figure 11. Harmonic level of the seven level CMLI.

Fig. 11 displays the harmonic rates of the 7 stadium cascaded MLI. Cumulative harmonic distortion (CHD) is shown to be 9.55%. The 3rd order is about 2.0%, and the harmonics are around 2.3%.

7. Hardware Results and Discussion

MOSFETs to test the HCMLI proposal using the H-Bridge Multi-stage inverter with seven switching system equivalent voltages. A single DC supply is used to power the inverter. MOSFET feed is the door signals. The MOSFET pulses of the first H-bridge are aligned with the H-bridge positive leg pair pulse. This is shown in Fig. 12. A standard PIC 16F877A microcontroller performs the control algorithm and generates the PWM waveforms for the MOSFET drive inverter. By using Microcontroller, the control gate signals are giving to the each pair of MOSFET unit and the corresponding gate pulse trigger to the particular pair of MOSFET unit will conducting. Finally we are connecting positive leg pairs of MOSFET and negative leg pairs of MOSFET through a three phase transformer. We are getting a seven level output voltage wave form which is shown in Fig. 13

Voltage in volt

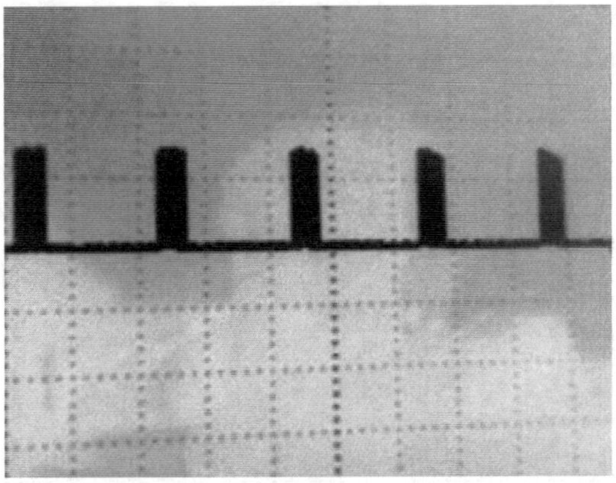

Time in ms

Figure 12. Pulse Width of MOSFET

Voltage in Volts

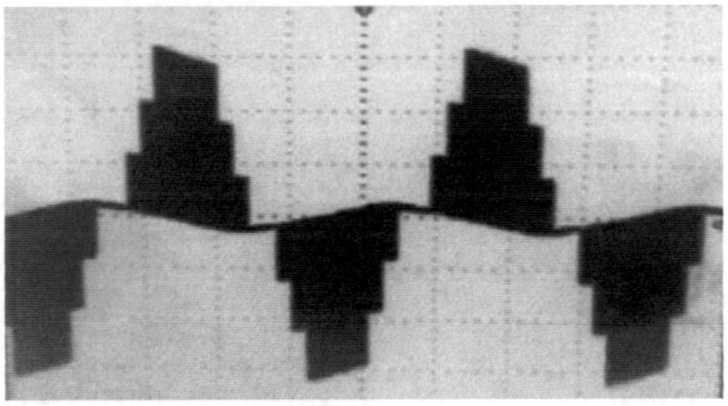

Time in second

Figure 13. Seven Level Output Voltage Wave Form

7. Conclusion

A multi-level waterfall inverter is the first cascaded hybrid inverter. For a hybrid multi-tiered converter, the 7-level wave shape is similar to the sinus wave. The seven-story cascading multi-level inverter created at the price was used to demonstrate the effects of simulations. Seven phases of cascading hybrid inverter load and voltage are given in the hardware implementation compared to the voltage of the output simulation. Regularly and in more harmonic intervals the various modern electronics are developed. Such frequencies are referred to and used to check their original elements, often without legislative power. Normal monitoring systems are also possible to detect the consequences of their variable control distortions in a certain way. The HCMI generates the desired waveforms with a single power source for each operation. Time is indifferent and the phase waveform had minimum total harmonic distortion for this experiment. The ratio of frequencies is indifferent. The harmonics in the fifth and seventh order are based on harmonic removal procedures. There is no algorithm for the fifth and seventh harmonics of this section. The simulation and testing were carried out to see how the program was operating. The studies show that only about 14.27 per cent of the THD is estimated. The third order is just 5.2% and the fourth order is 8.2%.

8. References

[1] J. N. Chiasson, B. Ozpineci, and L. M. Tolbert, "A five-level three-phase hybrid cascade multilevel inverter using a single DC source for a PM synchronous motor drive," in Proc. IEEE APEC, 2007, pp. 1504-1507.
[2] J.Vassallo, Nov. 2–6, 2003 "A power-equalized harmonic elimination scheme for utility-connected cascaded H-bridge multilevel converters", in Proc. IEEE Ind. Electron. Soc. Annu. Conf, pp. 1185–1190.
[3] J. R Wells., Jan. 2007 "Modulation-based harmonic elimination", IEEE Trans. Power Electron., vol. 22, no. 1, pp. 336–340.
[4] J. Rodriguez, P. W. Hammond, J. Pontt, R. Musalem, P. Lezana, and M. J. Escobar, "Operation of a medium-voltage drive under faulty conditions," IEEE Trans. Ind.Electron., vol. 52, no. 4, pp. 1080-1085, Aug. 2005.
[5] J.Rodriguez, " Multilevel voltage-source-converter topologies for industrial medium voltage drives", IEEE Trans. Ind. Electron., vol. 54, no. 6, pp. 2930–2945, Dec.2007.
[6] F. Z. Peng, L. Chen, and F. Zhang, "Simple topologies of PWM ac-ac converters," IEEE Power Electron. Letters, vol. 1, no. 1, pp. 10– 13, Mar. 2003.
[7] Z. Du, L. M. Tolbert, J. N. Chiasson, B. Ozpineci, "A cascade multilevel inverter using a single fuel cell DC source", Proc. IEEE APEC, vol. 1, pp. 426-430, Mar. 2006
[8] J. Rodriguez, B. Wu, S. Bernet, J. Pontt, and S. Kouro, "Multilevel voltage source-converter topologies for industrial mediumvoltage drives," IEEE Trans. Ind.Electron., vol. 54, no. 6, pp. 2930–2945, Dec. 2014.
[9] Xu, R.; Yu, Y.; Yang, R.; Wang, G.; Xu, D.; Li, B.; Sui, S. "A Novel Control Method for Transformerless H-Bridge Cascaded STATCOM with Star Configuration". IEEE Trans. Power Electron. 2015, 30, 1189–1202.
[10] Rahman, M.A.; Islam, M.R. "Modified High Frequency Phase Disposition Pulse Width Modulation for Modular Cascaded H-bridge Multilevel Converters". In Proceeding of the 2017 3rd International Conference on Electrical Information and Communication Technology (EICT 2017), Khulna, Bangladesh, 7–9 December 2017; pp. 1–5.
[11] Gadalla, A.S.; Yan, X.; Altahir, S.Y.; Hasabelrasul, H. "Evaluating the capacity of power and energy balance for cascaded H-bridge multilevel inverter using different PWM techniques". IET J. Eng. 2017, 1713–1718.

Software development of electric scooter

D G Sesa Balaji [1,2], S Shabeer Ahamed [1], Dr D Kirubakaran, [1]

[1] Electrical and Electronics Engineering, St. Joseph's Institute of Technology, Chennai

gomapraba95@gmail.com

Abstract. The Electric scooter is one which can be used to overcome the various problems faced by mankind in terms of the increase in price rates of petrol and diesel and also help us reduce the pollution in society. This project is a development of software for the low-cost electric scooter which uses a BLDC Motor as the primary motor as they have a feedback circuit and helps in identifying the problems when faced. We use 48V LifepO4 Battery (Lithium Ferro Phosphate) and developed charger which is compatible with E-scooter with 40kmph speed and 50 km mileage. We use a DC-DC converter to power up the E-scooter accessories which is sized at 1000W and 48V battery. The main accessories in E-scooter are display, headlights, horns all are 12V supplied, controlled by the software programmed, we have also designed a step-down converter. Then we have a solar panel which produces 125W and it reduces the voltage up-to 56V of constant supply DC. We have mainly designed software for the accessories and also for the proper operation of the BLDC Motor. This is done by the process of providing the PWM output to the power electronic device

1. INTRODUCTION

The main purpose of developing an electric scooter is to overcome the various problems facing the environment and also the universe. The major part of our project lies in the process of developing software for the proper running of the Electric scooter and also for the proper working of the main accessories. For the running of the Brushless DC Motor, we need a driver circuit that drives the Motor and we have used a Power Electronic Device namely Power MOSFET's which can drive the 48V rated Brushless DC Motor. The MCU will get the supply voltage value and the current battery voltage value and depending upon the comparison value the MCU drives the Driver circuit using the pulse produced by the PWM technique and charges the battery of the e-scooter. As per the algorithm designed, the three-phase voltage of the Brushless DC Motor is controlled with the help of the Pulse Width Modulation Technique. Using the PWM process an average output voltage is provided to the driver circuit which in this case is a MOSFET. The driver circuit is necessary because we won't be able to produce the maximum voltage to control the BLDC Motor and thus we require a driver circuit

The driver circuit is mainly used for speed control of the BLDC Motor. To develop and software for the headlights because it is essential to work it out in a software environment and we have used Proteus 8 Professional for the simulation. For the process of developing the code which is to be dumped into the E-scooter, we have used MPLAB X IDE which uses XC8 Compiler which converts the high-level language into low-level language. The microchip laboratory provides us with various data about the PIC Microcontroller which is used for the control of E-scooter. The PIC microcontroller has 40 pins in total and has various input and output pins which can be used for the accessories used in the Electric scooter. The PWM input to the Brushless DC Motor is provided via the PIC8 Microcontroller which is done using the CCP/PWM Module which also the part of the

MCU. The Code of the BLDC Motor is one that has the stator and rotor which is different from many other traditional motors as it has stator Coils and the rotor is made up of the permanent magnets and the permanent magnets on the rotor. The major reason we go for BLDC Motor with Hall effect sensors will turn on the supply for any two phases at a time depending upon the rotor position. The BLDC motor has various advantages such as speed, torque characteristics which are much better compared to other motors and higher torque, higher efficiency, silent operation, and various speed ranges. The stator winding is of two types namely trapezoidal motors and sinusoidal motors. Depending on various factors, the motor with the 48V rating is chosen and they are proceeded to use in the electric vehicles. The rotor of the BLDC Motor is made up of permanent magnets with the south and north poles designed alternatively. We use the electronic way of controlling the commutation process of the motor which also helps in the process of maintaining the effectiveness of the motor and also is used to rectify any mistake if a problem arises this can be done using the software code developed. The working of headlights is simulated using the LED in the Proteus 8 Professional environment and the battery indication is done using a bar graph in the same environment

In this paper, we provide the overview of the electric scooter designed in Section 2 along with the various novelties in the electric scooter such as the charging on the drive, etc; Section 3 gives information about the way the software is being developed for the proper running the Scooter and also discusses the components used to achieve the same. Section 4 gives information about the major component in the Brushless DC Motor. Section 5, gives an overview of the various tools used to design the software. Section 6 gives the Hardware implementation of the electric scooter set-up. Finally, Section 7 gives the conclusion obtained from the software development of an electric scooter

2. VARIOUS COMPONENTS OF THE ELECTRIC SCOOTER

In this section, the main components of the low-cost electric scooter is given an overview

2.1 DC-DC CONVERTER:

The DC-DC converter is designed in the process of making an electric scooter because it is required to achieve the working of the accessories by providing them supplies. The supply voltage of these main accessories is 12V but we have a battery which is 48V and it is not possible to power them with that voltage. This is the main reason why we prepare this DC-DC Converter. We have designed a Step-Down converter. There are various converters used in the research topologies but the perfect converter for the low-cost electric scooter is the buck converter as the output power is about 36W

2.2 BATTERY CHARGER:

The Battery charger is developed which is used to control the 48V based BLDC Motor controller, Throttle sensors, Mid-drive motor, and charger stations. The battery which we have used is Lithium Phosphate which has a cell made up of 3.5V, 6Ah capacity. The major reason to use the Lithium Phosphate batteries is that it will not catch fire even if punctured and is the safest among all Lithium Batteries. The typical estimated life of the Lithium-Ion battery is about 300 to 500 charge-discharge cycles. The life expectancy of the 48V Lifep04 battery charger is 7 yrs. We have added an important novelty that is not available in the industry is the BMS shows the Dead Battery Indication and also the BMS parameter indication

2.3 SOLAR DESIGN:

As of now, there are only AC mains charging option is available for Electric scooter. But, our Electric scooter is one that uses the solar panel which helps in charging the system while driving. To provide novelty to the Electric Scooter designed we have introduced "Charging On Drive". The panel produces 125W and its peak time supply open-circuit voltage of 70V with a charge controller and it reduces the voltage up-to 56V for the constant supply DC. The disadvantage with the charge on a drive is it may take about 3 to 4 hours to charge the vehicle (i.e.) the battery

2.4 SOFTWARE DEVELOPMENT:

Currently, the software industry has developed so much that various methods of software development have arrived to help us overcome the tedious and confusing software coding for electric scooters. Also, to designing software for a particular hardware is difficult with the traditional process. Thus, we have used modern technology to overcome the problems using the Microchip Laboratory environment for dumping the code into the PIC Microcontroller which is the MCU in the electric scooter and it's checked for the proper working in the Proteus 8 Professional environment by simulation. We use a Microcontroller Unit (MCU) along with a driver Circuit and a converter which provides the output to the battery management system of the e-scooter. The main purpose of the MCU is to get the supply voltage for the BLDC Motor and to compare that voltage with the output from the battery voltage obtained from the battery management system of the e-scooter. The voltage and the current value are obtained from the V & I Sensing which is attached to the Battery Management System

3. BLOCK DIAGRAM OF PROPOSED SYSTEM

The Proposed system is that we can see that there is an LCD Display which will provide all the update regarding the electric scooter condition. Now we get the supply voltage from the device which is the motor and we also get the battery voltage regarding the battery status. Now the output obtained from the MCU which is called the Microcontroller unit is provided to the Driver circuit which is nothing but the PWM value. We normally know that the PWM value is nothing but an average voltage value which can be obtained as one which is the maximum and the minimum voltage is 0V and 5V respectively. The average voltage obtained from the PWM output is about 2.5 V which cannot be used to drive the Brushless DC Motor which has a rating of about 48V. The driver circuit includes the Power MOSFETS which can drive the circuit smoothly and also helps in getting the position of the rotor and now we can get the speed control and other things in the system. Thus, the need for a driver circuit is very important for the proper running of the electric scooter. Usually, the driver circuit that is being used includes the MOSFET or the IGBT but we are using MOSFET because of its ability to be very efficient whilst having a good and efficient structure. The MOSFET also has a great driver speed

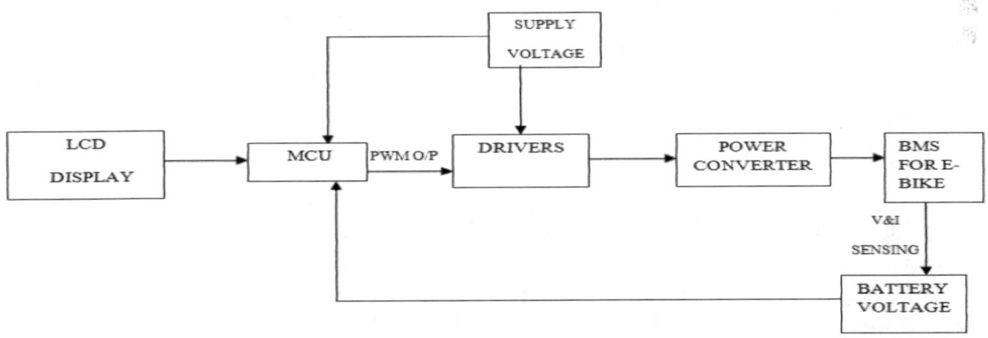

Figure 1 –The MCU gets the voltage value and battery voltage value compares them and drives the entire electric scooter.

low switching loss. This is a major advantage compared to IGBT. Now the MOSFET is various in number and can be used to run a BLDC Motor depending on the two of the three Phases switched On. Then, the PWM output is there which is the pin that gets the PWM values. The PIC microcontroller can also be used for various other processes such as developing the charging circuit for electric vehicles and also for various controller purposes

The PIC Microcontroller is very useful. Also for the working of the main accessories and the other spare parts, we need to code them using the PIC Microcontroller and this can be done in such a way we need to switch on the light when necessary and to use the horn when necessary which has to be coded using the Microchip Laboratory environment. Also, we display the battery charged condition and the amount of charge remaining in the Electric scooter using the code developed by the software components used. This can be represented using the bar graph which indicates the percentage remaining in the battery in terms of percentage and in case of charging needed we can use the inbuilt 125W solar panel in the electric scooter. The battery indication is very important because we need to know when to charge the battery and when not or else it may damage the battery

4. HALL EFFECT SENSOR

The Hall effect sensor is a major reason in the wide use of Brushless DC Motor because in other DC Motors there is commutation which is not controlled electronically but in BLDC Motor it is controlled electronically. For the BLDC Motor to rotate the stator coils of the motor has to be energized in a sequence. To understand which winding will be energized following the sequence we need to know the position of the rotor. This is done easily in a BLDC motor as it has the Hall effect sensors embedded in the stator part of the motor. The BLDC motor has the sensors present in the non-driving end of the motor. When the magnetic poles of the rotor come near the sensors we get a high signal or low signal which helps us in determining which pole is passing through or nearer to the sensor embedded in the stator of the motor. Based on the Hall sensor signals, the commutation sequence of the BLDC motor is determined easily

5. DESCRIPTION OF TOOLS

In this section, the various tools to design the software has been discussed,

5.1 PROTEUS 8 PROFESSIONAL

The Proteus 8 Professional environment is a software environment used for simulation of electronic devices and their automation. These are usually used for the design of PCB's by the design engineers. But, it can also be used to simulate the PIC microcontroller within the Proteus environment. We have used a 40 Pin PIC Microcontroller and some of the main simulations done using the PIC in Proteus are software implementation of the headlights and the horns as well as the pulse production of the PWM circuit which is necessary for the proper control of the BLDC Motor. Thus, Proteus is one of the modern methods of controlling the software part of the low-cost electric scooter

5.2 MPLAB X IDE

MPLAB X IDE is a software program that is used to develop applications for Microchip microcontrollers and digital signal controllers. This development tool is called an Integrated Development environment, or IDE because it provides a single integrated environment to develop code for embedded microcontrollers. MPLAB X IDE is one that is like an interactive console between the IDE and programming language. Using this IDE, we as a user can dump codes into the Microcontroller unit in a language understandable by the devices for proper operation of the low-cost electric scooter. The MCU is called the target microcontroller and does all the work for the functioning of the electric scooter. MPLAB X IDE is one of the most used tools for performing the software development of various devices in the field of the microcontroller. The MPLAB X IDE is one where we give the code to the system in terms of the C Language Program and the XC8 compiler converts it to Assembly Language

6. IMPLEMENTATION

The above diagram is the design implementation of our whole hardware setup but before that, we decided to implement it in the software to check whether it is producing the intended results. The PIC

IOP Conf. Series: Materials Science and Engineering **937** (2020) 012039 doi:10.1088/1757-899X/937/1/012039

Microcontroller plays a major role in the project as you can see from the diagram it has a lot of inputs and outputs and the required PWM inputs are connected to the CCP1 Pin and is called as the pin 16. The CCP2 Pin in the PIC Microcontroller is also used for the PWM outputs and it can be used to obtain the increase the voltage provided to the BLDC Motor and we can vary the PWM output (ie) the average output voltage is controlled and in that way, the speed is also controlled. We also see that the LED Pin is connected to the device which is the PIC Microcontroller in the Pin 14 and thus we can produce the night vision for the electric scooter during night times. The LEDs will be turned ON when the vehicle is turned and thus we can reduce the button required to turn on the Light or also called the headlights

The buzzer which is to be implemented in the device because we require and horn for the electric scooter is connected to the PIC Microcontroller and the PIN to which it is connected 4. This enables the user to indicate that to other vehicles on the road so that we can avoid accidents. The hall sensors are three in number and the feedback regarding the rotor position is provided to the PIC Microcontroller and thus the required phases are turned on to obtain the required positive and the negative torque and to obtain the maximum torque and thus the BLDC Motor is operated at its maximum rated speed which is about 5000 RPM in its full load

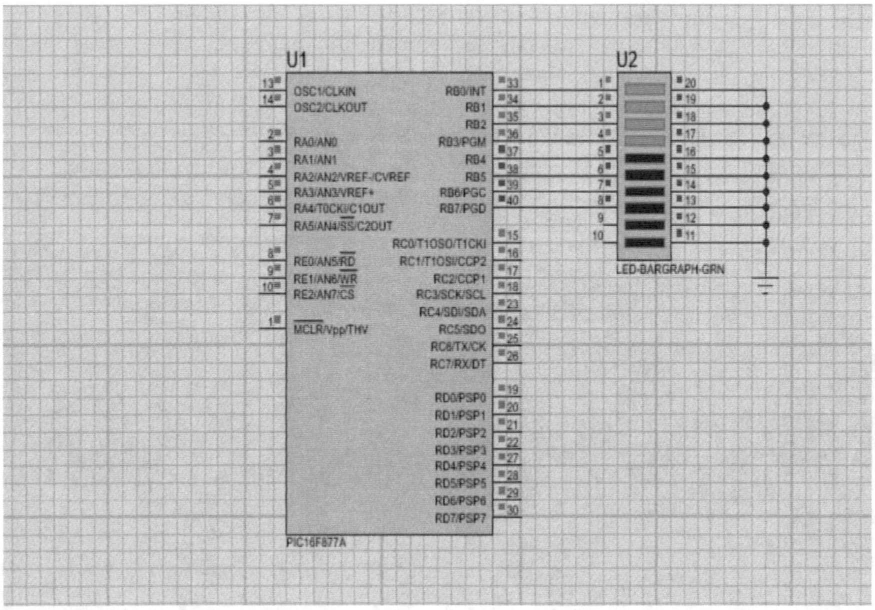

Figure 2 - It represents the simulation performed to indicate the charge status in the Electric scooter. It has about 10 indication bars and each indication step shows the percentage of the charge present in the electric scooter.

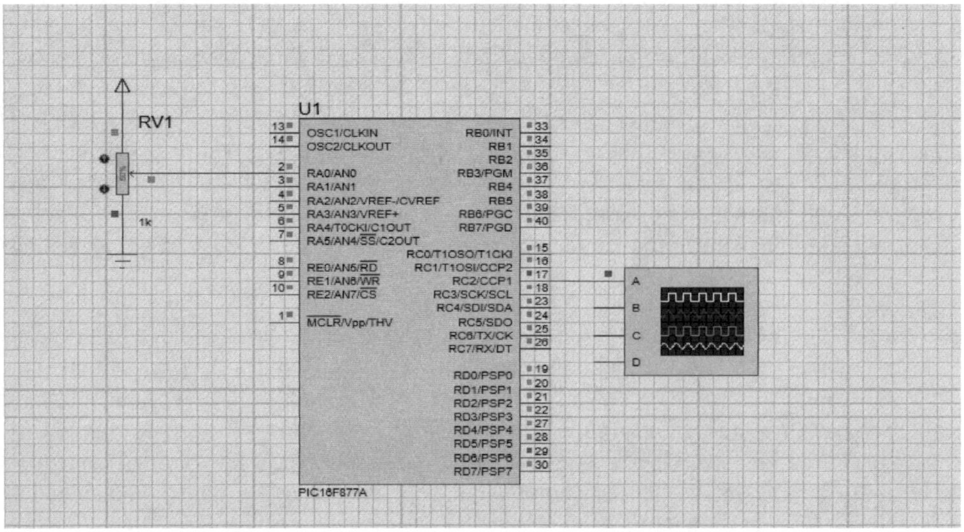

Figure 3 - It represents the PWM simulation process. The PWM output voltage value given to the system is that of about 5V but it is not enough to drive a Brushless DC Motor thus we have installed a Power Electronic Device which is used to drive the 48V Motor. The PWM input has an average voltage of about 2.5V and this can be achieved by varying the potentiometer present in the simulation. The device is connected to the pin 18 because it is a specially made pin PWM working within the PIC microcontroller.

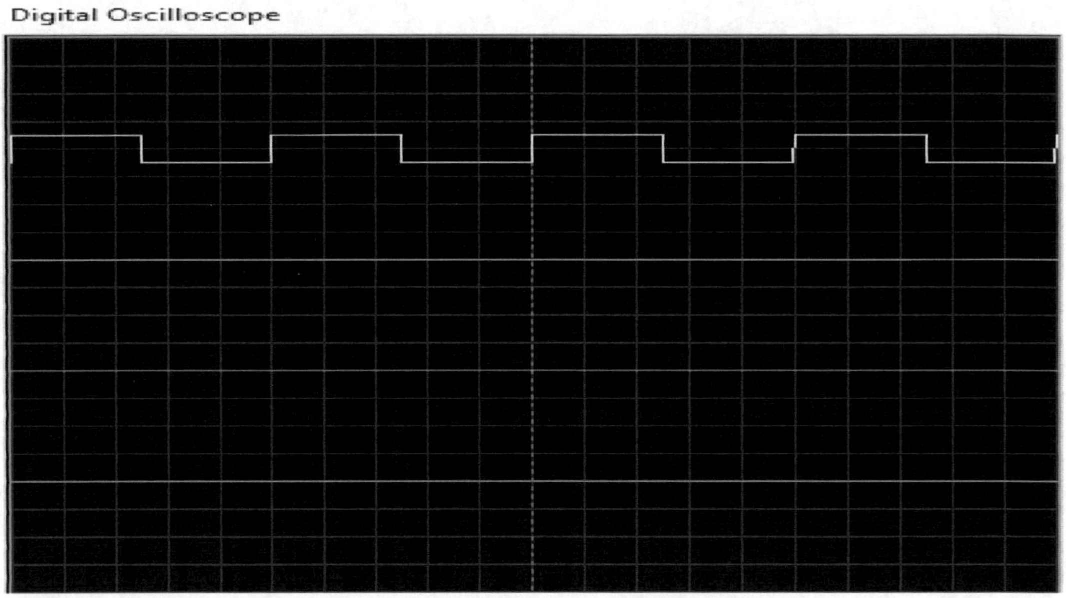

Figure 4 – The Digital Oscilloscope output figure is about 2.5V which is the average output voltage which we get when we have 50% duty cycle. The channel A is the yellow colour which gives square wave output as the pin from PIC Microcontroller is connected to channel A. The other lines are flat.

Figure 5 - It represents the main parts of the electric scooter. The charge controller(a) at the top and the 48V battery(b) at the bottom and the Brushless DC Motor(c) on the left corner with the charging pin.

7. CONCLUSION

The results obtained help to understand that the above-designed software development is much efficient and simple for the BLDC Motor control. This method of developing software can also be implemented in the control of various other devices when used in the development of the electric scooter and as well as the Brushless DC Motor control. The major portion in the proper design of the electric scooter is done using the selection power driver circuit. Though various power converters are present, the Power MOSFET's is the most efficient and used in various applications including electric scooters. Also, we have developed a charger controller system in the scooter which shows the defects when there is one which is indicated by the sparkling of the red light. This is a new method of identifying errors within the system. Also, our system has a solar panel th helps us to charge the system while driving instead of searching for a charging station nearby and reduces the time needed for the charging of the system which could be useful in many ways. We have also developed software to control the BLDC Motor Driver that helps us to get an indication of whether it is running properly or not with the help of the feedback sensors connected to the system driver in the motor

8. REFERENCES

[1] *Determining MOSFET Driver Needs for Motor Drive Applications* Author: Jamie Dunn Microchip Technology Inc. [2014]

[2] *Brushless DC (BLDC) Motor Fundamentals* Author: Padmaraja,Yedamale Microchip Technology [2016]

[3] *MPLAB XC8 C Compiler Datasheet* Author: Jamie Richard [2016][4]PIC16F87XA Datasheet Author: Nathan Richard [2017]

IOP Conf. Series: Materials Science and Engineering **937** (2020) 012039 doi:10.1088/1757-899X/937/1/012039

[4] Zhao, Yanbin , and Zaimin Zhong. "*Development on Automotive Electronic Real-Time Control Software based code Generation Technology "Computer-Aided Engineering (2015)"*

[5] Edwards, Stephen, et al. "*Design of Embedded Systems: Formal Models, Validation, and Synthesis." Proc., IEEE 85P3 (2015-2016).*

[6] *Infineon Technologies AG. "Tricore-Electric motor control" Infineon Application Notes* (2017)

[7] *Codes were taken from microchip official website* (2019)

[8] Stephen, et al. "*Design of Embedded Systems: Validation, and Synthesis." PROC., OF THE IEEE* 85.3(2015-2016).

[9] *PIC16F87XA Data Sheet* Author: Nathan Richard [2017] *"PROC., of the PIC Microcontroller in the Industry under IEEE Standards".*

Comparative Study of Recurrent and Non Recurrent Neural Network Based Approach for Modeling of PEM Fuel Cell Powered Electric Vehicle

K Gomathi[1], M Karthik[2] and S Usha[2]

[1]Department of Electrical and Electronics Engineering, Velalar College of Engineering and Technology, Erode, India

[2]Department of Electrical and Electronics Engineering, Kongu Engineering College, Perundurai, Erode, India

Email: gomathikalainesan@gmail.com

Abstract. In this paper, a non recurrent Focused Time Delay Neural Network (FTDN) and a recurrent nonlinear autoregressive network with exogenous inputs (NARX) Neural Network are employed as a black box prediction model for substituting the complex conventional model of PEM 5kW Proton Exchange Membrane (PEM) Fuel Cell system. A comparative assessment is performed between the recurrent and non recurrent neural network on certain performance measures to identify an optimal network for modeling the PEM Fuel Cell System in PEM Fuel Cell powered electric vehicle application. From the simulation result, it is observed that the recurrent NARX network is showed excellent prediction ability in terms of minimizing the Mean Square Error (MSE) value with faster convergence. The optimized network is tested with intermittent data for examining its adaptability and validated with experimental benchmark data for proving its reliability. Hence the optimum network is integrated with converters and vehicle dynamic system to develop a fuel cell based electric vehicle system. The performance of the proposed vehicle is tested with US06 drive cycle pattern for justifying its reliability.

1. Introduction

Hydrogen related studies especially generation from renewable sources and utilization with improved fuel economy has grasped great attention of researchers worldwide in the recent years [1]. As a clean energy source, PEM Fuel cell with its high efficiency, modularity, flexibility and fast load response are well applicable factors to be employed in vehicular and house hold applications [2]. Meanwhile, it is a huge challenging task to represent a transient behavioral system using empirical or semi empirical model with fixed parameters because the system is subjected to temporal variation due to non observable and non predictable changes on internal conditions uncertainly [3] and [4]. Estimating or approximating the system parameters is not been well appropriate for all the conditions especially for the long term précised prediction. These modelling constraints is effectively resolved using a powerful data mapping and modeling tool termed as 'neural network modeling' or it is also termed as 'black-box models' which provide better approximation without characterizing the complex input and output relationships [5] .

Some of the related studies that spotlighted on the modeling of neural network controller using either non recurrent neural network or dynamic recurrent network or both are elucidated as follows.

Puranik et al. is implemented a modified version of NARMAX recurrent neural network model for a 500-W PEMFC stack for examining the long and short term transient dynamics [6]. The training data is extracted from the nonlinear State-Space simulation model of PEMFC system [7]. Yu et al. has [8] determined the optimal linear operating point of fuel cell for obtaining a maximum and stable power density with the combined effect of Taguchi approach and neural networks against different current densities. The proposed neural network is trained for establishing the relationship between various control factors . Hayder et al. has optimized the fuel cell performance for different pressure and load by the influence of PID Controller whose gain is tuned with PSO technique [9]. Load tracking ability of neural network equipped with Fuzzy Logic Controller is examined by Mammar et al. [10] and it is mainly for residential application and subjected to abrupt load changes of active and reactive power demand thereby the flow of hydrogen for the load change can be effectively controlled.

This paper is focused on précised prediction of a time varying, uncertain and multivariable PEM Fuel Cell system using system identification approach as a substitution of complex conventional modeling in electric vehicular applications. For which, a tapped time delayed non recurrent network FTDN and recurrent network NARX are developed and compared with certain performance measures such as Mean Square Error (MSE), epochs and correlation (R).

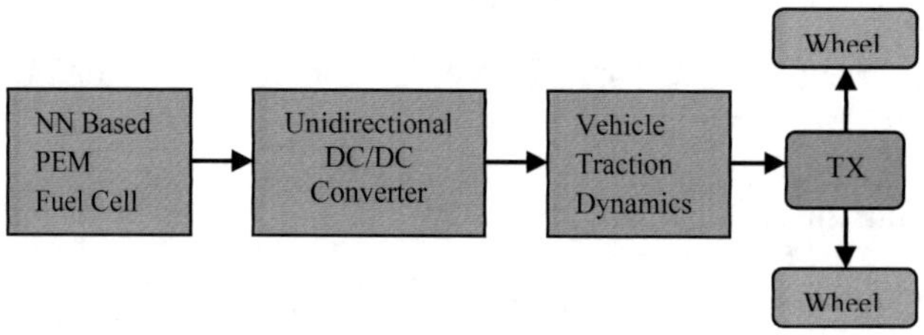

Figure 1. NN based PEM fuel cell powered electric vehicle system.

The optimum network is incorporated with DC – DC converter and transmission system to form a complete vehicle traction system. The overall layout of the neural network based PEM fuel cell powered electric vehicle system is presented in figure 1.

2. Modeling of ANN based PEM fuel cell powered electric vehicle system

The major aspects involved in the modeling of the ANN based PEM fuel cell powered electric vehicle system are the modeling of energy sources, the power electronic circuitry and the transmission systems. The main use of source in the vehicle traction system is to provide the supply for the load demand. The fuel cell energy source preferred here is the Proton Exchange Membrane (PEM) Fuel Cell because of its low operating temperatures, sustained operation at a high current density, low weight, compactness, the potential for low cost and volume, longer stack life, fast start-ups and suitability for discontinuous operation is prioritized among other types [2].

The Artificial Neural Network (ANN) is substituted as a black box for predicting the PEM Fuel Cell response that provide appropriate mapping between the input and the output of the PEM fuel cell without acquiring any empirical relationship inside the PEM Fuel cell [11]. The fluctuated output response from the source is regulated with the power electronics DC-DC converter unit. Since the fuel cell is only provided the supply to the load demand and lack of utilizing the regenerative power during braking process a unidirectional DC-DC converter unit is used.

The Dynamic behavior of the electric vehicle is modeled with the consideration of all the forces applied to the vehicle. The net force acting at the wheel of the vehicle for its propulsion is the combination of two forces that are tractive and resistive force. The tractive force applied along the direction of the movement of the vehicle while the resistive force applied in the opposite direction of the movement of the vehicle.

The four major components acting against the vehicle are the combination of aero dynamics force, rolling resistance force, inertial force and gradient force [12]. A simplified equivalent model of the vehicle dynamics can be used to estimate the tractive requirement of the vehicle drive-train that comprising of all the relative forces is showed in figure2. The vehicle specification to be used for the modeling of the proposed electric vehicle model is taken from the literature [13].

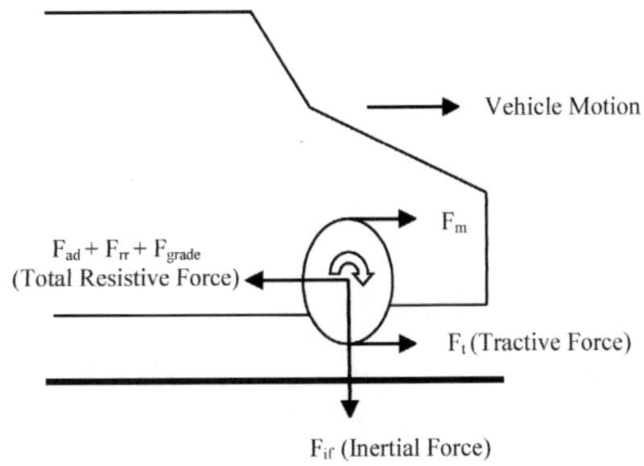

Figure 2. Simplified view of vehicle dynamics model.

3. Non Recurrent and Recurrent Neural Network

The Artificial Neural Network (ANN) is not required the knowledge of various empirical set of equation inside the fuel cell system or any processing parameter assumptions to reduce the overall complexity of the model. This approach not only reducing the complexity involved in the mathematical modeling but also reducing the time required to develop the model. The general block diagram representation of neural network approach for the prediction of output response is shown in figure 3.

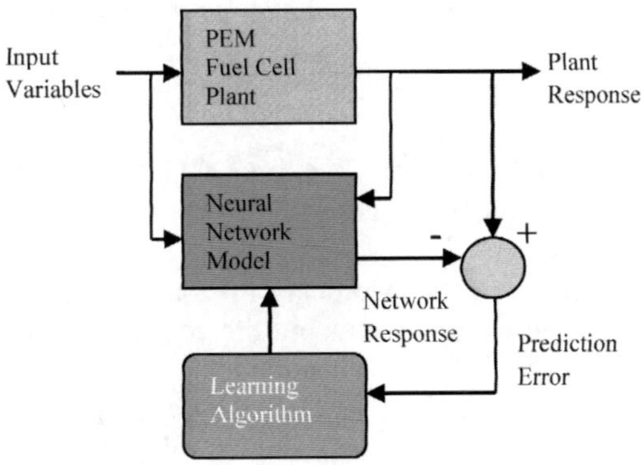

Figure 3. ANN modeling approach.

The polarization dynamics of PEM Fuel Cell is described the input output parameter for the NN modeling of PEM Fuel Cell. The effect of current density, hydrogen partial pressure and oxygen partial pressure on cell voltage and stack power prediction is investigated by keeping Stack temperature at 72 degree and Membrane Water Content at 14 as constant values. The data required to training the network is extracted from the Ballard-Mark-V 35-cell 5 kW PEM Fuel Cell experimental result from literature [14] and its specification is given in table 1. Among which 60% data set is used for training and remaining 40% data set is used for testing and validation. The raw extracted data are differed in their units and numerical data range and it is pre-processed to obtain a normalized data within a specific range -1 to 1 or 0 to 1 for improving the accuracy of subsequent numeric computation. In this paper, the raw data are normalized within -1 to 1 value and it is again denormalized to its original unit and range. The parameters used for modeling the dynamic non recurrent and recurrent neural networks are shown in table 2.

Table 1. Specifications of Ballard-Mark-V PEMFC system [15].

Component	Value
Stack temperature	72
Active area of the cell	232 cm^2
Anode pressure	3 atm
Cathode pressure	3 atm
Number of cells in the stack	35
Anode volume	0.005 m^3
Cathode volume	0.01m^3
Membrane dry density	0.002kg/cm^3
Membrane dry Eq. weight	1.1 kg/mol
Membrane thickness	0.0178 cm

Table 2. Network parameter selections.

Inputs: Outputs		3:2
Hidden layer		1
Hidden layer Neurons		20
Delay		1:2
Epochs		1000
Activation function	Hidden	Tangential sigmoid
	Output	Linear
Training Algorithm		Levenberg-Marquardt [19]

3.1. FTDN Network

A non recurrent Focused Time Delay Neural Network (FTDN) is also a dynamic network whose dynamic exist as a time delay variant only at a input layer and remaining parts resembles a static multilayer feed forward network. The FTDN Network is suitable for time series prediction and provided one step-ahead prediction. The tapped delay lines at input layer is enhanced the prediction ability than a static feed forward network by reducing the conflicts of propagating the input vector to the input nodes and the network provide faster training than other conventional dynamic networks because the tapped delay lines appears only at the input layer and the network does not contain any feedback loops or any adjustable parameter in between the network layers. Hence it does not require dynamic back propagation to compute the network gradient. This makes the convergence at the faster rate than other dynamic networks. The mathematical model of FTDN Network is showed in figure 4 and their representations are given in equation (1), (2), (3) and (4).

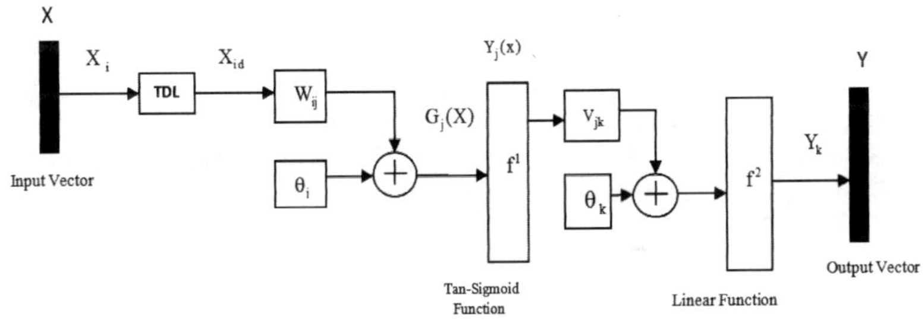

Figure 4. Mathematical model of FTDN network.

The input vector (X_i) after tapped delay (X_{id}) with its corresponding synaptic weight links (W_{ij}) and bias factor (θ_j) are furnished as a net input value ($G_j(X)$) and disseminated from the input layer towards the hidden layer is given in equation (1). The propagated input ($G_j(X)$) is functioned with tangential sigmoid activation function at the hidden layer is given in equation (2). Then the intermittent response ($Y_j(x)$) with corresponding synaptic weight links (V_{jk}) and bias factor (θ_k) are aggregated as $Y_k(X)$ and propagated towards the output layer for linear activation to yield a network output response (Y_k) are given in equation (3) and (4).

$$G_j(X) = (\sum_{i=0}^{n} X_{id} \times W_{ij}) + \theta_j \tag{1}$$

$$Y_j(x) = f^1 (G_j(X)) \tag{2}$$

$$Y_k(X) = \left(\sum_{j=0}^{p} Y_j(x) \times V_{jk} \right) + \theta_k \tag{3}$$

$$Y_k = f^2 \left(Y_k(X) \right) \tag{4}$$

3.2. NARX Network

The dynamic recurrent nonlinear autoregressive network with exogenous inputs (NARX) is constructed with feedback connection from output to the input layers of the network. In this network, the output signal from the network is regressed on past values of the output signal and present values of an independent (exogenous) input signal [16] and [17]. The dynamic recurrent NARX neural network mathematical model is showed in figure 5. The network is triggered by providing the input vector to the n input nodes in the network input layer that disseminated towards the successive layer with appropriate delays, weight links and bias between the layers.

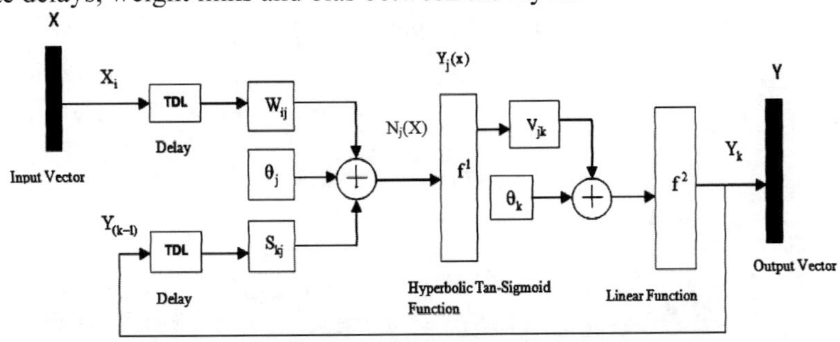

Figure 5. Mathematical model of NARX network.

The mathematical representation of these computations is represented in equation (5), (6), (7) and (8). The input vector (X_i) after tapped delay (X_{id}) with its corresponding synaptic weight links (W_{ij}), past recurrent output vector ($Y_{(k-1)}$) after tapped delay ($Y_{d(k-1)}$) from output layer with corresponding synaptic weight links (S_{kj}) and bias factor (θ_j) are aggregated as a net input value ($N_j(X)$) that propagated towards the hidden layer is given in equation (5). These values are functioned with tangential sigmoid activation function at the hidden layer is given in equation (6). Then the intermittent response (Y_j (x)) with corresponding synaptic weight links (V_{jk}) and bias factor (θ_k) are aggregated as $Y_k(X)$ and propagated towards the output layer for linear activation to yield a network output response (Y_k) are given in equation (7) and (8).

$$N_j(X) = \sum_{i=0}^{n} X_{id} \times W_{ij} + \sum_{k=0}^{p} Y_{d(k-1)} \times S_{kj} + \theta_j \quad (5)$$

$$Y_j(x) = f^1 (N_j(X)) \quad (6)$$

$$Y_k(X) = \left(\sum_{j=0}^{p} Y_j(x) \times V_{jk} \right) + \theta_k \quad (7)$$

$$Y_k = f^2 \left(Y_k(X) \right) \quad (8)$$

4. Simulation Result of Neural Network Modeling

The prediction of the proposed non recurrent and recurrent in terms of Mean Square Error (MSE), epochs and correlation (R) are investigated and compared for the optimum network identification that can be substituted as a PEM Fuel Cell black box model. The prediction performance of the FTDN Network for the training, validation and testing data set in terms of Mean Square Error (MSE) is showed in figure 6.

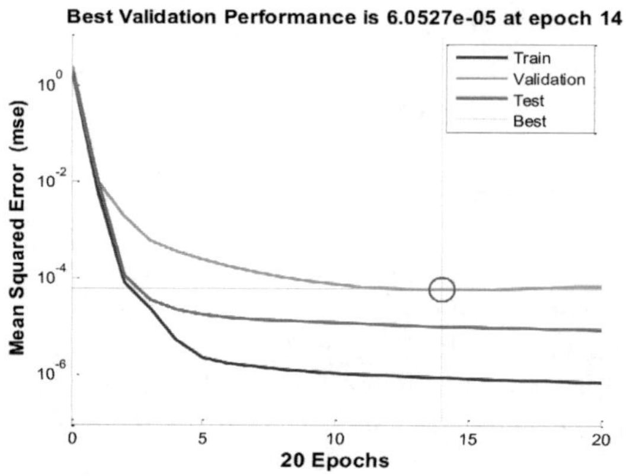

Figure 6. Prediction response of FTDN network.

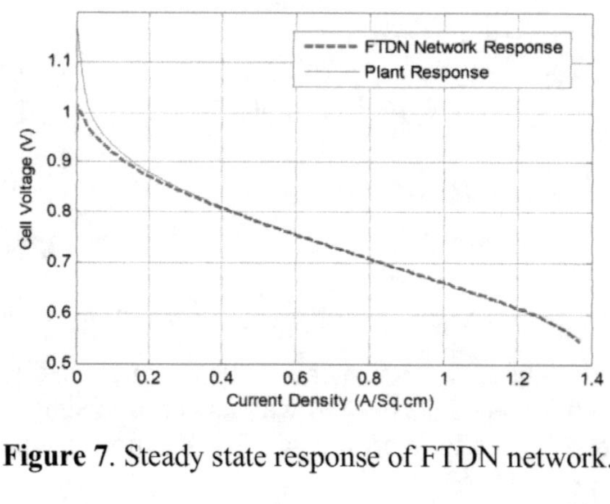

Figure 7. Steady state response of FTDN network.

The best validation performance is 6.0527e-05 attained at 14 epochs. The steady state V-I characteristics of NN based PEM Fuel Cell is validated with the experimental data from literature [18] and it is depicted in figure 7.

The prediction performance of the NARX Network for the training, validation and testing data set in terms of Mean Square Error (MSE) is showed in figure 8. The best validation performance is 8.4675e-07 attained at 11 epochs. The recurrent NARX network is showed far excellent result with minimum prediction error at a fast convergence rate than non recurrent FTDN network. The fitness of optimum NARX network prediction response with real experimental response in terms of Correlation factor values (R) for the training and validation data are showed in figure 9. The predicted response is showed good congruence with real measured value with R=0.99999 for the training data set and R=0.99957 for the validation data set.

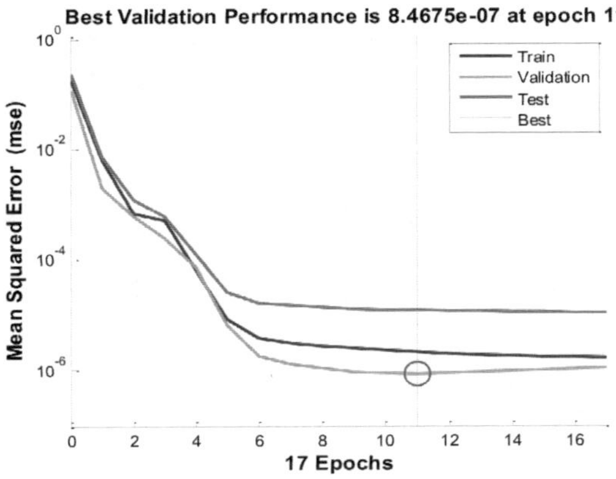

Figure 8. Prediction Response of NARX Network.

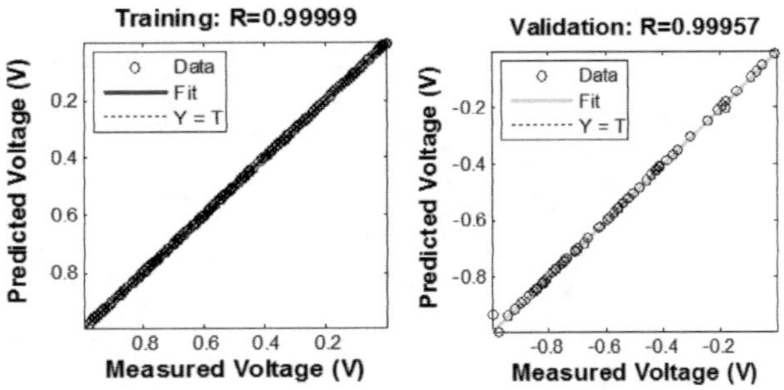

Figure 9. Correlation factor values of NARX Network.

The behavioral analysis of the optimum NARX network is carried out and validated the response with the experimental plant response from the literature [18]. In the steady state response analysis, cell voltage against current density and stack power against current density curves are investigated and depicted in figure 10 and 11. While in the transient response analysis, dynamic time varying stack voltage response and stack power response with respect to time are analyzed and depicted in figure 12 and 13.

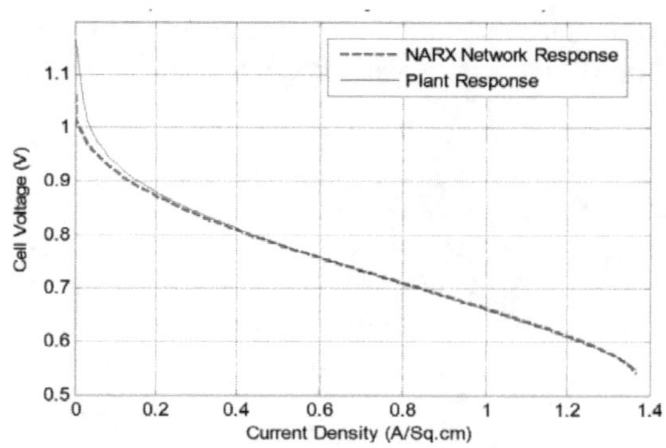

Figure 10. Polarization curve of NARX network.

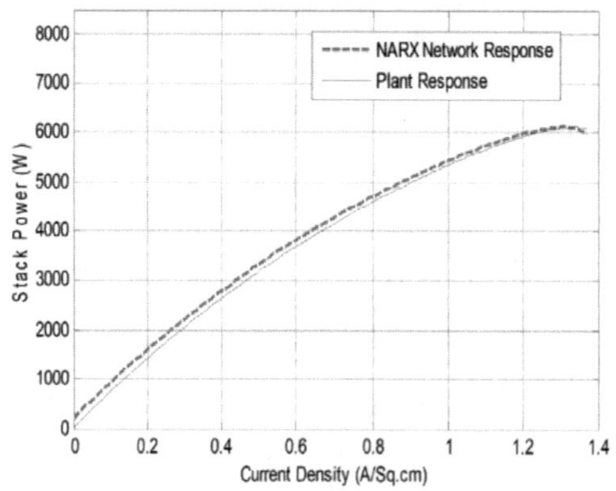

Figure 11. Stack power estimation curve of NARX network.

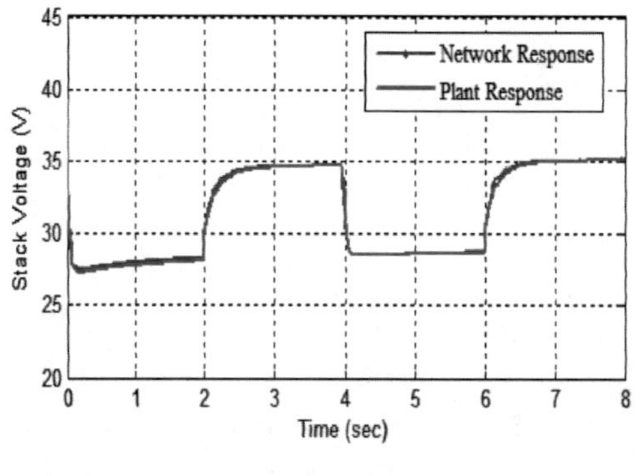

Figure 12. Stack voltage estimation curve of NARX network.

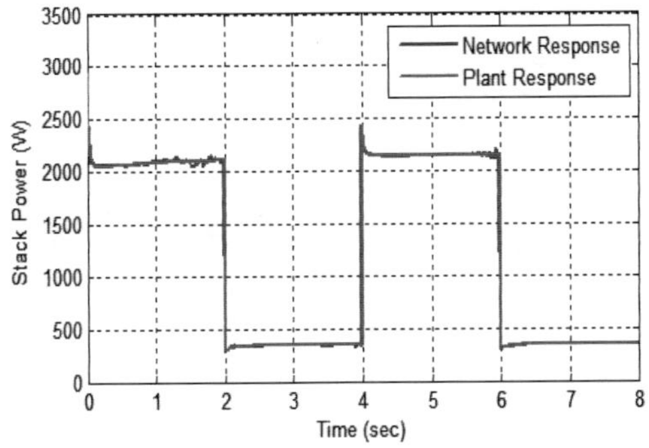

Figure 13. Stack power estimation curve of NARX network.

Both steady state and dynamic response of proposed ANN based PEM Fuel Cell extremely congruence with the experimental plant response. Hence the dynamic recurrent NARX network is declared as an optimum network to be used for further development in NN Based PEM Fuel Cell Electric Vehicle Model.

5. Results and Discussion

The MATLAB/Simulink platform is used to develop the ANN Based PEM Fuel Cell Electric Vehicle Model in which the optimum NARX network is integrated with DC – DC converter and vehicle dynamics and its Simulink Model is showed in figure 14. An unregulated DC output voltage form ANN based PEM Fuel Cell is regulated by means of power electronic circuitry of DC-DC converter and it propagated toward the vehicle transmission system for delivering the requested power at the wheel of the vehicle. Since PEM Fuel cell is a unidirectional energy source and lack of utilizing regenerative braking power, a unidirectional DC-DC converter is used in this paper.

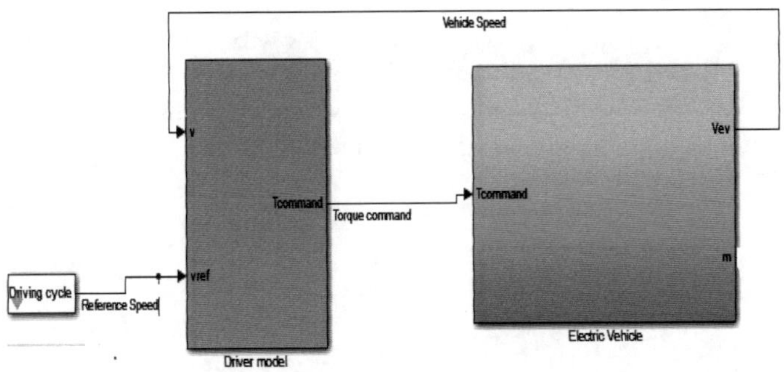

Figure 14. Simulink model of the proposed electric vehicle model.

A standardized driving pattern that described by means of a speed time table is used as reference drive cycle to analyze and compare the proposed vehicle performances. Among several driving patterns, US06 drive cycle with the maximum range of 35km/hr is used in this paper for analyzing the performance of the proposed vehicle and is showed in figure 15. The vehicle runs a distance of around 26 km/hr in 1200seconds for US06 drive cycle is shown in figure 16.

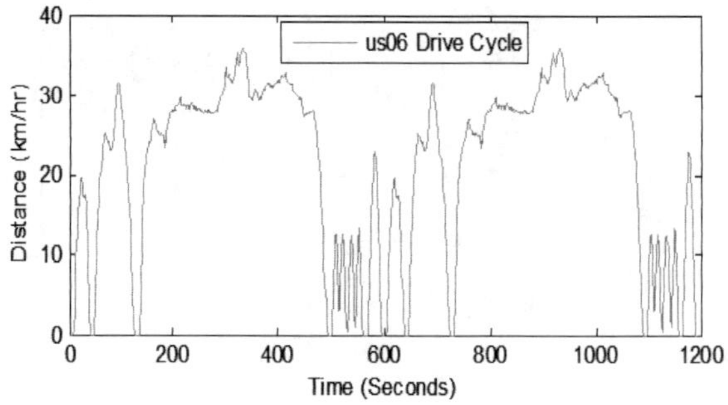

Figure 15. US06 drive cycle pattern.

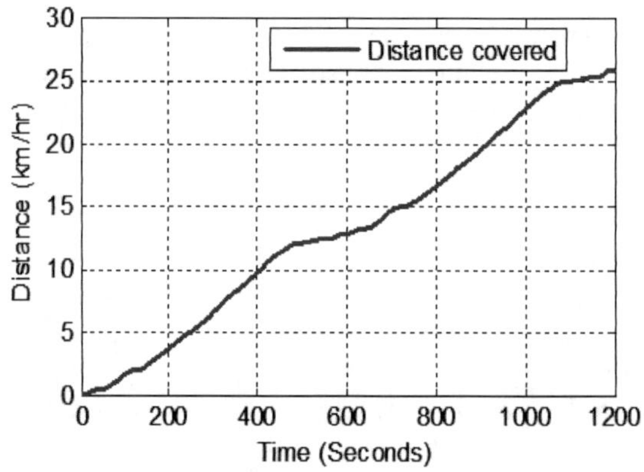

Figure 16. Distance coverage for US06 drive cycle.

The power required at the wheel of the vehicle for the US06 drive cycle is compared with power delivered from the NN Based PEM Fuel Cell is showed in figure 17. It is observed that, the proposed vehicle model is delivered the acceleration power demanded at the wheel of the vehicle but it is lacked in utilizing the negative regenerative braking power in an effective manner.

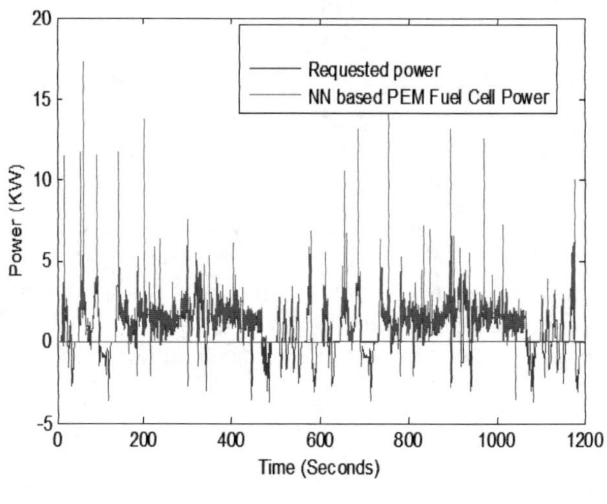

Figure 17. Comparison of requested power for US06 drive cycle with available NN based PEM fuel cell power.

6. Conclusion

Two dynamic networks such as a non recurrent Focused Time Delay Neural Network (FTDN) and a recurrent NARX Neural Network are developed and compared for identifying an optimum network to be employed for establishing a box NN based PEM Fuel Cell model. It is observed from the simulation result that the NARX Neural Network is showed improved prediction performance with minimum Mean Square Error (MSE) value of 8.4675e-07 at 11 epochs for the validation data than the FTDN network. The static and dynamic behavior analyses of optimum are carried out and validated the networks response with experimental benchmark data. ANN based PEM Fuel Cell powered electric vehicle model is developed by incorporating the optimum NARX Neural Network as a energy source with the unidirectional DC-DC converter units and vehicle dynamic system. The performance of the proposed vehicle model is investigated with US06 drive cycle pattern in terms of power requirement, power availability from the energy source and distance coverage. It is observed that, the proposed vehicle model has delivered the acceleration power but it is lacked in utilizing the negative regenerative braking power in an effective manner. In future, it is planned to develop a hybrid electric vehicle model for utilizing the braking power and to extend the life of energy source with improved drivability.

References

[1] Dincer I and Zamfirescu C 2014 Advanced power generation systems In: Boston: Elsevier 143-98.

[2] Pukrushpan J T, Stefanopoulou A G and Peng H 2004 Control of fuel cell breathing *IEEE Control System Magazine* **24(2)** 30–46.

[3] Larminie J and Dicks A 2000 *Fuel Cell Systems Explained* New York: Wiley 29–52.

[4] Pukrushpan J T, Peng H and Stefanopoulou A G 2002 Simulation and analysis of transient fuel cell system performance based on a dynamic reactant flow model *ASME International Mechanical Engineering Congress and Exposition* **36290(1)** 637-48.

[5] Pasricha S and Shaw S R 2006 A dynamic PEM fuel cell model *IEEE Tranactions on. Energy Conversion* **21(2)** 484–90.

[6] Nørgaard M, Ravn O, Poulsen N K and Hansen L K 2003 *Neural Networks for Modelling and Control of Dynamic Systems* Great Britain, Springer.

[7] Puranik S V, Keyhani A and Khorrami F 2010 Neural network modeling of proton exchange membrane fuel cell *IEEE Tranactions on. Energy Conversion* **25 (2)** 474–83.

[8] Puranik S V, Keyhani A and Khorrami F 2010 State-Space Modeling of Proton Exchange Membrane Fuel Cell *IEEE Tranactions on. Energy Conversion* **25(3)** 804-13.

[9] Wei-Lung Yu, Sheng-Ju Wub and Sheau-Wen Shiah 2010 Experimental analysis of dynamic characteristics on the PEM fuel cell stack by using Taguchi approach with neural networks *International Journal of Hydrogen Energy* **35(20)** 11138-47.

[10] Hayder Sabah, Abad Al-Amir, Hayder Abed Dahd and Eiman Ali Eh. Sheet 2015 Modeling and Control of Fuel Cell Using Artificial Neural Networks *Journal of Engineering* **21(12)** 124-38.

[11] Mammar K and Chaker A 2012 Neural Network-Based Modeling of PEM fuel cell and Controller Synthesis of a stand-alone system for residential application International *Journal of Computer Science Issues* **9(1)** 244-53.

[12] Demuth H B, Beale M 1998 *Users' Guide for the Neural Network Toolbox for Matlab* The Mathworks, Natica, MA.

[13] Robert Bosch Gmbh 2002 *Bosch-Automotive Handbook*, German.

[14] Lin B 2000 Conceptual design and modeling of a fuel cell scooter for urban Asia *Journal of Power Sources* **86(1-2)** 202-13.

[15] Lee C H, Yang J T 2011 Modeling of the Ballard-Mark-V proton exchange membrane fuel cell with power converters for applications in autonomous underwater vehicles *Journal of Power Sources* **196(8)** 3810-23.

[16] Murugesan K, Senniappan V 2013 Investigation of Water Management Dynamics on the Performance of a Ballard-Mark-V Proton Exchange Membrane Fuel Cell Stack System International *Journal of Electrochemical Science.,* **8** 7885-904.

[17] Chen S,Wang X X, and Harris C J 2008 NARX-Based Nonlinear System Identification Using Orthogonal Least Squares Basis Hunting *IEEE Transactions on Control Systems Technology* **16(1)** 78-84.

[18] Lin T, Horne B G, Tino P, Giles C L 1996 Learning long-term dependencies in NARX recurrent neural networks *IEEE Transactions on Neural Networks* **7(6)** 1329-38.

[19] Mann R F, Amphlett J C, Hooper M A I, Jensen H M, Peppley B A, and Roberge P R 2000 Development and application of a generalized steady-state electrochemical model for a PEM fuel cell *Journal of Power Sources* **86(1-2)** 173–80.

[20] Hagan M T and Menhaj M B 2008 Training Feedforward Networks with the Marquardt Algorithm *IEEE Transactions on Neural Network* **5(6)** 78-84.

Radio frequency identification system based automated public distribution system

K. Dhayalini [1] and P.Vigneshwaran [2]
[1] Professor, Department of EEE
[2] Assistant Professor, Department of EEE
[1&2] K. Ramakrishnan College of Engineering, Trichy, Tamilnadu, India

Email: [1] dhaya2k@gmail.com and [2] vigneshwaran@krce.ac.in

Abstract— This paper is designed at an aim of improving public distribution system by the process of minimizing manual work. In our nation, Ration card is important for every family for the purpose of identification and getting goods from the government. Today, public distribution system (PDS) having corruption and there is no transparency between government and users. This paper focuses on distribution of goods in PDS system automatically and creates a transparency between the customers and government.

Keywords— Radio Frequency(RF), Radio Frequency Identification (RFID), Public Distribution System(PDS)

1. INTRODUCTION

Nowadays ration card has become a major aspect for every household and its major purpose is for obtaining the details of members belonging to the family by acting as an address proof and it is also essential for getting gas connection. In order to buy things like kerosene, dal, flour, oil and getting other benefits this card is essential. Two major disadvantages are being identified in this system. The foremost is inaccuracy in materials, second is smuggling of materials. This project proposes an efficient method to overcome the difficulties of old system. As payment bills are computerized it avoids collaboration and malfunctions. Transparency can be maintained through this system. Performance speed is high, more congestion can be avoided by using public distribution system. Time delay can be avoided.

2.EXISTING SYSTEM DESCRIPTION

The main objective of a government is to control the distribution of goods and services to the common people. This aspect is even more essential when there is an outweighing of the demand items with respect to supply availability. When there is a war or when economies slide the process of rationing will be done by governments. The government will be in a need of controlling goods supply when there is an occurrence of black market. This is very much essential for enhancement of security network. Besides this, the poor people are suffering to meet their needs of consumption by getting their supply only by ration shops. For this purpose there are state-level ministries under the department of food and civil supply for the purpose of regulating ration shops that are present under their jurisdictions, and it is this ministry that is providing license for private traders. The process of issuing these ration cards is done by state governments in a proper manner and the allocation of goods are also done by them.

This will vary from one commodity to another and the authority for determining the prices lies with the state government. Public distribution system (i.e.) ration distribution is an area where there is a high possibility for illegal smuggling.The citizens of the country posses ration cards for buying the goods from the ration shops. While getting materials through the rations, first there is a necessity to give the ration card and a signature will be put by the official in the card on the basis of the materials that are to be allotted for that particular person. Through this process the materials are allocated to the stipulated person by using the assistance of humans. Because of using this system, the customer will get inaccurate materials .This type of public distribution as a whole entirely will not be able to satisfy all the needs of the common people because of excess population of the country. In addition to this , there is also a possibility for the materials to be robbed by entering mismatched or wrong entries into the official register which may also happen without bringing this news to the knowledge of the ration card holder .

3.OVERVIEW OF PROPOSED SYSTEM

In this system, RFID (Radio Frequency Identification) system is proposed in place of manual entry in the record books. The customer needs to place the RFID card to the RF (Radio Frequency) reader and it reads the signal. This information is send to the Microcontroller. The controller checks the customer codes and then microcontroller is interfaced with a computer and GSM module. These blocks are used for displaying the quantities and the messages are being sent to the user.

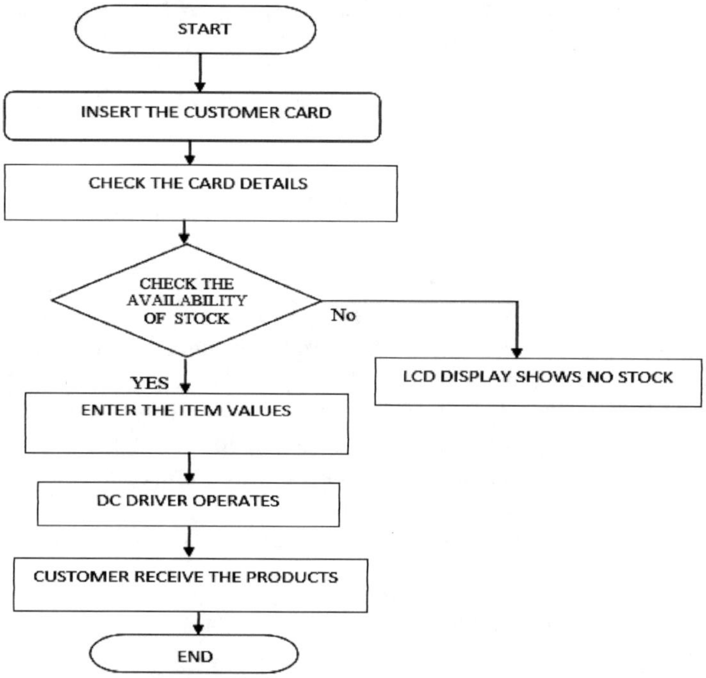

Figure 1: Flow Chart for PDS System

IOP Conf. Series: Materials Science and Engineering **937** (2020) 012041 doi:10.1088/1757-899X/937/1/012041

4. Block Diagram of Public Disribution System

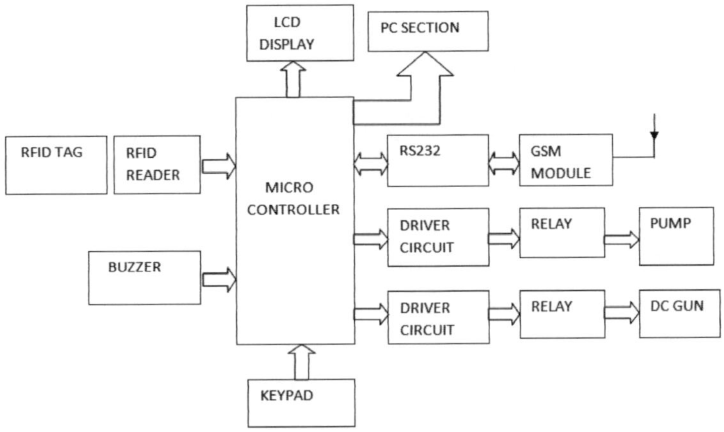

Figure 2: Overall Block Diagram of Automatic Public Distribution System

5. Block Diagram of RFID Section

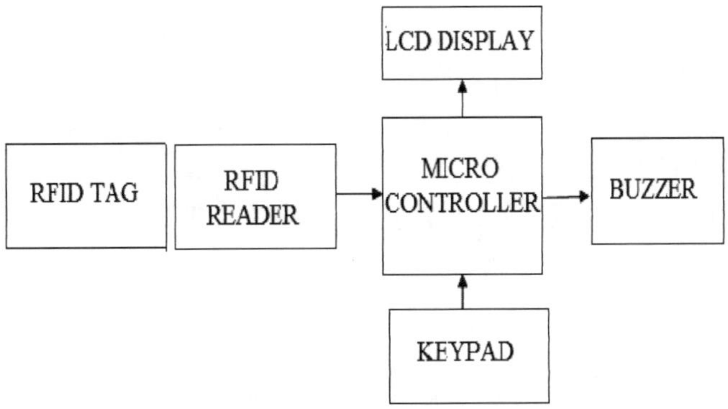

Figure 3: Block diagram of RFID Section

RFID is an advanced technology which has special characteristics such as wireless card. This card will have an inbuilt embedded chip and in addition to this it also has an loop antenna. A twelve digit number which is denoting the card number will be represented by this inbuilt chip. In this representation, for the purpose of library card we are using a more advanced RFID card. With the help of this technology, each and every individual will have an unique RFID card. For the transfer of data from the RFID reader, it is being interfaced with a suitable microcontroller. For this purpose we need a reprogrammable controller. Hence a flash type reprogrammable microcontroller is used. The card number is to be pre-programmed into the microcontroller.

Whenever the RFID card is placed in front of the reader by the person, the reader starts its operation by reading the number present in the card. After this the data is being sent to microcontroller. Then the person will be in need to press the PIN number i.e., personal identification number which is given as an input through key pad. All these data that are transferred reaches the microcontroller. Data like the card number and PIN number that is being entered by the person in the keypad will be compared with the number that is being already stored in the network. If all these data are found to be valid then the database of the concerned person will open. The database network will posses all the details like quantity of material to be issued, amount that is being already bought and allocated details. In contrast, there will be a display of message called "Authentication Fail" if any of the things becomes as invalid like PIN number. This will be displayed on the LCD screen. There will also be an additional indication like alarm sound which will be activated by the driver circuit. Hence if there is an alarm sound it can easily be concluded that there is authentication failure.

6. DESCRIPTION OF MECHANICAL SECTION

After RFID section process, the user can enter variety and quantity of the product based on product availability of particular customer. By the use of driver circuit and relay, the entered quantity can be delivered. The entered product is a solid material it can get from DC Gun, if it is a liquid material it can get from pump of the mechanical section.

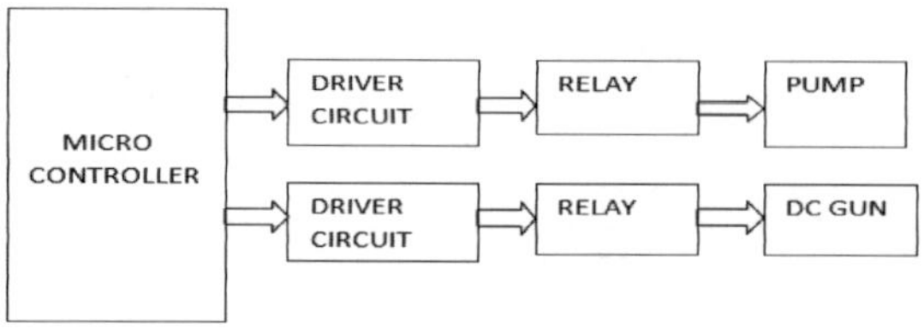

Figure 4: Block diagram of Mechanical section

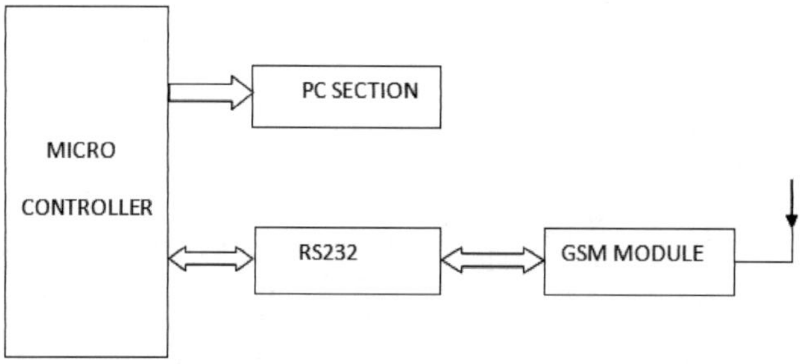

Figure 5: Block diagram of Interfacing Section

In mechanical section, RS232 cable is used for interfacing GSM module with the microcontroller. VB tool is used for interfacing PC with the microcontroller.

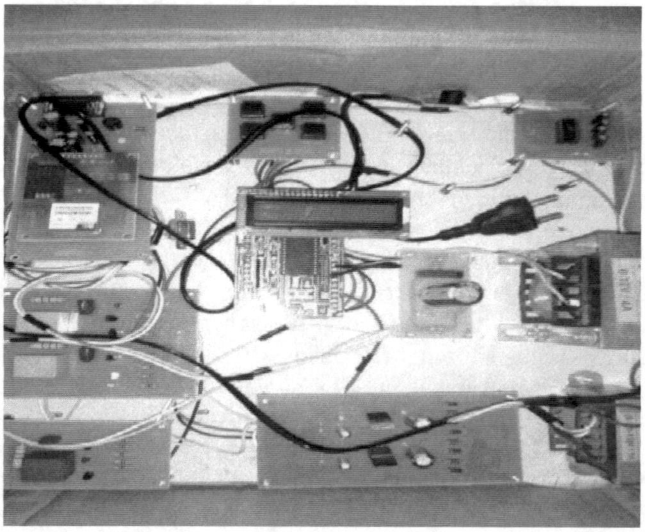

Figure 6: Experimental setup of PDS section

Figure 7: Mechanical Setup of PDS System

7.Conclusion

In the previous technology, the ration is not properly distributed to the consumers. This drawback is over come by using smart cards. We can avoid malfunctioning and corruption under going in ration shops. In the old ration system all the data's are entered manually. In modern public system based on smart card reader we overcome the above problems. This provides simple and accurate information about PDS to the public. In future, the PDS can be interfaced with pc and GSM for updating the details of the Customer.

REFERENCES

[1] Yogesh Kumar Sharma, Dr K B Shiva Kumar, Srinidhi G A and Dr Manoj Kumar. (2014)- "Multi-Modality Biometric Assisted smart Card Based Ration Distribution System".
[2] Rajesh C.Pingleand and P.B. Borole.(2013)-"Automatic Rationing for public Distribution System (PDS) using RFID and GSM Module to Prevent Irregularities".
[3] K.SRinivasaRavi, G.H.Varum, T.Vamsi ,P.Pratyusha. (2013) -"RFID Based Security System".
[4] Dhanashri Pingale, SonaliPatil, Nishigandha Gadakh, ReenaAvhad, and Gundal. (2013)- "Web Enabled Ration Distribution and Corruption Controlling System.
[5] M.Elizabeth Sherine and Shinu Sadeyone (2014)- "NFC based stock maintenance and billing system with auto alert to customers".
[6] Prasanna balaji.R, and Manikandan.T. (2013)- "Automatic Public Distribution System".
[7] Gyanendra K Verma and Pawan Tirupathi. (2010)- "A Digital Security System with Door Lock System Using RFID Technology".
[8] KashinathWakade, Pankaj Chidrawar, Dinesh Aitwade. (2015)- "Smart Ration Distribution and Controlling".
[9] K.Govinda and Yannick Ngabirano. (2012)-"Secure data Storage in Cloud Environment Using Biometrics".

IOP Conf. Series: Materials Science and Engineering **937** (2020) 012042 doi:10.1088/1757-899X/937/1/012042

Experimental analysis of PQ parameter estimation of VFD drives

Nabanita Dutta[1], Palanisamy Kaliannan[1], Umashankar Subramaniam[2]
[1]Department of Energy and Power Electronics, School of Electrical Engineering, VIT Vellore, Vellore 632014, India; nabajhilikbarbi@gmail.com, kpalanisamy@vit.ac.in,
[2]College of Engineering, Prince Sultan University Riyadh, Saudi Arabia, shankarums@gmail.com

Abstract. Controlling the AC motor speed and torque changing voltage and frequency for electrical and mechanical system the adjustable speed drive system is required which is known as Variable Frequency Drive. The usage of VFD is gradually increasing every day. The 25% of the world's energy is consumed by the motor load for which the usage of VFD is responsible. The cost of VFD can be reduced and size also reduced to help the system more compatible within a very short span of space. It is based on semiconductor topologies, driver circuit, control circuit and both software and hardware devices are also used. Now power quality is the important role plays for the energy efficiency of the system. The researchers are working on various issues of power quality like voltage sag, swell, system unbalance etc. which are required to be analysed for the improvement of the efficiency of the system. In this research the VFD based induction motor faces various power quality issues which are described. The paper is concentrating on various power quality issues which affects the performance of VFD drives through various parameters. The various power quality parameters are predicted for the change in loadings with the help of MATLAB curve fitting tool. With different load variations voltage and current unbalance have been measured. The process has been done though experimental setup with the help of power quality analyser Fluke 302 and both the controller technique like VVC+ and V/F methods have been used.
Keywords: VFD, Curve fitting, Power Quality, Induction Motor, Control circuit

1. Introduction

The requirement of energy is also increasing as world also progressing. Rather than 2016 to 2030 the energy consumption also increases 30% [1]. The optimal utilization of the energy is required with the increasing demand of the energy as the resources are exhausting day by day [2].

Energy efficiency improvement is the key matter to keep the environment sustainable and the main resources of the energy comes from electric power and water. As the electric power is very important for energy supplying case and every household sectors, industries. As the electric power is the main concern therefore power quality term comes into picture. Power supply is good or bad that can be measured by power quality analysis. The differentiation is based on various parameters likewise voltage, current, frequency, harmonics and system unbalance. The change in these parameters with respect to standards shows the status of power quality [13].

The factors are various types which leads to deterioration of power quality. In this case variable frequency drives plays a significant role [14]. The power electronics modules play a significant role in the usage of VFDs and behave as a nonlinear load and draws the harmonic currents from the supply and pollution is created in the power system. This also leads to electromagnetic interfaces and in three phase system nonlinearity causes the system unbalance and significant neutral currents. As a result there is a poor power quality and reduced efficiency can be observed [15-25]. Besides harmonics power quality also deals with voltage sag, swell, power factor, system unbalance, electromagnetic interfaces, frequency variation etc. [26-35]. These parameters should be monitored in proper way and should be

maintained for the energy efficient system. Active power filters are fully commercialised and provides a unique solution for various power quality problems [5].
The various power quality parameters have to be under specified standards. These standards are governed by IEEE and IEC. IEEE Standard-519 gives the various limits for current and voltages for a power system [6]. EN 50160 gives the voltage limits for a public distribution network [7] [8]. These standards should be maintained for healthy grid conditions [9].
In this work power quality parameter estimation of VFD (variable frequency drive) is done with respect to different loadings of motor. The estimation is done using MATLAB curve fitting tool [10]. A three phase induction motor has been analysed with respect to different loadings as no load, half load and full load and parameter for other loadings are estimated [11]. The various PQ parameters are studied with the varying load and detailed analysis with respect to comparison of control techniques is made in this work. Moreover comparison between the speed control methods is also carried out [12]. The various sections of this paper consists of VFD, experimental setup, test procedure and data acquisition, standards, followed by detailed analysis and conclusions.

2. VFD (Variable Frequency Drive)

Variable frequency drive also known as adjustable speed drive system which are used for electrical and mechanical system to control the AC motor torque speed characteristics by changing voltage and frequency. VFD is very helpful for large motor system and compressor application. Few decades the usage of VFD has been increased in alarming rate, and it makes the industry energy efficient as 25% of the energy is used by motor application. VFD system helps to reduce the size of the system and it helps to improve the performance of the semiconductor devices, control circuit for both software and hardware, semiconductor devices, drive topologies. VFD is made of AC-AC and DC-DC topologies.

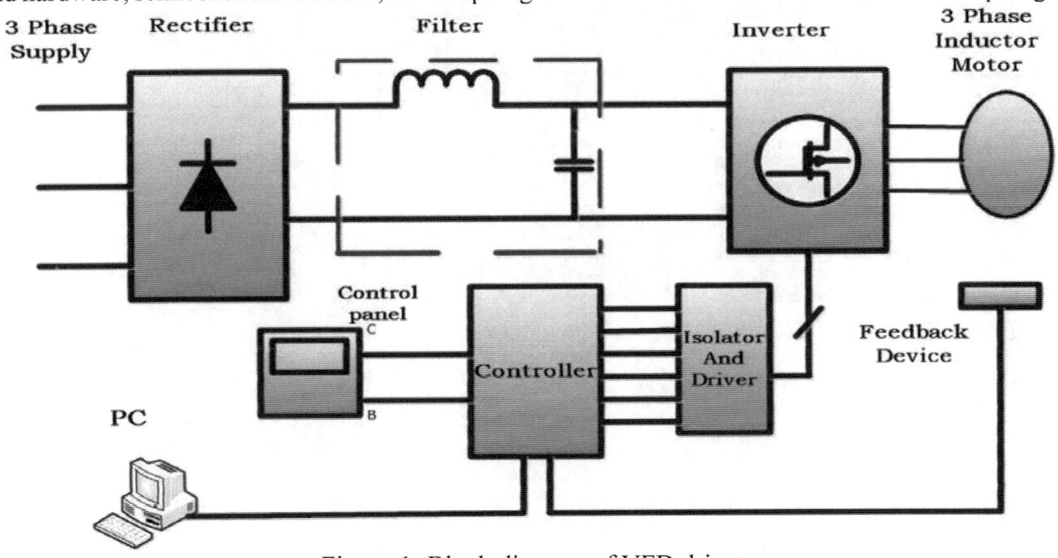

Figure 1: Block diagram of VFD drives

3. Experimental Setup

The circuit diagram of the experimental setup used for the analysis is shown in Figure 2. The circuit diagram consists of supply, measuring clamps, and VFD which is connected to the motor. The various electrical parameters are measured at the input to the VFD with a power quality analyser as shown in Figure 2. In this experiment the analysis of voltage, current, frequency unbalance, voltage unbalance, power consumed, power factor and efficiency factor, power factor and efficiency factor, energy consumed and harmonic spectrum etc.

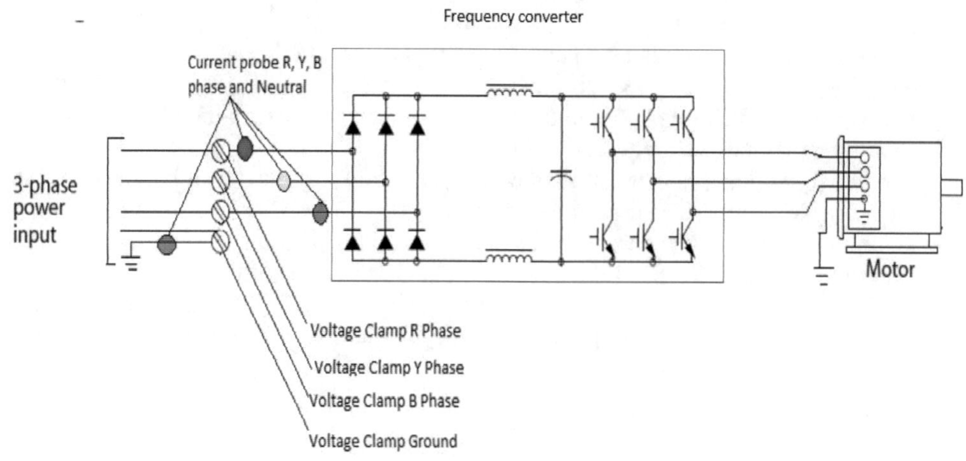

Figure. 2. Circuit diagram of the experimental setup used

Pictorially the test bench setup looks like Figure 3. The various parts are shown and labelled in Fig. 3.

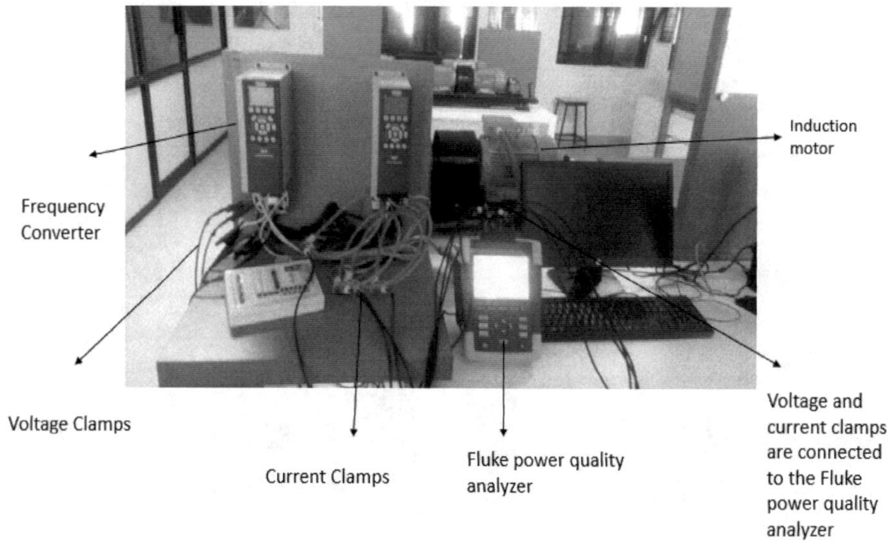

Fig. 3. Pictorial view of the Test Bench

4. Test Procedure and Data Acquisition

The below steps represent the step by step procedure involved in the analysis of frequency converter fed IM. First step is one end of the voltage and current clamps are connected on the input side of the converter and the other end is connected to the fluke meter. Second step is the frequency converter will be initialized and required settings should be done. Third step is the machine should be turned on and loaded. Fourth step is fluke power analyser is turned on and data acquisition has been started and recorded for 5 minutes. Fifth step is data is stored in fluke power analyser SD card. Sixth step is to connect the PC and SD card also connected through PC. Seventh step is to use power log software to use the fluke power analyser to analysis the data. Eighth step is to obtain the data and interfaces to be analysed.

The various PQ parameter's standard values are given in Table 1.

IOP Conf. Series: Materials Science and Engineering **937** (2020) 012042 doi:10.1088/1757-899X/937/1/012042

Table 1. Standards

S.NO	Parameters	Value
1	Voltage variation	<±10%
2	THD	Voltage (<8%) and current (<5%)
3	Unbalance	<±3%
4	Frequency variation	< ± 0.1%

5. Estimation of Power Quality parameter and comparison of control methods

The various power quality parameters are predicted for the change in loadings with the help of MATLAB curve fitting tool. The different plots are estimated for two control methods namely VVC+ and V/F method which are used for induction motor speed control.

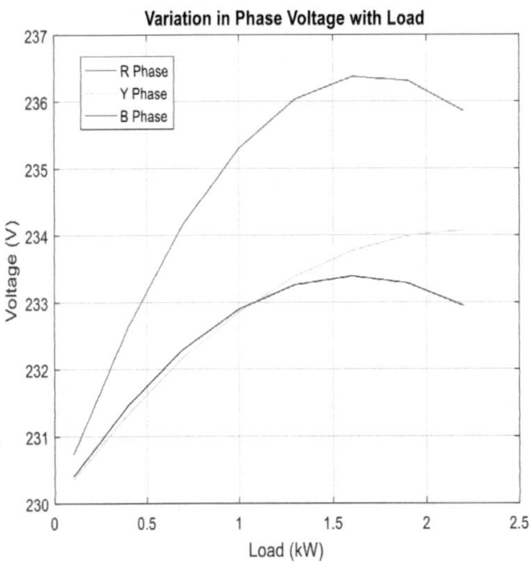

Figure.4 Variation of Phase Voltage (VVC+)

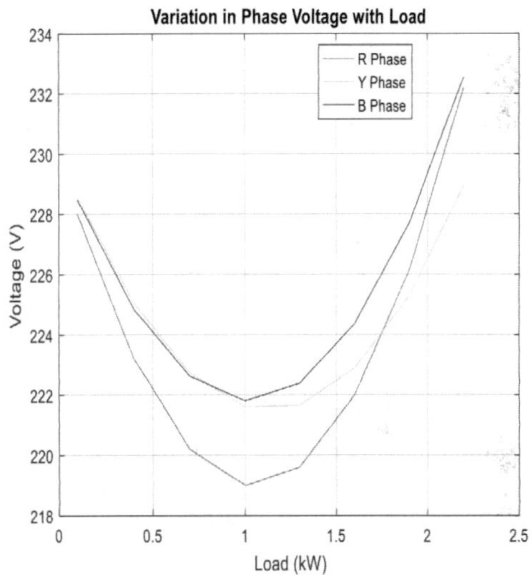

Figure.5 Variation of Phase Voltage (V/F)

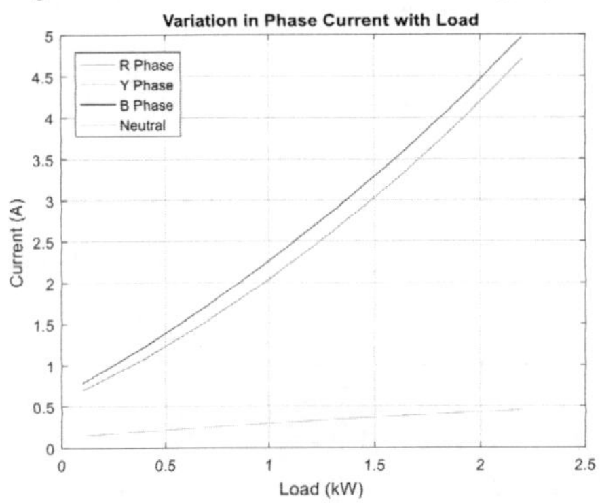

Figure. 6 Variation of Phase Current (VVC+)

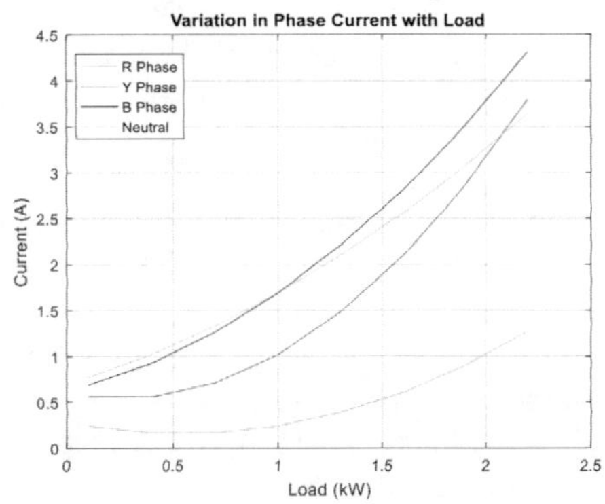

Figure. 7 Variation of Phase Current (V/F)

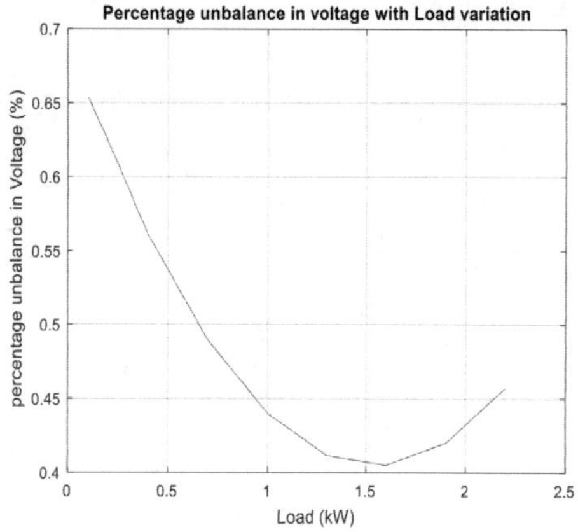

Figure.8 Voltage unbalance v/s load variation (VVC+)

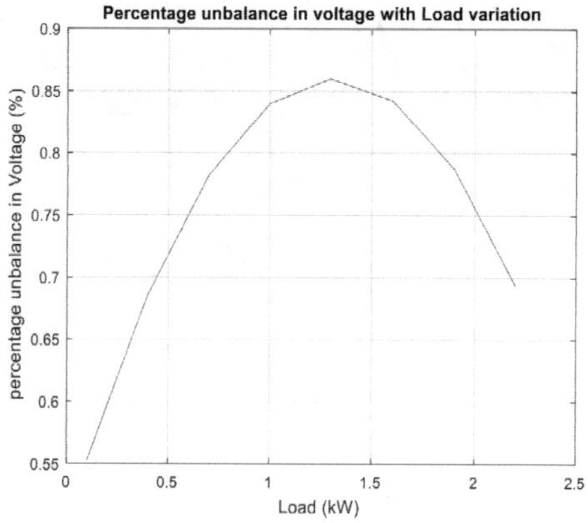

Figure 9 Voltage unbalance v/s load variation (V/F)

5.1. Phase Voltage

Using plotting tool of MATLAB, the phase voltage is estimated for VVC+ as shown in Fig. 4 and for V/F as shown in Fig. 5. The voltage increases in the case of VVC+ and decreases in the case of V/F. As the load further increases in the case of VVC+ the voltage comes to a saturation point while in V/F case the voltage increases.

5.2. Line Current

Line current variation for VVC+ is shown in Fig. 6 and for V/F is shown in Fig. 7. It can be seen that there is an increase in the value of line current with the increase in the loading for both the cases of VVC+ and V/F respectively. The difference is in the magnitude of neutral current. The neutral current is more with the increase in the loading for the case of V/F as compared to VVC+.

5.3. Voltage Unbalance

The unbalance in voltage for the case of VVC+ is shown in Fig. 8 and for the case of V/F is shown in Fig. 9. For the case of VVC+ the unbalance in voltage decreases with the increase in the load while in the case of V/F, the voltage unbalance increases with the increase in the load. After 1.5 kW, the nature changes, for VVC+ the unbalance increases while for V/F the unbalance decreases.

5.4. Current Unbalance

Unbalance in current for VVC+ is shown in Fig. 10 and unbalance in current for V/F is shown in fig. 11. It can be seen that with the increase in loading current unbalance for both the cases decreases. For the case of VVC+, the decrease is sudden while for V/F the decrease is gradual. Moreover, for VVC+, when the loading is reaching near the full load, there is an increase in current unbalance.

5.5. Frequency

Frequency variation is shown in fig. 12 for VVC+ and for V/F is shown in fig.13. For VVC+, the frequency first increases and then decreases. Similar graph is obtained for V/F also. The difference is that in the case of V/F method, the frequency never jumps below 50 Hz while for the case of VVC+, the frequency comes below 50 Hz with the increase in loading.

5.6. Active Power

Fig. 14 and Fig. 15 shows the active power variation for VVC+ and V/F respectively. In both the cases with the increase in loading of the motor the active power requirement increases as shown in fig. 14 and fig. 15.

IOP Conf. Series: Materials Science and Engineering **937** (2020) 012042 doi:10.1088/1757-899X/937/1/012042

5.7. *Reactive Power*

Fig. 16 and Fig. 17 shows the reactive power requirement for VVC+ and V/F control strategies respectively. The reactive power requirement in the case of VVC+ is more as compared to V/F. For same loading in both the cases of VVC+ and V/F, there is less reactive power requirement in the case of V/F which is also the cause of improved power factor in the case of V/F speed control method

5.8. *Apparent Power*

Apparent power demand for VVC+ and V/F method are shown in fig. 18 and fig. 19 respectively. It can be seen that apparent power is more in the case of VVC+ as compared to V/F.

5.9. *Power factor*

Power factor for both the control methods are shown in fig. 20 and fig. 21 respectively. It is noticed that power factor profile for V/F method is superior as compared to the VVC+ method.

5.10. *Voltage THD*

Voltage THD profile is shown in fig. 22 and fig. 23 respectively for VVC+ and V/F speed control methods. It can be seen that the THD in voltage increases in the case of VVC+ with the increase in load while in the case of V/F method, the voltage THD decreases with the increase in load.

5.11. *Current THD*

Current THD for VVC+ and V/F control methods are shown in fig. 24 and fig. 25 respectively. It can be noticed that current THD are greater in magnitude in the case of V/F method as compared to VVC+. Moreover, neutral current have huge THD in V/F method as compared to VVC+ method. For both the cases with the increase in load, there is an increase in current THD and then there is a decrease with the further increase in load. Only for the case of neutral current in V/F method, there is a increase in THD with an increase in load.

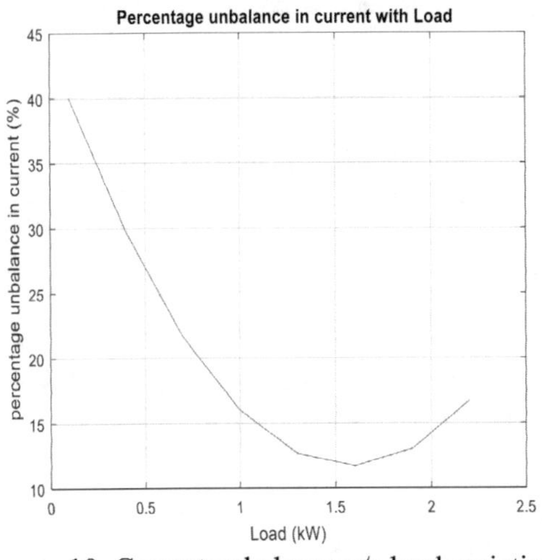

Figure. 10 Current unbalance v/s load variation (VVC+)

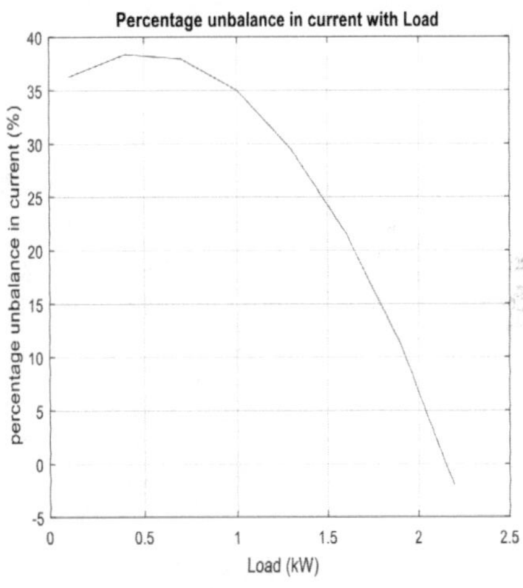

Figure. 11 Current unbalance v/s load variation (V/F)

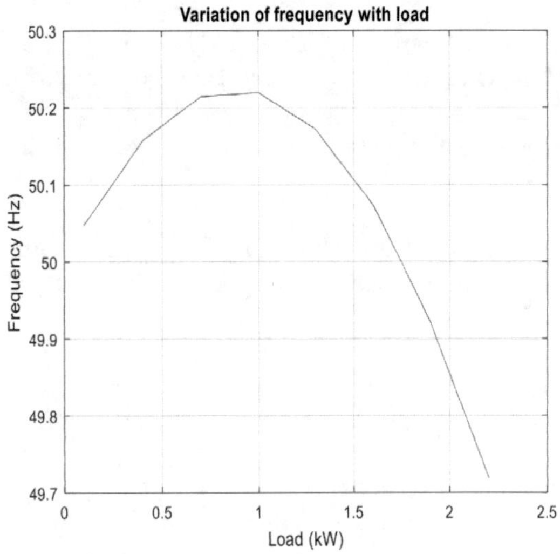

Figure.12 Frequency v/s load variation (VVC+)

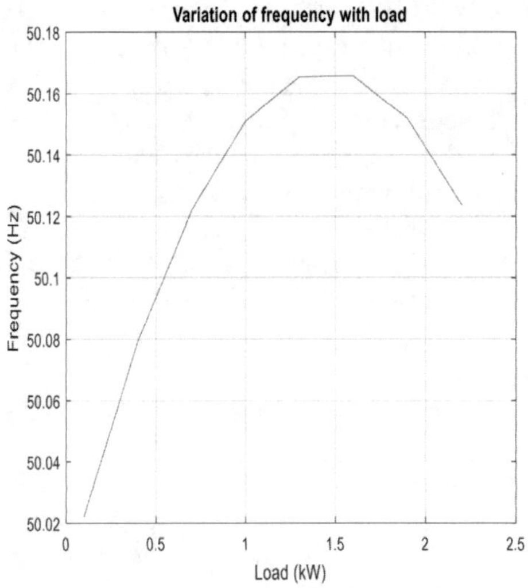

Figure. 13 Frequency v/s load variation (V/F)

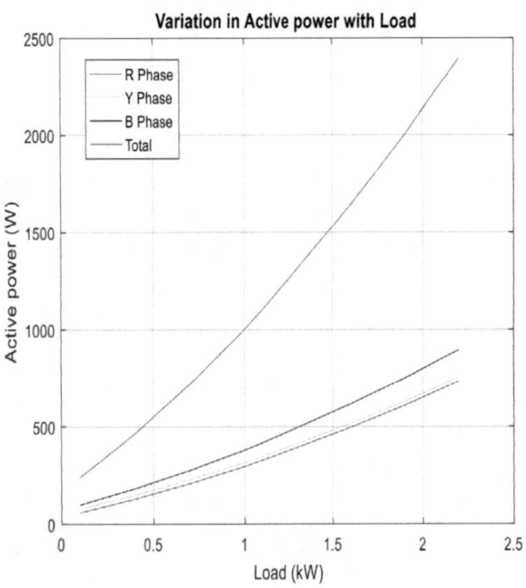

Figure. 14 Active Power v/s load variation (VVC+)

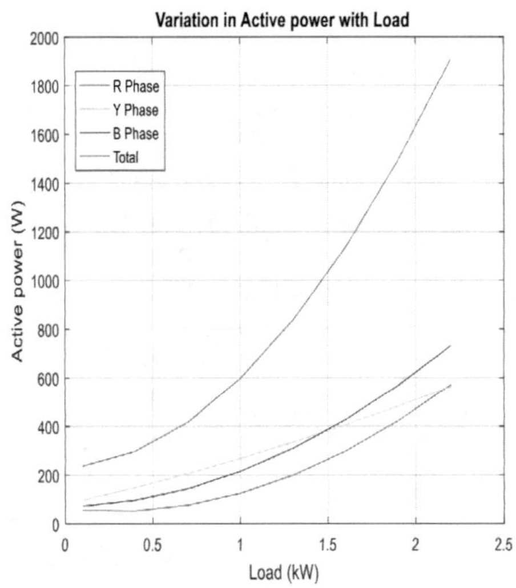

Figure. 15 Active Power v/s load variation (V/F)

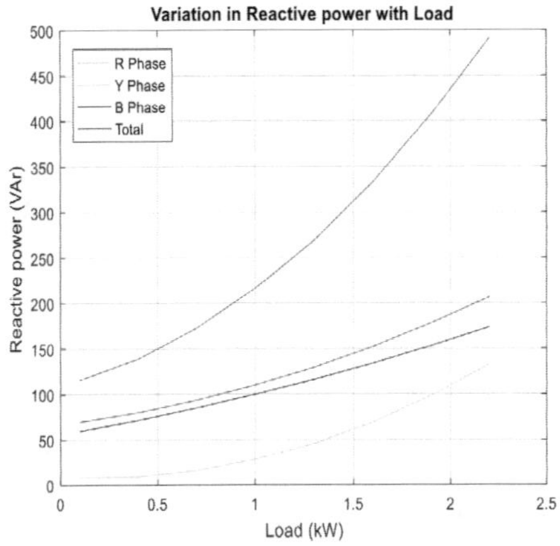

Figure. 16 Reactive power v/s load variation (VVC+)

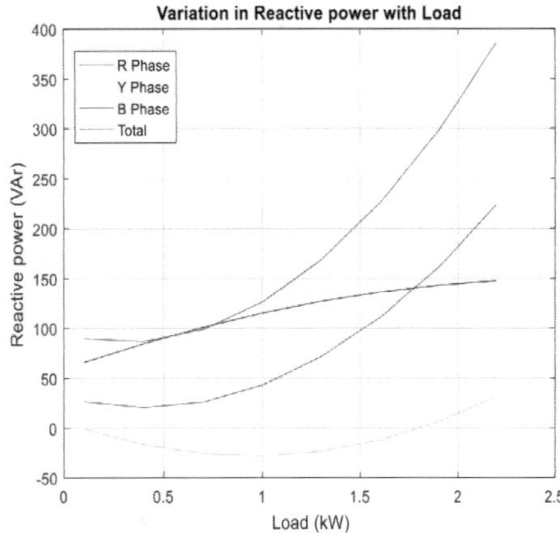

Figure. 17 Reactive power v/s load variation (V/F)

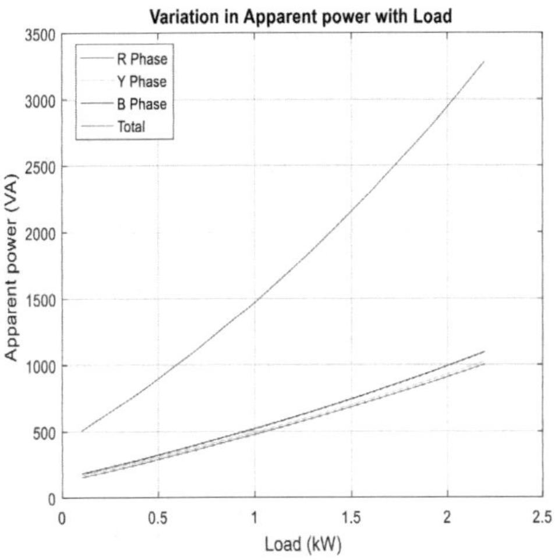

Figure. 18 Apparent power v/s load variation (VVC+)

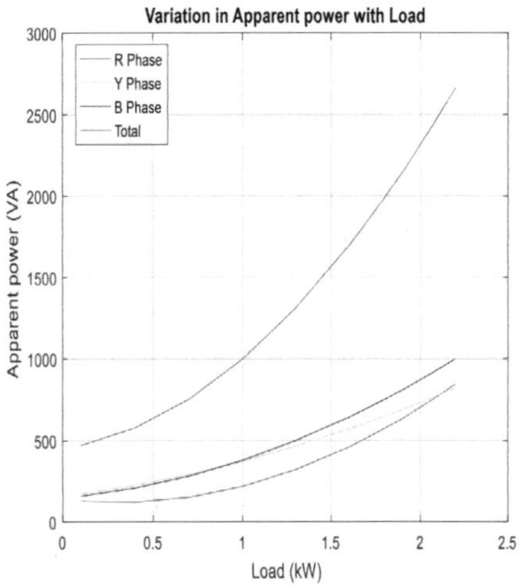

Figure. 19 Apparent power v/s load variation (V/F)

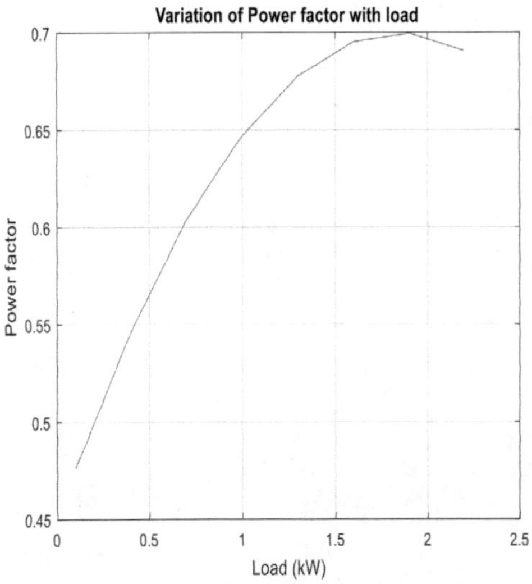

Figure. 20 Power Factor v/s load variation (VVC+)

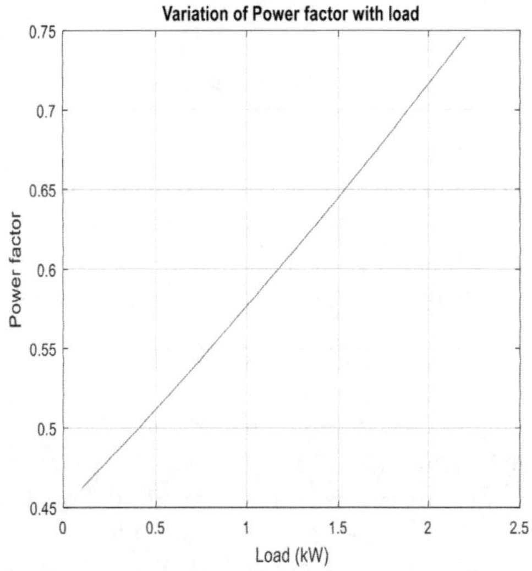

Figure. 21 Power Factor v/s load variation (V/F)

IOP Conf. Series: Materials Science and Engineering **937** (2020) 012042 doi:10.1088/1757-899X/937/1/012042

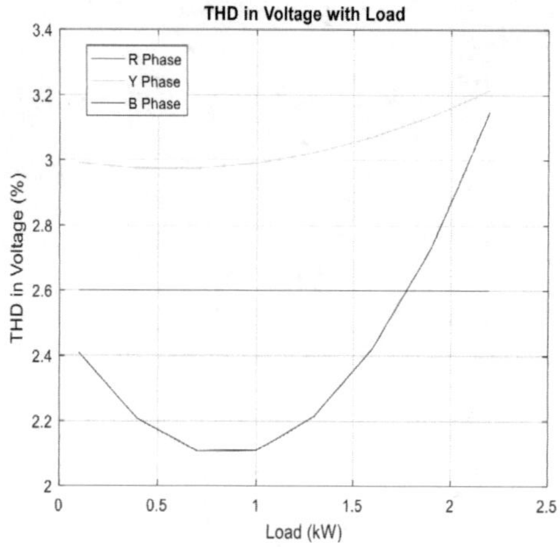

Figure. 22 Voltage THD v/s load variation (VVC+)

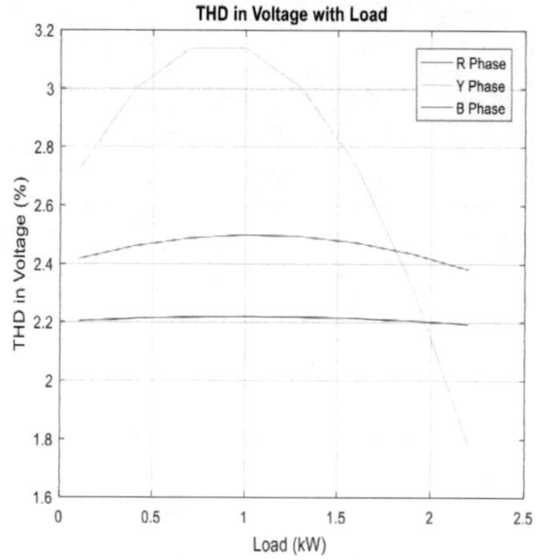

Figure. 23 Voltage THD v/s load variation (V/F)

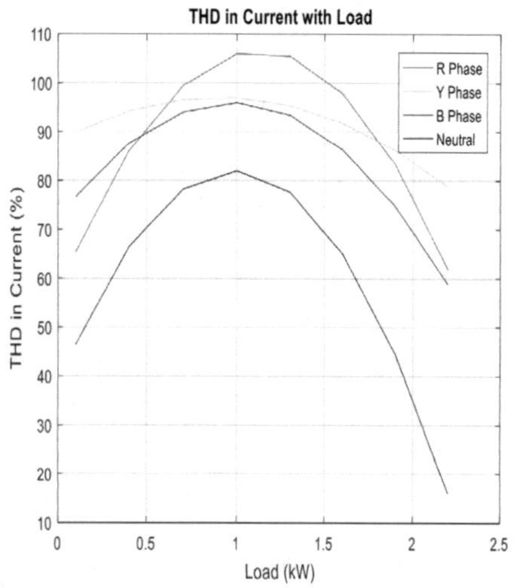

Figure. 24 Current THD v/s load variation (VVC+)

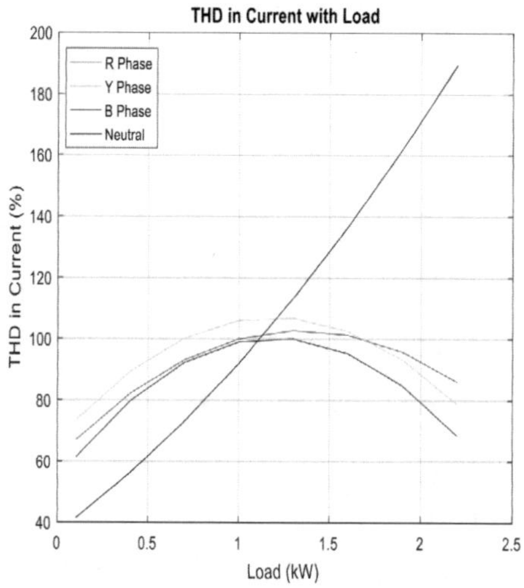

Figure. 25 Current THD v/s load variation (V/F)

6. Individual voltage harmonics

6.1. 5th Harmonic

The 5th harmonic voltage distortion is shown in fig. 26 for VVC+ and in fig. 27 for V/F speed control methods. It can be seen that distortion profile for both the cases is similar except that Y-phase shows a constant nature for VVC+ while in V/F the magnitude varies. Moreover the magnitude of distortion is higher in V/F method.

6.2. 7th harmonic

Voltage distortion for 7th harmonic is shown in fig. 28 for VVC+ and in fig. 29 for V/F. Both the figures shows that with the increase in the loadings the distortion decreases except for Y phase in the case of VVC+ and R phase in the case of V/F.

6.3. 11th Harmonic

For 11th harmonic distortion, the plots are shown in fig. 38 for VVC+ and in fig. 39 for V/F. For both the cases, the distortion profile is same except that the magnitude of the neutral current distortion is higher in the case of V/F control method as compared to VVC+.

6.4. 13th Harmonic

Fig. 40 and fig. 41 shows the distortion profile for 13th harmonic current distortion in the case of VVC+ and V/F respectively. It is found out that the distortion is considerably reduced for both cases. In the case of V/F, the neutral current distortion has higher magnitude.

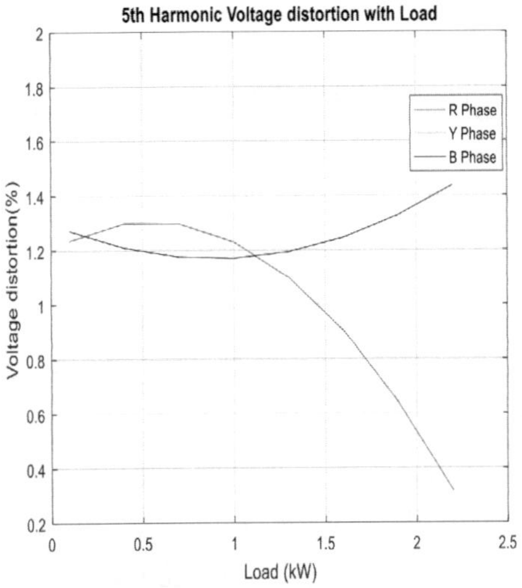

Figure. 26 5th harmonic voltage distortion v/s load variation (VVC+)

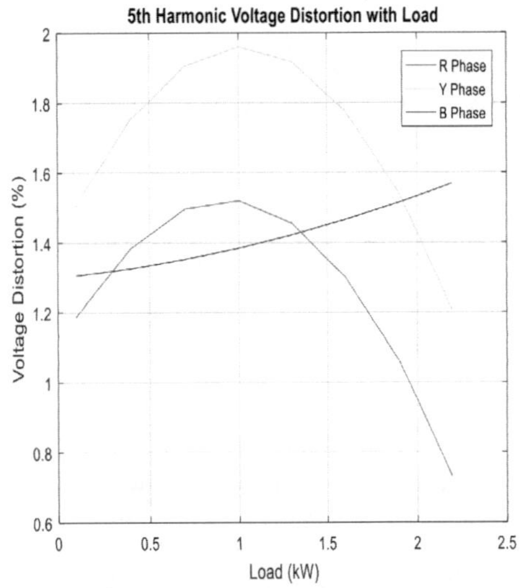

Figure. 27 5th harmonic voltage distortion v/s load variation (V/F)

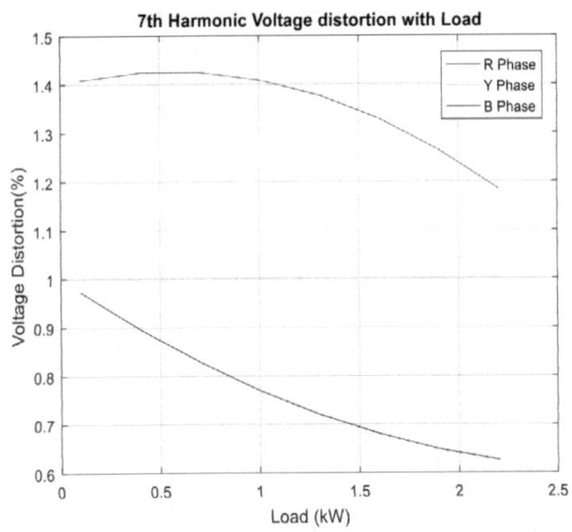

Figure. 28 7th harmonic voltage distortion v/s load variation (VVC+)

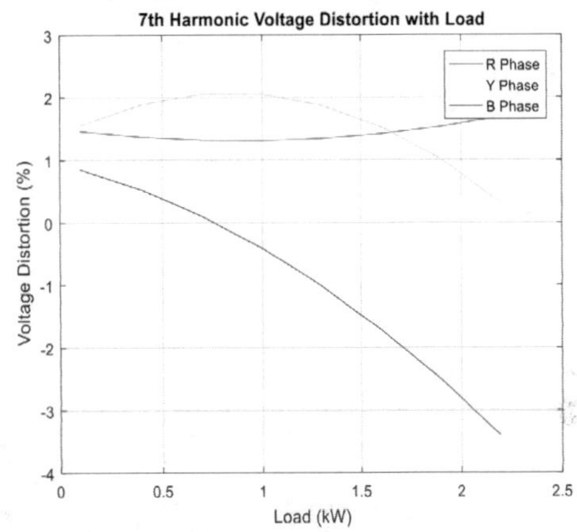

Figure. 29 7th harmonic voltage distortion v/s load variation (V/F)

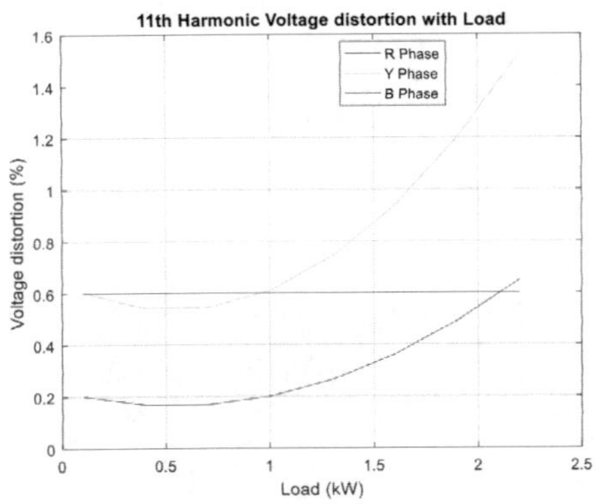

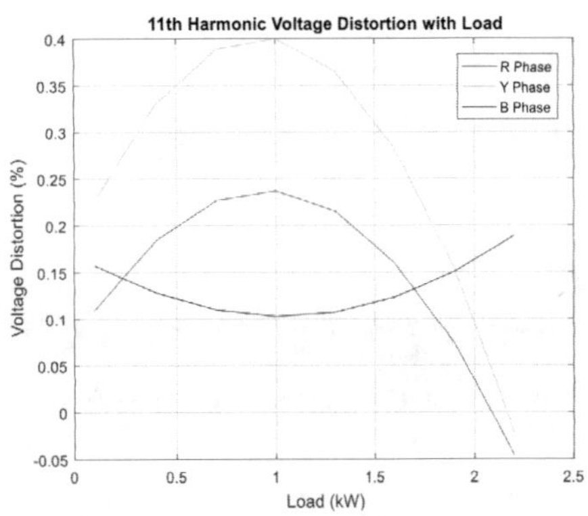

IOP Conf. Series: Materials Science and Engineering **937** (2020) 012042 doi:10.1088/1757-899X/937/1/012042

Figure. 30 11th harmonic voltage distortion v/s load variation (VVC+)

Figure. 31 11th harmonic voltage distortion v/s load variation (V/F)

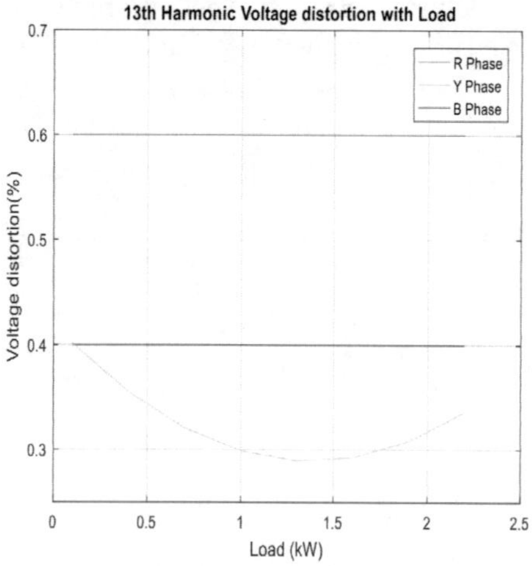

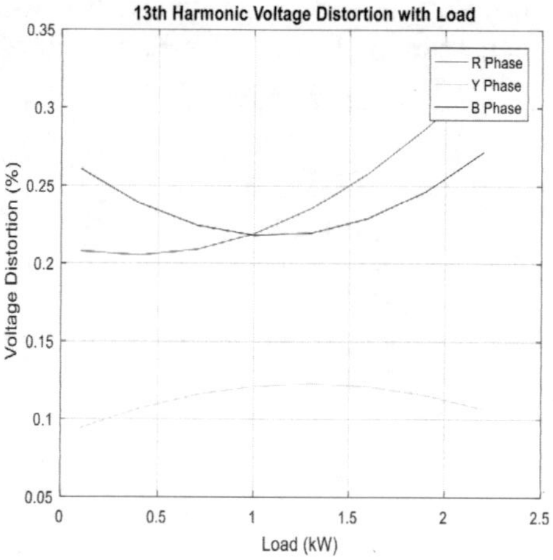

Fig. 32 13th harmonic voltage distortion v/s load variation (VVC+)

Fig. 33 13th harmonic voltage distortion v/s load variation (V/F)

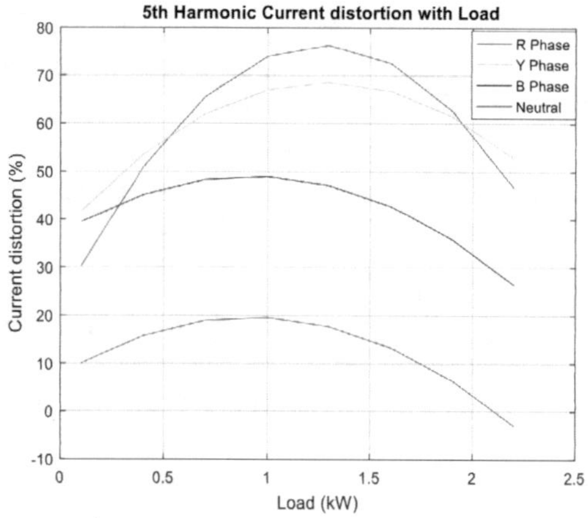

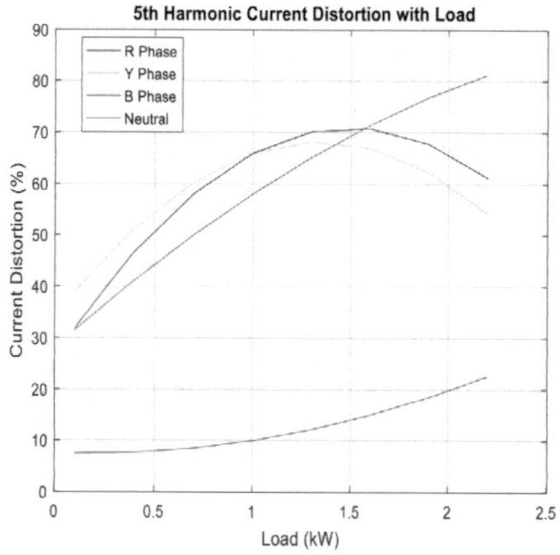

Fig. 34 5th harmonic current distortion v/s load variation (VVC+)

Fig. 35 7th harmonic current distortion v/s load variation (V/F)

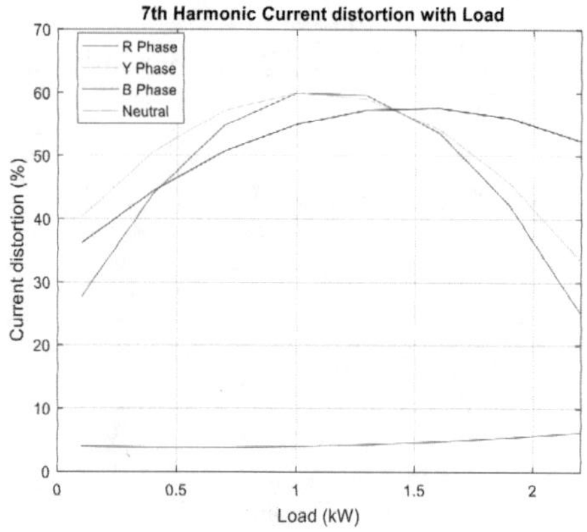

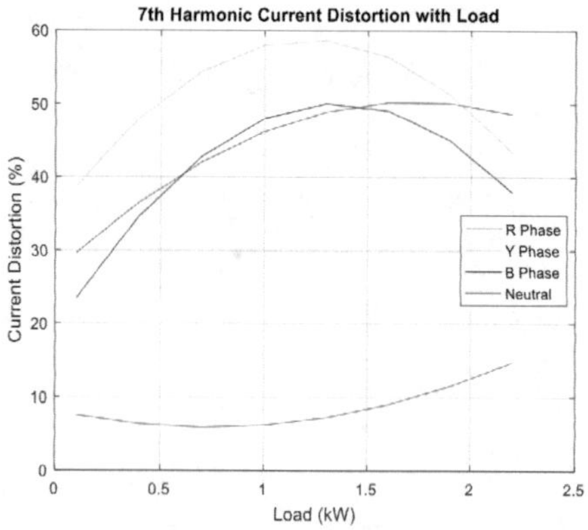

Fig. 36 7th harmonic current distortion v/s load variation (VVC+)

Fig. 37 7th harmonic current distortion v/s load variation (V/F)

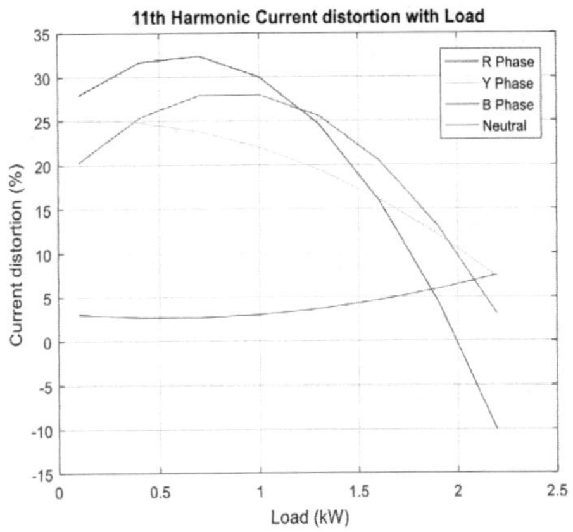

Fig. 38 11th harmonic current distortion v/s load variation (VVC+)

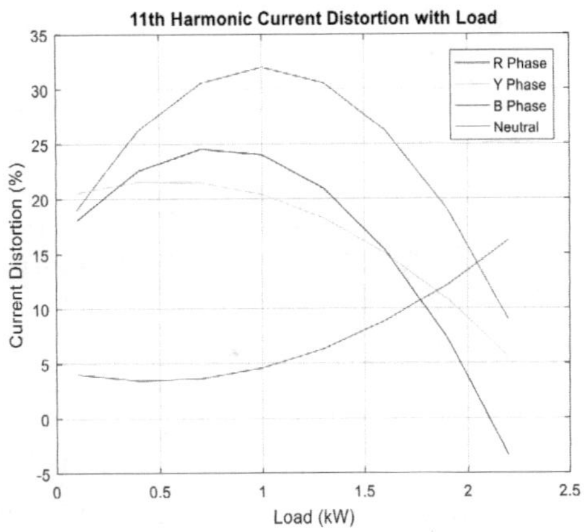

Fig. 39 11th harmonic current distortion v/s load variation (V/F)

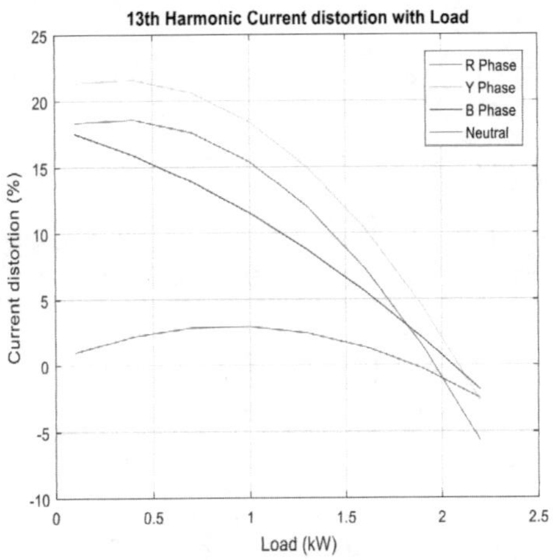

Fig. 40 13th harmonic current distortion v/s load variation (VVC+)

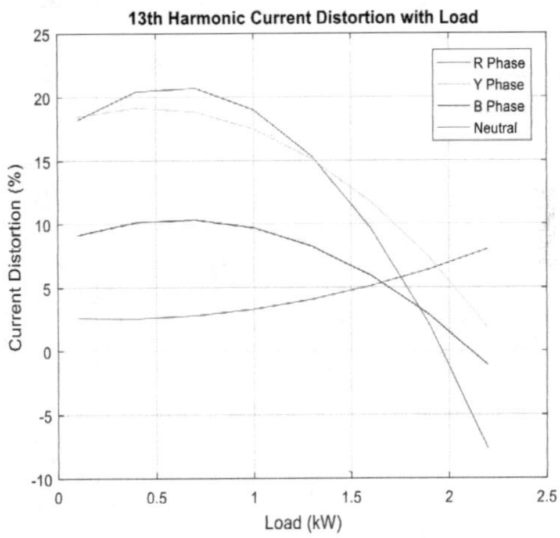

Fig. 41 13th harmonic current distortion v/s load variation (V/F)

7. Comparison Table

The various power quality parameters are tabulated in Table 2 with respect to both the control strategies used for speed control of induction motors. Table 3 and Table 4 shows the various harmonic distortion of voltage and current. The tabulation helps to compare the performance of VFD when fed through different control techniques.

Table 2. Comparison of various PQ parameters for VVC+ and V/F methods

Parameter	VVC+	V/F	Comment
Phase Voltage	Varies between 230V-237V	Varies between 218V-233V	Voltage profile is better in VVC+ as voltage variation is less in VVC+.
Line Current	Varies between 0.75A-5A Neutral current varies between 0.1A-0.5A.	Varies between 0.5A-4.4A Neutral current varies between 0.1A-1.3A	Neutral current is high in the case of V/F. VVC+ is better.
Voltage Unbalance	Varies between 0.4%-0.65%. Decreases with the increase in load.	Varies between 0.55%-0.65% . Increases with the load and then decreases.	Unbalance decreases in the case of VVC+, hence better.
Current Unbalance	Varies between 10%-40%.	Varies between 0.1%-36%. Even comes to lower values at higher loading.	V/F has lower values and is superior.
Frequency	Varies between 49.7Hz-50.22 Hz.	Varies between 50.02 Hz-50.17 Hz.	V/F is superior as the variation is less.
Active Power	Varies between 250W-2400W	Varies between 300W-1900W	
Reactive Power	Varies between 120VAr-490VAr	Varies between 20VAr-380VAr	V/F has less reactive power requirement and hence better.
Apparent Power	Varies between 500VA-3250VA	Varies between 500VA-2700VA	
Power Factor	Varies between 0.475-0.7	Varies between 0.470-0.75	There is a rapid improvement in power factor in case of VVC+, hence better.
Voltage THD	Varies between 2.1%-3.2%	Varies between 1.8%-3.15%.	
Current THD	Varies between 15%-105%	Varies between 40%-190%.	Neutral current THD is much higher in V/F. Thereby VVC+ is superior.

Table 3 Voltage distortion for different harmonics

Harmonic Number	VVC+	V/F	Comments
5	Varies between 0.1%-1.8%	Varies between 0.7%-1.9%	Both are similar
7	Varies between 0.65%-1.43%	Varies between 0.1%-2.1%	V/F has more distortion.
11	Varies between 0.19%-1.5%	Varies between 0.1%-0.37%	VVC+ has more distortion.
13	Varies between 0.25%-0.6%	Varies between 0.1%-0.3%	VVC+ has more distortion.

IOP Conf. Series: Materials Science and Engineering **937** (2020) 012042 doi:10.1088/1757-899X/937/1/012042

Table 4 Current distortion for different harmonics

Harmonic number	VVC+	V/F	Comments
5	Line current distortion variation-28%-75% Neutral current distortion variation- 0.1%-20%	Line current distortion variation- 32%-82% Neutral current distortion variation- 8%-22%	Neutral current has more distortion in the case of V/F.
7	Line current distortion variation- 27%-80% Neutral current distortion variation- 0.4%-0.7%	Line current distortion variation- 24%-58% Neutral current distortion variation- 5%-15%	Neutral current distortion is more in V/F.
11	Line current distortion variation-0.1%-33% Neutral current distortion variation- 3%-7%	Line current distortion variation- 0.1%-33% Neutral current distortion variation- 3%-17%	Neutral current distortion is more in V/F/
13	Line current distortion variation- 0.1%-22% Neutral current distortion variation- 0.1%-2.5%	Line current distortion variation-0.1%-22% Neutral current distortion variation- 2%-8%	Neutral current distortion is more in V/F.

8. Conclusion

In this work parameter prediction was carried out using curve fitting tool of MATLAB. Moreover the two control strategies used for the speed control of induction motor are compared. With reference to Table 2 to Table 4 we can draw the following conclusions. The VVC+ was found out to be superior as compared to V/F we talk about overall performance. For the power quality parameters like voltage, current, voltage unbalance, power factor and current THD, VVC+ is more preferred whereas for voltage THD, frequency profile, and unbalance in current V/F is superior. Overall if we consider, VVC+ is more superior as compared to V/F. Phase voltages varies for individual test cases which is obtained from the tabulation. There is variation in different test cases but change in voltage is limited to certain extend shown in table 4. The Y and R phase of voltage is followed by the B phase. The supply voltage is unbalanced due to variations in the phase voltages. When loaded the system voltage is greater than the no load system voltage. Unbalance in system voltage is due to the presence of voltage harmonics and unbalanced three phase supply loading. It can be observed that the percentage unbalance in voltage is getting decreased when the system is loaded. But change in voltage should be in acceptable rate as per table 4. Unbalance in current is because of current harmonics and unbalanced loading which draws unbalanced currents from the supply. If the supply voltage is unbalanced the currents drawn will also be unbalanced. As the load increases the real power required to drive the load increases, so this increase is drawn from the supply. Reactive power drawn to produce flux increases as the load also increases. As the unbalance in voltage and current reduces when machine is loaded which results in increase in system efficiency. The power factor of the machine increases when the machine is loaded which also leads to an increase in efficiency. Apparent power increases, real power increases at a higher rate when compared with reactive power rise. VFD has a rectifier on the front end which draws a non-sinusoidal source current (Current harmonics) which causes voltage harmonics. When induction machine is loaded the current drawn by the rectifier increases so THD of a loaded machine is more than the THD of the machine under no load.

Acknowledgement

Authors are like to thank the Danfoss Advance Drives laboratory of VIT Vellore for their modern setup and allow authors to complete the experiment and Danfoss Pvt Ltd, Chennai for their support in technical solution and Prince Sultan University, Riyadh, Saudi Arabia for technical guidance.

Reference:

[1]. Sitharthan R, Geethanjali M and Pandy TKS 2016 Adaptive protection scheme for smart microgrid with electronically coupled distributed generations *Alexandria Engineering Journal* **55(3)** 2539-2550

[2]. Fathima AH, and Palanisamy K 2014 Battery energy storage applications in wind integrated systems—a review *IEEE International Conference on Smart Electric Grid* 1-8

[3]. Prabaharan N and Palanisamy K 2015 Investigation of single-phase reduced switch count asymmetric multilevel inverter using advanced pulse width modulation technique *International Journal of Renewable Energy Research* **5(3)** 879-890.

[4]. Jerin ARA, Kaliannan P and Subramaniam U 2017 Improved fault ride through capability of DFIG based wind turbines using synchronous reference frame control based dynamic voltage restorer. *ISA transactions* **70** 465-474

[5]. Sitharthan, R, Sundarabalan CK, Devabalaji KR, Nataraj SK and Karthikeyan M 2018 Improved fault ride through capability of DFIG-wind turbines using customized dynamic voltage restorer *Sustainable cities and society* **39** 114-125

[6]. Prabaharan N and Palanisamy K 2016 A single-phase grid connected hybrid multilevel inverter for interfacing photo-voltaic system *Energy Procedia* **103** 250-255

[7]. Palanisamy K, Mishra JS, Raglend IJ and Kothari DP 2010 Instantaneous power theory based unified power quality conditioner (UPQC) *IEEE Joint International Conference on Power Electronics, Drives and Energy Systems* 1-5

[8]. Sitharthan R and Geethanjali M 2017 An adaptive Elman neural network with C-PSO learning algorithm-based pitch angle controller for DFIG based WECS Journal of Vibration and Control **23(5)** 716-730

[9]. Sitharthan R and Geethanjali M 2015 Application of the superconducting fault current limiter strategy to improve the fault ride-through capability of a doubly-fed induction generator–based wind energy conversion system *Simulation* **91(12)** 1081-1087

[10]. Sitharthan R, Karthikeyan M, Sundar DS and Rajasekaran S 2020 Adaptive hybrid intelligent MPPT controller to approximate effectual wind speed and optimal rotor speed of variable speed wind turbine *ISA transactions* **96** 479-489

[11]. Sitharthan R, Devabalaji KR and Jees A 2017 An Levenberg–Marquardt trained feed-forward back-propagation based intelligent pitch angle controller for wind generation system *Renewable Energy Focus* **22** 24-32

[12]. Sitharthan R, Sundarabalan CK, Devabalaji KR, Yuvaraj T and Mohamed Imran A 2019 Automated power management strategy for wind power generation system using pitch angle controller *Measurement and Control* **52(3-4)** 169-182

[13]. Sundar DS, Umamaheswari C, Sridarshini T, Karthikeyan M, Sitharthan R, Raja AS and Carrasco MF 2019 Compact four-port circulator based on 2D photonic crystals with a 90° rotation of the light wave for photonic integrated circuits applications *Laser Physics* **29(6)** 066201

[14]. Sitharthan R, Parthasarathy T, Sheeba Rani S and Ramya KC 2019. An improved radial basis function neural network control strategy-based maximum power point tracking controller for wind power generation system *Transactions of the Institute of Measurement and Control* **41(11)** 3158-3170

[15]. Rajesh M and Gnanasekar JM 2017 Path observation based physical routing protocol for wireless ad hoc networks *Wireless Personal Communications* **97(1)** 1267-1289
[16]. Palanisamy K, Varghese LJ, Raglend IJ and Kothari DP 2009. Comparison of intelligent techniques to solve economic load dispatch problem with line flow constraints *IEEE International Advance Computing Conference* 446-452
[17]. Sitharthan R, Ponnusamy M, Karthikeyan M and Sundar DS 2019 Analysis on smart material suitable for autogenous microelectronic application *Materials Research Express* **6(10)** 105709
[18]. Rajaram R, Palanisamy K, Ramasamy S and Ramanathan P 2014 Selective harmonic elimination in PWM inverter using fire fly and fireworks algorithm *International Journal of Innovative Research in Advanced Engineering* **1(8)** 55-62
[19]. Sitharthan R, Swaminathan JN and Parthasarathy T 2018 March. Exploration of wind energy in India: A short review *IEEE National Power Engineering Conference* 1-5
[20]. Karthikeyan M, Sitharthan R, Ali T and Roy B 2020 Compact multiband CPW fed monopole antenna with square ring and T-shaped strips *Microwave and Optical Technology Letters* **62(2)** 926-932
[21]. Sundar D Sridarshini T, Sitharthan R, Madurakavi Karthikeyan, Sivanantha Raja A, and Marcos Flores Carrasco 2019 Performance investigation of 16/32-channel DWDM PON and long-reach PON systems using an ASE noise source *In Advances in Optoelectronic Technology and Industry Development: Proceedings of the 12th International Symposium on Photonics and Optoelectronics* 93
[22]. Sitharthan R and Geethanjali M 2014 Wind Energy Utilization in India: A Review *Middle-East J. Sci. Res.* **22** 796–801 doi:10.5829/idosi.mejsr.2014.22.06.21944
[23]. Sitharthan R and Geethanjali M 2014 ANFIS based wind speed sensor-less MPPT controller for variable speed wind energy conversion systems *Australian Journal of Basic and Applied Sciences* **8**14-23
[24]. Jerin ARA, Kaliannan P, Subramaniam U and El Moursi MS 2018 Review on FRT solutions for improving transient stability in DFIG-WTs *IET Renewable Power Generation* **12(15)** 1786-1799
[25]. Prabaharan N, Jerin ARA, Palanisamy K and Umashankar S 2017 Integration of single-phase reduced switch multilevel inverter topology for grid connected photovoltaic system *Energy Procedia* **138** 1177-1183
[26]. Rameshkumar K, Indragandhi V, Palanisamy K and Arunkumari T 2017 Model predictive current control of single phase shunt active power filter *Energy Procedia* **117** 658-665
[27]. Fathima AH and Palanisamy K 2016 Energy storage systems for energy management of renewables in distributed generation systems *Energy Management of Distributed Generation Systems* 157
[28]. Rajesh M 2020 Streamlining Radio Network Organizing Enlargement Towards Microcellular Frameworks *Wireless Personal Communications* 1-13
[29]. Subbiah B, Obaidat MS, Sriram S, Manoharn R and Chandrasekaran SK 2020 Selection of intermediate routes for secure data communication systems using graph theory application and grey wolf optimisation algorithm in MANETs IET *Networks* doi:10.1049/iet-net.2020.0051
[30]. Singh RR and Chelliah TR 2017 Enforcement of cost-effective energy conservation on single-fed asynchronous machine using a novel switching strategy *Energy* **126** 179-191
[31]. Amalorpavaraj RAJ, Palanisamy K, Umashankar S and Thirumoorthy AD 2016 Power quality improvement of grid connected wind farms through voltage restoration using dynamic voltage restorer *International Journal of Renewable Energy Research* **6(1)** 53-60

[32]. Singh RR, Chelliah TR, Khare D and Ramesh US 2016 November. Energy saving strategy on electric propulsion system integrated with doubly fed asynchronous motors *IEEE Power India International Conference* 1-6

[33]. Singh RR, Mohan H and Chelliah TR 2016 November. Performance of doubly fed machines influenced to electrical perturbation in pumped storage plant-a comparative electromechanical analysis *IEEE 7th India International Conference on Power Electronics* 1-6

[34]. Sumithra M, Abraham A, Rajesh M and Nithiyanandam N 2019 Mining social media content to predict peer trust level in social networks *International Journal of Engineering and Advanced Technology* **6S3** 1959-1964

[35]. Venkatesh K, Ali MJS, Nithiyanandam N and Rajesh M 2019 Challenges and Research Disputes and Tools in Big Data Analytics *International Journal of Engineering and Advanced Technology* **6S3** 1949-1952

Seamless Engineering Process to Adapt IEC 61850 Standard for Substation Automation Through Requirements Specification

S U Khan[1], **A Kalam**[1,2]

[1] College of Engineering and Science, Victoria University, Footscray 3011 Victoria, Australia
[2] School of Electrical Engineering, Engineering Institute of Technology, Melbourne 3000 Victoria Australia

saad.khan@live.vu.edu.au

Abstract. Whatever demographic one belongs to in the technology adaption life cycle, for a seamless implementation of IEC 61850 Standard it is required to answer the key concern related to extent of change. This paper provides guidelines on the evaluation of amount of change by going through the requirement specification process, interface evaluation, and configuration tools. All this information is presented in a process form which makes it easier to quantify the inputs for organizational change management.

1. Introduction

The comprehension and integration of IEC 61850 standard for an inexperienced operator and maintainer can be challenging at times. The company's protection engineers have to work closely with the vendor offering IEC 61850 enabled equipment and systems during implementation of systems as complex as substation automation.

Although the vendors will perform the detailed implementation of the IEC 61850 object and service models, the substation engineers must be able to decide what settings should be established for a particular substation. Therefore, the company's workforce needs to upskill so that all the potential benefits of IEC 61850 can be earned while attaining the change management sweet spot when smart substation automation systems and devices are implemented for the grid. A Six gate process for requirements specification management can help to attain that sweet spot for IEC 61850 adoption.

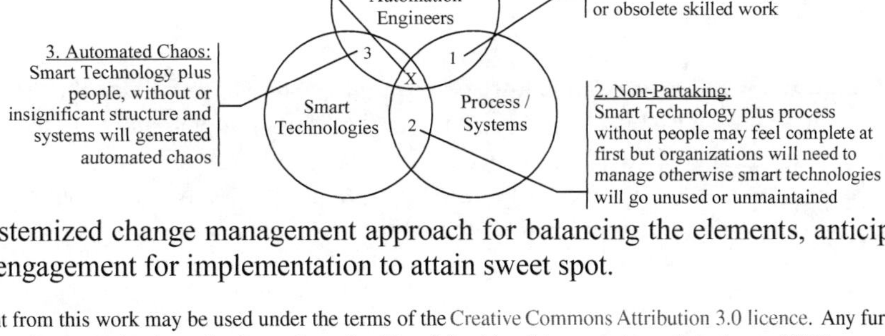

Figure 1 – A systemized change management approach for balancing the elements, anticipating value and facilitating engagement for implementation to attain sweet spot.

2. Six Phase-Gate Process:

A phase gate process deals with a project by dividing it into phases separated by decision points and at the end of each gate the management decides on continuation. Following are the six gate process for requirements specification of IEC 61850:

Stage - 1		Stage - 2		Stage - 3		Stage - 4		Stage - 5		Stage - 6	
Define functional requirements	Gate 1	Establish IEC 61850 logical nodes and its parameters	Gate 2	Specify IEC 61850 data exchanges within the station	Gate 3	Specify IEC 61850 data exchanges with external interfaces	Gate 4	Specify conformance testing	Gate 5	Specify IEC 61850 configuration tools	Gate 6

2.1. Stage/Gate 1 – Define Functional Requirements

By adapting a systems approach, functional requirements can drive the application architecture of substation automation system. A detailed functional requirement assessment can produce a comprehensive scope relevant to the organization. Following this step, the vendor independent outcomes in the form of technical requirements can be adopted by a sub-contractor (vendors/integrator) to execute the implementation of IEC 61850. Following are the basis of these requirements:

Substation Electrical Layout	Equipment Selection and Specification	Assessment of available or necessary data	Assessment of protection scenarios
Determine SCADA Requirements – Real time and control/set-point/parameter capabilities	I/O (Input/Output) List	Information security requirements	Network Management Requirements
Functional requirements around change in automation philosophy			

The functional requirement specification is more than a datasheet and should include the requirements of all stakeholders. This holistic approach will enable to reap the full advantage of smart protection system with the use IEC 61850. Following are some of the stakeholders involved:

- Planning Engineers
- Protection Engineers
- Maintenance
- Operations and Control Management
- Market operations

By answering "what questions" first rather jumping to "how" serves as the basis of functional requirement management. This brainstorming coupled by stakeholder management can be facilitated by modelling techniques such as depicted in figure 2. After progressive iterations of the model, a final model and specification can be created and at the first gate, a go- or no-go decision can be made. At this stage the effect on both conceptual and physical systems can be analyzed.

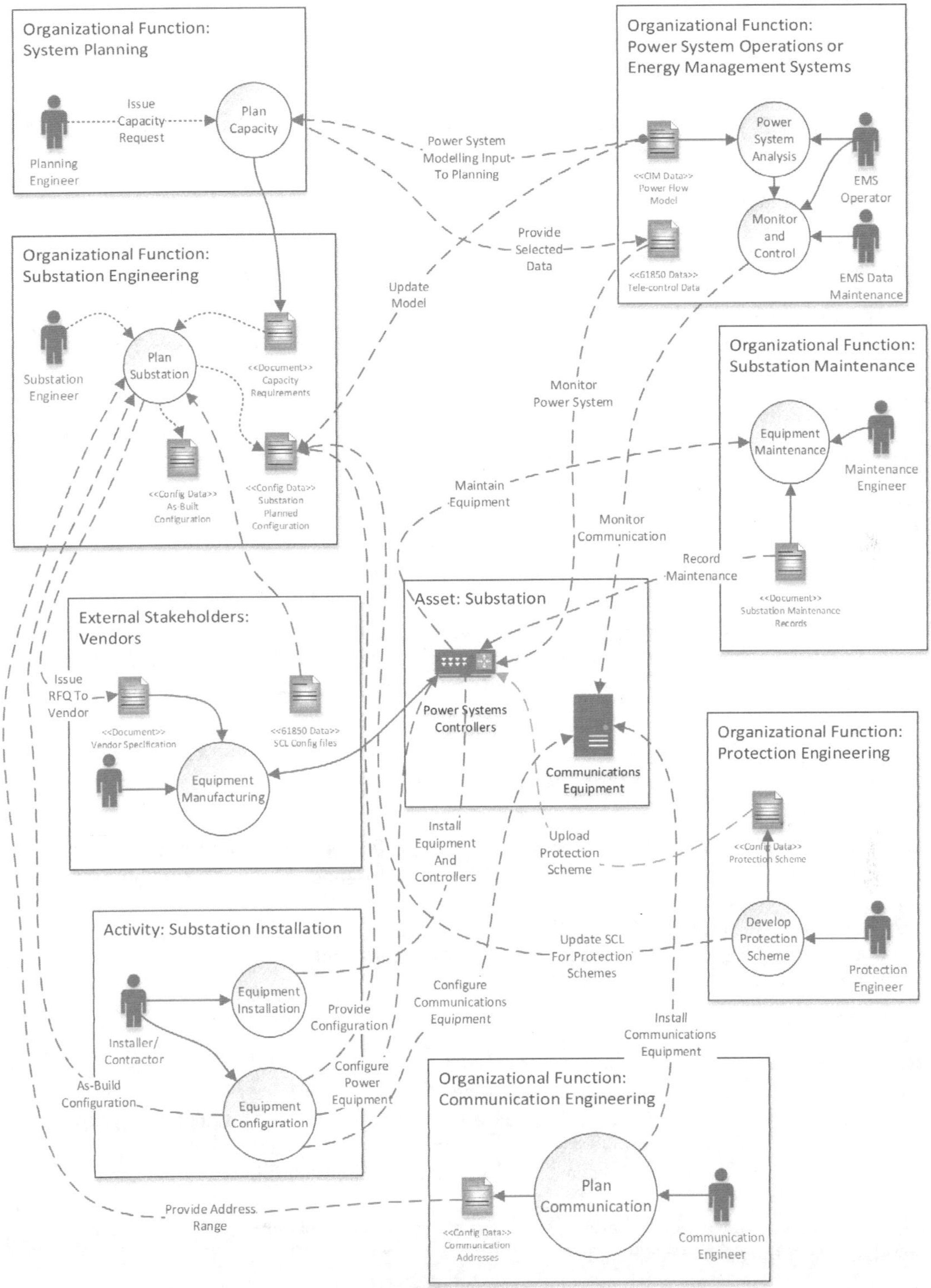

Figure 2 – A typical UML (Unified Modelling language) for substation automation using SCL (Substation Configuration Language) with the purpose of visualizing automation system along with its main actors, roles, actions and artefacts in order to document system information.

2.1.1. Stage/Gate 2 – Establish IEC 61850 logical nodes and its parameters
Based on the functional requirements, the data items are instantiated in to logical nodes.

2.1.2. Stage/Gate 3 – Specify IEC 61850 data exchanges within the station
The functional requirements in Stage 1 describe the types of data to be exchanged. Stage 3 defines following:

- what IEC 61850 data items are sent
- Where IEC 61850 data items are sent
- Under what conditions IEC 61850 data items are sent

In IEC 61850 terminology, these data exchange definitions are PICOMs (Pieces of information for communication), which are only for example and not normative. Thus, it is essential that actual data exchanges define following [6]:

- average and maximum transfer times
- the average and maximum response times
- the average and maximum size of messages
- security
- availability
- backup and/or redundancy

2.2. Stage/Gate 4 – Specify IEC 61850 data exchanges with external interfaces
The external interfaces for data exchange can be identified with system modelling. Further brainstorming on questions in following categories will finalize the interactions.

- Monitored data
- Controls and parameter settings

2.2.1. Stage/Gate 5 – Specify Conformance Testing
With IEC 61850-10, a standardized approach is available for interoperability. Interoperability is not only important for flexibility but also for reduced life cycle cost and obsolesce/future proofing.

2.2.2. Stage/Gate 6 – Specify IEC 61850 Configuration Tools
The decision to adopt a particular vendor specific tools is important as any change afterwards can have implications on people, systems and equipment.

3. Conclusion:
There is a degree of resistance in the industry to adopt the IEC 61850 standard for substation automation and a systematic approach involving change management can help derive the maximum benefit from an IEC 61850 enabled substation automation. The benefit offered to organizations in terms of cost and people in terms of skill development can convince users to implement smart substation automation solutions. A phase gate process can give confidence to decision makers by evaluation of change, reducing risk and cost.

It is concluded that a UML based visualization technique can help quantify change across all disciplines, departments and stakeholders.

4. References:
[1] Substation Automation Using IEC 61850, the International Power System Information Modelling Standard by Electric Power Research Institute (EPRI), Inc.
[2] IEC 61850-1.; Communication networks and systems for power utility automation –

Introduction and overview, IEC Technical Report, Edition 2.0, 2013-03
[3] IEC 61850-2.; Communication networks and systems in substations - Glossary, IEC Technical Specification, Edition 1.0, 2003-08
[4] IEC 61850-3.; Communication networks and systems for power utility automation – General requirements, IEC International Standard, Edition 2, 2013-12
[5] IEC 61859-4.; Communication networks and systems for power utility automation – System and project management, IEC International Standard, Edition 2, 2011-04
[6] IEC 61850-5.; Communication networks and systems in substations. – Communication requirements for functions and device models, IEC International Standard, Edition 2, 2013-01.
[7] IEC 61850-6.; Communication networks and systems for power utility automation.– Configuration description language for communication in electrical substations related to IEDs, IEC International Standard, Edition 2, 2009-12
[8] IEC 61850-7-1.; Communication networks and systems for power utility automation – Basic information and communication structure – Principles and models, IEC International Standard, Edition 2.0, 2011-07.
[9] IEC 61850-7-2.; Communication networks and systems for power utility automation – Basic information and communication structure – Abstract communication service interface (ASCI), IEC International Standard, Edition 2.0, 2010- 08.
[10] IEC 61850-7-3.; Communication networks and systems for power utility automation – Basic information and communication structure – Common data classes, IEC International Standard, Edition 2.0, 2010-08.
[11] IEC 61850-7-4.; Communication networks and systems for power utility automation – Basic communication structure – Compatible logical node classes and data object classes, IEC International Standard, Edition 2.0, 2010-03.
[12] IEC 61850-7-410.; Communication networks and systems for power utility automation – Basic communication structure – Hydroelectric power plants – Communication for monitoring and control, IEC International Standard, Edition 2.0, 2015-11.
[13] IEC 61850-7-420.; Communication networks and systems for power utility automation – Basic communication structure – Distributed energy resources logical nodes, IEC International Standard, Edition 1.0, 2009-10.
[14] IEC 61850-7-510.; Communication networks and systems for power utility automation – Basic communication structure – Hydroelectric power plants – Modelling concepts and guidelines, IEC Technical Report, Edition 1.0, 2012-03.
[15] IEC 61850-8-1.; Communication networks and systems for power utility automation – Specific communication service mapping (SCSM) – Mappings to MMS (ISO 9506-1 and ISO 9506-2) and to ISO/IEC 8802-3, International Standard, Edition 2.0, 2011-06.
[16] IEC 61850-9-2.; Communication networks and systems for power utility automation – Specific communication service mapping (SCSM) – Sampled values over ISO/IEC 8802-3, International Standard, Edition 2.0, 2011-09.
[17] IEC 61850-9-3.; Communication networks and systems for power utility automation – Precision time protocol profile for power utility automation, International Standard, Edition 1.0, 2015-05.
[18] IEC 61850-10.; Communication networks and systems for power utility automation – Conformance testing, International Standard, Edition 2.0, 2012-12.

IOP Conf. Series: Materials Science and Engineering **937** (2020) 012044 doi:10.1088/1757-899X/937/1/012044

IoT and Renewable Energy systems based Connected Chair for Health Monitoring

Dr.K.Dhayalini[1*], S.A.Priyadarshini[2], M.Abirami[3],

[1*]Professor and Head, Assistant Professor[2], Undergraduate Scholar[3]

Department of EEE, K.Ramakrishnan College of Engineering, Tiruchirappalli, Tamilnadu, India.

*Corresponding Author: dhaya2k@gmail.com

ABSTRACT

Health monitoring of a system involve in various activities done in order to make the system operating condition feasible with certain constrains such as current systems states which includes repair and maintenance of the observations. Remote health monitoring is a process in which patients with numerous conditions such a heart disease or diabetics are monitored based on the latest technology. Internet of things and renewable energy sources are the brain of this work based smart wheel chair for special people for autonomous movement and also gathering person's health information using smart sensors displayed in webpage. For this innovative work a digital single board computer for gathering sensor information and also transfer the status of the persons through inbuilt Wi-Fi, and also the smart chair movement dependent upon user's voice, which the sceneries are recognized by connected camera to raspberry pi. Renewable energy system such as PV array or wind energy systems can be replaced for supply of the entire health monitoring system and connected chair system.

Keywords: Raspberry pi, Internet of Things, sensor, Health monitoring, smart chair monitoring, renewable energy system.

INTRODUCTION:

Human being health is primary element for their survival. Growth of the society and growth of the individual primarily depends on the health of the human being. Health of the individual includes the fitness of mental and physical development which is an important parameter. Global health organization has issued lot of concern regarding individual health issues [1]. Due to population aging, chronic diseases existence and increasing cost of the medical expenses the health care of an individual plays a burden factor. Because of these challenging factors many researches are performing various innovative studies and solutions to invent new innovative solutions and new technologies to find a solution for the above said challenging issues. Patient disease management and detection plays an important innovative invention for the improved health quality of the individual [2].

Nowadays lot of devices such as wearable sensors are available commercially to take care of the health of the individuals. Even devices are available to take care of the fitness, personal health care and fitness awareness. In addition remote health monitoring also available in existence to monitor the fitness arena, long term recording of the health system and management and patient's physiological information. Technologies are developed to monitor the clinical healthiness of the individual and group of individual. Based on the trends of the current technology the routine physical examination and physiological monitoring and be done in present and future time domain using wearable sensors and other equipments in and inexpensive way [3]

Interaction of the things with internet or cloud involving various sensors, micro controller, microprocessor, transmitters and receivers is known as Internet of Things (IoT). The transceivers are meant for empowering the communication and the required introduce the necessary protocol stacks to built and interact with each other and with far of communication users or devices. Nowadays, Internet of things have become the part of the day to day life and happenings. All the medical and health monitoring applications make use of Internet of Thing in the day to day activity of everyday life. In view of these things IoT are developed in every physical object is connected to the Internet by employing sensor devices. The health care monitoring system using Internet of Things and its need are increasing to take care of the patients day by day to increase and strengthen their health care and reduce the cost of health monitoring of individual [4].

In recent years to facilitate the remote health monitoring during emergency situations Internet of Things (IoT) devices are widely used in various advanced healthcare systems. At the present many challenges are faced by the individual in the real world, which need to be solved and solution to be arrived to deal realistically. The manpower and time consumption are more important challenges in IoT. The Internet of things of rehabilitation of resources has become prominent and popular for the development of the smart IoT applications like smart home and smart devices and equipments. On comparing the traditional system with existing system the aim of smart rehabilitation is an effective treatment for providing solution aiming sufficient interaction and quick reconfiguration. To determine the probable use of the medical resources, equipments and sensors according to the patient's requirements and the diseases encountered by the patients. For interconnecting all the medical resources and devices of the rehabilitation systems the primary and important technology are included by having sensors and wearable devices. For wide range of application the network technologies are used and implemented to interact and communicate among themselves [5]. In this paper the status of the patient along with orientation of the wheel chair according to voice of the user.

The proposed framework manages the way toward controlling and monitoring wheel chair and also health information of user with the idea of IOT and Computer Vision. The working model of connected chair for health monitoring through IOT and computer vision is explained below.

Input block comprises of MEMS sensors, temperature sensors, USB camera, Raspberry Pi and analog to digital converter.

MEMS SENSOR (ADXL335):

MEMS sensor is to monitor the orientation of the wheel if the wheel starts skids it sends alert message to the person.

TEMPERATURE SENSOR:

Temperature sensor is to monitor the temperature of the person. If the temperature of the person rises it sends the message.

USB CAMERA:

Webcam is used for real time monitoring of the patients health.

ADCMCP3008:

Analog values are converted into digital values using the device analog to digital converters. The analog values measurement of temperature sensor and MEMS sensor are converted into digital values for giving as input.

BLOCK DIAGRAM:

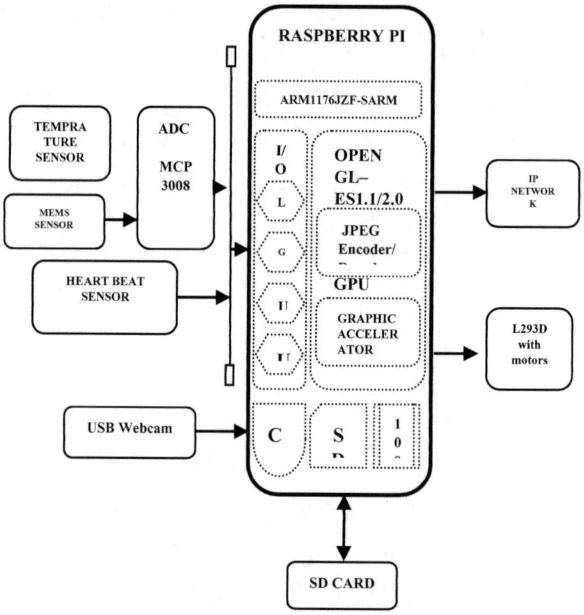

Figure 1: Block Diagram of existing system

RASPBERRY PI:

Raspberry pi is used to interface the Temperature sensor, MEMS sensor, L293D dc motor. Raspberry pi is good at software applications and can run multiple programs at a time. It stores the program in SD card.

OUTPUT BLOCK:

L293DMOTORDRIVER:

This motor driver IC is used for rotation of the wheel according to the voice input of the user. If the user wants to move the wheel chair by using the voice as input they can move forward, backward, right, left and stop.

WORKING DESCRIPTION

The Raspberry Pi is a three microchip reads the input from USB cam and build the wheel chair move in specific direction given by the user and conjointly the sensors conjointly monitored the user's physical parameters to observance.

In the existing work the wheelchair is controlled through IoT employing an android application which is connecting via Wi-Fi module and through button system which is fixed on the hands of the chair. The dependency of the user will decrease due to this application. The wheel chair is employed with MEMS sensor to sense and avoid the accidents which causes due to obstacles in real or remote location. The sensors on detecting the obstacle will act immediately and the commands recorded will be communicated to the heart of the hardware module microcontroller for further processing and further motion. The doctor can monitor the information's and data which are communicated by the health monitoring system and by which necessary action can be taken by the doctor to cure the respective patients.

Wheelchair is provided with four wheels in all the four directions, two rear wheels and two castor wheels. The two caster wheel are fixed in front of the wheel chair, and all the wheels are designed in such a way having the same diameter. The drive wheels are designed in rear side of the base of the vehicle in such a way that , chair is allowed to turn according to voice command enabled by the system and the wheels are engaged directly to a gear train that transmit torque form motor to wheels by two grooves in each wheel and nut.

HARDWARE IMPLEMENTATION

In the hardware design all the sensors are placed on the seat for visibility. The programming is done using Python Language through an open source LINUX Software.

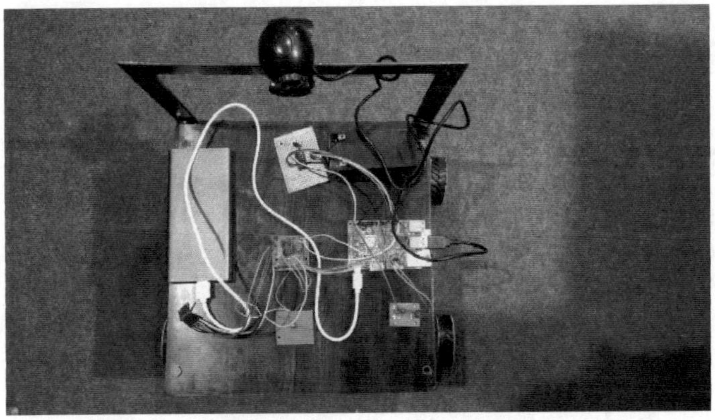

Figure 2. Hardware Design of the existing system

The following table shows the list of components and their corresponding specifications used to design the hardware kit.

Table 1. Specification table

NAME	SPECIFICATION
Raspberry Pi	MODEL 3 B
Motor Driver	L293D
Mems Sensor	ADXL335
ADC	MCP3008
Temperature Sensor	LM35

IOP Conf. Series: Materials Science and Engineering **937** (2020) 012044 doi:10.1088/1757-899X/937/1/012044

The Python language is used to program using Linux Software which is an open source. All the sensors are individually programmed according to the consumers' requirement. The pins used here are called the GPIO pins. The sensors after programming is connected to the Raspberry Pi.

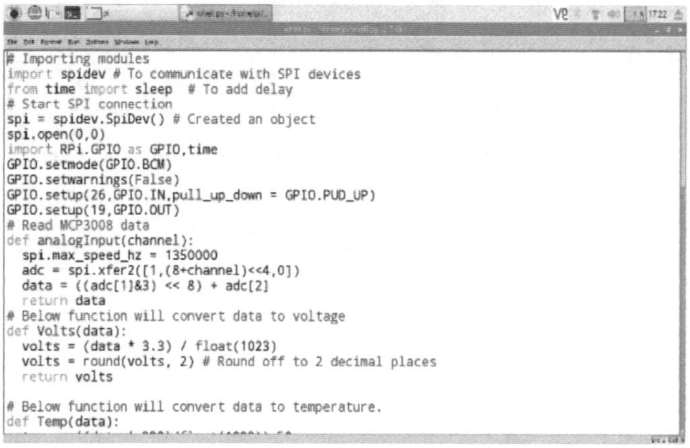

Figure 3. Sensors Output

EXPERIMENTAL RESULTS:

The outputs of the sensors are taken from the Raspberry Pi. This output is made to be displayed in the browser for doctors' use. For security purpose the output is viewed in the browser only when the IP address of the particular patient is entered. Since IP address is used it can be viewed it Laptop, mobile phone or any gadget.

Figure 4. Camera Output

CONCLUSION

A unique model of wheel chair for future IoT-based health monitoring systems was designed and which can be applied to both general systems and systems that monitor specific conditions. This designed model can be used in hospitals as indoor application effectively for monitoring the health condition of patients admitted from the doctors place itself. The doctor can answer emergencies without intimation of helper. The alert system helps to provide immediate service to the patient. Accidents can drastically be reduced through this system. The proposed system of patient health monitoring can be highly used in emergency situations as it can be daily monitored, recorded and stored as a database. The IOT device is combined with the computer vision so that the database can be shared in all the hospitals for the intensive care and treatment. Renewable energy system such as PV array or wind energy systems can be replaced for supply of the entire health monitoring system and connected chair system.

REFERENCE:

[1] Zankhana Mehul Kalarthi, "A REVIEW PAPER ON SMART HEALTH CARE SYSTEM USING INTERNET OF THINGS", International Journal of Research in Engineering and Technology (IJRET), Volume5, Mar-2016.

[2] Raga Lavima1, Mr. G. Subhramanya Sarma "AN IOT BASED INTELLIGENT MEDICINE", International Journal of Computer Science and Mobile Computing (IJCSMC), Vol. 4, Issue. 10, October 2015.

[3] Moen Hassanalieragh, Alex Page, TolgaSoyata, Gaurav Sharma, Mehmet Aktas, Gonzalo MateosBurak Kantarci, Silvana Andreescu, "Health Monitoring and Management Using Internet-of-Things (IoT) Sensing with Cloud-based Processing: Opportunities and Challenges", IEEE International Conference on Services Computing, 2015.

[4] Ravi Kishore Kodali, Govindaswamy, Boppana Lakshmi., "An Implementation of IoT for Healthcare", ResearchGate Publications, December 2015.

[5] P.Elanthiraiyan & Dr.S.Babu, "Smart Medicine and Physical Health System Using IoT", International Journal of Computer Science and Mobile Computing (IJCSMC), Vol. 4, March 2015

[6] S. M. Riazul Islam, "The Internet of Things for Health Care: A Comprehensive Survey", Institute of Electrical and Electronics Engineers (IEEE), June 2015

[7] Shubhamsagar Nayak, Upasana, Prateek Gupta and Atul B. Wani, "IOT BASED- SMART WHEELCHAIR WITH HEALTH MONITORING SYSTEM", International Journal of Recent Innovation in Engineering and Research, Volume: 02, April 2017.

[8] Shubhamsagar Nayak1, Prateek Gupta2, Upasana3, Prof.Atul B. Wani4, "Wheel Chair with Health Monitoring System Using IoT", International Research Journal of Engineering and Technology (IRJET), Vol 4, May2017.

[9] Ashutosh Gupta, Tathagata Ghosh, Pradeep Kumar, Shruthi.S, Bhawna, "Development of Android Based Powered Intelligent Wheelchair for Quadriplegic Persons", IOP Conf. Series: Materials Science and Engineering, 2017.

[10] M. Li, S. Yu, Y. Zhen, K. Ren, and W. Lou," Scalable and Secure Sharing of Personal Health Records in Cloud Computing using Attribute-based Encryption" IEEE Trans. on parallel and distributed system, vol. 24(1), UT, Mar. 2012, pp.131-143.

[11] J.S. Khan, V. Aulakh, and A. Bosworth, "What It Takes: Characteristics of the Ideal Personal Health Record", Health Aff (Millwood), vol. 28(2), pp. 369-376.

[12] D. Slamanig, and C. Stingle," Privacy Aspects of eHealth," IEEE 3rd Int. Conf. on Availability, Reliability and Security, Barcelona, Mar. 2008, pp. 1226-1233.

[13] A. Ukil, S. Bandyopadhyay, and A. Pal,"IoT-Privacy: To Be Private or Not To Be Private," IEEE INFOCOM, Toronto ON, Apr. 2014, pp.123-124.

Extended Boost DC-DC-AC Converter for Electric Vehicle Applications

R.Uthirasamy[1], J.Karpagam[2], Umashankar Subramaniam[3], V.J.Vijayalakshmi[4]

[1,2,4]Department of Electrical and Electronics Engineering, KPR Institute of Engineering and Technology, Coimbatore 641 407, India
[3]Renewable Energy Lab, Department of Communications and Networks Engineering, College of Engineering, Prince Sultan University, Riyadh, Saudi Arabia

Corresponding Author E-mail: rusamy83@gmail.com

Abstract. In this paper, extended boost DC-DC-AC converter is proposed for Electric Vehicle (EV) applications. The extended boost DC-DC power converter overcomes the limitations of traditional voltage source and impedance source power converters. In addition to the DC-DC converters, charge storage systems are also evenly plays a pre-dominant role in EVs. This paper provides a comprehensible idea concerning the right selection of power converters, energy storage systems and electric drives systems. A wide research was made on electric vehicle models in the design phase, but most reached the prototype model with better efficiency. The electrical systems and combustion engines found in these vehicles are compared in order to forecast the evolution trend in terms of specifications and performance of the whole vehicle and of each system. The entire system is developed and simulated using SIMULINK tools. A prototype model of single phase extended boost DC-DC converter is developed and its results are validated.

IndexTerms—Battery systems, extended boost DC-DC power converters, electric vehicles.

1. Introduction

The Electric vehicles (EVs) put forward a massive potential to significantly dwindle carbon emissions in the present scenario from internal combustion engines and the generation companies. Plug-In Electric Vehicles (PHEVs) may perhaps activate a stepwise electrification of the complete transport sector. In recent times quite a lot of electric vehicles have been introduced and used all throughout the world. But only some vehicles has been designed and manufactured for commercial use. Some vehicles are still only in the design phase. The foremost important is regarding the long range vehicle which contributes to the major of carbon emission. Based on the predictions from the existing data the future development was made in the EVs. For low range vehicles the manufacturers prefer pure electric configuration while for the high range vehicles hybrid configuration is preferred [1]-[3]. A comparison is made between the IC engines and EVs to anticipate the performance of the system.

Reducing the level of greenhouse gas emissions attributable to the transportation sector is a major part of the environmental protection policies in many developed and developing economies. The transportation sector and individual transport in meticulous is highly dependent on fossil fuels. Due to the oil crisis most of the countries are moving towards electric vehicles [4]-[6]. From the survey it is identified that electrifying the entire transport sector diminish the energy consumption to one fifth of the present utilization. Although EVs have been around for quite some time, their limited range has hindered their wide spread use [9]. Over the past decade key technologies have progressed so that mass-market viable plug-in electric vehicles (PEVs). In this paper, ultra capacitor based AC drive is developed for plug-in electric vehicles.

This paper presents an ultra capacitor based extended boost power converter fed AC Motor Drive for EV application. The main motivation of this work is to implement the modified Z-source or impedance source inverter to activate it in high ambient temperature conditions. The enhancements made in the modified Z-source inverter will mitigate the limitations of conventional DC-DC converter topology and will get better operation in the soft switching region. The projected configuration functions as a forward boost/reverse buck or forward buck/reverse boost configuration where the functioning type is preselected by the analog or digital controller.

In this project, ultra capacitors are proposed as the storage system. Through the literature it is understood that the performance of ultra capacitor are better than the lead-acid batteries. This project provides a clean and emission free environment. Plug-in concept is introduced to charge the electric vehicle as like that of fuels [17]-[21].

In the article; Section I shows the introduction of EV system. Section II shows the introduction of extended boost power converter system for EV application. Storage system of extended boost power converters is addressed in Section III. Simulation model of the modified boost impedance source converter are presented in Section IV. Results and discussions of proposed converter are presented in Section V. Section VI concludes the development of extended boost power converters. The general structure of EV systems is represented in Figure. 1.

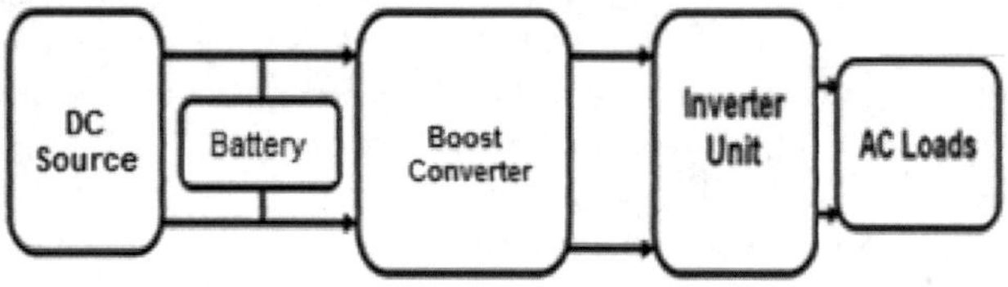

Figure. 1 Block diagram of conventional electric vehicle system

2. Extended boost power converters

Theoretically, boost converter can produce infinite gain, on the other side; it is understood that the derating of the system performance is due to presence of parasitic elements. By considering all the performance parameters, the increase of duty cycle above 0.9possibly will increase power losses and the system will lead to malfunction [10]-[16].The voltage stresses across the inverter switches are high in conventional power converters. Without compromising the performance parameters and boost capability of conventional power converters, certain modifications made in normal boost network named "Extended Boost power Inverter". The block diagram of proposed extended boost power converter based EV is shown in Figure. 2. Operation of shoot through state decides the setup voltage in the passive components in the modified impedance source extended boost power converter. By controlling both the stages of the converter the input voltage is get boosted to obtain the desirable AC output voltage for EV application.

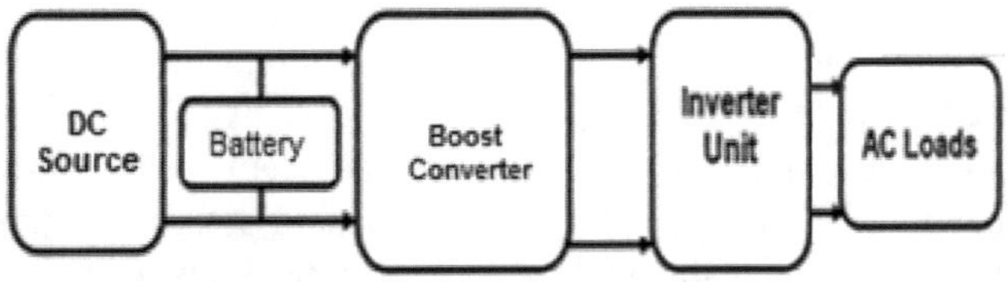

Figure. 2 Block diagram of extended boost converter based electric vehicle system

The proposed impedance source inverter has two main operating states as like that of traditional Z-Source converter topology. Impedance source based maximum boost power topology can be operated in boost and buck mode. Modified Z-source converter with continuous current converter is shown in the Figure 3(a). In this topology additional switching and passive components are added with first extension [7].

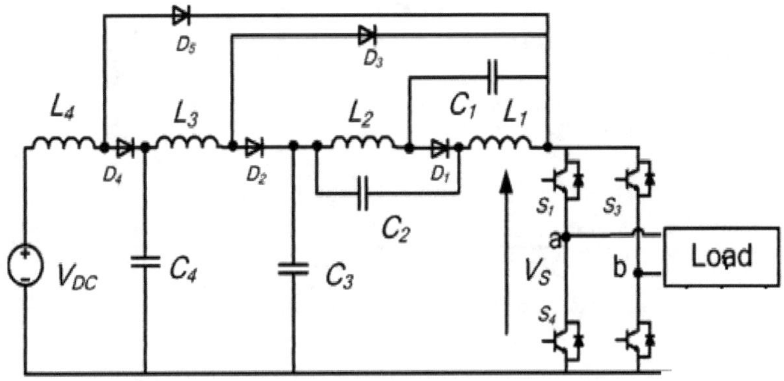

Figure. 3(a) Equivalent circuit of diode assisted power converter

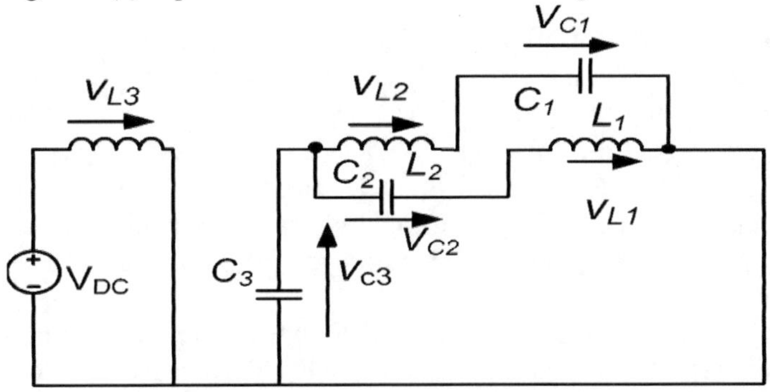

Figure. 3(b) Diode assisted power converter circuit for shoot-through state

Modified impedance source extended boost configuration consists of three power semiconductor diodes, three passive components such as inductors capacitors. Figure. 3(b) illustrates the equivalent-circuit for the voltage boosting stage/state called shoot-through state. The voltage equations are as follows:

$$V_{dc} + V_{L3} = V_{C3} \quad (1)$$

$$V_{L1} = V_{C1} \quad (2)$$

$$V_S = V_{C3} + V_{C2} + V_{L1} \quad (3)$$

Power flow from the source to the inductor or the capacitor to the inductor is happened when one of the passive components like capacitor is getting discharged.

By taking into account of the nominal voltage across the magnetic components such as inductors is zero for the voltage boosting stage/state of shoot-through duty ratio as D_s and voltage step down stage of non-shoot-through duty ratio as D_A where $D_A + D_S = 1$. Let us assign $B = 1/[(1-2D_S)(1-D_S)]$, the voltage boost factor in the DC side, Then the maximum AC supply can increased by increasing the factor of $1/(1-D_S)$. Likewise, the stable condition equations are derived for the discontinuous-current mode of modified inverter. It is achievable that the same voltage boost factor as that of the continuous current topology.

3. Review of storage systems

Due the high power density and high energy per unit mass of Lithium-ion batteries are widely used in electronics applications [8]. Majority of the components of lithium-ion batteries are recycled. By sacrificing the cost parameter of Lithium-ion batteries, it is adopted for plug-in hybrid electric vehicles. The charging characteristics of Lithium ion battery is shown in Figure. 4.

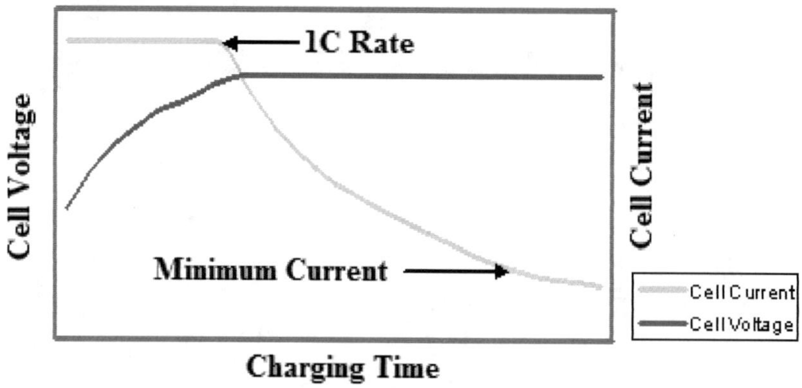

Figure. 4 Charging characteristics of Lithium ion battery

Nickel-metal hydride batteries proffer realistic explicated energy and precise control capabilities and are adopted characteristically in processors of supercomputers and therapeutic apparatus. Nickel-metal hydride batteries encompass a great deal for longer existence in charging and discharging cycle than lead-acid batteries. Nickel-metal hydride batteries are now adopted for electric vehicle applications and are extensively adopted for hybrid electric vehicles. Apart from battery cost, the well known challenges with conventional nickel-metal hydride batteries are high self-discharge and thermal creation for high temperatures, and it needs to control hydrogen loss. The characteristics of Nickel ion battery is represented in Figure. 5.

Lead-acid batteries can be premeditated to be high authority in terms of power, energy and are reasonably priced, secure, and consistent. On the other side, low specific energy, pitiable cold-temperature concert, and diminutive schedule and cycle life hamper their utilization. Highly developed lead-acid batteries are being manufactured and commercialized for electric drive vehicles for additional systems. The charge characteristics of Lead-acid battery are shown in Figure. 6.

Ultra capacitors from miracle substance grapheme charges batteries in 4 minutes. The new storage technology uses a ultra capacitor, rather than a conventional storage system, A ultra capacitor stores energy on the surfaces of materials in the form of tacit electricity as like that of super capacitor.

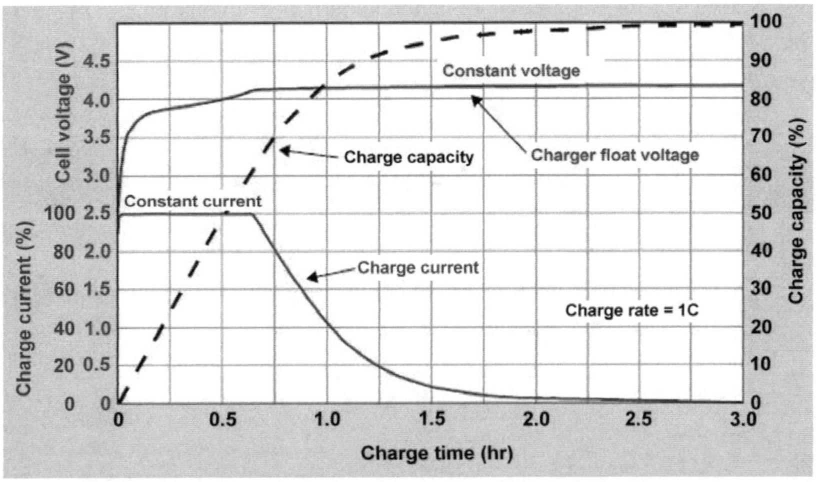

Figure. 5 Charging characteristics of Nickel ion battery

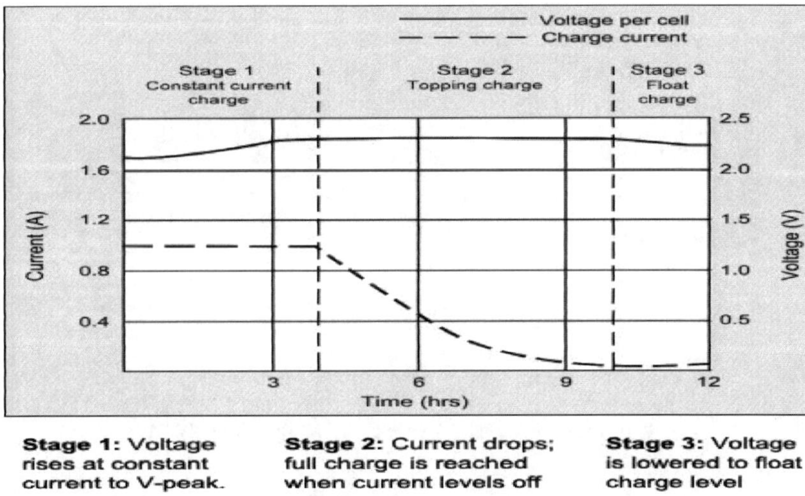

Figure. 6 Charging characteristics of Lead-acid battery

4. Simulation of extended boost power converter

The energy stored in the battery is converted into AC voltage with the help of inverters. In this project the proposed modified boost extended impedance source converter is used to convert Direct Current to Alternating Current and also to step up the voltage of 140V to the desired voltage of 220V (RMS). The projected modified impedance source extend boost topologies are considered for superior boost and subordinate voltage stress across the passive components like capacitors. The secure condition manoeuvre is executed to analyze the maximum stepping up capability of the modified Z-source inverter system. In this project maximum boost control topology is used for PWM pulse generation the generated gate pulses are given to the inverter switches as shown in the Figure. 7.

Figure. 8 shows the simulation of extended boost converter, which gets the input from the battery and boosted to the desired level by the switching operation of inverter switches. Based on the switching strategy of inverter the voltage levels get boosted. Obtained output voltage is given to the load system.

IOP Conf. Series: Materials Science and Engineering **937** (2020) 012045 doi:10.1088/1757-899X/937/1/012045

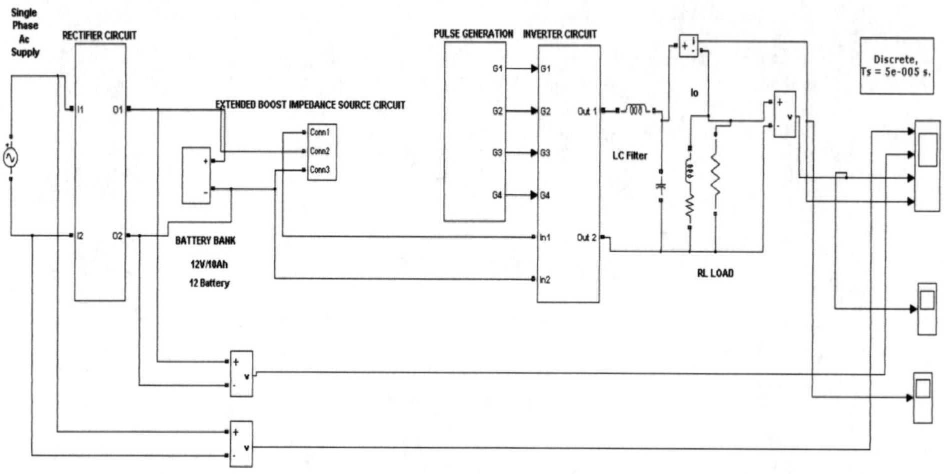

Figure.7 Extended boost power converter based AC load System

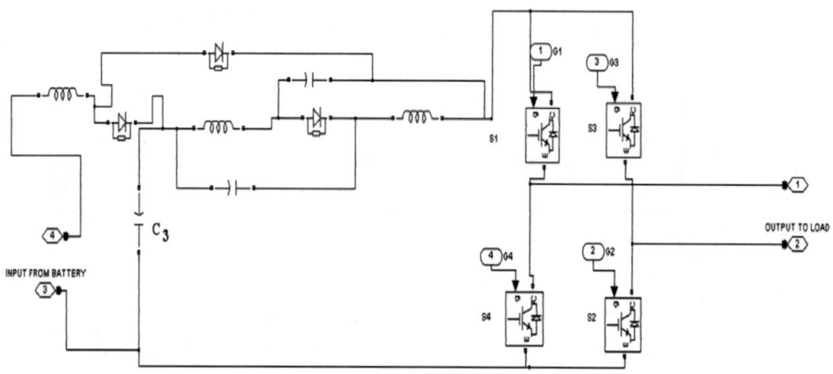

Figure. 8 Simulation of extended boost converter

5. Results and discussion of extended boost power converter

5.1. Simulation results and analysis

The triggering pulses are engendered using a maximum boost control techniques. The reference signals are superimposed with the triangular carrier wave to synthesise the triggering scheme with the main shoot through states. When the triangular carrier waves are greater than the superior envelope V_P, or subordinate than the bottom packet V_n the track turns into shoot through state or else it activates as a conventional inverter. Figure. 9 shows the engendered switching schemes for the switches with boost have power over another method with modulation index of 0.79.

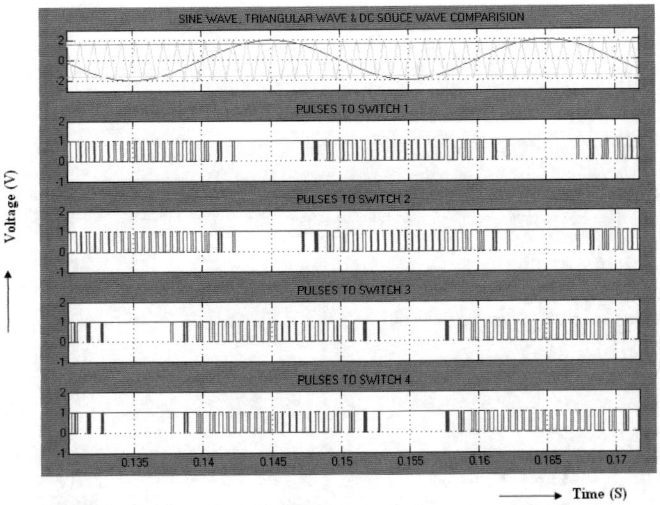

Figure. 9 Gate pulses to extended step-up converter

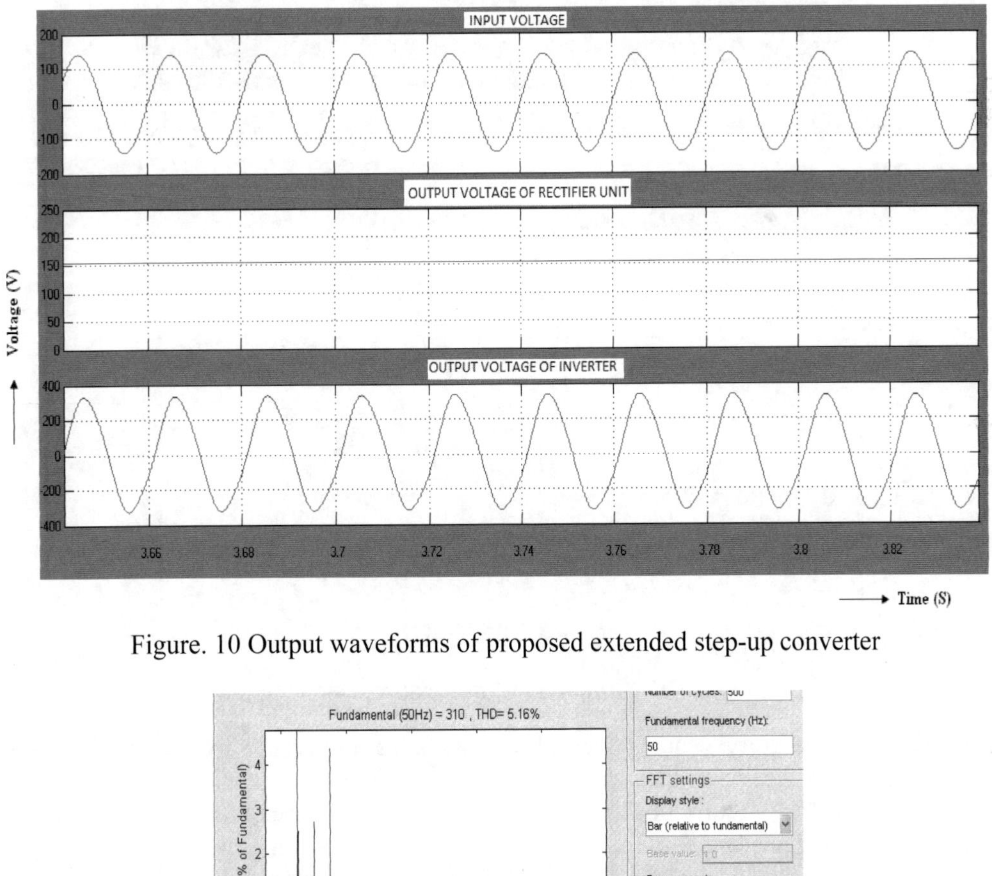

Figure. 10 Output waveforms of proposed extended step-up converter

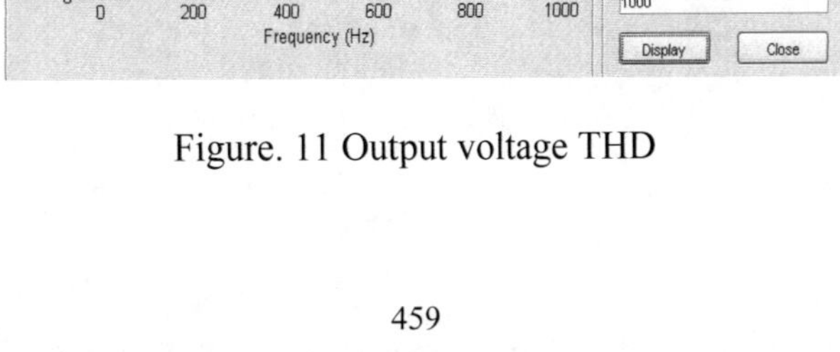

Figure. 11 Output voltage THD

Figure. 10 shows the output voltage and current waveforms of proposed extended boost impedance source inverter if the input voltage is 144 V (V_m).Figure. 11 shows output voltage THD analysis of modified Z-source inverter with minimum THD of 5.16%.

5.2. Experimental results and analysis

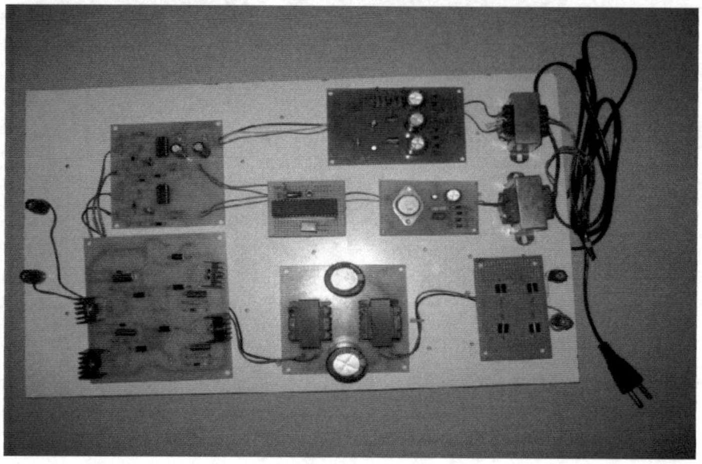

Figure. 12Experimental setup of maximum boost impedance source system

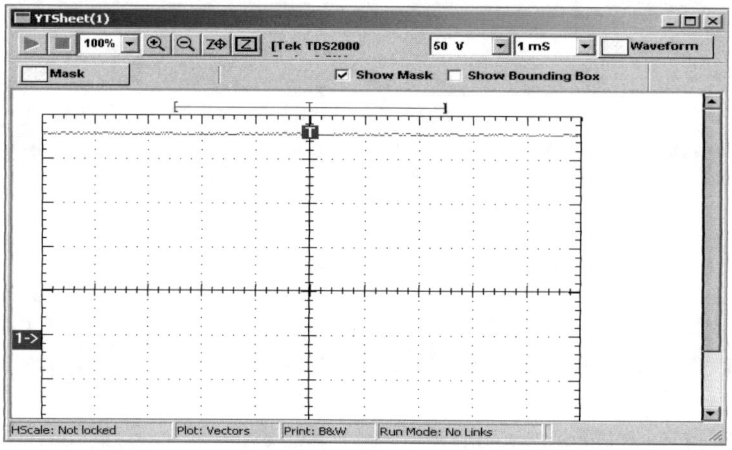

Figure. 13 Maximum boost impedance source output

Figure. 12 shows the experimental setup of maximum boost impedance source inverter system for EV applications. Figure. 13 shows the maximum boost impedance source output, which is about 200V DC from the input of 15V. Figure. 14 shows the output waveform of impedance source inverter which is about 180V, 50 Hz. Impedance source inverter converts 200V DC supply (which is obtained from maximium boost impedance source) into 180V AC supply. The inverter output is suitable for domestic applications.

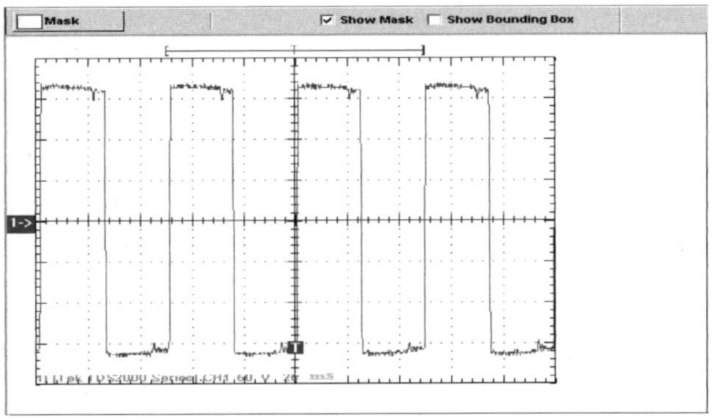

Figure. 14 Inverter output voltage

6. Conclusion

The proposed modified impedance source system was simulated using MATLAB/SIMULINK tools their results are validated. From the achieved results it is clearly inferred that the voltage anxiety in the inverter modules are reduced from end to end the using auxiliary resonant soft switching system. Thus electric vehicles are more eco-friendly than any other vehicles. The need for electric vehicles will be increasing more in the future generation. The efficient utilization of the storage system is improved through the proposed extended boost power converters.

References

[1] Ambühl, D, Sundström, O. Sciaretta, A. and Guzzella, L,"Explicit Optimal Control Policy and its Practical Application for Hybrid Electric Power trains", *Control Engineering Practice, vol.18, pp. 1429-1439, 2010.*

[2] Bakker, S, "The Car Industry and the Blow-Out of the Hydrogen Hype", Energy Policy, vol.38, pp. 6540-6544, 2010.

[3] Bellur, D. M. and Kazimierczuk, M. K,"DC-DC Converters for Electric Vehicle Applications", Proceedings of Electrical Insulation Conference and Electrical Manufacturing Expo, pp. 286-293, 2007.

[4] Bento, N, "Dynamic Competition between Plug-in Hybrid and Hydrogen Fuel Cell Vehicles for Personal Transportation" International Journal of Hydrogen Energy, vol.35, pp. 11271-11283, 2010.

[5] Chen, K, Bouscayrol, A, Berthon, A, Delarue, P,Hissel, D. and Trigui, R."Global Modeling of Different Vehicles",IEEE Vehicular Technology Magazine, pp. 80-89, June 2009.

[6] Clement-Nyns, K, Haesen, E. and Driesen, J, "The Impact of Vehicle-to-Grid on the Distribution Grid", Electric Power System Research, vol.81, pp. 185-192, 2011.

[7] Chandana Jayampathi Gajanayake, Fang Lin LuoHoay Beng Gooi, Ping LamSo and Lip Kian Siow "Extended-Boost Z-Source Inverters" IEEE Trans. Power Electron. vol. 25, no. 10, pp. 2642–2652, Oct. 2010.

[8] Burke, A. F, "Batteries and Ultra capacitors for Electric, Hybrid, and Fuel Cell Vehicles", Proceedings of the IEEE, vol.95, no.4, pp. 806-820, April 2007.

[9] Chan, C. C, "The State of the Art of Electric, Hybrid, and Fuel Cell Vehicles. Proceedings of the IEEE, vol.95, no.4,pp. 704-718, April 2007.

[10] Gajanayake.C.J, D.M.Vilathgamuwa, and P.C.Loh "Development of a comprehensive model and a multiloop controller for Z-source inverter DG systems," IEEE Trans. Ind. Electron., vol. 54, no. 4, pp. 2352–2359, Aug.2007.

[11] Loh.P.G, F. Gao, and F. Blaabjerg, "Topological and modulation design of three-level Z-source inverters," IEEE Trans. Power Electron., vol. 23, no. 5, pp. 2268–2277, Sep. 2008.

[12] Miaosen.S, W. Jin, A. Joseph, P. Fang Zheng, L. M. Tolbert, and D. J. Adams,"Constant boost control of the Z-source inverter to minimize current ripple and voltage stress," IEEE Trans. Ind. Appl., vol. 42, no. 3, pp. 770–778, May/Jun. 2006.

[13] Peng.F.Z, "Z-source inverter," IEEE Trans. Ind. Appl., vol. 39, no. 2, pp. 504–510.

[14] Peng.F.Z, M. Shen, and Z. Qian,] "Maximum boost control of the Z source inverter," IEEE Trans. Power Electron., vol. 20, no. 4, pp. 833–838, Jul./Aug. 2005.

[15] Poh Chiang Loh, Member, Feng Gao, Student Member, Pee-Chin Tan, Blaabjerg, Fellow, "Three-Level AC–DC–AC Z-Source Converter Using Reduced Passive Component Count," IEEE Trans. Power Electron., vol. 24, no.7, July 2009.

[16] Zhi Jian Zhou, Xing Zhang, Po Xu, and Weixiang X. Shen, "Single-Phase Uninterruptible Power Supply Based on Z-Source Inverter," IEEE Trans. Ind. electron., vol. 55, no. 8, Aug 2008.

[17] Lucena, S. E. De Marcelino, M. A. and Grandinetti, F. J, "Low-Cost PWM Speed Controller for an Electric Mini-Baja Type Vehicle", Journal of the Brazilian Society of Mechanical Sciences and Engineering, vol.28, no.1, pp. 21-25, 2007.

[18] Ohn, H.; Yu, S. and Min, K. "Spark Timing and Fuel Injection Strategy for Combustion Stability on HEV Powertrain", Control Engineering Practice, vol.18, pp. 1272-1284, 2010.

[19] Dakshina M. Bellur and Marian K. Kazimierczuk, "DC-DC Converters for Electric Vehicle applications", IEEE Electrical Insulation Conference and Electrical Manufacturing Expo, 2007.

[20] Fang Z. Peng, Hui Li, Gui-Jia Su, and Jack S. Lawle, "A New ZVS Bidirectional DCDC Converter for Fuel Cell and Battery Application", IEEE Trans. Power Electron., vol. 19, no.1, 2004.

[21] M. Kwon, S. Oh, and S. Choi, "High gain soft-switched Bidirectional DC-DC Converter for Eco-Friendly Vehicles", IEEE Trans. Power Electron. vol. 29, no. 4, pp. 1659-1666, Apr. 2014.

Battery modelling and state of charge estimation methods for Energy Management in Electric Vehicle-A review

Srinivas Singirikonda, Y P Obulesu

School of Electrical Engineering, Vellore Institute of Technology, Vellore-632014, Tamilnadu,India.

E-mail: srinivas.singirikonda2018@vitstudent.ac.in yp.obulesu@vit.ac.in

Abstract. To reduce global warming, the Electric Vehicles (EV) are more attracting Worldwide for replacement of conventional IC engine vehicle but the main problem is driving range and the cost of EV is very high compared to a conventional vehicle. The driving range is mainly depending on the type of battery and size of the battery pack used in EV, for long driving range more number of batteries are required which automatically increase the weight and cost of EV. An effective battery management system will increase battery life and driving range of the EV with less number of batteries. In battery management system of EV the battery is major component but battery is costly and managing power of the battery is very much essential in EV technology. Majority of the issues can be solved by developing advanced battery management system (BMS) in EV such as, Battery modelling, accurate battery state of charge and state of health estimation, which can provide an exact driving range of EV and charging/discharging strategies work more effectively. This review paper mainly focuses on different battery modelling techniques and existing battery SOC estimation methods, issues and challenges.

1. Introduction

The BMS manage the electric energy storage system in EV. BMS is used to monitor and protect each cell in battery pack continuously and it's often interfaced with other devices in EV. The main function of the BMS is to provide the over voltage and under voltage protection, over current protection, state of charge (SOC) and state of health (SOH) estimation in each cell of battery, fault detection, cell balancing and thermal management. The secondary function of BMS is to current limit calculations, digital or analog interfaces with application, cell health monitoring, isolation fault monitoring.

BMS will improve the lifespan battery by providing over charge and over discharge protection to each cell in battery pack. Driving range and cost of EV mainly depends on the battery capacity. The capacity of the battery is depending on the battery chemistry, ambient temperature, aging effects, control algorithm used in the BMS and maintenance. The control algorithms are mainly used for estimate the SOC, SOH and thermal management. There are different control algorithms are proposed in the literature for battery parameters estimation, mainly they are classified as electrochemical battery impedance spectroscopy (ESI), Equivalent circuit models (ECM) and electrochemical model (EM), each technique have its own advantages and disadvantages which are clearly explained in the following chapters.

For Different types of batteries anode and cathode materials have different property and Every battery have its own electrochemical and dynamic characteristics. Most of equivalent circuit model hysteresis effect of the battery was not consider. So, it's very difficult to model the equivalent circuit of it. It is necessary to design suitable model of the battery which can work with any load conditions.

In literature different battery models are there, they can classify as a thermal, physical, electrochemical, equivalent circuit model and so on. From that equivalent circuit model (ECM) is more using for estimation of SOC due to its simple structure.

2. EV Battery modelling

Now a days mostly the Lithium-ion batteries (LIB) are using in EV applications due its energy density, charging and low self-discharging characteristics. The Different LIB model-based methods are explained in the literature.

2.1. Rint ECM

The Rint ECM, as shown in the figure 1, it is simple model for practical implementation but output equation leads to uncertainties in state estimation. The equation 1 is the output voltage (V_L) is the sum of the open circuit voltage V_{oc} and internal resistance R_0, where V_{oc} and R_0 are the function of SOC, SOH and temperature. The I_L is load current which is positive during discharging and negative during discharging of the cell.

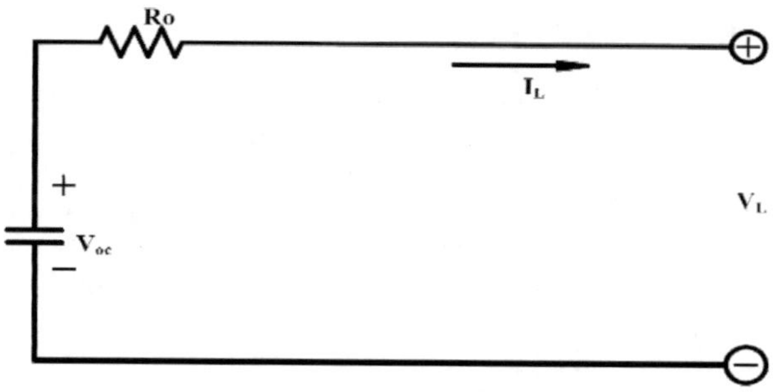

Figure 1. The Rint ECM.

$$V_L = V_{OC} - I_L R_0 \tag{1}$$

2.2. Thevenin ECM

Thevenin ECM is shown in figure 2. It is an Improved model of R_{int} with Parallel RC network in series. It Most used in practical applications but it has accuracy problem during charging and Discharging process. The internal resistance of this model includes R_0 and R_{th} namely called as ohmic and polarization resistances. The C_{Th} describes the transient response of the ECM. The behavior of the output voltage is shown in equation 2 and equation 3.

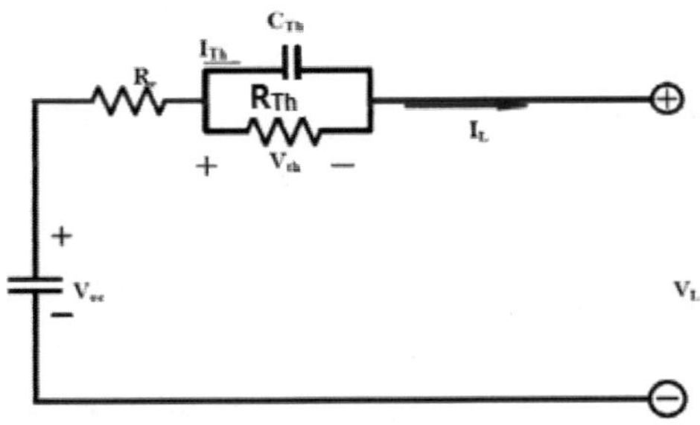

Figure 2. Thevenin ECM.

$$\dot{V}_{Th} = -\frac{V_{Th}}{R_{Th}C_{Th}} + \frac{I_L}{C_{Th}} \quad (2)$$

$$V_L = V_{OC} - V_{Th} - I_L R_0 \quad (3)$$

2.3. Dual polarization ECM

Dual polarization ECM is shown in the figure 3. It is an improved model of the Thevenin's model. It has more accurate during charging and discharging. The output voltage equation is shown in 4 to 6.

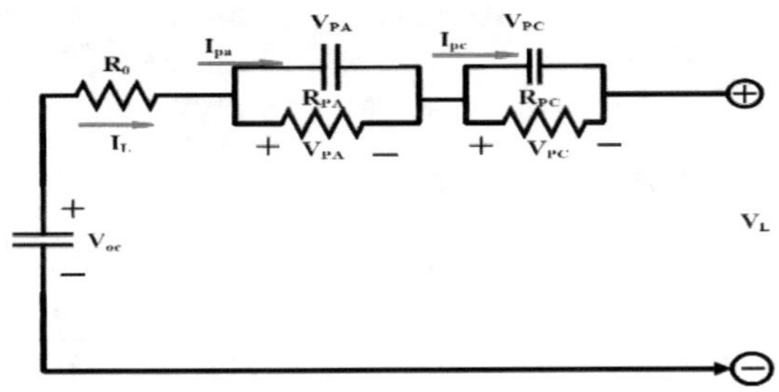

Figure 3. Dual polarization ECM.

$$\dot{V}_{pa} = -\frac{V_{pa}}{R_{pa}C_{pa}} + \frac{I_L}{C_{pa}} \quad (4)$$

$$\dot{V}_{pc} = -\frac{V_{pc}}{R_{pc}C_{pc}} + \frac{I_L}{C_{pc}} \quad (5)$$

$$V_L = V_{OC} - V_{pa} - V_{pc} - I_L R_0 \quad (6)$$

2.4. PNGV (partnership for new generation of vehicle) ECM

PNGV ECM is shown in the figure 4. It an improved model of the Thevenin model by adding a capacitor in series with the source voltage of the Thevenin model. Data saturation and accumulation over time of load current problem can be overcome using PNGV model. Equation 7 to 9 describe the output voltage the model.

IOP Conf. Series: Materials Science and Engineering **937** (2020) 012046 doi:10.1088/1757-899X/937/1/012046

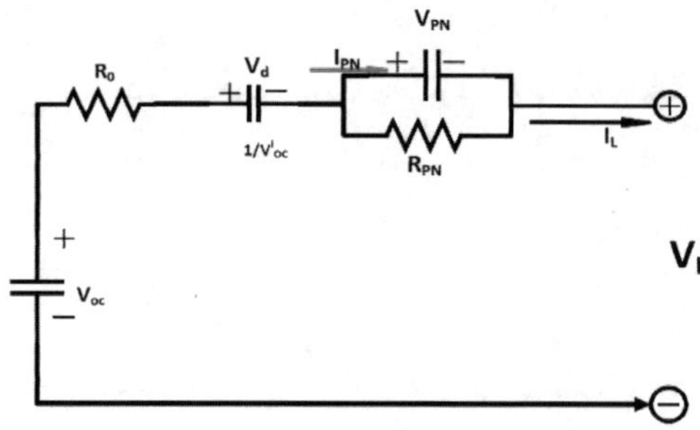

Figure 4. PNGV ECM.

$$\dot{\boldsymbol{V}}_{\boldsymbol{d}} = \boldsymbol{V'}_{\boldsymbol{oc}}\, \boldsymbol{I}_{\boldsymbol{L}} \tag{7}$$
$$\ddot{V}_{pn} = -\frac{V_{pn}}{R_{pn}C_{pn}} + \frac{I_L}{C_{pn}} \tag{8}$$
$$V_L = V_{oc} - V_d - V_{pn} - I_L R_0 \tag{9}$$

2.5. GNL (General non-linear) ECM

GNL ECM is shown in the figure 5. It is combination model of the PNGV and Dual polarization with additional parallel sours resistance R_S. This model is used in non-linear battery modelling due its less self-discharging features. The equation 10 to 13 describe the output voltage of model.

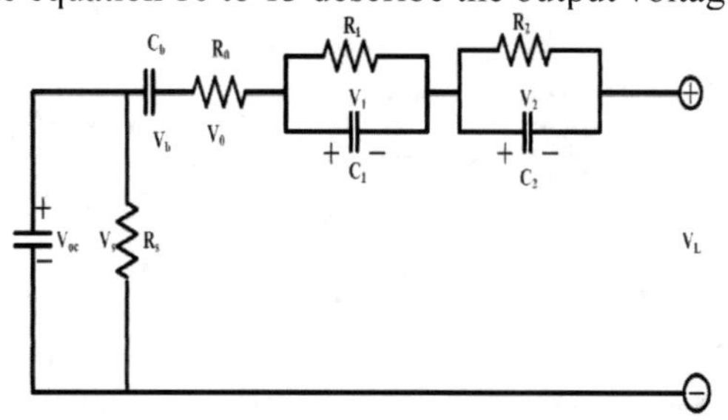

Figure 5. GNL ECM.

$$\dot{V_b} = V_{oc}' \tag{10}$$
$$\dot{V_1} = \frac{1}{C_1} - \frac{V_b}{C_1R_s}\left(\frac{V_1}{C_1R_s} + \frac{V_1}{C_1R_1}\right) - \frac{V_2}{C_1R_s} + \frac{V_{oc}}{C_1R_s} \tag{11}$$
$$\dot{V_2} = \frac{1}{C_2} - \frac{V_b}{C_2R_s}\left(\frac{V_2}{C_1R_s} + \frac{V_2}{C_2R_2}\right) - \frac{V_1}{C_2R_s} + \frac{V_{oc}}{C_2R_s} \tag{12}$$
$$V_L = V_{oc} - V_b - V_1 - V_2 - I_L R_0 \tag{13}$$

2.6. RC Equivalent model

The RC equivalent model is shown in the figure 6. It will give the dynamic voltage performance of the battery. Equation 14 and 15 describe the output voltage of the model.

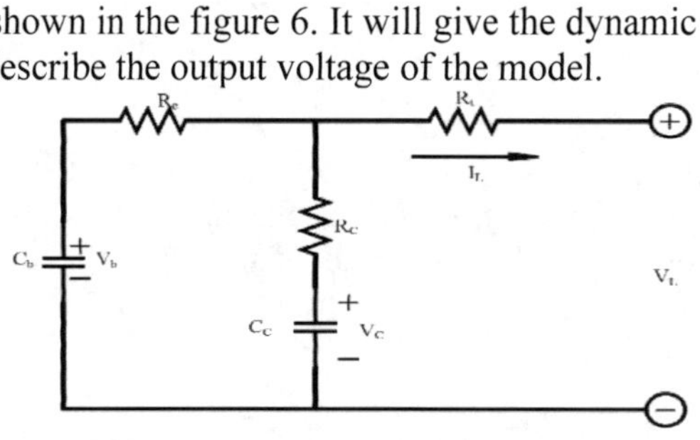

Figure 6. RC Equivalent model.

$$\begin{bmatrix} \dot{V}_b \\ \dot{V}_c \end{bmatrix} = \begin{bmatrix} \frac{-1}{C_b(R_e+R_c)} & \frac{1}{C_b(R_e+R_c)} \\ \frac{1}{C_C(R_e+R_c)} & \frac{-1}{C_C(R_e+R_c)} \end{bmatrix} \begin{bmatrix} V_b \\ V_C \end{bmatrix} + \begin{bmatrix} \frac{-Rc}{C_b(R_e+R_c)} \\ \frac{-R_e}{C_C(R_e+R_c)} \end{bmatrix} [I_L] \quad (14)$$

$$[V_L] = \begin{bmatrix} \frac{R_C}{(R_e+R_c)} & \frac{R_e}{(R_e+R_c)} \end{bmatrix} \begin{bmatrix} V_b \\ V_C \end{bmatrix} + \begin{bmatrix} -R_t & -\frac{R_eR_C}{(R_e+R_c)} \end{bmatrix} [I_L] \quad (15)$$

2.7. *Fractional order ECM*

The fractional order ECM is shown in the figure 7. It is used to reduce the computational complexity and to improve the efficiency of outdated existing ECMs faults for assurance of optimal trend off among ECMs. The equations 16 to 18 describe output voltage of the model.

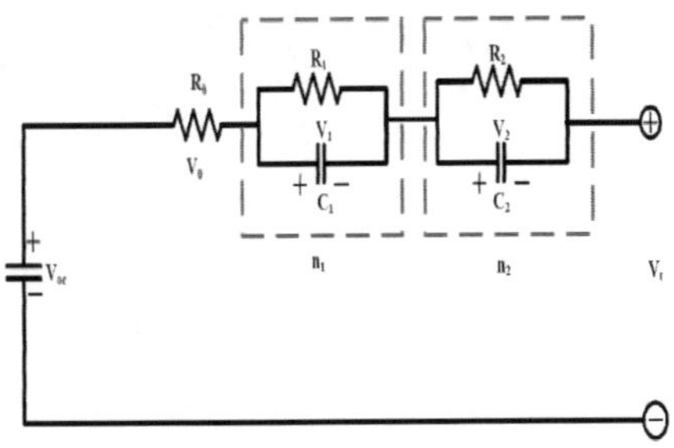

Figure 7. Fractional order ECM

$$D^{n_1}\dot{V}_1 = \frac{1}{C_1} - \frac{V_1}{C_1R_1} \quad (16)$$

$$D^{n_2}\dot{V}_2 = \frac{1}{C_2} - \frac{V_2}{C_2R_2} \quad (17)$$

$$V_t = V_{oc} - V_1 - V_2 - IR_0 \quad (18)$$

We can conclude that the dual polarization model has best dynamic performance and less losses compare to other model-based method.

3. STATE OF CHARGE ESTIMATION METHODS.

The SOC estimation is main challenge in EV batteries, we can't measure the SOC of the battery directly, a specific algorithm required for it, which describe the remaining capacity of it. The accurate SOC estimation is paly major role in the battery life and performance improvement. This estimation algorithms continuously monitor all internal parameters of the battery and provide the protection from battery from over charging and over discharging. The SOC and SOH of battery is defined as equation 19 and 20. The exiting SOC and SOH estimation methods are reviewed in the following sections.

$$State\ of\ Charge = \frac{remaining\ capacity\ of\ the\ battery}{Total\ capacity\ of\ the\ battery} \quad (19)$$

$$State\ of\ Health = \frac{Maximum\ charging\ capacity\ of\ the\ battery}{Rated\ capacity\ of\ the\ battery} \quad (20)$$

The evolution of the SOC estimation methods is shown in the figure 8. There are various methods available in the literature for battery SOC Estimation. Each method having some merits and demerits. Mainly the SOC estimation methods are classified into Conventional Methods, Adaptive filter based, artificial intelligent based algorithms, Non-linear Observer algorithms, Hybrid methods. each method is further classified into sub method which are explained as follows.

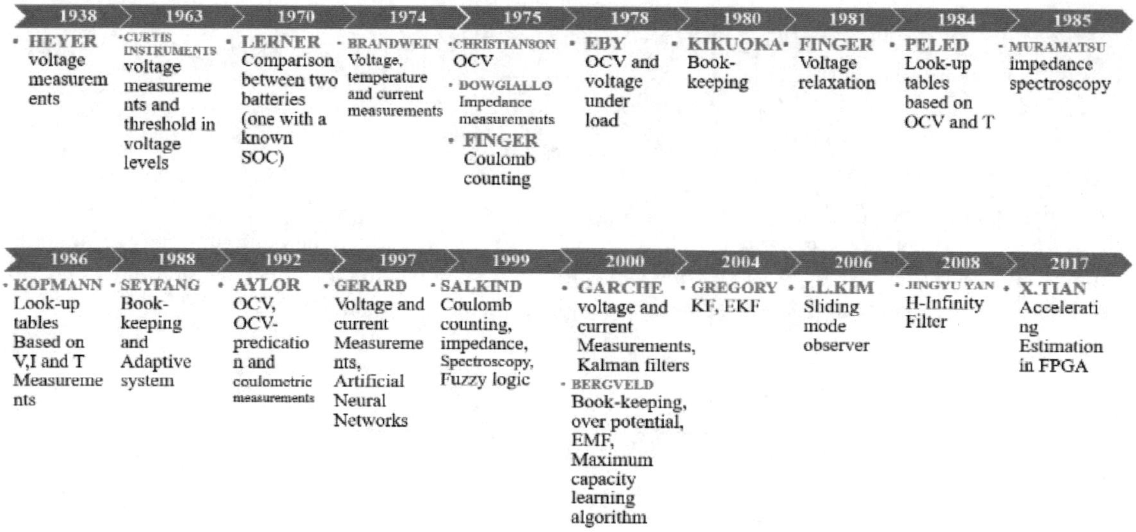

Figure 8. Battery SOC estimation methods evolution.

3.1. Conventional Methods

3.1.1. Ampere-Hour Counting Method (AHM) / Coulomb Counting (CC) Estimation Method.

The ampere-hour counting method is most commonly used conventional method for SOC estimation of batteries. It is accurate method for short term calculations and the main advantage is it is very simple, concise method it can be implement easily in practice. This method is using in the applications of industries, Portable Electronics and EVs. In [1] J. Rivera-Barrera *et al* proposed the ampere-hour counting method (AHM) / coulomb counting (CC) SOC Estimation method. It will measure the charging and discharging current over a time of period to estimate the remaining capacity of the battery. Mathematically that can be define as equation 21.

$$SOC = SOC(t_0) + \frac{1}{C_n}\int_{t_0}^{t_0+t} I_{bat}\,(dT) \times 100 \tag{21}$$

Where SOC(t0) is the Initial SOC, I_{bat} is the Charging or Discharging Current and Cn is the Nominal Capacity.

However, this CC method is suffering from cumulative error due to its initial value estimation problem. So, to implement this method initial vales must be Known.

3.1.2. Modified Coulomb Counting Method

In [2] I. Baccouche *et al* from the piecewise relationship of SOC-OCV(open circuit voltage), an improved coulomb counting algorithm is proposed, which can increase the accuracy of the SOC estimation in LIB compare to coulomb counting algorithm. In [3] Venkatesh Prasad K. S *et al* proposed a modified coulomb counting (CC) method by using a peukert mathematical model to overcome inaccuracy problem in CC method. The modified CC SOC estimation equation is given by equation 22.

$$SOC_t = \left(1 - \frac{1}{Q_\tau}\int_0^{\tau} \eta_d\, i_\tau dT\right) \times 100 \tag{22}$$

Where SOC_t is corrected SOC, η_d is coefficient of efficiency.

3.1.3. The open circuit voltage method

The OCV algorithm is to require less number of computations for SOC estimation. However, the SOC estimation time increasing along with the age of the battery and estimation time is changing along with temperature change. This OCV method is not suitable for online SOC estimation because the battery required long resting time to reach balance condition and to get accurate SOC. This method is not possible in practical because its measuring OCV that means estimation of battery SOC is not available while Battery is Charging and Discharging. Direct OCV method can be useful for low power consumption application, it will provide the Comparatively high Accuracy. In [4] M.A.Hannan *et al* proposed a EMF method for SOC and capacity of the battery at time by OCV relaxation process to reduce the SOC estimation time of the battery and to eliminate the effect of the over voltage.

3.1.4. Electrochemical Method

In electrochemical battery we can directly calculate the SOC from its positive and negative Electrodes. Theoretically this method can give the accurate values of the SOC and it is suitable for battery offline parameters analysis only. However, this method consisting of partial difference equations of multi parameters. So, it is very complex process and difficult to solve the online SOC estimation and it has very poor accuracy in real time application because of poor parameters fitting. In [5] Nima Lotfi *et al* proposed a reduced-order electrochemical model based on the Luenberger and recursive least square algorithms, without any partial differential equations and with less mathematical complexity. it is observed that the algorithm reaches their actual values within 1min. In [6] Rui Xiong *et al* proposed electrochemical genetic algorithm for SOC estimation of lithium-ion battery, which is also estimate the SOH battery based on battery degradation characteristics.

3.1.5. Model Based Method

The model-based method widely adopted for SOC estimation of the batteries in the BMS of EV, for accurate online SOC Estimation in batteries first we need to develop the battery equivalent model, which can describe the actual dynamic behaviour of the battery. There are different model-based methods available for battery modelling like electrochemical or white box model. The most commonly used other model is Linear equivalent circuit model [7-9]. However, some of the characteristics of the battery cannot be designed in the simulation like hysteresis and Warburg effects. To overcome that problem Mathematical model with Hysteresis is used in simulation for better SOC estimation.

3.2. Adaptive filter algorithm methods

Adaptive filtering algorithms are used to estimate the SOC of battery by reducing noise influence on the battery model and these algorithms also improve the accuracy and robustness of the battery.

3.2.1. Kalman filter

This algorithm is used as state estimation tool in linear systems, it is also used as dynamic state of charge estimator in batteries. This computational method is useful to predict the past, present and future states of parameters [10]. The main drawback of the Kalman filter is it requires large computing capacity, suitable battery model and problem of determining initial parameters [11].

3.2.2. Extended Kalman filter

The EKF can overcome the non-linear estimation problem by linearizing the state equations [12]. In [13] Ruifeng Zhang *et al* the partial derivatives and first order Taylor series expansion are used in EKF to linearize the battery model. The state-space model is linearized at every time instance, which compares the estimated value with its measured batteries terminal voltage to correct the estimation parameters for SOC. However, if the system is highly non-linear, linearization error may occur due to the lack of accuracy in the first order Taylor series under a highly non-linear condition [14]. In [15] Jia, J *et al* Proposed multi-rate strong tracking EKF, this algorithm was improving the battery parameters tracking stability and SOC estimation accuracy 55.34% compare to EKF algorithm.

3.2.3. Fading Kalman Filter
In [16] KaiChin Lim *et al* proposed a Fading Kalman filter (FKF) to estimate the OCV and SOC. FKF algorithm uses a fading factor due to that factor its can capable to compensate for any modelling error. FKF algorithm can avoid the large SOC estimation errors in batteries, which may occur with a conventional KF. This algorithm implementation is easy and SOC estimation accuracy is more in the real time applications compare to KF and EKF. In [17] Yunfei Zhao *et al* proposed an Adaptive fading EKF algorithm, it is combination of the EKF and FKF algorithm, which can reduce the noise problems and improve the convergent speed and provide more accurate SOC estimation results of the battery.

3.2.4. Unscented Kalman filtering
The Unscented Kalman filter (UKF) is extended version of the EKF. The performance of the EKF is not poor for more than second order model. This problem can be overcome by using UKF. The UKF is designed by using state space model for higher order nonlinear applications, which can also reduce the sampling noise [18].

3.2.5. Sigma-Point Kalman Filter
Sigma-point Kalman filter (SPKF) algorithm is subjected to numeric approximation and algorithm selects the sets of sigma points, which is completely similar to the value of mean and covariance of the model being developed. The advantages of using SPKF are that it has an identical calculation complexity to EKF without considering Jacobian matrices [19-20]. The SPKF can proved more accurate with less memory and less computational calculations [21].

3.2.6. H∞ Filter
This model is simple in design and which high robust under certain conditions. The accuracy of SOC estimation will reduce due to the thermal and aging effects. In order to improve the accuracy and robustness in [22] cheng chen *et al* proposed a multiscale dual H infinity filter, the results are shown that it have more accurate SOC estimation and robustness compare to KF algorithms.

However, these KF family algorithms are not suitable for high complexity system because of its high mathematical computational cost. These methods consist of complex matrix operations which can make the numerical instability. The SOC estimation sensor accuracy and battery model are the major concerns for SOC estimation. For the performance analysis KF's variants are mainly depends on measurement noise covariance and required Prior knowledge of the model.

3.3. Adaptive Learning Algorithm

3.3.1. Artificial Neural Network (ANN) based algorithms
In [23] F. Zhao, Y *et al* proposed a recursive NN algorithm to estimate the SOC of the lithium-ion battery, the multi-channel extended convolution NN is used to extract the battery parameters. The accuracy of this algorithm is 11.3% more compared to ampere hour count method. In [24] Jian Chen *et al* proposed a radial basis function NN algorithm along with nonlinear observer for online SOC estimate of the battery.

3.3.2. Extreme Machine Learning (EML) based algorithms
To reduce the mathematical complex calculation and when we combine any two algorithms more complex and expensive in the implementation. In order to overcome that problem in [25] M. S. Hossain Lipu et al proposed an improved extreme learning machine algorithm for battery SOC estimation and gravitational search algorithm which improve the speed and reduce the computational complexity of the EML algorithm by finding the optimal value hidden layer neurons. This algorithm is robust and does not require the battery internal mathematical model and knowledge.

3.3.3. Support Vector Machine (SVM) based algorithms

In [26] Juan Carlos Álvarez Antón *et al,* proposed SVM algorithm for battery SOC estimation. The algorithm will extract the battery model parameters from its charging and discharging test cycles, cell thermal temperature, voltage and current as independent variables. This algorithm dose not required any matrix inversion. So, which will reduce the mathematical computational complexity.

3.3.4. Genetic Algorithm (GA) based algorithms

In [27] Shen, Y *et al* proposed an improved chaos genetic algorithm for SOC estimation of the battery with less computational complexity and high initial stability, ampere hour approach and adaptive switch mechanism is advised to predict the available capacity of the battery. In [28] Jiahuan Lu y*et al* proposed the online approach of GA-based estimation method that can improve the State of power (SOP) estimation accuracy. Compared with the traditional Taylor method the online approach of GA-based estimation method can improve the 7.2% more accuracy.

3.3.5. Fuzzy logic-based algorithms

In [29] Yan Ma *et al* proposed a two-stage bidirectional equalization model with fuzzy logic control scheme to improve the inconsistency of series connected batteries. Compared to mean difference algorithm, the fuzzy logic control scheme can be reducing the standard deviation of final SOC, equalization time by 18.5% and 23% respectively and can improve the energy efficiency 5.54%.

3.4. Non-Linear Observers

3.4.1. The Luenberger Observer

Online equivalent circuit model are used to estimate SOC and power capacity of battery, but the due to the noise corruption accuracy will reduce, to overcome these problem [30] Z. Wei *et al* proposed online model identification method based on adaptive forgetting recursive total least squares to compensate the noise effect and combined this method with Luenberger observer to estimate the SOC of the battery, the simulation results are shown that the estimation accuracy was improved and robust in noise corruption. In [31] Hu. X *et al* proposed online SOC estimation method for Lithium-ion battery using Luenberger observer. This observer method will reduce difference between actual and estimated voltage error, the stochastic gradient method is used for adaptive adjustment of the observer gain.

3.4.2. Sliding mode observer

In order to increase stability by compensating non-linear dynamic uncertainties in the battery model, in [32] Kim I.L first introduced the SMO in battery SOC estimation in HEV. The SMO control the convergence of high charging and discharging values by using Lyapunov inequality equations. the author shown that the test results with less than 3% SOC estimation error. In [33] Ning B *et al* proposed adaptive SMO for accurate online SOC estimation in the adaptive parameters-based battery model. This observer mothed will improve the accuracy of SOC estimation by using the dynamic self-adjusting switching gain in response to the tracking errors. The result shown that the error between actual and estimated SOC is less than 2%. In [34] Y. Feng et al proposed a three-terminal sliding-mode observer (TSMOs) for battery SOC and SOH estimation, each observer in the TSMOs will estimate one parameter in battery variables. This estimation algorithm doesn't contain any low pass filters, which will increase the accuracy and robustness, reduce cost.

3.5. Hybrid methods

Based on the analysis of number of SOC estimation methods, each method or algorithm have its own merits and demerits. By combing these algorithms, we can get advantages both algorithms, which can improve the performance and accuracy as well as reduce cost. In [35] J. Wang et al proposed hybrid coulomb counting and Kalman filter algorithm, the main limitation of coulomb count counting method is that, in order to estimate the battery SCO, the initial values of the battery parameters are needed, to solve this issue, the Kalman filter is used in this process, which can correct the initial values of the battery parameters and be used as input to the coulomb counting algorithm. In [36] Bizhong Xia et al have proposed the Levenberg-Marquardt hidden layer (LMMWNN) and PSO algorithm for SOC estimation of battery. The piecewise network and seven-point linear smoothing method are used to optimize the LMMWNN. The proposed LMMWNN method tested on New European Driving cycle and results showed that it has better robustness, noise performance compared to the Levenberg-Marquardt wavelet neural network algorithm. In [37] Chen, X et al, proposed a hybrid BPNN-EKF algorithm for SOC estimation method of thevenin equivalent model battery pack. The model bias and parameter uncertainties are considered as model uncertainties. In this hybrid model bias correction method was used to reduce model uncertainties, A BPNN based bias functioning method is used to reduce the polarization and temperature effects. Finally, the EKF is used for SOC estimation of the battery equivalent model. In [38] Yuejiu Zheng et al, has proposed a hybrid EKF method and PSO algorithm for SOC inconsistency estimation for RC second-order equivalent model. The POS algorithm is used to identify the battery parameters and EKF is used SOC estimation. In [39] Yanqing Shen, proposed a hybrid adaptive chaos GA(CGA) based on EKF for SOC estimation of LIBs. The EKF is used for local linear approximation and CGA is used for global optimization search for battery parameter estimation. This method quickly evaluates SOC estimation with high accuracy and it has high robustness without being affected by uncertain initial values. In [40] Ju wang et al, proposed recursive least square and adaptive H∞ filter joint estimation framework. The proposed dual estimator setup analyzed with wide temperature range (-10°C to 25°C), results showed that the proposed hybrid method have batter SOC estimation performance even at very low temperature i.e error is less than 3.5% at -10 °C, and less than 2% at 0 °C and 25°C. In [41] Shehab El Din et al, proposed adaptive equivalent circuit model with artificial neural network (ANN) controller with hybrid unscented Kalman filter (UKF) and Autocovariance least square (ALS) method. In the proposed hybrid method, ANN is used to identify the optimal parameters of battery model, UKF is used for fast covariance and ALS used to measure the noise covariance which improves the SOC estimation accuracy. In [42] Minghui Hu et al, proposed a mixed swarm based co-operative particle swarm optimization method, to improve the accuracy and better robustness with different driving range conditions for LIB. the proposed method can adequately coordinate the dynamic balance between local and global optimization for the optimal solution with fast accuracy. In [43] N. Guo et al Proposed fuzzy weighted algorithm, which is a combined model of genetic algorithm (GA) -back propagation neural network (BPNN) algorithm and ampere integration method (AIM) for battery SOC estimation. The weights of fuzzy controller are controlled by the GA-BPNN and AIM, for large voltage variation the weights of the fuzzy is propositional to the GA-BPNN and for small voltage variations proportional to the AIM. This algorithm can have advantage of both algorithms based on the situation and state of voltage and SOC levels. However, this algorithm will depend on battery working environmental conditions. In [44] F. Yang *et al*, proposed hybrid long-short-term memory (LSTM)- recurrent neural network (RNN) and (unscented Kalman filter) UKF algorithm. The battery current, voltage and temperature are taken as input Parmenter to the LSTM-RNN and SOC estimation as output, the UKF is used as noise filter at output, the algorithm is model free and data driven, tested at different temperature from 0^0C to 50^0C and estimation error was less than 1%.

CONCLUSION

In this paper an over view of battery Equivalent circuit models, SOC and SOH estimation algorithms has been given. The major challenges and constrains in battery modeling, SOC and SOH estimation are discussed. Performance of different battery model, SOC and SOH estimation algorithms are compared in point of accuracy, speed and mathematical complexity. Our future work is focus on the development of new hydride estimation algorithms, which can provide more accurate and high-speed results with less complexity in development of smart BMS for EVs.

References

[1] J. Rivera-Barrera, N. Muñoz-Galeano, and H. Sarmiento-Maldonado, SoC Estimation for Lithium-ion Batteries: Review and Future Challenges, vol. 6, no. 4. 2017.

[2] I. Baccouche, S. Jemmali, A. Mlayah, B. Manai, and N. E. Ben Amara, "Implementation of an Improved Coulomb-Counting Algorithm Based on a Piecewise SOC-OCV Relationship for SOC Estimation of Li-Ion Battery," Int. J. Renew. Energy Res., vol. 8, no. 1, pp. 178–187, 2018.

[3] Venkatesh Prasad K. S. and B. P. Divakar. "Real Time Estimation of SoC and SoH of Batteries". international journal of renewable energy research, Vol.8, No.1, March, 2018.

[4] M. A. Hannan, M. S. H. Lipu, A. Hussain, and A. Mohamed, "A review of lithium-ion battery state of charge estimation and management system in electric vehicle applications: Challenges and recommendations," Renew. Sustain. Energy Rev., vol. 78, no. May, pp. 834–854, 2017.

[5] Nima Lotfi ; Robert G. Landers ; Jie Li ; Jonghyun Park. "Reduced-Order Electrochemical Model-Based SOC Observer With Output Model Uncertainty Estimation". IEEE Transactions on Control Systems Technology, Year: 2017 , Volume: 25 , Issue: 4, Page s: 1217 – 1230.

[6] Rui Xiong, Linlin Li, Zhirun Li, Quanqing Yu, Hao Mu, An electrochemical model based degradation state identification method of Lithium-ion battery for all-climate electric vehicles application, Applied Energy, Volume 219, 2018, Pages 264-275.

[7] Y. Zheng, M. Ouyang, X. Han, L. Lu, and J. Li, "Investigating the error sources of the online state of charge estimation methods for lithium-ion batteries in electric vehicles," J. Power Sources, vol. 377, no. November 2017, pp. 161–188, 2018.

[8] J. Rivera-Barrera, N. Muñoz-Galeano, and H. Sarmiento-Maldonado, SoC Estimation for Lithium-ion Batteries: Review and Future Challenges, vol. 6, no. 4. 2017.

[9] Scott J. Moura, Nalin A. Chaturvedi, Miroslav Krstic "Adaptive Partial Differential Equation Observer for Battery Estimation Via an Electrochemical Model," J. Dyn. Syst. Meas. Control, vol. 136, pp. 1–11, 2014.

[10] Zhihao Yu , Ruituo Huai and Linjing Xiao. "State-of-Charge Estimation for Lithium-Ion Batteries Using a Kalman Filter Based on Local Linearization ". Energies 2015, 8, 7854-7873.

[11] Sabine Piller, Marion Perrin, Andreas Jossen, Methods for state-of-charge determination and their applications, Journal of Power Sources, Volume 96, Issue 1, 2001, Pages 113-120.

[12] Pawel Malysz ; Ran Gu ; Jin Ye ; Hong Yang ; Ali Emadi."State-of-charge and state-of-health estimation with state constraints and current sensor bias correction for electrified powertrain vehicle batteries". IET Electrical Systems in Transportation, Year: 2016 , Volume: 6 , Issue: 2 Page s: 136 – 144.

[13] Ruifeng Zhang , Bizhong Xia , Baohua Li , Libo Cao , Yongzhi Lai , Weiwei Zheng ,Huawen Wang and Wei Wang . " State of the Art of Lithium-Ion Battery SOC Estimation for Electrical Vehicles". Energies 2018, 11, 1820.

[14] Hannan, M.A.; Lipu, M.S.H.; Hussain, A.; Mohamed, A. A review of lithium-ion battery state of charge. estimation and management system in electric vehicle applications: Challenges and recommendations. Renew. Sustain. Energy Rev. 2017, 78, 834–854.

[15] Jia, J.; Lin, P.; Chin, C.S.; Toh, W.D.; Gao, Z.; Lyu, H.; Cham, Y.T.; Mesbahi, E. Multirate strong tracking extended Kalman filter and its implementation on lithium iron phosphate (LiFePO4) battery system. In Proceedings of the IEEE International Conference on Power Electronics and Drive Systems, Sydney, NSW, Australia, 9–12 June 2015; pp. 640–645.

[16] KaiChin Lim, Hany Ayad Bastawrous, Van-Huan Duong, Khay Wai See, Peng Zhang, Shi Xue Dou, Fading Kalman filter-based real-time state of charge estimation in LiFePO4 battery-powered electric vehicles, Applied Energy, Volume 169, 2016, Pages 40-48.

[17] Yunfei Zhao, Jun Xu, Xiao Wang, Xuesong Mei, The Adaptive Fading Extended Kalman Filter SOC Estimation Method for Lithium-ion Batteries, Energy Procedia, Volume 145, 2018, Pages 357-362.

[18] Weida Wang ; Xiantao Wang ; Changle Xiang ; Chao Wei ; Yulong Zhao. Unscented Kalman Filter Based Battery SOC Estimation and Peak Power PredictionMethod for Power Distribution of Hybrid Electric Vehicles. IEEE Access, Year: 2018 , Volume: 6, Page s: 35957 – 35965.
[19] Gregory L. Plett, Sigma-point Kalman filtering for battery management systems of LiPB-based HEV battery packs: Part 1: Introduction and state estimation, Journal of Power Sources,Volume 161, Issue 2, 2006,Pages 1356-1368.
[20] Gregory L. Plett, Sigma-point Kalman filtering for battery management systems of LiPB-based HEV battery packs: Part 2: Simultaneous state and parameter estimation,Journal of Power Sources, Volume 161, Issue 2, 2006, Pages 1369-1384.
[21] He, Z.; Liu, Y.; Gao, M.;Wang, C. A joint model and SOC estimation method for lithium battery based on the sigma point KF. In Proceedings of the Transportation Electrification Conference and Expo, Dearborn, MI, USA, 18–20 June 2012; pp. 1–5.
[22] Cheng Chen ; Rui Xiong ; Weixiang Shen. "A Lithium-Ion Battery-in-the-Loop Approach to Test and Validate Multiscale Dual H Infinity Filters for State-of-Charge and Capacity Estimation". IEEE Transactions on Power Electronics. Volume: 33 , Issue: 1,Page s: 332 – 342, Year: 2018.
[23] F. Zhao, Y. Li, X. Wang, L. Bai and T. Liu, "Lithium-Ion Batteries State of Charge Prediction of Electric Vehicles Using RNNs-CNNs Neural Networks," in IEEE Access, vol. 8, pp. 98168-98180, 2020, doi: 10.1109/ACCESS.2020.2996225.
[24] Jian Chen ; Quan Ouyang ; Chenfeng Xu ; Hongye Su. "Neural Network-Based State of Charge Observer Design for Lithium-Ion Batteries". IEEE Transactions on Control Systems Technology. Volume: 26 , Issue: 1, Page s: 313 – 320, Year: 2018.
[25] M. S. Hossain Lipu, M. A. Hannan, A. Hussain, M. H. Saad, A. Ayob and M. N. Uddin, "Extreme Learning Machine Model for State-of-Charge Estimation of Lithium-Ion Battery Using Gravitational Search Algorithm," in *IEEE Transactions on Industry Applications*, vol. 55, no. 4, pp. 4225-4234, July-Aug. 2019, doi: 10.1109/TIA.2019.2902532.
[26] Juan Carlos Álvarez Antón; Paulino José García Nieto; Cecilio Blanco Viejo; José Antonio Vilán Vilán."Support Vector Machines Used to Estimate the Battery State of Charge". IEEE Transactions on Power Electronic, Year: 2013, Volume: 28, Issue: 12, Page s: 5919 – 5926.
[27] Shen, Y. (2018). Improved chaos genetic algorithm based state of charge determination for lithium batteries in electric vehicles. Energy, 152, 576–585.
[28] Jiahuan Lu ; Zeyu Chen ; Ying Yang ; Ming Lv, Online Estimation of State of Power for Lithium-Ion Batteries in Electric Vehicles Using Genetic Algorithm. IEEE Access, Year: 2018, Volume: 6, Page s: 20868 – 20880.
[29] Yan Ma ; Peng Duan ; Yanshuai Sun ; Hong Chen. Equalization of Lithium-Ion Battery Pack Based on Fuzzy Logic Control in Electric Vehicle. IEEE Transactions on Industrial Electronics Year: 2018 , Volume: 65 , Issue: 8, Page s: 6762 – 6771.
[30] Z. Wei, J. Zhao, R. Xiong, G. Dong, J. Pou and K. J. Tseng, "Online Estimation of Power Capacity With Noise Effect Attenuation for Lithium-Ion Battery," in IEEE Transactions on Industrial Electronics, vol. 66, no. 7, pp. 5724-5735, July 2019, doi:10.1109/TIE.2018.2878122.
[31] Hu, X.; Sun, F.; Zou, Y. estimation of state of charge of a lithium-ion battery pack for electric vehicles using an adaptive Luenberger observer. Energies 2010, 3, 1586–1603.
[32] Il-Song Kim. Nonlinear State of Charge Estimator for Hybrid Electric Vehicle Battery. IEEE Transactions on Power Electronics, Year: 2008 , Volume: 23 , Issue: 4, Page s: 2027 – 2034.
[33] Ning, B., Cao, B., Wang, B., & Zou, Z. (2018). Adaptive sliding mode observers for lithium-ion battery state estimation based on parameters identified online. Energy, 153, 732–742.
[34] Y. Feng, C. Xue, Q. Han, F. Han and J. Du, "Robust Estimation for State-of-Charge and State-of-Health of Lithium-Ion Batteries Using Integral-Type Terminal Sliding-Mode Observers," in IEEE Transactions on Industrial Electronics, vol. 67, no. 5, pp. 4013-4023, May 2020, doi: 10.1109/TIE.2019.2916389.
[35] J. Wang, B. Cao, Q. Chen, and F. Wang, "Combined state of charge estimator for electric

vehicle battery pack," Control Engineering Practice, vol. 15, no. 12, pp. 1569–1576, 2007.

[36] B. Xia et al., "State of charge estimation of lithium-ion batteries using optimized Levenberg-Marquardt wavelet neural network," Energy, vol. 153, pp. 694–705, 2018.

[37] X. Chen, H. Lei, and R. Xiong, "A Bias Correction Based State-of-Charge Estimation Method for Multi-cell Battery Pack under Different Working Conditions," IEEE Access, vol. 6, pp. 78184–78192, 2018.

[38] Y. Zheng, W. Gao, M. Ouyang, L. Lu, L. Zhou, and X. Han, "State-of-charge inconsistency estimation of lithium-ion battery pack using mean-difference model and extended Kalman filter," J. Power Sources, vol. 383, pp. 50–58, 2018.

[39] Y. Shen, "A chaos genetic algorithm based extended Kalman filter for the available capacity evaluation of lithium-ion batteries," Electrochim. Acta, vol. 264, pp. 400–409, 2018.

[40] J. Wang, R. Xiong, L. Li, and Y. Fang, "A comparative analysis and validation for double-filters-based state of charge estimators using battery-in-the-loop approach," Appl. Energy, vol. 229, no. 5, pp. 648–659, 2018.

[41] M. Shehab El Din, A. A. Hussein, and M. F. Abdel-Hafez, "Improved battery SOC estimation accuracy using a modified UKF with an adaptive cell model under real EV operating conditions," IEEE Trans. Transp. Electrif., vol. 4, no. 2, pp. 408–417, 2018.

[42] M. Hu, Y. Li, S. Li, C. Fu, D. Qin, and Z. Li, "Lithium-ion battery modeling and parameter identification based on fractional theory," Energy, vol. 165, pp. 153–163, 2018.

[43] N. Guo, Y. Fang, Z. Tian and S. Cao, "Research on SOC fuzzy weighted algorithm based on GA-BP neural network and ampere integral method," in The Journal of Engineering, vol. 2019, no. 15, pp. 576-580, 3 2019, doi: 10.1049/joe.2018.9385.

[44] Fangfang Yang, Shaohui Zhang, Weihua Li, Qiang Miao, State-of-charge estimation of lithium-ion batteries using LSTM and UKF, Energy, Volume 201, 2020, 117664, ISSN 0360-5442, https://doi.org/10.1016/j.energy.2020.117664.

Design and Simulation of Wind Turbine Blades

[1] Swapnil Kumar, [2] Dr. Sitaram Dash

[1] Student, School of Mechanical Engineering, Vellore Institute of Technology, Vellore.
[2] Professor, School of Mechanical Engineering, Vellore Institute of Technology, Vellore.

Email : swapnilsrivastava1997@gmail.com

Abstract. A Wind Turbine Blade has been modelled in Solidworks and its simulation implemented in ANSYS frameworks. Structural, fatigue and Computational Fluid Dynamics (CFD) analysis has been carried out to determine the performance of this blade. The analysis incorporates a blade velocity of 100 m/sec. High Lift to Drag Parameter has been considered for obtaining optimal performance of this blade. The lift-to-drag ratio happens to be a key design parameter for benchmarking such blades. k-ε turbulent model was adopted for our CFD based simulation. Fine **meshing** was carried out for improving accuracy of results.

Keywords: Computational Fluid Dynamics, Meshing, Structural, Velocity, Lift, Drag

1. Introduction

T Wolff and R J Seume (2016) have recently presented a paper on "Modeling the transient aerodynamic effects during the motion of a flexible trailing edge". In their research paper, they have developed an empirical model that takes into the account the transient effects while considering the trailing edge motion for the wind turbine. They have developed a wind turbine simulation considering all the load parameters. Main source of load is aerodynamic load that generally gets affected by relative velocity which occurs on the blade section. One of the solutions proposed in earlier studies by Troldborg and Wolf et. al. was a flexible trailing edge in proximity of the blade tip. They have conducted the simulations using 2D RANS for obtaining an optimal trailing edge. Due to this the stress variation was reduced on the blade roots concomitant with the dynamic forces acting in the vicinity of the trailing edges. Wind tunnel tests were carried out by Pechlivanoglou et. Al and Bak et. al. (2010) are reported in their paper "Active aerodynamic control of wind turbine blades with high deflection flexible flaps". The research findings illustrated direct variations of lift of wind turbine airfoil with trailing edge geometry. According to Troldborg, who has carried out the study on EllipSys 2D solver, it has been inferred that curved trailing edge should have relative length between 5% - 10% of chord length. Wolf et. al. on the other hand, have suggested that a 20% relative length should be considered optimal. In this Bergami (2013) has presented a paper on "Adaptive Trailing Edge Flaps for Active Load Alleviation in a Smart Rotor Configuration". He has developed an aerodynamic model while considering both steady state and dynamic effects. He has evaluated two different algorithms using his model in the aeroelastic

IOP Conf. Series: Materials Science and Engineering **937** (2020) 012047 doi:10.1088/1757-899X/937/1/012047

simulation code HAWC2. His simulation predicted a load alleviation occurring up to 30%. It has been found that modelled lift coefficient was consistent with the CFD results only for the cases where angle of attack is less than 4 degree. ($\alpha < 4°$). Achilles M. Boulamatsis et. al. (2018) have mentioned about regulating varying blade tip sweep using an active control. To minimize the aerodynamic and structural loads, the huge mass of the blades and relatively less energy production have always been considered as major parameters. Sanaa et.al have presented a paper on "Aerodynamics and Structural analysis of wind turbine blades". Finite element analysis involving static structural and aerodynamic analyses of Horizontal Axis Wind Turbine (HAWT) have been carried out in the research content. This research content provides us a clear view on design and simulation of wind turbine blades. The aerodynamic models have been coupled with a nonlinear formulation. From the paper, it has been concluded that reliability of wind turbine blades has been increased through the development of the air foil structure to calculate an optimum blade shape. It has been concluded that air foils lift and drag performance depicts the angles of twist and chord lengths in order to improvise aerodynamic performance. Swapnil et al (2020) have presented a paper on performance enhancement of wind turbine in which efficiency of Wind Turbine that has been optimized by changing the design parameters of wind turbine blade. Reduction of performance error was kept under consideration. Structural and vibrational simulation of wind turbine blades have been carried out in the research content. High lift to drag parameter has been optimized in the research content for better efficiency of wind turbine blades. Summer School held at Peter the Great St. Petersburg Polytechnic University, Russia, where the author (Swapnil) participated, has been a major source of technical information regarding horizontal axis wind turbine. Turbine rotor rpms are dictated by the generator synchronism speed for arriving at appropriate angle of attack at varying wind speeds. Wind Turbine blades are turned around their feathering axis and regulating the pitch (or blade) angle θ. The same mechanism allows to feather the blades, minimizing the aerodynamic load at extreme wind speeds. Variable speed turbines achieve best performance by allowing the rotor to work at the tip-speed-ratio λ of maximum Cp. They are also much quieter because rotation is slower at smaller wind speeds. Airfoils used in HAWT are generally cambered. The angle of attack is a function of angular position θ and the tip speed ratio λ.

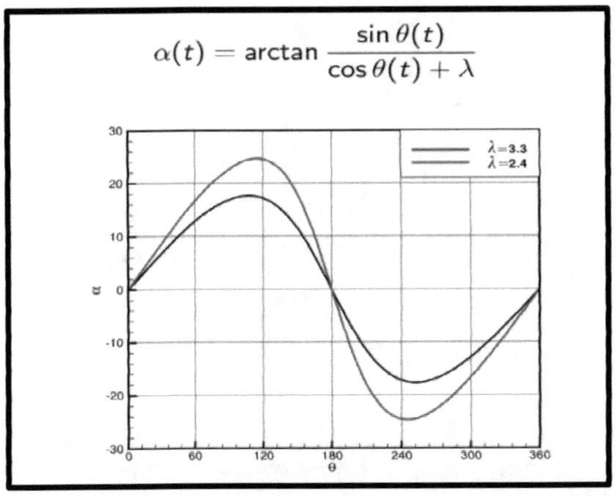

Figure 1. Angle of Attack with respect to angular position and tip speed ratio

2. Mathematical Model

Equation of Continuity and Navier-Stokes equation are the governing equations used in the frame of reference of the rotating blade.

Conservation of mass:

$$\frac{\partial \rho}{\partial t} + \nabla . \rho v_r = 0 \quad \textbf{(1)}$$

Conservation of Momentum (Navier-Stokes):

$$\nabla . (\rho v_r) + \rho (2\omega \times v_r \times \omega \times \omega \times r) = -\nabla p + \nabla . \tau_r \quad \textbf{(2)}$$

Where v(r) is the relative velocity and ω is the angular velocity

Turbulence Model: The k-epsilon Realizable

K-epsilon (k-ε) turbulence model has been used for Computational Fluid Dynamics (CFD) simulation.

For turbulent kinetic energy:

$$\frac{\partial(\rho k)}{\partial t} + \frac{\partial(\rho k u_i)}{\partial x_i} = \frac{\partial}{\partial x_j}\left[\frac{\mu_t}{\sigma_k}\frac{\partial k}{\partial x_j}\right] + 2\mu_t E_{ij} E_{ij} - \rho\varepsilon \quad \textbf{(3)}$$

For dissipation:

$$\frac{\partial(\rho \varepsilon)}{\partial t} + \frac{\partial(\rho \varepsilon u_i)}{\partial x_i} = \frac{\partial}{\partial x_j}\left[\frac{\mu_t}{\sigma_\varepsilon}\frac{\partial \varepsilon}{\partial x_j}\right] + C_{1\varepsilon}\frac{\varepsilon}{k} 2\mu_t E_{ij} E_{ij} - C_{2\varepsilon}\rho\frac{\varepsilon^2}{k} \quad \textbf{(4)}$$

3. Design Model

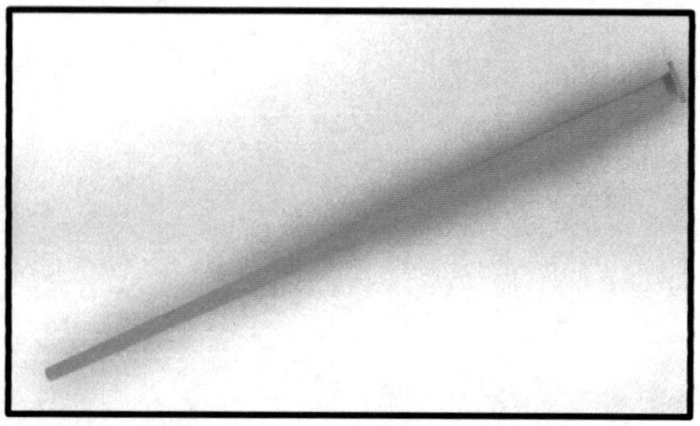

Figure 2. Design Model – Wind Turbine blade

IOP Conf. Series: Materials Science and Engineering **937** (2020) 012047 doi:10.1088/1757-899X/937/1/012047

4. Material Selection

Silicon material has been considered as the test material for wind turbine blade.

Table 1: Material properties of Silicon

S.No	Parameter	Values
1	Density (Kg/m^3)	2330
2	Young Modulus (GPa)	112.4
3	Poisson Ratio	0.28
4	Shear Modulus (GPa)	49

5. Meshing & Analysis

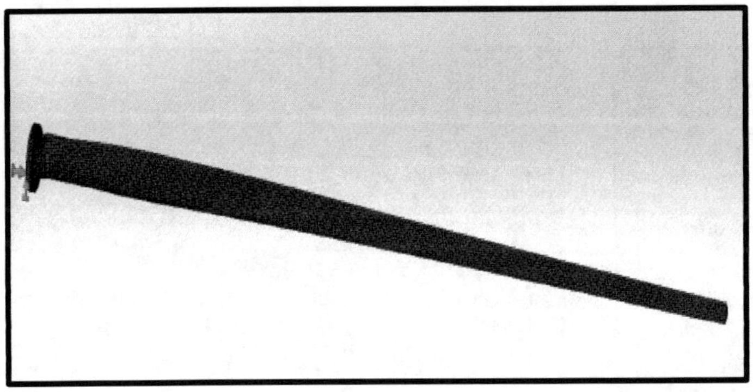

Figure 3. Meshing

Table 2: Meshing results

S.No	Parameter	Values
1	Nodes	101850
2	Elements	60770
3	Mesh Quality	High
4	Mesh Type	Solid

Arc length of wind turbine blade is 1.3 m.

Structural and Computational Fluid Dynamics analysis has been carried out by considering I.5 atm pressure and 100 m/sec velocity.

IOP Conf. Series: Materials Science and Engineering **937** (2020) 012047 doi:10.1088/1757-899X/937/1/012047

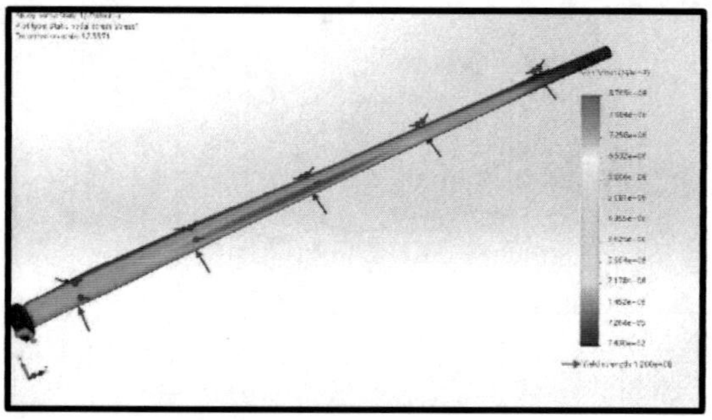

a) Equivalent Stress

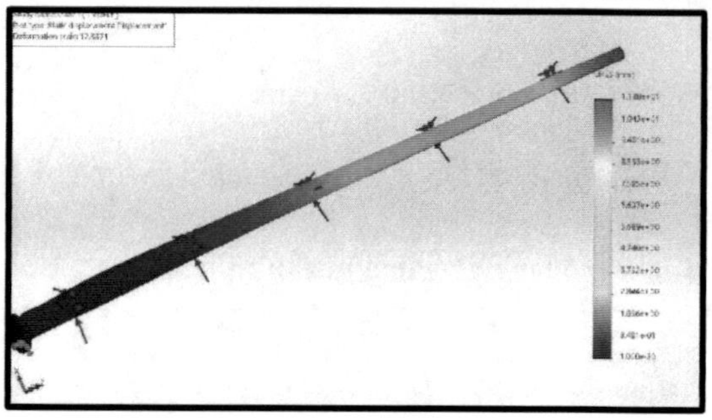

b) Displacement

Figure 4. Equivalent Stress and Displacement behavior for wind turbine blade

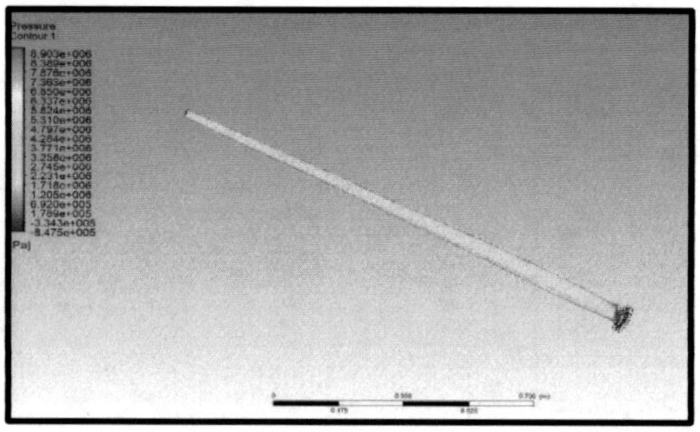

c) Inlet Pressure

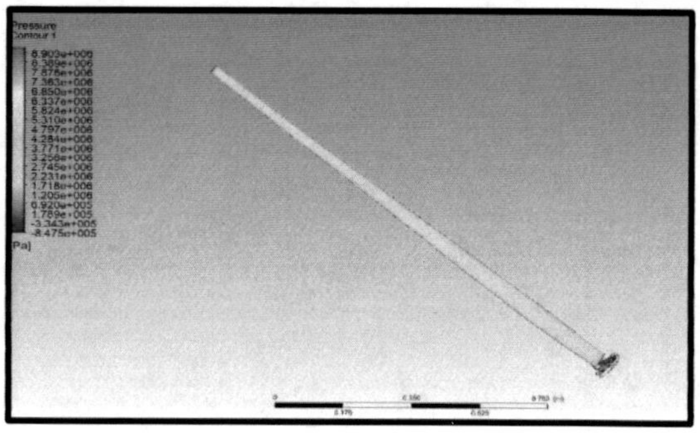

d) Outlet Pressure

Figure 5. Inlet and Outlet pressure behavior for wind turbine blade

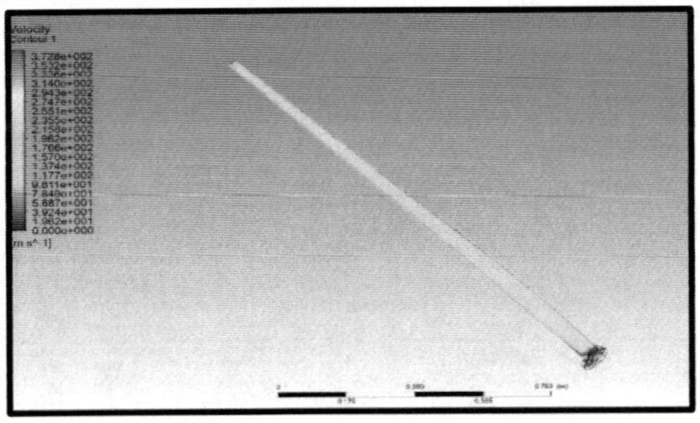

e) Inlet Velocity

f) Outlet velocity

Figure 6. Inlet and Outlet Velocity behavior for wind turbine blade

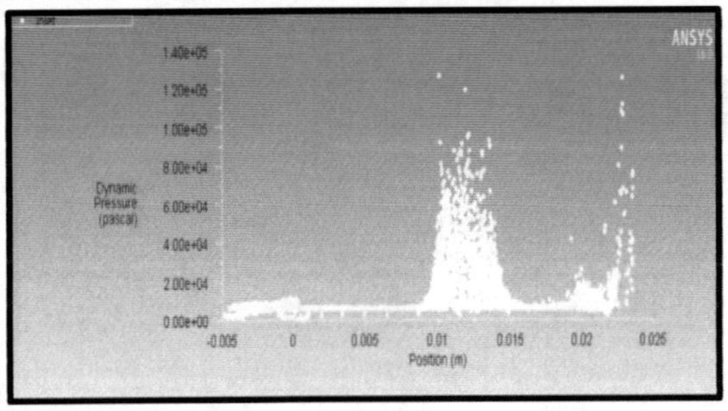

g) Dynamic Pressure for Inlet

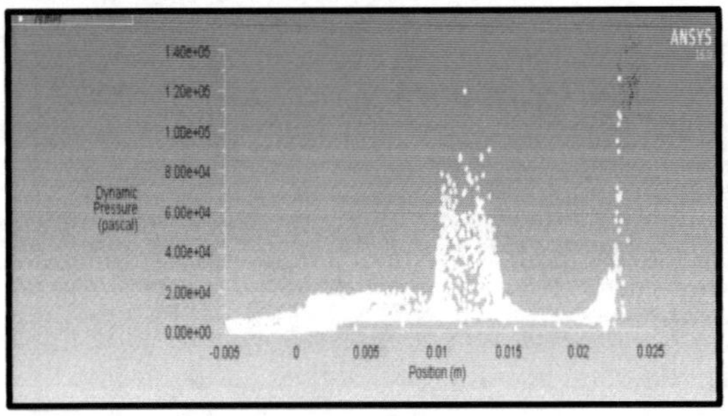

h) Dynamic Pressure for Outlet

Figure 7. Dynamic pressure behavior of Inlet and Outlet for wind turbine blade

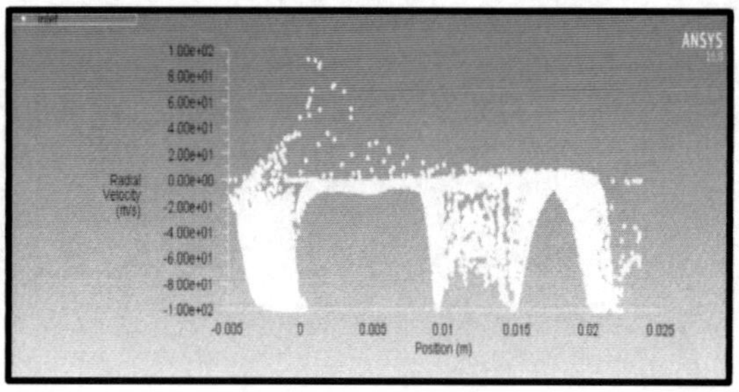

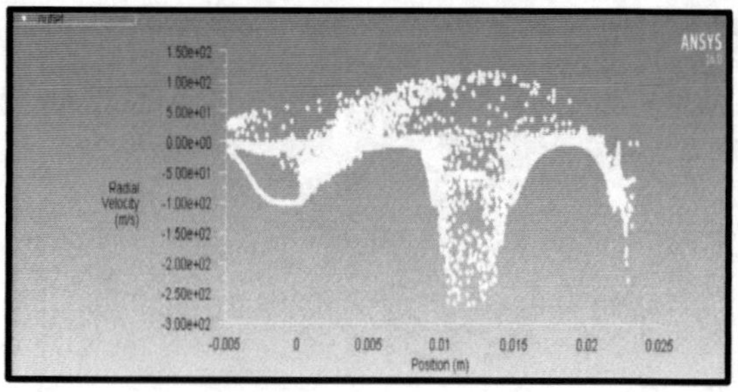

i) Radial Velocity for Inlet

j) Radial Velocity for Outlet

Figure 7. Radial Velocity behavior of Inlet and Outlet for wind turbine blade

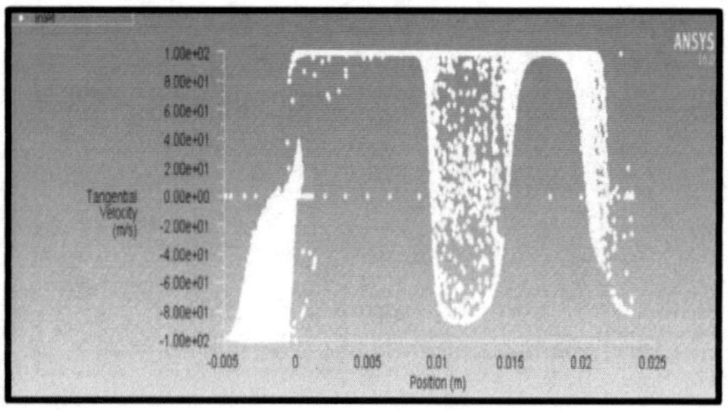

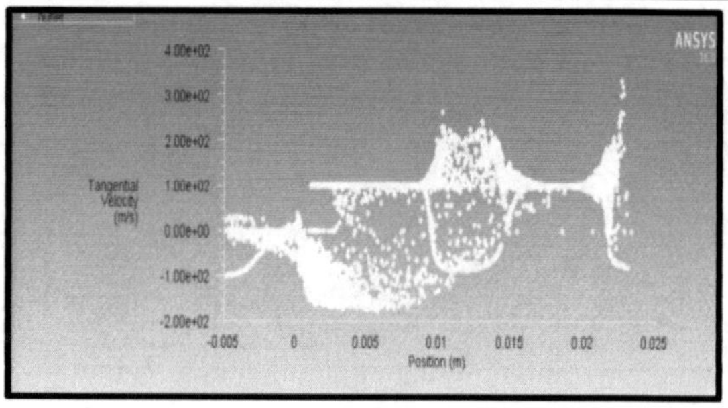

k) Tangential Velocity for Inlet

l) Tangential Velocity for Outlet

Figure 7. Tangential Velocity behavior of Inlet and Outlet for wind turbine blade

6. Results & Discussion

Designing of CAD modelling has been carried out in Solidworks. Structural and Computational Fluid Dynamics (CFD) analyses have been carried in Solidworks and ANSYS, respectively. Stress and deformation contours are observed for wind turbine blades. Inlet and Outlet velocities and pressure behaviour could be deciphered using k-ε turbulent model. Inlet velocity has been considered as 100 m/sec as an input parameter. Fine meshing was considered for improving simulation accuracy. Second Order has been considered for ANSYS fluent simulation.

7. Conclusion

Fine Meshing was carried out for simulation of deformation response in wind turbine blades. Maximum equivalent stress induced in the wind turbine blades was deduced as 8.71e+06 from our studies. Inlet and Outlet velocities and pressure contours were exhibited in CFD simulation. High Lift to Drag Parameter was considered for optimum performance of Wind Turbine blades. Blades have been designed for Horizontal Axis Wind Turbines (HAWT). Dynamic Pressure, Radial Velocity and Tangential Velocity vs position plots were derived by executing ANSYS Fluent.

References

[1] T Wolff and J R Seume (2016) J. Phys. Modeling the transient aerodynamic effects during the motion of a flexible trailing edge .

[2] Sanaa El Mouhsine, Karim Oukassou, Mohammed Marouan Ichenial, Bousselham Kharbouch, Abderrahmane Hajraoui (2018) Aerodynamics and structural analysis of wind turbine blades, Elsevier, Proceedings in Manufacturing.

[3] Pechlivanoglou G K, Wagner J, Nayeri C N and Paschereit C O 2010 Active aerodynamic control of wind turbine blades with high deflection flexible flaps 48th AIAA Aerospace Sciences Meeting (Orlando, Florida) AIAA 2010-644.

[4] Bergami L 2013 Adaptive Trailing Edge Flaps for Active Load Alleviation in a Smart Rotor Configuration DTU Wind Energy PhD-0020(EN)

[5] Swapnil (2020) Perfromance Enhancement of wind turbine blades, International Journal of Engineering and Advanced Technology.

[6] Summer School, Peter the Great St.Petersburg Polytechnic University, Russia.

Modelling and control of electric scooter driven by induction motor

Aishwarya M[1*], R M Brisilla[2]
School of Electrical Engineering
Vellore Institute of Technology, Vellore, TamilNadu
*Email: aishu25mannan@gmail.com

Abstract. This paper presents the modelling and control of electric scooter driven by single-phase Induction motor drive. Single-phase induction motor is employed for this design due to lower cost and little maintenance, although a complex inverter is needed for the vehicle drive system. This paper proposes an integral sliding motor controller and pole placement controller to improve the overall performance of the electric scooter system. The stability of the system using the proposed control strategies are validated with and without disturbances. The proposed controller is validated through MATLAB simulations.

Keywords: Electric Scooter, Single-phase Induction Motor, Speed control, Integral sliding mode controller, and pole placement control

1. Introduction

Depletion of fossil fuels, advancements in power converter technologies, and improved battery charging and management technologies has brought electric vehicles back. Electric vehicles are proven to be one of the promising solutions for the global environmental problem caused by air pollution [1]. The usage of a two-wheeled vehicle is predominant among city commuters due to the increase in personal transportation. Besides, these vehicles have a small footprint [2]. The traditional internal combustion engines in these vehicles produce local emissions and also poor in efficiency.

Electric scooters are becoming popular due to their higher efficiency with zero local emissions. The wide diffusion of electric vehicles demands battery charging infrastructure with fast charging capability [3] and optimal assignment of chargers [4]. Also, the range of electric vehicles depends on the capacity of the batteries. The average usage of urban vehicles is 25 km and it may go up to 40 km and hence extra battery packs can be added to increase the range which increases the cost. Instead, energy management is a more reasonable solution [5]. In [6], a contactless energy transfer system is proposed for electric scooters whereas in [7] dynamic battery swapping technique is proposed as a battery management technique. Due to the advancements in power converters, electric motors such as brushless DC motor, switched reluctance motor and induction motor are used in electric vehicles. Selection of proper electrical propulsion system with optimal energy management and optimal control strategies are important for an electric vehicle. In [8] design of an in-wheel permanent magnet, brushless DC motor is proposed for electric scooter whereas in [9] axial-flux switched reluctance motor for e-scooter is presented. In [10] speed limiter is designed for an electrically assisted kick scooter.

Induction motors have been increasingly applied in electric vehicles due to their remarkable advantages such as simple structure, high reliability, low cost, and high ruggedness. The dynamic model of induction motor is nonlinear and nonlinear control techniques such as direct torque control [11, 12, 13], backstepping, sliding mode control [14] are designed to achieve the high-performance control in electric vehicles. In [15] Lagrangian model of induction motor is used for the torque control in electric vehicles whereas in [16] speed controller is designed using model reference adaptive control. In [17] field-oriented control of multiphase induction motor is proposed. In [18], fault-tolerant cruise control of the induction motor is presented. This paper is organized as follows. Section 2 briefs about the equivalent circuit of induction motor. Section 3 details the design criteria considered for electric scooter.

Section 4 briefs about mathematical modelling of electric vehicle system while section 5 elaborates the design of control techniques. Section 6 depicts the simulation results and finally, section 7 gives some concluding remarks.

2. Motor design

The per-phase equivalent circuit of an induction motor concerning the stator side is given in 'figure 1'. The speed of the stator revolving flux is given by $Ns=\frac{120f}{P}f$; where f is the frequency in Hz and P is the number of poles.

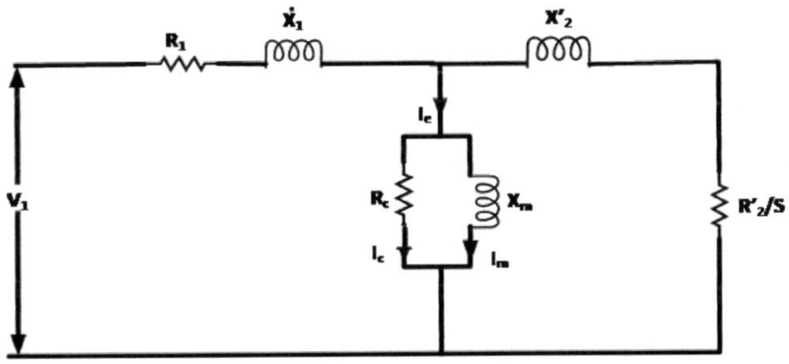

Figure 1. Per-phase Equivalent circuit of Induction Motor

Where,

R_1= Resistance of stator

X_1=Stator leakage reactance

I_1=Stator Current

R_C=Shunt branch resistance

X_m=Magnetizing reactance

I_e=Per-phase no-load current

I_2=Rotor Current

I_2'=Rotor current per phase referred to stator

X_2'=Standstill rotor reactance referred to stator

R_2'=Rotor resistance referred to stator

V_1=Stator voltage

E_1=Stator induced emf

S=Slip

This paper details the development of controllers to regulate the speed of induction motor used in an electric scooter. A single phase 4 pole induction motor which is commonly used in low power applications is considered for the design of the electric scooter. The system is modelled using a state-space approach. The design parameters of the electric scooter are as follows; Total mass is 80kg, Maximum speed withstand is 20Km/hr, acceleration time is 0.5s. the effective wheel diameter is 0.2m.

IOP Conf. Series: Materials Science and Engineering **937** (2020) 012048 doi:10.1088/1757-899X/937/1/012048

The motor is fed by an inverter of 230 V ac supply. The motor parameters and the electric scooter parameter required for the EV is given in Table 1.

Table 1. Parameters for EV system

Parameters	Value
V_{RMS}-inverter Voltage	230V
Hz-Supply frequency	50Hz
P-Number of poles	4
J-Moment of inertia	0.01 N-ms^2/rad
Motor rating	0.26kW
M-Mass	80kg
r_{eff}-Effective wheel radius	0.1m
a-Acceleration constant	9.8m/s^2
R_a-Resistance	3.9Ω
L_a-Inductance	0.285H
C_d-Drag coefficient	0.2
ρ-Air density	0.129
A_r-Cross-sectional area of the vehicle	0.3m^2

3. Design criteria's for Dynamic vehicle model

During the normal operating condition, the electric scooter has to overcome the disturbances such as gradient, friction, and wind (aerodynamic drag) resistance[4]. The system modelling is done by considering all the disturbances and the exploded view of an electric scooter as shown in 'figure 2'

3.1. Friction force

The friction force is an opposing force existing between the wheel and road that restricts the vehicle movement. This force can be calculated as,

$$F_f = Mgf_r$$

$$F_r = 0.01\left(1 + \frac{v}{447.09}\right)\ rad/s$$

Where M is the mass and g is the acceleration gravity.

3.2. Gradient force

When the vehicle climbs up/down a hill only the gradient force will exist. The direction of movement is projected on the component of vehicle weight.

$$F_g = Mgsin(\alpha)$$

Where M is the mass, g is the acceleration gravity and α is the road gradient.

IOP Conf. Series: Materials Science and Engineering **937** (2020) 012048 doi:10.1088/1757-899X/937/1/012048

3.3. Aerodynamic drag force

Aerodynamic drag force is one of the speed limiting factors of a vehicle and it depends on the shape of the vehicle body or its speed.

$$F_a = 0.5C_D A_r \rho v^2$$

Where ρ is the mass density of air, A_r is the cross-sectional area of the vehicle, C_D is the Drag coefficient.

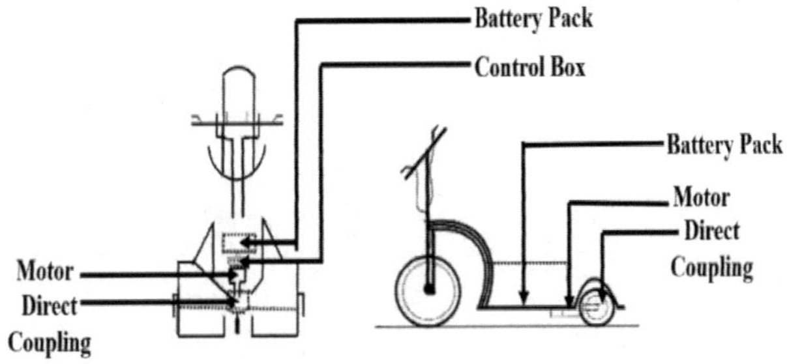

Figure 2. E-scooter

4. Mathematical modelling of an electric vehicle system

For designing the EV system under dynamic environmental condition, the following equations are considered:

The mathematical modelling of IM is obtained by the physical equations governing the system [5]

The electrical equation of the machine is,

$$V_a = I_a R_a + JX_a I_a + E_b \tag{1}$$

The induced back emf E_b,

$$E_b \propto \frac{d\theta}{dt} \tag{2}$$

Applying Laplace transforms for (1) and (2),

$$V_a(s) = I_a(s)(R_a + JX_a) + E_b(s) \tag{3}$$

$$E_b = k_t s\theta_s \tag{4}$$

$I_a(s)$ is derived by substituting (4) in (3),

$$V_a(s) = I_a(s)Z_s + k_t s\theta_s$$

$$I_a(s) = \frac{V_a(s) - k_t s\theta_s}{Z_s} \tag{5}$$

Electrical torque of IM is given by,

$$T_e = \frac{kE^2R}{R^2 + x^2} \tag{6}$$

Taking Laplace of equation (6),

$$T_e = E^2\sin\theta \tag{7}$$

Mechanical torque is given by,

$$T_m = J\ddot{\theta} + B\theta \tag{8}$$

Where $B = r_{eff}(F_f + F_g + F_a)$whereas, $\theta = s$; $\ddot{\theta} = s^2$

Taking the Laplace transform of equation (8),

$$T_m = Js^2\theta(s) + r_{eff}(F_f + F_g + F_a)s\theta(s)$$

$$T_m = (Js^2 + Bs)\theta(s) \tag{9}$$

Combining both electrical and mechanical torque equations, the speed equation of EV is derived and it is detailed as follows

$$T_e = T_m = E^2\theta(s) = (Js^2 + Bs)\theta(s)$$

$$T_m = \left((Js^2 + Bs) - E^2\right)\theta(s) \tag{10}$$

$$\frac{\theta_s}{T_m} = \frac{1}{Js^2 + Bs - E^2} \tag{11}$$

Also, the torque equation is represented as,
$T \propto \varphi I_a \cos\theta$
we know that,

$$T = k\varphi I_a \tag{12}$$

Following the torque determination, the transfer function of EV is obtained by considering

Taking the Laplace transform of equation (12),

$$T = k_a\theta(s)I_a(s) \tag{13}$$

Applying equation (5) & (10) in (13)

$$T = ka\left(\frac{V_a(s) - k_t s\theta(s)}{Z(s)}\right) = (Js^2 + Bs - E^2)\theta(s)$$

$$k_a V_a(s) = (Js^2 + Bs - E^2)\theta(s)Z(s) + k_t k_a s\theta(s) \tag{14}$$

$$\frac{V_a(s)}{\theta(s)} = \frac{J^2Bs - E^2\theta(s)Z(s) + k_t k_a s}{k_a} \tag{15}$$

$$\frac{\theta(S)}{V_a(S)} = \frac{K_a}{(JS^2 + BS - E^2)Z(S) + k_a k_t S}$$

$$\frac{\theta(S)}{V_a(S)} = \frac{K_a}{(JS^2 + BS - E^2)(R_a + L_a S) + k_a k_t S} \tag{16}$$

$$\frac{\theta(S)}{V_a(S)} = \frac{K_a}{\left(JS^2 + r_{eff}\left(F_f + F_a + F_g\right)S - E^2\right)(R_a + L_aS) + k_ak_tS} \tag{17}$$

$$\frac{\theta(S)}{V_a(S)} = \frac{K_a}{\left(JL_aS^3 + L_ar_{eff}\left(F_f + F_a + F_g\right) + R_aJS^2\right) + \left(R_ar_{eff}\left(F_f + F_g + F_a\right) + k_ak_t\right)S} \tag{18}$$

$$\frac{\theta(S)}{V_a(S)} = \frac{K_a}{(JL_aS^3 + a1S^2 + a2S)} \tag{19}$$

Equation (19) gives the transfer function model of an electric scooter. The differential equation governing the system is obtained from (19) as follows

$$(JL_aS^3 + a1S^2 + a2S)\theta(S) = K_aV_a(S) \tag{20}$$

$$JL_as^3\theta(s) + a_1s^2\theta(s) + a_2s\theta(s) = k_aV_a(s)$$

$$JL_a\dddot{\theta} + a_1\ddot{\theta} + a_2\dot{\theta} = k_aV_a(s) \tag{21}$$

Defining, $\dddot{\theta} = \dot{Z}_2;\ \ddot{\theta} = \dot{Z}_1;\ \dot{\theta} = Z_1$

The second-order state equations governing the system (19) is given by

$$\dot{Z}_1 = Z_2\ ;\ \ \dot{Z}_2 = \frac{-a_1Z_2 - a_2Z_1 + k_a}{JL_a} \tag{22}$$

Where, $\dot{Z}_2 = \text{speed}\ ;\ \dot{Z}_1 = \text{acceleration}$

The second-order system equations in standard form are given as,

$$\begin{bmatrix}\dot{z}_1\\ \dot{z}_2\end{bmatrix} = \begin{bmatrix}0 & 1\\ \frac{-a_2}{JL_a} & \frac{-a_1}{JL_a}\end{bmatrix}\begin{bmatrix}z_1\\ z_2\end{bmatrix} + \begin{bmatrix}0\\ \frac{k_a}{JL_a}\end{bmatrix}u \tag{23}$$

5. Control Techniques

This section details the control design procedure for the speed regulation of electric scooter using pole placement and integral sliding mode control.

5.1. Pole placement controller

The pole placement technique is based on classical control theory [19], in which poles of the closed-loop system are placed at the desired location to obtain stability and desired performance.

$$\dot{x} = A_x + B_u\,, x(0) = x_0$$

Using equation (23), the system matrices A and B are obtained based on the parameters in Table (1) as follows,

$$[A] = \begin{bmatrix} 0 & 1 \\ -3.333 & -13.684 \end{bmatrix}$$

$$[B] = \begin{bmatrix} 0 \\ 0.5415 \end{bmatrix}$$

Using pole placement controller the controller gains K_1, K_2 are obtained as $K_1 = 86.181$ and $K_2 = 2.430$.

Hence, Va=-86.181*(z1-z1des)-2.4303*(z2) (24)

5.2. Sliding mode control

The sliding mode control (SMC) is one of the robust nonlinear control techniques commonly used in uncertain systems due to its ease in design and the ability to discard uncertainties parameter. It can alter the dynamics of a nonlinear system utilizing a switching control. It is designed in two steps; 1. Design of a surface called a sliding surface; 2. Design of a discontinuous control. The design of the surface is such that once the state trajectory is considered on it, the controlled system will exhibit the desired performance[20]. The discontinuous control forces the system to reach the surface in a finite time. Once the system trajectory reaches the surface, it slides along the surface to reach the origin as shown in 'figure 3'.

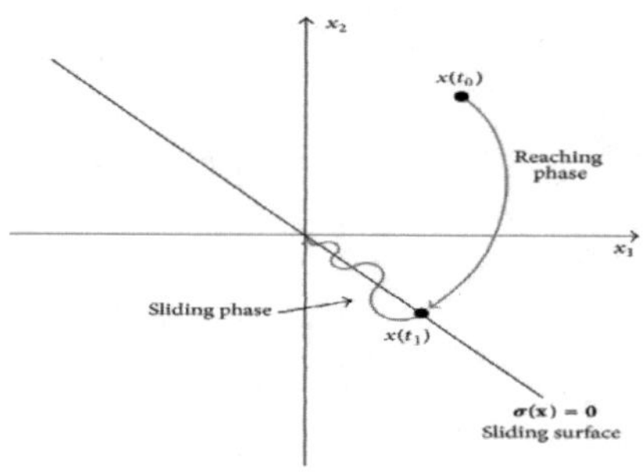

Figure 3. State trajectory of a second-order system under SMC[21]

5.2.1 ISMC design for E-scooter. ISMC is an advanced version of SMC which gives a fast dynamic response to the system during external disturbances [22]-[25]. The ISM control method can ensure less chattering and also provide compensation for both matched and unmatched disturbances.

The sliding surface is defined as

$$s = c_1 x_1 + c_2 x_2 + c_3 \int (x_1 - x_{1ref})$$

Where, $x_1 = \text{Speed}$; $x_2 = \text{acceleration}$;

$$\dot{s} = c_1 \dot{x}_1 + c_2 \dot{x}_2 + c_3 (x_1 - x_{1ref})$$

Where, $\dot{x}_1 = z_1$; $\dot{x}_2 = z_2$

This c_1, c_2, c_3 are the Surface constant and gain "k" are presented in the "Integral Sliding mode controller" for both with and without disturbance. The values for without disturbance c1 is 1 and the with disturbance c_1 is 3, then c_2 and c_3 values are represented as 0.5 and 0.001. The gain k value for without disturbance is 500 while with disturbance condition to achieve the expected speed to increase the gain value as 1000.

The second-order equation,

$$-\mathrm{ksign}(s) = c_1 z_1 + c_2\left(-\frac{a_2}{JL_a}z_1 - \frac{a_1}{JL_a}z_2 + \frac{k_a}{JL_a}V_a\right) + c3(x_1 - x_{1ref})$$

$$V_a = \frac{-\mathrm{ksign}(s) - c_1 z_1 - c_2\left(-\frac{a_2}{JL_a}z_1 - \frac{a_1}{JL_a}z_2\right) - c_3 * (z_2 - z_{2ref})}{c_2\frac{k_a}{JL_a}} \tag{25}$$

Whereas, $\dot{Z}_2 = \mathrm{speed}\,;\ \dot{Z}_1 = \mathrm{acceleration}$

6. Simulation results

The proposed control techniques are validated by numerical simulations in MATLAB. The results obtained are given in this section. The simulation results for pole placement controller are given in 'figure 4' and 'figure 5'. The simulation results for sliding mode controller are given in 'figure 6' and 'figure 7'. The reference speed is chosen as 20km/hr. 'Figure 5' and 'figure 7' shows that, speed can be maintained to 19.1Km/hr for pole placement and 19.8Km/hr for ISMC.

6.1 Pole placement Controller

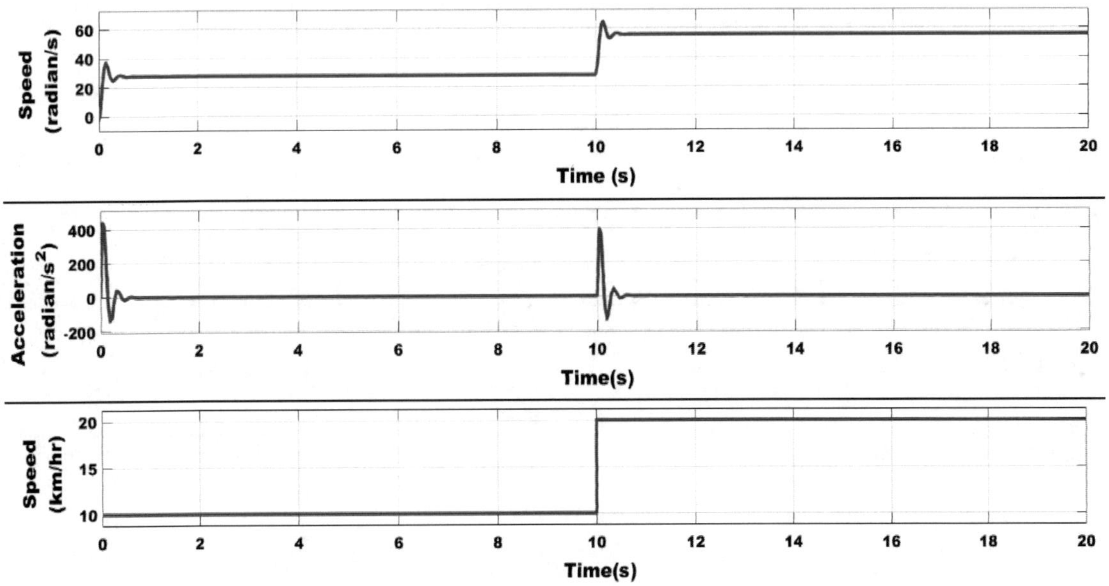

Figure 4. Without disturbance

IOP Conf. Series: Materials Science and Engineering **937** (2020) 012048 doi:10.1088/1757-899X/937/1/012048

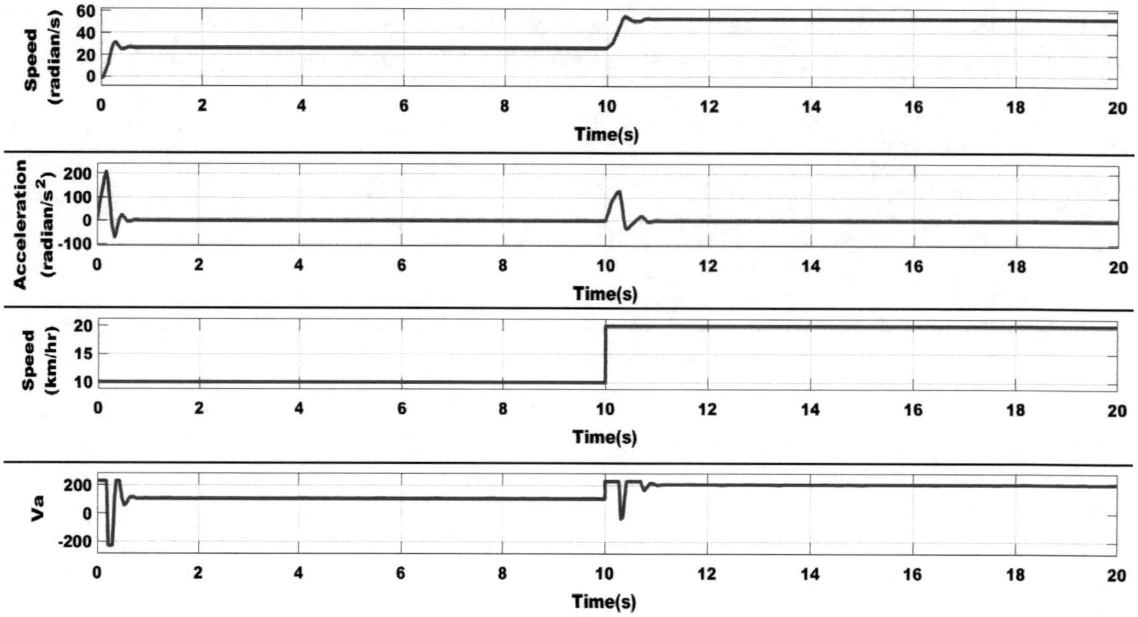

Figure 5. With disturbance

6.2 ISMC Controller

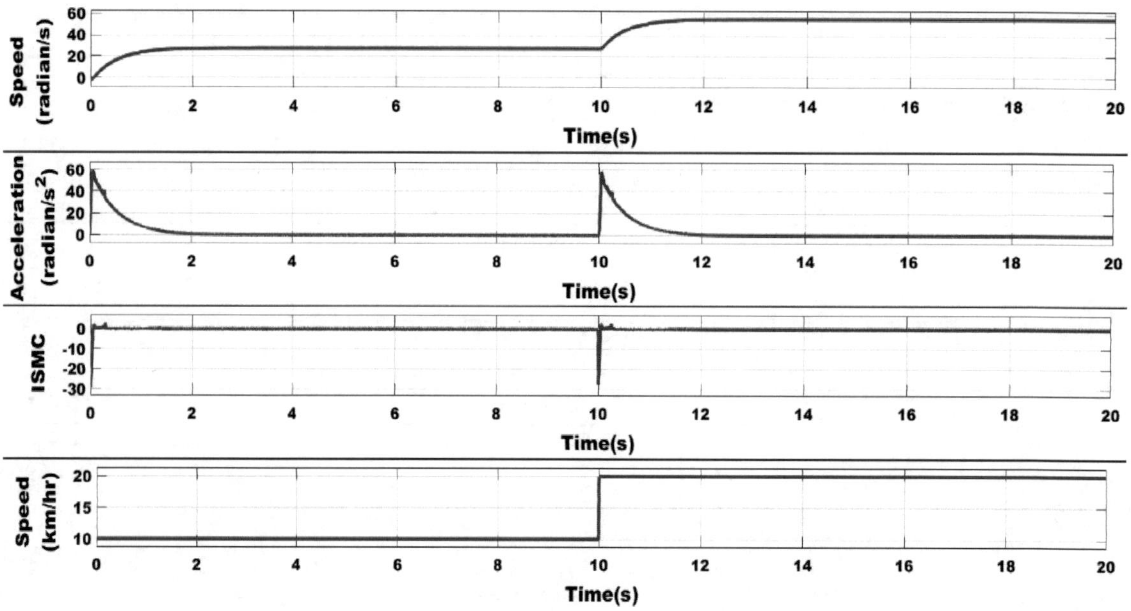

Figure 6. Without disturbance

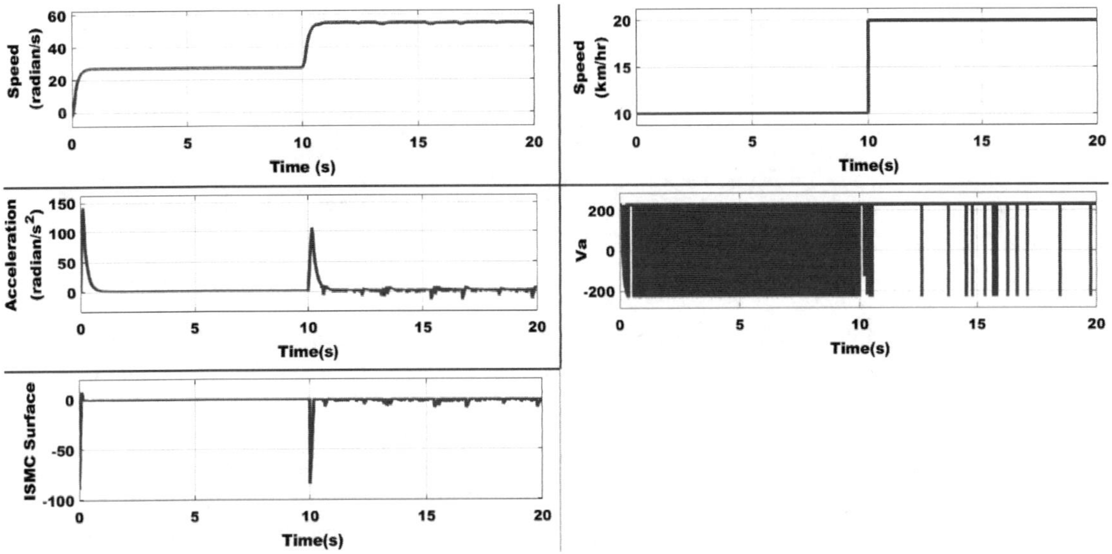

Figure 7. With disturbance

Comparative analyses of the results show that ISMC (Integral sliding mode controller) gives better performance when compared to the pole placement system.

7. Conclusion

In this paper, modeling and control of electric vehicle driven by a single-phase induction motor has been presented. Simulations have been performed and results validate the effectiveness of proposed nonlinear control. The analysis shows ISMC system design provides a comparatively better response than the pole placement controller. Also, it is observed that the ISMC design offers a perfect speed range even under external disturbances. Hence, it is evident from the simulation results that the Single-phase induction motor can be utilized to drive the electric scooter with good performances.

Reference

[1] Chan C C An overview of electric vehicle technology 1993 *Proc. of the IEEE* **81** 1202.

[2] Lee S H, Kim T H and Ahn J W *Electric powertrain system for E-scooter with rear two in-wheeled propulsion motors* 2013 *Int. Conf. on Electrical Machines and Systems* (Busan: South Korea) p 1491.

[3] Kaneko K, Omori H, Kimura N, Morizane T and Nakaoka M *A novel type of EDLC electric motor driven scooter with pulse super-rapied charger* 2015 *Int. Conf. on Electrical Drives and Power Electronics* (Tatranská Lomnica: Slovakia) p 476.

[4] Masoud M, Elhenawy M, Almannaa M H, Liu S Q, Glaser S and Rakotonirainy A *Optimal assignment of e-scooter to chargers* 2019 *Intelligent Transportation Systems Conf.* (Auckland: New Zealand) p 4204.

[5] Alli G, Formentin S and Savaresi S M *A range-bounding strategy for electri scooters* 2012 *Int. Electric Vehicle Conf.* (Green ville: USA) p 1.

[6] Kumar K, Mude K N, Santra S B, Dutta S and Banerjee S *Analysis and design of contactless energy trafer system for e-scooter* 2017 *14th IEEE India Council Int.Conf.* (Roorkee: India) p 1.

[7] Tai N C, Wu D and Jang H W *Extending the range of electric scooters from fixed-point battery stations by using a dynamic battery swapping mobile application* 2018 *Int. Conf. on Applied System Innovation* (Chiba: Japan) p 389.

[8] Ravi N, Ekram S and Mahajan D *Design and development of a in-wheel brushless DC motor drive for an electric scooter* 2006 *Int. Conf. on Power Electronic, Drives and Energy Systems* (New Delhi: India) p 1.

[9] Andrada P, Martínez E, Blanque B, Torrent M, Perat J I and Sanchez J *A New axial flux switched reluctance motor for E-scooter* 2016 *Int. Conf. on Electrical Systems for Aircraft, Railway, Ship Propulsion and Road Vehicles & Int. Transportation Electrification Conf.* (Toulouse: France) p 1.

[10] Brankovic A, Berretta D, Formentin S, Corno M and Savaresi S M *Modeling and speed limitation control of an electric kick scooter* 2015 *European Control Conf.* (Linz: Austria) p 697.

[11] User Y and Tabanli N *Development of induction motor torque control algorithm for electric vehicles on inclined road* 2017 *Int. Conf. on Electromechanical and Power Systems* (Iasi: Republic of Moldova) p 472.

[12] Raj A and Sharma *R Improved direct torque control for induction motor in electric vehicle application* 2018 *IEEE 8th Power India Int. Conf.* (Kurukshetra: India) p 1.

[13] Tabbache B, Kheloui A and Benbouzid M.E.H *Design and control of the induction motor propulsion of an electric vehicle* 2010 *Vehicle Power and Propulsion Conf.* (Lille: France) p 1.

[14] Jezernik K *Robust induction motor control for electric vehicles* 1996 *Proc. of 4th Int. Workshop on Advanced Motion Control-AMC'96-MIE.* (Mie: Japan) p 436.

[15] Pei W *Lagrangian modelling and passivity based control of induction motors for electric vehicles* 2018 *IFAC-Papers OnLine* **51** 499.

[16] Pal A, Kumar R and Das S *Sensorless speed control of induction motor driven ekectric vehicle using model reference adaptive controller* 2016 *Energy Procedia* **90** 540.

[17] Iffouzar K, Amrouche B, Cherif T O, Benkhoris M F, Aouzellag D and Ghedamsi K Improved direct field oriented control of multiphase induction motor used in hybrid electric vehicle application 2017 *Int. J.of Hydrogen Energy* **42** 19296.

[18] Marino R, Scalzi S, Tomei P and Verrelli C M *Fault-tolerant cruise control of electric vehicles with induction motors* 2013 *Control Engineering Practice* **21** 860.

[19] Brasch F and Pearson J R Pole placement using dynamic compensators 1970 *IEEE Trans. on Automatic Control* **15** 34.

[20] Brisilla R M and Sankaranarayanan V *Extended State Observer Based Sliding Mode Control of Permanent Magnet DC motor* 2015 *Annual IEEE India Conf.* (**New Delhi: India**) p 1.

[21] Edwards, Christopher and Sarah S 1998 *Sliding mode control: theory and applications* CRC Press

[22] Abidi K, Xu J X and Yu X On the discrete time integral sliding mode control 2007 IEEE Trans. Automatic Control **52** 709.

[23] Liang Y W, Ting L W and Lin L G Study of reliable control via an integral-type sliding mode control scheme 2012 *IEEE Trans. Ind. Electronics* **59** 3062.

[24] Xu J X, Guo Z Q and Lee T H Design and implementation of integral sliding-mode control on an underactuated two-wheeled mobile robot 2014 *IEEE Trans. Ind. Electronics* **61** 3671.

[25] Zhang J, Liu X, Xia Y, Zuo Z and Wang Y Disturbance Observer Based Integral Sliding Mode Control for Systems with Mismatched Disturbances 2016 *IEEE Trans. on Ind. Electronics* **63** 7040.

IOP Conf. Series: Materials Science and Engineering **937** (2020) 012049 doi:10.1088/1757-899X/937/1/012049

An Integrated IoT LabVIEW Based Fuzzy-PLC Controller for Automation of Boiler

S Boobalan [1], S Gobinath[2], A Dharmalingam[3], A Jayaprakash [4], M Poovarasan [5], K Poovendan [6]

Assistant Professor, Department Electrical and Electronics Engineering, Sri Krishna College of Engineering and Technology[1,2]

UG Scholar, Department Electrical and Electronics Engineering, Sri Krishna College of Engineering and Technology[3,4,5,6]

Abstract. The proposed prototype, to develop and automate and controlling of steam outlet from the boiler unit in the Power Plants using Fuzzy logic-based PLC Controller. The set point of the steam outlet could be decided by the Fuzzy Logic Controller. Automation is a growing technology in almost all fields and It reduces the man power and time complexity in all the way to the thermal power plant also. It reduces the risk of making mistakes and avoid any danger. The details of the load is collected from P.T and then given to the LabVIEW. The data about the load requirements, peak load data requirements ae once fed into the LabVIEW. The LabVIEW sends signals to the PLC and steam outlet is controlled by Servo Motor. The steam outlet set point is measured and controlled by fuzzy logic with the help to PLC controller. Based on alternator current and voltage signal the fuzzy logic will decide the steam outlet set point. If in case the load requirement is higher or lower, the changes in the coal input and the water requirement are changed by means Fuzzy logic control. The steam valve sensor closes or opens the valve by servo motor using the variable frequency drive that is connected to the PLC. Depending upon the load, to the steam turbine. The PLC is connected to the ethernet by means in LabVIEW, in which the data are collected are given to the authorized person in the control room. If in case required the data can be used to store the data in the cloud for future use.

Index Terms— Boiler Automation, PLC Controller, HMI, LabVIEW, Fuzzy Logic Controller, IOT, ThingSpeak.

I. INTRODUCTION

The market for automated systems that can work more proficiently has increased to deliver better quality products. In the case of a steam boiler factory, proper production control and proper temperature monitoring are frequently needed. Therefore, manual plant management turned out to be a laborious one for the job. The suggested concept design of boiler automation was introduced to achieve the goals identified with the automation control system: creation of processes, enhancement of management level and high-performance method.

The boiler automation systems are very necessary to keep human employees from making mistakes. A variety of reports on the boiler automation system[1] have been performed since. Automation was performed in the manufacturing field using different techniques. T. Karuppiah and Dr. Azha built a framework for power plant boiler automation by using microcontroller and GSM technology[2]. The temperature of the device in their proposed design is regulated by alternator production. The central water tank has an ultrasonic sensor for monitoring water pressure. In every boiler pipe there is a certain temperature sensor which controls the machine temperature. The GSM module is used to give details to the user. Udaykumar S. Kulkarni & Venkat N. Ghodke have suggested a boiler automation system utilizing FPGA & GSM[3]. A principal water tank is attached to many boilers in their planned configuration. Water level sensor & temperature controls are used for monitoring main tank water level & boiler temperature. GSM cell phone is used to provide present temperature knowledge in every boiler. VHDL (VHSIC Hardware Definition Language) is used and introduced in Field Programmable Gate Array (FPGA) equipment. Aaron James et. Al . suggested a boiler automation framework for a power station utilizing a distributed control system (DCS)[4]. The device is typically used up to about as a water tube boiler. Three hundred bar or around as a strong steam boiler. Thirty windows. With the assistance of DCS, the steam production from the steam header covers the required temperature by means of a good temperature sensor. The roles have not yet been explicitly established for this framework, however.

This research addresses the concept and testing of an integrated programmable logical controller (PLC) for the fuzzy logic system. The chosen system is part of the IQ-F MITSUBISHI sequence, while code in Ladder Diagram has been created. The fuzzy controller is of the form Mamdani and is used for servomotor operation. The validation of the built software module and its viability to handle this form of control algorithm are contrasted with a LabVIEW fuzzy controller. Fuzzy means "not visible, distinct or true, blurred." A kind of depiction of information relevant to notation that can not be strictly specified but relies on its meaning. A fugitive logic controller imitates the way information is managed by human operators.

A Fuzzy adds a series of control laws that can vary and sometimes contradict each other to the simulation. A sequence of answer rules developed by the professional operator[5] are used by the fluid logic control[6]. The quantizer moves the knowledge from the sensor to a way which can be used by the fluid logic system.

LabVIEW uses a interactive programming framework to imagine all facets of the design, including equipment, calculations and debugging. This show simplifies the incorporation of each vendor's measuring equipment, reflects a dynamic diagram structure, integrates data processing algorithms and creates unique user interfaces. Classic Control proved long enough to do machine control function, but it is based on an exact mathematical model of the program and is not a basic mathematical operation.

The fuzzy logic is able to simulate incorrect or imprecise structures, as opposed to conventional logic systems. The fluoridated logic strategy presents a easier, cheaper and more efficient alternative with strong benefits over modern technology. At present, LabVIEW simplifies the implementations of technology, process management, analysis, manufacturing usage and calculation. Thanks to the fact that LabVIEW lacks the programming language's simplicity coupled with built-in monitoring, measurement and control tools[8][9]. Through utilizing the interactive LabVIEW system to communicate with real-world signals, data are processed for relevant evidence and findings are exchanged. Take LabVIEW then to build a control framework that appends to fudge logic and which is more flexible than normal traditional regulation for device parameter variance, and the advantages of fogging are [10][11].

2.BLOCK DIAGRAM

The following diagram is the main block diagram of proposed An Integrated IoT LabVIEW Based Fuzzy- PLC Controller for Automation of Boiler

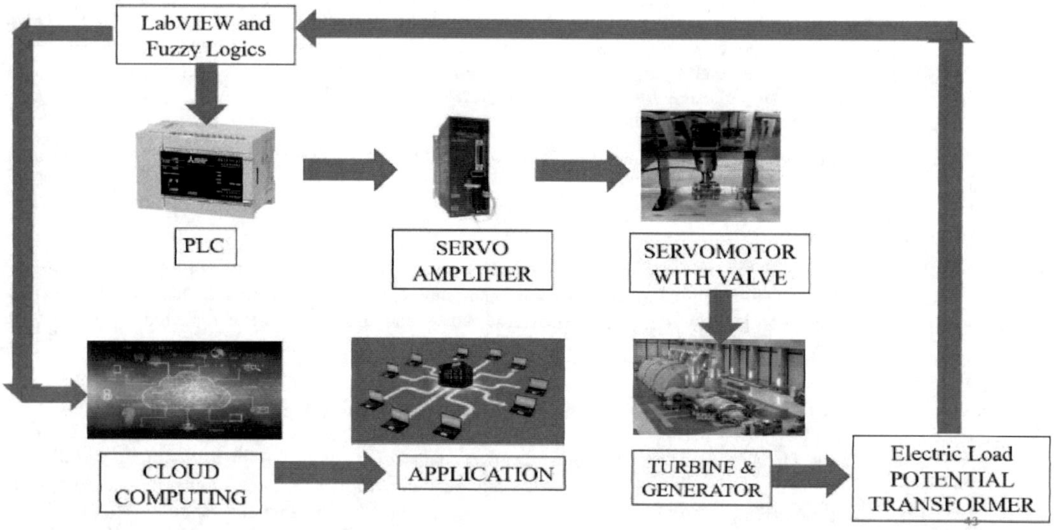

Fig. 1. Block diagram of proposed An Integrated IoT LabVIEW Based Fuzzy- PLC Controller for Automation of Boiler

In the proposed system LabVIEW plays a important role as measurements i.e Data Acquisition, Analysis, Monitoring and Controlling. For automating the steam valve in requirement with the load demand, it need to get an output from the generator. So the Potential Transformer is connected at the Generator side, to measure the current load demand. Then the P.T is connected as input to the LabVIEW.

The Fuzzy-Logic is built on the LabVIEW. The Fuzzy-logics is like adding an thinking ability to the PLC. The Fuzzy-logics keeps on monitoring the P.T input. When the Load demand rises the Fuzzy-logics decides how much angle to open the steam valve to rotate and when demand reduces the steam valve is closed in based on the demand. This change in angle is sent to the PLC through OPC Server. OPC stands for Open Flatform Communication. The PLC receives data through the OPC from LabVIEW, and it is mapped on the Position control of the Ladder Logics. So the PLC sent data to the servo amplifier and turns the motor to the required angle for the steam flow to steam turbine.

To monitor the changes of the steam valve, IOT is introduced on the LabVIEW to monitor the process remotly. The LabVIEW

IOP Conf. Series: Materials Science and Engineering **937** (2020) 012049 doi:10.1088/1757-899X/937/1/012049

sends information such as VOLTAGE, CURRENT, POWER, STEAM VALVE ANGLE, RPM(of the generator) to the cloud database. This database from the cloud can be monitored on various platforms.

In currently major industries are using human manpower and their knowledge-based supervising of change it setpoint in manually. In this proposed prototype project has the option to it automatically using artificial network. Further the setpoint also chosen by the artificially by sing intelligent technique using LabVIEW.

3. DESIGN AND IMPLEMENTATION

The construction of a Fuzzy logic controller is an essential phase in boiler automation and the required process variable must be defined before beginning automation or constructing a Fuzzy logic controller. Sometimes "linguistic words" are used when defining the value of the factors in the input equation. For eg, error and actual location are two language words that can be used as input variables easily specified. To order to monitor the device, the blurry output must be used. Over recent decades, the usage in the fields of power electronics, control systems and positioning systems of multiple software technologies such as hybrid [1, 2] or Artificial Intelligence has increasingly increased. A broad data base and knowledge of control methods are required to explain the conceptual framework to facilitate decision taking in order to apply such implementation methods. The fugitive logic [3, 4] is one of these approaches, which is important to strengthen the classical logic and is essentially a assertion that can be interpreted as a logical value of 1 or 0.

The LabView Fuzzy Logic Toolkit is put in the LabVIEW development framework software section. The Fuzzy Logic Toolkit is a GUI, a digital user interface, in true LabView form. Fuzzy logic is an expert framework and process management mechanism rule-based decision taking tool. The philosophy of Fuzzy varies from the standard Boolean logic since Fuzzy theory allows partial membership. The Fuzzy Logic utilizes descriptive, vocabulary definitions for the management of systems. The data are collected by a blurry sensor in order to handle or manipulate a specific device in real time. The outputs of the fuzzy controller are easily performed through analog hardware DAQ to monitor the operation in real time. To interactively build a fuzzy framework using the Fuzzy Program Builder. For the programming of fuzzy programs, using Fuzzy Logic VIs.

The following diagram shows the fuzzy logic controller, which is designed in LabVIEW.

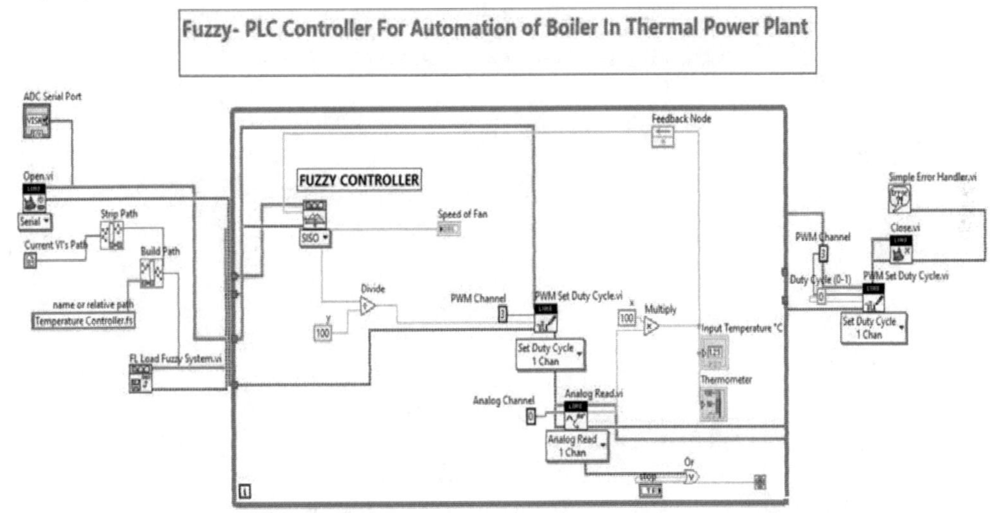

Fig. 2. Building the Fuzzy-Logic Control using NI LabVIEW

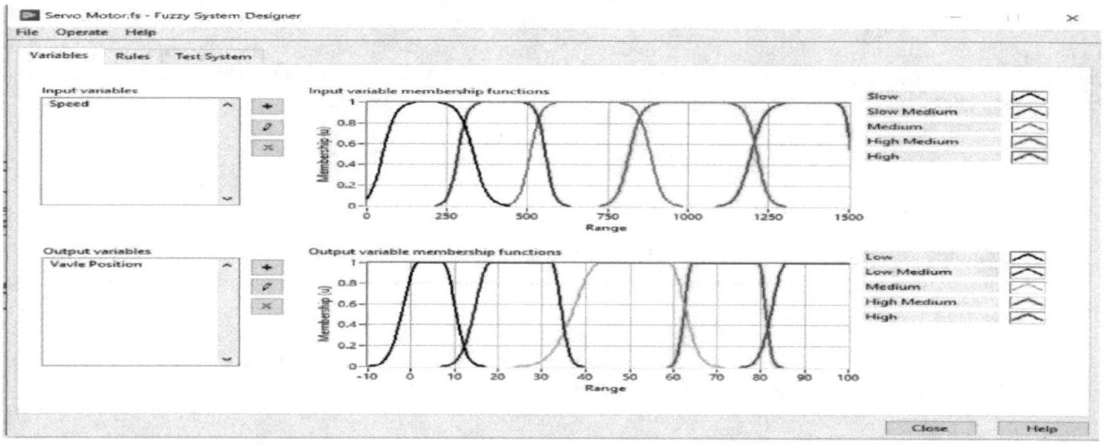

Fig. 3. Fuzzy system Designer

The above diagram shows the fuzzy membership function for speed of alternator and valve position. As per the realtime process variable the sigmoidal membership function is chosen as best one. Through this membership function, the realworld process easily captured because of the boiler process.

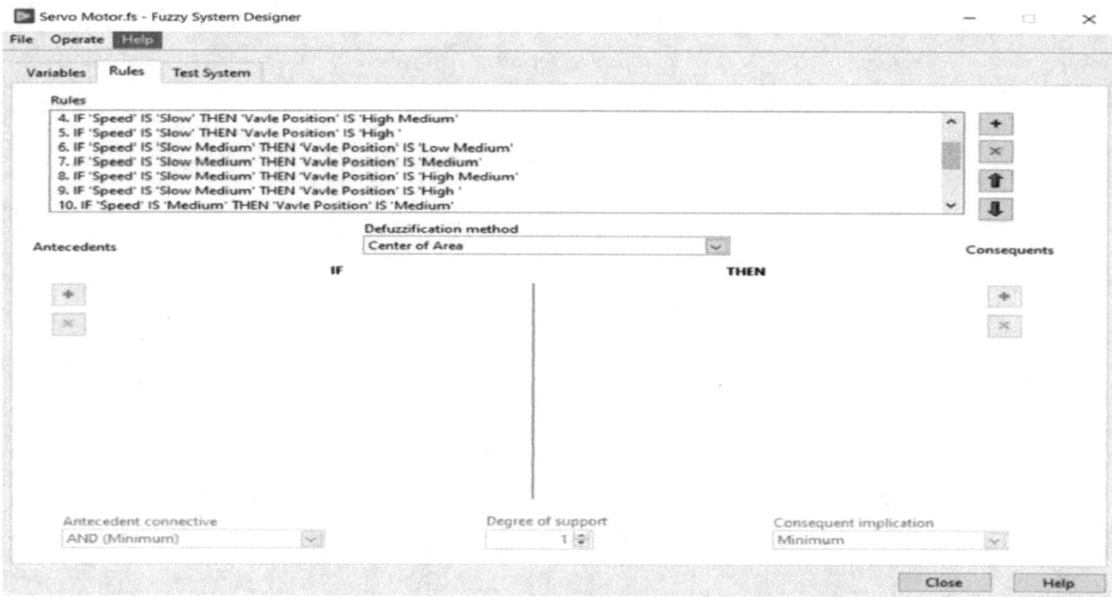

Fig. 4. Fuzzy system Rules

The above picture is the Realtime snapshot of the fuzzy rules system designer. There are 14 rules has been framed based the operating procedure of the boiler. If the rules are designed in such manner that the system is going to yield good result. So, the rules are very most important thing to control the process. There are 14 rules here is used to manipulate the input and output variable and it is used to control the output using fuzzy controller.

Fig. 5 shows the Testing system of the boiler automation. Once the rules were fixed and framed in fuzzy designer system, the system itself has the ability to simulate the output process. By using this advance feature, this will leads to help to design a controller in a better manner. Before applying to the real-time process, it could help to simulate the entire controlled process and error correction. By using this feature, we could able to adjust and fine tune the rules and membership function values and factors. So it will leads to improve the controller output efficiency. The following Fig. 5. Shows the Test system Design.

IOP Conf. Series: Materials Science and Engineering **937** (2020) 012049 doi:10.1088/1757-899X/937/1/012049

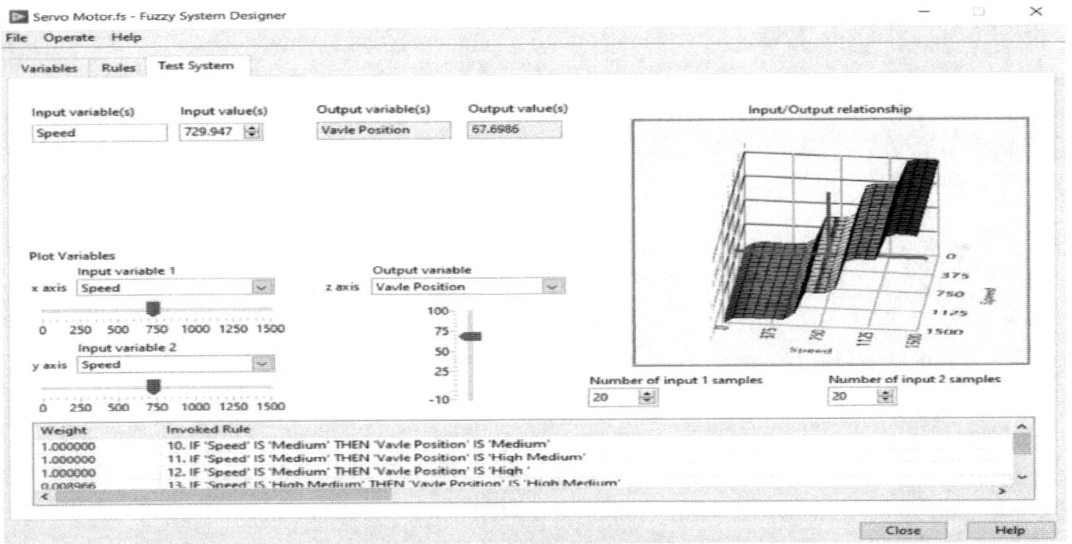

Fig. 5. Test System of Fuzzy Rules Based System

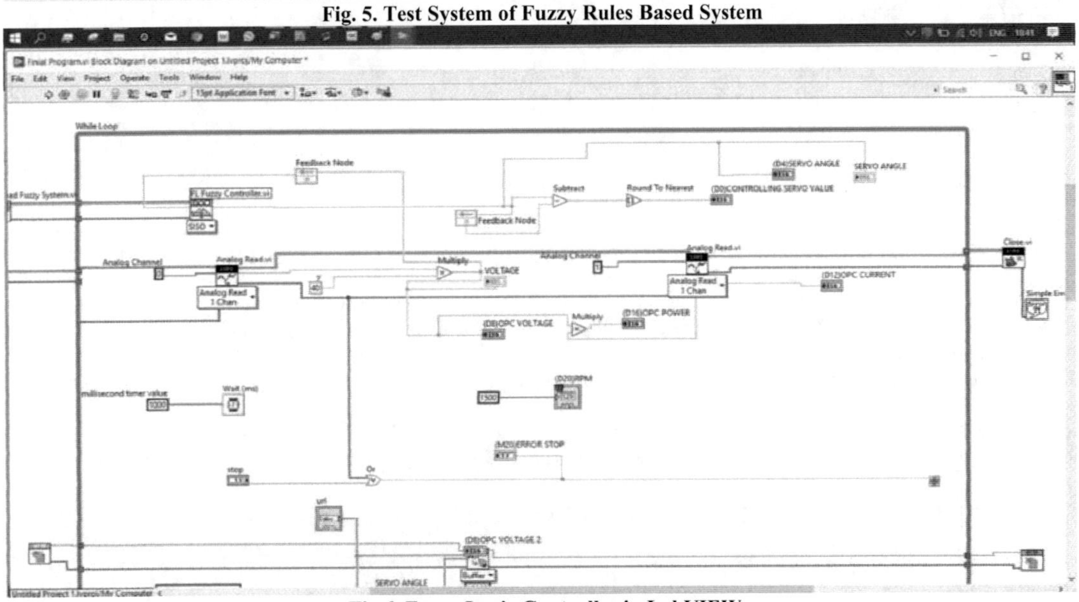

Fig 6. Fuzzy Logic Controller in LabVIEW

The above picture shows the entire Fuzzy Logic Controller system, which is designed in LabVIEW.

4.PLC PROGRAM

PLC-Programmable Logic Controller is the field controller is used to control the steam valve with the help of servo motor. So PLC is need to get the input control signal LabVIEW. This is the important step is required to control the entire process. Before connecting PLC to labVIEW, PLC itself to get programmed to monitor the its input card and output card signal status. For that the PLC needs to program itself to monitor those signals.

The proposed prototype project we have used Mitsubishi FX5CPU PLC Controller. The PLC is configured for the entire process using GX Works 3 Environment.

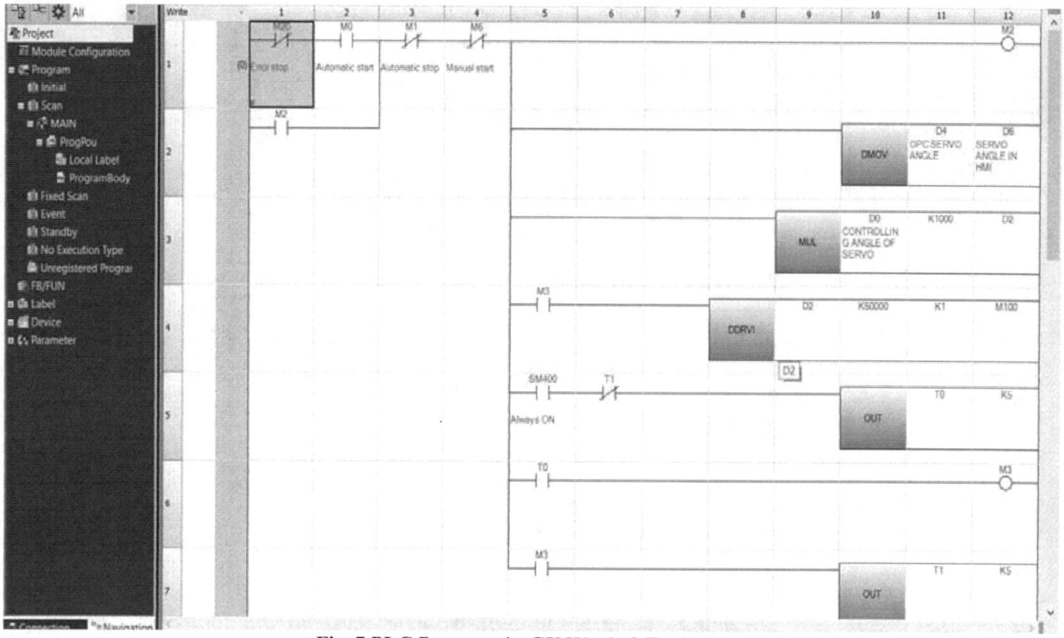

Fig. 7 PLC Program in GX Works3 Environment

5.HUMAN MACHINE INTERFACE

HMI- Human Machine Interface is the system to operate and monitor the process easily. The GOT 2000 series Mitsubishi is used here to monitor the input and output variable. There is a separate program is needed to program the HMI system. Before Monitoring the variable such as Sped of the alternator, Temperature, Steam valve position, it requires the register need to allocate it first. Then it could map with the PLC Program, which is already programmed in PLC Controller. The following fig8. Shows the HMI Front panel Screen.Fig. 8. HMI Program in GOT 2000 Mitsubishi HMI

Based on the operating condition, it could have the two cases like manual mode and automatic mode. In automatic mode selection the operator who is handling the boiler system, who couldn't able to change the process variable. The system itself is going to process it based the rules, which is already fed into the system in LabVIEW. But in maual mode the operator, who could able to change the process variable based on the real-time condition. The following fig.9 Shows the /manual mode operation.

Fig. 9. Manual Mode Selection Window in HMI -1
Fig. 10 . Manual Mode Selection Window in HMI-2

Based on the real-time condition the operator could able to change the process variable based on the operation. This is the additional feature included in the proposed project prototype.

6. OPC: OPEN PLATFORM COMMUNICATION

OPC: Open Platform Communication is the software which is used to communicate the two different hardware's. NI OPC is used in this project to communicate the and sharing the information between LabVIEW and Mitsubishi PLC in a single platform. **Fig.10.**

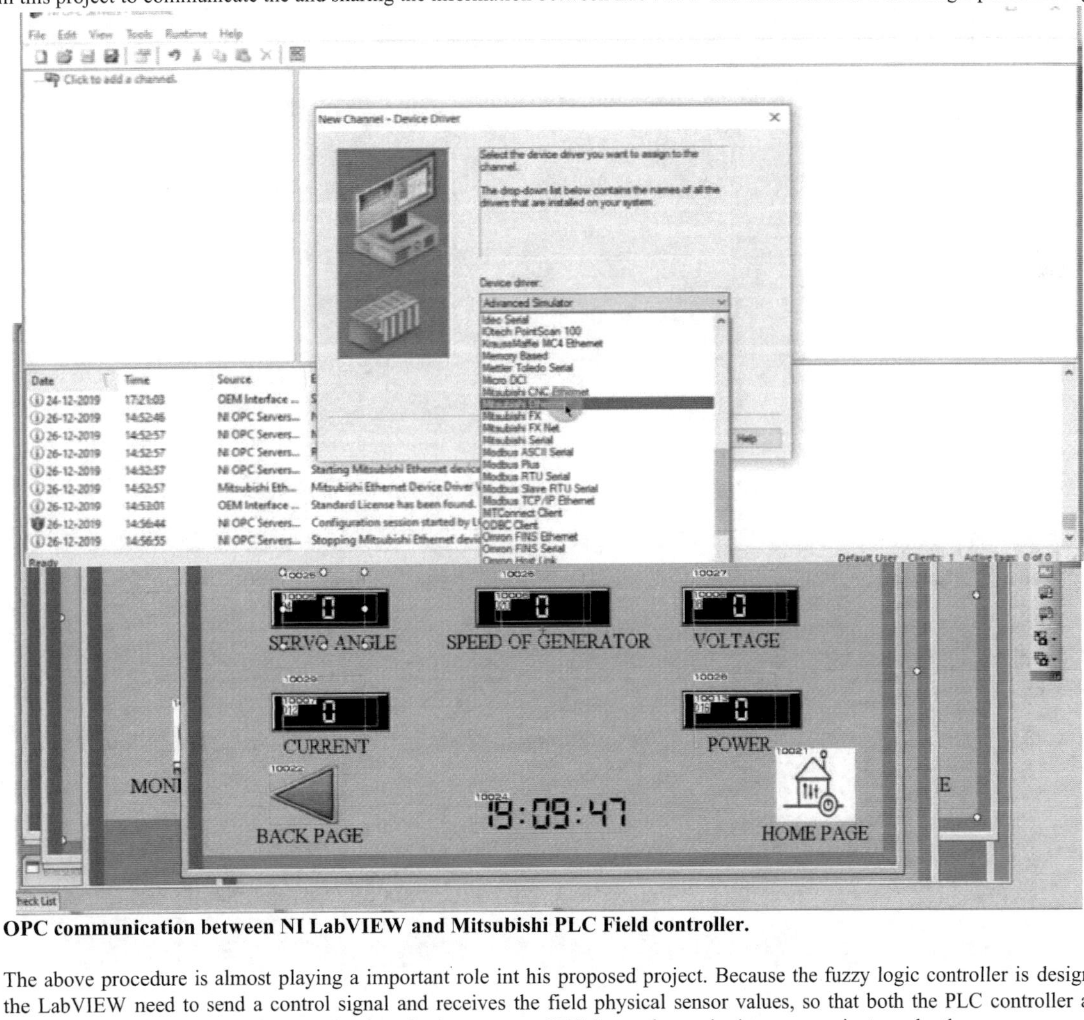

OPC communication between NI LabVIEW and Mitsubishi PLC Field controller.

The above procedure is almost playing a important role int his proposed project. Because the fuzzy logic controller is designed the LabVIEW need to send a control signal and receives the field physical sensor values, so that both the PLC controller and LabVIEW is need to communicate each other. So that purpose OPC Server is required to communicate each other.

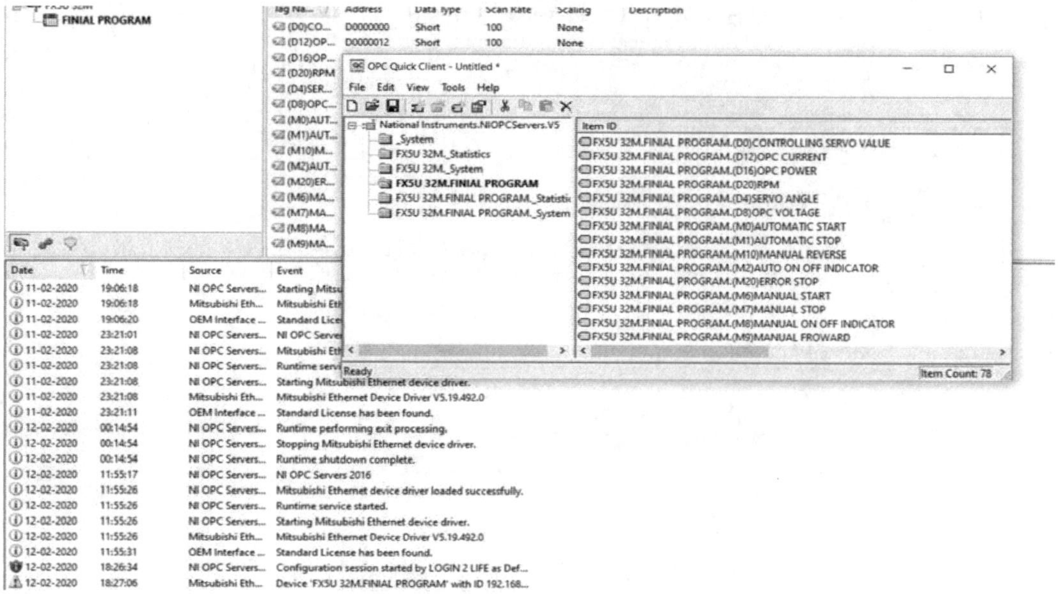

Fig.10. OPC Server Configuration

7. RESULTS AND DISCUSSIONS

The following sections are the real-time graphs which is really designed a fuzzy logic controller and the output response are drawn using IOT Sever.

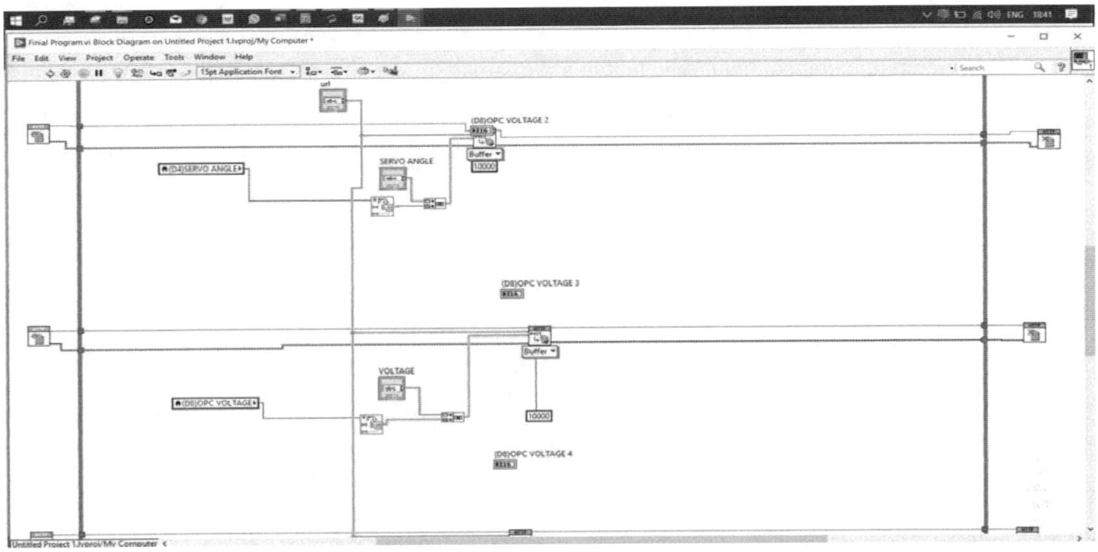

Fig.11. IOT Interface using LabVIEW using Thinspeak Server

ThingSpeak Server is used here to monitor the input and output process variable of the boiler system. These values are Alternator speed and Temperature, Alternator voltage, Current, etc. These process variables are measured using corresponding sensors using LabVIEW and through the LabVIEW the data's were sent to ThingSpeak Server for IOT enabled system.

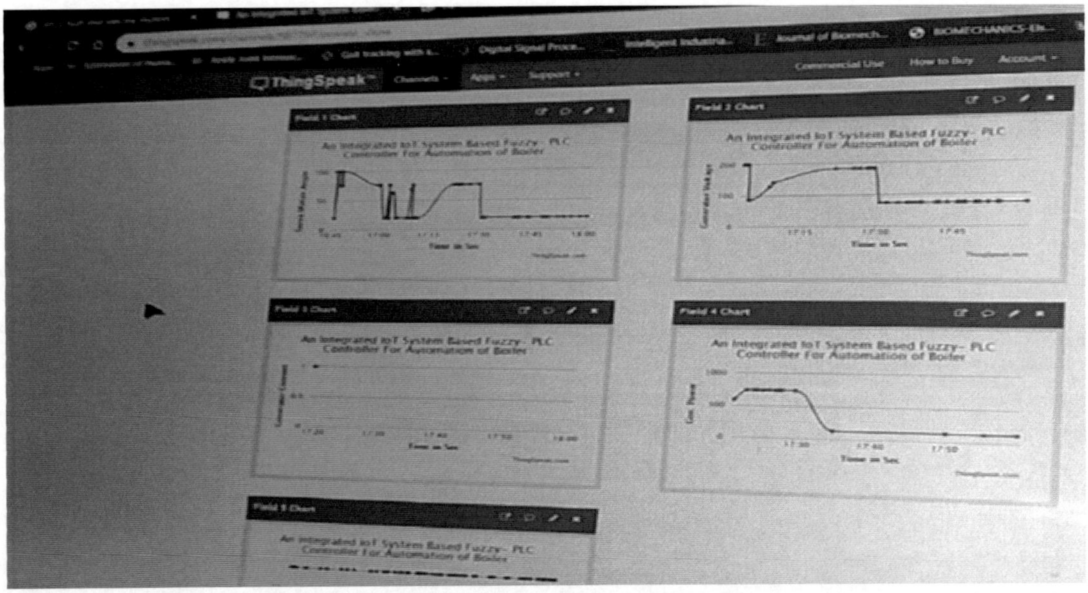

Fig.11. IOT Server Data Monitoring system

Thus, the above Fig.11 shows the real time datas of the alternator. By using this we could able to monitor the health and process of the boiler in remote. This syste yields the good profit at end case. Because we could able to do predict the problem as well as we could able to increase the health of the alternator.

8. CONCLUSION

In this proposed prototype we have automated the steam valve in the Thermal Power Plants by using servo motor coupled with valve, based on demand of load using Fuzzy-Logics build on LabVIEW. Here the pneumatic control valve is replaced by using servo motor. The servo motor is used here to achieve accurate position control of the steam valve in requirement with the load demand.

9. FUTURE WORKS

Even though this system has more capable to achieve good result. Further Complete Automation of Boiler including Water level monitoring, coal intake, temperature control needs to introducing the Artificial Intelligence (AI) for the controlling of steam valve. And Controlling of the Boiler and the Steam valve through cloud computing. Temperature control of boiler using thermal camera for better study of Boiler.

10. REFERENCES

1. PENG, G.L. and LIU, Z.G., 2006. Boiler control system with PLC [J]. Electric Power Automation Equipment
2. Conrad, D.A. and Muldoon, D.P., 1995. Design and application of a PC/PLC utility boiler control system. In INSTRUM CONTROL AUTOM POWER IND, PROC (Vol. 38, pp. 351-362).
3. Xu, L., Jiang, Y., Wang, J., Gao, T. and De, X., 2008, July. Design of fuzzy-PID controller based on prediction model and its realization in PLC. In 2008 Chinese Control and Decision Conference (pp. 3808-3811).
4. Weibin, C. and Qingjian, M., 2010, October. Based on PLC temperature PID-Fuzzy control system design and simulation. In 2010 International Conference on Information, Networking and Automation (ICINA) Vol. 2, pp. V2-417.
5. Shuisheng, C. and Qingjian, M., 2011. Based on PLC Temperature PID-fuzzy Control System Design and Simulation [J]. Journal of Agricultural Mechanization Research
6. YU, H.Y., JING, F., SONG, Q.K. and CHEN, S.S., 2003. Fuzzy-control in the Combustion System of Burning Coal Boiler
7. Thepsatorn, P., Numsomran, A., Tipsuwanporn, V. and Teanthong, T., 2006, October. DC motor speed control using fuzzy logic based on LabVIEW. In 2006 SICE-ICASE International Joint Conference pp. 3617-3620.
8. Pei, F. and Yang, W.S., 2004. Design and Application of Fuzzy Logic Controller Based on LabVIEW. Control and Instruments In Chemical Industry, 31(1), pp.41-43.
9. Jin, M.L. and Ho, M.C., 2009. Labview-based fuzzy controller design of a lighting control system. Journal of marine science and technology, 17(2), pp.116-121.
10. Zhao, B.C., An, L. and Liu, X.H., 2006. Design and simulation of fuzzy logic controller based on LabVIEW. Control Engineering of China, 12, pp.49-52.
11. Gowthaman, E., Saravanan, S., Naveenbalu, K., Aravind, S. and Naveen, S., 2016, November. Performance analysis of hybrid fuzzy-PID controller action on boiler drum level control. In 2016 Online International Conference on Green Engineering and Technologies (IC-GET) (pp. 1-6).
12. Kar, P.P. and Saikia, P., 2013. Performance of different control strategies for boiler drum level control using LabVIEW
13. Wang, J., Song, A.J. and Zhu, Q.X., 2011. Adaptive PID temperature controller based on fuzzy logic. In Applied Mechanics and Materials (Vol. 55, pp. 1455-1458).

A Review on Power Electronics and Drives in Electric Propulsion System

Saurabh Shivam[1], Rohan Singh[1], R. Raja Singh[2], Sitharthan R[1]

[1]Department of Electrical and Electronics Engineering,Vellore Institute of Technology Vellore, India.
[2]Department of Energy and power electronic,Vellore Institute of Technology Vellore, India.
Correspondence: saurabh.shivam0143@gmail.com

Abstract. This era is the part of the fast-growing world and it is thriving more in the field of advancement of technology. Power electronics and drives are one of the major fields that is getting updated every day. Due to increased power generation capacity, improved power qualities, increased automation, and improved capabilities, the electric propulsion has replaced mechanical propulsion. Similarly, the variable speed motors have replaced transverse motor starters to save energy and reduce load instability. With the detailed structure and working of the marine electric propulsion system, this paper focus on reviewing about the various power electronic devices and converter configurations that are used in marine applications, specifically in the propulsion system. Moreover, the detailed assessment of different motor and drive schemes employed in propulsion system is carried out. Further, the review will shift its centre towards controlling drives, which includes various control strategies and control techniques.

1. Introduction

In earlier times the mechanism that was in operation for propulsion of marine ships was mechanical. Although due to advancement of sophisticated circuits it has been replaced by electrical ships which has electrical propulsion system [2]. Power electronics has played a great role on the efficiency and uses of almost every electrical device. There are three major phases in an electric ship: (i) power source (ii) distribution centre and (iii) load centre [3]. Power electronics switches and devices are used to connect power source and load centers through bus which eventually distributes the power to every part of ship (which includes propulsion system). Due to increased power generation capacity, improved power qualities, increased automation, and improved capabilities, marine electric propulsion has replaced the mechanical propulsion ships and similarly the variable speed motors have replaced transverse motor starters to save energy and reduce load instability [6].

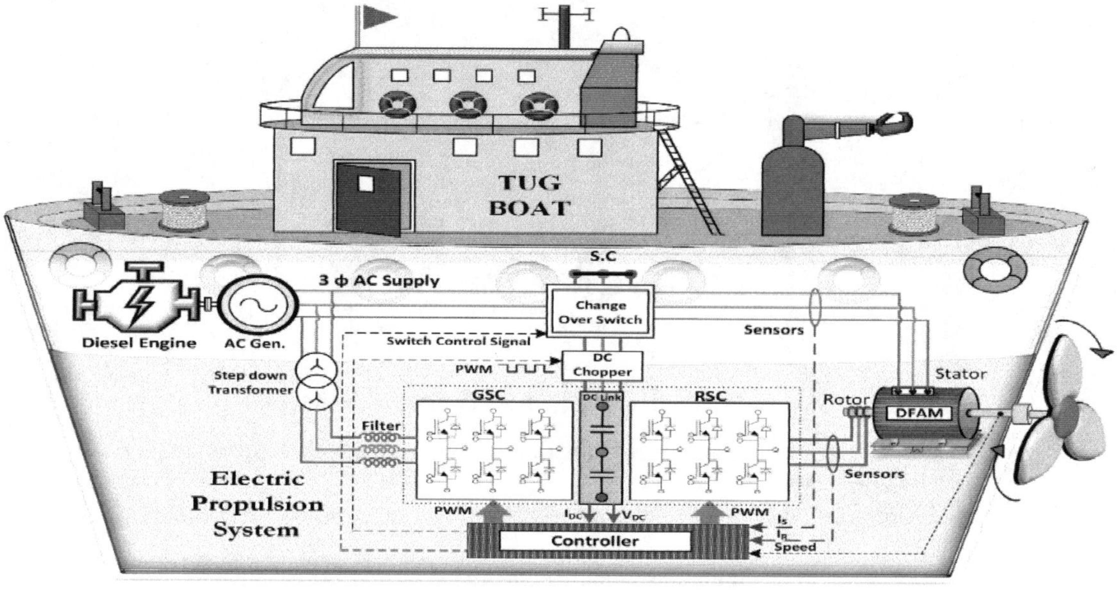

Figure 1. Overall schematic of electrical propulsion system [28].

The above diagram (figure 1) represents the schematic representation of electrical propulsion system. Here the diesel engine which acts as a prime mover rotates the generator. This rotating generator produces electricity and supply it to the transformer (step down transformer) and the load (stator of the motor). The transformer then supplies power to the motor through a series of power converters. These power converters are connected in series with the help of a dc link. The control strategy applied for the power converters is pulse width modulation. Controllers are also used in the circuit for controlling the speed of the propellers and the current fed into the rotor and stator.

The below diagram (figure 2) basically gives us the overview of the electric propulsion system. In the generation part, AC generator generates power and gives it to the bus bar. In the distribution part, power is distributed to the motor from the bus bar. In the propulsion part, electric motor rotates the propellers which produces propulsion. Here the motor is not directly connected to the propellers but is connected to the propelling shaft and the gearbox. The gearbox and propelling shaft are then connected to the propellers.

Traditional marine mechanical propulsion system uses main engines for propelling the shaft and each shaft is powered by its own dedicated main engine with particular provisions. Except when difficult combination of gearboxes cannot be used for shafts interconnect.

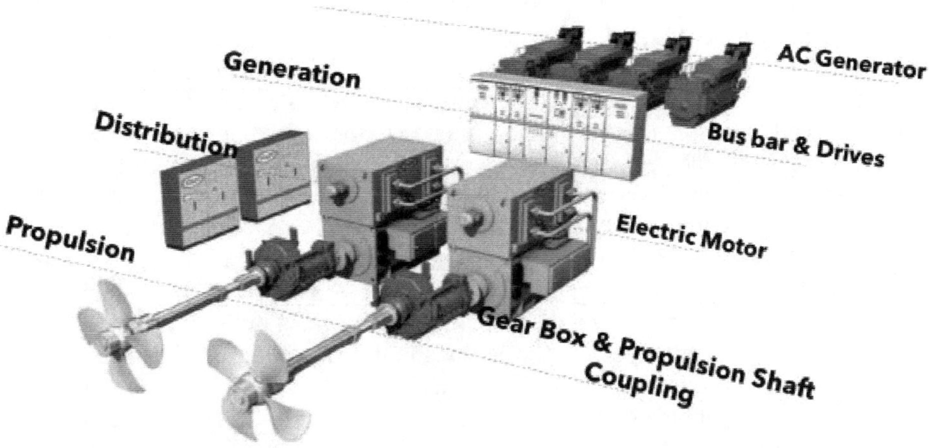

Figure 2. Schematic representation of Electrical propulsion system [28].

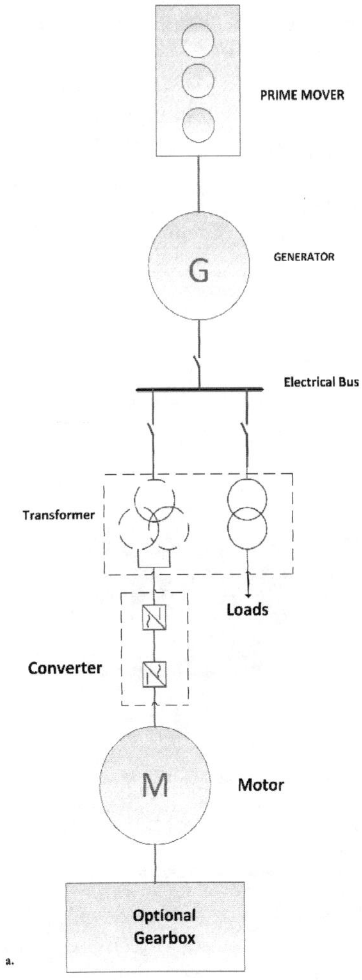

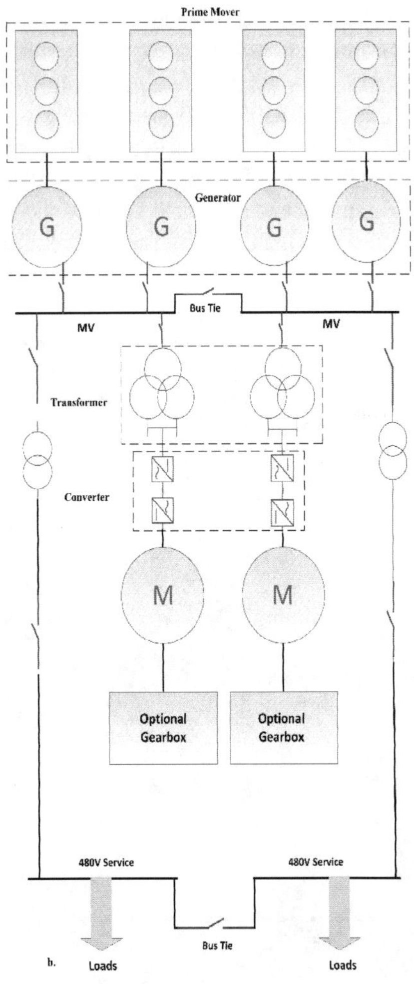

Figure 3. (a) Single sourced propulsion system. **(b).** Multi-sourced propulsion system with redundancy.

In fig (a), the prime mover produces the useful work and rotates the generator which in turn produces electricity for the marine electric propulsion. This prime mover can be a diesel engine which uses diesel and produces useful work or it can be a gas turbine too. Now this running generator provides electricity to the electric bus as shown in the fig (a).This electric bus provides electricity to the loads as well as the transformer which gets connected to the motor through a series of two converters as shown in fig (a).The motor is then connected to the gearbox which helps in the propulsion of the electric propulsion system. Fig (b) shows the multi sourced marine electric propulsion system with redundancy. In this single line diagram four prime movers are used which are then connected to four generators. Each prime mover rotates one generator. All the rotating generators produce electricity and supply it to the electric bus. Two electric busses are used for four generators (each bus for two generators combined) as shown in the fig (b). Each bus powers the load and the transformers. The transformers are connected to the motors through converters. These motors are connected to the gearbox which helps in the propulsion. This circuit is more redundant than that of fig (a) because here the electric supply to the loads and the transformer is not dependent only on a single bus. Even if one bus fails or has some faults in it the other bus can compensate for it by establishing a connection through the bus tie line. But in case of fig (a) the supply to the transformer and loads has no alternative. Therefore, circuit in fig (b) has redundancy.

Electric propulsion (EP) is a class of space propulsion that uses electrical energy to accelerate a propeller by various possible electrical and / or magnetic means [8]. The use of electrical energy improves the propulsion performance of EP propellants compared to conventional chemical propellants [10]. The major reasons to adapt to electric ship (electric propulsion system) is that astern operation and ship manoeuvring is possible without recourse to controllable pitch propellers or complicated reversing gearboxes [18]. Addition to electrical propulsion the electrical generators need not be limited to supplying propulsion power alone – they can also be arranged to provide the electrical power required for the ship's domestic and other use[24].For the electric boat related applications cost, weight, loss and size (whichever is has the priority) for different applications in the equipment is a major limitation on the acceptance of power electronics on boats[26].

2. Power electronic switches used in marine electric propulsion system

For the marine electric propulsion system to work smoothly and hassle free, power electronics switches plays a major role. All the major on-off circuits and control circuits are dependent on power electronic switches. Major types of power electronic switches are discussed ahead:

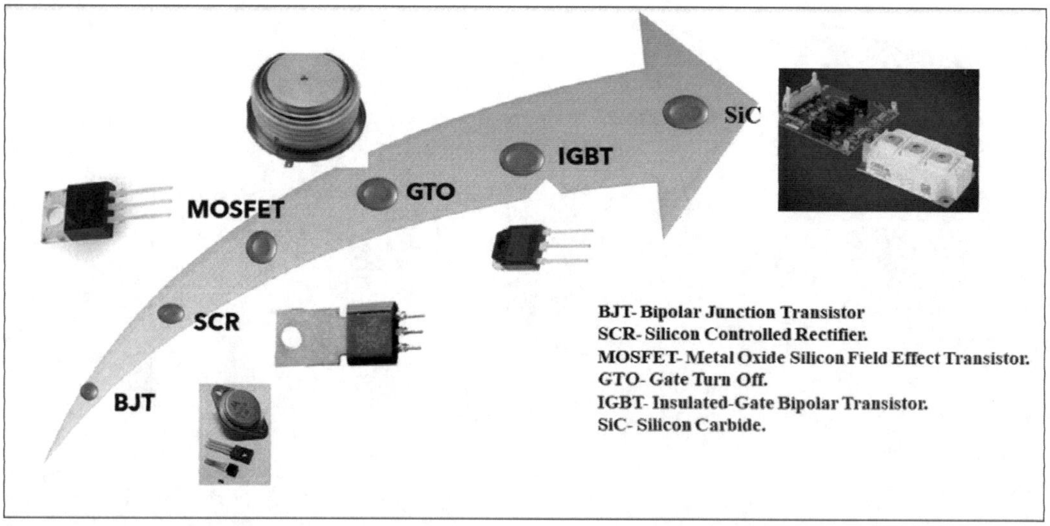

Figure 4. Evolution of switches in Marine Propulsion System(MPS).

IOP Conf. Series: Materials Science and Engineering **937** (2020) 012050 doi:10.1088/1757-899X/937/1/012050

2.1. Thyristor
The thyristor is locked with an impulse switch on the gate. ON low-voltage conductivity drops forward and does not turn off with pulse removal. Change needs a negative opening impulse. Normal used thyristors, line converters, and cycle converters are built without their own disconnect [26] capability, in which case the thyristor goes from its locked conduction state to a non-conductive condition, even if the current is zero in various directions. The approach results in a very effective solution and unity with the highest scores. Regarding high power converters are as follows.

- Gate Turn off (GTO)
- Integrated gate commutated thyristor (IGCT)
- MOS turn-off thyristor (MTO)
- Emitter turn-off thyristor (ETO)
- Super GTO (SGTO)[29]

2.2. Transistor
A transistor conducts a gate signal in its forward direction when the gate signal is turned off and on. A shutter has manual ignition; If the signal is less than what is needed for a full ignition, it will work as long as the anode remains at a partial cathode voltage [26]. This current limiting feature is an essential asset to transistors: it buys several microseconds of time, enough for protection purposes [29]. Nowadays mainly MOSFET and IGBT switches are being modified and used in most of the circuits due to its controlled cut-off maintaining features.

2.3. Power-Diode
A power diode is a two-terminal unit, where one is an anode and the other is a cathode. If the voltage of the anode is greater than the voltage of the cathode, then the diode is biased forward and the forward current flows through the diode [24]. Through this position the current rises gradually through the power diode, and the voltage increases as well. If the diode is biased in reverse, only a small current pass through the diode-this is called the leakage current [26]. In certain instances, this can be overlooked. The current is fairly insignificant in this role before some degree of reverse voltage occurs-this is called the voltage of breakdown.

2.4. SiC
Silicon carbide (SiC) chips can be used in principle to manufacture any of the devices mentioned above. The software offers advantages high voltage capacity (20 kV) of much lower switching losses (score per device) and high temperature [26]. Nonetheless, the hardware is expensive, the amount of small chips is huge. It must be connected in parallel, and the commercially available devices are around 1200 V and one hundred amps [29]. Although SiC devices are expensive, they have a very high impact on drive size, loss, weight, cooling requirements, and high PWM frequency potential [29].

3. Motor
The following table represents different types of motors that can be used in marine electric propulsion system. This table shows us a comparative study on the evolution of motors mainly dependent upon five factors: range of the motor, starting torque required by the motors , rating of converter associated with the motor and the efficiency of the motor, after the comparative study of different types of motor, DFA motor is found to be the best fit for usage in marine electric propulsion system due to the fact that it has partial power converter rating for the converter associated with it. This means that DFA motor can run at full capacity without burning up the power converter that is associated with it.

IOP Conf. Series: Materials Science and Engineering **937** (2020) 012050 doi:10.1088/1757-899X/937/1/012050

Table 1. Comparison table of different types of motors.

Motors	Range	Starting torque	Rating of converter associated	Power rating	Efficiency
DC Motor	Wide-speed	High	Full rating of power converter		70 to 80 %
Synchronous Motor	No range (Sync speed)	Zero	Full rating of power converter	High power rating	High efficiency
Induction Motor	Runs at approx. synchronous speed	High starting torque	Full rating of power converter		85-97%
PMS Motor	Wide speed range	No starting torque	Full rating of power converter	High power rating	High efficiency
HTS Motor	Wide speed range	Good starting torque	Full rating of power converter	High power rating	High efficiency
DFA Motor	Wide speed range	High starting torque	Partial rating of power converter	High power rating	High efficiency

4. Power converter configuration in marine electric propulsion system.

The electronic power converter converts electrical energy in one way to electrical energy in a different way: AC to DC, AC to AC, DC to DC, DC to AC. Converters can adjust the power generated by source to an increasing voltage distribution; and load converters. The adjustment of voltage distribution depends according to the type of power supply that requires preparation. Resources can be configured to work at their best punctures [18]. The distribution and sharing of loads between active sources upgraded. Conversion point of use ensures loads are getting to their needed strength. Nearly every source and charge need today a conversion tool. Power and control circuits were special in the past and hardwired, unchangeable. Computer configurations are now made with Power Electronic Building Block (PEBB) can be used to produce variations during the design process miscellaneous products. In addition, these, with digital power products can be designed to operate the electrical selected functionality. The end result is therefore special and immutable like the corresponding wired converters. Today, digital control networks, a given PEBB configuration, can see their control algorithms modified to be delivered in the field [21]. This system can be used for many different functions. Different types of converters used in drives in electrical propulsion system (EPS) are Voltage Source Inverter (VSI) type converters for AC motors normally for asynchronous motors, cyclo-converters for AC motors normally for synchronous motors, Current Source Inverter type (CSI) converters for AC motors normally for synchronous motors and DC converters, or SCR (Silicon Controlled Rectifier) for DC motors[30].

Modern marine power electronic converters are Alstom series connected converters, voltage clamped converters and isolated voltage converters[30].

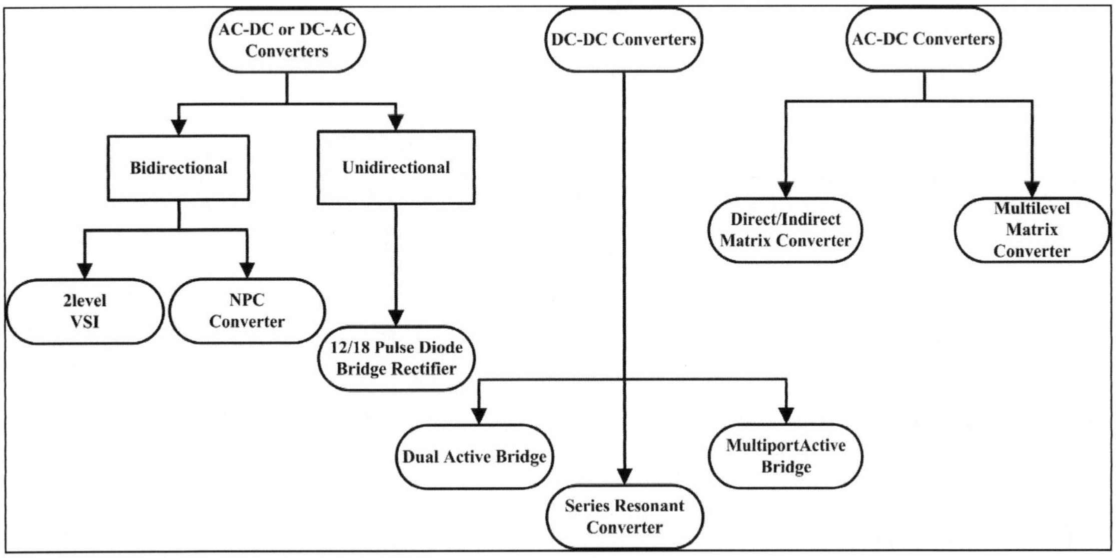

Figure 5. Power electronics converter classification tree.

4.1. Alstom Series Connected Converters

It is a multilevel inverter. It uses switches connected in series with flying capacitors connected in parallel to those set of series connected switches, the flying capacitor is used to cope up with the short circuits at the load or overvoltage problems. The Alstom series connected converter increases the power rating and maintains high quality of power and voltage without derating the motor.

These converters permit a rise in power rating and voltage while not sacrificing the purity of the voltage waveform and also the subsequent motor de-rating. Semiconductors need not be stratified and may be place in series with none preselection [21]. Finally, energy levels are minimized beneath fault conditions. The benefits of this device are no longer needed to be established for high-power variable speed drives. In the years to return, it will inevitably become the topology of alternative all told sectors of energy conversion [30].

4.2. Voltage Clamped Converters

DC–DC device with reduced reverse recovery current and switch-voltage stress, with an influence quality of low voltage and high current [18]. It will choose the switch with lower property voltage for assuaging the switch conductivity loss in order to use of voltage-clamped technique. In this case the extra capacitor is employed for providing the reverse current path of the electrical device to lift the utility rate of the core [30]. further snubber circuits for engrossing the voltage spikes within the diodes aren't needed to additional hamper the manufacture price.

IOP Conf. Series: Materials Science and Engineering **937** (2020) 012050 doi:10.1088/1757-899X/937/1/012050

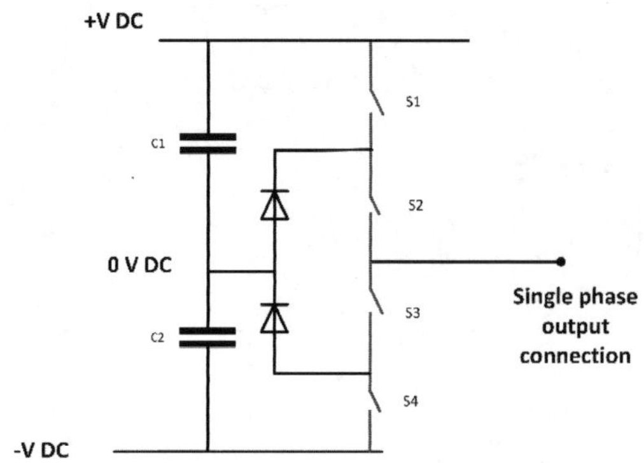

Figure 6. Voltage clamped converter.

4.3. Resonant Converters

Recently, resonant DC / DC converters have sparked considerable interest in analysis by tracking the progress of their applications. This increase in its industrial application has led to many efforts to improve smooth switching, smooth waveforms, high octane density, and high-power options for resonant converters. Its quality for high-frequency use and its ability to mitigate switching losses were attracted to industrial applications compared to typical hard-switched drives. Numerous studies have been carried out on the exchange of DC-DC converters to confirm that they meet the strictest criteria for the application of the philosophy of natural energy. The possibility of minimizing the switching losses and conductivity in the switching mode by increasing the switching frequency makes them very attractive. However, many switch topologies will achieve high octane transfer [1][2], but the problem is that the power switches (transistors), diodes, and passive energy storage components (capacitors and inductors) contained in the converter capacity structure, which affects its power. The idea is that an auxiliary circuit around the main power switch creates a resonant flow of current in the same circuit that forces the current or voltage of the switch to be around zero at all times switching interval [21]. In this way, the switching loss is essentially eliminated because the volt-amp product is also around zero. once this control is around zero, the circuit is assumed to resonate in current, on the contrary, once the voltage is controlled around zero, the circuit is assumed to resonate in voltage [30].

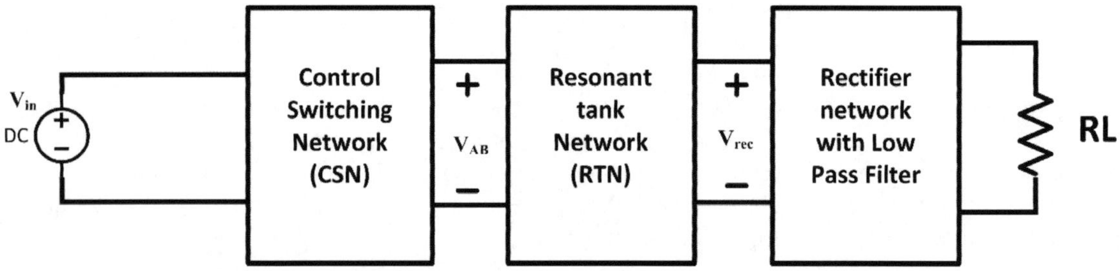

Figure 7. Resonant converter.

4.4. Isolated Voltage Converter

In terms of accuracy in output voltage regulation, isolated DC / DC converters generally fall into three categories: regulated, unregulated, and semi-regulated. In a regulated isolated DC / DC converter, the input voltage, load current, and shutdown temperature have an effect on the output voltage accuracy of unregulated isolated DC / DC devices[30]. For some applications that require specific output voltage and tight regulation, this is generally unacceptable and regulated isolated DC / DC devices must be used.

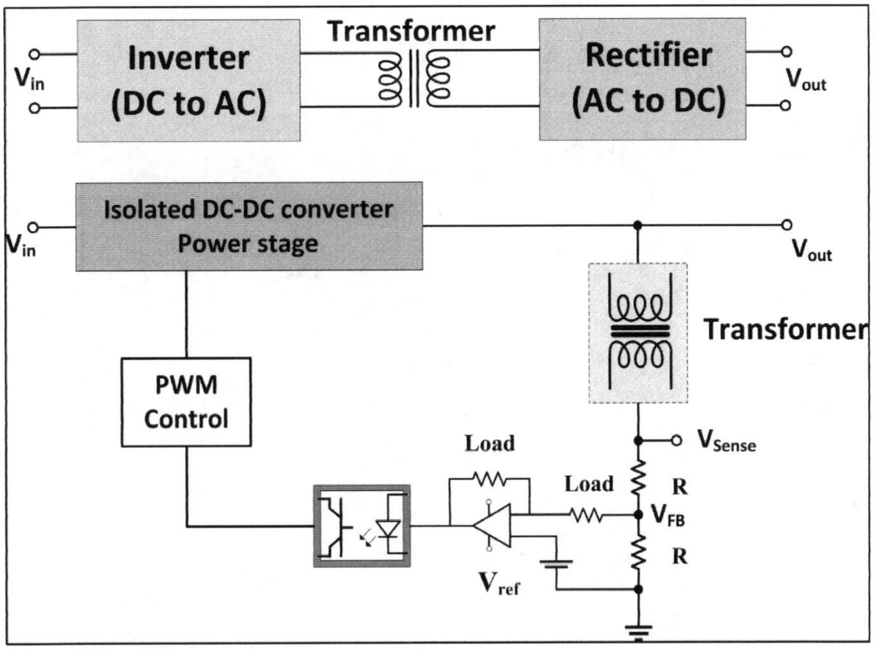

Figure 8. Isolated voltage converter.

4.5. Matrix Converter

A matrix converter is used to convert input voltage direct into an arbitrary AC voltage instead of converting that input voltage as DC voltage (as in case of inverters). This matrix converter has high efficiency, long life span, smaller harmonics and has high potential for completing the demands of today's market[21]. The matrix converter arranges the switches in matrix order and controls them to convert input voltage direct into an arbitrary AC voltage. Since there is no conversion into DC component therefore energy storage devices are not required [30].

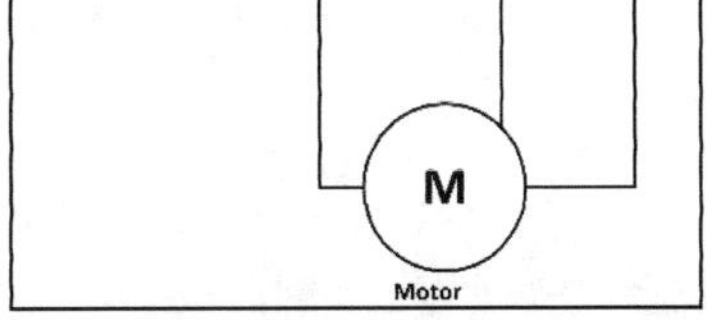

Figure 9. Matrix converter.

The below table gives us a comparative study on the three different types of enhanced power converters. The comparison parameters for the power converters are mainly: number of switches used in the power converter, different types of configurations in which the power converters can be used and the most important parameter that is the MVA capacity of the power converters.

Table 2. Comparison table of modern converters.

Power Converters	No. of Switches	Modes of Output	Configuration	MVA Capacity	Year [citation]
Alstom Series Connected Converters	10	3	Transformers, Multilevel-Inverter, Motor-filter	25 MVA	2002 [30]
Isolated Voltage Converter	8	1	Refer diagram	10-11 MVA	2002 [30]
Voltage Clamped Converters	12	2	Refer diagram	5 MVA	2002 [30]

5. Control strategies for on board micro-grid system.

5.1. Primary Control

The primary control focus of a dc microgrid are dc voltage guideline and burden sharing. If the breakers are shut, the AC/DC converters associated with the generators are taking care of a solitary HV transport, in this way, the voltage of the transport and the power sharing of the two generators must be managed [21]. A few methodologies can be embraced to accomplish these objectives, and they can be separated into concentrated and decentralized methodologies. The best decentralized methodology depends on the hang control, looking like the controllers embraced in the inheritance framework. At the point when two DC/DC converters are working equal, their yield voltage is controlled to de-wrinkle relatively to the current [31]. This usage of the hang control impersonates the activity of a perfect voltage source with an arrangement resistor. On the off chance that the yield current builds, the voltage diminishes, taking into account the equal association of numerous voltage-controlled converters. The hang coefficient can likewise be named the virtual resistor for the voltage lease hang control. Different methodologies are conceivable, similar to the current/voltage hang (where the converter is worked with a present control conspire) or by utilizing the power rather than the current (power/voltage hang or volt- age/power hang) The decentralized controllers consider a basic execution; in any case, consistent state voltage blunders or soundness issues may emerge [29]. A voltage blunder shows up when the converter is preparing the most extreme current Imax diminishing the estimation of the virtual resistor R_v decreases the mistake however makes the power sharing less precise because of estimation blunders[30].

5.2. Secondary Controls

To settle the consistent state blunder because of the hang control, auxiliary circles are utilized. The thought is that a focal controller forms the estimations at every hub and afterward sends refreshed references for the hang control to re-establish the ostensible voltage condition [18]. This technique looks like the optional control of the heritage network, and it needs a low-transfer speed correspondence connect to play out the estimation and speak with the neighbourhood controllers. Techniques that don't require a correspondence interface however depend on improvement strategies are likewise professional presented to this point [21].

In an external adjustment circle is received to forestall the framework in-strength on account of little yield channels. An eye-witness based framework is along these lines proposed to play out the adjustment activity with-out the need of extra sensors [31].

5.3. Storage Management System

This control mechanism is most essential part of any system. Power required for all the equipment other than traction power comes from this system. It basically deals with the battery state estimation i.e. state of charge, state of health, state of life, and state of power estimation for the batteries. Overall supply of energy for all the systems at regular basis as well at the emergency state depends upon the wellness of the batteries that stores energy. All the supporting element gets power from this for operation.

5.4. Multi-objective global optimization

The heart of all control types. This keeps an eye on rest of the other control techniques for the smooth working of the equipment under desirable range. It's working ladder is based on the importance of equipment on the priority list for a particular work. It's feature for controlling the targets depending on the phase of operation makes it very compatible for installation to get better control over the equipment. But due to computational complexity it's establishments are affected.

Table 3. Comparison table for different control type.

Control type	Control targets	Features	Issues	Year [citation]
Primary	Load current and voltage DC or AC bus voltage	Local implementation of the inner controls Possibility to use solutions tested in other applications	Customization to the ship environment required.	2019 [21]
Secondary	Voltage restoration Stability improvement Power flow regulation	Improvement of the distributed control performance. Global network management.	Communication link must be established. Susceptible to single point failure. Several sensors needed.	2019 [21]
Storage Management System	State of charge Peak shaving Voltage control	Energy storage scheduling depending on the technology.	Many control targets makes the optimization difficult. Fault isolation required because of the risks.	2019 [21]

Multi-objective global optimization	All variables within the specified limits, load prioritization and storage management.	Better performance than distributed control. Possibility to adapt the control targets depending on the phases of operation.	Mismatch between predicted and actual mission profile. Computational complexity.	2019 [21]

6. Energy management strategies

Power age in transport microgrids are commanded by coordinated generators, which are controllable. Notwithstanding, the heaps are exceptionally powerful and, in specific cases, may contain quick changes. In this way, so as to oversee complexities and accomplished to control goals, the various levelled structure can be received for transport microgrids also. In addition, as clarified beneath, the current control advancements in transport power frameworks can likewise be portrayed in accordance with the various levelled control system. In various levelled control, the essential level goal is to accomplish load sharing among the power sources. The optional level control objective is to make sure about transport signals at their ostensible qualities. The tertiary level control is utilized to accomplish ideal activity with deliberate goals [32]. In this plan, the higher the degree of control, the more-slow the guideline it gives. Besides, the extent of the control augments as the level increments.

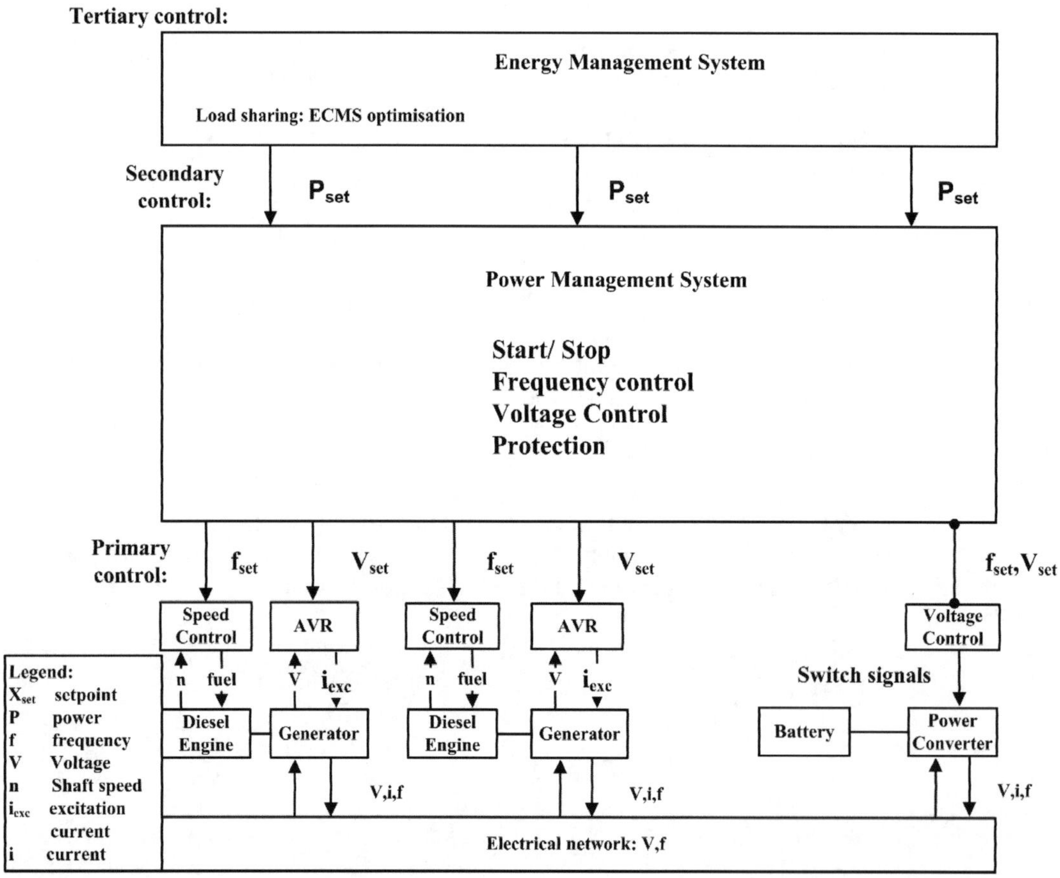

Figure 10. Different levels of control strategies and control techniques in MPS.

IOP Conf. Series: Materials Science and Engineering **937** (2020) 012050 doi:10.1088/1757-899X/937/1/012050

6.1. Isochronous Control

Most marine power scaffolds are AC flow scaffolds where repetition and tension are the two key parameters that must be met within certain limits. Since simultaneous generators for diesel engines or gas turbines are the sources normally used in this setting, the senator's controller manages the repeat while the automatic voltage controller (AVR) directs the voltage. The senator, supplied in the form of a hydromechanical or electronic control, correctly directs the fuel to the engine and therefore manages the speed (ω) of the rotor.[3] Delivered as an electronic controller, the AVR controls the current applied to the generator field turn, which manages the performance voltage[4].

Under isochronous control, generator voltage and repetition performance are maintained at fixed focal points, regardless of changes in the heap. Unlike different level control structures, isochronous control for a single generator falls into the area of essential and auxiliary levels, as it manages the transport voltage and repetition in fixed foci and at the same time provides the required force [18].

The previously mentioned isochronous control functions admirably for a solitary generator. On the off chance that at least two generators are associated with a similar power transport, one of the motors may attempt to take the whole burden while the others probably won't take the heap. This prompts dangers and may bring about power outage. In this way, so as to settle this issue, correspondence between the representative controllers, as a heap sharing line or a correspondence connect, for example, Controller Area Network (CAN) transport or Field transport, is required with the assistance of the correspondence interface, every motor can be set to take a particular portion of the heap without going into boundaries or insecurities[31]. In this setup, the Power Management System (PMS), decides the power reference for every motor, and can be ascribed to the tertiary degree of the various levelled control structure [19].

Despite the fact that isochronous burden sharing is equipped for controlling voltage and recurrence at set focuses, it has not gotten the well-known decision in transport microgrids, primarily due the cruelty of nature in ships, which unfavourably influences correspondences [8]. Additionally, so as to execute isochronous force sharing, all the senator controllers ought to be perfect, and a large portion of the cases should originate from a similar maker, which may not be conceivable at times [31]. Despite the fact that arrangements, for example, propelled generator oversight have been created to forestall power outage in flawed circumstances, isochronous control is as yet not the mainstream decision with regards to huge shipboard force frameworks.

6.2. Droop Control

Contrasted with isochronous control, hang control is the mainstream decision for power partaking in multi-generator shipboard power frameworks, as it doesn't require correspondence between the representative controllers. As opposed to the fixed recurrence and fixed voltage activity in the isochronous control, hang control lets the recurrence and voltage differ with respect to the dynamic, (P), and receptive force, (Q), requests of the heap. Where the speed and voltage references are diminished directly as the dynamic and responsive power requests increment.

The droop control matches with the intrinsic P/f droop nature of simultaneous machines where stacks on the electrical side hinders the rotor and thus, speed drops [3][4][5]. The senator infuses more fuel in light of the speed drop and subsequently, the generator set gets steady at another recurrence, which is lower than the ostensible recurrence, fundamental frequency (f_o). The recurrence as well as the voltage settles at another incentive similarly when there is an adjustment in the responsive power request [18]. In a multi generator framework, every senator detects the speed drop, supplies more capacity to the matrix lastly settles at another recurrence. The measure of intensity provided by each generator set relies upon the hang settings of the genset. In the event that the settings are equivalent, all the generators set similarly share the heap. Additionally, the hang control can also be applied for power converter-based frameworks. Also, it may be, independent of the framework, hang based power sharing falls inside the extent of the essential reaction in the progressive control plot.

In a multi-generator framework, the auxiliary control can be accustomed to take the recurrence back to the ostensible worth, by adding a counter-balance to the speed reference, ω ref. This is known as the optional reaction that can be credited to the level 2 control of the various levelled control plot. Further changes can be made to the hang controller by changing the hang gain, kp, which changes the force levels of every motor to their most ideal levels under the given condition. This tertiary reaction has a place with the level 3 control of the progressive plan. Like the previously mentioned speed, hang controllers work dependent on the responsive force request. As indicated by the guidelines of the majority of boat arrangement social orders, the proportionality of burden sharing needs to biggest generator should be inside the scope of ±15% of the evaluated dynamic force and ±10% of the appraised receptive intensity of the biggest generator. As opposed to isochronous control, hang control requires just neighbourhood estimations of voltage and recurrence, and in this manner, it permits different generators to share the heap without chasing and without the need between unit correspondence. This makes hang control strong, exceptionally solid, and adaptable in including/expelling the generators of various power evaluations to the network [7][8]. By the by, since the dynamic and receptive force provided by generators rely upon recurrence and voltage deviations , huge burdens bring about expanded deviations, and this is an innate exchange off of hang control.

7. Result

The space required for installation of electrical propulsion machinery is very less and compact as compared to conventional system. There is no direct connection of propeller shaft and prime mover, and hence transmission of severe stresses such as torsional and vibration is restricted. Environmental benefits from lower fuel consumption and emissions.

Whereas the efficiency of the electrical plant is less than that of a conventional system. The installation cost of electrical propulsion plant is much higher. Different and improved training for ship's crew as the system is completely different from mechanical system and involves major automation.

8. Conclusion

At present, electric propulsion is applied mainly in following type of ships: Cruise vessels, ferries, DP drilling vessels, thruster assisted moored floating production facilities, shuttle tankers, cable layers, pipe layers, icebreakers and other ice going vessels, supply vessels, and war ships. There is also a significant on-going research and evaluation of using electric propulsion in new vessel designs for existing and new application areas. The development of offshore electrical power systems has been extremely strong in the past two decades and this is due to advances in topology and the construction of independent electronic switches and power converters. This paper compared various converters used in marine propulsion and saw various power switching devices and also various control strategies. Electric propulsion systems are majorly dependent on power electronics sources and devices which is depicted above in this paper. Large and small ships will greatly bee benefited from continuous improvements in devices and controls in the area of power electronics. Space and weight are always a very important on board a ship, so current developments will result in devices that significantly reduce the cost, size, weight, and losses of power electronics.

References

[1] Sitharthan R, Geethanjali M and Pandy TKS 2016 Adaptive protection scheme for smart microgrid with electronically coupled distributed generations *Alexandria Engineering Journal* **55(3)** 2539-2550

[2] Fathima AH, and Palanisamy K 2014 Battery energy storage applications in wind integrated systems—a review *IEEE International Conference on Smart Electric Grid* 1-8

[3] Prabaharan N and Palanisamy K 2015 Investigation of single-phase reduced switch count asymmetric multilevel inverter using advanced pulse width modulation technique

International Journal of Renewable Energy Research **5(3)** 879-890.

[4] Jerin ARA, Kaliannan P and Subramaniam U 2017 Improved fault ride through capability of DFIG based wind turbines using synchronous reference frame control based dynamic voltage restorer. *ISA transactions* **70** 465-474

[5] Sitharthan, R, Sundarabalan CK, Devabalaji KR, Nataraj SK and Karthikeyan M 2018 Improved fault ride through capability of DFIG-wind turbines using customized dynamic voltage restorer *Sustainable cities and society* **39** 114-125

[6] Prabaharan N and Palanisamy K 2016 A single-phase grid connected hybrid multilevel inverter for interfacing photo-voltaic system *Energy Procedia* **103** 250-255

[7] Palanisamy K, Mishra JS, Raglend IJ and Kothari DP 2010 Instantaneous power theory based unified power quality conditioner (UPQC) *IEEE Joint International Conference on Power Electronics, Drives and Energy Systems* 1-5

[8] Sitharthan R and Geethanjali M 2017 An adaptive Elman neural network with C-PSO learning algorithm-based pitch angle controller for DFIG based WECS Journal of Vibration and Control **23(5)** 716-730

[9] Sitharthan R and Geethanjali M 2015 Application of the superconducting fault current limiter strategy to improve the fault ride-through capability of a doubly-fed induction generator–based wind energy conversion system *Simulation* **91(12)** 1081-1087

[10] Sitharthan R, Karthikeyan M, Sundar DS and Rajasekaran S 2020 Adaptive hybrid intelligent MPPT controller to approximate effectual wind speed and optimal rotor speed of variable speed wind turbine *ISA transactions* **96** 479-489

[11] Sitharthan R, Devabalaji KR and Jees A 2017 An Levenberg–Marquardt trained feed-forward back-propagation based intelligent pitch angle controller for wind generation system *Renewable Energy Focus* **22** 24-32

[12] Sitharthan R, Sundarabalan CK, Devabalaji KR, Yuvaraj T and Mohamed Imran A 2019 Automated power management strategy for wind power generation system using pitch angle controller *Measurement and Control* **52(3-4)** 169-182

[13] Sundar DS, Umamaheswari C, Sridarshini T, Karthikeyan M, Sitharthan R, Raja AS and Carrasco MF 2019 Compact four-port circulator based on 2D photonic crystals with a 90° rotation of the light wave for photonic integrated circuits applications *Laser Physics* **29(6)** 066201

[14] Sitharthan R, Parthasarathy T, Sheeba Rani S and Ramya KC 2019. An improved radial basis function neural network control strategy-based maximum power point tracking controller for wind power generation system *Transactions of the Institute of Measurement and Control* **41(11)** 3158-3170

[15] Rajesh M and Gnanasekar JM 2017 Path observation based physical routing protocol for wireless ad hoc networks *Wireless Personal Communications* **97(1)** 1267-1289

[16] Palanisamy K, Varghese LJ, Raglend IJ and Kothari DP 2009. Comparison of intelligent techniques to solve economic load dispatch problem with line flow constraints *IEEE International Advance Computing Conference* 446-452

[17] Sitharthan R, Ponnusamy M, Karthikeyan M and Sundar DS 2019 Analysis on smart material suitable for autogenous microelectronic application *Materials Research Express* **6(10)** 105709

[18] Rajaram R, Palanisamy K, Ramasamy S and Ramanathan P 2014 Selective harmonic elimination in PWM inverter using fire fly and fireworks algorithm *International Journal of Innovative Research in Advanced Engineering* **1(8)** 55-62

[19] Sitharthan R, Swaminathan JN and Parthasarathy T 2018 March. Exploration of wind energy in India: A short review *IEEE National Power Engineering Conference* 1-5

[20] Karthikeyan M, Sitharthan R, Ali T and Roy B 2020 Compact multiband CPW fed monopole antenna with square ring and T-shaped strips *Microwave and Optical Technology Letters* **62(2)** 926-932

[21] Sundar D Sridarshini T, Sitharthan R, Madurakavi Karthikeyan, Sivanantha Raja A, and Marcos Flores Carrasco 2019 Performance investigation of 16/32-channel DWDM PON and long-

reach PON systems using an ASE noise source *In Advances in Optoelectronic Technology and Industry Development: Proceedings of the 12th International Symposium on Photonics and Optoelectronics* 93

[22] Sitharthan R and Geethanjali M 2014 Wind Energy Utilization in India: A Review *Middle-East J. Sci. Res.* **22** 796–801 doi:10.5829/idosi.mejsr.2014.22.06.21944

[23] Sitharthan R and Geethanjali M 2014 ANFIS based wind speed sensor-less MPPT controller for variable speed wind energy conversion systems *Australian Journal of Basic and Applied Sciences* **8**14-23

[24] Jerin ARA, Kaliannan P, Subramaniam U and El Moursi MS 2018 Review on FRT solutions for improving transient stability in DFIG-WTs *IET Renewable Power Generation* **12(15)** 1786-1799

[25] Prabaharan N, Jerin ARA, Palanisamy K and Umashankar S 2017 Integration of single-phase reduced switch multilevel inverter topology for grid connected photovoltaic system *Energy Procedia* **138** 1177-1183

[26] Rameshkumar K, Indragandhi V, Palanisamy K and Arunkumari T 2017 Model predictive current control of single phase shunt active power filter *Energy Procedia* **117** 658-665

[27] Fathima AH and Palanisamy K 2016 Energy storage systems for energy management of renewables in distributed generation systems *Energy Management of Distributed Generation Systems* 157

[28] Rajesh M 2020 Streamlining Radio Network Organizing Enlargement Towards Microcellular Frameworks *Wireless Personal Communications* 1-13

[29] Subbiah B, Obaidat MS, Sriram S, Manoharn R and Chandrasekaran SK 2020 Selection of intermediate routes for secure data communication systems using graph theory application and grey wolf optimisation algorithm in MANETs IET *Networks* doi:10.1049/iet-net.2020.0051

[30] Singh RR and Chelliah TR 2017 Enforcement of cost-effective energy conservation on single-fed asynchronous machine using a novel switching strategy *Energy* **126** 179-191

[31] Amalorpavaraj RAJ, Palanisamy K, Umashankar S and Thirumoorthy AD 2016 Power quality improvement of grid connected wind farms through voltage restoration using dynamic voltage restorer *International Journal of Renewable Energy Research* **6(1)** 53-60

[32] Singh RR, Chelliah TR, Khare D and Ramesh US 2016 November. Energy saving strategy on electric propulsion system integrated with doubly fed asynchronous motors *IEEE Power India International Conference* 1-6

[33] Singh RR, Mohan H and Chelliah TR 2016 November. Performance of doubly fed machines influenced to electrical perturbation in pumped storage plant-a comparative electromechanical analysis *IEEE 7th India International Conference on Power Electronics* 1-6

[34] Sumithra M, Abraham A, Rajesh M and Nithiyanandam N 2019 Mining social media content to predict peer trust level in social networks *International Journal of Engineering and Advanced Technology* **6S3** 1959-1964

[35] Venkatesh K, Ali MJS, Nithiyanandam N and Rajesh M 2019 Challenges and Research Disputes and Tools in Big Data Analytics *International Journal of Engineering and Advanced Technology* **6S3** 1949-1952

Classification and prediction of social attributes By K-Nearest Neighbor Algorithm with Socially-aware wireless networking- A study

Sujatha krishanmoorthy,Han Bingxu,Zhang Huimin,Chai Yunsheng,Cheng Nuo,

Assistant professor ,Students

School of Science and Technology

Wenzhou-Kean University Wenzhou, China
sujatha@wku.edu.cn.zhanghui@kean.edu, chaiyu@kean.edu,chengn@kean.edu

Corresponding Author: Sujatha.ssps@gmail.com,sujatha@wku.edu.cn

Abstract. In this paper, we analyze the data which was collected from SIGCOMM 2009, where 76 users were taken part in and traces of Bluetooth encounters, opportunistic messaging, and social profiles of them were collected. The experiment started by conducting a the data sorting and data cleaning. In data processing, mainly three main types of characteristic data in the dataset: message, proximity and participant were chosen. After that, Hierarchical cluster analysis was applied on the processed data, which can score the selected segments by similarity measurement, and then form and visually describe the hierarchical structure of these selected clusters. T can either be manually set or a machine language algorithm can be used and here the manually number of categories are set to be 3 after applying the fviz_nbclust function for optimal clustering analysis. As this was a study article five different calculation methods were analyzed to accomplish hierarchical clustering: single linkage, complete linkage, median distance method, UPGMA and Centroid Clustering. The comparison is deployed iwith a simulation tool and the results are plotted .After the clustering algorithms, the participants are divided into three categories, and for each clustering algorithm the divisions are different. In this regard, k-nearest neighbor classification algorithm is used to calculate the error rate of comparison between the original data and the predicted data. To conclude, Our method results in specific label classification and then analyze the strength of nodes and the tightness between nodes via the KNN algorithm.

1. Introduction

Zhiwen, et al. [1] have defined the socially aware computing. Social perception computing is to perceive and identify the behavior of social individuals in real time, analyze and excavate the characteristics and rules of group social interaction, assist individual social behaviors, and support the interaction .communication and collaboration of the community through the large-scale multi-type sensor devices increasingly deployed in human living space. The core of social perception computing lies in the word "perception", which has two meanings: first perceiving the real world, then perceiving and responding to it. Social perception computing emphasizes the use of advanced computer science and technology to perceive individual behaviors and group interactions in the real world, understand human social activity patterns, and provide intelligent assistance and support for individual and group interactions. [1]. In the Internet era, especially the emergence of Web 2.0, researchers have been provided with massive data sources for analyzing online behaviors and social networks. Socially aware computing has been booming, and social software and applications have been emerging, such as netizens' behavior pattern mining and network public opinion analysis. Therefore, Socially aware computing can be used for human behavior and interaction analysis, individual behavior recognition, and analysis of behavior characteristics and rules [2-3]. Based on the data of individual behavior, mobility and social interaction, the features of group social interaction are analyzed by means of social network analysis, machine learning and data mining.Therefore, in this paper has taken advantage of the abundant socially interaction data resources from the internet and classify and predict network users by utilizing the K-Nearest Neighbor Algorithm.

2. Collation and Classification of Data

2.1. A Data cleaning and Sorting

These data were collected in two field experiments using mobiclique smart phones in two large-scale network conferences. They show who each node (personal ID) is associated with, interest classification of each node, regional characteristics, participant characteristics and sent information characteristics.[4] After the data is collected, it needs to be sorted out first. Put the original disorderly data into excel table, which is conducive to observation and calculation. We use R language to clean the data. Data cleansing plays an important role in data management and data analysis. At the same time, data cleansing, cleansing or appending can effectively help the subsequent processing of data, because it can correct or delete inaccurate and damaged data.[5] First, the data is transformed into CSV file type, and then imported into R language. Through STR function, the structure of each group of data is obtained, including the data type and data length of each column vector. And get the network graph for the connection between nodes. A total of 76 nodes, for the ring layout and hierarchical layout drawing corresponding to the network diagram. These network diagrams can let us clearly see the connections of these nodes, and clearly show that there are complex but close connections between each node. This is good for data classification. The network diagram of the ring layout clearly shows that there are different connections between the 76 nodes. The connections between some nodes are closer and more complex, but the connections between some nodes are relatively sparse. This strongly suggests that there are different categories between these relationships, so we can classify these data. In data processing, we choose three main types of characteristic data: message, proximity and participant. We use statistical nodes to audit the quality of data. From the audit results, we find that there are missing data, and we process the missing data. We choose to delete these missing nodes.

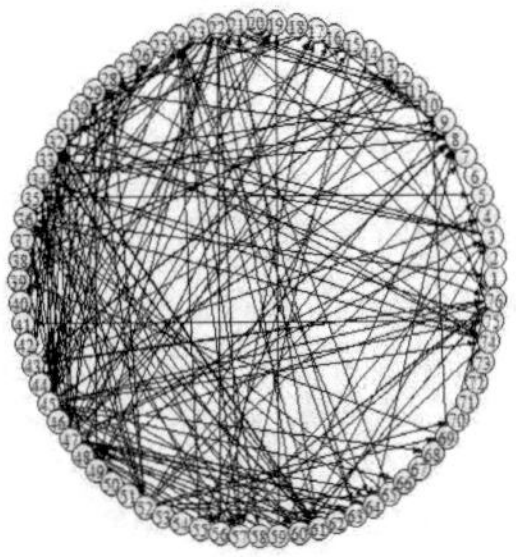

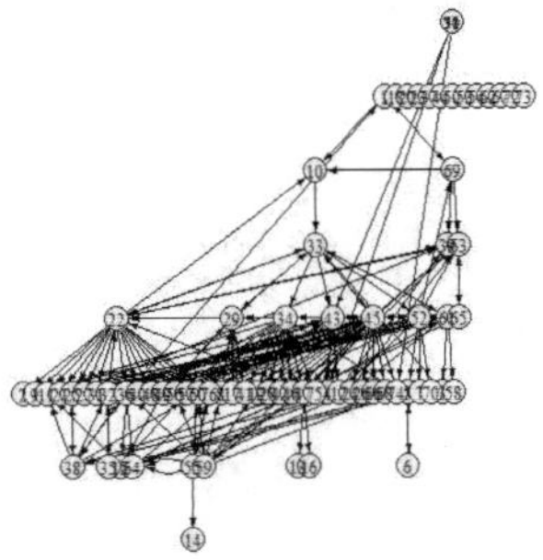

. Figure 1: Ring layout Figure 2: Hierarchical layout

2.2 Hierarchical cluster analysis

After the preliminary data cleaning and sorting, three kinds of main data information can be selected as the eigenvalues, and then the data classification can be completed. Cluster analysis is to divide a group of data into several categories according to similarity and difference. Its purpose is to make the similarity between data belonging to the same category as large as possible, and the similarity between data in different categories as small as possible. We choose hierarchical clustering to analyze, and try to classify them according to the messages, proximity and participants. Hierarchical clustering analysis can score the selected segments by similarity measurement, and then form and visually describe the

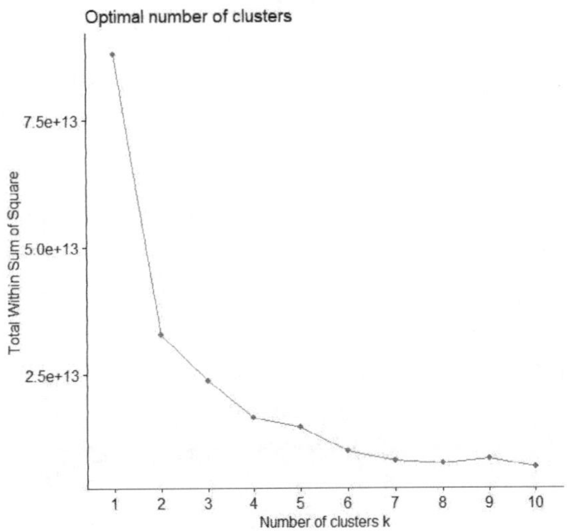

hierarchical structure of these selected clusters.

Figure 3: hierarchical clustering line graph

IOP Conf. Series: Materials Science and Engineering **937** (2020) 012051 doi:10.1088/1757-899X/937/1/012051

This approach allows us to understand the relationship between fragments more easily. [6] However, hierarchical clustering cannot automatically determine the number of categories, it is necessary to define the number of categories of the final clustering in a user-defined way. We can use fviz_nbclust function for optimal clustering analysis.

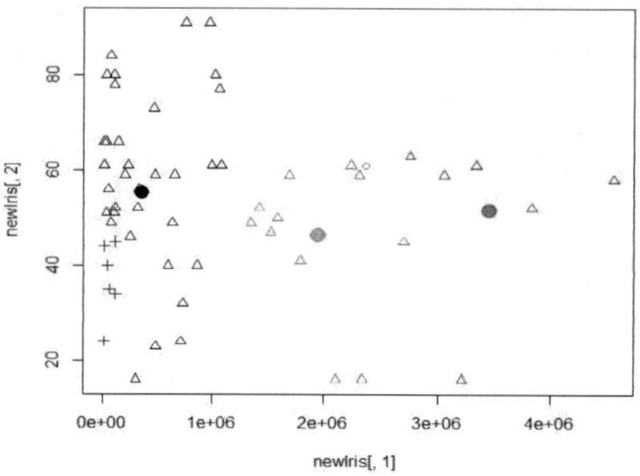

Figure 4: hierarchical clustering

In the resulting line graph, this clearly shows that the data classification takes a big turn in 2 and 4. But when we grade these data, two categories may be too few and four categories may be too many. Therefore, we finally decided to divide it into three categories. In the preliminary clustering graph, the data are clearly divided into three categories, and only a few outliers exist.

We can clearly feel that these data distribution have hierarchical clustering, and further prove that we can divide into three categories. These data will be finally hierarchical clustering through five different calculation methods: single linkage, complete linkage, median distance method, UPGMA and Centroid Clustering. According to these five methods,

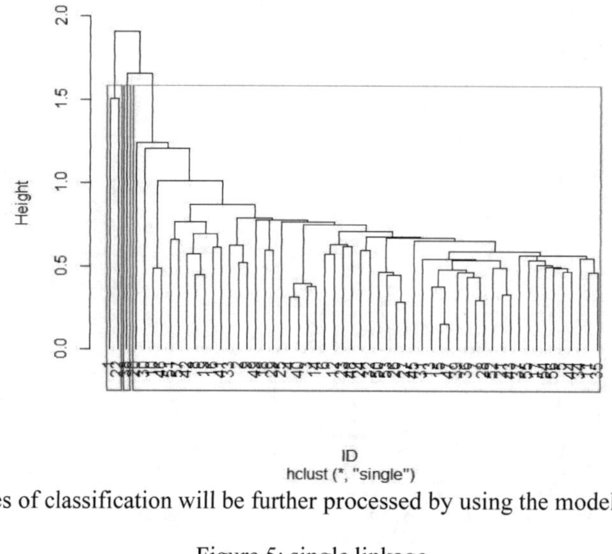

different types of classification will be further processed by using the model.

Figure 5: single linkage

The first is the single linkage. Single linkage can process data of various shapes and densities, but it is sensitive to noise. A single linkage considers local density to ensure that the distance involves the core of each group.[7] The single linkage clustering method is to find out the non-diagonal elements of the original m × m distance matrix, to merge the classification objects GP and GQ into a new class GR, and then according to the calculation formula Calculate the distance between the original classes and the new classes, so as to get a new (m-1) Order distance matrix; then select the smallest DIJ from the new distance matrix, and merge GI and GJ into the new class; then calculate the distance between the new classes and the new classes, so on until the classification pairs are classified into one class. Through this calculation method, figure 5 can be obtained. Figure 5 clearly shows that tags 1 and 22 are divided into one category, tag 38 is a separate category, and all remaining data tags are a separate category. Then through the artificial custom division, set a, B, C three different levels of categories for the next step model test.

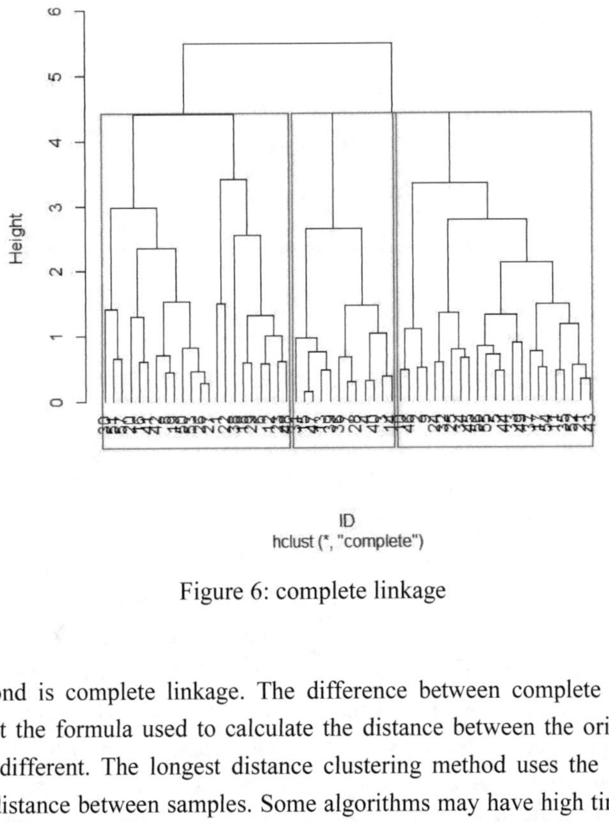

Figure 6: complete linkage

The second is complete linkage. The difference between complete linkage and single linkage is that the formula used to calculate the distance between the original class and the new class is different. The longest distance clustering method uses the longest distance to measure the distance between samples. Some algorithms may have high time complexity. But complete linkage algorithms can provide the best possible result in the manageable time. Compared with the solution of single linkage hierarchical clustering which can find individual clusters with high precision, complete linkage can find the optimal solution according to the linkage between data.[8] In this way, we can get Figure 6. Figure 6 clearly shows that all data can be divided into two categories and one subclass. In addition, we set three different levels of categories A, B and C for the next step of model test through custom division.

The third method is median distance method. Intermediate distance method is a kind of system clustering method. It is a compromise between the minimum distance method and the maximum distance method. For the distance between classes, the minimum distance between two classes of samples and the maximum distance between two classes of samples shall not be taken, but a certain intermediate distance shall be taken. This method uses the center theorem to find the center of a class, and then finds the minimum distance between centers in different classes. According to the classification diagram obtained in Fig. 7, it can be clearly seen that labels 10 and 46 are in the same category and labels 1 and 22 are in the same category. This is similar to the previous single linkage.

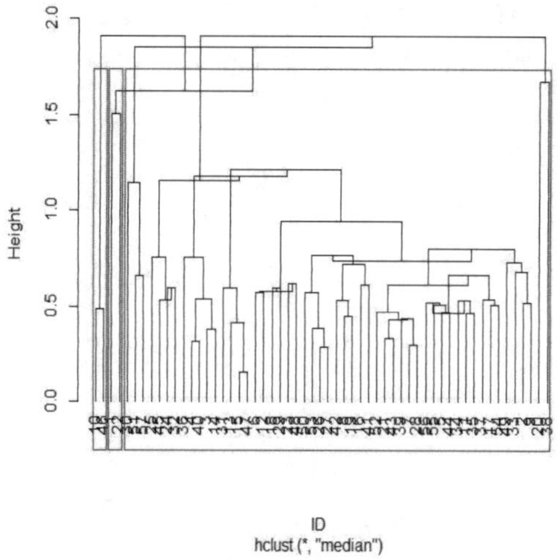

Figure 7: Median distance method

The fourth is UPGMA (unweighted pair group method with arithmetic mean), and the classification result is shown in Figure 8. Average link analysis is usually called UPGMA (unweighted pair group method with arithmetic mean). Because big data sets are becoming more and more popular in the field, and UPGMA is easy to understand conceptually and very fast in practice, it can be said that it is the most popular hierarchical clustering algorithm, and it is still widely used and very popular.[9] The fifth is Centroid Clustering, and the classification result is shown in Figure 9. This method is simple and efficient, which is suitable for both the tasks to be solved and the internal clustering module.[10] There are one or two labels in the two categories as a separate category, which may be caused by the influence of outliers. In Figure 8, tags 1 and 22 are a separate category, and the remaining data are classified into two categories. In Figure 9, label 30 is a separate category, and label 1 and 22 are a separate category. This is similar to the previous classification results.

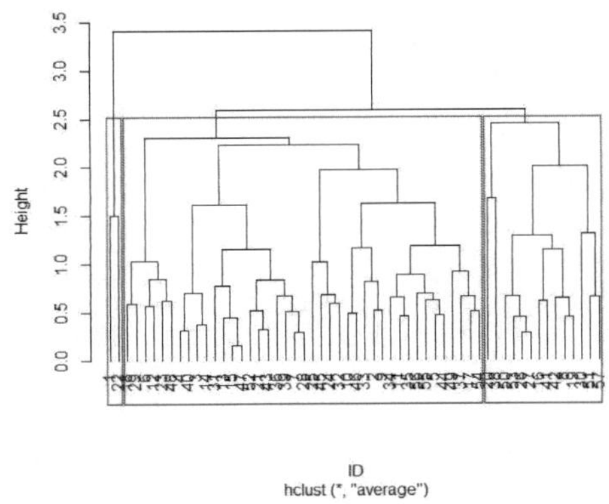

Figure 8: UPGMA

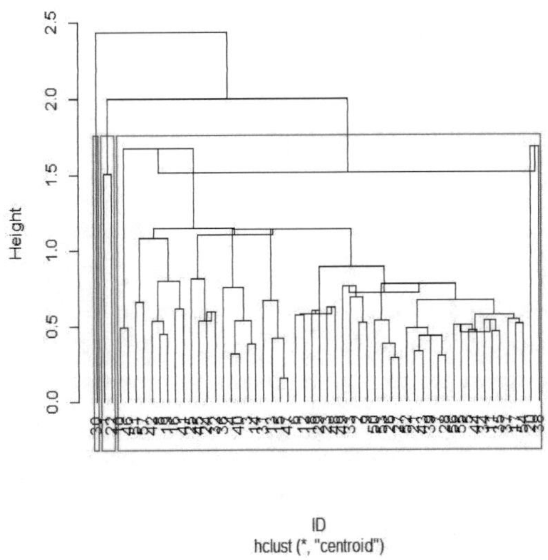

Figure 9: Centroid Clustering

After getting these detailed hierarchical clustering, hence after customizing the three categories of selection labels according to the attributes of the data analysis and prediction can be performed . Compared with these clustering analysis, these classification charts clearly show that for tags 1 and 22, they have high similarity and are classified into the same category. This means that in the follow-up research, we can better mine the relevance between the two tags. It may even serve as the basis for classification so that when new data is obtained, it can be classified according to these two labels. At the same time, the result graph of clustering based on the longest distance method shows that there is no single or very small amount of data that is clustered into a single group, which may be related to the characteristics of classification, so it needs to be further tested and observed in the analysis and prediction.

3. USING K-NEAREST NEIGHBOR ALGORITHM FOR ANALYSIS AND PREDICTION

After five clustering algorithms, the data set grouping will be obtained, which is composed of many similar objects. Through clustering algorithm to find the internal structure of data, all data instances are organized into some similar groups. The data instances in the same cluster are the same as each other, and the instances in different clusters are different from each other. When using five groups of algorithms, the original data will be divided into three groups. In this regard, k-nearest neighbor classification algorithm is used to calculate the error rate of comparison between the original data and the predicted data. [11] Combined with the advantages of KNN, the visualization results are constructed and the error rate prediction results are analyzed according to the data. For each node variable, we use the data of N nodes in the dataset. For the given data, the value of n is set to 76. The specific attributes of 76 nodes are copied out to five copies, which are labeled with five groups of feature

classification labels analyzed by clustering algorithm before. According to each copy, the text file is generated, parsed by python, and a two-dimensional diffusion graph is drawn by using Matplotlib. Part of the data of the text file is used as the test sample to calculate the error rate. If the error rate is within the acceptable range, k-nearest neighbor algorithm can be run to classify. From the beginning of k = 1, gradually increase, use the test data to analyze the accuracy, so as to choose the best K. This can provide a reference for communication operators to implement the D2D payload balance and scheduling. [12] At the same time, it has some reference significance for ICN. The data traverses the network according to the match between its name and the user's interest in this kind of content, which is independent of the network location and realizes the efficient and rich content transfer. [13]

3.1 Limitations

In data processing, we choose three main types of feature data: messages, proximity and participants. In the analysis of all the overall data, there are some limitations. After using clustering algorithm for analysis, the label results can be verified by k-NN algorithm, and the label category of node attributes can be obtained accurately, so as to infer the strength and weakness of information nodes. In this respect, the strength of information nodes and the compactness of adjacent nodes need more research to prove. When KNN model is used to analyze data, it has the limitation of parameter selection and heavy learning time. [14] Despite these limitations, we are still trying to analyze and predict data results in social networks and communications, which will be discussed below.

3.2 Analysis Process and Results

In the initial analysis, we get and separate the original data set. The value of each node is extracted from the participants data table, and the device_average data is calculated from the proximity data table. Moreover, the corresponding created messages data of each node is extracted from the messages data table. These data are collected and imported into a data table. After the analysis of five clustering algorithms, five different copies of data table are obtained. In these copies, each node has a label attribute generated by clustering algorithm. These data are parsed by python, and two-dimensional diffusion graph is drawn by Matplotlib.

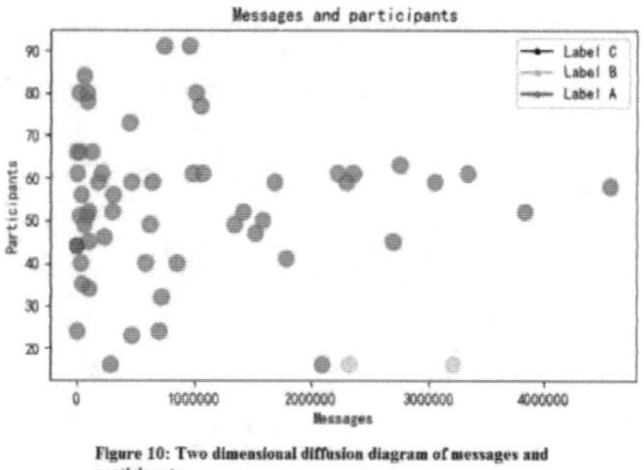

Figure 10: Two dimensional diffusion diagram of messages and participants

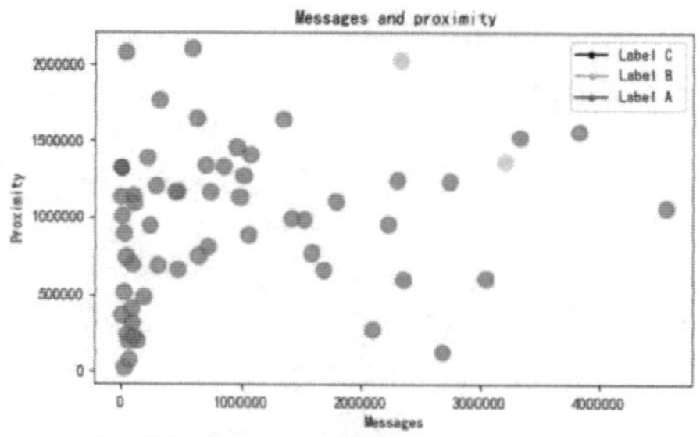

Figure 11: Two dimensional diffusion diagram of messages and proximity

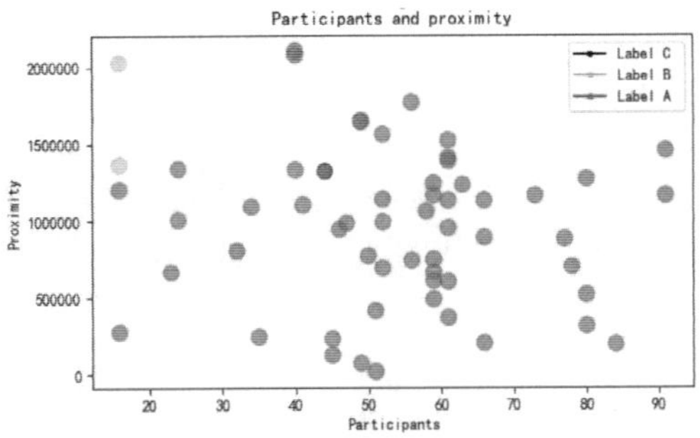

Figure 12: Two dimensional diffusion diagram of participants and proximity

We can get figure 10,11,12 according to the data copy of classification analysis using the shortest distance method. Through the visualization of data, we can clearly feel the decentralized relationship and diffusion form between the two attribute tags. The data analyzed by other four clustering algorithms are also analyzed in the same way. Through comparison, from the perspective of visual results, their distribution pattern has some similarity. In the equation, the attribute with the largest number difference has the greatest impact on the calculation results, that is to say, messages will have a far greater impact on the calculation results than the other two features. The reason for this is that the number of messages is larger than the other two eigenvalues. If these three-feature data are regarded as equally important, we need to balance their feature weights to prevent any of them from seriously affecting the analysis and prediction results. When dealing with the eigenvalues of different value ranges, we use the method of numerical normalization to process the value ranges from 0 to 1. We use the following formula to normalize the three characteristic attributes:

NewValue = (OldValue - Min) / (Max - Min)

The normalized data is obtained by subtracting the minimum value from the original value and dividing it by the difference between the maximum value and the minimum value. Moreover, the range of data and the minimum value of data are calculated, and the data is preprocessed. Next, 60% of the selected node attribute data is selected as the training sample to train the classifier, and the rest 40% data is used to test the classifier to detect the accuracy of the classifier. This 40% can be selected randomly from the data to prevent affecting its randomness. We chose the option of including three labels in the measurement to prevent the situation that there are only one or two labels in the test sample. The error rate is 13.043478% by setting the K value to 1 and cross checking the experiment. At the same time, we analyze and test the other four copies, and get their error rate as follows: 34.782609%, 21.739130% (K=3), 17.391304% (K=3), 4.347826%. The change of error rate is detected by changing the value of function, inner variable and classifier K. It can be seen from the results of the validation classifier that the error rate of KNN is 4.347826%. Compared with other classification results, it is a result with the lowest cross validation error rate. Through the analysis, we can get that this model has a low error rate for the data classified by barycenter method, which proves the accuracy of the algorithm.

Therefore, the new input data can be predicted, and the expected results can be obtained. According to the prediction results, we can get the specific label classification, and then analyze the strength of nodes and the tightness between nodes.

4. Deployment and application

The purpose of this work is to analyze the high-intensity information nodes and their close relationship by establishing a model. From the expected results with the analysis high-intensity information nodes can be determined. These nodes can be applied in many fields.

4.1 Positioning with nodes.

Positioning with information nodes is very important for wireless networks. Based on the received signal strength, the position estimation of the source node is generally carried out by converting the received signal strength measurements of several known positions into the distance from the source node. [15] The process is as follows: the transmitting node's transmitting signal strength is known, the receiving node calculates the transmission loss of the signal according to the received signal strength, transforms the transmission loss into distance by using theoretical and empirical model, and then calculates the node's position by using the existing algorithm. Through the model, the high-strength nodes are determined, and the distance to the source node is calculated by detecting its strength measurement value, which can help to locate. For some network operators, calculating the location of high-strength nodes and ranging according to their strength can help them determine the location of the signal tower and maximize the communication resources as much as possible. In addition, node location is widely used in military, transportation, industry, and environmental fields. For example, military reconnaissance, radar, intelligent transportation, environmental monitoring, emergency rescue, etc.

4.2 Provides a reference for D2D to achieve payload balance and scheduling.

D2D (device to device) communication is a new technology, which allows terminals to directly communicate by reusing cell resources under the control of the system. It does not rely on the central node, but on the edge nodes of the network, to realize self-organization and peer-to-peer collaboration of resource discovery and sharing. D2D has the potential to improve system performance, enhance user experience and expand the application prospect of cellular communication. D2D adopts the authorized frequency band of telecom operators, with controllable interference environment and high data transmission reliability. In the edge computing network of D2D, the task will choose to compute at the sensor node or on the edge server according to its own attributes and the load balance state of the network. [16] When the edge server is overloaded, it needs to schedule tasks to other relatively idle servers to achieve load balancing between the edge servers. This model infers the load of the task on the node by judging the strength of the node, so as to judge whether the node is overloaded. This provides a reference for D2D to achieve load balancing and scheduling.

4.3 Provide a new link prediction algorithm.

Link prediction is an effective method of data mining and prediction analysis in social networks. Link prediction is mainly divided into two types: one is to predict the future links according to the current network status. Second, it can identify existing but unknown links.

[17] Link prediction plays an important role in some fields, such as information retrieval, system recommendation, etc. In the future, through the model to determine the node strength, take the public neighbor node as the main research object, to study the link strength between each public neighbor node and two neighbor nodes, so as to predict the nodes. The link strength of the common neighbor node and the prediction node will affect the link prediction. Compared with other algorithms for the similarity between nodes, using the link strength of nodes to predict the link has better test results while maintaining the same time complexity.

5. Conclusion

In conclusion, this paper analyzes and processes the sigcom data. Then, five different calculation methods are used to analyze the processed data: single link, full link, median distance, UPGMA and centroid clustering. After scoring each data segment, the hierarchical structure of clustering is described intuitively by image. After hierarchical clustering analysis, k-nearest neighbor algorithm is used to classify and predict network users. After getting the prediction data, compare with the original data, and get the error rate. The result shows that the model has low error rate for the data classified by barycenter method, which proves the correctness of the algorithm. Through this model, we can determine some high-intensity information nodes. In the subsequent work development, the strength of these nodes and their close relationship can be determined. These identified high-intensity information nodes also have important applications in some fields. For example, in the fields of military, industry, communication, and transportation, nodes are used to locate; for D2D, it provides a reference for load balancing and scheduling; for link prediction, it provides a new algorithm.[18-22] There are many different application were the clustering can be applied .

References

[1] yu zhiwen, yu zhiyong, zhou xingshe. Social perception computing: concepts, problems and research progress [J]. Journal of computer science, 2012,35 (01) : 16-26.

[2] ChenL, NugentCD CookD, YuZ. Knowledge-driven activity recognition in intelligent environments. Pervasive and Mobile Computing 2011 7 (3): 285 - 286

[3] MadanA, CaneelR PentlandA. GroupMedia: Distributed multimodal interfaces //Proceedings of the 6th International Conference on Multimodal Interfaces (ICMI. State College, Pennsylvania, USA, 2004: 309-316)

[4] A.-K. Pietiläinen, E. Oliver, J. LeBrun, G. Varghese, and C. Diot, "MobiClique: middleware for mobile social networking," in Proceedings of the 2nd ACM workshop on Online social networks, 2009, pp. 49-54.

[5] Chu, X., Ilyas, I. F., Krishnan, S., & Wang, J. (2016, June). Data cleaning: Overview and emerging challenges. In Proceedings of the 2016 International Conference on Management

of Data (pp. 2201-2206).

[6] Heckerman, D. E., Bradley, P. S., Chickering, D. M., & Meek, C. A. (2008). U.S. Patent No. 7,333,998. Washington, DC: U.S. Patent and Trademark Office.

[7] Ros, F., & Guillaume, S. (2019). A hierarchical clustering algorithm and an improvement of the single linkage criterion to deal with noise. Expert Systems with Applications, 128, 96-108.

[8] Mamun, A. A., Aseltine, R., & Rajasekaran, S. (2016). Efficient record linkage algorithms using complete linkage clustering. PloS one, 11(4).

[9] Moulton, V., Spillner, A., & Wu, T. (2018). UPGMA and the normalized equidistant minimum evolution problem. Theoretical Computer Science, 721, 1-15.

[10] Huang, G., Larochelle, H., & Lacoste-Julien, S. (2019). Centroid networks for few-shot clustering and unsupervised few-shot classification. arXiv preprint arXiv:1902.08605, 3(7).

[11] J. Shi and L. Yang, "A Climate Classification of China through k-Nearest-Neighbor and Sparse Subspace Representation," Journal of Climate, vol. 33, no. 1, pp. 243-262, 2020.

[12] Y. Cao, C. Long, T. Jiang, and S. Mao, "Share communication and computation resources on mobile devices: A social awareness perspective," IEEE Wireless Communications, vol. 23, no. 4, pp. 52-59, 2016.

[13] W. Moreira and P. Mendes, "Social-aware forwarding in opportunistic wireless networks: Content awareness or obliviousness?," in Proceeding of IEEE international symposium on a world of wireless, mobile and multimedia networks 2014, 2014: IEEE, pp. 1-6.

[14] G. Alimjan, T. Sun, Y. Liang, H. Jumahun, and Y. Guan, "A new technique for remote sensing image classification based on combinatorial algorithm of SVM and KNN," International Journal of Pattern Recognition and Artificial Intelligence, vol. 32, no. 07, p. 1859012, 2018.

[15] H. Wen, Z. Fang, M. Wang. (2017). "A source node location algorithm based on received signal strength." JOURNAL OF GEODESY AND GEODYNAMICS, vol. 37 Issue (4): 380-384

[16] R Wang, x.nie, d. Wu, h. Li. (2020). "Social attribute aware task scheduling strategy for edge computing. "Journal of electronics & information technology, vol. 42no. 1. P271-278

[17] Y. Guisheng, y. Wansi and d. Yuxin, "A new link prediction algorithm:

[18] Node link strength algorithm," 2014 ieee symposium on computer applications and communications, weihai, 2014, pp. 5-9, doi: 10.1109/scac.2014.8.

[19] Punithavathani, D.S., Sujatha, K. & Jain, J.M. Surveillance of anomaly and misuse in critical networks to counter insider threats using computational intelligence. Cluster Comput 18, 435–451 (2015). https://doi.org/10.1007/s10586-014-0403.

[20] Lakshmanaprabu, S. K., Mohanty, S. N., Krishnamoorthy, S., Uthayakumar, J., & Shankar, K. (2019). Online clinical decision support system using optimal deep neural networks. Applied Soft Computing, 81, 105487.

[21] Anbarasan, M., Muthu, B., Sivaparthipan, C. B., Sundarasekar, R., Kadry, S., Krishnamoorthy, S., & Dasel, A. A. (2020). Detection of flood disaster system based on IoT, big data and

convolutional deep neural network. Computer Communications, 150, 150-157.

[22] Appathurai, Ahilan, et al. "A study on ECG signal characterization and practical implementation of some ECG characterization techniques." Measurement 147 (2019): 106384.

IOP Conf. Series: Materials Science and Engineering **937** (2020) 012052 doi:10.1088/1757-899X/937/1/012052

Performance Improvement of ZSI Based PMS Motor Drive System for Electric Vehicles

[1]B Karunamoorthy [2]J Ramprabu,

Department of electrical and Electronics Engineering, Kumaraguru college of Technology, Coimbatore, India
[2]Assistan professor II, Department of electrical and Electronics Engineering, Kumaraguru college of Technology, Coimbatore, India
jrameee@gmail.com

Abstract. In traction system the choice of motor is very important, multipurpose motors are available in the market. By considering the motor parameter like high power to weight ratio, high torque to current ratio with that high efficiency and robustness. Owing all the advantages the motor chosen is PMSM (Permanent magnet synchronous motor) for the variable speed modern drive which is most widely used in EV's (Electric vehicle and in Hybrid electric vehicle). Without weakening the field by controlling the DC-link voltage the PMSM machine can run at high speed by implementing ZSI to the drive system. Z source inverter working principle and modulation method is discussed in this study. By considering the Limitation of Maximum voltage and available line current a new Vector control scheme is developed for PMSM drive voltage boosting. At last the proposed system implementation effectiveness and reliability is tested by several simulation and experimental results.

1.INTRODUCTION

Increase in population is proportionally increased in internal combustion engine Automobile. This leads demand in natural resources and drastic increase in the price of petroleum products. By concerning the global problem all the IC engines should be replaced with Electric Motor which turns our attention to the research area Electric Vehicle (EV). The EV's are Eco friendly and no emission to the environment. In Idle speed it will not consume energy and also it will not produce any sound at the time of running.

A drive system is necessary for the Electric vehicle to control the speed from stand still to the maximum speed to propel the vehicle. Multiple motor and their drive system are available in market out of which Permanent magnet synchronous motor is very popular and attractive because of their reliability, high power density and efficiency [5] [10]. To operate the motor in high-speed region, field weakening control scheme is utilized since the PMSM machine have a short constant-power region due to their limited field weakening capability [5], still it requires some more current to reduce the magnetic flux of the machine. In reference [6] [11], Dc-Link voltage is changed with boost converter connected in series with the PWM inverter for the rated speed of PMSM drive system. Z source inverter existing PWM technique total harmonic distortion is bigger due to unsymmetrical shoot through is discussed in [1]. Different modulation strategies impedance source network to generate DC to AC voltage is discussed in [2], MPC model predictive control discuss about the fast torque control of PMSM machine [4], Fuzzy logic based vector control led PMSM machine speed control using DS1103 discussed in this paper [3].

A competitive alternative topology is introduced in single stage power converter and Z-source inverter due to the defects in tradition voltage inverter because of the complexity of the circuit and and control, the two stage system requires more space and proposed single stage ZSI is discussed in [7] [8] [12]. Both the buck and boost capabilities advantages adopted in the different applications are discussed in [7] and the same is discussed with Ac induction motor drive in [12]. A birectional Z source inverter with the steady state operating principle and voltage boosting modified vector control scheme of permanent magnet synchronous motor drive is presented in the paper.

Z source inverter both direction topologies has discussed in [20]. The discontinuous induction current causes unstable and un controllable ZSI, it can be overcome by operating the ZSI in Low load power factor or with small inductance [28], The system bidirectional power flow is discussed in [21]. The proposed circuit ZSI is used to feed current to PMSM machine as current fed Z-source DC/DC converter under the condition of reverse power flow. The complicated operation and increase in number of switches and their mode of operation are compared to [20]. Z source inverter, Pulse width modulation traditional inverter and DC/DC boosted PWM inverter are compared in [22] [29]. As per the comparison the ZSI Conversion efficiency of inverter and motor system efficiency is better when compared to the earlier two existing inverter system.In this survey PMSM drive system is controlled by Bi-directional Z source inverter as a single stage converter instead of a two stage converter (DC/DC converter bidirectional and VSI). In the next section, The working principle and pulse width modulation method and the equivalent circuit of the Bi-directional Z source inverter are discussed. Limitations of maximum line current and voltages of the Permanent magnet synchronous motor are discussed in the next section. In the last section the experimental results and simulations results for the implemented drive system is shown.

2. BIDIRECTIONAL Z-SOURCE INVERTER

The proposed drive system is shown in the figure which consists of Permanent magnet synchronous motor connected with conventional voltage source inverter with a impedance network. The power to the circuit is fed by the Battery pack connected to it. The bidirectional power flow system ability is enabled by introducing switch S7. A PN junction diode is connected Anti parallel to S7 which eliminates the discontinuous inductor current caused by undesirable mode of operation. To achieve the required propertied two similar inductors and capacitors are connected in specific manner as shown in Fig.1

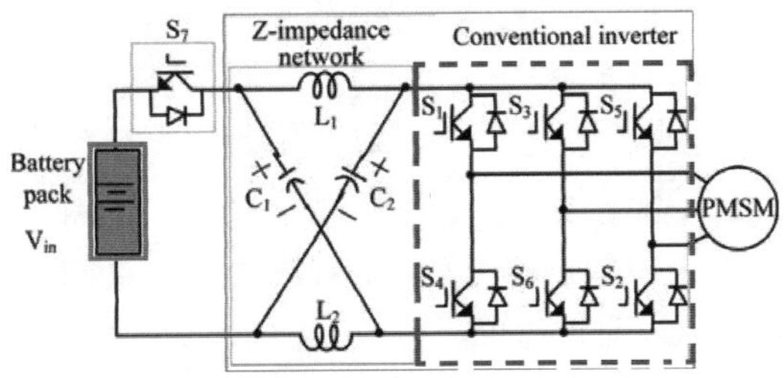

Fig. 1 Bidirectional ZSI for PMSM drive

To maintain the output voltage the active states should be kept as same and partially replace the traditional zero states or the zero state to be shoot-through entirely based on voltage level required to boost base on our need. Each operating state characteristics are described separately as shown below

From the equivalent circuit based on symmetry we have $V_{c1} = V_{c3} = V_{c2}$, $v_{l2} = v_{l1} = v_{l3}$ (1)

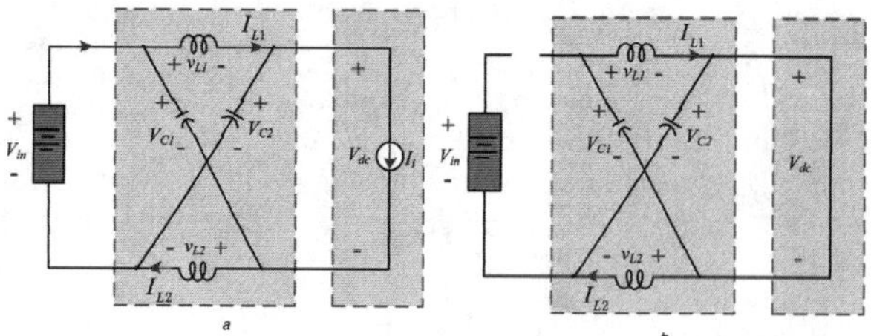

Fig. 2 ZSI Equivalent circuits

2.1 Bidirectional ZSI PWM control strategy

Traditional voltage source converter uses various Pulse width modulation control methods at present. Traditional PWM methods will not use the shoot through state so to control the ZSI DC boost PWM methods need to be modified. The method we require is controllability in wide range AC output voltage and less harmonic. A technique that bridges PWM and Space vector control is space vector pulse width modulation (SVPWM) technique. Four PWM control methods are existing, Maximum constant boost control method is discussed in [23]. Simple boost control method is discussed in [15], Modified space vector control method discussed in [31] and maximum boost control method is discussed in [30]. In this paper for attaining variable speed application shoot through is control by using MSVPWM method.

While inverter is operated in shoot through state switch S7 has to be kept in OFF in the mode analysis of ZSI. Switch position remains in ON state when the inverter is operated in its non-shoot through states. Boost capacitor is used in non shoot through sate when inverter is operated in regenerating mode. In the different mode of operation whenever reverse current is monitored the switch S7 is closed permanently since because the battery receives constant reverse current.

3.PMSM DRIVE SYSTEM DESIGN

3.1Current and Voltage Range

The rotor reference frame steady state voltage equation of PMSM is,

$$T_e = P\frac{3[i_{sq}\lambda_{pm} + i_{sq}(L_d - L_q)i_{sd}]}{2} \quad (2)$$

$$\begin{bmatrix} V_{sd} \\ V_{sq} \end{bmatrix} = \begin{bmatrix} L_d P + R & -L_q\omega_e \\ L_d\omega_e & L_q P + R \end{bmatrix} \begin{bmatrix} i_{sd} \\ i_{sq} \end{bmatrix} + \begin{bmatrix} 0 \\ \lambda_{pm}\omega_e \end{bmatrix} \quad (3)$$

Where, d- and q-axis current sare the i_{sd} and i_{sq} currents ; d- and q-axis voltages are v_{sd} and $v_{sq\ voltages}$; d- and q-axis inductances are the L_d and L_q; Stator resistance is the R ; Electrical angular frequency is ω_e ; number of pole pairs is P and electromagnetic torque T_e d/dt is the derivative operator of p; flux linkage is PM respectively

$v_{s\,max}$ maximum available voltage and $i_{s\,max}$ maximum line current amplitude of the motor with that we can get the following equation

$$v_{sq}^2 + v_{sd}^2 \le v_{s\,max}^2 (4)$$

$$i_{sq}^2 + i_{sd}^2 \le i_{s\,max}^2 \ (5)$$

In high speed operation by neglecting the armature resistance drop, in steady state the derivative operator P becomes zero by substituting (4) and (5) into (2), With that we can obtain an voltage constraint equivalent as follow,

$$\left(i_{sq}L_q\right)^2 + \left(\lambda_{pm}+L_d i_{sd}\right)^2 \leq \frac{v_{s\,max}^2}{\omega_e} \quad (6)$$

3.2 Maximum Torque per Ampere Operation(MTPA)

Motor parameters depends on the MTPA trajectories, MTPA operation can be done and IPM motor has a saliency (L_q, L_d) With the reluctance torque

The following equation obtains the operating point as follows, The operating point can be obtained by the following equations simplified is,

$$i_{sq} \leq \sqrt{i_{s\,max}^2 - i_{sd}^2} \quad (7)$$

$$i_{sd} = -i_{sq}^2 \frac{\left(L_q - L_d\right)}{\lambda_{pm}} \quad (8)$$

3.3 Current Range of the battery

To run a motor above its rated speed with constant power and also to operate below rated speed with the same power when current of the battery is limited. When motor draws the rated current that value corresponds to the DC-link Current. I_{BTmax} [32-35].

Current value is limited by considering the parameters and characteristics of the battery. The below equation defines the maximum torque output of the machine.

$$T_{max} = \frac{I_{BT\,max}V_{in}}{\eta\omega_r} \quad (9)$$

Constant torque can be achieved only if the current of the battery is not limited; if the current of the battery is limited then I_{DC} DC-Link current is as follows

$$I_{DC} = I_{BTmax}\eta\left(\frac{V_{in}}{V_{DC}}\right) \quad (10)$$

Therefore the VDC DC-link voltage increases as the I_{DC} decreases. In other words, current I_{BTmax},. for the constant battery the current limited circle shrinks as the speed increases.

4.CONTROL DESIGN

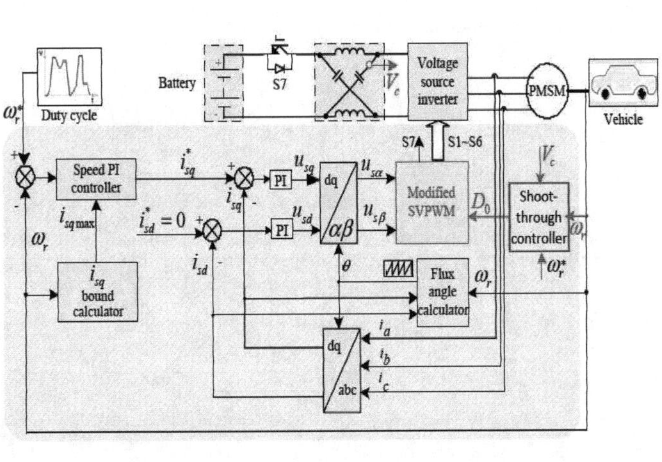

Fig.3 Control scheme PMSM with Bi-directional ZSI

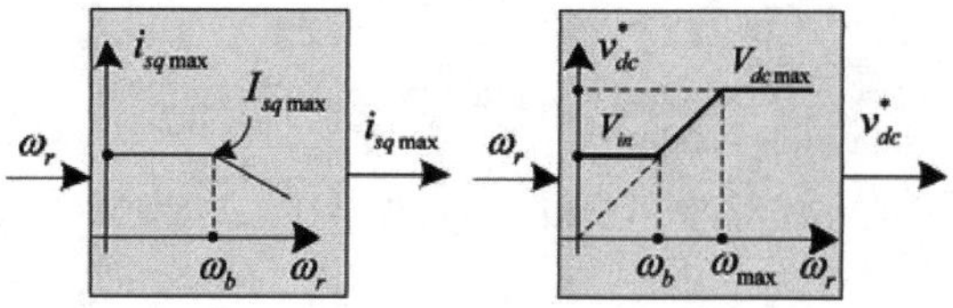

Fig. 4 Functional diagrams a) i_{sq} current calculator b) voltage command generator DC-link

Fig.3 show the functional diagram of i_{sq} bound calculator .The q-axis current is the output maximum amplitude and rotor speed is the input parameter. Maximum line current of the motor is limited using speed PI controller in the zero d-axis control modes such that current of the battery will not exceed the limit. Fig. 4(b). Shows the command block of the DC-link, ω_r rotor speed is less than the rated speed ω_b, Z source inverter will operate without the DC-link voltage command and boost which are equal to the applied input voltage. In the high speed operation the rated input voltage should apply higher. Feedback signal cannot be selected because of the DC-link voltage in the shoot through state of Z source inverter. The duty ratio D_o shoot through time by controlling it V_c capacitor voltage of the Z-source can be increased. So in this paper, v_c is selected as feedback signal to control v_{dc} indirectly. In order to overcome the nonlinear problem between v_{dc} and v_c, a linear capacitor voltage controller [9] is adopted. The task of shoot through controller in Fig.3 is to generate D_o.

5. Simulation Results

Saber software is used to take the simulation result .Bidirectional ZSI based proposed PMSM drive system for an electric vehicle by considering the PMSM parameters like, battery, Z-source network, Transmission system are listed in the Table 1,

Table 1 System Parameters

Component	Parameter	Value
Transmission system	Ratio Efficiency	1:2 0.91
PMSM	rated speed rated torque d-axis inductance (L_d) q-axis inductance (L_q) flux of field rated output power armature resistance (R) pole number	51 rad/s 339 Nm 9 mH 8.5 mH 0.7 Wb 17 kW 0.4375 ohm 8
Battery	rated capacity rated voltage (EBT)	100 A h 410 V

IOP Conf. Series: Materials Science and Engineering **937** (2020) 012052 doi:10.1088/1757-899X/937/1/012052

Vehicle	aerodynamic drag coefficient	0.3
	gross mass	990 kg
	rolling resistance coefficient	0.015
	radius of wheels	0.288
	frontal area	1.7 m^2
Z-source network	Capacitor	300 μF
	inductor	36.5 μH

The electric vehicle proposed drive system dynamics and effectiveness are simulated extensively through saber. The table 1 list all the parameters. The different operating mode of the system is detail shown in Fig. 5. ZSI driven PMSM characteristics and simulation results are shown in Fig.7. The rated torque is achieved in the time interval of 0.4-3.9 s in the acceleration mode. 3.9-4.2s time interval requires to reach the rated speed in cruise mode. 4.2-11s requires to decrease in torque in acceleration mode. 11-12s, time interval to reach the maximum speed in cruise control mode. 12-18s required to reach maximum speed to rated speed in regenerative braking mode. 20-23.5s time intervals required for regenerative braking mode with the rated torque. 18-20s time intervals with the rated speed requires in cruise operation mode.Here, a dip is observed in the battery current and the q-axis current waveforms at 3.9 s. This is because the vehicle's resistance decreasing significantly during the time interval (3.9–4.2 s), where the vehicle is in cruise operation mode. The shoot-through time duty ratio, boost factor, DC-link voltage and capacitor voltage of PMSM driven by ZSI against speed characteristics presented in Fig. 7.

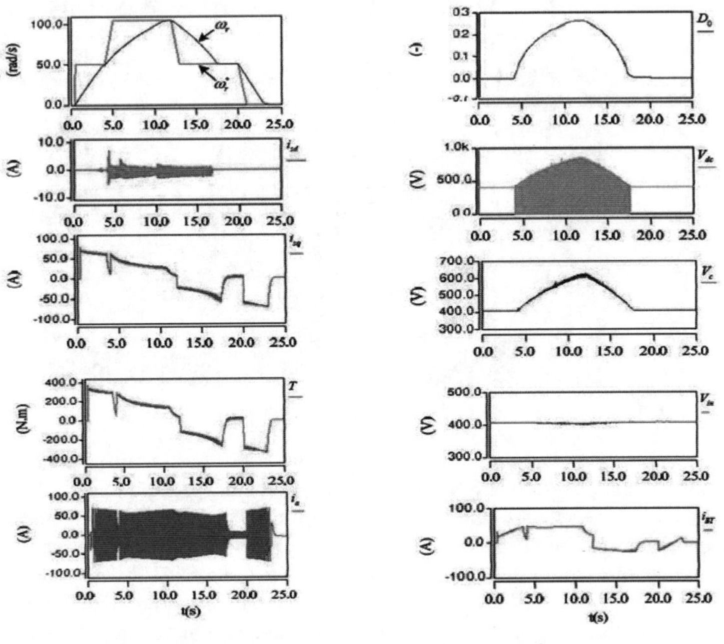

Fig. 5. PMSM driven by ZSI Transient waveforms

Fig. 6. ZSI operation Characteristic

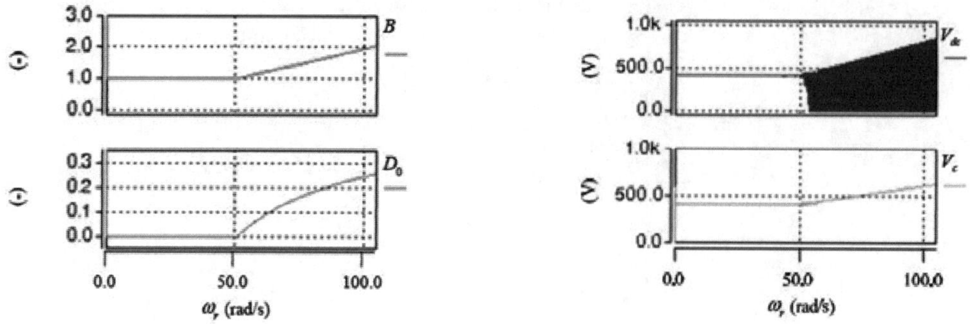

Fig. 7 (a) speed characteristics vs shoot-through and Boost factor time duty ratio (b) speed characteristics vs capacitor voltage versus and DC-link voltage and

Fig.5 - 7, one finds the following mentioned below,

- Above the rated speed the d- axis current kept zero. Hence there is no additional current required to reduce the flux of the motor.
- Since i_{BT} is negative bidirectional power flow is possible in regenerative braking mode.
- Input voltage and DC-link voltage are equal at rated speed while ZSI working in non-shoot through mode.

Fig. 8 ZSI experimental set-up

Using the table 1 parameters the bidirectional Z source inverter prototype for PMSM drive is designed and implemented in the laboratory set up. For the switching frequency generation the TMS320F2808 board has chosen to generate 10KHZ system frequency for testing the proposed control methods and realization. Code generation is done with code composer studio for the real time control. DAC 12 bit resolution is used to find the wave forms and inspect in oscilloscope. System functional block is shown in the Fig.7. Laboratory setup is shown in Fig.8. Bidirectional ZSI switches (S1–S7) and intelligent power module PM800HSA120 is used. Fig. 9 shows the d-axis current, stator current, speed command and motor speed, DC-link voltage and battery voltage; motor speed and capacitor voltage short-through time duty ratio; respectively, which are quite consistent with the simulation results.

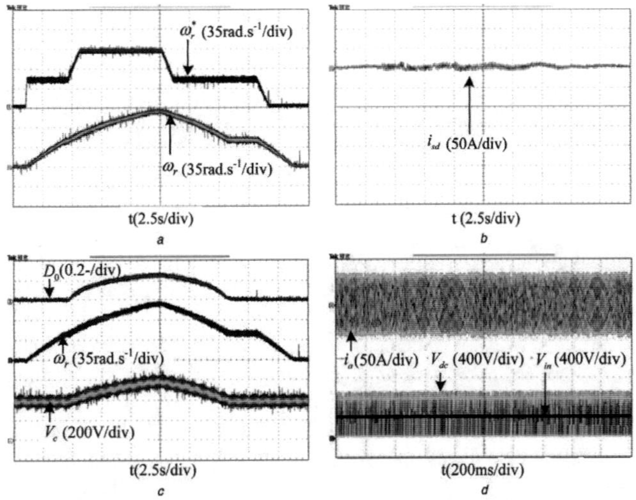

Fig. 9 Experimental result of characteristics for PMSM driven by ZSI (a) motor speed and Speed command (b) current d-axis (c) capacitor voltage, motor speed and Short-through time duty ratio (d) battery voltage, Stator current, DC-link voltage and

6. Conclusion

The bidirectional Z source inverter is presented in the paper. The main focus is to increase the speed range of permanent magnet synchronous motor by reducing the amplitude of the current in the high speed region, this is achieved by gating simultaneously both the lower and upper switches of the same phase leg, when the speed of the machine goes higher than the rated speed. The shoot through of the circuit no longer disturbs the system and hence the reliability goes higher. It is a stable low cost single stage highly efficient system. The experimental result is compared with the simulated result and the simulated results were validated. In the proposed system the D-axis current is totally eliminated in the constant power region, obviously in the flux weakening operation the copper loss of the motor will be reduced.

References

[1]. Sitharthan R, Geethanjali M and Pandy TKS 2016 Adaptive protection scheme for smart microgrid with electronically coupled distributed generations *Alexandria Engineering Journal* **55(3)** 2539-2550

[2]. Fathima AH, and Palanisamy K 2014 Battery energy storage applications in wind integrated systems—a review *IEEE International Conference on Smart Electric Grid* 1-8

[3]. Prabaharan N and Palanisamy K 2015 Investigation of single-phase reduced switch count asymmetric multilevel inverter using advanced pulse width modulation technique *International Journal of Renewable Energy Research* **5(3)** 879-890.

[4]. Jerin ARA, Kaliannan P and Subramaniam U 2017 Improved fault ride through capability of DFIG based wind turbines using synchronous reference frame control based dynamic voltage restorer. *ISA transactions* **70** 465-474

[5]. Sitharthan, R, Sundarabalan CK, Devabalaji KR, Nataraj SK and Karthikeyan M 2018 Improved fault ride through capability of DFIG-wind turbines using customized dynamic voltage restorer *Sustainable cities and society* **39** 114-125

[6]. Prabaharan N and Palanisamy K 2016 A single-phase grid connected hybrid multilevel inverter for interfacing photo-voltaic system *Energy Procedia* **103** 250-255

[7]. Palanisamy K, Mishra JS, Raglend IJ and Kothari DP 2010 Instantaneous power theory based unified power quality conditioner (UPQC) *IEEE Joint International Conference on Power Electronics, Drives and Energy Systems* 1-5

[8]. Sitharthan R and Geethanjali M 2017 An adaptive Elman neural network with C-PSO learning algorithm-based pitch angle controller for DFIG based WECS Journal of Vibration and Control **23(5)** 716-730
[9]. Sitharthan R and Geethanjali M 2015 Application of the superconducting fault current limiter strategy to improve the fault ride-through capability of a doubly-fed induction generator–based wind energy conversion system *Simulation* **91(12)** 1081-1087
[10]. Sitharthan R, Karthikeyan M, Sundar DS and Rajasekaran S 2020 Adaptive hybrid intelligent MPPT controller to approximate effectual wind speed and optimal rotor speed of variable speed wind turbine *ISA transactions* **96** 479-489
[11]. Sitharthan R, Devabalaji KR and Jees A 2017 An Levenberg–Marquardt trained feed-forward back-propagation based intelligent pitch angle controller for wind generation system *Renewable Energy Focus* **22** 24-32
[12]. Sitharthan R, Sundarabalan CK, Devabalaji KR, Yuvaraj T and Mohamed Imran A 2019 Automated power management strategy for wind power generation system using pitch angle controller *Measurement and Control* **52(3-4)** 169-182
[13]. Sundar DS, Umamaheswari C, Sridarshini T, Karthikeyan M, Sitharthan R, Raja AS and Carrasco MF 2019 Compact four-port circulator based on 2D photonic crystals with a 90° rotation of the light wave for photonic integrated circuits applications *Laser Physics* **29(6)** 066201
[14]. Sitharthan R, Parthasarathy T, Sheeba Rani S and Ramya KC 2019. An improved radial basis function neural network control strategy-based maximum power point tracking controller for wind power generation system *Transactions of the Institute of Measurement and Control* **41(11)** 3158-3170
[15]. Rajesh M and Gnanasekar JM 2017 Path observation based physical routing protocol for wireless ad hoc networks *Wireless Personal Communications* **97(1)** 1267-1289
[16]. Palanisamy K, Varghese LJ, Raglend IJ and Kothari DP 2009. Comparison of intelligent techniques to solve economic load dispatch problem with line flow constraints *IEEE International Advance Computing Conference* 446-452
[17]. Sitharthan R, Ponnusamy M, Karthikeyan M and Sundar DS 2019 Analysis on smart material suitable for autogenous microelectronic application *Materials Research Express* **6(10)** 105709
[18]. Rajaram R, Palanisamy K, Ramasamy S and Ramanathan P 2014 Selective harmonic elimination in PWM inverter using fire fly and fireworks algorithm *International Journal of Innovative Research in Advanced Engineering* **1(8)** 55-62
[19]. Sitharthan R, Swaminathan JN and Parthasarathy T 2018 March. Exploration of wind energy in India: A short review *IEEE National Power Engineering Conference* 1-5
[20]. Karthikeyan M, Sitharthan R, Ali T and Roy B 2020 Compact multiband CPW fed monopole antenna with square ring and T-shaped strips *Microwave and Optical Technology Letters* **62(2)** 926-932
[21]. Sundar D Sridarshini T, Sitharthan R, Madurakavi Karthikeyan, Sivanantha Raja A, and Marcos Flores Carrasco 2019 Performance investigation of 16/32-channel DWDM PON and long-reach PON systems using an ASE noise source *In Advances in Optoelectronic Technology and Industry Development: Proceedings of the 12th International Symposium on Photonics and Optoelectronics* 93
[22]. Sitharthan R and Geethanjali M 2014 Wind Energy Utilization in India: A Review *Middle-East J. Sci. Res.* **22** 796–801 doi:10.5829/idosi.mejsr.2014.22.06.21944
[23]. Sitharthan R and Geethanjali M 2014 ANFIS based wind speed sensor-less MPPT controller for variable speed wind energy conversion systems *Australian Journal of Basic and Applied Sciences* **8**14-23
[24]. Jerin ARA, Kaliannan P, Subramaniam U and El Moursi MS 2018 Review on FRT solutions for improving transient stability in DFIG-WTs *IET Renewable Power Generation* **12(15)** 1786-1799
[25]. Prabaharan N, Jerin ARA, Palanisamy K and Umashankar S 2017 Integration of single-phase reduced switch multilevel inverter topology for grid connected photovoltaic system *Energy Procedia* **138** 1177-1183
[26]. Rameshkumar K, Indragandhi V, Palanisamy K and Arunkumari T 2017 Model predictive current control of single phase shunt active power filter *Energy Procedia* **117** 658-665
[27]. Fathima AH and Palanisamy K 2016 Energy storage systems for energy management of renewables in distributed generation systems *Energy Management of Distributed Generation Systems* 157
[28]. Rajesh M 2020 Streamlining Radio Network Organizing Enlargement Towards Microcellular Frameworks *Wireless Personal Communications* 1-13
[29]. Subbiah B, Obaidat MS, Sriram S, Manoharn R and Chandrasekaran SK 2020 Selection of intermediate routes for secure data communication systems using graph theory application and grey wolf optimisation algorithm in MANETs IET *Networks* doi:10.1049/iet-net.2020.0051

[30]. Singh RR and Chelliah TR 2017 Enforcement of cost-effective energy conservation on single-fed asynchronous machine using a novel switching strategy *Energy* **126** 179-191

[31]. Amalorpavaraj RAJ, Palanisamy K, Umashankar S and Thirumoorthy AD 2016 Power quality improvement of grid connected wind farms through voltage restoration using dynamic voltage restorer *International Journal of Renewable Energy Research* **6(1)** 53-60

[32]. Singh RR, Chelliah TR, Khare D and Ramesh US 2016 November. Energy saving strategy on electric propulsion system integrated with doubly fed asynchronous motors *IEEE Power India International Conference* 1-6

[33]. Singh RR, Mohan H and Chelliah TR 2016 November. Performance of doubly fed machines influenced to electrical perturbation in pumped storage plant-a comparative electromechanical analysis *IEEE 7th India International Conference on Power Electronics* 1-6

[34]. Sumithra M, Abraham A, Rajesh M and Nithiyanandam N 2019 Mining social media content to predict peer trust level in social networks *International Journal of Engineering and Advanced Technology* **6S3** 1959-1964

[35]. Venkatesh K, Ali MJS, Nithiyanandam N and Rajesh M 2019 Challenges and Research Disputes and Tools in Big Data Analytics *International Journal of Engineering and Advanced Technology* **6S3** 1949-1952

Visualization and Classification of Diseases using Deep Learning Based Convolution Neural Network

Janeera.D.A[1], P.Amudhavalli[2], Sherubha.P[3], Sasirekha.S.P[4], P.Anantha Christu Raj[5], Sitharthan R[6]

[1]Assistant Professor, Sri Krishna College of Engineering and Technology, Coimbatore,
[2]Assistant Professor, Department of Computer Applications, B S AbdurRahman Crescent Institute of Science and Technology, Chennai,
[3]Independent Researcher, Coimbatore, ORCID: 0000-0002-5945-2402,
[4] Research Scholar, Karpagam Academy of Higher Education, Coimbatore,
[5]Assistant Professor, Robotics Engineering, Karunya Institute of Technology and Sciences, Coimbatore,
[6]School of Electrical Engineering, Vellore Institute of Technology, Vellore
janeeera@skcet.ac.in

Abstract. In the contemporary phase of big data, data visualization is one of the challenging segment of the discovery process. For a classifier, the primary goal is to identify the hidden levels of big data. The performance of classifiers depends on the feature space, number of classes and size of the data. To improve the reliability, efficiency and accuracy of the classifiers, new algorithms are required for analysis. This paper enables classification and visualization of information on diseases using a Deep Learning based Convolution Neural Network classifier. For feature selection and handling massive data in analysis of multivariate data is performed using particle swarm optimization (PSO) and principal component analysis (PCA) techniques. Real-world datasets are utilized for demonstration of the proposed learning algorithm. Deep learning classifiers are scientifically higher and offers better performance when compared to other classifiers according to the comparative study.

1. Introduction

Modern data science uses deep learning and big data most often due to their adaptability towards a wide range of applications. Various sources and sensors produce massive amount of data which is challenging to analyse and handle using the traditional data processing tools. The conventional tools cannot analyse the Exabyte or even larger daily data that is generated by the industries. In order to extract significant information from large quantity of data and learn from it, an advanced technique called machine learning is used [1]. Accurate data representation decides the performance of these techniques. Diminished execution is prompted by machine learning techniques that provides poor representation of data. Distinguishing the data patterns and obtaining a complete perspective of data can be done using visualization. While dealing with massive data, visualization with decent heterogeneity and diversity is a major issue. In applications pertaining to data science, visualization and big data analytics play a major role.

Deep learning is inspired by the human brain functionality and performs data representation and classification by using supervised and unsupervised approaches. In applications pertaining to voice, speech and face recognition, deep learning techniques are exploited by several search engines.

Recognition of significant information from the original data patterns and deficient precision in grouping are caused by the insignificant features. The proposed approach enables classification and visualization of information by means of deep learning based CNN. Deep learning techniques along with the CNN extension provides improved performance benefits in several applications. Fully connected layers, pooling, rectifier logic unit (ReLU), manifold convolution are the key engagements in multi-hidden layer perceptrons (MLPs) based CNN [2]. The classification and visualization results are displayed after examination of the dataset by CNN. This work proposes four major contributions namely – boosting the classifier performance by diminishing the dimensionality, minimization of computational time, improvement in the classifier accuracy and comparison of the proposed model with existing models.

2. Literature Review

A simple sentence classification model using CNN is proposed by Kim et al. using static and task-specific vectors for upgrading the conventional CNN model [3]. For question classification and sentimental analysis, a simple CNN model with single convolution layer works extremely well. In case of large volume of traffic data, prediction and visualization model presented by Ari et al. can be used [4]. Humongous traffic data and pictures are analysed and processed using FIMT-DD in this work. Within a specific sensor point, the traffic stream can be envisioned using the proposed prediction framework.

Together with pooling layers, feature maps and images are generated as a convolution of image filter sets in convolutional layers of CNN. Estimation of a class is performed with an output label on the network through the repetitive processing layers and convolution for extraction of feature maps in image classification [5]. For the purpose of solving the classification problems for any given training datasets, features are generated by the hidden layers where the filter parameters and weights are optimized by CNN rather than using the traditional hand-crafted machine learning features and techniques. The classification error can be minimized using approaches like gradient descent and back propagation for optimizing the network parameters [6].

Multiple learning techniques and PCA is used for dimensionality reduction in huge samples by Farrelly et al. for extraction of global and local features from the datasets [7]. When compared to the complete datasets, high dimensional data involves lower dimensional space and are considered as universal sets in the modern data analytics. In order to guide information towards this lower dimensional space and enable accurate recognition of this space, several dimensionality reduction strategies are established [8]. In problems regarding classification, the effectiveness of dimensionality reduction techniques is demonstrated.

The large amount of features in datasets were identified and examined by [9-19]. For efficient classification, it was identified hat not all the features were valuable. The performance of the classifier is reduced by the repetitive and unwanted features. Hence the performance improvement is done by implementing feature selection for choosing the features that contain significant attributes rather than using all the available features [11-18]. Particle Swarm Optimization is used for feature selection and to improve the classifier performance.

3. Proposed Methodology

Prediction and visualization of disease data using three different techniques namely pre-processing of information, selection of prominent features and classification is analyzed in this section. Figure 1 represents the architectural model of the proposed deep learning based CNN.

3.1. Data Preprocessing

While handling big data, it is essential to represent the data with lowest possible proportions without misplacement or losing the originality of the information. For this purpose, dimensionality reduction technique is used. Principal component analysis technique is used for reducing the big data dimensionality. The underlying features are combined to producing new features using principal component analysis [19-26]. Mapping of the given dataset and its occurrences from one dimensional space to another such that the second dimension is larger than the first is performed by PCA. Considering the first dimension as g and second as p, g enables the generation of a new set of principal components p. These principle component PCk can be represented using the following expression.

$$PC_k = x_1Y_1 + x_2Y_2 + \cdots + x_gY_g$$

Where x_i is the numerical coefficient of the original feature Y_i.

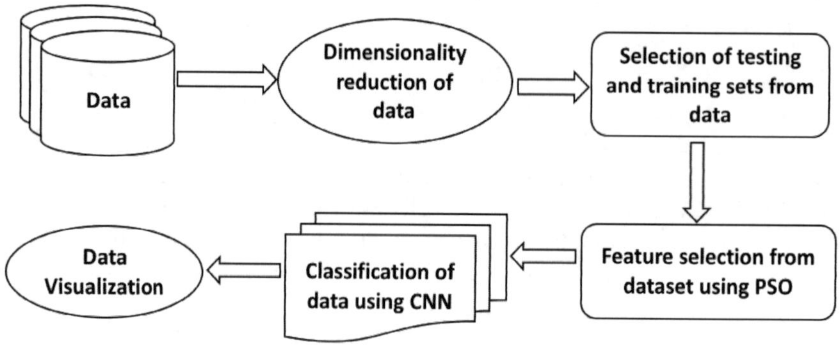

Figure 1: Architecture of the proposed model

3.2. Feature Selection

Feature selection assists in selection of a subset of unique features for improving the performance of the classifier. In text visualization and classification, feature selection is a substantially challenging task. [12] Multiple features are available in such data with high dimensionality which makes visualization and classification of the content more complex. Particle Swarm Optimization with swarm intelligence is used in this paper for extracting unique and suitable features from the feature subset [27-35]. Algorithms relating to fish schooling or bird flocking can also be used for stimulating the features. The steps involved in PSO algorithm is as represented below. The steps form a loop and iterates unless the completion state is met.

Input: Dataset
Output: Optimized subset of features

Step 1: Speed, position and population factors are initiated by the user
Step 2: The individual particle fitness is surveyed (Pbest)
Step 3: The most raised individual fitness is screened (Gbest)
Step 4: The speed of location of Pbest and Gbest is changed
Step 5: Particle position is refreshed
Step 6: If the condition is met, end, else go to Step 3

Step 1 and 2 are not very significant. The objective task receives positive responses for the fitness assessment. Position, universal ideal fitness and entity are updated in contrast with the fitness values assessed recently with respect to the individual and past global best fitness values. It is critical to replace positions and the best fitness values. The PSO technique is streamlined by refreshing the position and speed of the operational data volume. The performance of the classifier is improved by the algorithm by decreasing the classifier complexity through selection of optimal feature subset.

Gradient ascent technique called activation maximization is used by the CNN for visualization of features learnt on observation of respective neuron activation [14]. This technique is also called as feature visualization. The input image gradient is calculated in alignment to the specific neuron's mean output value by feeding the neural network with a random noise image as input [15]. Gradients can be added repetitively to the input image and the image can be optimized based on the high activation neuron direction for feature visualization of the captured neuron.

3.3. Classification

The performance of the classifier is improved by the decent portrayal of information. In this paper, we use CNN for visualization and classification of information. Feature map layer, classification layer and several other layers are available in the Convolutional Neural Network [16]. Fully connected layers, pooling and specific convolutional layers are the major categories of layers in CNN. The future portrayal of input is accomplished and learnt using the convolutional layer. This layer receives and acknowledges the data from the input layer during the operation of CNN. The convolutional layer is accountable for the channel maps of the convolution tasks to be of smaller number and of similar sizes [17]. The future layer sizes are deteriorated when the yield from convolution layer is directed to the sampling layer. The number of connections between the layers of convolution are reduced by the pooling layer thereby reducing the computational load. The CNN performance is improved by using a suitable activation function for a specific task. Figure 2 represents the various components and layers of CNN.

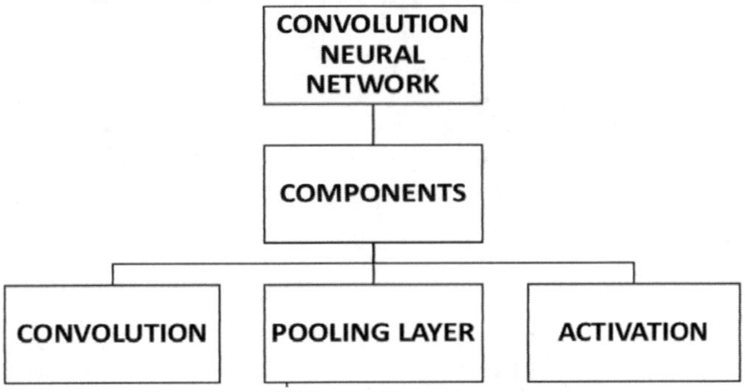

Figure 2: CNN and its components

4. Proposed Methodology

Classification and visualization of the disease data is the foremost aim of the proposed technique. Random weights are used for initialization of the network. Segmentation of the images are performed based on the pixel values. Based on the experimentation condition and type of images, the time taken per epoch is calculated and optimized. Traditionally the learned features are visualized with naïve ways for extraction of the output of hidden or intermediate layers. The calculation is halted at the layer of interest on passing the image to CNN. The assessment of the proposed classifier through various experimental analysis is discussed in this section.

4.1. Experimental Setup

The datasets of three diseases namely liver, migraine and mifem (myocardial infarction) are used in this approach for performance analysis of the proposed methodology. Along with CNN, PSO and PCA techniques have also been used in this paper for visualization and classification purposes. ANN and KNN models are compared with the proposed CNN model. Statistical measurements such as precision, G-mean, accuracy and specificity are analysed and compared for evaluation.

4.2. Performance Metrics

The efficiency of machine learning models are determined by true and false positive and negative rates. Specificity and accuracy evaluation using these parameters are calculated by the following formulae.

$$\text{Specificity} = TN/\,(TN+FP)$$
$$\text{Accuracy} = (TP+TN)/\,(TP+FP+TN+FN)$$

4.3. Experimental Results

For the given datasets, various strategies are implemented and the classification accuracy is analysed. K nearest neighbour (KNN), Artificial Neural Network (ANN) and proposed deep learning enabled Convolution Neural Network (CNN) classifiers are compared for samples datasets of liver, migraine and mifem as represented in figure 3.

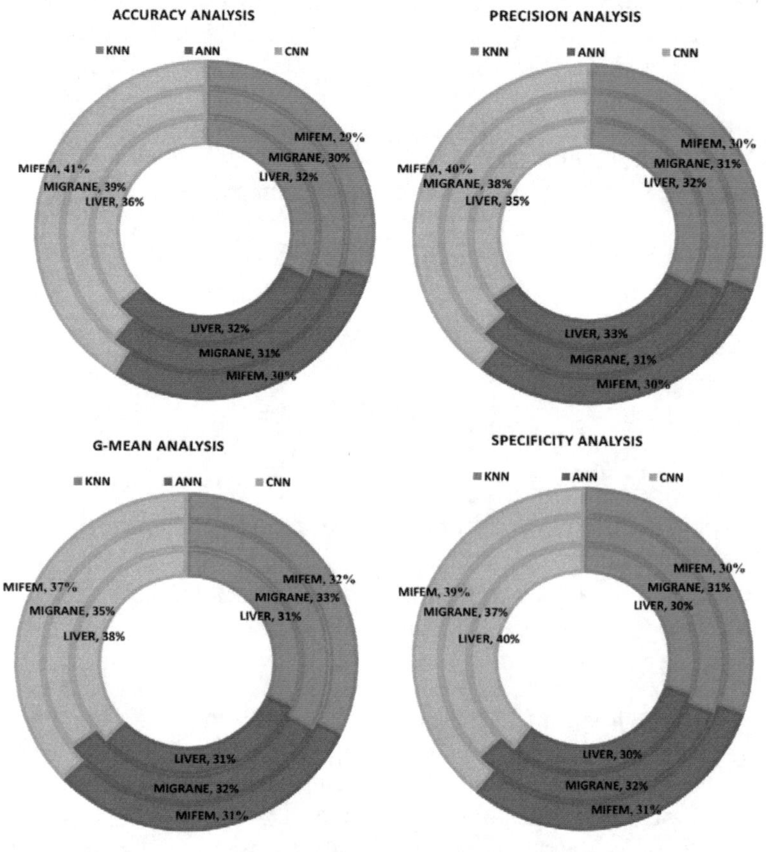

Figure 3: Performance Analysis of proposed approach

The accuracy of classification, precision, G-Mean and specificity analysis shows improved performance of the proposed deep learning based CNN classifier when compared to other classifiers. The proposed methodology is analysed statistically for classification of diseases as shown in table 1.

Table 1: Analysis of operational indicators of the proposed classifier

Sl. No.	Operational Indicators	Dataset	Obtained Value
1	Accuracy	Liver	0.8
		Migraine	0.85
		Mifem	0.9
2	Precision	Liver	0.78
		Migraine	0.87
		Mifem	0.95

3	G-mean	Liver	0.87
		Migraine	0.84
		Mifem	0.91
4	Specificity	Liver	0.76
		Migraine	0.87
		Mifem	0.94

The images of the diseased region can be understood in a much clearer manner using feature visualization techniques in the deep learning architectures and models. Reference based visualization, Grad-CAM visualization, Gradient-based visualization and Perturbation-based visualization techniques are used for this purpose. Saliency map and heat maps are some of the commonly used visualization maps. Heat maps often fail to detect some of the diseased regions from the input images. In order to overcome this issue, guided back propagation technique is used in saliency maps such that all diseased regions are identified accurately. Hence saliency maps provide better results when compared to the heat maps. Classification of specific diseases are done using semantic dictionary and feature visualization on a wide range of applications. The attention maps can be interpreted for accurate understanding of the diseased region in the images.

5. Conclusion and Future Scope

In recent days, text, speech, image and voice processing is performed using CNNs (Convolutional Neural Networks). Despite the extraordinary progress made my CNNs in the experimental assessments, various issues that requires examination still exist. In order to train the information, CNN requires huge computing power as well as large datasets. This paper offers an efficient solution for classification and visualization of disease datasets using CNN classifier. Liver, migraine and mifem (myocardial infarction) datasets can be analysed and evaluated using the proposed technique. The experimental results demonstrate that the proposed classifier offers improved efficiency in comparison with the other existing classifiers.

References

[1]. Sitharthan R, Geethanjali M and Pandy TKS 2016 Adaptive protection scheme for smart microgrid with electronically coupled distributed generations *Alexandria Engineering Journal* **55(3)** 2539-2550

[2]. Fathima AH, and Palanisamy K 2014 Battery energy storage applications in wind integrated systems—a review *IEEE International Conference on Smart Electric Grid* 1-8

[3]. Prabaharan N and Palanisamy K 2015 Investigation of single-phase reduced switch count asymmetric multilevel inverter using advanced pulse width modulation technique *International Journal of Renewable Energy Research* **5(3)** 879-890.

[4]. Jerin ARA, Kaliannan P and Subramaniam U 2017 Improved fault ride through capability of DFIG based wind turbines using synchronous reference frame control based dynamic voltage restorer. *ISA transactions* **70** 465-474

[5]. Sitharthan, R, Sundarabalan CK, Devabalaji KR, Nataraj SK and Karthikeyan M 2018 Improved fault ride through capability of DFIG-wind turbines using customized dynamic voltage restorer *Sustainable cities and society* **39** 114-125

[6]. Prabaharan N and Palanisamy K 2016 A single-phase grid connected hybrid multilevel inverter for interfacing photo-voltaic system *Energy Procedia* **103** 250-255

[7]. Palanisamy K, Mishra JS, Raglend IJ and Kothari DP 2010 Instantaneous power theory based unified power quality conditioner (UPQC) *IEEE Joint International Conference on Power Electronics, Drives and Energy Systems* 1-5
[8]. Sitharthan R and Geethanjali M 2017 An adaptive Elman neural network with C-PSO learning algorithm-based pitch angle controller for DFIG based WECS Journal of Vibration and Control **23(5)** 716-730
[9]. Sitharthan R and Geethanjali M 2015 Application of the superconducting fault current limiter strategy to improve the fault ride-through capability of a doubly-fed induction generator–based wind energy conversion system *Simulation* **91(12)** 1081-1087
[10]. Sitharthan R, Karthikeyan M, Sundar DS and Rajasekaran S 2020 Adaptive hybrid intelligent MPPT controller to approximate effectual wind speed and optimal rotor speed of variable speed wind turbine *ISA transactions* **96** 479-489
[11]. Sitharthan R, Devabalaji KR and Jees A 2017 An Levenberg–Marquardt trained feed-forward back-propagation based intelligent pitch angle controller for wind generation system *Renewable Energy Focus* **22** 24-32
[12]. Sitharthan R, Sundarabalan CK, Devabalaji KR, Yuvaraj T and Mohamed Imran A 2019 Automated power management strategy for wind power generation system using pitch angle controller *Measurement and Control* **52(3-4)** 169-182
[13]. Sundar DS, Umamaheswari C, Sridarshini T, Karthikeyan M, Sitharthan R, Raja AS and Carrasco MF 2019 Compact four-port circulator based on 2D photonic crystals with a 90° rotation of the light wave for photonic integrated circuits applications *Laser Physics* **29(6)** 066201
[14]. Sitharthan R, Parthasarathy T, Sheeba Rani S and Ramya KC 2019. An improved radial basis function neural network control strategy-based maximum power point tracking controller for wind power generation system *Transactions of the Institute of Measurement and Control* **41(11)** 3158-3170
[15]. Rajesh M and Gnanasekar JM 2017 Path observation based physical routing protocol for wireless ad hoc networks *Wireless Personal Communications* **97(1)** 1267-1289
[16]. Palanisamy K, Varghese LJ, Raglend IJ and Kothari DP 2009. Comparison of intelligent techniques to solve economic load dispatch problem with line flow constraints *IEEE International Advance Computing Conference* 446-452
[17]. Sitharthan R, Ponnusamy M, Karthikeyan M and Sundar DS 2019 Analysis on smart material suitable for autogenous microelectronic application *Materials Research Express* **6(10)** 105709
[18]. Rajaram R, Palanisamy K, Ramasamy S and Ramanathan P 2014 Selective harmonic elimination in PWM inverter using fire fly and fireworks algorithm *International Journal of Innovative Research in Advanced Engineering* **1(8)** 55-62
[19]. Sitharthan R, Swaminathan JN and Parthasarathy T 2018 March. Exploration of wind energy in India: A short review *IEEE National Power Engineering Conference* 1-5
[20]. Karthikeyan M, Sitharthan R, Ali T and Roy B 2020 Compact multiband CPW fed monopole antenna with square ring and T-shaped strips *Microwave and Optical Technology Letters* **62(2)** 926-932

[21]. Sundar D Sridarshini T, Sitharthan R, Madurakavi Karthikeyan, Sivanantha Raja A, and Marcos Flores Carrasco 2019 Performance investigation of 16/32-channel DWDM PON and long-reach PON systems using an ASE noise source *In Advances in Optoelectronic Technology and Industry Development: Proceedings of the 12th International Symposium on Photonics and Optoelectronics* 93

[22]. Sitharthan R and Geethanjali M 2014 Wind Energy Utilization in India: A Review *Middle-East J. Sci. Res.* **22** 796–801 doi:10.5829/idosi.mejsr.2014.22.06.21944

[23]. Sitharthan R and Geethanjali M 2014 ANFIS based wind speed sensor-less MPPT controller for variable speed wind energy conversion systems *Australian Journal of Basic and Applied Sciences* **8**14-23

[24]. Jerin ARA, Kaliannan P, Subramaniam U and El Moursi MS 2018 Review on FRT solutions for improving transient stability in DFIG-WTs *IET Renewable Power Generation* **12(15)** 1786-1799

[25]. Prabaharan N, Jerin ARA, Palanisamy K and Umashankar S 2017 Integration of single-phase reduced switch multilevel inverter topology for grid connected photovoltaic system *Energy Procedia* **138** 1177-1183

[26]. Rameshkumar K, Indragandhi V, Palanisamy K and Arunkumari T 2017 Model predictive current control of single phase shunt active power filter *Energy Procedia* **117** 658-665

[27]. Fathima AH and Palanisamy K 2016 Energy storage systems for energy management of renewables in distributed generation systems *Energy Management of Distributed Generation Systems* 157

[28]. Rajesh M 2020 Streamlining Radio Network Organizing Enlargement Towards Microcellular Frameworks *Wireless Personal Communications* 1-13

[29]. Subbiah B, Obaidat MS, Sriram S, Manoharn R and Chandrasekaran SK 2020 Selection of intermediate routes for secure data communication systems using graph theory application and grey wolf optimisation algorithm in MANETs IET *Networks* doi:10.1049/iet-net.2020.0051

[30]. Singh RR and Chelliah TR 2017 Enforcement of cost-effective energy conservation on single-fed asynchronous machine using a novel switching strategy *Energy* **126** 179-191

[31]. Amalorpavaraj RAJ, Palanisamy K, Umashankar S and Thirumoorthy AD 2016 Power quality improvement of grid connected wind farms through voltage restoration using dynamic voltage restorer *International Journal of Renewable Energy Research* **6(1)** 53-60

[32]. Singh RR, Chelliah TR, Khare D and Ramesh US 2016 November. Energy saving strategy on electric propulsion system integrated with doubly fed asynchronous motors *IEEE Power India International Conference* 1-6

[33]. Singh RR, Mohan H and Chelliah TR 2016 November. Performance of doubly fed machines influenced to electrical perturbation in pumped storage plant-a comparative electromechanical analysis *IEEE 7th India International Conference on Power Electronics* 1-6

[34]. Sumithra M, Abraham A, Rajesh M and Nithiyanandam N 2019 Mining social media content to predict peer trust level in social networks *International Journal of Engineering and Advanced Technology* **6S3** 1959-1964

[35]. Venkatesh K, Ali MJS, Nithiyanandam N and Rajesh M 2019 Challenges and Research Disputes and Tools in Big Data Analytics *International Journal of Engineering and Advanced Technology* **6S3** 1949-1952

A Study on Optimization of Network latency and Pocket loss Rate

Sujatha krishanmoorthy [1], Zhangyu Wei [1],Yihuai Zhang[1], and Hao Jin[1],Hao Dong[1]

[1]Assistant professor , Students (1064082,1064213,1063845,1063739) Department of Computer science, Wenzhou-Kean University Wenzhou, China.

Sujatha@wku.edu.cn, weiz@kean.edu,
zhanyihu@kean.edu,jinhao@kean.edu,dongh@kean.edu

corresponding author's email:sujatha.ssps @gmail.com

Abstract. Network delay refers to the transmission of various data in the network medium through network protocol, such as TCP / IP. If the amount of information is too large and there is no limit at the same time, the excessive network traffic will lead to the slow response of equipment, resulting in network delay. [2] The main causes of network delay are transmission delay, propagation delay, packet switching, queuing, packet drop and processing. [1]Therefore, in order to study the optimization of network delay, the causes of network delay was studied in detail.Some of the methods were deployed to learn the testing network latency, delay measurement algorithm, for example, Cristian's Algorithm, network analysis by establishing a packet processing model,and TCP BBR algorithm and analyze the optimization of network delay through the simulation of BBR. Finally, by comparing the data before and after optimization, the results of the optimization of the network delay test data by the BBR algorithm are obtained.

1. Introduction

With the rapid increase in the number of Internet users, the scale, and the dependence on the network, people are increasingly demanding high performance, low latency, and mode of the network. However, because the network not only carries huge data services, but also multimedia voice, video, financial, and other real-time services, and due to some issues such as insufficient server performance, system design defects and etc., a large number of network delays have seriously affected the user experience.

Unreasonable server system design and low computing power are an important cause of network delays. In the network, the server provides users with resources (such as CPU, network, data, communication, etc.). If the server performance is poor (there is software on the network that can detect the server performance), it will take a long time to perform related operations. In other words, the time spent on network transmission is negligible. Therefore, according to the barrel principle, if the bottleneck of network delay is the server, even if the network performance is increased, the delay will not change. At the same time, system design also plays an important role in the overall performance of the server.

The transmission of the application data unit (ADU) on the network is delayed compared to the application layer ADU. After the application layer protocol invokes the services provided by the lower layer protocol, a protocol data unit (PDU) is formed. Since the protocol payload lengths of different layers are different, the data packet size is not necessarily the same. PDU is transmitted on the network as an independent data unit. The main delays at this time are caused by packaging, transmission, propagation, transmission, and processing.

Routing delays include domain name request delay, routing delay, connection establishment delay, and connection release delay. First, due to the special nature of the host address, it is difficult to remember. Now accessing the network is basically accessed through the domain name, so the host address must be resolved before communication. The delay caused by that process called the domain name request delay, which is related to the speed of the network and server load. Secondly, after finding the address, you also need to select the optimal path according to the relevant situation. This process is called routing delay, which includes the processing of the gateway routing table and the address resolution delay. The gateway address resolution is done by ARP at the local machine, so the delay in routing delay is relatively slight. Finally, the process of user connection establishment and connection release is related to network and server load.

According to queuing theory, when the utilization rate of a channel increases, the delay caused by the channel increases rapidly. When there is little traffic on the network, the delay caused by the network is not large. However, in the case of increasing network traffic because the packets need to be queued for processing at the network node (router or node switch), the delay caused by the network will increase. If we let D0 denote the delay when the network is idle and D denote the current delay of the network, then under appropriate assumptions, the following simple formula can be used to express the relationship between D, D0 and utilization U: $D=D0/(1-U)$. When the utilization rate of the network reaches 1/2 of its capacity, the delay will be doubled. It is particularly worth noting that when the network utilization is close to the maximum of 1, the network delay tends to infinity. Therefore, we must have the concept that excessive channel or network utilization will cause very large delays.

In this paper, the contribution is to study about a reasonable network delay measurement method. BBR Congestion Control algorithm was implemented to optimize network latency and packet loss rate.

2. Methodology

Our simulation follows a quantitative experimental study method during the whole procedure. Before deciding our topic, we conducted a survey about some usual network knowledge. The questionnaires are conducted to our friends and relatives, without gender, culture or age limits. Questions we designed follow quantitative method and are all open-ended, since we want to collect participants' description and opinion rather than answers such as "Yes", "No" or "how many times". For the question "what do you think of latency", the participants' feed backs are 100% negative and that's why we choose this topic. Then we start from theoretical research.

IOP Conf. Series: Materials Science and Engineering **937** (2020) 012054 doi:10.1088/1757-899X/937/1/012054

First of all, we conclude what is latency and what causes latency. Some of the answers come from textbook, and I will generalize them with the help of academic library sources. Network latency can be defined by the time taken for a request to travel from the sender to the receiver and for the receiver to process that request. Though requests and data travel through fiber cable with a speed of light, when the distance goes further, delays must exist since you cannot beat the laws of physics. Maybe the speed of light makes latency paltry when distance is small enough, the repeaters or amplifiers also introduce additional delays. When the requests need to travel through Atlantic submarine cable, latency becomes influential to your life.

Bandwidth determines the maximum rate of data transfer across a given path. Since there are typically many clients per server, the requests to be processed on a server will be greater than each individual client, and sometimes the traffic on such a network exceeds the capacity of calculation and causes congestion. In that situation, a server administrator may control the number of requests a server responds to within a specified period of time. When a server reaches the limit, new requests will be offloaded without responds. This also causes more latency. Other factors such as transmission medium and storage delays also impact your network latency time. Storage delays can occur when a packet is stored or accessed resulting in a delay caused by intermediate devices like switches and bridges.

Hence we get a list of the factors that impact network latency: distance, transition delays, storage delays, bandwidth limitation, transmission medium. Other factors may be added with our study goes deeper. After finishing quantitative description of the factors, the next step we search for methods point at each of the factors listed.

Before testing algorithms, we need tools to measure network latency for an immediate observation. RTT, also known as Round Trip Time, is the length of time it takes for a signal to be sent plus the length of time it takes for an acknowledgement of that signal to be received. It is also known as the ping time for public. RTT has an impact on service response time and it depends on OWD, One-Way Delay, which is the length of time that a packet takes from one point to another across the network. Our analysis of the following tests is all based on the two measurements.

Simulate and analyze methods. Here we choose TCP-BBR Congestion Control Algorithm and Cristian's Algorithm as our research goal. The BBR algorithm is chosen because the simulation process can be finished on a simple computer, and Linux network names paces provide lightweight network emulation, including processes. Though bandwidths available is low, the Linux's Token-Bucket Filter (TBF) helps to limit rate and set the buffer size. Traffic from the receivers to the senders is not subject to rate limiting since we only send data from the senders and the returning acknowledgment stream does not exceed the bottleneck bandwidth. Since many users make no sense what do 10ms delay and 100ms delay mean, and their feelings are just "the feedback is quick" or "the waiting time is too long", we perform qualitative tests which can offer a direct assessment of the new algorithms. In the next part we will talk more about those algorithms.

2.1. Network latency measurement method

This part, it will introduce two network latency measurement methods, they are One-way delay measurements method and Round-trip delay measurement method. In fact, the network performance measurement method is divided into active measurement and passive measurement according to the measurement method. Active measurement is to inject probe packets into the network under test. The probe packets record the transmission in the network. The terminal analyzes the probe packets to obtain network performance data. The passive measurement method uses monitoring equipment or tools to analyze the network characteristics by capturing the data packets transmitted in the network under test. According to the definition of the network delay metric, the delay of any layer can be attributed to one-way delay or round-trip delay, but the protocol is different. Therefore, the following test methods can be applied to the delay measurement above the network layer.

2.1.1 One-way delay measurement method

Firstly, selecting the protocol used for measurement, such as TCP, UDP, OWDP (One Way Delay Protocol), IPMP (Internet Protocol Measurement Protocol, IP layer measurement protocol), and determine the details of the selected protocol, Such as TCP, UDP port number, timeout waiting time and measurement packet size. After determining these parameters, proceed as follows:

(1) Negotiate source and destination synchronization: That is, the clocks between the source and destination must be accurately synchronized. (2) At the source end, determining the IP address of the destination end and fill test packets according to the structure of the selected measurement protocol. The vacant part should be filled with random bits to avoid that the measured delay is lower than the delay after the compression technique is used in the path. Set the time stamp in the test package, and then send the filled test package to the destination host. (3) At the destination, preparing to receive the packet. If the packet arrives within a reasonable time, the estimated one-way delay can be calculated by immediately subtracting the receiving time from the sending time in the packet. This value is only valid under the premise that the source and destination clocks are synchronized. If the packet cannot arrive within a reasonable time, the one-way delay value should be set to undefined.

There are two main methods to solve the clock synchronization problem: software synchronization and hardware synchronization. Synchronization based on GPS belongs to hardware synchronization; synchronization based on Network Time Protocol and synchronization based on algorithm estimation belong to software synchronization. Time synchronization on the Internet is mostly achieved by using an NTP server.

2.1.2. Round-trip delay measurement method

The preparation work is the same as the one-way delay measurement method. Then, the measurement is performed in the following steps:

(1) At the source end, determine the IP address of the destination end and fill the test packet according to the structure of the selected measurement protocol. In order to avoid that the measured delay is lower than the expected result due to the compression technology on the path, the vacant part should be filled with random bits; and the test packet must be set with an ID number so that after receiving the response packet, the source host can judge that it is the response of sending out the test packet by itself. (2) At the source, set the sending timestamp in the test package, and then send the filled test package to the destination host. The timestamp can be placed in the package or outside the package, as long as it contains an appropriate identifier so that the received timestamp can be compared with the sent timestamps. (3) At the destination, prepare to receive test packets. If the test packet reaches the destination host, a response packet should be quickly generated and sent to the source.

(4) At the source, prepare to receive the corresponding response packet. According to the ID number to determine whether it is the response of the test package. If the response packet arrives within the timeout waiting time, immediately subtract the receiving time from the sending time in the response

IOP Conf. Series: Materials Science and Engineering **937** (2020) 012054 doi:10.1088/1757-899X/937/1/012054

packet to calculate the round trip delay value. If the response packet cannot arrive within the timeout period, the round trip delay value is set to undefined

2.2 Cristian's Algorithm

Delay:

request sent time: T_1

reply received time: T1

Then, we can Assume network delays are symmetric.

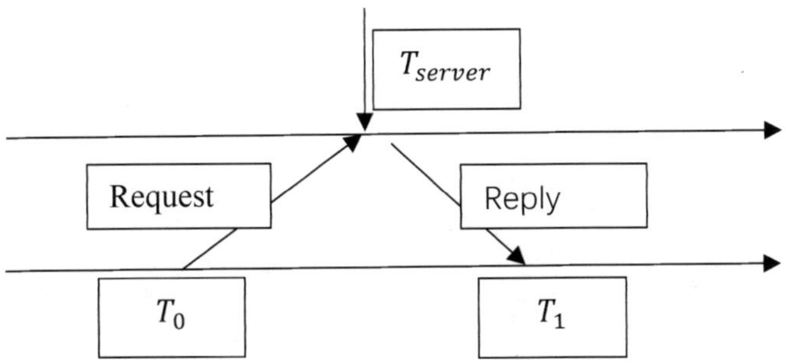

The Client Time:

$$T_{new} = T_{server} + \frac{T_1 - T_0}{2}$$

Accuracy Time:

minimum message transit time: T_{min}

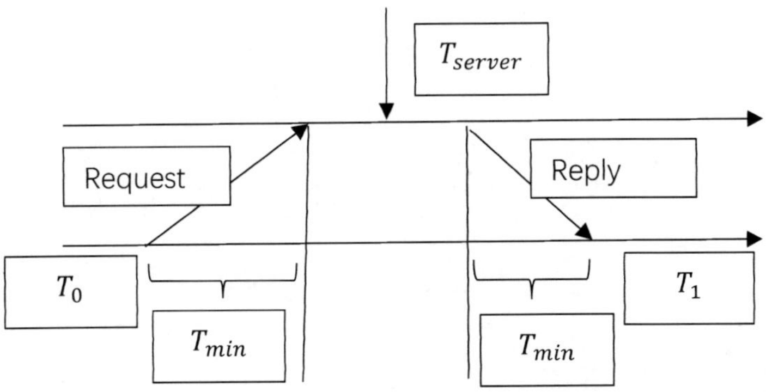

$$Range = T_1 - T_0 - 2T_{min}$$

Therefore, the accuracy time $= \pm \frac{T_1 - T_0}{2} - T_{min}$

2.2.1 Processing Packet Modeling

From the graph, when the packet arrives, metadata's behavior will be matched with flow meter. If there is no reasonable flow meter that contain L7 metadata, ->sentinto SDN controller -> be inspected by DPI based application-aware module. If malicious behaviors are identified, packet will be discarded.

Therefore, it can prevent malicious programs stepping into the networks. Otherwise, a new flow meter which contains behaviour type and priority of this packet will be distributed into OpenFlowswitch.

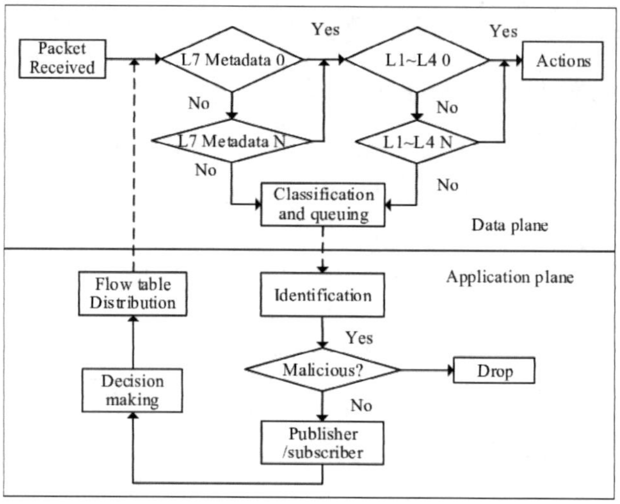

Then, we can get the Network Latency Analysis that

"n" is network nodes

"K" kinds of new traffic flow

Latency time = T_{LTC}

T_{LTC}(Lc)=$\sum_{i=1}^{n}(\frac{N-K}{N}t_1'(i) + \frac{K}{N}t_3'(i))$

"m" unidentified packets in new packets

"M" identified packets in total

T_{LTC}(Lc)=$\sum_{i=1}^{n}(\frac{N-K-M+m}{N}t_1'(i) + \frac{M-m}{N}t_3'(i) + \frac{K-m}{N}t_3'(i) + \frac{m}{N}t_4'(i))$

Whereby, t'1(i)=t1(i) + t2(i), t'3(i)=t3(i) + t4(i)

3. CP-BBR Congestion Control Algorithm.

BBR is a new TCP congestion control algorithm proposed by Google. It uses round-trip delay and bandwidth to control the congestion window. It no longer uses packet loss as a signal of congestion, but instead detects congestion based on RTT like Vegas. Compared with packet loss, detecting RTT can detect congestion earlier, and then take measures to slow down. In the BBR algorithm, the RTT and bandwidth are periodically detected to discover the network congestion.

The BBR congestion control algorithm has four stages, they are STARTUP, DRAIN, PROBE_BW, PROBE_RTT. Compared with other congestion control algorithms, BBR does not have the traditional slow start phase.

[1]STARTUP can be regarded as equivalent to slow start. In the STARTUP phase, gain will take a relatively large value. The default is 2/ln2, which is about 2.885. In this phase, the amount of data injected into the network will increase rapidly from a small value until the bandwidth no longer increases.

[2]In the DRAIN phase, a smaller gain value will be set, the default is ln2/2, which is about 0.3466. After too many packets injected into the network before this phase are emptied, it will enter the PROBE_BW phase.

[3]Under normal circumstances, the majority of the BBR time will be in the PROBE_BW phase. In this phase, the sender will tentatively inject more data packets into the network to detect whether there is excess bandwidth. The specific method is to increase the gain to 1.25 in the first RTT time to detect whether there is excess bandwidth. In the second RTT, reduce the gain to 0.75 to empty the excess data packets, then set the gain in 6 RTT time Is 1, continue to send data.

[4]In this way, every 8 RTTs are used as a period to detect the bandwidth. In general, without data queuing, the measured RTT value is the smallest. BBR will consume 200ms to detect the minimum RTT value every 10 seconds of transmission. This is the PROBE_RTT stage. At this stage, the window is reduced to 4 MSS, and the smallest RTT can be detected, which is also a process of bandwidth politeness, which allows other TCP connections to have the opportunity to occupy bandwidth.

3.1 Components of BBR Congestion Control Algorithm

Calculation of immediate rate

Calculate an instantaneous bandwidth, which is the benchmark for all calculations of BBR. BBR will calculate the pacing rate and cwnd (Congestion Windows) based on the current instantaneous bandwidth and the pipe status it is in. The breakthrough improvement of the bandwidth calculation method is the reason why BBR is simple and efficient. The calculation scheme is calculated according to a scalar quantity, and no longer pays attention to the meaning of the data. During the operation of BBR, the system will track the largest instantaneous bandwidth so far.

RTT tracking

The reason why BBR can obtain a very high bandwidth utilization rate is because it can detect the maximum bandwidth and the minimum value of RTT very safely and boldly. The BDP calculated in this way is the maximum capacity of the TCP pipeline so far. Therefore, we think BBR is to achieve this maximum capacity! This goal ultimately drives the calculation of cwnd. During BBR operation, the system will track the minimum RTT so far.

Maintenance of the BBR pipe state machine

The BBR algorithm specifically defines four states according to the congestion behavior of the Internet, namely STARTUP, DRAIN, PROBE_BW, PROBE_RTT.

Results output -pacing rate and cwnd

First of all, it must be said that the output of BBR is not just a cwnd, but more important is the pacing rate. In the traditional sense, cwnd is the only output of the TCP congestion control algorithm, but it only specifies how much data the current TCP can send at most. It does not specify how to send so much data. In the implementation of Linux

Use of other external mechanisms fq, rack.

The reason why BBR can run efficiently and so simple is because many mechanisms are not implemented by itself, but use external existing mechanisms

3.2 BBR Simulation

The experience use the MiniNet Simulator in VMware, then using the BBR algorithm, the bandwidth setting is 10Mbit/s, the link delay is 250ms, and the amount of data sent in the first 20 seconds.

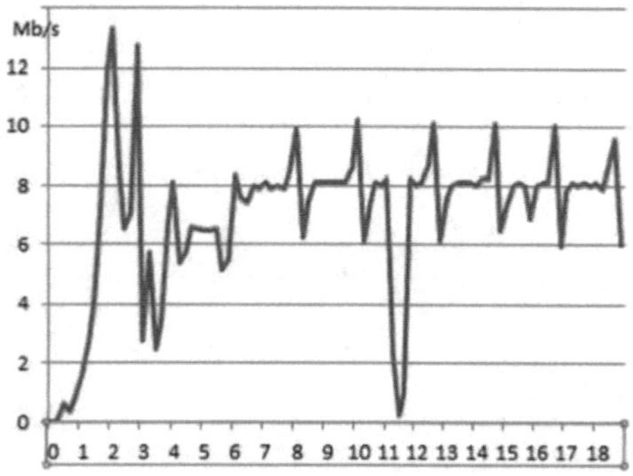

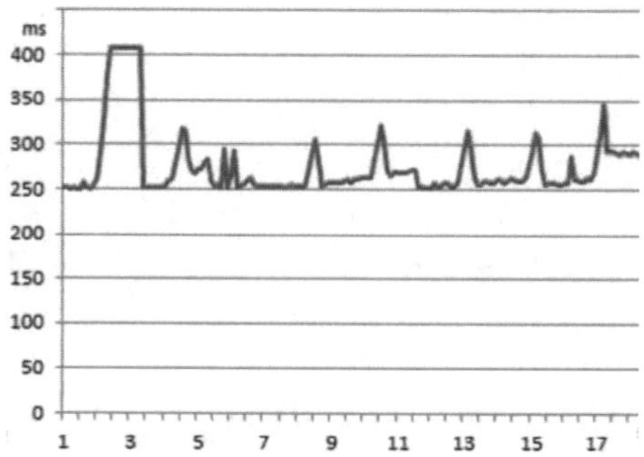

Figure.1 BBR Sending Data Figure2. RRT value of BBR Algorithm

It can be seen from the figure that the initial data transmission volume grows rapidly, enters a stable state at the 6th second, and there is a possibility of a larger window for multiple detection. At about 11 seconds, the transmission volume decreases rapidly and enters the RTT detection phase. When the amount of data injected into the network is large, the RTT becomes large. When the amount of data is small, the RTT becomes small, but it is not less than the delay of the link.According to above, as a new congestion control algorithm, BBR is mainly reflected in the adjustment of the transmission rate based on delay in TCP transmission. Since BBR is not sensitive to packet loss, it has positive significance on some networks with high bit error rate.

4. Conclusion

Due to server performance and design defects and other reasons, network delay will seriously affect the user's experience. By reducing network delay, that is, reducing the time of network transmission, users can hardly feel the existence of the transmission process, so as to achieve a good experience effect. Aiming at this goal, we optimize two algorithms, TCP and BBR. After testing the network delay before and after the algorithm optimization, we get the conclusion that there is a big difference between before and after the algorithm optimization, and then there is a more obvious delay reduction and stability improvement.

References

[1]. Scholz, D., Jaeger, B., Schwaighofer, L., Raumer, D., Geyer, F., & Carle, G. (2018, May). Towards a deeper understanding of tcp bbr congestion control. In *2018 IFIP Networking Conference (IFIP Networking) and Workshops* (pp. 1-9). IEEE.

[2]. Li, G., Dong, M., Ota, K., Wu, J., Li, J., & Ye, T. (2016, December). Deep packet inspection based application-aware traffic control for software defined networks. In 2016 IEEE Global Communications Conference (GLOBECOM) (pp. 1-6). IEEE.

[3]. Fetzer, C., & Cristian, F. (1995, June). An optimal internal clock synchronization algorithm. In COMPASS'95 Proceedings of the Tenth Annual Conference on Computer Assurance Systems Integrity, Software Safety and Process Security' (pp. 187-196). IEEE.

[4]. Cardwell, N., Cheng, Y., Gunn, C. S., Yeganeh, S. H., & Jacobson, V. (2017). BBR: congestion-based congestion control. Communications of the ACM, 60(2), 58-66.

[5]. Tools, P. (2020). Interpreting Latency and Packet Loss | PingPlotter. Retrieved 26 May 2020, from https://www.pingplotter.com/wisdom/article/latency-packet-loss.html

[6]. Yi, X. (2020). What causes network latency? How to reduce (optimize) network latency?. Retrieved 26 May 2020, from https://zhuanlan.zhihu.com/p/142790243

[7]. Singh, Parminder, et al. "Soft-computing-based false alarm reduction for hierarchical data of intrusion detection system." International Journal of Distributed Sensor Networks 15.10 (2019): 1550147719883132.

[8]. Sujatha, Krishnamoorthy. "Automatic epilepsy detection using hybrid decomposition with multi class support vector method." Multimedia Tools and Applications (2019): 1-20.

[9]. Krishnamoorthy, Dr, et al. "A Novel Approach towards Lowest Energy Usage Routing Protocol (LEuRP) in Wireless Mobile Ad-Hoc Network (WMANet)." International Journal of Pure and Applied Mathematics 116.12 (2017): 21-29.

IOP Conf. Series: Materials Science and Engineering **937** (2020) 012055 doi:10.1088/1757-899X/937/1/012055

Design & Stability Analysis of Vehicle Suspension System

R Saranraj[1], J Subajayasri[2], J Persey[3], R Nikila[4], SP Balasubramanian[5]

[1]Assistant Professor/ EEE Department, K. Ramakrishnan college of Engineering, Trichy, India.

[2,3,4] IV Year/EEE Department, K. Ramakrishnan college of Engineering, Trichy, India.

[5] III Year/EEE Department, K. Ramakrishnan college of Engineering, Trichy, India.

Abstract- This exploration depends upon the attestation of the parameters and fragmentary sales PID controllers anticipated vehicle suspension framework. From the beginning, without considering dynamic suspension, introduction of uninvolved suspension layout for various wheel load were recorded and introduced on utilizing an exchange furthest reaches of the framework. As the examination subsequent to assessing the exhibition of old style PID controller, fragmentary request PID controller was applied that improves presentation of old style controller. Thus, variables of the controller were additionally acquired on utilizing a similar improvement calculations. In this paper, street had been demonstrated with sinusoidal (street includes slope) or arbitrary changes, i.e. a saw tooth signal. Execution results were shown that introduction of fragmentary solicitation PID controller is inconceivably improved than normal PID controller.

.

Keywords: **Fragmentary request PID, suspension framework, quarter-vehicle.**

1. Introduction

Suspension structures were presented between vehicle body and wheel to hold an unexpected vibration because of road condition. Road dealing with capacity of vehicle was key factor for security and comfort. Need for suspension frameworks can be abridged as below

- To assimilate the vibration because of the defective states
- solace of travellers
- To moves braking power to the haggle the respectability among haggle body.

Like air stages, wheeled vehicles having three fundamental wavering power as introduced in Fig. 1. Saltation was the movement in vehicle from high to low caused from light harsh street on rapid conditions. Swing was the development on front and end of vehicle due to over whelming unpleasant street at moderately moderate speed (quick speed isn't recommended). Rollover was development of right and left of vehicle, caused due to turns. Every one of these developments relates to the power upon the suspension. As it were, the situation of wheel causes these motions to deal with the conditions and assurance a safe and solace travel.
The models of the suspension can be separated into three structures regarding the control point of view: inactive, semi-dynamic and dynamic suspension frameworks. Uninvolved suspension

IOP Conf. Series: Materials Science and Engineering **937** (2020) 012055 doi:10.1088/1757-899X/937/1/012055

frameworks a pre-planned and module gadgets with the end goal that practically all parameters are resolved at the creation stage.

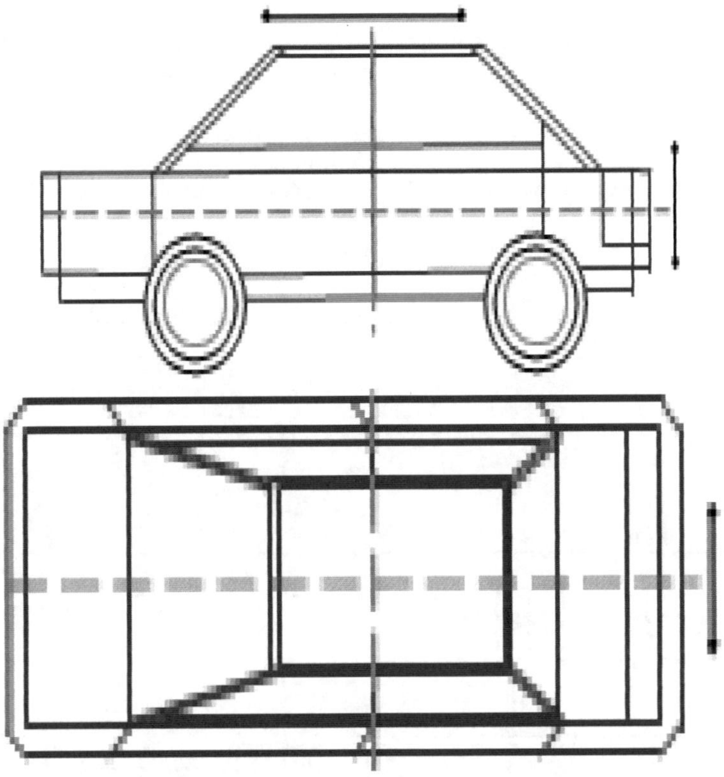

Figure 1. Wheeled Vehicle Oscillation Behaviour

In the above diagram the three oscillation behavior are marked as 1 (rapid swing), 2(saltation) & 3 (roll over). Semi dynamic suspension frameworks having variable damping power and coefficients [1]. The upside of these frameworks are their moderately minimal effort, since they by and large contain less number of gadgets (additionally don't have interconnected gadgets, for example, actuator, pneumatics and power through pressure identified with suspension framework).

A functioning suspension framework is an issue condition and the improved form of the uninvolved frameworks upon controllable actuators. These frame work can prepared to gracefully essentialness when it is needed [3].

Suspension structures were confined to two classes as Depended and Independent Systems. Depended frameworks demonstrates its truly associated frameworks and don't lean toward as front suspension frameworks (at present day vehicles) for quite a long time because of their weight and comparing wavering. However, they are commonly favored as back suspension frameworks. The notable autonomous suspension framework was "McPherson" created in 1947 [4]. As suspension frameworks were partitioned two sections and are correspondingly isolated into three gatherings such as Quarter-vehicle, half-vehicle and full-vehicle models. In Quarter vehicle model only one wheel of vehicle is taken into consideration [5]. Half-vehicle model is used as subordinate suspension systems [6]. Full-vehicle model is on overall model that consists of front and back structures [7].

2.Mathematical Approach for the Problem

In mathematical approach the difficult condition, numerical depiction of the issue, and the advancement calculations utilized for tackling control issue were examined.

Problem Definition

This paper focusses on structure controller for altering distinction between street conditions and body position. In this way, at first the impact of change at degree of wheel to vehicle position is clarified on latent suspension model. At that point, controller intended for dynamic framework. The parameters at fig. 2 from base to top is specified as "u" that refers changing position at wheel base. Wheel is considered as an unspring mass "M2" with a spring "K2" Spring and damper (with coefficient "K1" 'and "b") are put in the middle of the haggle body with mass M1.Mathematical equation of framework in fig 2 is as

$$M_1 \ddot{x} = b(\dot{y} - \dot{x}) + k_1(y - x) + \frac{1}{A}\dot{Q} \quad (1)$$

$$M_2 \ddot{y} = b(\dot{y} - \dot{x}) + k_1(y - x) - k_2(u - x) - 1/A \quad (2)$$

$$\dot{Q} = -c_f Q + k_f i \quad (3)$$

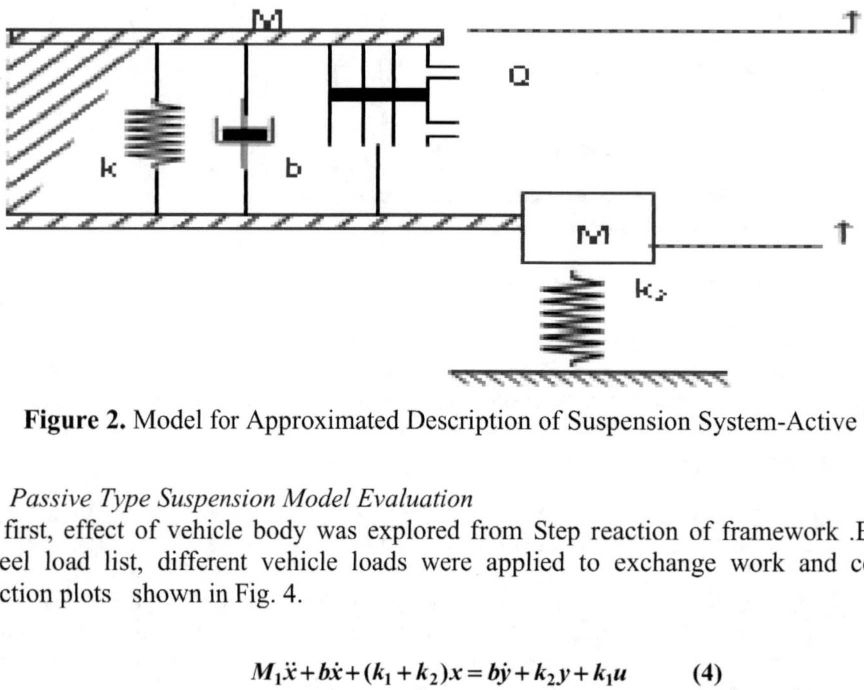

Figure 2. Model for Approximated Description of Suspension System-Active Type

Passive Type Suspension Model Evaluation

At first, effect of vehicle body was explored from Step reaction of framework .By considering wheel load list, different vehicle loads were applied to exchange work and comparing step reaction plots shown in Fig. 4.

$$M_1\ddot{x} + b\dot{x} + (k_1 + k_2)x = b\dot{y} + k_2 y + k_1 u \quad (4)$$

$$M_2\ddot{y} + b\dot{y} + k_2 y = b\dot{x} + k_2 x \quad (5)$$

In Fig. 4. just heaviness of vehicle body (M1) is changed with step reactions were introduced. As body size expands, exhibition markers for transient reaction additionally diminishes, as in Table 1.

IOP Conf. Series: Materials Science and Engineering **937** (2020) 012055 doi:10.1088/1757-899X/937/1/012055

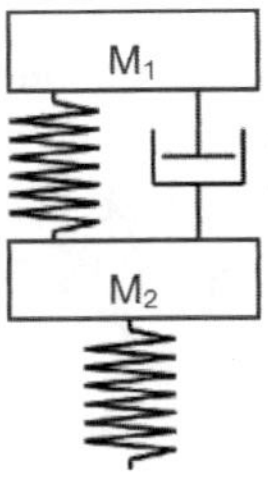

Figure 2. Model for Approximated Description of Suspension System-Passive Type

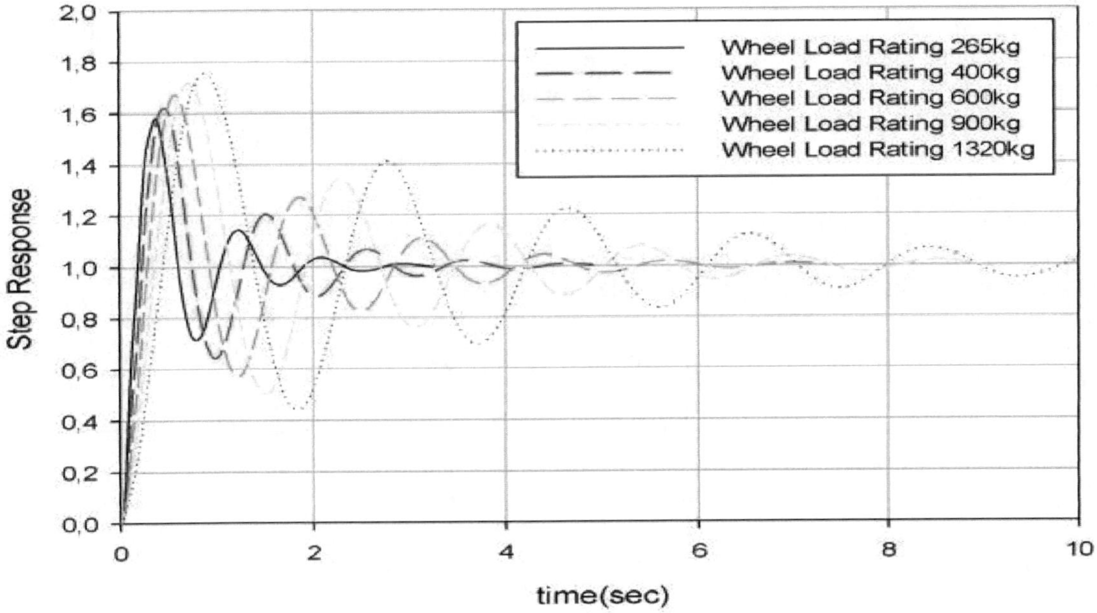

Figure 2. Step Response of Changing Load Rating for Passive Type Suspension System

Table 1. Time Domain Specifications values for the Step Response of Changing Load Rating for Passive Type Suspension System

Load (kg):	265kg	400kg	600kg	900kg	1320kg
T_P	0,3707	0,4747	0,5694	0,6988	0,8465
T_R	0,1277	0,1616	0,2013	0,2589	0,3188
M_P	0,5804	0,6221	0,6752	0,7185	0,7533
T_S	2,2213	3,6701	5,2078	7,9126	11,4748

3.Stability Analysis

Indeed, PID controller were intended of dynamic suspension framework, a traditional root locus configuration is employed with additional "K" is made and results are displayed in graphical manner. Fig.5.shows root locus structure of main relatively controlled suspension framework. Fig. 5a. shows an estimation for addition of fundamentally stable framework, increase in nature which makes framework insecure. Fig. 5b. shows an ideal damping proportion esteem comparing gain . At the end of the day, with just relative increase it isn"t conceivable to get a steady framework with an ideal shut circle shafts as exhibited in Fig.5c. Thus, right now, controllers are structured and parameters were enhanced. In following segment, PID controller (both traditional and partial request) clarified with advancement calculations.

PID and Fractional-order PID Controller

Three parameters should have been improved for the ideal execution. To sum things up, corresponding parameter (KP) diminishes the ascent time and consistent state mistake. Be that as it may, for a generally enormous consistent state blunder fundamental term (KI) is expected to wipe out this mistake. The inconvenience of an essential term is expansion at the overshoot. In this manner, sometimes the expansion on peak overshoot and settling time is moderately little. In different situations, subordinate term is expected diminish overshoot and settling time of the transient reaction. Table 2 gives rundown impacts.

$$G_c\,(S) = K_P + \frac{K_I}{S} + K_D\,S \qquad (6)$$

The commitment of parameters isn't acknowledged fragmentary request PID controller (FRPID). Nonetheless, FRPID controller has a bigger number of degrees adaptable than traditional PID controller that expands adaptability of controller and conceivable to all more likely to modify framework.

$$G_{CF}\,(S) = K'_P + \frac{K'_I}{S^{\lambda'}} + K'_D\,S^{\mu'} \qquad (7)$$

For an alternate estimation of μ and λ, fragmentary request PID controller becomes traditional PID , where for μ"=1 and λ' =1 relate to old style PID, for μ"=0 and λ" =1 compare to PI , for μ"=1 and λ" =0 compare to PD , and for λ"=0 and μ"=0 compare to corresponding controller.

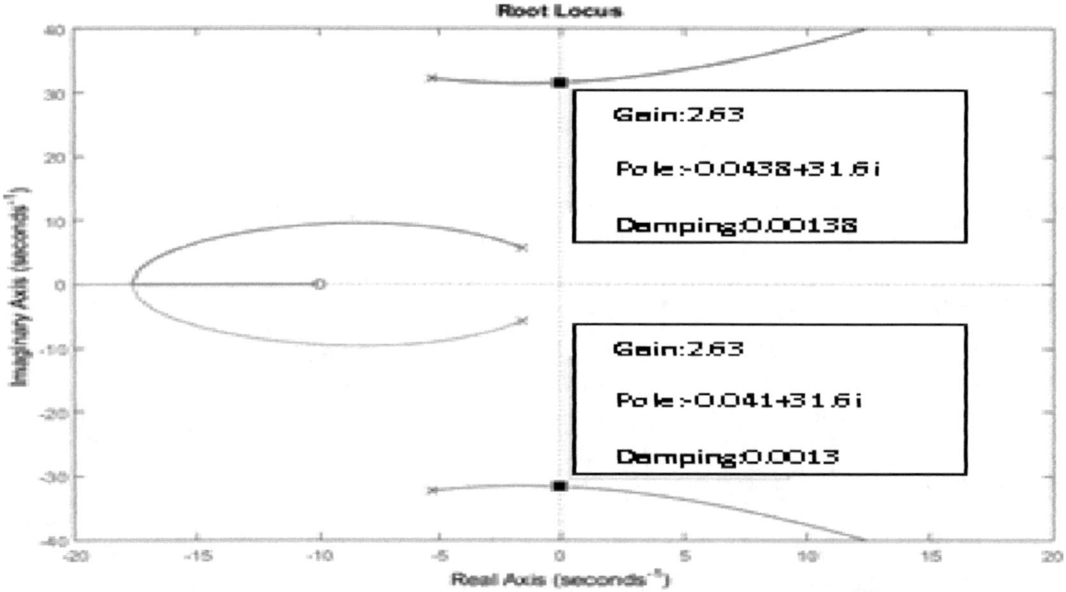

Figure.5.a Root Locus Plot : Passive Type Suspension System for critically stable condition.

IOP Conf. Series: Materials Science and Engineering **937** (2020) 012055 doi:10.1088/1757-899X/937/1/012055

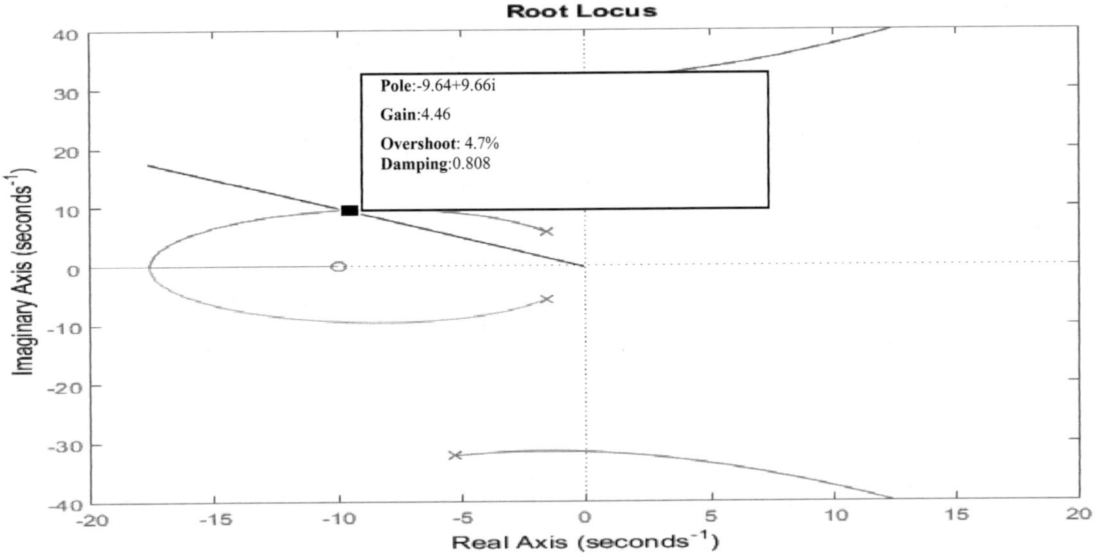

Figure.5.b. Passive Type Suspension System of desired damping ratio line

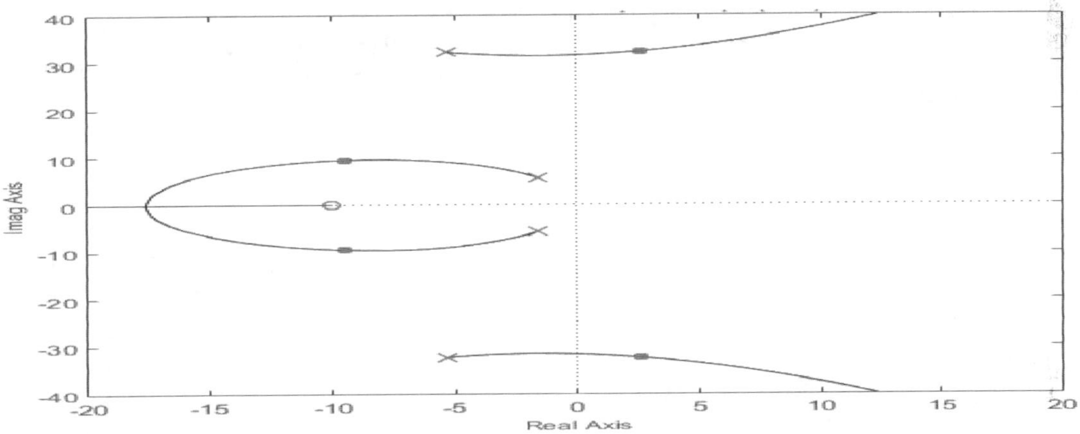

Figure.5.c. Root Locus Plot: Passive Type Suspension System corresponding unstable system

Table 2. Time Domain specifications of PID Parameters

Closed Loop response	T_R	M_P	T_S	E_{ss}
K_P	Low	High	-	Low
K_I	Low	High	High	Eliminate
K_D	-	High	Low	-

4.Conclusion

The presentation of the inactive suspension framework is researched concerning the distinctive wheel load file. An active type is replaced with an optimal PID controllers. At that point to show the significance of controller which need for moderately unpredictable calculation, Straight forward relative coefficient was applied to solidarity input framework with the guide of root locus, an exhibition of framework talked about. In an accompanying areas, first classical PID controller was intended on dynamic framework. At that point to improve presentation of PID controller, a fragmentary request PID (FRPID) controller was applied with general framework rather than classical PID controller. The best outcomes from FRPID controller with little overshoot greater undershoot at the slope fall second. In this paper, dynamic suspension framework issue is explained with ideal PID controllers.

References

[1] M. Canale, M. Milanese, and C. Novara, "Semi-active suspension control using fast model-predictive techniques," *Ieee Transactions on Control Systems Technology,* vol. 14, pp. 1034-1046, Nov 2006.
[2] J. T. Cao, P. Li, and H. H. Liu, "An Interval Fuzzy Controller for Vehicle Active Suspension Systems," *Ieee Transactions on Intelligent Transportation Systems,* vol. 11, pp. 885-895, Dec 2010.
[3] A. Hac, "Optimal Linear Preview Control of Active Vehicle Suspension," *Proceedings of the 29th Ieee Conference on Decision and Control, Vols 1-6,* pp. 2779- 2784, 1990.
[4] S. Frik and M. Hiller, "Kinematics and Dynamics of a Mcpherson Front Wheel Suspension with Elastic Rear Transverse Pivot Bearing," *Zeitschrift Fur Angewandte Mathematik Und Mechanik,* vol. 69, pp. T398-T399, 1989.
[5] H. M. Soliman, A. Benzaouia, and H. Yousef, "Saturated robust control with regional pole placement and application to car active suspension," *Journal of Vibration and Control,* vol. 22, pp. 258-269, Jan 2016.
[6] J. Marzbanrad, G. Ahmadi, H. Zohoor, and Y. Hojjat, "Stochastic optimal preview control of a vehicle suspension," *Journal of Sound and Vibration,* vol. 275, pp. 973-990, Aug 23 2004.
[7] O.Tolga Altinoz , A. Egemen Yilmaz,"Optimal PID design for control of Active Car suspension System"IJITCS 2018

IOP Conf. Series: Materials Science and Engineering **937** (2020) 012056 doi:10.1088/1757-899X/937/1/012056

A Control Methodology of Doubly Fed Induction Generator for Wind Energy Generation

Pandikumar M

Department of Electrical and Electronics Engineering, SSN College of Engineering, Kalvakkam-603110, Tamilnadu, INDIA

pandikumarm@ssn.edu.in

Abstract. This paper presents the Direct Power Control (DPC) strategy for a Doubly-Fed Induction Generator (DFIG) using Matrix Converter (MC). Wind turbines are subjected to irregular wind speeds and thus the power generated is varying in nature. The Active and Reactive power flow to the grid from the DFIG connected to the wind turbines can be controlled by connecting a Power Electronic converter at its rotor terminals. By this method, the Induction Machine can be operated in generating mode even at sub-synchronous speed, by controlling the slip power, fed-to or absorbed-from the grid. The DPC Scheme is a high performance, dynamic power control method for DFIG. In DPC, at every instant of switching, the vectors for the power converter are selected based on the instantaneous active and reactive power error command from a hysteresis band, rather than a predetermined pattern as in conventional PWM drives. This work presents the Matrix converter based DPC for DFIG where the conventional AC to DC to AC converter at the rotor terminal has been replaced by a three phase MC.

1. Introduction

Wind energy is a popular renewable energy source. Doubly-fed induction generator (DFIG) wind turbines are widely used in the Wind Energy Conversion Systems (WECS). DFIGs are variable speed generators with more advantages compared to other solutions. They are used more and more in wind turbine applications due to easy controllability, high energy efficiency and improved power quality. Fixed speed generators and induction generators had the disadvantage of having low power efficiencies at most speeds [1-5]. To improve the efficiency, power electronic converters driven by control algorithms are commonly connected to the generators. Voltage Source Inverter (VSI) is used to convert the voltage magnitude and frequency to synchronize with the grid values. Power converters in a DFIG system only deal with rotor power control. Due to this factor the electronics' costs are kept low, about approximately 20-25% of the total generator power making the system economically efficient. This is due to the fact that the rotor voltage is lower than the stator voltage. This implies that the converter is designed to suit the rotor parameter and making the system more economical than using a full power rated converter in a series configuration [6-10].

To observe the system and the flow of active and reactive power, a dynamic model of the machine is needed and has been developed [11-16]. For a normal generation regime, the energy obtained by processing the wind speed as an input is fed into the network by both the stator and the rotor. This work presents a modified WECS with the AC to DC to AC converter replaced by a three phase Matrix

Converter. The control of the corresponding active and reactive power supply to the grid can be controlled by various control techniques of the Induction Machine such as V/Hz control or Vector control or Direct Power Control (DPC) [17-25]. This work presents a novel DPC scheme which replaces the conventional AC-DC-AC Converter with a three phase Matrix Converter. The complete schematic of the WECS has been given in figure 1.

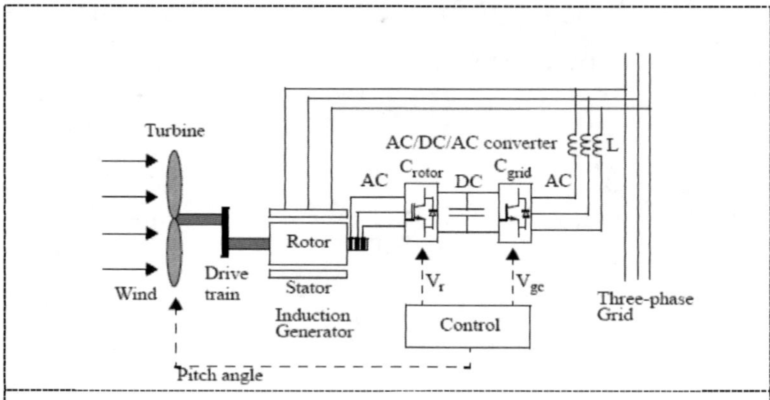

Figure 1. Complete schematic of the WECS.

2. System Design

The DFIG is grid connected and draws or supplies active and reactive power based on the availability of energy in the wind.

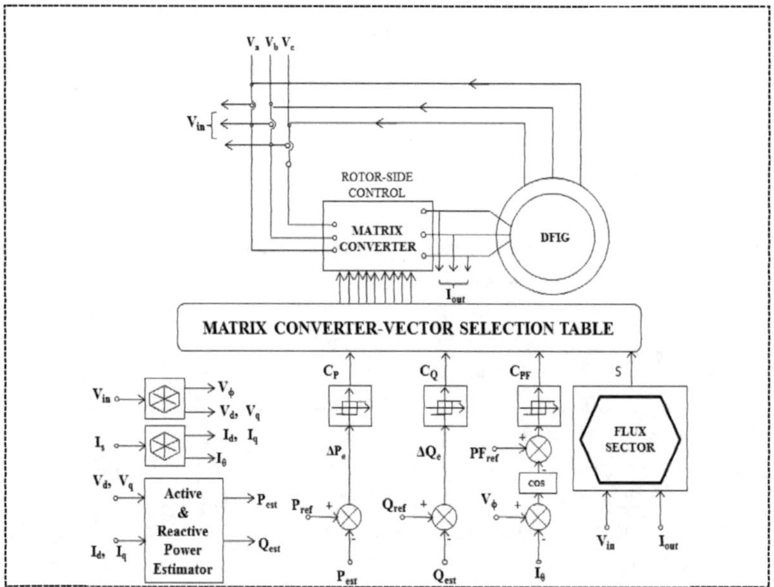

Figure 2. Block Diagram of Direct Power Control of DFIG.

Direct Real and Reactive Power control strategy for a Doubly Fed Induction Generator driven by a Wind turbine, using Matrix Converter at the rotor side for the supplying the slip power as shown in Fig 2.

IOP Conf. Series: Materials Science and Engineering **937** (2020) 012056 doi:10.1088/1757-899X/937/1/012056

3. Concept of Direct Torque / Power Control

The evolution of the variable speed AC drive technology is driven by the desire to achieve characteristics as that of the DC motor, such as fast torque response, easier speed control to accuracy, while using inexpensive, robust and maintenance-free Induction machine drives. Direct Torque Control (DTC) has found broad industrial applications . It is an optimised vector control strategy of Induction Machine, where the machine variable (torque and speed) are controlled using a power converter according to the dynamic loading conditions [26-30]. DTC eliminates complex frame transformations thus making it faster and involving selection of appropriate switching vector of the power converter according to the loading condition. In spite of its high dynamic response, sensor-less, transformation-less operations and robustness, the DTC has some disadvantages: Control of torque and flux at low speeds are difficult; higher current and torque ripples result in higher machine losses and noise. To retain the advantages of DC drives and eliminate the problems of DTC, a hybrid Space Vector Modulation (SVM) based DTC method was proposed. Introduction of an AC to AC converter called Matrix Converter (MC) as shown in Fig 2 with n x m switches arranged in a matrix form, paved way for easier control of input current and consequently the input power factor. This DTC scheme has been the basis for the Direct Power Control (DPC) of DFIG used in WECS. The technique involves direct control of rotor active and reactive power of DFIG by selection of appropriate switching vectors of the Power Converter (Matrix Converter). The control of the Power in rotor side in-turn controls the Active and Reactive Power that the DFIG supplies to the Grid.
The change in input voltage immediately changes the stator flux. This stator flux change can be brought about using a power converter, with which the input voltage vector can be determined by the space vector position, value of reference Active and Reactive Power values.
The errors in active and reactive power are fed to 3-level hysteresis band comparators. The output of the active power hysteresis band in figure 3 is CP = -1 || 0 || +1, which represents the action to be taken in the power converter for the change desired to occur. The other hysteresis band as in figure 4 gives a signal CQ = -1 || 0 || +1, which indicates the error in flux and the switching change to be done in the power converter for bringing about the change in stator flux and torque.

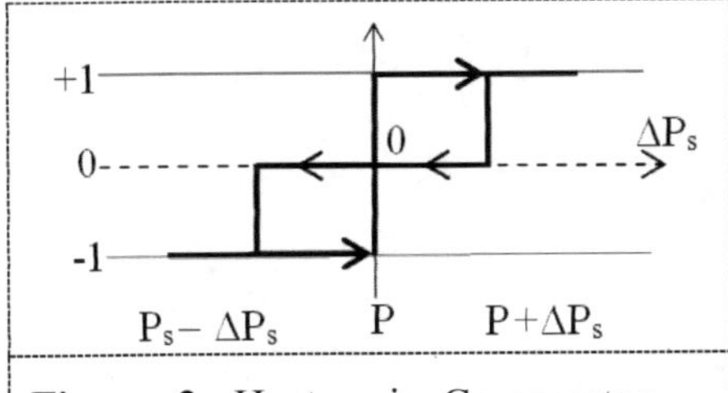

Figure 3. Hysteresis Comparator - Active Power, CP.

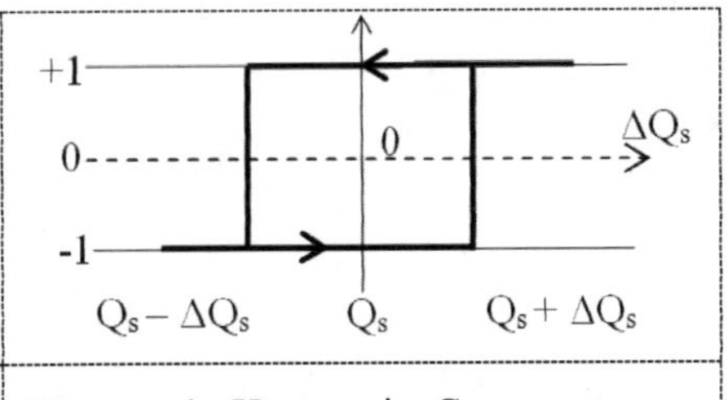

Figure 4. Hysteresis Comparator - Reactive Power, CQ.

4. Direct Power Control of DFIG

The basic concept of direct control of active and reactive powers can be appreciated from the phasor diagrams based on the equivalent circuit of the doubly-fed machine as shown in figure 5. From the phasor diagram in figure6, it is noted that the stator current Is has to be controlled to control the stator active power Ps and the stator reactive power Qs. This is achieved in turn by controlling the rotor currents Ir in conventional field oriented control strategy. The effect of injection of these rotor currents on the air-gap and rotor fluxes can be derived by subtracting and adding the respective leakage fluxes.
The reactive power drawn from the grid by the stator can be reduced by increasing the magnitude of the rotor flux and vice-versa.
It can be concluded that

i. The stator active power can be controlled by controlling the angular position of the rotor flux vector.

ii. The stator reactive power can be controlled by controlling the magnitude of the rotor flux vector.

DPC is a technique in which, at every instant of switching, the vectors are selected on the basis of the instantaneous active and reactive power hysteresis command, rather than a predetermined pattern as in conventional PWM flux vector drives. The actual control of power by directly changing the input voltage can be expressed as below using equation (1) to (17), representing the behavior of Induction Machine. Neglecting the copper loss in stator of the IM, the Stator voltage of an induction machine is given by,.

$$V_s = \frac{d\overrightarrow{\psi_s}}{dt} \tag{1}$$

Where $\overrightarrow{\psi_s}$ is the stator flux,

Assuming flux changes in a considerable amount of time, the voltage is given by,

$$V_s = \frac{\Delta\overrightarrow{\psi_s}}{\Delta t} \tag{2}$$

Thus, the stator flux can be directly controlled changing the input voltage of the motor for a given amount of time t, as in equation (3)

$$V_s \, \Delta t = \Delta\psi_s \tag{3}$$

$$\overrightarrow{\psi_s} = \psi_{qs}^{s} - \psi_{ds}^{s} \tag{4}$$

The stator current, $\overrightarrow{I_s}$ is given by,

$$\overrightarrow{I_s^r} = i_{qs}^{s} - i_{ds}^{s} \tag{5}$$

Expressing Stator flux and the Rotor flux in complex form,

$$\overrightarrow{\psi_s^r} = L_s\overrightarrow{I_s^r} - L_m\overrightarrow{I_r^r} \tag{6}$$

$$\overrightarrow{\psi_r^r} = L_r\overrightarrow{I_r^r} - L_m\overrightarrow{I_s^r} \tag{7}$$

Equations (6) and (7) can be equated and reframed as follows. In Equation (6), eliminating the term I_r, Stator flux

$$\overrightarrow{\psi_s^r} = \frac{L_m}{L_r}\overrightarrow{\psi_r^r} - L_a\overrightarrow{I_s^r} \tag{8}$$

Where $L_a = (L_sL_r - L_m^2) / L_r$.

The corresponding stator current is given by equation (9)

$$\overrightarrow{I_s^r} = \frac{1}{L_a}\overrightarrow{\psi_s^r} - \frac{L_m}{L_rL_a}\overrightarrow{\psi_r^r} \tag{9}$$

$$P_s = \frac{3}{2}V_s^r . I_s^r \tag{10}$$

$$Q_s = -\frac{3}{2}V_s^r x \, I_s^r \tag{11}$$

$$\overrightarrow{V_s^r} = \frac{d\overrightarrow{\psi_s^r}}{dt} + R_s\overrightarrow{I_s^r} + j\omega_r\psi_s^r \tag{12}$$

Neglecting copper losses in equation (12) and substituting equation (12) and (9) in equation (10) and (11),

$$P_s = \frac{3}{2} * \frac{L_m}{L_rL_a}\, \omega 1 * |\psi_s^r|\, |\psi_r^r| \sin\theta \tag{13}$$

$$Q_s = \frac{3}{2} * \frac{\omega 1}{L_a} |\psi_s^r| \left(\frac{L_m}{L_r} |\psi_r^r| \cos\theta - |\psi_s^r|\right) \tag{14}$$

$$\frac{dP_s}{dt} = \frac{3}{2} * \frac{L_m}{L_r L_a} \omega 1 * |\psi_s^r| \frac{d|\psi_r^r| \sin\theta}{dt} \tag{15}$$

$$\frac{dQ_s}{dt} = \frac{3}{2} * \frac{L_m}{L_r L_a} \omega 1 * |\psi_s^r| \frac{d|\psi_r^r| \cos\theta}{dt} \tag{16}$$

According to equation 15 and 16, it can be seen that fast reactive and active power changes can be achieved by changing $d|\psi_r^r|\cos\theta$ and $d|\psi_r^r|\sin\theta$ respectively. From figure 5 & 6, $|\psi_r^r|\sin\theta$, $|\psi_r^r|\cos\theta$ represent the components of the rotor flux ψ_r^r at the perpendicular and the same direction of the stator flux respectively. This indicates that, if the change of the rotor flux is at the stator flux direction, i.e., $|\psi_r^r|\cos\theta$, reactive power Qs is changed. Alternatively, if the change of the rotor flux is at 90° to the stator flux direction, i.e., $|\psi_r^r|\sin\theta$, active power Ps is changed.

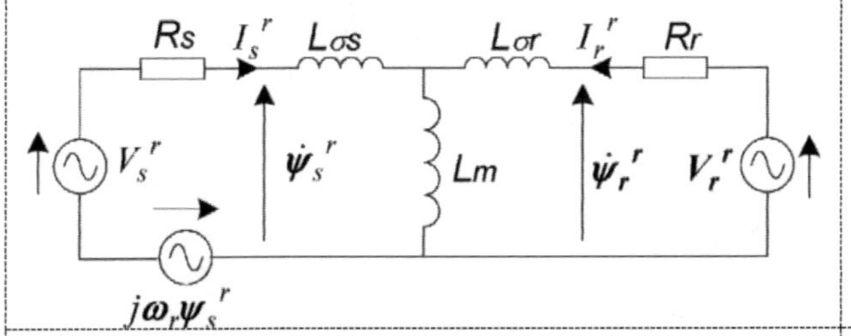

Figure 5. Equivalent Circuit of DFIG.

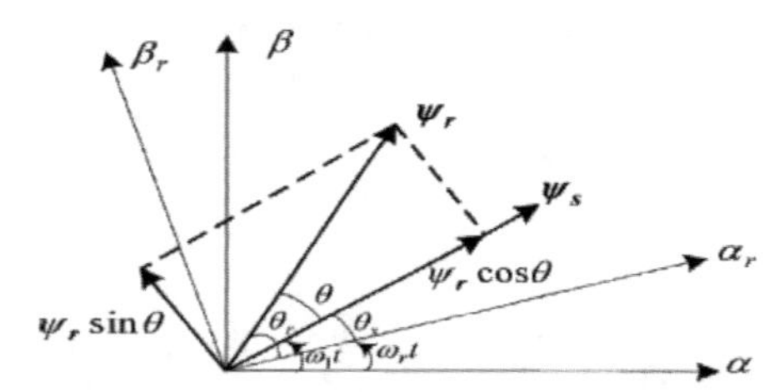

Figure 6. Stator and Rotor Flux Components.

The initial position of the rotor flux and its amplitude do not directly affect the active and reactive power changes

$$\overrightarrow{V_r^r} = \frac{d\overrightarrow{\psi_r^r}}{dt} + R_r \overrightarrow{I_r^r} \tag{17}$$

Neglecting the effect of rotor resistance Rr, equation (17) indicates that the variation of rotor flux is determined by the applied rotor voltage. The rotor flux moves at the direction of applied rotor voltage vector and its speed is proportional to the amplitude of the voltage vector applied. Therefore, by selecting appropriate voltage vectors, the movement of the rotor flux can be controlled. The selection of the voltage vector also depends on the location of the flux linkage and the a_r–$ß_r$ plane is divided into six regions (I–VI) as shown in figure 7.

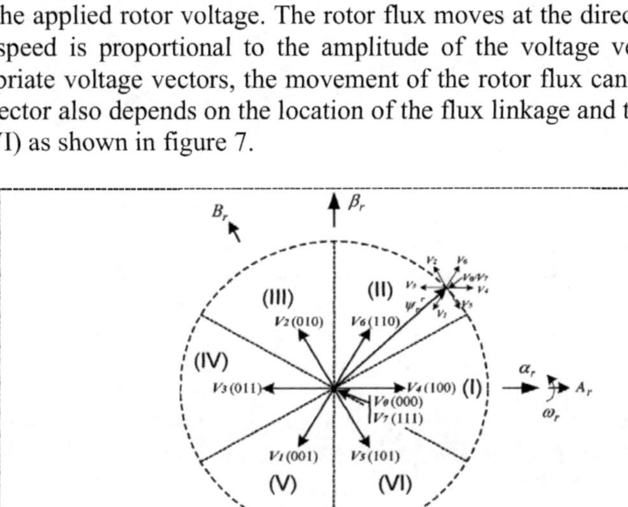

Figure 7. Space Vectors and their Locus.

IOP Conf. Series: Materials Science and Engineering **937** (2020) 012056 doi:10.1088/1757-899X/937/1/012056

5. Implementation

The selection of switching vectors for triggering the MC for DPC can be directly done but selected from the switching table of VSI in table 2. The active power error CP and the reactive power error CQ are obtained from the active power hysteresis and reactive power hysteresis bands respectively.

Table 1. Matrix Converter Switching Vectors.

Vector	Switches On			Output Voltage	Vector Position
abb(+1)	S_{aA}	S_{bB}	S_{bC}	$2/3V_{ab}$	0
baa(-1)	S_{bA}	S_{aB}	S_{aC}	$-2/3V_{ab}$	0
bcc(+2)	S_{bA}	S_{cB}	S_{cC}	$2/3\ V_{bc}$	0
cbb(-2)	S_{cA}	S_{bB}	S_{bC}	$-2/3\ V_{bc}$	0
caa(+3)	S_{cA}	S_{aB}	S_{aC}	$2/3\ V_{ca}$	0
acc(-3)	S_{aA}	S_{cB}	S_{cC}	$-2/3\ V_{ca}$	0
bab(+4)	S_{bA}	S_{aB}	S_{bC}	$2/3V_{ab}$	120
aba(-4)	S_{aA}	S_{bB}	S_{aC}	$-2/3V_{ab}$	120
cbc(+5)	S_{cA}	S_{bB}	S_{cC}	$2/3V_{bc}$	120
bcb(-5)	S_{bA}	S_{cB}	S_{bC}	$-2/3V_{bc}$	120
aca(+6)	S_{aA}	S_{cB}	S_{aC}	$2/3V_{ca}$	120
cac(-6)	S_{cA}	S_{aB}	S_{cC}	$-2/3V_{ca}$	240
bba(+7)	S_{bA}	S_{bB}	S_{aC}	$2/3V_{ab}$	240
aab(-7)	S_{aA}	S_{aB}	S_{bC}	$-2/3V_{ab}$	240
ccb(+8)	S_{cA}	S_{cB}	S_{bC}	$2/3V_{bc}$	240
bbc(-8)	S_{bA}	S_{bB}	S_{cC}	$-2/3V_{bc}$	240
aac(+9)	S_{aA}	S_{aB}	S_{cC}	$2/3V_{ca}$	240
cca(-9)	S_{cA}	S_{cB}	S_{aC}	$-2/3V_{ab}$	240
aaa(10)	S_{aA}	S_{aB}	S_{aC}	0	0
bbb(11)	S_{bA}	S_{bB}	S_{bC}	0	0
ccc(12)	S_{cA}	S_{cB}	S_{cC}	0	0

The switching table 2 for the inverter based DPC has been formed based on the flux Sector(S), CP and CQ. The input voltage sector (Vs) is calculated and is utilized for the MC-DPC vector selection. Also, the input power factor control is performed in this method and a separate hysteresis band is used for the same. A two-level hysteresis band fed by the error of the power factor and giving output variable namely CPF with values -1 || +1 for increasing or decreasing the power factor has been used to an appropriate switching vector for the MC for the PF control.

The MC switching table is selected from this VSI based DPC, CPF, and the input voltage sector number and has been selected from table 1 and has been tabulated in table 3. The simulation results are shown in figure 8 to figure 15.

IOP Conf. Series: Materials Science and Engineering **937** (2020) 012056 doi:10.1088/1757-899X/937/1/012056

Table 2. Optimal Switching Table.

Sector(S)		1	2	3	4	5	6
C_F	C_T						
	-1	2	3	4	5	6	1
-1	**0**	9	7	8	9	7	8
	1	6	1	2	3	4	5
	-1	3	4	5	6	1	2
1	**0**	9	7	8	9	7	8
	1	5	6	1	2	3	4

Table 3. Matrix Converter Based DTC.

Input Voltage Sector (Vs)		1	2	3	4	5	6
C_{PF}	VSI Vector						
	V1	1	-3	2	-1	3	2
	V2	-7	9	-8	7	-9	-8
-1	**V3**	4	-6	5	-4	6	5
	V4	-1	3	-2	1	-3	-2
	V5	7	-9	8	-7	9	8
	V6	4	6	-5	4	6	-5
	V1	-3	2	-1	3	2	1
	V2	9	-8	7	-9	-8	-7
1	**V3**	-6	5	-4	6	5	4
	V4	3	-2	1	-3	-2	-1
	V5	-9	8	-7	9	8	7
	V6	6	-5	4	6	-5	-4

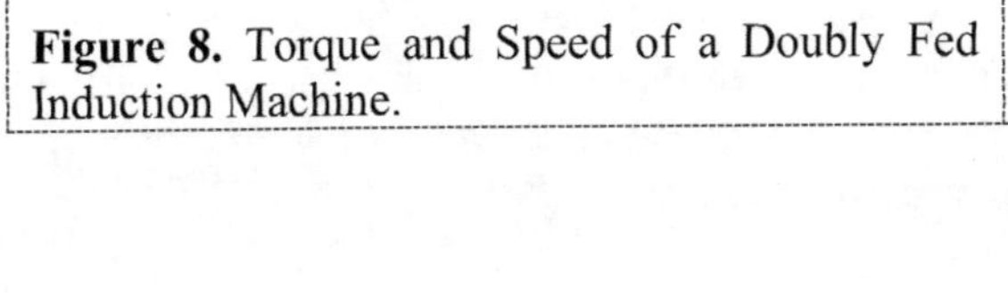

Figure 8. Torque and Speed of a Doubly Fed Induction Machine.

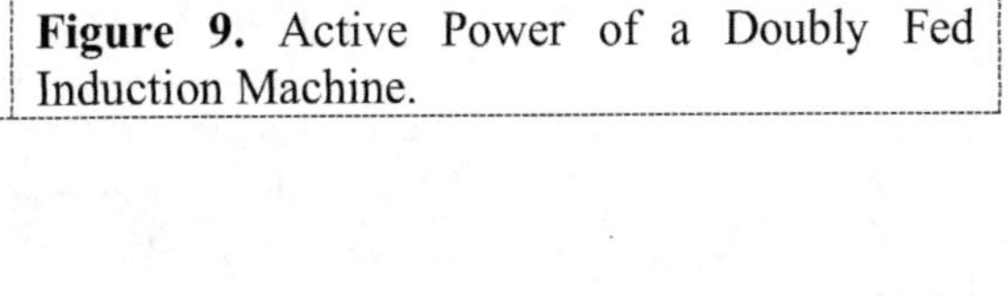

Figure 9. Active Power of a Doubly Fed Induction Machine.

IOP Conf. Series: Materials Science and Engineering **937** (2020) 012056 doi:10.1088/1757-899X/937/1/012056

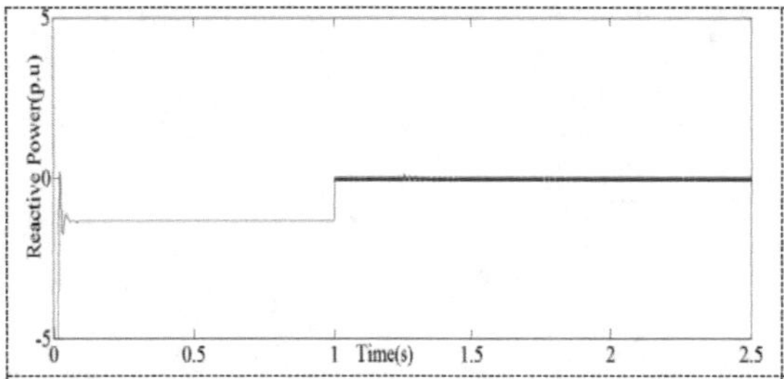

Figure 10. Reactive Power of a Doubly Fed Induction Machine.

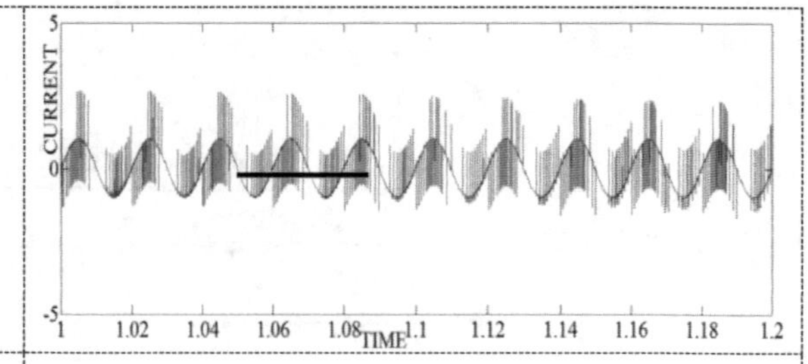

Figure 11. Unfiltered Grid current.

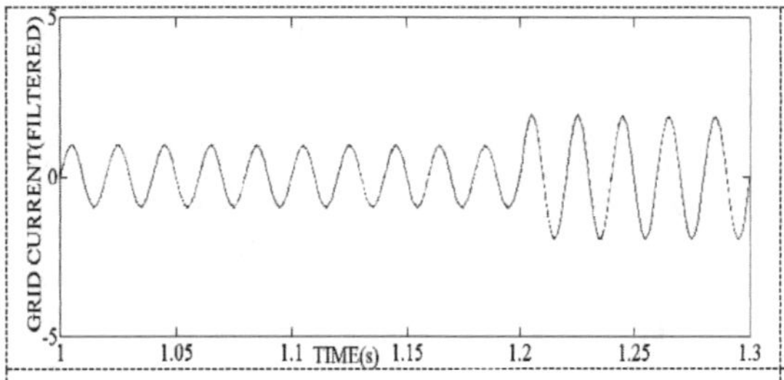

Figure 12. Filtered Grid current

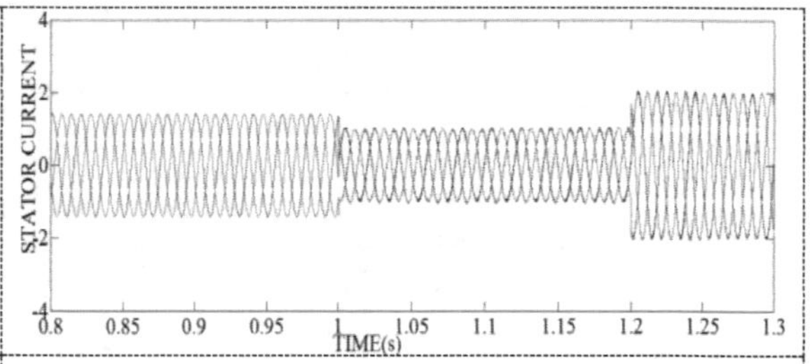

Figure 13. Stator Current of a Doubly Fed Induction Machine.

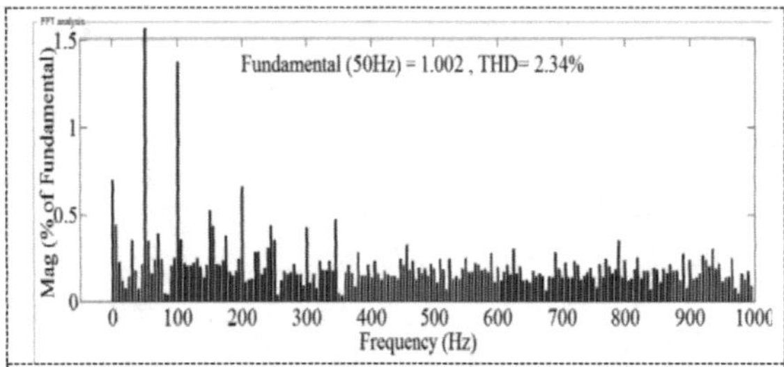

Figure 14. THD of GRID Current

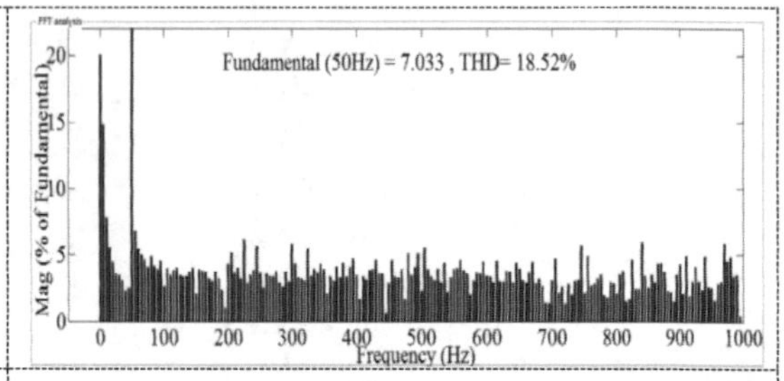

Figure 15. THD of Rotor Side Converter Voltage.

6. Conclusion

Matrix Converter provides a very handy solution for compact applications eliminating the use of bulky capacitors as there is no DC link between the AC-DC-AC converters. The major drawbacks associated with the Matrix Converters are Common Mode Voltage; High frequency switching harmonics; complexity of switching algorithm; complexity of commutation circuit. MCs are normally triggered using standing vectors- those that are pulsating in magnitude at a particular position in space. The MC's output voltage, synthesized from the standing space vectors, produces a high frequency Common Mode Voltage (CMV). The CMV produces bearing current causing the discharge of the stored energy in the parasitic capacitance through the low impedance path provided by the bearing leading to its failure and the subsequent failure of the machine shaft.

References

[1] Bolik S M 2004 *Modelling and analysis of variable speed wind turbines with induction generator during grid fault* Institute of Energy Technology Aalborg University

[2] Venturini M and Alesina A 1980 The generalized transformer: A new bidirectional sinusoidal waveform frequency converter with continuously adjustable input power factor, *Proc. IEEE PESC'80* pp 242–252

[3] Alesina A and Venturini M 1989 Analysis and Design of Optimum-Amplitude Nine-Switch Direct AC-AC Converters *IEEE Transactions on Power Electronics* vol 4 pp 101-112

[4] Blaabjerg F, Casadei D, Klumpner C, and Matteini M 2002 Comparison of two current modulation strategies for matrix converters under unbalanced input voltage conditions, *IEEE Trans. On Industrial Electronics* vol 49 pp 289–296

[5] Wheeler P W, Rodriguez J, Empringham L, Weinstein A and Clare J C 2002 Matrix Converters: a technology review, *IEEE Trans. On Industrial Electronics* vol 49 pp 276 – 288

[6] Hanjucha and Prasad N Enjeti 2003 An approach to reduce Common mode voltage in Matrix converters *IEEE trans. on Industry Applications* vol 39 pp 1151-1159

[7] Cha H J 2004 *Analysis and design of matrix converters for adjustable speed drives and distributed power sources* (Doctoral dissertation, Texas A&M University)

[8] Gao F and Iravani M R 2007 Dynamic model of a space vector modulated matrix converter *IEEE transactions on power delivery* vol 22 pp 1696-1705

[9] Gupta R K 2010 *Rotating-space-vector-modulated matrix-converter-based advanced drive topologies with enhanced features* (Doctoral dissertation, University of Minnesota)

[10] Fletcher J and Yang J 2010 Introduction to doubly-fed induction generator for wind power applications *Paths to sustainable energy* pp 259-278

[11] Barakati S M 2008 *Modeling and controller design of a wind energy conversion system including a matrix converter*

[12] Harris B J 2009 *Matrix converter technology in doubly-fed induction generators for wind generators*

[13] Bose B K 2002 *Modern power electronics and AC drives* **123** Upper Saddle River (NJ: Prentice hall)

[14] Datta R and Ranganathan V T 2001 Direct power control of grid-connected wound rotor induction machine without rotor position sensors *IEEE Transactions on Power Electronics* 16 pp 390-399

[15] Xu L and Cartwright P 2006 Direct active and reactive power control of DFIG for wind energy generation *IEEE Transactions on energy conversion* 21 pp 750-758

[16] Robyns B, Francois B, Degobert P and Hautier J P 2012 Vector control of induction machines *In Vector Control of Induction Machines* (London: Springer) pp 75-121

[17] Prabaharan N and Palanisamy K 2016 A single-phase grid connected hybrid multilevel inverter for interfacing photo-voltaic system *Energy Procedia* **103** 250-255

[18] Palanisamy K, Mishra JS, Raglend IJ and Kothari DP 2010 Instantaneous power theory based unified power quality conditioner (UPQC) *IEEE Joint International Conference on Power Electronics, Drives and Energy Systems* 1-5

[19] Sitharthan R and Geethanjali M 2017 An adaptive Elman neural network with C-PSO learning algorithm-based pitch angle controller for DFIG based WECS Journal of Vibration and Control **23(5)** 716-730

[20] Sitharthan R and Geethanjali M 2015 Application of the superconducting fault current limiter strategy to improve the fault ride-through capability of a doubly-fed induction generator–based wind energy conversion system *Simulation* **91(12)** 1081-1087

[21] Sitharthan R, Karthikeyan M, Sundar DS and Rajasekaran S 2020 Adaptive hybrid intelligent MPPT controller to approximate effectual wind speed and optimal rotor speed of variable speed wind turbine *ISA transactions* **96** 479-489

[22] Sitharthan R, Devabalaji KR and Jees A 2017 An Levenberg–Marquardt trained feed-forward back-propagation based intelligent pitch angle controller for wind generation system *Renewable Energy Focus* **22** 24-32

[23] Sitharthan R, Sundarabalan CK, Devabalaji KR, Yuvaraj T and Mohamed Imran A 2019 Automated power management strategy for wind power generation system using pitch angle controller *Measurement and Control* **52(3-4)** 169-182

[24] Sundar DS, Umamaheswari C, Sridarshini T, Karthikeyan M, Sitharthan R, Raja AS and Carrasco MF 2019 Compact four-port circulator based on 2D photonic crystals with a 90° rotation of the light wave for photonic integrated circuits applications *Laser Physics* **29(6)** 066201

[25] Sitharthan R, Parthasarathy T, Sheeba Rani S and Ramya KC 2019. An improved radial basis function neural network control strategy-based maximum power point tracking controller for wind power generation system *Transactions of the Institute of Measurement and Control* **41(11)** 3158-3170

[26] Rajesh M and Gnanasekar JM 2017 Path observation based physical routing protocol for wireless ad hoc networks *Wireless Personal Communications* **97(1)** 1267-1289

[27] Palanisamy K, Varghese LJ, Raglend IJ and Kothari DP 2009. Comparison of intelligent techniques to solve economic load dispatch problem with line flow constraints *IEEE International Advance Computing Conference* 446-452

[28] Sitharthan R, Ponnusamy M, Karthikeyan M and Sundar DS 2019 Analysis on smart material suitable for autogenous microelectronic application *Materials Research Express* **6(10)** 105709

[29] Rajaram R, Palanisamy K, Ramasamy S and Ramanathan P 2014 Selective harmonic elimination in PWM inverter using fire fly and fireworks algorithm *International Journal of Innovative Research in Advanced Engineering* **1(8)** 55-62

[30] Sitharthan R, Swaminathan JN and Parthasarathy T 2018 March. Exploration of wind energy in India: A short review *IEEE National Power Engineering Conference* 1-5

Estimation of Lifetime and Reliability of Input Rectifier in Variable Frequency Drives

[1]Ramesh. P, [1]Govarthanan. R, [2]Dr. Palanisamy K, [3]Dr. Paramasivam S
[1]Danfoss Industries Pvt Ltd, Chennai, India, ramesh@danfoss.com, govi@danfoss.com
[2]Vellore Institute of Technology, Vellore, India, kpalanisamy@vit.ac.in
[3]ESAB India Ltd, Chennai, India, param.s@esab.co.in

Abstract. This research paper investigates the causes of power module failures in front-end rectifier of variable frequency drive (VFD) and their lifetime estimation. The front-end rectifier plays an important role in the electrical system for energy conversion. However, from the industrial survey it is found that they are sensitive to harsh operating conditions and are prone to damage. Hence, it becomes inevitable to analyse the causes of their failure and lifetime. Due to prolonged usage, sudden variations in supply and load, the power modules will be subjected to stresses such as overvoltage, voltage variations due to sag, phase unbalance, over current, temperature, harmonics and pre-distortions. The major causes of failure are due to several stresses such as solder joint break, bond wire lift-off, cracks in power modules are discussed. In this paper, the lifetime of the power module is estimated for a given mission profile data. The junction temperature profile is obtained from the VFD Front-end rectifier simulation model. The number of temperature cycles from the temperature profile is estimated using cycle counting methods. Lifetime estimation and damage modelling is done using Coffin-Manson rule and Palmgren Miner rules. All these modelling and simulations are done for a specific VFD power size using MATLAB and PLECS.

***Keywords*:** Lifetime estimation, reliability, Power modules, Cycle counting algorithm, VFD, Coffin-Manson rule, Palmgren Miner rule.

1. Introduction

Power converters contributes significant role in energy conversion systems which are widely used in various industrial applications. Reliability of Power semiconductor devices used in these converters is an important criterion for determination of their lifetime. Power converters are deployed in severe operating environment which are prone to various power quality disturbances such as voltage sag and swell, voltage unbalance, harmonic distortion, transients, rapidly changing temperature cycles and grid impedance variations. Power semiconductor devices reliability are mainly dependent on junction temperature variations subjected to thermal stresses due its operating environment, thereby reducing their lifetime. VFD lifetime is recommended based on the reliability of the DC filter capacitors and the Power modules. Front-end rectifiers were not given high priority for reliability in the past. However, in modern day scenario, the failures in the Front-end diodes and SCRs in VFD are getting increased. Input rectifier diodes/SCRs are subjected to various such disturbances and hence electrical, mechanical and thermal stresses vary considerably. Very few research works have been done in these areas.

In small power VFDs, there used to be an integrated Power module which consists of both rectifier and inverter sections designed in the same package. Sometimes, these modules will also have brake chopper built in. So, in this case, compared to the reliability of the front-end components, the IGBT inverters will be poor. However in drives rated beyond certain power range, this structure is not feasible.

IOP Conf. Series: Materials Science and Engineering **937** (2020) 012057 doi:10.1088/1757-899X/937/1/012057

The Front-end rectifier/Diode rectifiers will be kept in separate packages. In such case, it is important to know the reliability of this diode/SCR bridge also.

To understand in detail the major causes of concerns during power module usage, both dynamic power and standalone thermal tests are required to be done. Detached thermal cycling tests cause stress within the interconnection between base plate and substrate. Amid dynamic power cycles, there will be stress on bond wires resulting in concerns over solder joints in between the substrate and chip. Amid brief cycles, stress is observed on bonding wires and during long cycles, the chip-solder layer encounters increased stress. Investigations are carried out to understand issues due to stress in temperature such as average temperature, swing in temperature, junction temperature, temperature cycle duration and its impact on power module reliability and lifetime. Many investigations are carried out in bonding technology improvements and lifetime expectation models for checking the unwavering quality of the power module. The unwavering quality and lifetime of the power module is vital concern because it decides the reliability of the whole system [1]. The lifetime of the modules depends on the field of application. Power converter operates in extreme conditions subjected to longer duration of high currents, voltages and temperature swings.

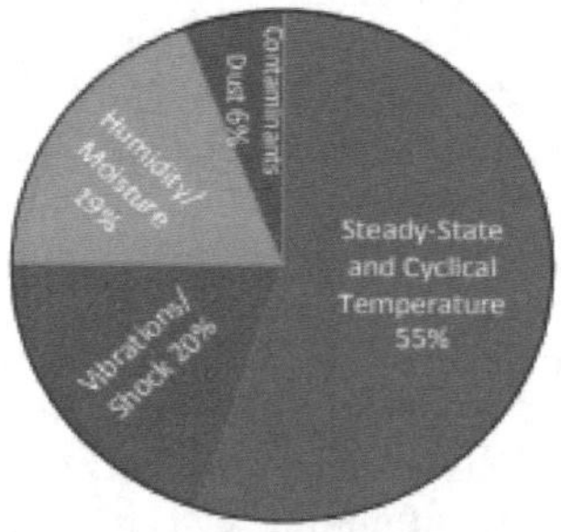

Figure. 1. Stress distribution sources

Analysis of failure mechanism is important criterion to estimate lifetime of power modules. Fig. 1 shows the source of stress distribution in power module. Major contribution of power module failure is random in nature, resulting in thermal stress caused due to increased junction temperature during passive thermal and active power cycling. Solder joints and bonding wires are the weakest interconnections within power module prone to frequent damage. These failures are mainly attributed to various thermal expansion coefficients of different adjacent layers within power module. Fig. 2 shows the failure mode distribution in power module.

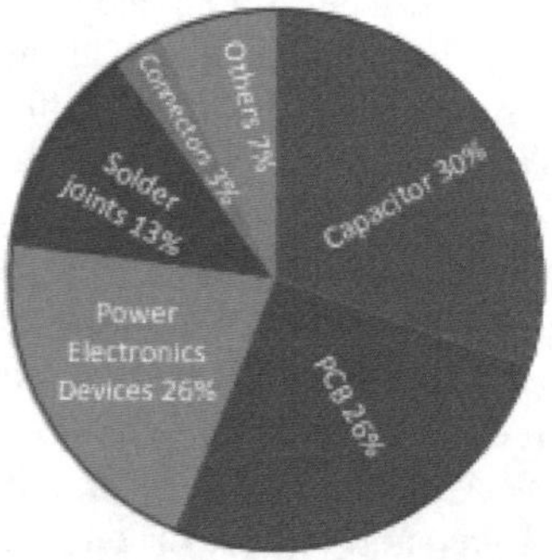

Figure. 2. Failure mode distribution

Lifetime prediction and failure mechanisms of power converters are well understood by creating various electrical and thermal cyclic models, subjected to accelerated life cycle tests carried over entire operating cycles. Analytical models are most suited to estimate lifetime over several cycles until failure, subjected to various factors such as current through bond wires, swing in temperature and the average value of

temperature. Cycle counting algorithms are used to determine the number of cycles and lifetime prediction is suitably modelled.
Three phase front-end rectifier, analysis of failure mechanisms, power modules failure modes, lifetime estimation through various simulation models and analytical methods are detailed in various sections of this investigation. Front-end rectifier in modelled using MATLAB® and PLECS® and the simulation results are discussed in detail.

2. Three phase Front-end rectifier

Three phase rectifiers are commonly used front-end topology in six pulse Variable Frequency Drives. The major parts of VFD are three phase front-end rectifier, control and regulation section, intermediate dc link and inverter as shown in Fig. 3. Front-end rectifiers are the most vulnerable part of VFD in the overall electrical system. Rectifiers must deal with high power and power quality disturbances. Hence, they are subjected to severe damages, stresses due to overvoltage, current and thermal effects. Thus, the reliability of VFD mainly depends on the rectifier reliability. The system reliability can be increased by replacing rectifier components before the components fail. In order to replace at the right moment, the lifetime of the device must be estimated. The estimation involves various testing. The tests must be carried out under different loading conditions considering the harsh environment in which rectifier is in operation.

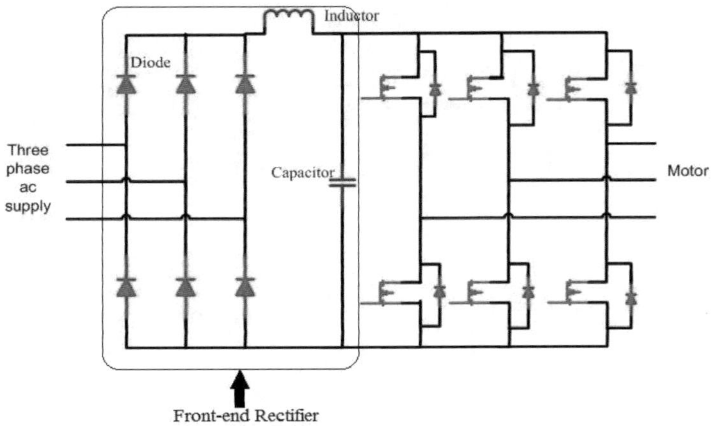

Figure. 3. Front-end rectifier components in VFD

3. Power modules

The foremost dependable operation of front-end rectifiers can be accomplished by the unwavering quality of diode and SCR power modules. Power module as shown in Fig. 4 comprises of distinctive layers each made of diverse materials. Silicon is the main constituent of the device chip and is soldered onto coordinate copper fortified ceramic substrate. This substrate serves as an insulation separator to chip from the metal base plate and scatters heat onto the cooling system framework. For giving electrical contact to the outside circuit, the chip is connected to aluminium bond wires. Chip top is joined with aluminium bond wire. All these materials have diverse CTEs (Coefficients of Warm Development) [1, 2, 3, 5]. The foremost untrustworthy parts in power module are the bonding wires and solder joints. At different loading conditions, power module is subjected to stresses due to increased temperature and variations in voltage and current, thereby resulting in junction temperature increase.

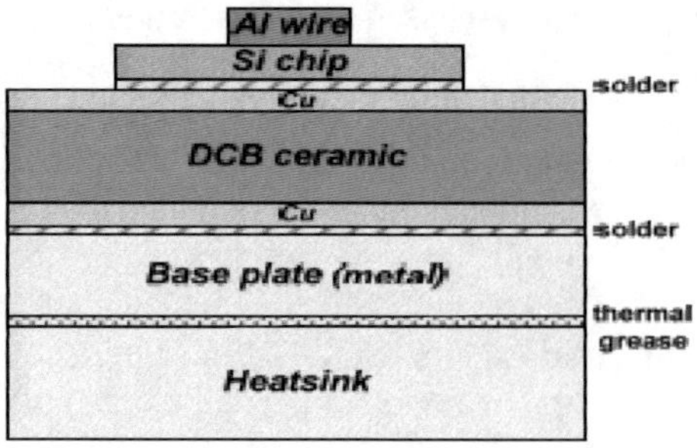

Figure. 4. Layers of Power Module

Dissipation of heat along with change in CTE, tends to lift-off distinct layers within the module thereby accelerating failure of the power module. Various accelerated life tests are conducted to ascertain the reliability of the module under several stress factors. Thereby, module is subjected to uneven temperature distribution and variations. Lifetime of module with its inherent design components can be assessed by subjecting to different tests thereby obtaining the junction temperature and examining the observations through different algorithms. Table.1 shows the co-efficient of thermal expansion of different materials used within the power module.

Table. 1. Coefficient of thermal expansion of different materials in Power module

S No.	**Coefficient of Thermal Expansion**	
	Material	*Value*
1	Al	$23.5*10^{-6}K^{-1}$
2	Al2O3	$6.8*10^{-6}K^{-1}$
3	AlN	$4.7*10^{-6}K^{-1}$
4	Cu	$17.5*10^{-6}K^{-1}$
5	Si	$2.6*10^{-6}K^{-1}$

4. Mission Profile

The temperature swing on the module is the important factor which has significant effect on the lifetime. From several research analysis it is understood that for every 10 °C temperature rise of the module, the failure probability is almost doubled [2, 3, 9, 12]. Hence monitoring temperature of the module gives good insight on the reliability enhancement. Mission profile variations impacts further temperature swings on the module.

Mission profile is the representative voltage pattern applied to the input of on the power module which has significant impact on device performance and junction temperature, thereby module lifetime [17]. Mission profile variations impacts loading stress of the module. Device strength depends on their physical properties. Stress on the device is directly related to device strength, which attributes device failure. These factors directly impact rectifier reliability and lifetime. Mission profile of the module in real time application ensures good lifetime estimation and reliability. The module ought to have way better electro-thermal characteristics. Mission profile transformation to thermal loading of rectifier level ensures reasonable lifetime estimation.

Power quality disturbances in the system has significant contribution in mission profile creation [17]. Mission profile consists of several input parameters such as power supply and environmental conditions including their variations. Various types of voltage sags have been accounted in Mission profile modelling. Voltage sag and types, their classification, occurrence reasons and effects on the system are discussed as below.

4.1. *Voltage Sag*

Voltage sag is the foremost critical power quality issue because it causes antagonistic impacts on the connected equipment's in the utility side [11, 16]. Voltage sag could be a diminish in RMS voltage for brief period caused by different occasions – power system framework faults, sudden energization of the transformer, connected load variations and high-power motor starts. Voltage sag basically happens due to fault current streaming through the power system onto the fault location. Voltage list is well defined by its magnitude, recurrence period and frequency. System impedance of the power system determines sag magnitude.

As per IEEE standard 1159-1995, voltage sag is characterized as a diminish between 0.1 and 0.9 p.u. in root mean square (RMS) voltage at the power frequency for duration length of 0.5 cycles to 1 minute [16].

4.1.1. Voltage sag classification

Voltage sags are classified depending upon the combination of above factors. Table. 2 provides information about the classification of sags.

Table. 2. Classification of voltage sag

Classification of sag	**Factors**		
	Fault type	***Transformer connection***	***Load connection***
Sag A	Three-phase to ground	Δ/Y_g	Y_g connection
Sag B	Single-phase to ground	Y/Y	Y_g connection
Sag C	Single-phase to ground	Δ/Y	Y_g connection
Sag D	Phase to phase	Δ/Y	Y_g connection
Sag E	Two-phase to ground	Y/Y	Y_g connection
Sag F	Two-phase to ground	Δ/Y	Y_g connection
Sag G	Two-phase to ground	Δ/Δ	Y_g connection

Type Sag A incorporates equal voltage drop in all three phases and there's equal 120° phase displacement as this happens due to symmetrical fault. Type Sag B has no voltage drops in Y and B phase wherein there's a drop-in phase R with 120° displacement in phase. Type Sag C sort incorporates voltage drop in R and B phases with no voltage drop in phase Y. As there is no symmetrical fault, the phase displacement between the phases isn't 120°. In type Sag D, the voltage drop is more in R phase compared to Y and B phases. Fault is unsymmetrical and phase displacement isn't 120°. In type Sag E, there's voltage drop in R and Y phases with no voltage drop in B phase, also there's 120° phase displacement among the phases.

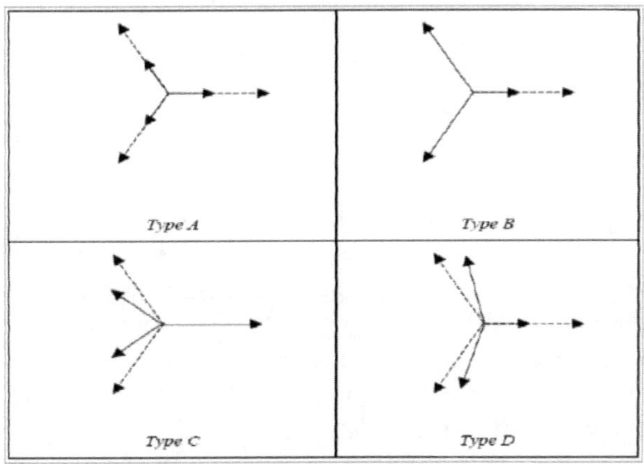

Figure. 5. Voltage sag types and one or three phase faults

Type Sag F has more voltage drop in R phase as compared with Y and B phase, resulting in the phase displacement not as 120°. Less voltage in phases Y and B as compared to phase R and displacement in phase is not 120° [16] (Fig. 5, 6).

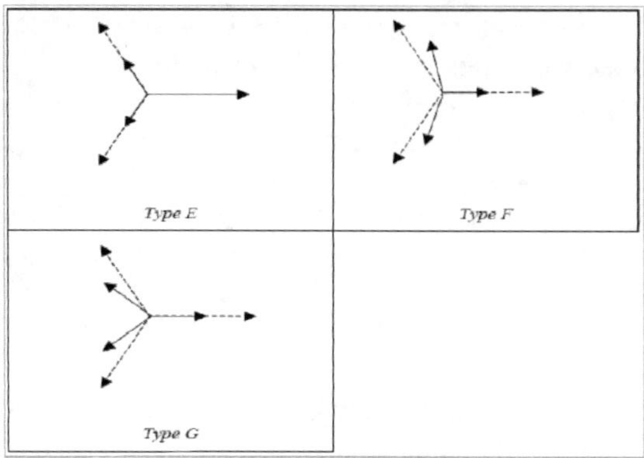

Figure. 6. Voltage sag types and two-phase faults

5. Lifetime estimation

VFD Front - end rectifier module and its lifetime are estimated [15] through proper methodology. Right models provide more accurate results. In this investigation, the following steps are followed.

1. System modelling
2. Thermal modelling
3. Lifetime modelling
4. Damage modelling

Fig.7 shows the flow chart of the lifetime estimation model and interconnecting information between different blocks.

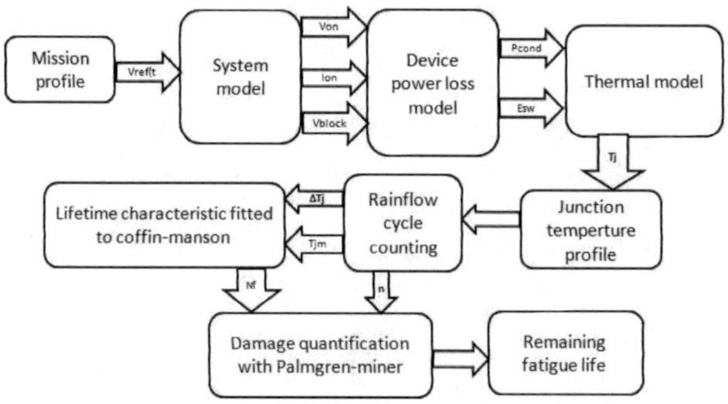

Figure.7. Power module lifetime estimation flow chart

5.1. System modelling

Prior to analyzing the characteristics of the module, circuit is first modelled, its working and characteristics are studied in detail, information are recorded. The system model will be made through use of reliable simulation tool or a mathematical model. The model should provide the resultant electrical data of the Power module based on the operating conditions. The input to the system model is the mission profile data. The loss contributing electrical parameters will be extracted from the system model. A repetitive voltage profile pattern is applied as the mission profile input to the circuit and variations in the module are observed and recorded.

5.2. *Thermal modelling*

Electro-thermal model is generally designed for establishing junction temperature of the module, switching power losses and conduction losses. Thermal characteristics of the module components can be portrayed by RC models [6, 14,16]. Thermal resistance defines the device ability to resist the heat flow and thermal capacitance defines the capability of the device to absorb or store heat. RC networks like Foster or Cauer (Fig. 8) are being used. Device mathematical modelling through Foster network does not give enough information on the internal structure. However, Cauer model provides physical meaning and needs more information on power module internal structure which is difficult to obtain. Hence the thermal modelling is designed using Foster RC network (Fig.9).

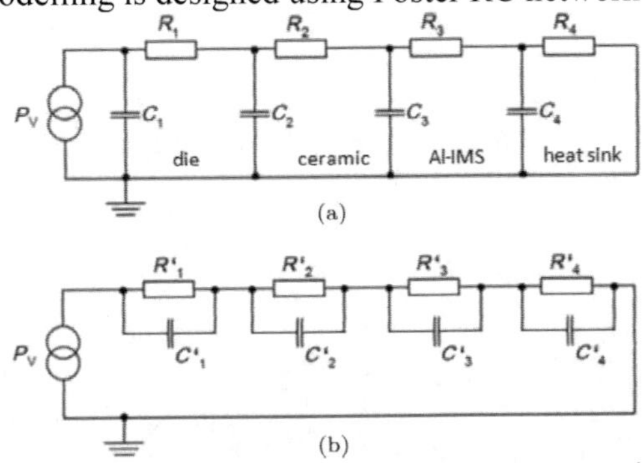

Figure 8. Thermal equivalent of RC Networks (a) Foster network (b) Cauer network

Thermal modelling hence deploys RC network equivalent and thermal coefficients available in the manufacturer's datasheet. Thermal resistances R_{thi} is available in the datasheet and hence thermal capacitance is derived from known thermal time constants τ_i and using the formula as below in equation (1),

$$C_{thi} = \frac{\tau_i}{R_{thi}} \tag{1}$$

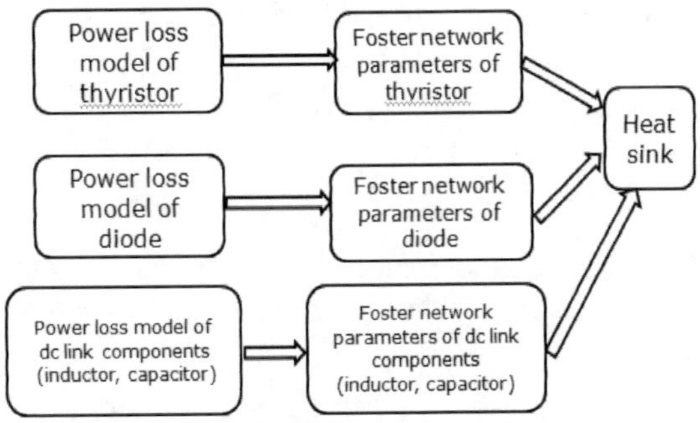

Figure.9. Thermal modelling

5.3. *Lifetime modelling*

Junction temperature is the most critical criteria in lifetime estimation of the power device. Junction temperature profile is derived from the thermal modelling and is generally irregular. Hence lifetime estimation is done by using suitable cycle counting method along with simulation models which determines fatigue of solders and bonding wires. The extent of damage on the device due to temperature stress is well measured using Palmgren miner rule. Several lifetime models are discussed as below [15].

Various counting methods such as level crossing, simple range, peak and rain flow are in general use. Amongst the popular counting algorithms, rain flow counting is the most commonly used method [7, 8, 15]. Rain flow counting has the unique advantage of reducing irregular temperature profile to simple and regular temperature intervals.

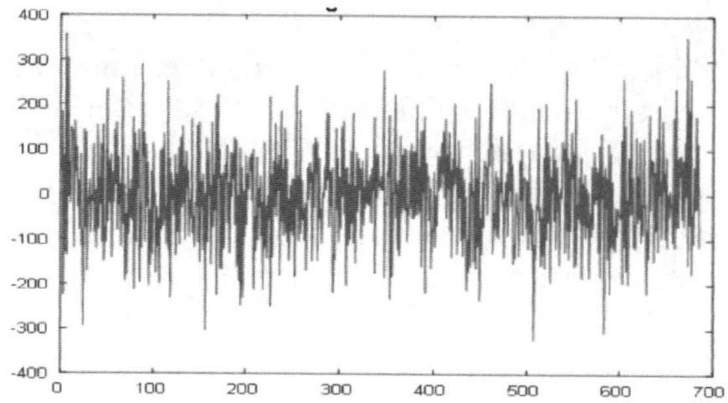

Figure 10. Irregular waveform

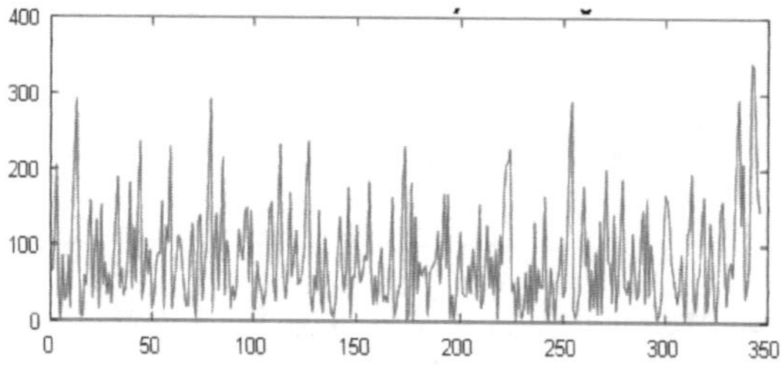

Figure 11. Simplified waveform - Rain flow cycle counting method

5.3.1. *Rain flow cycle counting*

Among the different counting techniques inside Rainflow method, three-point counting algorithm is being used. This algorithm thus eliminates shorter patterns through regular profile pattern checking (Fig 10, 11). Three consecutive points are being checked to determine completion of cycle [8, 13, 14, 17]. The amplitude and mean value of thermal cycle is calculated by equation (2) and equation (3) respectively.

$$T_{ja} = \frac{|T_{jmax} - T_{jmin}|}{2} \tag{2}$$

$$T_{jm} = \frac{T_{jmax} + T_{jmin}}{2} \tag{3}$$

By knowing the maximum junction temperature T_{jmax} and minimum junction temperature T_{jmin}, the temperature rise can be estimated as shown in equation (4).

$$\Delta T_j = |T_{jmax} - T_{jmin}| \tag{4}$$

5.3.2. *Different Lifetime models*

Lifetime models [1, 8, 15] are designed in general based on the module thermal characteristics. Various Lifetime models are discussed as below:

- *Coffin Manson model:* Coffin-Manson is used in general to estimate the fatigue caused by solder joints and bonding wires due to temperature changes in every cycle, expressed by equations (5, 6).

$$N_f = A * \Delta T_j^{\alpha} * \exp\left(\frac{E_a}{k_b * T_{jm}}\right) \tag{5}$$

wherein,

N_f – *number of cycles due before failure*

α – *Expression Constant* $[K^{-\alpha}]$

ΔT_j – *range of temperature [K]*

A – *general constant*

k_b – *Boltzmann constant [J/K]*

E_a – *Activation energy [J]*

T_{jm} – *Average temperature[K] calculated by expression*

$$T_{jm} = T_{jmin} + \frac{\Delta T_j}{2} \tag{6}$$

- *Norris-Landzberg model:* Norris-Landsberg model estimates number of cycles to failure taking into consideration temperature swing and the frequency by equation (7).

$$N_f = A * f^{-n_2}(\Delta T_j^{-n_1}) * \exp\left(\frac{E_a}{k_b * T_{jm}}\right) \tag{7}$$

- *Bayerer model:* Bayerer model is multi-parametric which considers temperature swing ΔT_j, heating time t_{on}, applied DC current I, diameter of the bond D, blocking voltage V and junction temperature T_j. to determine number of cycles to failure by equation (8).

$$N_f = K(\Delta T_j^{-\beta_1}) * \exp\left(\frac{\beta_2}{T_j * 273K}\right) t_{on}^{\beta_3} * I^{\beta_4} * V^{\beta_5} * D^{\beta_6} \tag{8}$$

K and β are the constants derived through reliability test experiments. Among the above models, the simplest and commonly used model is Coffin-Manson model.

5.4. *Damage modelling*

Damage modelling is calculated through estimation of accumulated stress on the device. Palmgren miner rule is used to evaluate extent of damage in each cycle and thus calculated as in equation (9).

$$D = \frac{n[i]}{N_f[i]} \tag{9}$$

n_i - *the number of identified temperature cycles*

j - *the number of cycles and half-cycles distinguished from the history*

N_{fi} - *the number of cycles to failure computed with Coffin-Manson law*

However total damage is expressed as sum of damage caused every cycle as shown in equation (10). The device is expected to be failed [1, 8, 17] whenever total sum equals to 1.

$$D = \sum_{i=1}^{j} D[i] \tag{10}$$

IOP Conf. Series: Materials Science and Engineering **937** (2020) 012057 doi:10.1088/1757-899X/937/1/012057

5.5. Lifetime Estimation

Device lifetime can be estimated by understanding in detail the interval and total damage over the duration of entire cycle as shown in equation (11).

$$L = \frac{T_0}{D} \tag{11}$$

T_0 – cycle interval
D – Total damage over the entire cycle

Hence, the lifetime of VFD front-end rectifier is estimated through suitable design of system and electro-thermal model. These models are subjected to various mission profiles to analyse various device parameters mainly junction temperature. From the junction temperature profile lifetime is established by using Rain flow counting strategy, Coffin-Manson law and Palmgren miner rule.

6. Simulation results

VFD front-end rectifier is suitably modelled using MATLAB and PLECS software (Fig. 12). Thermal model of the front-end rectifier in specific is done using PLECS (Fig.13). PLECS software has built-in options for entering device data obtained from the manufacturer datasheets within the framework of multi-dimensional lookup tables and thereby losses are estimated.

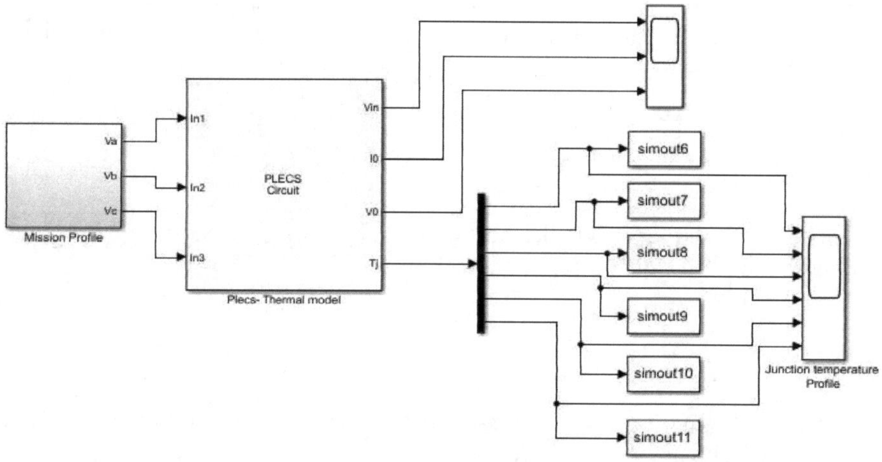

Figure. 12. Front-end rectifier MATLAB-PLECS model for lifetime estimation

Table 3. Specifications of the components used in the circuit

S.No	Component	Specification
1	Front-end Diode	MCD 162-16io1
2	VFD DC inductor (L)	181μH
3	VFD DC capacitor (C)	4700μF

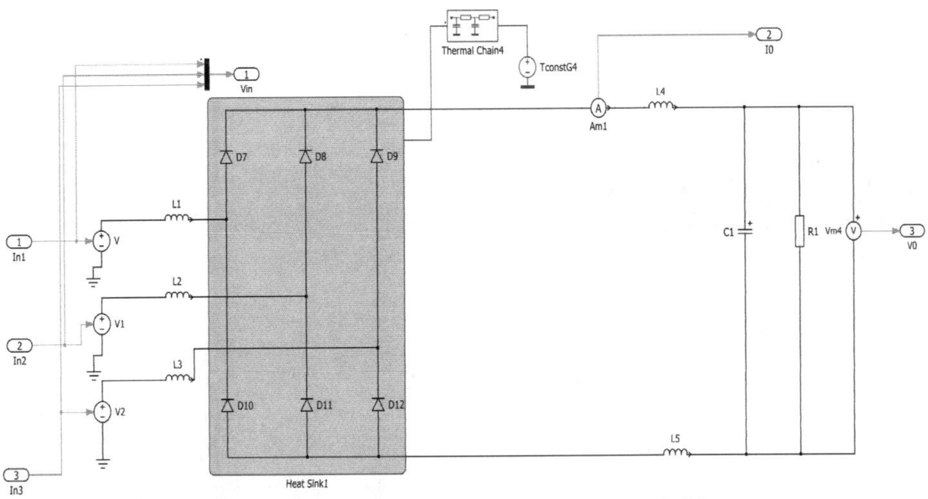

Figure. 13. Front-end rectifier Electro-thermal modelling: PLECS

The simulation is run for $T_0 = 1000$ s to analyse the damage accumulated for a considerable period and to analyse the model on the occurrence of voltage sag for few cycles at various instants of time. The specifications of the VFD and its components used in the simulation circuit are mentioned in Table. 3.

7. Mission profile

Three phase 415V, 50 Hz is fed as input supply to the system under investigation. Grid impedance plays significant role in establishing the stability and lifetime of the system. Grid impedance can be established from the short circuit ratio of the system. Short circuit ratio (SCR) is defined as the ratio of RMS current produced in the system to the nominal current of the system.

Short circuit ratio determines the voltage distortion and the stability of the grid. Short circuit ratio is taken into consideration to determine the distortion in the voltage. Higher impedance at the input minimizes the distortion in the supply. Lower the SCR, source impedance will be high. The test is done on front-end rectifier of a VFD. The nominal current of the drive for 415 V, 50 Hz input supply is found to be 37.56 A.

The source inductance is calculated for two different short circuit ratios. The source impedance is calculated and with available data system is modelled, simulation is carried out with mission profile as input supply without any disturbance (Fig. 14) and with voltage sag (Fig.15) for SCR as 20 and 120.

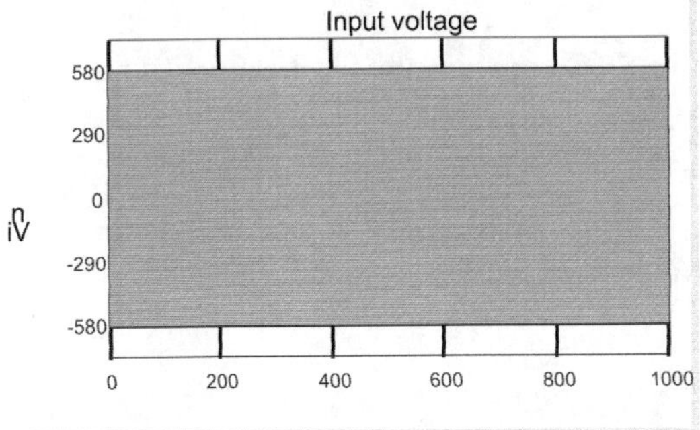

Figure. 14. Input supply without any disturbance

The voltage sag is generated at various instants throughout the loading profile to the system. Amplitude of the voltage sag is about 0.1 to 0.9 p.u. of the nominal voltage for a duration of few cycles to few minutes during its occurrence. During the simulation, voltage sag is generated at few instants for a duration of 60s, 50s, 40s, 20s and 40s with reduction in the nominal voltage wherein the voltages are 0.5 p.u., 0.2 p.u., 0.7 p.u., 0.4 p.u. and 0.8 p.u. of the nominal voltage respectively.

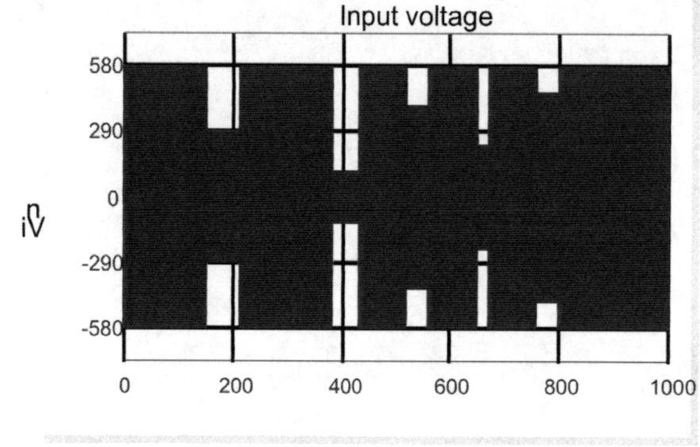

Figure. 15. Input supply with voltage sag

8. System model results

VFD front-end rectifier is suitably modelled using MATLAB and PLECS software. Front-end rectifier diode electrical parameters are updated in the lookup table of PLECS. This data corresponds to diode MCD 162-16io1 considered in this investigation.

9. Thermal model results

Thermal modelling of the device is developed using Foster RC network. However, thermal resistance values are obtained from diode manufacturer datasheet (Table.4). Thermal capacitance values are

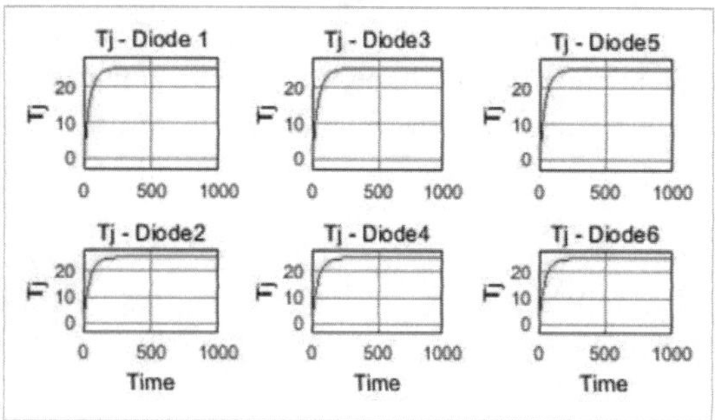

Fig.16 Thermal results for Rsce=20 without Sag

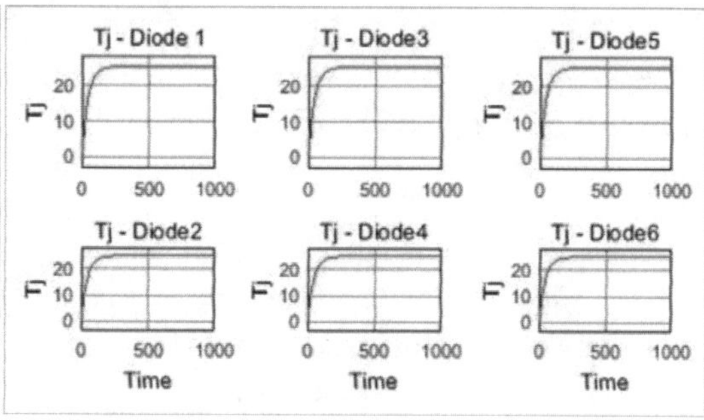

Fig.17 Thermal results for Rsce=120 without Sag

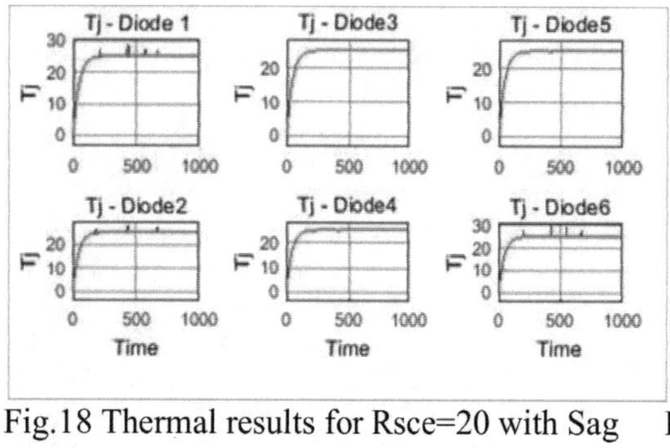

Fig.18 Thermal results for Rsce=20 with Sag

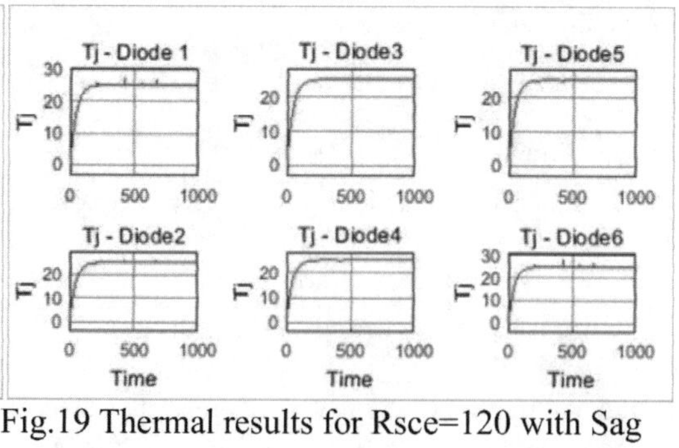

Fig.19 Thermal results for Rsce=120 with Sag

obtained from the thermal time constant values τ_i using the equation (12) as below,

$$C_{thi} = \frac{\tau_i}{R_{thi}} \tag{12}$$

C_{thi} – *thermal capacitance*
R_{thi} – *thermal resistance*
τ_i – *thermal constant*

The associated thermal and electrical parameters of semiconductor devices considered in this investigation are interconnected through means of heat sink. Heat sink is further associated to foster RC network and ambient (Table. 5). Ambient is considered to be maintained at 25°C and equivalent thermal parameters of the diode and heatsink are obtained from the datasheet.

Table 4. Thermal parameters for diode : Foster RC network

Thermal resistance R_{th} (K/W)	**Thermal constant τ_i (s)**
0.0072	0.001
0.0188	0.080
0.1290	0.200
0.07	14.29

Table 5. Thermal parameters for Heat sink: Foster RC network

Thermal resistance R_{th} (K/W)	**Thermal constant τ_i (s)**
0.0045	0.0045
0.0013	0.39
0.0057	7.1649

Additionally, thermal capacitance is calculated using equation (12) and entered in PLECS lookup table.

10. Lifetime and Damage Modelling results

Three phase 415V, 50 Hz is fed as input supply to the system and is simulated for time period of T_0 = 1000s with *SCR* = 20 and 120 respectively , with and without the event of voltage sag. VFD front-end diode junction temperature profile is determined for all cases (Fig 16, 17, 18, 19). It is observed from the graphs that there is considerable stress on diodes due to the occurrence of voltage sag and maximum short circuit ratio.

Outcome of the simulation and modelling is the irregular temperature profile which is subjected to rainflow counting algorithm for extracting optimized regular profile, determining the number of cycles and mean temperature. These values are then applied to Coffin-Manson lifetime model for obtaining the intensity of damage occurred and thereby determining lifetime of the device as shown in equation (5), where T_{jm} is calculated using equation (6). Damage is evaluated for the entire test cycle duration. Total damage of the system is estimated by the summation of individual damages in each cycle as shown in equation (10).

Simulation is carried out with the mission profile as input voltage supply without any disturbance and with the input supply with voltage sag and for short circuit ratio = 20 and 120. Damage on all diodes is obtained and the results are plotted in a graph (Fig 20, 21). The results show that the diodes experience more stress and damage on the occurrence of voltage sag which is more during higher short circuit ratio.

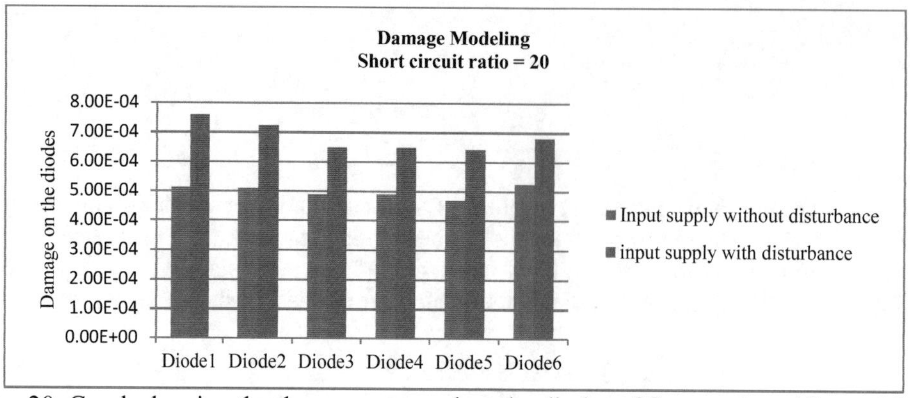

Figure. 20. Graph showing the damage occurred on the diodes of front-end rectifier at SCR=20

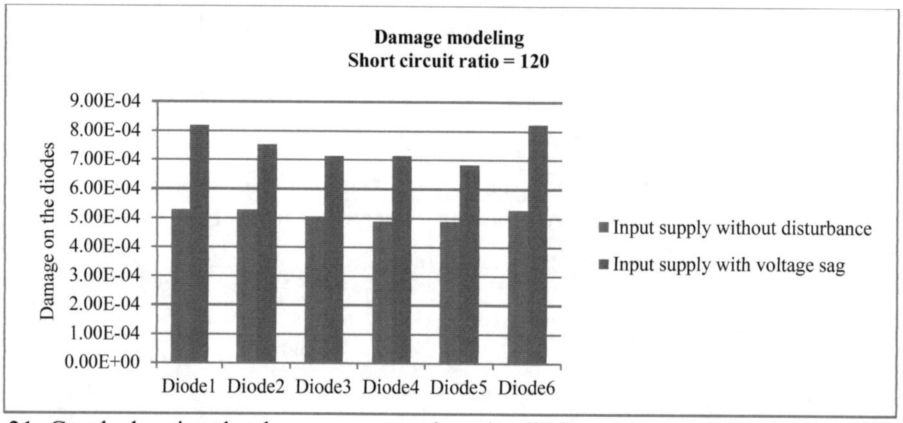

Figure. 21. Graph showing the damage occurred on the diodes of front-end rectifier at SCR=120

From the data from simulation, the lifetime of the diodes is plotted in graph for comparing the lifetime in case of input supply without any disturbance and input supply with occurrence of sag (Fig. 22,23)

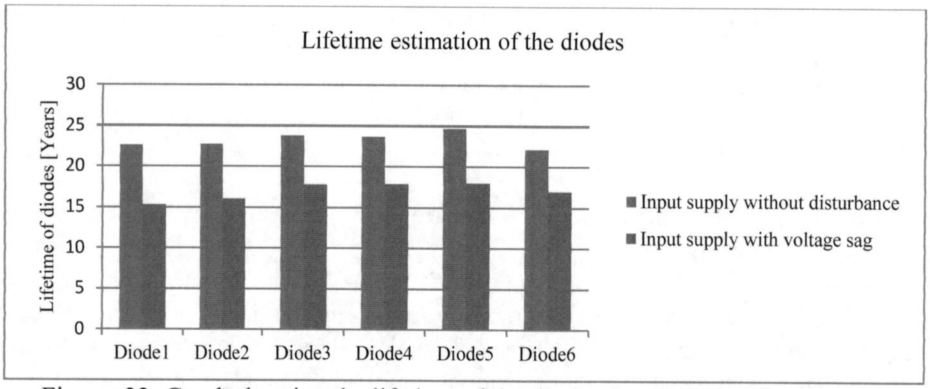

Figure. 22. Graph showing the lifetime of the diodes for short circuit ratio = 20

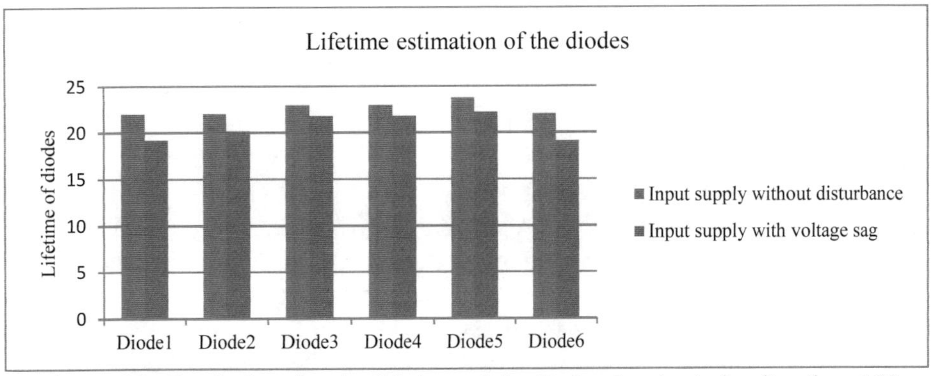

Figure. 23. Graph showing the lifetime of the diodes for short circuit ratio = 120

Thus, lifetime of the diodes in the front-end rectifier is calculated and as seen from the data and graphs, the lifetime of the device is reduced during the occurrence of voltage sag. And in case of minimum short circuit ratio, the impedance at the source is high which reduces the ripple in the supply. But in case of maximum short circuit ratio, the impedance at the source is less and the ripples in the supply are more and this too becomes a reason for the reduction in lifetime of the diodes.

11. Conclusion

VFD Front-end rectifier failure modes and causes are discussed in this investigation. Appropriate system modelling and simulation are carried out. Results are thus obtained through suitable simulation and analytical models. Device junction temperature profile evaluated is then considered for lifetime estimation as thermal stress contributes significant cause of failure. Lifetime estimation is calculated by suitable application of device junction temperature waveform through rain flow counting method, Coffin Manson model and Palmgren miner rule. Thus, with these three algorithms, the lifetime of the device is calculated. With this estimation it is possible to recommend an appropriate replacement time of the front- end rectifier. Lifetime of the device under voltage sag and at different short circuit ratio are also analysed. These effects also have additional impact on the device lifetime. Poor power quality and high short circuit ratio reduces the lifetime of device. Thus, the device can be tested with worst case scenario and the lifetime can be obtained while operating under real time harsh conditions.

Reference

[1]. Sitharthan R, Geethanjali M and Pandy TKS 2016 Adaptive protection scheme for smart microgrid with electronically coupled distributed generations *Alexandria Engineering Journal* **55(3)** 2539-2550

[2]. Fathima AH, and Palanisamy K 2014 Battery energy storage applications in wind integrated systems—a review *IEEE International Conference on Smart Electric Grid* 1-8

[3]. Prabaharan N and Palanisamy K 2015 Investigation of single-phase reduced switch count asymmetric multilevel inverter using advanced pulse width modulation technique *International Journal of Renewable Energy Research* **5(3)** 879-890.

[4]. Jerin ARA, Kaliannan P and Subramaniam U 2017 Improved fault ride through capability of DFIG based wind turbines using synchronous reference frame control based dynamic voltage restorer. *ISA transactions* **70** 465-474

[5]. Sitharthan, R, Sundarabalan CK, Devabalaji KR, Nataraj SK and Karthikeyan M 2018 Improved fault ride through capability of DFIG-wind turbines using customized dynamic voltage restorer *Sustainable cities and society* **39** 114-125

[6]. Prabaharan N and Palanisamy K 2016 A single-phase grid connected hybrid multilevel inverter for interfacing photo-voltaic system *Energy Procedia* **103** 250-255

[7]. Palanisamy K, Mishra JS, Raglend IJ and Kothari DP 2010 Instantaneous power theory based unified power quality conditioner (UPQC) *IEEE Joint International Conference on Power Electronics, Drives and Energy Systems* 1-5

[8]. Sitharthan R and Geethanjali M 2017 An adaptive Elman neural network with C-PSO learning algorithm-based pitch angle controller for DFIG based WECS Journal of Vibration and Control **23(5)** 716-730

[9]. Sitharthan R and Geethanjali M 2015 Application of the superconducting fault current limiter strategy to improve the fault ride-through capability of a doubly-fed induction generator–based wind energy conversion system *Simulation* **91(12)** 1081-1087

[10]. Sitharthan R, Karthikeyan M, Sundar DS and Rajasekaran S 2020 Adaptive hybrid intelligent MPPT controller to approximate effectual wind speed and optimal rotor speed of variable speed wind turbine *ISA transactions* **96** 479-489

[11]. Sitharthan R, Devabalaji KR and Jees A 2017 An Levenberg–Marquardt trained feed-forward back-propagation based intelligent pitch angle controller for wind generation system *Renewable Energy Focus* **22** 24-32

[12]. Sitharthan R, Sundarabalan CK, Devabalaji KR, Yuvaraj T and Mohamed Imran A 2019 Automated power management strategy for wind power generation system using pitch angle controller *Measurement and Control* **52(3-4)** 169-182

[13]. Sundar DS, Umamaheswari C, Sridarshini T, Karthikeyan M, Sitharthan R, Raja AS and Carrasco MF 2019 Compact four-port circulator based on 2D photonic crystals with a 90° rotation of the light wave for photonic integrated circuits applications *Laser Physics* **29(6)** 066201

[14]. Sitharthan R, Parthasarathy T, Sheeba Rani S and Ramya KC 2019. An improved radial basis function neural network control strategy-based maximum power point tracking controller for wind power generation system *Transactions of the Institute of Measurement and Control* **41(11)** 3158-3170

[15]. Rajesh M and Gnanasekar JM 2017 Path observation based physical routing protocol for wireless ad hoc networks *Wireless Personal Communications* **97(1)** 1267-1289

[16]. Palanisamy K, Varghese LJ, Raglend IJ and Kothari DP 2009. Comparison of intelligent techniques to solve economic load dispatch problem with line flow constraints *IEEE International Advance Computing Conference* 446-452

[17]. Sitharthan R, Ponnusamy M, Karthikeyan M and Sundar DS 2019 Analysis on smart material suitable for autogenous microelectronic application *Materials Research Express* **6(10)** 105709

SVM Based Reduction of Input Current Harmonics in Three Phase Rectifier

T.Kesavan[1] , S. Sheebarani Gnanamalar [2] , S.Sivaranjani [3], V.Gomathy[4], G.Radhakrishnan[5] and R. Sitharthan[6]

[2,4,5] Prof, Department of Electrical and Electronics Engineering,
[1,3]Asst.Professor, Department of Electrical and Electronics Engineering,
Sri Krishna College of Engineering and Technology
Coimbatore, Tamilnadu, India
[6]School of Electrical Engineering, Vellore Institute of Technology, Vellore.

Email:[1]t.kesavan87@gmail.com , [2]sheebaranis@skcet.ac.in , [3] sivaranjanis@skcet.ac.in, [4]gomathyv@skcet.ac.in and [5] radhakrishnang@skcet.ac.in

Abstract. Current Scenario of power system engineering field is facing problem of harmonics in voltage and current. In power converter harmonics are mainly disturb the stability of voltage and current in uncertainty of load and transient. This paper proposes space vector modulation based analysis of input current in 3 phase converter. Harmonics in input current in converter is a big challenge of power Electronics research. In power converters non-linear load are one of the reason for increasing the harmonics in electronic circuits. Modified space vector modulation is proposed for reducing the harmonics in 3 phase converter and also control the total harmonic distortion. Matlab simulink software is used to simulate and analyse the 3 phase converter and SVM is modelled for above converter. In this propose method THD of input current harmonics has reduced below 5 percentages and compared with IEEE standard.

Keywords:-- **Harmonic current, Harmonic distortion, PWM , SVM.**

1. Introduction

Increasing a stage of non linear load in our electrical field harmonics are placed in would it an current. Harmonics are defenders the multiple integral of fundamental frequency In linear load the voltage and current are linearly very good that means both are directly proportional to the each other. Example of linear load is a direct current serious and front motors. In non linear load the voltage and current are inversely proportional and not buried linearly. Induction motor and synchronous motor and special electrical machines are examples of non linear load. in non linearity electrical circuits having problems in frequency. A power Electronics converter operation depends on switching variations of power electronic devices. Due to transient of electronic circuits .harmonious may be developed changes or load variation of circuit as developing the harmonics.

In advance the life the world is started to run essay automatic smart and digital. Games and controllers are parts of automatic welding. Man linear load or necessarily for real time automatic condition power electronic converter for controlled need to control the non linear load especially for Electrical drives. Generally conventional electrical motors may be used in open loop system without automatic. Non-conventional Motors power electronic commentator are converted with electrical Motors search with the exact speed control. Harmonics are developing the converter circuit due to the speed variation in power electronics circuits. As per IEEE standard total harmonic distortion of voltage should be e less than 5 percentages and current less than 10 percentages but real time due to usage non linear loads and also based on induction principle motors, it produces maximum harmonics in voltage and current. Some of major harmonics producers from AC drives in induction motor, synchronous motor, Bldc motor, switched reluctance motor and PMSM motor.

Figure 1 shows the proposed concept block diagram of three phase inverter with SVM technique. In this concept output current are taken as feedback from 3 phase converter and is compared with reference current. Error can be detected and it is converted into two current regulators, the duty of current regulator is producer architecture access current and direct Axis current. In this time direct answer is current or considered as zero. Proportional integral controllers are used as current regulator for developing the current Iqs based on the error from detector. In generally proportional controller may be used to get quick settling time of a signal and integral controller are prescribed for reducing peak overshoot of the signal. Voltage regulator develops the voltage of quadrature Axis and direct axis depend upon the harmonic calculation. Three phase voltages of ABC are converted into vds and vqs by DQ transformation. Based the basic concept of SPM produces the gate pulses for three phase converter. In this type getting calls itself hormones can reduces to converter. Three phase rectifier can operating by giving firing pulse from space vector modulation. Finally three phase converter competitions the output voltage with fewer amounts of harmonics and total harmonic distortion of current can be read easily below 5 percentages with attribute standard.

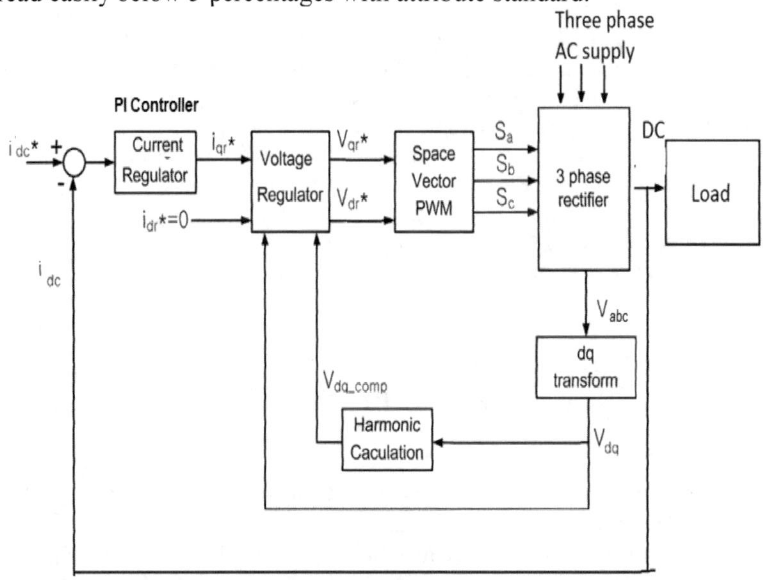

Figure 1. Block diagram of the Proposed PWM rectifier

2. Space Vector Pulse Width Modulation

Space Vector Modulation was developed in the year of 1980s. In conventional pulse width modulation techniques, reference signals compare with carrier signals modulation pwm signals can produced. Some of conventional methods are sinusoidal pulse width modulation multiple pwm, single pwm, trapezoidal pwm and harmonic pwm. In SVM consist of eight vector for states during the time period of TS and eighth factor are called as zero state or null vector because there is no power old flowing from source to load during the zero state operation so in which state all the switches or off condition. In other states in anyone switch will be turn on so the voltage are flowing from input to load the switching States from 2 to 7 are operating States. Figure 2 shows the switching state of proposal space vector modulation. The total time period of modulation technique is 360 degree and each modulation consists of minimum 30degree. Fear power converter each States consists of vectors and switches are controlled by bye svm get pulses. Output voltage can be improved in this method and harmonics are maximum reduced [13].

$$V_k = \frac{2}{3} V_{dc} * e^{j(k-1)\frac{\pi}{3}}, k = 1,2,...,6. \tag{1}$$

$$T_1 = M * \frac{Sin(60^0 - \alpha)}{Sin60^0} * T_s \tag{2.1}$$

$$T_2 = M * \frac{Sin\alpha}{Sin60^0} * T_s \tag{2.2}$$

$$T_Z = T_s - T_1 - T_2 \tag{2.3}$$

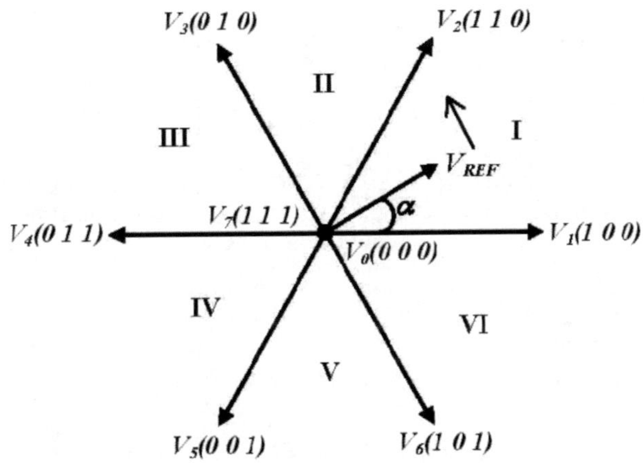

Figure 2. Switching states and corresponding voltage vectors of a three phase VSI. I, II, III, IV, V, VI are the sectors.

Thus, SVPWM uses V_0, V_k, V_{k+1}, V_7, V_{k+1}, V_k, V_0 sequence for odd sectors and V_0, V_{k+1}, V_k, V_7, V_k, V_{k+1}, V_0 sequence for even sectors respectively.

3. Harmonics Reduction Technique of Input Currents

Stationary frame of current can be converted into synchronous form is shown in equation 3.

$$\begin{bmatrix} i_d^{s*} \\ i_q^{s*} \end{bmatrix} = \begin{bmatrix} \cos\theta & -\sin\theta \\ \sin\theta & \cos\theta \end{bmatrix} \begin{bmatrix} i_d^{e*} \\ i_q^{e*} \end{bmatrix}. \tag{3}$$

$$\begin{bmatrix} i_d^{s*} \\ i_q^{s*} \end{bmatrix} = \begin{bmatrix} \cos(\theta+\Delta\theta) & -\sin(\theta+\Delta\theta) \\ \sin(\theta+\Delta\theta) & \cos(\theta+\Delta\theta) \end{bmatrix} \begin{bmatrix} i_d^{e**} \\ i_q^{e**} \end{bmatrix} = \begin{bmatrix} \cos\theta & -\sin\theta \\ \sin\theta & \cos\theta \end{bmatrix} \begin{bmatrix} i_d^{e*} \\ i_q^{e*} \end{bmatrix}. \tag{4}$$

By using the references $i^{e**}{}_{d}$, $i^{e**}{}_{q}$ and the phase angle, the corrected current references $i^{e**}{}_{d}$, $i^{e**}{}_{q}$ can be derived by (6).

$$\begin{bmatrix} i_d^{e**} \\ i_q^{e**} \end{bmatrix} = \begin{bmatrix} \cos(\theta+\Delta\theta) & -\sin(\theta+\Delta\theta) \\ \sin(\theta+\Delta\theta) & \cos(\theta+\Delta\theta) \end{bmatrix}^{-1} \begin{bmatrix} i_d^{s*} \\ i_q^{s*} \end{bmatrix}$$
$$= \begin{bmatrix} \cos(\theta+\Delta\theta) & \sin(\theta+\Delta\theta) \\ -\sin(\theta+\Delta\theta) & \cos(\theta+\Delta\theta) \end{bmatrix} \begin{bmatrix} i_d^{s*} \\ i_q^{s*} \end{bmatrix}$$
$$= \begin{bmatrix} \cos(\theta+\Delta\theta) & \sin(\theta+\Delta\theta) \\ -\sin(\theta+\Delta\theta) & \cos(\theta+\Delta\theta) \end{bmatrix} \begin{bmatrix} \cos\theta & -\sin\theta \\ \sin\theta & \cos\theta \end{bmatrix} \begin{bmatrix} i_d^{e*} \\ i_q^{e*} \end{bmatrix}$$
$$= \begin{bmatrix} \cos\Delta\theta & \sin\Delta\theta \\ -\sin\Delta\theta & \cos\Delta\theta \end{bmatrix} \begin{bmatrix} i_d^{e*} \\ i_q^{e*} \end{bmatrix}. \quad (5)$$

$$[R(\theta)] \begin{bmatrix} \cos(\theta+\Delta\theta) \\ \sin(\theta+\Delta\theta) \end{bmatrix} = \begin{bmatrix} \cos\theta & \sin\theta \\ -\sin\theta & \cos\theta \end{bmatrix} \begin{bmatrix} \cos(\theta+\Delta\theta) \\ \sin(\theta+\Delta\theta) \end{bmatrix}$$
$$= \begin{bmatrix} \cos\theta\cos(\theta+\Delta\theta) + \sin\theta\sin(\theta+\Delta\theta) \\ -\sin\theta\cos(\theta+\Delta\theta) + \cos\theta\sin(\theta+\Delta\theta) \end{bmatrix}$$
$$= \begin{bmatrix} \cos\Delta\theta \\ \sin\Delta\theta \end{bmatrix}. \quad (6)$$

The distortion components sinΔΘ and cosΔΘ necessary for calculating (4) can be obtained by (5).To extract such distortion components, the phase angles Θ and Θ + ΔΘ should be known

4. Simulations and Results

Figure.3 Shows MATLAB simulation of Three phase Rectifier under the distortion of Input Voltage. Here Source voltage is distorted by series connected reactance. MATLAB /Simulink diagram has been run up to 0.1 sec and fundamental frequency of supply is 50 Hz.

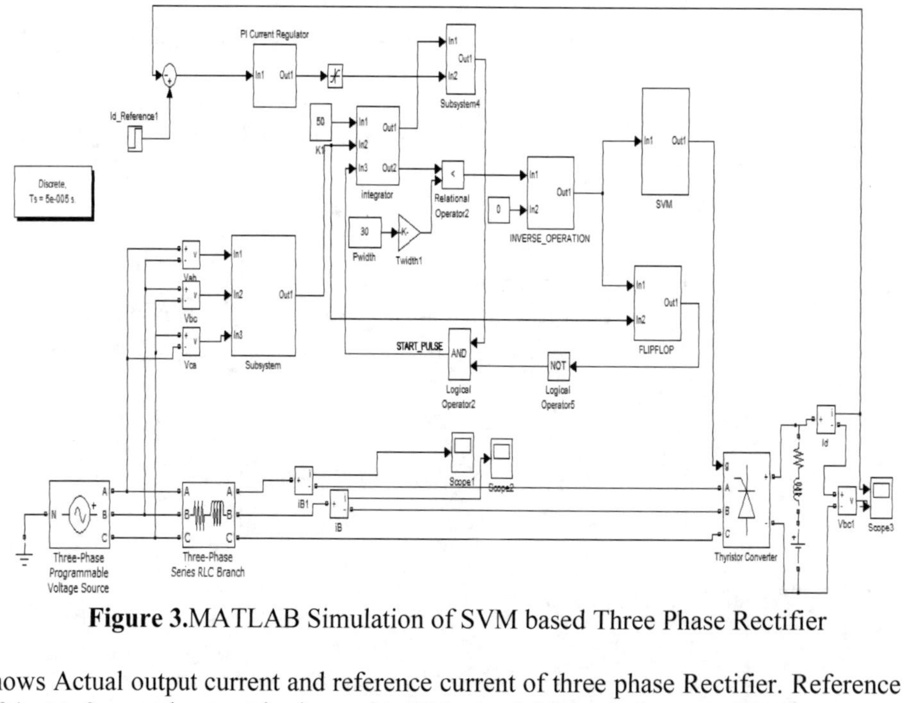

Figure 3.MATLAB Simulation of SVM based Three Phase Rectifier

Figure 4 shows Actual output current and reference current of three phase Rectifier. Reference current is settled 10A at t=0sec and current is changed to 25A at t=0.06sec as shown in Figure

IOP Conf. Series: Materials Science and Engineering **937** (2020) 012058 doi:10.1088/1757-899X/937/1/012058

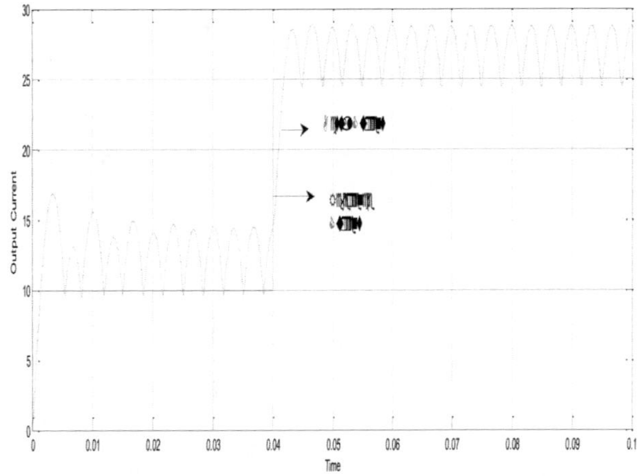

Figure 4. Actual Output current and reference current of three phase Rectifier.

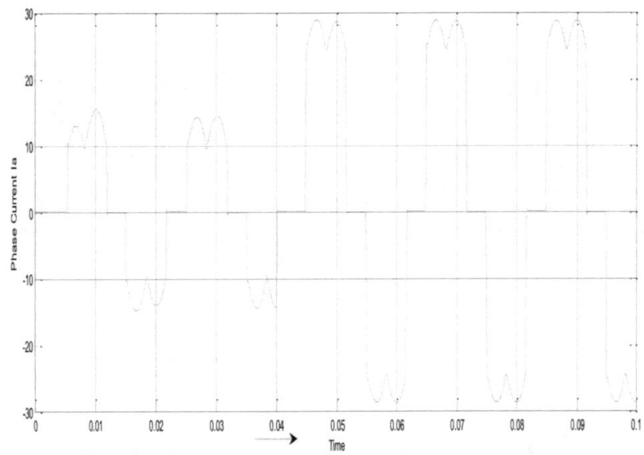

Figure 5. Input Phase Current Ia of Three phase Rectifier

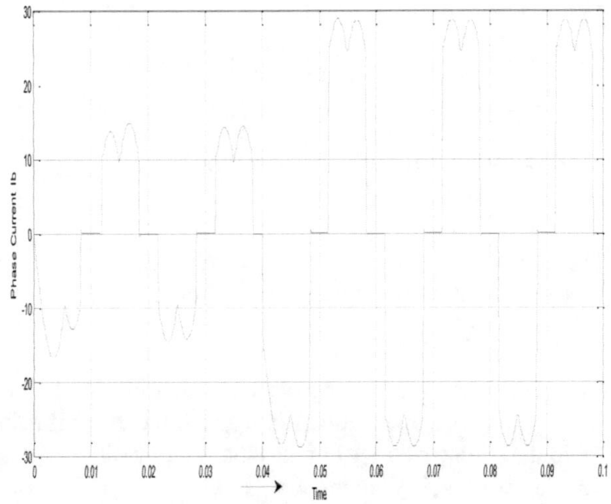

Figure 6. Input Phase Current Ib of Three phase Rectifier

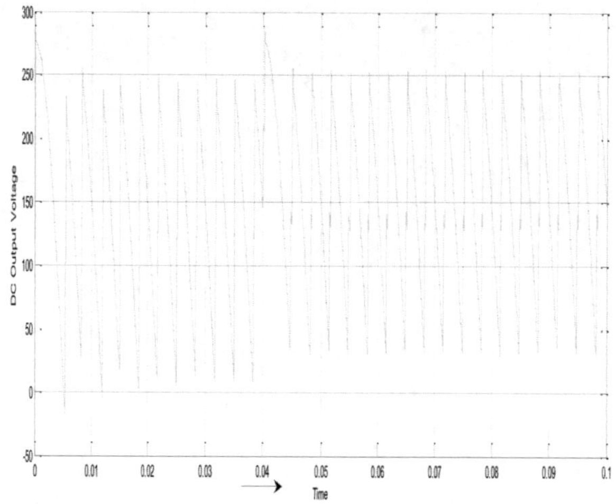

Figure 7.Output DC Voltage of Three phase Rectifier

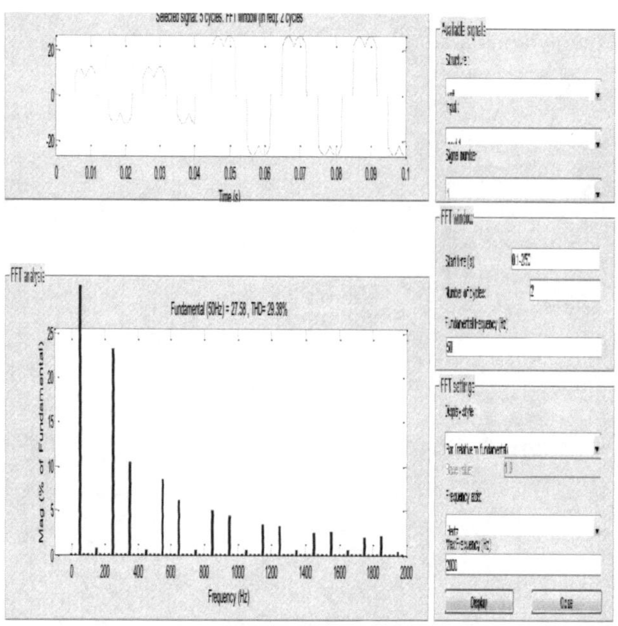

Figure 8.THD Analysis of Input Phase Current Ia of Three phase Rectifier

Figure 8 shows the THD Analysis of Input Phase Current Ia of Three phase Rectifier. By using proposed method, getting the THD 2.23% of Input phase current.

5. Conclusion

In this paper input phase current harmonics of three phase rectifier has been analysed by MATLAB simulation.THD of input current under source voltage distorted has been minimized without using any additional filter circuit. Output current of rectifier has been controlled by space vector Modulation should have a brief caption describing it and, if necessary, a key to interpret the various lines and symbols on the figure.

References

[1]. Jerin ARA, Kaliannan P and Subramaniam U 2017 Improved fault ride through capability of DFIG based wind turbines using synchronous reference frame control based dynamic voltage restorer. *ISA transactions* **70** 465-474

[2]. Sitharthan, R, Sundarabalan CK, Devabalaji KR, Nataraj SK and Karthikeyan M 2018 Improved fault ride through capability of DFIG-wind turbines using customized dynamic voltage restorer *Sustainable cities and society* **39** 114-125

[3]. Prabaharan N and Palanisamy K 2016 A single-phase grid connected hybrid multilevel inverter for interfacing photo-voltaic system *Energy Procedia* **103** 250-255

[4]. Palanisamy K, Mishra JS, Raglend IJ and Kothari DP 2010 Instantaneous power theory based unified power quality conditioner (UPQC) *IEEE Joint International Conference on Power Electronics, Drives and Energy Systems* 1-5

[5]. Sitharthan R and Geethanjali M 2017 An adaptive Elman neural network with C-PSO learning algorithm-based pitch angle controller for DFIG based WECS Journal of Vibration and Control **23(5)** 716-730

[6]. Sitharthan R and Geethanjali M 2015 Application of the superconducting fault current limiter strategy to improve the fault ride-through capability of a doubly-fed induction generator–based wind energy conversion system *Simulation* **91(12)** 1081-1087

[7]. Sitharthan R, Karthikeyan M, Sundar DS and Rajasekaran S 2020 Adaptive hybrid intelligent MPPT controller to approximate effectual wind speed and optimal rotor speed of variable speed wind turbine *ISA transactions* **96** 479-489

[8]. Sitharthan R, Devabalaji KR and Jees A 2017 An Levenberg–Marquardt trained feed-forward back-propagation based intelligent pitch angle controller for wind generation system *Renewable Energy Focus* **22** 24-32

[9]. Sitharthan R, Sundarabalan CK, Devabalaji KR, Yuvaraj T and Mohamed Imran A 2019 Automated power management strategy for wind power generation system using pitch angle controller *Measurement and Control* **52(3-4)** 169-182

[10]. Sundar DS, Umamaheswari C, Sridarshini T, Karthikeyan M, Sitharthan R, Raja AS and Carrasco MF 2019 Compact four-port circulator based on 2D photonic crystals with a 90° rotation of the light wave for photonic integrated circuits applications *Laser Physics* **29(6)** 066201

[11]. Sitharthan R, Parthasarathy T, Sheeba Rani S and Ramya KC 2019. An improved radial basis function neural network control strategy-based maximum power point tracking controller for wind power generation system *Transactions of the Institute of Measurement and Control* **41(11)** 3158-3170

[12]. Rajesh M and Gnanasekar JM 2017 Path observation based physical routing protocol for wireless ad hoc networks *Wireless Personal Communications* **97(1)** 1267-1289

[13]. Palanisamy K, Varghese LJ, Raglend IJ and Kothari DP 2009. Comparison of intelligent techniques to solve economic load dispatch problem with line flow constraints *IEEE International Advance Computing Conference* 446-452

[14]. Sitharthan R, Ponnusamy M, Karthikeyan M and Sundar DS 2019 Analysis on smart material suitable for autogenous microelectronic application *Materials Research Express* **6(10)** 105709

[15]. Rajaram R, Palanisamy K, Ramasamy S and Ramanathan P 2014 Selective harmonic elimination in PWM inverter using fire fly and fireworks algorithm *International Journal of Innovative Research in Advanced Engineering* **1(8)** 55-62

[16]. Sitharthan R, Swaminathan JN and Parthasarathy T 2018 March. Exploration of wind energy in India: A short review *IEEE National Power Engineering Conference* 1-5

[17]. Karthikeyan M, Sitharthan R, Ali T and Roy B 2020 Compact multiband CPW fed monopole antenna with square ring and T-shaped strips *Microwave and Optical Technology Letters* **62(2)** 926-932

[18]. Sundar D Sridarshini T, Sitharthan R, Madurakavi Karthikeyan, Sivanantha Raja A, and Marcos Flores Carrasco 2019 Performance investigation of 16/32-channel DWDM PON and long-reach PON systems using an ASE noise source *In Advances in Optoelectronic Technology and Industry Development: Proceedings of the 12th International Symposium on Photonics and Optoelectronics* 93

[19]. Sitharthan R and Geethanjali M 2014 Wind Energy Utilization in India: A Review *Middle-East J. Sci. Res.* **22** 796–801 doi:10.5829/idosi.mejsr.2014.22.06.21944

[20]. Sitharthan R and Geethanjali M 2014 ANFIS based wind speed sensor-less MPPT controller for variable speed wind energy conversion systems *Australian Journal of Basic and Applied Sciences* **8**14-23

AUTHOR INDEX